FIBER OPTIC SENSORS

Edited by **Moh. Yasin,**
Sulaiman W. Harun and **Hamzah Arof**

INTECHOPEN.COM

Fiber Optic Sensors

Edited by Moh. Yasin, Sulaiman W. Harun and Hamzah Arof

CBS, Edition 2016

Published by InTech

Janeza Trdine 9, 51000 Rijeka, Croatia

Publishing Process Manager Martina Blecic

Technical Editor Teodora Smiljanic

Cover InTech Design Team

Additional hard copies can be obtained from orders@intechopen.com

Fiber Optic Sensors, Edited by Moh. Yasin, Sulaiman W. Harun and Hamzah Arof
p. cm.
ISBN 978-953-307-922-6

INTECH

open science | open minds

Contents

Preface

Fiber optic is made of a plastic or glass core surrounded by cladding material. The difference in reflective index between these two components allows light to be guided inside the core with the principle of total internal reflection. The optical fiber and opto-electronics technologies are progressing rapidly due to innovations in telecommunications, semiconductor and consumer electronics sectors. The revolution in communication industries significantly reduces the prices of optical components and stimulates the development of optical fiber sensors. These sensors use optical fiber either as the sensing element ("intrinsic sensors"), or as a means of relaying signals from a remote sensor to the signal processor ("extrinsic sensors"). In the future, it is expected that optical fiber sensors will replace most of the conventional devices for the measurement of various physical, chemical and biological parameters such as temperature, pressure, strain, position, rotation, acceleration, electric, magnetic fields, acoustics, vibration, strain, humidity, viscosity, PH, glucose, gases, pollutants and many more.

The field of optical fiber sensors is expected to expand and develop, influenced by new applications of the latest technologies. In this way, the subject continuous to mature and reach into new areas of engineering. This book reviews the recent topics on optical fiber sensors. Chapter 1 presents an overview of fiber optic sensors and their applications. The chapter discusses a review based on rare-earth doped fiber and new materials such as conducting polymer. Chapter 2 focuses on optical fiber sensors for volatile organic compound (VOC) detection. Fiber Bragg grating (FBG) and distributed Brillouin fiber sensors are the most popular sensing techniques for structural health detection. Recent progress in the use of these distributed sensors for structural health monitoring in energy infrastructures in China are discussed in Chapter 3. Chapter 4 presents an overview of intrinsic optical fiber sensors. Chapter 5 discusses a theoretical analysis and characterization of Brillouin gratings in optical fibers. Two applications of Brillouin grating are also given in this chapter. The first application is for the distributed birefringence measurement in polarisation maintaining fiber (PMF), and the second is for simultaneous measurement of temperature and strain.

Chapter 6 reviews the principle of two frequency symmetrical radiation (TFSR) synthesis and its applications in fiber optic structural monitoring. A variety of TFSR multiplexed sensing functions can be provided by the TFSR technique. In this chapter, various sensor systems are also introduced based on optical reflectometry, distributed lateral stress location, multiplexed FBG and Fabry-Perot interferometer. Chapter 7 presents a thorough theoretical study of the optical fiber evanescent field sensors. A new method to estimate the sensitivity of the sensor is then proposed and verified experimentally. Chapter 8 presents a theoretical study of tapered optical fibers (TOFs) of different forms. The description starts with the rigorous analytical approach for conventional dielectric TOFs, and ends with the dispersion features as well as the relative power distribution for different low-order modes. The results are compared with those of conventional optical fibers in terms of dispersion characteristics, and it is found that the normalized frequency parameter is reduced for the TOFs. A microcsphere coupling system is presented in Chapter 9. The main aim of this chapter is to demonstrate the practical thermal sensing application based on the robust fiber-integrated microsphere coupling structure. Chapter 10 investigates the optical effects connected with coherent, polarized light propagation through a step-index fiber.

The use of long period fiber gratings (LPFGs) as sensors is thoroughly explained in chapters 11 to 13. Chapter 11 reviews the fabrication methods, the theory behind the operation and applications of LPFGs. The application of LPFG produced by arc discharges in temperature and curvature sensors is explained in Chapter 12. Chapter 13 focuses on the possible application of long-period gratings technology in environmental monitoring, particularly in the measurement of surrounding refractive index or salinity. Chapter 14 provides an overview of the state-of-the-art, birefringence concepts, and new developments of high-birefringence fiber loop mirror configurations that can be used as sensing elements. Recently, plastic optical fiber sensors represent an emerging alternative for various applications in engineering. Chapters 15 to 17 present the development of plastic optical fiber-based sensors, which offer many unique features that could be exploited to achieve cost-effective sensing systems. The performance of various fiber optic displacement sensors is investigated theoretically and experimentally in Chapter 15. Chapter 16 presents the potential of POF sensing technique as an attractive option for various applications in civil engineering such as for monitoring strain, deflection, liquid level, vibration and detection of cracks. Chapter 17 demonstrates a facile method to develop POF pH sensors with a tip-based sensing element prepared by a sol-gel process, and consisting of phenol red indicator entrapped in a polymer-silica organic-inorganic hybrid material. Chapter 18 discusses the mechanical property and strain transferring mechanism of optical fiber sensors.

Chapter 19 describes the construction of the optical fiber based system for efficient detection of bioluminescence at the optical fiber end. The sensitivity of Adenosine triphosphate (ATP) detection is investigated by using an avalanche photon diode (APD). ATP is a reliable indicator of biochemical reaction or life activity, since ATP is considered to be the universal currency of biological energy for all living things. Chapter 20 demonstrates fiber optic thermometers for hyperthermia therapy. This optical technique is normally used when electrical insulation and electromagnetic immunity are necessary. In chapter 21, high voltage optical fiber sensor systems with compensation for optical power fluctuations are demonstrated using a white light interferometry approach.

Dr Moh. Yasin,
Dept. of Physics, Faculty of Science,
Airlangga Univ. Surabaya,
Indonesia

Prof. Sulaiman W. Harun,
Dept. of Electrical Engineering,
Faculty of Engineering, Univ. of Malaya,
Malaysia

Dr Hamzah Arof,
Dept. of Electrical Engineering,
Faculty of Engineering, Univ. of Malaya,
Malaysia

1

Optical Fiber Sensors: An Overview

Jesus Castrellon-Uribe
Center for Research in Engineering and Applied Sciences, CIICAp
Autonomous University of Morelos State, UAEM
México

1. Introduction

Fiber optic sensor technology has been under development for the past 40 years and has resulted in the production of various devices, including fiber optic gyroscopes; sensors of temperature, pressure, and vibration; and chemical probes. Fiber optic sensors offer a number of advantages, such as increased sensitivity compared to existing techniques and geometric versatility, which permits configuration into arbitrary shapes. Because fiber optic sensors are dielectric devices, they can be used in high voltage, high temperature, or corrosive environments. In addition, these sensors are compatible with communications systems and have the capacity to carry out remote sensing. Recently, investigation in the field has focused on the development of new materials with non-linear optical properties for important potential applications in photonics. Examples of these materials are the conjugated semiconducting polymers that combine optical properties with the electronic properties of semiconductors. In addition, these conducting polymers have photoluminescent and electroluminescent properties, making them attractive for applications in optoelectronics.

This chapter presents an overview of fiber optic sensors and their applications. It also describes new optical materials that are being investigated for the development of chemical optical sensors. The chapter is organized into five sections (including conclusions) to provide a clear and logical sequence of topics. The first section briefly reviews optical fiber fundamentals, including basic concepts, optical fiber structure, and their general characteristics. The propagation of light in optical fibers, which involves Snell's law, the critical angle, and the total internal reflection, is also discussed. The second section offers an extensive introduction to fiber optic sensors, including their characteristics, functional classification, modulation methods, and principal applications. The third section discusses fluorescent optical sensors that employ rare-earth-doped fibers, such as erbium (Er^{3+}), neodymium (Nd^{3+}), ytterbium (Yb^{3+}), praseodymium (Pr^{3+}), samarium (Sm^{3+}), europium (Eu^{3+}), holmium (Ho^{3+}), and erbium/ytterbium (Er/Yb). A review of the performance of rare-earth-doped fiber sensors and their applications in remote temperature measurement is also presented, taking into account the sensing material, the temperature range, and its temperature sensitivity. The next section provides an overview of new materials with optical properties and evaluates their potential as optical fiber sensors. Conducting polymers, such as polypyrrole (PPy), polyaniline (PANI), polythiophene (PTh), and their derivatives, are discussed as potential optical sensors because of their interesting electrical, chemical, and optical properties. The final section provides the conclusions of the chapter.

The chapter ends with a bibliography on the topic that offers the reader an extensive selection of scientific references on optical fiber sensors.

2. Optical fiber basics

The optical fiber has represented a revolution in the world of telecommunications mainly because of its capacity to transmit large quantities of information, including video and data. Erbium-doped fibers can be used as optical amplifiers to extend the distance of transmission. The investigations in this field have permitted the expansion of the spectrum of applications of optical fibers, leading to the development of new devices, such as fiber lasers and optical fiber sensors, which are the subject of this chapter.

An optical fiber is an optical waveguide in the shape of a filament and is generally made of glass (although it can also be made of plastic materials). An optical fiber is composed of three parts: the core, the cladding, and the coating or buffer. Fibers can be produced in a range of sizes; a common cladding diameter is 125 μm, whereas the core typically ranges from 10 to 50 μm. The basic structure of an optical fiber is shown in Figure 1.

The core is a cylindrical rod of dielectric material and is generally made of glass. Light propagates mainly along the core of the fiber. The cladding layer is made of a dielectric material with an index of refraction, n_2, that is less than that of the core material, n_1. The cladding is generally made of glass or plastic. The cladding decreases the loss of light from the core into the surrounding air, decreases scattering loss at the surface of the core, protects the fiber from absorbing surface contaminants, and adds mechanical strength. The coating or buffer is a layer of plastic used to protect the optical fiber from physical damage. The core and the cladding provide the conditions necessary to permit an optical signal to be guided along the optical fiber.

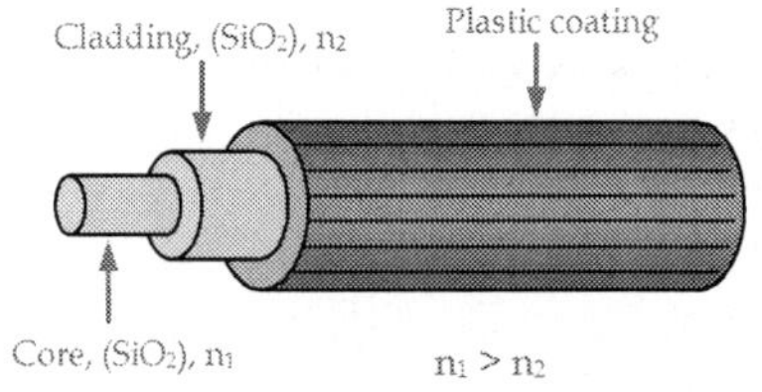

Fig. 1. Schematic of a single fiber optic structure.

The principle of transmission of light along optical fibers is based on *total internal reflection,* which is related to a light beam incident on the boundary between two materials with different refractive indices, as illustrated in Figure 2. When light is incident from a medium with a high index (n_1) to one with a lower index (n_2), the transmitted beam always emerges at an angle, ϕ_2, that is greater than the incident angle, ϕ_1 (see Fig. 2a). If we increase the measure of ϕ_1, there will come a point where ϕ_2 is 90°; at this point, the value of the angle of incidence is known as the critical angle, ϕ_c (see Fig. 2b). If the angle of incidence is greater than ϕ_c, there is no refraction of the light, and all of the rays (radiation) become *totally internally reflected* toward the material with the refractive index n_1 (see Fig. 2c).

For a ray to be effectively *"trapped"* within the fiber core, it must strike the core/cladding interface at an angle, ϕ, that is greater than the critical angle, ϕ_c. This critical angle is related to the refractive indices of the core n_1 and the cladding n_2 by Snell's law ($n_1 \sin \phi_1 = n_2 \sin \phi_2$) and can be calculated as $\phi_c = \arcsin (n_2/n_1)$. This requirement means that any ray entering the fiber with an incidence angle, ϕ_0, between 0 and $\pm\ \theta$ will be internally reflected along the fiber core. This angle θ is known as the acceptance angle and is related to the numerical aperture (NA) of an optical fiber as follows: $NA = n_0 \sin \theta = (n_1^2 - n_2^2)^{1/2}$, where n_0 is the refractive index of the medium surrounding the optical fiber.

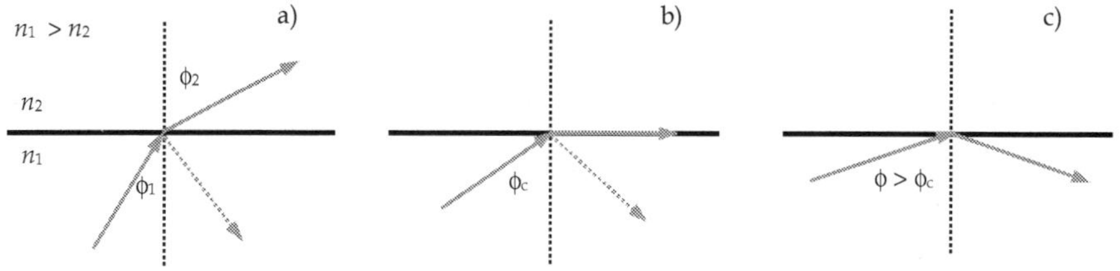

Fig. 2. Representation of the *critical angle* and *total internal reflection (TIR)* between two different materials.

Two types of fibers are commonly used: *step-index fibers* and *graded-index fibers*. In the first case, the refractive index of the core is uniform throughout and undergoes an abrupt change (or step) at the cladding boundary. In the second case, the core refractive index is made to vary as a function of the radial distance from the center of the fiber. Both types of fibers can be further divided intro *single-mode* and *multimode fibers*. A single-mode fiber sustains only one mode of propagation, whereas multimode fibers contain many hundreds of modes.

One of the principal characteristics of an optical fiber is its attenuation as a function of wavelength. The systems of optical communications operate in the band centered at 1550 nm because, in this region, the optical signal travelling by an optical fiber suffers from the lowest attenuation. This region is the named the third window of communications. Currently, new materials are being investigated for the production of optical fibers that further diminish the attenuation of the signal for applications in communications.

The main advantages of optical fiber technology are low attenuation, wide bandwidth, reduced weight and size, and immunity to electromagnetic interference (EMI). A more extensive description of the characteristics and properties of optical fibers can be found in the following references (Ghatak & Thyagarajan, 2000; Keiser, 1991).

Today, the investigation and development of optical-fiber devices encompasses optical amplifiers (Erbium Doped Fiber Amplifiers, EDFAs), fiber lasers, and optical fiber sensors.

3. Optical fiber sensors

Currently, the research and development of fiber-optic sensor devices has extended their applications to diverse technological fields, including the medical, chemical, and telecommunications industries. Optical fiber sensors have been developed to measure a wide variety of physical properties, such as chemical changes, strain, electric and magnetic fields, temperature, pressure, rotation, displacement (position), radiation, flow, liquid level, vibrations, light intensity, and color. Fiber-optic sensors are devices that can performance in harsh environments where conventional electrical and electronic sensors have difficulties.

In comparison with the other types of sensors, optical fiber sensors exhibit a number of advantages; they

- Are non-electrical devices
- Require small cable sizes and weights
- Enable small sensor sizes
- Allow access into normally inaccessible areas
- Often do not require contact
- Permit remote sensing
- Offer immunity to radio frequency interference (RFI) and electromagnetic interference (EMI)
- Do not contaminate their surroundings and are not subject to corrosion
- Provide high sensitivity, resolution and dynamic range
- Offer sensitivity to multiple environmental parameters
- Can be interfaced with data communication systems

Optical fiber sensors are dielectric devices that are generally chemically inert. They do not require electric cables for their performance and are technically ideal for working in hostile media or corrosive environments for applications in remote sensing.

The basic components of an optical fiber sensor are an optical source, a transducer, and a receiver, as is observed in the schema of Figure 3. Lasers, diodes, and/or LEDs are often used as the optical source in these sensing devices. An optical fiber (single or multimode), doped fibers, and/or bulk materials are employed as the transducer (sensor heart). At the output of the sensor system, a photodetector is used to detect the variation in the optical signal that is caused by the physical perturbation of the system. In the optical fiber sensors systems, the optical parameters that can be modulated are the amplitude, phase, color (spectral signal), and state of polarization. The optical modulation methods of the sensors involve the following:

The **amplitude change** is related to the transmission, absorption, reflection, or scattering of the optical signal. Currently, Fiber Bragg Gratings (FBG) and Long Period Fiber Gratings (LPFG) are employed as the sensor heads in optical fiber sensors systems. The optical parameters that can be modulated for these sensors are the wavelength, transmission, reflection, and refraction index, which are associated with the perturbation environment.

The **phase change** is associated with the optical frequency and wavelength variation.

The **change in color** is proportional to the changes in the absorption, transmission, reflection, or luminescence of the optical signal, whereas the polarization is related to the strain birefringence.

The *transmission concept* is normally associated with the interruption of a light beam that is travelling via the optical fiber. The sensors that are based on *reflection* employ two bundles

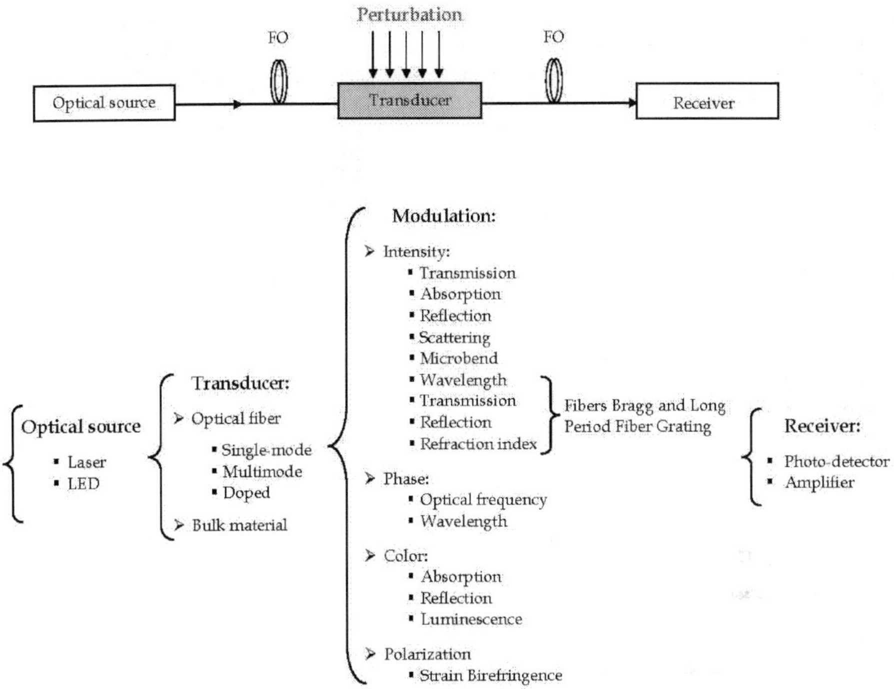

Fig. 3. Basic components of an optical fiber sensor.

of fibers or a pair of single fibers. One bundle of fibers transmits light to a reflecting target; the other bundle traps reflected light and transmits it to a detector. The variation in the intensity detected with a photodetector is directly proportional to the perturbation environment. In a sensor that is based on *microbending*, small amounts of light are lost through the wall of the fiber if the fiber is bent. If the fiber is bent due to a physical perturbation (e.g., pressure), then the amount of received light is related to the value of the physical parameter.

In addition, the optical fiber can be doped in the core with a chemical. Then the *absorption concept* is related to the absorbance spectrum of the chemical (dopant) incorporated in the fiber. According to the characteristics of the dopant, some peaks or bands of the absorption are dependent on some physical parameters, such as temperature. A similar approach can be considered for *scattering*.

Similar to the absorption concept, *luminescence* can be achieved by doping the fiber or some glass material with a chemical. In this kind of sensor, a light source can be used to stimulate a fluorescence signal, which is affected by some external physical parameter. In the same way, the fiber can be stimulated by outside radiation, and the fluorescence signal can be detected as a measure of the level of incident radiation. Similarly, a change in the *luminescence wavelength* can be transduced in a change of *color* as a function of a perturbing environment. *Refractive index changes* in the core of an optical fiber (e.g., fiber grating) due to a perturbing environment can change the optical frequency and, consequently, the amount of received light (transmitted or reflected) on the photodetector. The combination of some of these concepts can be used with some of the mechanisms of modulation to improve or to complement the sensor required for covering a specific need.

Optical fiber sensors can be divided into two basic categories: intensity-modulated sensors and phase-modulated sensors.

Intensity-modulated sensors: This class of sensors detects the variation of the light intensity that is proportional to the perturbing environment. The concepts associated with intensity modulation include transmission, reflection, and microbending. For this, a reflective or transmissive target can be incorporated in the fiber. Other mechanisms that can be used independently or in conjunction with the three primary concepts include absorption, scattering, fluorescence, and polarization. Intensity-modulated sensors normally require more light to function than phase-modulated sensors; as a result, they employ large core multimode fibers or bundles of fibers.

Phase-modulated sensors: This type of sensor compares the phase of the light in a sensing fiber to a reference fiber in a device known as an interferometer. Generally, these sensors employ a coherent laser light source and two single-mode fibers. The light is split and injected into the reference and sensing fibers. If the light in the sensing fiber is exposed to the perturbing environment, a phase shift occurs between them. The phase shift is detected by the interferometer. There are four interferometric configurations used in optical sensors: the Mach-Zehnder, Michelson, Fabry-Perot, and Sagnac. The Mach-Zehnder interferometer configuration is the most widely used for acoustic sensing. Phase-modulated sensors are much more accurate than intensity-modulated sensors.

Generally, fiber optic sensors can be conveniently classified according to the manner in which the optical fiber is used. These sensors can then be functionally classified into intrinsic and extrinsic sensors.

Intrinsic fiber-optic sensor: These sensors directly employ an optical fiber as the sensitive material (sensor head) and also as the medium to transport the optical signal with information of the perturbation environment to be measured. They operate through the direct modulation of the light guided into the optical fiber. The light does not leave the fiber, except at the detection end (the output) of the sensor. In intrinsic sensors, the variable of interest (physical perturbation) must modify the characteristics of the optical fiber to modify the properties of the light carried by the fiber (see Fig. 4a). These sensors can use interferometric configurations, Fiber Bragg Grating (FBG), Long Period Fiber Grating (LPFG), or special fibers (doped fibers) designed to be sensitive to specific perturbations.

Extrinsic or hybrid fiber-optic sensor: In an extrinsic sensor, the optical fiber is simply used to guide the light to and from a location at which an optical sensor head is located. The sensor head is external to the optical fiber and is usually based on miniature optical components, which are designed to modulate the properties of light in response to changes in the environment with respect to physical perturbations of interest. Thus, in this configuration, one fiber transmits optical energy to the sensor head. Then this light is appropriately modulated and is coupled back via a second fiber, which guides it to the optical detector. This is the principle of an intensity-based optical transmission sensor.

Alternatively, the modulated light may be coupled back into the same fiber by reflection or scattering and then guided back to the detection system (see Fig. 4b).

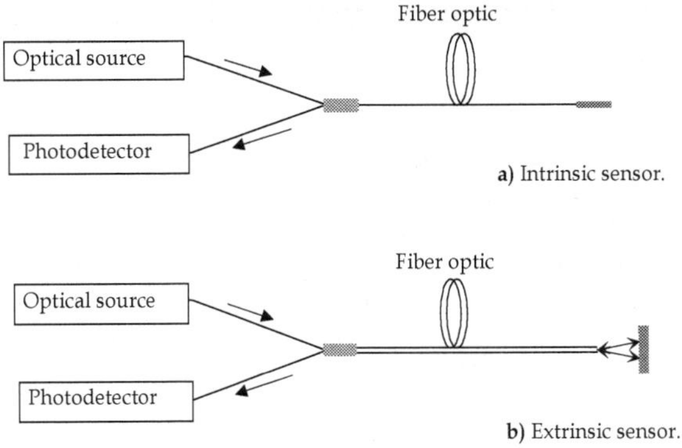

Fig. 4. Arrangements of an optical fiber sensor: a) intrinsic and b) extrinsic sensor.

Optical fiber sensors, whether intrinsic or extrinsic, operate by the modulation of one (or more) of the following characteristics of the guided light: the intensity, wavelength or frequency, state of polarization, and phase.

Today, fiber optic sensors have become essential devices for process control in measurement systems, finding countless applications in, for example, factory automation, the automotive industry, telecommunications, computers and robotics, environmental monitoring, health care, and agriculture. An extensive review of fiber optic sensors and their applications can be found in the following bibliography (Culshaw, 2004; Krohn, 1999; Lopez-Higuera, 2002; Othonos & Kalli, 1999; Rai, 2007; Udd, 1991; Yu et al., 2008).

New challenges in diverse technological fields requiring the monitoring, control, and security of processes are continuously arising. New optical sensor systems, for example, have been implemented for the monitoring of corrosion processes as an alternative to electrochemical sensor systems. The corrosion in metallic structures is a serious problem that involves security, maintenance or replacement costs, and the occasional interruption of the machine, which affects diverse processes in the industry.

Typically, the corrosion rate in a metallic sample is evaluated through measuring its weight-loss or by electrochemical techniques. Alternatively, one of the most well known optical techniques employed for corrosion monitoring is based on holographic interferometry (Habib, 1993, 1995). The main constraint of these techniques arises when measurements need to be taken *in situ* under different laboratory-controlled conditions. Therefore, it is important to investigate new alternatives for measurements. Recently, optical sensor systems based on the change in intensity have been proposed for the measurement of corrosion (Castrellon-Uribe et al., 2008; Dong S, 2005a, 2005b). The main advantages of this optical technique include its insensitivity to the intensity variations of the optical source signal, which helps to avoid errors in measurements; the simple detection system of the signal with the corrosion information; and the possibility of developing a fiber optic sensor to carry out measurements of corrosion *in situ*.

4. Rare-earth-doped optical fiber sensors

A rare-earth-doped optical fiber (laser fiber) undergoes the processes of absorption and spontaneous and stimulated emission of radiation when it is excited with photons of a particular energy. An investigation of these processes was conducted to improve the development of an erbium-doped fiber amplifier (EDFA) with the goal of extending the distance of transmission in optical communication systems (Desurvire, 1994; Digonnet, 2001). The investigation of nonlinear processes in laser fibers has allowed for the development of new optical fiber lasers by up-conversion (Mejia et al., 2002; Talavera & Mejia, 2005). In addition, laser fibers have been investigated to develop new temperature sensors because their properties of emission and absorption are dependent on temperature (Berthou & Jorgensen, 1990; Farries et al., 1986; Krug et al., 1991).

In general, radiative methods of temperature measurement are highly advantageous because they do not require physical contact or temperature equilibrium between different objects with distinct thermal masses. Frequently, the temperature can only be measured indirectly at a distance from the object to be measured. Fiber optic sensors have proven to be very efficient due to their small thermal mass, their ability to transmit light efficiently, and their mechanical flexibility, which allows for access to small remote volumes.

A number of optical fiber-based temperature sensors have been developed using approaches based on fluorescence. The techniques most commonly used are based on the fluorescence lifetime (FL) and the fluorescence intensity ratio (FIR). These techniques generally use rare-earth-doped optical fibers as the sensing medium. In these materials, the fluorescence signal is induced by widely available light sources (CW or pulsed) in a variety of wavelengths. A simple photodetector can be used to measure the variation in the intensity of the fluorescence signal as a function of temperature.

The fluorescence intensity generated from two closely spaced energy levels of an ensemble of ions doped in a host material depends on a number of parameters, including the host material, the particular energy level of interest, the dimensions of the material doped with the ion, the concentration level (doping), and the excitation method employed. The separation of the energy levels should be of the order of the thermal energy (a few kT, where kT is ~200 cm^{-1} at room temperature). There are a number of materials that have pairs of energy levels that are separated by energy differences such that they may be considered to be thermally coupled; hence, they could potentially be used in conjunction with the FIR method for temperature sensing. In particular, rare-earth-doped materials have been extensively investigated in the development of new fluorescent sensors of temperature.

The **fluorescence lifetime (FL)** of an energy level of a material is a measure of the rate of reduction in the intensity of fluorescence after the source of excitation has been removed. This rate of decay has been shown to depend strongly on temperature for the energy levels of many materials; therefore, it can be used as a measure of temperature. This technique has been investigated using a relatively large number of sensing materials in a variety of forms, including phosphors, bulk samples, and doped optical fibers. (Grattan & Zhang, 1995; Rai & S.B. Rai, 2007)

The **fluorescence intensity ratio (FIR)** technique involves utilizing the fluorescence intensities from two closely spaced energy levels for monitoring the temperature. In this technique, the fluorescence intensities from these levels to a common final (lower) level are monitored at the desired wavelength. The temperature dependent ratio of these intensities is independent of the source intensity because the emitted intensities are proportional to the population of each energy level involved. Therefore, the fluorescence intensity ratio, R, from two thermally coupled energy levels may be given as (Maurice et al., 1995)

$$R = \frac{N_2}{N_1} = \frac{I_2}{I_1} = B \exp\left[-\frac{\Delta E}{kT} \right] \tag{1}$$

An extensive review of rare-earth doped optical fiber sensors based on the fluorescence-intensity ratio technique is given in the references at the end of the chapter (Castrellon-Uribe, 1999, 2002a, 2002b, 2005, 2010; Dos Santos et al., 1999; Imai & Hokazono, 1997; Maurice, 1994, 1995a, 1995b, 1997a, 1997b; Wade 1997, 1998, 1999a, 1999b).

There are several advantages of using thermally coupled levels over using two non-coupled levels when the fluorescence intensity ratio method is utilized:

- The theory of the relative changes in the fluorescence intensity originating from thermally coupled levels is reasonably well understood, and thus, their behavior can be easily predicted.
- The population of the individual thermally coupled levels is directly proportional to the total population. Therefore, any changes in the total population due to changes in excitation power, for example, will affect the individual levels to the same extent. This helps to reduce the dependence of the measurement technique on the excitation power, which avoids errors in the measurements.
- For relatively closely spaced energy levels, the fluorescence wavelengths will be relatively close, which helps to reduce any wavelength-dependent effects caused by the fiber bends.

In the sensor systems, it is important to know the rate at which the fluorescence intensity ratio changes as a result of a change in temperature. This parameter is known as the sensitivity, *S(R)*, which is given by

$$S(R) = \frac{1}{R}\frac{dR}{dT} = \frac{\Delta E}{kT^2} \tag{2}$$

From Equation 2, it is clear that when using a pair of energy levels with a larger energy difference, the sensitivity of the fluorescence intensity ratio is increased. It is important to notice that the largest energy difference is limited by the occurrence of thermalization. As the energy difference becomes larger, the population and hence the fluorescence intensity from the upper of the two thermalizing levels will decrease, which may introduce problems when measuring very low light levels.

Additionally, there are other factors that limit the feasibility of using a material as a sensor. These factors include costs and availability, the temperature range for which the material can be used, and the fluorescence yield of the particular level of interest. The materials that have been found to meet the above requirements are the triply ionized rare-earth ions.

In the implementation of temperature sensors, the energy levels do not only have to be thermally coupled, but they should also meet other requirements that depend largely on the host matrix into which the active ions are doped. When considering a silica-based glass host, for example, the energy levels should meet the following requirements:

- The first condition is that the pair of energy levels should be thermally coupled, and as a result, Equation 1 can be applied. The energy level separation should be smaller than 2000 cm^{-1} (the separation should not be too large); otherwise, the upper level would have a very small population for the temperature range of interest.
- The separation between the energy levels must be more than 200 cm^{-1} to avoid substantial overlap of the two fluorescence wavelengths.
- To obtain sufficient fluorescence intensity from the pair of upper levels, the radiative transitions must dominate the non-radiative transitions. The non-radiative transition rate decreases with the increase of the energy gap to the next lower energy level. Therefore, it is preferable that the two thermalizing levels lie at least 3000 cm^{-1} above the next lowest energy level.
- For commonly available detectors (such as silica photodiodes) to be utilized in the sensor system, the energy levels should have radiative transitions (fluorescence) with energies between 6000 and 25000 cm^{-1} corresponding to wavelengths of 1.66 μm and 0.4 μm, respectively.
- For practical sensors, the fluorescence signal must be excited by commercially available light sources, such as laser diodes (LD) or light-emitting diodes (LEDs).

A review of the literature shows that there are only a few rare-earth ions with a pair of energy levels that meet all of these above requirements. Therefore, the rare-earth ions that can be used as sensing materials for temperature measurements are praseodymium (Pr^{3+}), neodymium (Nd^{3+}), samarium (Sm^{3+}), europium (Eu^{3+}), holmium (Ho^{3+}), erbium (Er^{3+}), and ytterbium (Yb^{3+}), which can be doped into a wide variety of glass or crystal hosts. The energy levels of the rare-earth ions, as well as their fluorescence transitions of particular interest, can be found in the literature for a variety of host materials. The performance characteristics of rare-earth-doped fibers used as temperature sensors that employ the fluorescence-intensity ratio technique are provided in Table 1.

There are a number of experimental arrangements employed in the fluorescence intensity ratio technique (FIR) for sensing temperature; the basic elements used in the technique are described as follows. To investigate the photo-thermal properties of these rare earth ions in different hosts, the samples can be excited by a pump source (a laser or pig-tailed diode) that excites the fluorescence from a pair of energy levels of interest. Then the samples can be cooled and/or heated, and their temperature can be detected independently using a thermocouple or a similar device in close proximity to the sample. Next, an optical spectrum analyzer (OSA) can be used for recording the fluorescence spectrum and calculating the intensity ratio as a function of the temperature of the sample from the data obtained. A photodetector and bandpass filters also can be used to measure the fluorescence intensity changes as a function of temperature in the sample.

In most practical cases, compact optical fiber sensors with a high signal-to-noise ratio (SNR) and sensitivity are desirable. To evaluate these parameters, an erbium-doped fiber was analyzed as a temperature sensor in terms of the standard radiometric figures of merit to evaluate its ability to detect thermally generated radiation (Castrellon-Uribe, 1999, 2002). Afterward, the performance of the erbium-doped fiber as a temperature sensor was shown

Sensing material	Transition	Energy gap[1]	λpump	Fluorescence Intensity Ratio Technique (FIR)	Temperature Range	Sensitivity at 20°C	Ref.
Er^{3+}: silica fiber	$(^2H_{11/2}\,;\,^4S_{3/2}) \rightarrow {}^4I_{15/2}$	~800 cm^{-1}	800 nm	530 nm/555nm	23 - 600°C	1.3 %/°C	a
Er^{3+}: silica fiber	$(^2H_{11/2}\,;\,^4S_{3/2}) \rightarrow {}^4I_{11/2}$	~800 cm^{-1}	800 nm	1.13 nm/1.24nm (lines)	25 - 600°C	0.8 %/°C	b
Er^{3+} and Er/Yb: chalcogenide glasses	$(^2H_{11/2}\,;\,^4S_{3/2}) \rightarrow {}^4I_{15/2}$	~800 cm^{-1}	1.540 μm and 1.064 μm	530 nm/555 nm	20 - 220°C	1.02 %/°C, 0.52 %/°C at 220°C	c
Nd^{3+}: silica fiber	$(^4F_{3/2}\,;\,^4F_{5/2}) \rightarrow {}^4I_{9/2}$	~1000 cm^{-1}	802 nm	[820-840 nm]/[895-915 nm]	-50 - 500°C	1.68%/°C	d
Pr^{3+}: silica fiber	$[(^3P_1 + {}^1I_6)\,;\,^3P_0] \rightarrow {}^1G_4$	~580 cm^{-1}	488 nm	[820 - 850 nm]/[890 - 900 nm]	22 - 250 °C	0.39 %/°C	e
Pr^{3+}: aluminosilica fiber	$[(^3P_1 + {}^1I_6)\,;\,^3P_0] \rightarrow {}^1G_4$	~580 cm^{-1}	488 nm	[820 - 850 nm]/[890 - 900 nm]	-185 - 257 °C	0.14 %/°C	f
Pr^{3+}: ZBLAN glass	$[(^3P_1 + {}^1I_6)\,;\,^3P_0] \rightarrow {}^1G_4$	~580 cm^{-1}	450 nm (LED)	877 nm/906 nm	-45 - 255°C	0.48%/°C	g
Yb^{3+}: silica fiber	$^2F_{5/2} \rightarrow {}^2F_{7/2}$	Stark sublevels	810 nm	910 nm/1030 nm	20 - 600 °C	0.95 %/°C	h
Er^{3+} doped fiber	$^4I_{13/2} \rightarrow {}^4I_{15/2}$	Stark sublevels	1.48 μm	1552 nm/1530 nm	-50 - 90°C	0.7 %/°C	i
Sm^{3+}: silica fiber	$(^4F_{3/2}\,;\,^4G_{5/2}) \rightarrow {}^6H_{5/2}$	~1000 cm^{-1}	476.5 nm	[520-535 nm]/[564-574 nm]	~22 - 475 °C	1.85 %/°C	j
Eu^{3+}: silica fiber	$(^5D_1\,;\,^5D_0) \rightarrow {}^7F_2$	~1750 cm^{-1}	465 nm	[552-560 nm]/[607-630 nm]	-172 - 400 °C	0.178 %/°C	k
Dy^{3+}: silica fiber	$(^4I_{15/2}\,,\,^4F_{9/2}) \rightarrow {}^6H_{13/2}$	~1000 cm^{-1}	477 nm	[530-542.5 nm]/[560-590 nm]	21.5 - 250 °C	1.05 %/°C	l
Er^{3+}: silica fiber	$(^2H_{11/2}\,;\,^4S_{3/2}) \rightarrow {}^4I_{15/2}$	~800 cm^{-1}	975 nm	[525-535 nm]/[555-565 nm]	20 - 200°C	3.5 %/°C	m

[1] Energy gap of thermally-coupled energy levels.

a (Maurice et al., 1995); b (Maurice et al., 1995); c (Dos Santos et al., 1999); d (Wade et al., 1999); e, f, j (Wade, 1999); g (Maurice et al., 1997); h (Maurice et al., 1997); i (Imai & Hokazono, 1997); k (Wade et al., 1998); l (Wade et al., 1997); m (Castrellon-Uribe & Garcia-Torales, 2010).

Table 1. Summary of the performance of rare-earth-doped fibers and materials as temperature-sensing elements based on the fluorescence intensity ratio technique.

experimentally. In the fluorescent sensor, a detection system was incorporated to interpret the temperature information encoded in the measured fluorescence spectrum. The detection system incorporated two optic channels to select the fluorescence spectral bands emitted from levels $^2H_{11/2}$ and $^4S_{3/2}$ of the erbium-doped fiber (Castrellon-Uribe, 2002, 2005).

Recently, this new method based on the analysis of radiometric figures of merit, such as the SNR, the noise equivalent power (NEP), sensitivity, and the temperature resolution (ΔT_{min}), was applied to evaluate the performance of rare-earth-doped fiber sensors (Castrellon-Uribe & Garcia-Torales, 2010). To select the optimum sensor for the monitoring of temperature *in situ*, this radiometric analysis allowed the selection of the limits of detection for these fluorescent sensors. In that work, the performance of an erbium-doped fiber as a remote temperature sensor employing the fluorescence intensity-ratio technique was analyzed. In this case, the green fluorescence signal was generated by up-conversion processes in the erbium-doped fiber pumped by a pigtail laser diode at 975 nm. A summary of the main results obtained in this investigation are presented as follows.

When an erbium-doped fiber was pumped with a photon energy of $2.028\text{x}10^{-19}$ J (λ=980 nm), the $^4I_{11/2}$ erbium level was excited through ground state absorption (GSA), and the $^4I_{13/2}$ metastable level was quasi-instantaneously populated due to non-radiative transitions. At the $^4I_{13/2}$ level, an emission to the ground state was observed around 1530 nm (near-IR). The $^4I_{11/2}$ level absorbed the pump photons and excited the $^4F_{7/2}$ level through excited state absorption (ESA). The latter process populated the $^2H_{11/2}$ and $^4S_{3/2}$ levels, which were responsible for emissions around 530 nm and 545 nm, respectively (see Fig. 5). The latter levels were said to be in quasi-thermal equilibrium because of the small energy gap between them (about 800 cm^{-1} = 1.59×10^{-20} J) in contrast to the relatively large energy difference between them and the next lowest level (about 3000 cm^{-1} = 5.9636×10^{-20} J). In silica, a fast thermal coupling between these two levels has been studied theoretically and observed experimentally (Berthou & Jorgensen, 1990; Krug et al., 1991; Maurice, 1994, 1995).

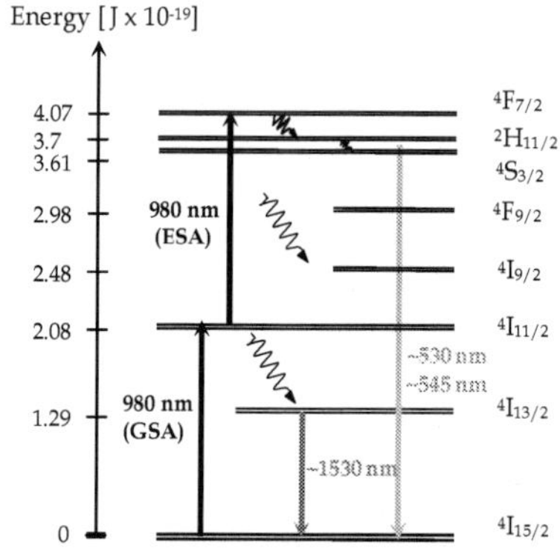

Fig. 5. Erbium energy levels diagram illustrating the excited state absorption (ESA) and the up-conversion fluorescence process. (Castrellon-Uribe & Garcia-Torales, 2010).

The ratio, R, of the intensities, I, radiating from two respective levels ($^2H_{11/2}$ and $^4S_{3/2}$) was proportional to their frequency ratio (ν), their emission cross-section ratio (σ), and the population distribution:

$$R = \frac{I(\Delta\lambda, T; {}^2H_{11/2})}{I(\Delta\lambda, T; {}^4S_{3/2})} = \frac{v({}^2H_{11/2})}{v({}^4S_{3/2})} \times \frac{\sigma({}^2H_{11/2})}{\sigma({}^4S_{3/2})} \exp\left[-\frac{\Delta E}{k \times T}\right] \quad (3)$$

Figure 6 shows the experimental setup that was used to evaluate the performance of the erbium-doped silica fiber sensor for remote temperature measurements. A pigtail laser diode with an emission at 975 nm (near-IR) was employed to excite the fluorescence of an erbium-doped (960-ppm) fiber with a length of 20 cm and a core diameter of 3.2 μm, which was located inside an enclosure whose temperature, T, was additionally monitored with a thermocouple. The green fluorescence power measured was 50 μW at 20°C for 60 mW of pump power when considering a pump power coupling efficiency to the fiber core of about 30%. A dichroic mirror transmitted the pumping infrared laser radiation and reflected the green fluorescence radiation. In the detection system, a dichroic mirror and wavelength division multiplexing (WDM) was used to separate the different spectral lines of the fluorescence-spectrum toward the two optical channels of the sensor. Interference filters with a 10-nm transmission spectral width centered on the maximum peak of transmission were employed to isolate the fluorescence spectral bands of the beam in each channel. A transducer was placed in each channel to interpret the temperature information encoded in

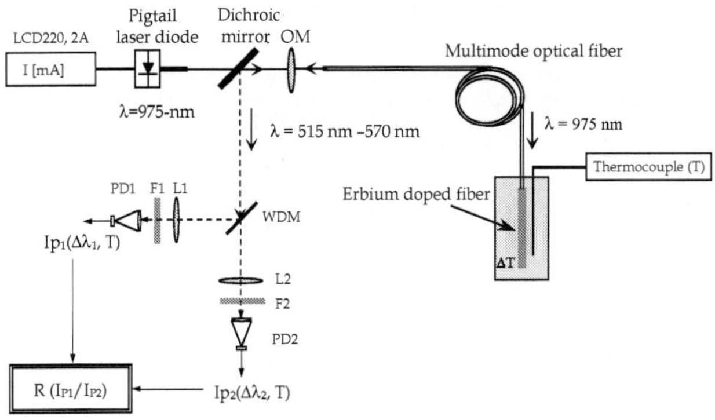

Fig. 6. Experimental setup of the erbium-doped silica fiber sensor for remote temperature measurements, employing the up-conversion fluorescence intensity ratio technique. (Castrellon-Uribe & Garcia-Torales, 2010).

the optical signal. Finally, the integrated radiation over the different wavelength intervals was detected and divided to give the spectral band power ratio. The detection system converted the measured fluorescence spectrum of the two thermally coupled energy levels ($^2H_{11/2}$ and $^4S_{3/2}$) of the erbium-doped fiber into temperature information.

Figure 7a shows the normalized fluorescence spectrum of the erbium-doped silica fiber as a function of the wavelength in the temperature interval from 20°C to 200°C. The power of the fluorescence spectrum centered at 530 nm ($^2H_{11/2}$ transition) increased with temperature, while the fluorescence spectrum centered at 545 nm ($^4S_{3/2}$ transition) decreased over the same temperature interval (see Fig. 7a). Figure 7b shows the measured power ratio (photocurrent-ratio measured in the detection system) as a function of temperature for the different fluorescence spectral bands integrated over the 10-nm width determined by the interference filters. The power ratios for a number of possible different fluorescence spectral bands considered for use in the erbium-doped fiber as remote temperature sensors were analyzed. The power ratio varied roughly linearly with the temperature in the interval from 20°C to 200°C with different slopes and a nearly linear increase in the y-intercepts (see Fig. 7b).

Afterward, the sensitivity of the sensor, *S(R)*, was evaluated as the ratio of the change in intensity integrated over the spectral bands, $\Delta R(I_1/I_2)$, to an increase in its temperature signal input, ΔT_{fiber}. The expression used to evaluate the sensitivity of the sensor was as follows:

$$S(R) = \frac{\Delta R\left[\frac{I_{p1}(\Delta\lambda_1, T)}{I_{p2}(\Delta\lambda_2, T)}\right]}{\Delta T_{fiber}} \quad [1/°C] \tag{4}$$

where $I_{p1}(\Delta\lambda_1, T)$ is the photocurrent of the channel 1 ($^2H_{11/2}$ transition) for the different spectral bands as a function of the temperature, and $I_{p2}(\Delta\lambda_2, T)$ is the photocurrent of the

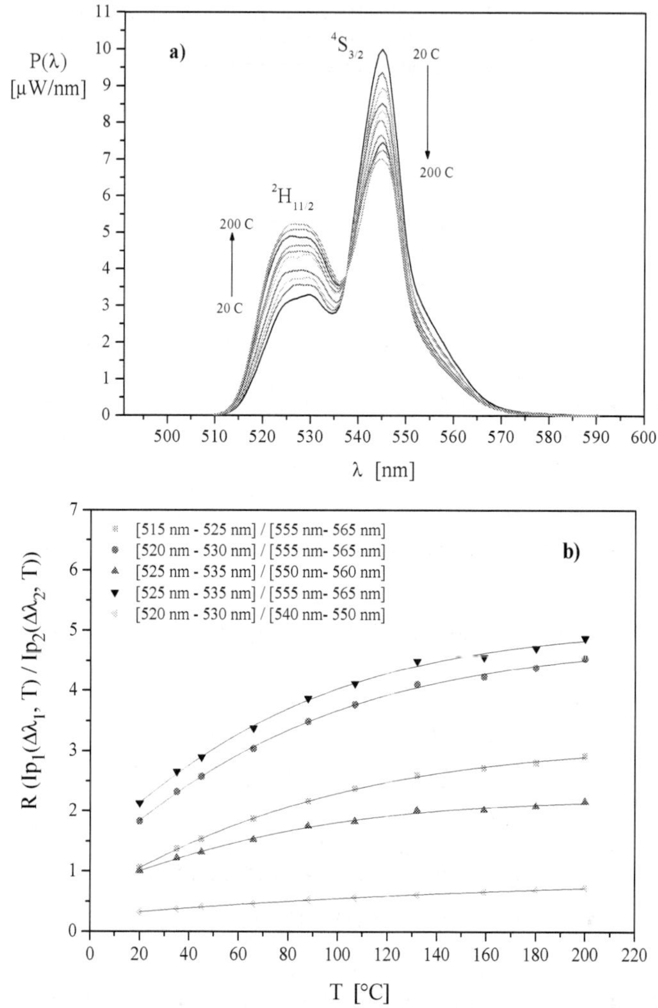

Fig. 7. a) Measured fluorescence spectrum of the erbium-doped silica fiber for different temperature values and b) measured power ratio for different fluorescence spectral bands, as a function of temperature. (Castrellon-Uribe & Garcia-Torales, 2010).

channel 2 ($^4S_{3/2}$ transition) for the different spectral bands as a function of the temperature. ΔT_{fiber} is the temperature change in the erbium-doped fiber.

The sensor sensitivity with the spectral bands [525 nm - 535 nm] / [555 nm - 565 nm] and [520 nm - 530 nm] / [555 nm - 565 nm] changed from approximately 35x10^{-3}/°C to 9x10^{-3}/°C and from approximately 33x10^{-3}/°C to 8x10^{-3}/°C, respectively. In addition, the sensitivities for the spectral intervals [515 nm - 525 nm] / [555 nm - 565 nm] and [525 nm -

535 nm] / [550 nm - 560 nm] changed from about $21x10^{-3}/°C$ to $6x10^{-3}/°C$ and from about $15x10^{-3}/°C$ to $4x10^{-3}/°C$, respectively. It was concluded that the sensor sensitivity exponentially decreases with an increase in the temperature.

Nevertheless, considering that the main characteristics for the best performance of any fiber optic sensor are a high SNR and excellent sensitivity, the authors also proposed to use the ratio of powers of spectral bands [520 nm - 530 nm] / [540 nm - 550 nm] with sensitivities from approximately $4x10^{-3}/°C$ to $2x10^{-3}/°C$ in the temperature interval of 20°C - 200°C. These spectral bands exhibited smaller sensitivities and power ratio slopes than the others. However, they had a very high SNR and responsivity because these spectral bands corresponded with the maximum peaks of fluorescence for the $^{2}H_{11/2}$ and $^{4}S_{3/2}$ transitions (channels of the sensor).

Finally, the optimal spectral bands proposed to use in the sensor were [520 nm - 530 nm] and [525 nm - 535 nm] ($^{2}H_{11/2}$ transition) of the erbium-doped fiber with signal-to-noise ratios of 110 dB and 111 dB, respectively, at 20°C; while for the spectral bands [540 nm – 550 nm] and [555 nm – 565 nm] ($^{4}S_{3/2}$ transition) of the erbium-doped fiber, the signal-to-noise ratios were 120 dB and 104 dB, respectively, at 20°C. The highest sensitivity obtained for the sensor was from approximately $35x10^{-3}/°C$ to $10x10^{-3}/°C$ for the temperature interval of 20°C - 200°C. Therefore, radiometric analysis is a powerful tool for predicting and comparing the performance of fiber optic sensors, and it allows one to determine the optimum sensor for specific applications.

5. New electro-optical materials for applications in chemical sensing

The development of new materials with non-linear optical properties (NLO) has been one of the main objectives of research and development in the field during the past few decades, due to their important applications mainly in photonics (Nalwa, 2001). The organic second-nonlinear optical materials have been widely investigated because of their great potential applications in optoelectronic devices and optical information processing, and many new NLO materials have been prepared and researched (Dalton, 1995; Yesodha et al., 2004). Generally, the organic materials are composed of a polymeric matrix in which the chromophores are distributed and produce the non-linear optical properties.

Polymers are normally used in electrical and electronic applications as insulators due mainly to the intrinsic property of covalent bonding present in most commodity plastics. These polymers with localized electrons are incapable of providing electrons as charge carriers or a path for other charge carriers to move along the chain. However, polymers are also widely exploited because of their special characteristics, such as low density, mechanical strength, ease of fabrication, flexibility in design, stability, resistance to corrosion, and low cost. Thanks to the investigations conducted by Shirakawa, Heeger, and Mac Diarmid since 1997 (prizewinners of the 2002 Nobel Prize in Chemistry), these polymers can also be synthesized in their conductive form (Shirakawa, 1977a, 1977b). Therefore, conjugated semiconducting polymers are a novel class of materials that combine optical properties with the electronic properties of semiconductors.

Since the early 1980s, conducting polymers, such as polypyrrole (PPy), polyaniline (Pani), polythiophene (PTh), and their derivatives, have been investigated due to their chemical, electrical, and optical properties (Skotheim & Reynolds, 2007). The conducting polymers are easy to synthesize through chemical or electrochemical processes, and their molecular chain structures can be conveniently modified by copolymerization or structural derivations. One of the most important characteristics of these conducting polymers is their capacity to be oxidized or reduced when they are in contact with positive or negative ions. The change from the conductive state (oxidized) to the non-conductive (reduced) state of the polymer is reversible and is associated with its *redox* property. In addition, the conducting polymers combine interesting optical and electrical properties, such as photoluminescence (PL) and electroluminescence (EL), making them attractive for applications in optoelectronics.

Luminescence: Luminescence is defined as the de-excitation of an atom or molecule by the emission of photons. According to the origin of the excitation, the luminescent process can be photoluminescence, electroluminescence, chemo-luminescence, bioluminescence, or incandescence. Fluorescence is a photoluminescence in which the molecular absorption of a photon triggers the emission of a photon with a longer wavelength (less energetic). The luminescence can be classified according to the duration of the emission after the excitation. When the excitation is suspended, an exponential decay of the emitted light occurs. The luminescent process is called fluorescence when the time of decay of the emission has a duration on the order of 10^{-3} s or less; for decay times greater than this value, the process is called phosphorescence (Lakowicz, 2006). The conjugated polymers based on the luminescence can be used for several applications, particularly in chemical sensors (Lange et al., 2008; Liu et al., 2009).

Electroluminescence: The electroluminescent conjugated polymers are materials that emit light when they are excited by the flow of an electric current. Conjugated polymers are particularly versatile because their physical properties, such as color and emission efficiency, can be fine-tuned by the manipulation of their chemical structures. The research on these new fluorescent materials has contributed to the development of organic light-emitting diodes (Akcelrud, 2003; Friend et al., 1999; Kraft et al., 1998). Organic thin-film electroluminescence devices were developed in the 1980s by Tang and Van Slyke (Tang & Van Slyke, 1987) and Saito and Tsutsui et al. (Adachi et al., 1988).

In recent years, research has been focused on thiophene-based polymers due to their structural versatility, solubility upon functionalization, and environmental stability (Chan & Ng, 1998). The polythiophenes are electroluminescent and photoluminescent materials, and their electro-optical properties are of considerable interest due to their potential applications, particularly as fluorescent chemical sensors based on fluorescence quenching (Li et al., 2005; Marti, 2009a, 2009b; Somanathan & Radhakrishnan, 2005; Tang et al., 2006).

Generally, the polythiophenes are excited with UV radiation, and the fluorescence signal is observed in the visible region of the electromagnetic spectrum. Fluorescence quenching refers to any process that decreases the fluorescence intensity of a sample. There are a wide variety of quenching processes; they include excited state reactions, molecular rearrangements, ground state complex formation, and energy transfer (Lakowicz, 2006).

In these conducting polymers, the quenching efficiency increases with an increasing tendency of the polymer to associate with the quencher in solution. This association can occur either through the formation of a non-luminescent complex between the polymer and the quencher (static quenching) or through collisions between the photo-luminescent macromolecule and the quencher (dynamic quenching).

In general, conjugated polymers have become an important class of materials employed in a wide variety of applications, including light-emitting diodes (LEDs) (Adachi et al., 1988; Akcelrud, 2003; Friend et al., 1999; Kraft et al., 1998; Tang & Van Slyke, 1987), light-emitting electrochemical cells (LECs) (Pei et al., 1995), plastic lasers (Hide et al., 1997), solar cells (Gunes et al., 2007), field effect transistors (FETs) (Sirringhaus, 2005), and more recently, chemical or biological sensors (Achyuthan et al., 2005; Castrellon-Uribe et al., 2009; Liu & Bazan, 2004; McQuade et al., 2000; Pinto & Schanze, 2002; Thomas et al., 2007).

Particularly, conducting polymers, such as polypyrrole, polyaniline, polythiophene, and their derivatives, have been investigated and used as the sensitive materials for developing gas sensors (Ameer & Adeloju, 2005; Bai & Shi, 2007; Maksymiuk, 2006; Nicho et al., 2001; Rahman et al., 2008).

Sensors composed of conducting polymers have important characteristics, such as high sensitivities and short response times. Conducting polymers are easy to synthesize through chemical or electrochemical processes, and their molecular chain structure can be conveniently modified by copolymerization or structural derivations. Furthermore, conducting polymers have good mechanical properties, which allow for the facile fabrication of sensors. Chemical sensors are devices that allow the continuous and reversible measurement of chemical parameters. The conducting polymers are conjugated macromolecules that exhibit electrical and optical property changes when they are protonated/deprotonated by certain chemical agents. In recent years, conducting polymers, such as polypyrrole and polyaniline (PANI), have been proposed as chemical sensors based on the changes in their electric conductivity when they are exposed to ammonia (Agbor et al., 1995; Brie et al., 1996; Koul & Chandra, 2005).

Recently, sensors of polyaniline films that are based on the change in their optical absorption have been investigated for the measurement of ammonia (Jin et al., 2001; Lee et al., 2003; Nicho et al., 2001). In these chemical sensors, the optical absorption at a wavelength of about 630 nm changes with an increasing ammonia concentration. Nevertheless, these sensing materials are not suitable to carry out remote measurements with optical fibers because the attenuation of the multimode fibers is 9 dB/km at 600 nm and 1 dB/km at 980 nm (Keiser, 1991).

Recently, a study of the optical response of polyaniline films that had been exposed to low concentrations of aqueous ammonia was reported (Castrellon-Uribe et al., 2009). The synthesis of the PANI films was carried out by the chemical bath method. In that work, polyaniline films were exposed to different concentrations of aqueous ammonia (10–4000 ppm), and their optical transmittances were measured in the wavelength interval of [350-1100 nm] to determine their optical sensitivities. In addition, an optical sensor system was developed based on the power ratio of transmittance for monitoring low concentrations of aqueous ammonia; it employed a polyaniline film, a pigtailed laser diode at 975 nm, photodetectors, and a multimode optical fiber.

Figure 8 shows the laser sensor system based on the optical power of transmittance for the optical detection of ammonia with PANI films. Generally, optic sensors that are based on change in intensity are susceptible to the variation of the optical signal of the source, causing errors in the measurement. Thus, the authors proposed the use of the optical power ratio technique to carry out the remote optical detection of ammonia.

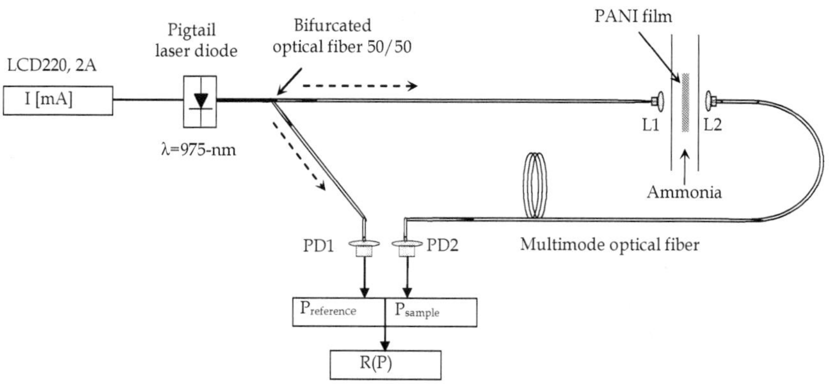

Fig. 8. Laser sensor system for the remote optical detection of ammonia with PANI films employing the optical transmittance ratio technique. (Castrellon-Uribe et al., 2009).

To evaluate the optical response of the PANI films, the sample was exposed to 4000 ppm of aqueous ammonia; immediately, the PANI film showed a chromatic change (from green to blue) when in contact with the ammonia. The PANI (EB) samples (blue color) were then treated with 0.2 M hydrochloric acid to chemically return them to their (ES) state (green color). When the PANI (ES) film was exposed to a basic solution, such as ammonia, it underwent a deprotonation process and was converted to an emeraldine base (EB) state with a blue color. In contrast, if the reaction medium was acidic, such as with hydrochloric acid, the polymer was in a protonated state, known as the emeraldine salt (ES), which had a green color. The optical transmittance of the PANI film in the (ES) and (EB) states is observed in Figure 9a. Afterward, the optical transmittance of the PANI film was measured to determine its optical sensitivity to different concentrations of ammonia (see Fig. 9b). In the visible region (VISR), the signal of transmittance showed a gradual shift in wavelength with increasing ammonia concentrations. The PANI (ES) film exposed to different concentrations of aqueous ammonia presented a better optical response at the wavelength centered at 975 nm (NIR), as observed in Figure 9b.

The response time and the recovery time of the PANI film when in contact with the ammonia and its regeneration in hydrochloric acid were also investigated. The response time and the recovery time of the PANI film exposed to a basic solution (such as ammonia) and an acid medium (such as hydrochloric acid) were less than 10 sg at room temperature. The response of the PANI film when exposed to aqueous ammonia, as well as its recovery when regenerated in hydrochloric acid, was immediate, as shown in Figure 10a. Finally, the calibration curve of the optical sensor system was obtained from the change in the power ratio of transmittance ratio, ($P = P_{Sample}/P_{ref}$), at different concentrations of ammonia as is observed in Figure 10b.

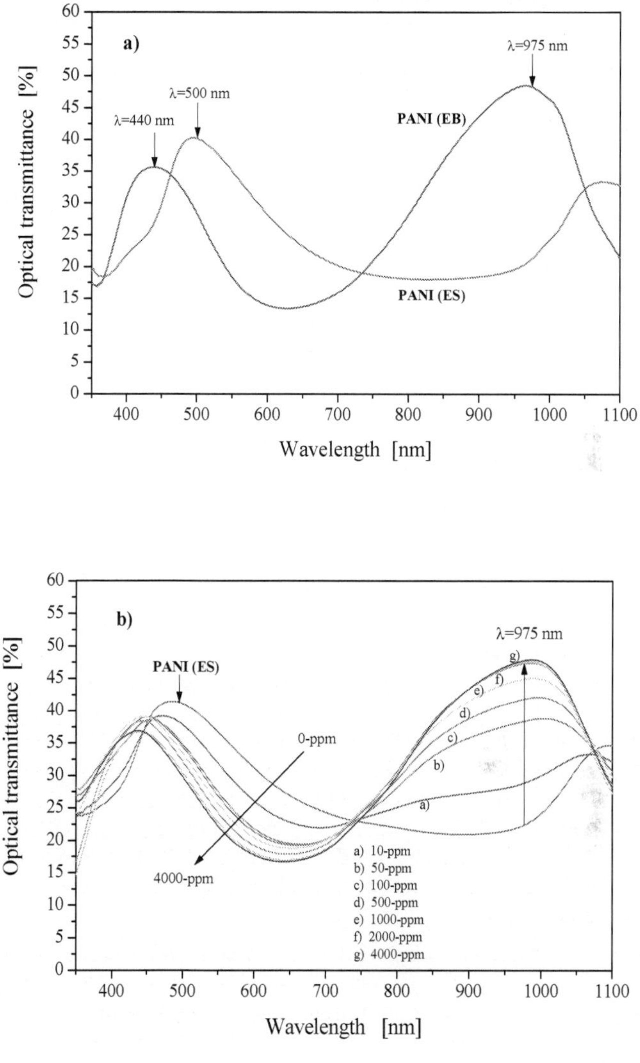

Fig. 9. a) Optical transmittance of the PANI film upon protonation and deprotonation reactions and b) optical response of the PANI film exposed to different concentrations of aqueous ammonia as a function of wavelength. (Castrellon-Uribe et al., 2009).

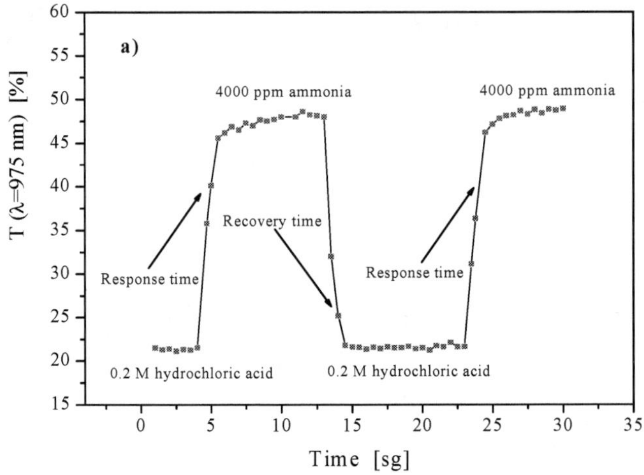

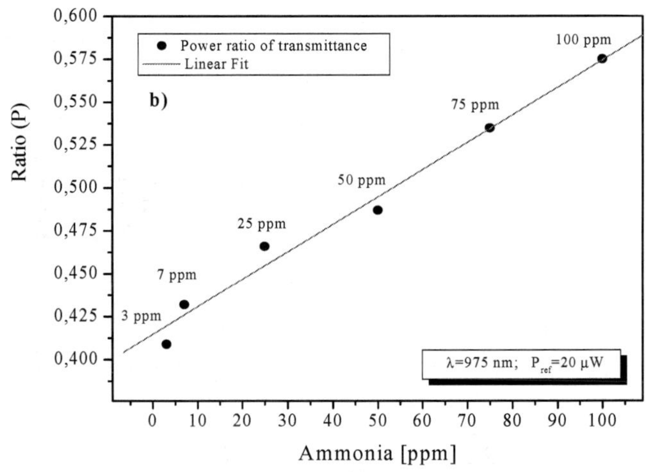

Fig. 10. a) Response time and recovery time curve of a polyaniline film measured at 975 nm and b) calibration curve of the power ratio of transmittance of the optical sensor system employing a pigtail laser diode at 975 nm. (Castrellon-Uribe et al., 2009).

6. Conclusions

In conclusion, the main advantages of the optical sensor system proposed for monitoring ammonia with PANI films are the following: its insensitivity to the intensity variations of the optical source signal, which helps to avoid errors in measurements; its simple detection system of the signal with the ammonia information; and the possibility of utilizing a light-emitting diode (LED) as the optical source instead of a laser diode. Therefore, the feasibility of employing polyaniline polymers in the development of intrinsic optical fiber sensors for the remote optical detection of ammonia was shown.

The development and commercialization of optical fiber sensors has increased in recent years. The area of application of optical fiber sensors is now well identified, and its extension toward sensor systems optoelectronics has contributed to a wide range of applications in diverse fields. However, the continuous technological progress in diverse fields establishes new challenges for the development and instrumentation of reliable optical fiber sensor systems and devices with high performance. The investigation and development of new materials that combine electrical and optical properties, such as the conductive polymers, open the possibility of new optoelectronic devices, such as sensor systems and their implementation with optical fibers (Cao & Duan, 2005; Castrellon-Uribe et al., 2009; Christie et al., 2003; Scorsone et al., 2003).

7. References

Achyuthan, K.E. Bergstedt, T.S. Chen, L. Jones, R.M. Kumaraswamy, S. Kushon, S.A. Ley, K.D. Lu, L. McBranch, D. Mukundan, H. Rininsland, F. Shi, X. Xia, W. & Whitten, D.G.(2005). Fluorescence super quenching of conjugated polyelectrolytes: applications for biosensing and drug discovery. *Journal of Materials Chemistry,* Vol. 15, (April 2005), pp. 2648–2656, ISSN 0959-9428.

Adachi, C. Tokito, S. Tsutsui, T. & Saito, S. (1988). Electroluminescence in organic films with three-layer structure. *Jpn. Journal of Applied Physics,* Vol. 28, L269-L271, ISSN 0021-4922.

Agbor, N. Petty, M. & Monkman, A. (1995). Polyaniline thin films for gas sensing. *Sensors & Actuators B: Chemical,* Vol. 28, No. 3, (Oct. 1995), pp. 173–179, ISSN 0925-4005.

Akcelrud, L. (2003). Electroluminescent polymers. *Progress in Polymer Science,* Vol. 28, (July 2002), pp. 875–962, ISSN 0079-6700.

Ameer, Q. & Adeloju, S.B. (2005). Polypyrrole-based electronic noses for environmental and industrial analysis. *Sensors and Actuators B: Chemical,* Vol. 106, No. 2, (May 2005), pp. 541-552, ISSN 0925-4005.

Bai, H. & Shi, G.(2007). Gas Sensors Based on Conducting Polymers. *Sensors,* Vol. 7, No. 3, (March 2007), pp. 267-307, ISSN 1424-8220.

Berthou, H. & Jorgensen, C. K. (1990). Optical-fiber temperature sensor based on upconversion-excited state fluorescence. *Optics Letters,* Vol. 15, No. 19, (October 1990), pg. 1100–1102, ISSN 0146-9592.

Brie, M. Turcu, R. Neamtu, C. & Pruneanu, S. (1996). The effect of initial conductivity and doping anions on gas sensitivity of conducting polypyrrole films to NH_3. *Sensors and Actuators B: Chemical,* Vol. 37, No. 3, (Dec. 1996), pp. 119–122, ISSN 0925-4005.

Cao, W. & Duan, Y. (2005). Optical fiber-based evanescent ammonia sensor. *Sensors and Actuators B: Chemical*, Vol. 110, No. 2, (October 2005), pp. 252–259, ISSN 0925-4005.

Castrellon, J. & Paez, G. (1999). Radiometric Figures-of-Merit of a Fiber Optic Temperature Sensor, *Proceedings of SPIE Infrared Spaceborne Remote Sensing VII*, Denver Colorado, Vol. 3759, pp. 410-421, (July1999).

Castrellon-Uribe, J. & Garcia-Torales, G. (2010). Remote temperature sensor based on the up-conversion fluorescence power ratio of an erbium-doped silica fiber pumped at 975 nm. *Fiber and Integrated Optics,* Vol. 29, No. 4, (July 2010), pp. 272-283, ISSN 0146-8030.

Castrellon-Uribe, J. (2005). Experimental results of the performance of a laser-fiber as a remote sensor of temperature. *Optics and Lasers in Engineering,* Vol. 43, No. 6, (June 2005), pp. 633-644, ISSN 0143-8166.

Castrellon-Uribe, J. Cuevas-Arteaga, C. & Trujillo-Estrada, A. (2008). Corrosion monitoring of stainless steel 304L in lithium bromide aqueous solution using transmittance optical detection technique. *Optics and Lasers in Engineering*, Vol. 46, No. 6, (June 2008), pp. 469-476, ISSN 0143-8166.

Castrellon-Uribe, J. Nicho, M. E. & Reyes-Merino, G. (2009). Remote optical detection of low concentrations of aqueous ammonia employing conductive polymers of polyaniline. *Sensors & Actuators: B: Chemical*, Vol. 141, No. 1, (June 2009), pp. 40-44, ISSN 0925-4005.

Castrellon-Uribe, J. Paez, G. & Strojnik, M. (2002). Radiometric analysis of a fiber optic temperature sensor. *Optical Engineering,* Vol. 41, No. 6, (June 2002), pp. 1255-1261, ISSN 0091-3286.

Castrellon-Uribe, J. Paez, G. & Strojnik, M. (2002). Remote temperature sensor employing erbium-doped silica fiber. *Infrared Physics & Technology,* Vol. 43, No. 3-5, (June 2002), pp. 219-222, ISSN 1350-4495.

Chan, H. S. O. & Ng, S. C. (1998). Synthesis, characterization and applications of thiophene-based functional polymers. *Progress in Polymers Science,* Vol. 23, No. 7, (November 1998), pp. 1167–1231, ISSN 0079-6700.

Christie, S. Scorsone, E. Persaud, K. & Kvasnik, F. (2003). Remote detection of gaseous ammonia using the near infrared transmission properties of polyaniline. *Sensors & Actuators: B: Chemical*, Vol. 90, No. 1-3, (April 2003), pp. 163–169, ISSN 0925-4005.

Culshaw, B. (2004). Optical fiber sensor technologies: Opportunities and–perhaps-pitfalls. *Journal of Lightwave Technology*, Vol. 22, No. 1, (Jan. 2004), pp. 39-50, ISSN 0733-8724.

Dalton, L.R. Harper, A.W. Ghosn, R. Steier, W. H. Ziari, M. Fetterman, H. Shi, Y. Mustacich, R.V. Jen, A. K.-Y. & Shea, K.J. (1995). Synthesis and Processing of Improved Organic Second-Order Nonlinear Optical Materials for Applications in Photonics. *Chemistry of Materials,* Vol. 7, No. 6, (June 1995), pp 1060–1081, ISSN 0897-4756.

Desurvire, E. (1994). *Erbium Doped Fiber Amplifiers: Principles and Applications* (1 ed.), John Wiley & Sons Inc, ISBN 0471589772, New York USA.

Digonnet, M. J. F. (2001). *Rare Earth Doped Fiber Lasers and Amplifiers,* (2 ed), Marcel Dekker Inc, ISBN 0824704584, New York USA.

Dong S, Liao Y, & Tian Q. (2005). Intensity-based optical fiber sensor for monitoring corrosion of aluminium alloys. *Applied Optics,* Vol. 44, No. 27, (September 2005), pp. 5773–7, ISSN 0003-6935.

Dong S, Liao Y, & Tian Q. (2005). Sensing of corrosion on aluminium surfaces by use of metallic optical fiber. *Applied Optics*, Vol. 44, No. 30, (October 2005), pp. 6334-6337, ISSN 0003-6935.

Dos Santos, P.V. de Araujo, M.T. Gouveia-Neto, A. S. Medeiros Neto, J.A. & Sombra, A.S.B. (1999). Optical thermometry through infrared excited upconversion fluorescence emission in Er^{3+} and Er^{3+}- Yb^{3+} -doped chalcogenide glasses. *IEEE Journal of Quantum Electronics*, Vol. 35, No. 2, (February 1999), pp. 395-399, ISSN 0018-9197.

Farries, M. C. Fernmann, M. E. Laming, R. I. Poole, S. B. Payne, D. N. & Leach, A.P. (1986). Distributed temperature sensor using Nd^{3+}-doped optical fibre. *Electronics Letters*, Vol. 22, No.8, (April 1986), pp. 418-419, ISSN 0013-5194.

Friend, R. H. Gymer, R. W Holmes, A. B. Burroughes, J. H. Marks, R. N. Taliani, C. Bradley, D. Dos Santos, D. A. Bredas, J. L. Logdlund M. & Salaneck, W. R. (1999). Electroluminescence in conjugated polymers. *Nature*, Vol. 397, (Jan 1999), pp. 121-128.

Ghatak, A. & Thyagarajan, K. (2000). *An Introduction to Fiber Optics*. (1 ed), Cambridge University Press, ISBN 0521571200, New York USA.

Grattan, K. T. V. & Zhang, Z. Y. (1995). *Fiber Optic Fluorescence Thermometry*. (1 ed), Chapman & Hall, ISBN 0412624702, London.

Gunes, S. Neugebauer, H. & Sariciftci, N.S. (2007). Conjugated polymer-based organic solar cells. *Chemical Reviews*, Vol. 107, No. 4, (April 2007), pp. 1324-1338, ISSN 0009-2665.

Habib, K. (1993). Model of holographic interferometry of anodic dissolution of metals in aqueous solution. *Optics and Laser in Engineering*, Vol. 18, No. 2, (August 2002), pp. 115-20, ISSN 0143-8166.

Habib, K. (1995). Non-destructive evaluation of metallic electrodes under corrosion fatigue conditions by holographic interferometry. *Optics and Laser in Engineering*, Vol. 23, No. 1, (January 2000), pp. 65-70. ISSN 0143-8166.

Hide, F. Diaz-Garcia, M.A. Schwartz, B.J. & Heeger, A.J. (1997). New developments in the photonic applications of conjugated polymers. *Accounts of Chemical Research*, Vol. 30, No. 10, (January 1997), pg. 430-436, ISSN 0001-4842.

Imai, Y. & Hokazono, T. (1997). Fluorescence-based temperature sensing using erbium-doped optical fibers with 1.48 μm pumping. *Optical Review*, Vol. 4, No. 1A, (January 1997), pp.117-120, ISSN 1340-6000.

Jin, Z. Su, Y. & Duan, Y. (2001). Development of a polyaniline-based optical ammonia sensor. *Sensors & Actuators B*, Vol. 72, No. 1, (Jan. 2001), pp. 75-79, ISSN 0925-4005.

Keiser, G. (1991). *Optical Fiber Communications*. (2 ed), McGraw-Hill, ISBN 0070336172, New York USA.

Koul, S. & Chandra, R. (2005). Mixed dopant conducting polyaniline reusable blend for the detection of aqueous ammonia. *Sensors and Actuators B: Chemical*, Vol. 104, No. 1, (January 2005), pp. 57-67, ISSN 0925-4005.

Kraft, A. Grimsdale, A. & Holmes, A. (1998). Electroluminescent Conjugated Polymers-Seeing Polymers in a New Light. *Angewandte Chemie International Edition*, Vol. 37, No. 4, pp. 402-428, ISSN 1433-7851

Krohn, D. A. (1999). *Fiber optic sensors, Fundamental and Applications*, (3 ed), Instrument Society of America, ISBN 9781556170102, USA.

Krug, P.A. Sceats, M.G. Atkins, G.R. Guy, S.C. & Poole, S.B. (1991). Intermediate excited-state absorption in erbium-doped fiber strongly pumped at 980 nm. *Optics Letters*, Vol. 16, No. 24, (December 1991), pp. 1976-1978, ISSN 0146-9592.

Lakowicz, J. R. (2006). *Principles of Fluorescence Spectroscopy* (3 ed), Springer, ISBN 0387312781, New York USA.

Lange, U. Roznyatovskaya, N.V. & Mirsky, V.M. (2008). Conducting polymers in chemical sensors and arrays. *Analytica chimica acta*, Vol. 614, No. I-26, (March 2008), pp. 1–26, ISSN 0003-2670.

Lee, Y.S. Joo, B.S. Choi, N.J. Lim, J.O. Huh, J.S. & Lee, D. D. (2003). Visible optical sensing of ammonia based on polyaniline film. *Sensors and Actuators B: Chemical*, Vol. 93, No. 1-3, (August 2003), pp. 148–152, ISSN 0925-4005.

Li, C. Numata, M. Takeuchi, M. & Shinkai, S. (2005). A sensitive colorimetric and fluorescent probe based on a polythiophene derivative for the detection of ATP. *Angewandte Chemie International Ed.*, Vol. 44, No. 39, (Sep. 2005), pp. 6371–6374, ISSN 1521-3773.

Liu, B. & Bazan, G.C. (2004). Homogeneous fluorescence-based DNA detection with water-soluble conjugated polymers. *Chemistry of Materials*, Vol. 16, No. 23, (September 2004), pp. 4467–4476, ISSN 0897-4756.

Liu, Y. Ogawa, K. & Schanze, K. S. (2009). Conjugated polyelectrolytes as fluorescent sensors. *Journal of Photochemistry and Photobiology C: Photochemistry Reviews*, Vol. 10, (October 2009), pp. 173–190, ISSN 1389-5567.

Lopez-Higuera, J. M. (2002). *Handbook of Optical Fibre Sensing Technology*, (1 Ed), John Wiley & Sons, ISBN 0471820539, New York USA.

Maiti, J. & Dolui, S.K. (2009). Photoluminescence properties of poly(thiophene-3yl-acetic acid 8-quinolinyl ester) in solution and in acid medium. *Journal of Luminescence*, Vol. 129, (January 2009), pp. 611–614, ISSN 0022-2313.

Maiti, J. Pokhrel, B. Boruah, R. & Dolui, S. K. (2009). Polythiophene based fluorescence sensors for acids and metal ions. *Sensors and Actuators B: Chemical*, Vol. 141, (July 2009), pp. 447–451, ISSN 0925-4005.

Maksymiuk, K. (2006). Chemical reactivity of polypyrrole and its relevance to polypyrrole based electrochemical sensors. *Electroanalysis*, Vol. 18, No. 16, (July 2006), pg. 1537-1551, ISSN 1521-4109.

Maurice, E. Monnom, G. Baxter, G.W. Wade, S.A. Petreski, B.P. & Collins, S.F. (1997). Blue light-emitting-diode-pumped point temperature sensor based on a fluorescence intensity ratio in Pr^{3+}: ZBLAN glass. *Optical Review*, Vol. 4, No. 1A, (January 1997), pp. 89-91, ISSN 1340-6000.

Maurice, E. Monnom, G. Dussardier, B. Saissy, A. Ostrowsky, D. B. & Baxter, G. W. (1994). Thermalization effects between upper levels of green fluorescence in Er-doped silica fibers. *Optics Letters*, Vol. 19, No. 13, (July 1994), pp. 990–992, ISSN 0146-9592.

Maurice, E. Monnom, G. Ostrowsky, D.B. & Baxter, G.W. (1995). High dynamic interval temperature point sensor using green fluorescence intensity ratio in erbium-doped silica fiber. *Journal of Lightwave Technology*, Vol. 13, No. 7, (July 1995), pp. 1349-1353, ISSN 0733-8724.

Maurice, E. Monnom, G. Ostrowsky, D.B. & Baxter, G.W. (1995). 1.2-μm transitions in erbium-doped fibers: the possibility of quasi-distributed temperature sensors. *Applied Optics*, Vol. 34, No. 21, (July 1995), pp. 4196-4199, ISSN 0003-6935.

Maurice, E. Wade, S.A. Collins, S.F. Monnom, G. & Baxter, G.W. (1997). Self-referenced point temperature sensor based on a fluorescence intensity ratio in Yb^{3+}-doped silica fiber. *Applied Optics*, Vol. 36, No. 31, (November 1997), pp. 8264-8269, ISSN 0003-6935.

McQuade, D.T. Pullen, A.E. & Swager, T.M. (2000). Conjugated polymer-based chemical sensors. *Chemical Reviews*, Vol. 100, No. 7, (Sep. 2000), pp. 2537–2574, ISSN 0009-2665.

Mejia, E.B. Senin, A.A. Talmadge, J.M. & Eden, J.G. (2002). Gain and saturation intensity of the green Ho:ZBLAN upconversion fiber amplifier and laser. *IEEE Photonics Technology Letters*, Vol. 14, No. 11 (November 2002), pp. 1500-1502, ISSN 1041-1135.

Nalwa, H. S. (2001). *Handbook of advanced electronic and photonic materials and devices*. (1 ed), Academic Press, ISBN 0125137583, New York USA.

Nicho, M.E. Trejo, M. García-Valenzuela, A. Saniger, J.M. Palacios, J. & Hu, H. (2001). Polyaniline composite coating interrogated by a nulling optical-transmittance bridge for sensing low concentrations of ammonia gas. *Sensors and Actuators B: Chemical*, Vol. 76, No. 1-3, (June 2001), pp. 18–24, ISSN 0925-4005.

Othonos, A. & Kalli, K. (1999). *Fiber Bragg Gratings: Fundamentals and Applications in Telecommunications and Sensing*, (1 ed), Hartech House INC, ISBN 0890063443, USA.

Pei, Q.B. Yu, G. Zhang, C. Yang, Y. & Heeger, A.J. (1995). Polymer light-emitting electrochemical-cells. *Science*, Vol. 269, No. 5227, pp. 1086–1088.

Pinto, M.R. & Schanze, K.S. (2002). Conjugated polyelectrolytes: synthesis and applications. *Synthesis*, Vol. 2002, No. 9, pg. 1293–1309, ISSN 0039-7881.

Rahman, M. A. Kumar, P. Park, D. & Shim, Y. (2008). *Electrochemical Sensors Based on Organic Conjugated Polymers. Sensors*, Vol. 8, No. 1, (Jan. 2008), pp. 118-141, ISSN 1424-8220.

Rai, V. K. & Rai, S.B. (2007). A comparative study of FIR and FL based temperature sensing schemes: an example of Pr^{3+}. *Applied Physics B: Lasers and optics*, Vol. 87, No. 2, (April 2007), pg. 323-325, ISSN 0946-2171.

Rai, V.K. (2007). Temperature sensors and optical sensors. *Applied Physics B: Lasers and Optics*, Vol. 88, No. 2, (July 2007), pp. 297-303, ISSN 0946-2171.

Scorsone, E. Christie, S. Persaud, K.C. Simon, P. & Kvasnik, F. (2003). Fibre-optic evanescent sensing of gaseous ammonia with two forms of a new near-infrared dye in comparison to phenol red. *Sensors and Actuators B: Chemical*, Vol. 90, No. 1-3, (April 2003), pp. 37–45, ISSN 0925-4005.

Shirakawa, H. Chiang, C. K. Fincher, C. R. Park, Y. W. Heeger, A. J. Louis, E. J. Gau, S. C. &Mac Diarmid, A. G. (1977). Electrical conductivity in doped poluyacetylene. *Physical Review Letters*, Vol. 39, No. 17, (October 1977), pp. 1098-1101, ISSN 0031-9007.

Shirakawa, H. Louis, E. Mac Diarmid, A. Chiang, C. & Heeger, A. (1977). Synthesis of Electrically Conducting Organic Polymers: Halogen Derivatives of Polyacetylene, $(CH)_x$. *Journal of the Chemical Society, Chemical Communications*, Vol. 477, pp. 578-580.

Sirringhaus, H. (2005). Device physics of solution-processed organic field-effect transistors. *Advanced Materials*, Vol. 17, No. 20, (October 2005), pp. 2411–2425, ISSN 1521-4095.

Skotheim, J.T.&Reynolds, J.R. (2007). Handbook of conducting polymers: Theory, synthesis, properties, and characterization, (3 ed), CRC Press, ISBN 1574446657. N. York USA.

Somanathan, N. & Radhakrishnan, S. (2005). Optical properties of functionalized polythiophenes. *International Journal of Modern Physics B*, Vol. 19, No. 32, (November 2005), pp. 4645-4676, ISSN 0217-9792.

Talavera, D.V. & Mejia, E.B. (2005). Blue up-conversion Tm^{3+}-doped fiber laser pumped by a multiline Raman source. *Journal of Applied Physical,* Vol. 97, No. 5, (February 2005), pp. 053102-1 - 053102-4, ISSN 0021-8979.

Tang, C. W. & Van Slyke, S. A. (1987). Organic electroluminescent diodes. *Applied Physics Letters*, Vol. 51, No. 12, (September 1987), pp. 913- 915, ISSN 0003-6951.

Tang, Y. He, F. Yu, M. Feng, F. An, L. Sun, H. Wang, S. Li, Y. & Zhu, D. (2006). A reversible and highly selective fluorescent sensor for mercury (II) using Poly(thiophene)s that contain thymine moieties. *Macromolecular Rapid Communications,* Vol. 27, No. 6, (March 2006), pp. 389–392, ISSN 1521-3927.

Thomas, S.W. Joly, G.D. & Swager, T.M. (2007). Chemical sensors based on amplifying fluorescent conjugated polymers. *Chemical Reviews*, Vol. 107, No.4, (March 2007), pp. 1339–1386, ISSN 0009-2665.

Udd, E. (1991). *Fiber Optic Sensors: An Introduction for engineers and scientists,* (New ed), Wiley-Interscience, ISBN 0471830070, USA.

Wade, S. A. (1999). Temperature measurement using rare earth doped fibre fluorescence. *Ph.D. Thesis*. Victoria University, Melbourne, Australia, 1999.

Wade, S. A. Bogdanov, V. K. Collins, S. F. & Baxter, G. W. (1998). Thermalisation of the $^5D_0$ and $^5D_1$ excited states in europium-doped optical fibre, *Proceedings of the 13th National Congress of the Australian Institute of Physics*, pp. 327-341, ISBN: 1863080716, Perth Australia, 1998.

Wade, S. A. Monnom, G. Collins, S. F. & Baxter, G.W. (1997). Thermalisation of the $^4F_{9/2}$ and $^4I_{15/2}$ levels of Dy^{3+}-doped silica fibre, *11th Australian Optical Society Conference*, Adelaide, pg.WP41.

Wade, S.A. Muscat, J.C. Collins, S.F. & Baxter, G.W. (1999). Nd^{3+} doped optical fiber temperature sensor using the fluorescence intensity ratio technique. *Review Scientific Instruments*, Vol. 70, No. 11, (Nov. 1999), pp. 4279-4282, ISSN 0034-6748

Yesodha, S. K. Pillai, C.K.S. & Tsutsumi, N. (2004). Stable polymeric materials for nonlinear optics: a review based on azobenzene systems. *Progress in Polymers Science*, Vol. 29, No. 1, (January 2004), pp. 45–74, ISSN 0079-6700.

Yu, F. T. S. Yin, S. Ruffin, P. B. (2008). *Fiber Optic Sensors,* (2 ed), CRC Press Taylor&Francis Group, ISBN 9781420053654, United State of America.

2

Optical Fiber Sensing Applications: Detection and Identification of Gases and Volatile Organic Compounds

Cesar Elosua, Candido Bariain and Ignacio R. Matias
Department of Electrical and Electronic Engineering,
Public University of Navarre
Spain

1. Introduction

Optical fiber has produced an authentic revolution in the field of communications. Nowadays, the principal guided data networks of the world use this media. The great success in optical communications raises the question of whether this technology could be also used in other fields, for instance, in the development of sensors. Optical fiber exhibits small dimensions, it is light weight and it is made of an inert and abundant dielectric material, vitreous silica. Even so, regarding to the sensors domain, electronic technology is much more advanced and mature: miniaturization allows mass production and hence, electronic sensors have affordable prices. Photonic devices currently have a justifiable cost in main infrastructures and networks, but they are still less competitive than electronic sensors in applications where there is a wide-spread deployment. Even though the increasing demand for optical devices is reducing its price, some researchers believe that, to be realistic, the success in the communication field does not have to be applicable in sensor technology (Leung, 2001). This idea is far from being pessimistic: applications whose requirements could be satisfied better by the intrinsic features of optical fiber than the ones of electronic devices, have to be identified: for instance, gyroscopes are Optical Fibre Sensors (OFSs) that have been successfully used to measure rotations (Lee, 2003).

Some important physical properties of optical fiber, such as its low size and light weight have been already mentioned, but there are other relevant features:

- The raw material is vitreous silica (SiO2), which is a dielectric. It means that the fiber is immune from external electromagnetic interference. On the contrary, electronic sensors handle with electric signals and are subject to the resulting noise and cross-talk. A factory with heavy machinery or a high tension installation are places where optical fiber sensors could be a good alternative to electronic devices.
- Optical fiber sensors are passive: there is no need for any biasing electric signal to operate. Passive devices have a great autonomy because they do not have to be electrically fed: only the light source needs that. Moreover, the just mentioned feature is interesting in applications where flammable gases or vapors or even explosives are present. Mines with explosive gases or eco-plants are some examples of environments where electric signals would be dangerous.

- Attenuation suffered by light when propagating through the fiber is very low, down to 0.2 dB/km at 1550 nm. It permits the measuring point and the receiver to be separated by several tens of kilometers without amplification. As a result, it can be possible to work remotely in hazardous environments, such as applications where highly toxic wastes have to be controlled or chemical solvents are shipped.
- The signals from many sensors (up to hundreds) can be guided through the same fiber using Wavelength Division Multiplexing (WDM) techniques (Barbosa et al., 2008), which are explained later in the current chapter. Even sensors measuring different parameters (temperature, humidity, pressure, strain) could be connected to the same optical fiber bus. These features are ideal in multi sensor applications, as for example, the structural control of buildings (Rao et al., 2006).
- A very interesting property that is being studied nowadays is Distributed Sensing: it is based on using the fiber itself as a sensing element, so measurements such as temperature can be determined with a spatial resolution below one meter (Diaz et al., 2008). Distributed sensing is not offered by any electronic sensor but only by optical fiber technology; it is well orientated to structural monitoring applications.

(A) (B) (C)

Fig. 1. Some potential market niches where OFSs could be used: (A) structural controlling of bridges; (B) eco plants where methanol is generated; (C) shipping of chemical products.

The most important features of OFSs have been listed, showing some potential applications where their high initial cost would be justified. One field where this technology shows great potential is the detection of Volatile Organic Compounds (VOCs). These substances are present in daily or industrial environments: cleaning products, toxic agents or odors are some examples of VOCs mixtures (Ampuero&Bosset, 2003, Hudon et al., 2000). Although electronic devices already exist for these tasks, they show practical drawbacks such as their large size or high weight (Goschnick et al., 2005).

This chapter is focused on optical fiber sensors used to handle with VOCs because this field can take advantage of optical fiber features. The second section of the chapter shows an overview about OFSs opportunities in VOCs applications; the distinct sensing architectures and construction methods are described in Section 2 as well. The third one covers the factors related to the development of the sensors, whereas some multiplexing networks are described in section 4. Data mining processes typically used to identify VOCs are detailed in section 5 and finally, an all fiber system able to identify beverages is described in section 6.

2. Optical fiber sensors to detect and identify VOCs

Applications related with VOCs detection cover several areas such as environmental monitoring, chemical industry, safety at work or food industry, just to mention a few. OFSs offer interesting features that solve some inherent problems of electronic sensors, as it was pointed out at the introduction. In this way, they have been studied and developed since almost three decades ago. This chapter will detail some potential market niches for them.

A volatile organic compound is, by definition, an organic compound that has a vapor pressure between 0.13 kPa and 101.3 kPa: in other words, it evaporates easily. Carbon is usually the base element of these compounds, although there are other ones without it that are considered VOCs as well. These substances are highly inflammable, easy to inhale, and in many cases, toxic (depending on the VOC, concentration and exposure time).

VOC	Dangerousness	Applications
Alcohols	Inflammables Irritation	Drinks Industry
Formaldehydes	Highly Toxic	Environmental / Safety at work
Methane	Explosive	Greenhouse effect / Cattle Health
Ketones	Industrial Processes	Inflammable / Irritant
COx	Asphyxia Risk	Greenhouse Effect / Safety at work
O2	Explosion risk	Air Quality
H2	Explosion risk	Storage and Leakage Detection

Table 1. Dangerousness and applications of some important VOCs.

There is a large number of potential applications based on identifying the VOCs present in these examples. The production of certain gases, such as methane, during the digestion of cattle (Culshaw et al., 1998) is a good health indicator, and furthermore, this gas affects directly greenhouse effect. In fermentation control applications, these sensors can be useful during the elaboration of alcoholic drinks (Santos et al., 2004), in olive oil production (Stella et al., 2000) or even in maturing stages of sausages (Garcıa et al., 2003). Finally, VOCs can be used also to check the condition of foods or drinks as milk, depending on the odors produced by them (Brudzewski et al., 2004). All these tasks can be performed by human operators, but, in cases where a continuous control is mandatory, the health of the workers could get affected, and their smell sense would become damaged.

On the other hand, most chemical solvents are volatile and are widely used to synthesize products present daily: lacquers, paints, dyes, cleaning products, cosmetics, pharmaceuticals, etc… The high toxicity of VOCs and their environmental impact have forced governments to legislate their emissions (in Spain, it is detailed in Real Decreto 117/2003 del 31 de enero). From this reason, it is necessary to control these vapors concentration continuously in the places where they are handled: a long exposure of several VOCs can produce health disorders. This phenomenon is present in many office buildings: it is named the Sick Building Syndrome (SBS) (Elosua et al., 2009).

Thanks to the features that these sensors offer and compared to their electronic counterpart, OFSs have a potential market niche in VOCs detection. Although there is an evident drawback in terms of technological maturity, there are some real applications where they

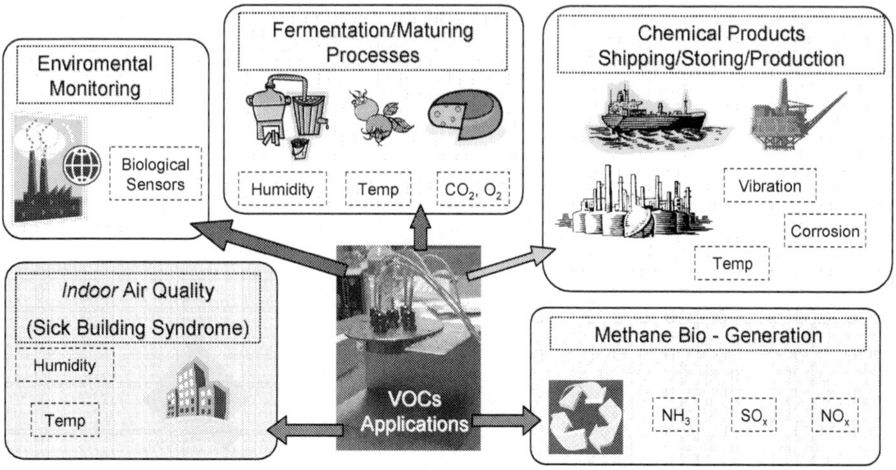

Fig. 2. Some potential market niches related with VOCs for optical fiber sensors. In each one, some other interesting physical or chemical magnitudes are pointed, so all of them could be monitored with optical fiber sensors using the same fiber network.

are successfully been used (Culshaw, 2005). Eventually, requirements of the market will determine whether the use of OFSs is generalized when working with VOCs.

2.1 Optical fiber sensors. General concepts

Briefly, a sensor is a device able to transduce a physical or chemical magnitude into a physical measurable signal; however, some authors assume that it covers the whole system, including transduction, transmission and data mining (Lopez-Higuera, 1998). Along the current work, the first definition is considered (referring just to the transducing process). Specifically, the magnitude to be measured is VOC concentration, so it has to be transduced into an optical beam (or beams) travelling through the fiber. The first classification of optical sensors can be made regarding to the parameter affected by transduction: so that, there are amplitude/intensity, phase or wavelength modulated sensors. Sometimes the signal of the sensor is conditioned by applying different modulations looking for making it more robust.

Some parameters, such as response/recovery times, should be always as short as possible, but other ones such as linearity, detection limit or selectivity will depend on the measuring conditions and the final requirements of the application. In the case of VOCs detection, applications can be divided in two main groups:

- Continuous monitoring of VOCs concentration: the concentration has to be known in real time, so the sensors have to be linear and show short response and recovery times. Otherwise, there would be a delay between the real value and the one obtained from the sensor. Moreover, cross correlation must be as low as possible and selectivity as high as possible.
- Individual solvents or VOCs mixtures identification: samples identification is achieved by specific fingerprints. These applications work in a qualitative rather than in a

quantitative way, so, linearity and selectivity are not so necessary. However, it is important that the sensor response was different for each VOC. In fact, mammals smelling sense consists of thousands receptors with overlapped selectivities: their responses conform a fingerprint for each odor (Burl et al., 2001).

2.2 Sensing topologies

An overview about the different techniques employed to detect VOCs with OFS will be exposed. In configurations based on spectrometry, each sensing device needs its own light source and detector; in other cases, no fiber is used, so the response of the sensors is first recorded by a CCD camera and then processed. This last solution is optimal for portable systems, but it is not useful in situations where remote sensing is required. In the case of the first configuration, the system is simple but shows a low scalability: the number of active devices grows with the number of sensors. On the contrary, optical fiber allows resources to be shared and also to take advantage of its remote sensing capability (Figure 3).

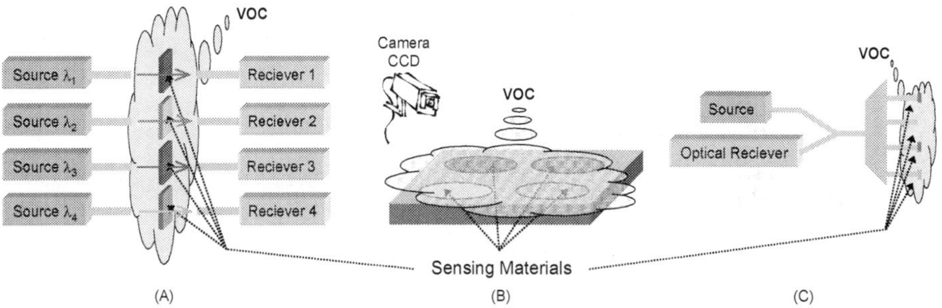

Fig. 3. Different VOCs sensors architectures: (A) one source/receiver per sensor; (B) configuration based on image processing, typically used in portable systems; (C) optical fiber multiplexing network, optimal for remote sensing applications.

There are many criteria to classify all the sensing architectures (Culshaw, 2000, Matias et al., 2006). Among all of them, the one based on where the transduction takes place will be considered in this chapter: in or onto the fiber (intrinsic sensors) or outside it (extrinsic sensors). This classification is followed by several authors (Elosua et al., 2006).

2.2.1 Extrinsic sensors

In this type of sensors, fiber acts as a transmission line. The sensing idea lies in guiding out the light from one fiber and then, coupling it back to another one after the signal interacts with the gas to detect. Extrinsic sensors are being used in real applications, multiplexing them in networks that monitor VOCs through wide extensions (Culshaw, 2004).

The way these sensors work is inspired by spectrometric techniques (Garriguesa et al., 2001). Every chemical compound has some specific absorption spectral lines: the material absorbs electromagnetic radiations at certain wavelengths. Therefore, each compound can be identified by the fingerprint formed by its spectral lines. If one of these lines falls into the transmission range of the optical fiber, the gas can be detected launching a light beam centered at this wavelength. If the target gas is present, it will absorb part of the optical signal, decreasing the power coupled to the receiver; otherwise, the signal will preserve its power level. These sensors are based on an intensity modulation, although frequency modulations have been used to make the measurements more robust (Chambers et al., 2004). The most successful application for this architecture is methane detection.

Another important feature is that a sensing unit can be used to detect different gases, optimizing resources and increasing the versatility of the whole system. It can be achieved by tuning the wavelength of the optical source to match spectral lines from different gases, which is known as Wave Spectral Modulation (WSM). This technique combines both wavelength and time division multiplexing (WDM and TDM): optical signals at different wavelengths are emitted at different time slots. Cross correlation with other gasses can be decreased by using modulations (Whitenett et al., 2004).

Other extrinsic configurations are based on guiding the optical signal from the fiber to a substrate doped with a material sensitive to the VOC to be detected (using the transmitted or reflected signal). Sometimes, if the distance between light source - substrate - receiver is short enough, no optical fiber is needed: with a certain incident angle, the reflected beam is directly coupled to the receiver. Some portable devices to detect formaldehydes have been fabricated following the option that uses no fiber (Kawamura et al., 2006).

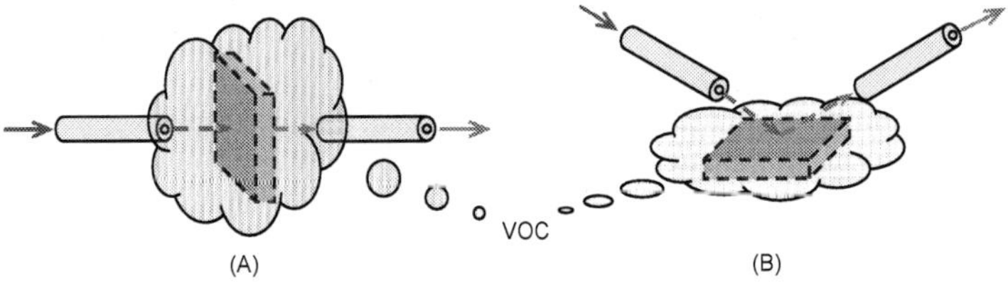

Fig. 4. Examples of extrinsic sensors with a sensing material: (A) the optical signal passing through the substrate with the sensing material is used; (B) the reflected signal is registered.

2.2.2 Intrinsic sensors

In this case, transduction takes place on the fiber, either because it is directly affected by VOCs or because a sensing material is fixed onto it. This category can be also organized in several subgroups, depending on the kind of transduction.

Transmission configurations

Most research efforts during the nineteen nineties were focused on this type of sensors (Khijwania&Gupta, 1999). The optical signal passes through the sensor: it is transmitted across it. The working principle is based on replacing a cladding segment by a material sensitive to the VOC to be detected. In this way, any change suffered by this new cladding will affect the light passing through that segment. For that reason, the sensitivity of the sensor depends on the optical power coupled to the evanescent field and its penetration depth into the modified cladding.

Light propagation through an optical fiber depends on the ratio between the refractive index of the core (n_1) and one of the cladding (n_2). The modified cladding has a variable refractive index n_{2m} and the last index to be taken into account is the one of the surrounding environment n_a. More rigorously, n_{2m} has a real part that affects the incident angle (used in Snell law) and an imaginary one that determines the optical absorption of the modified cladding. Depending on the ratio between these refraction indexes, the optical power transmitted through the sensor will vary (Figure 5), so this is the transduction principle.

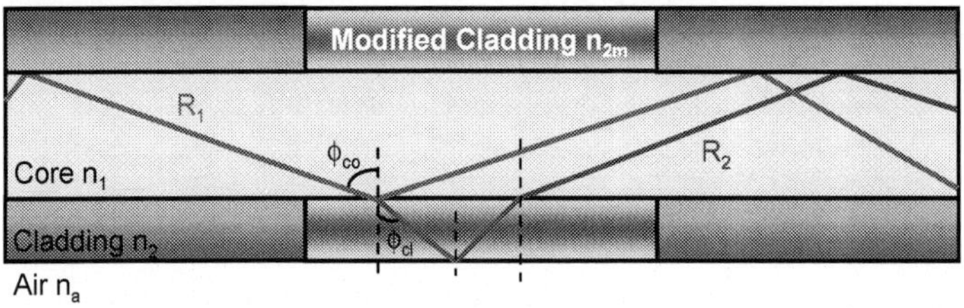

Fig. 5. Ray Theory applied to Evanescent Field Weak Guiding.

The modified cladding fiber can be bended or tapered in order to increase the final sensitivity of the sensor (Scorsone et al., 2004, Tao et al., 2006). Special fibers are also employed to implement sensors: Hollow Core Fibers (Smith et al., 2003) or Photonic Crystal Fibers (Frazao et al., 2008) have been used to develop sensors showing a high sensitivity. Even photonic devices typically used in telecommunication networks, as Bragg Grattings, can be used to implement sensors (Tao et al., 2004). Moreover, other kind of effects such as electromagnetic resonances, can take place when depositing metal oxides nanolayers. In this case, the transduction alters the frequency where the resonance occurs: in this manner, the response of the sensor is modulated in wavelength. Several devices are been developed nowadays, measuring also volatile organic compounds (Zamarreño et al., 2010).

Reflection configuration

In this category, sensors consist of an optical fiber pigtail cleaved ended, on whose extreme is fixed the sensing material. Thus, light travelling through the core reaches the end of the pigtail, interacts with the deposition and a part of the incident signal is reflected depending on the transduction. The sensor head looks like a chemical electrode; therefore in applications where the concentration of some chemical species has to be measured, it is called optrode (Wolfbeis et al., 1998).

The working principle is, a priori, simple: if any of the optical properties of the sensing material changes in presence of any VOC, it will cause a variation in the reflected signal (Giordano et al., 2005). In a quantitative way, the response of reflection sensors can be explained by interferometry (Lee et al., 1992). After the sensing layer is fixed, the interface between the fiber and it, together with the interface between the layer and the surrounding media, conform an interferometric cavity. If the refractive index of the sensing layer, its thickness or its optical absorbance changes in presence of VOCs, so will do the reflected power in the fiber - sensing layer interface. This idea is summarized in Figure 6 (Consales et al., 2009):

Other configurations

It is also possible to combine reflection and transmission topologies: it is known as the *hybrid configurations* (Mitsubayashi et al., 2003). Another possibility is based on fixing onto the fiber a luminescent material whose emission varies in presence of the target VOC. The most

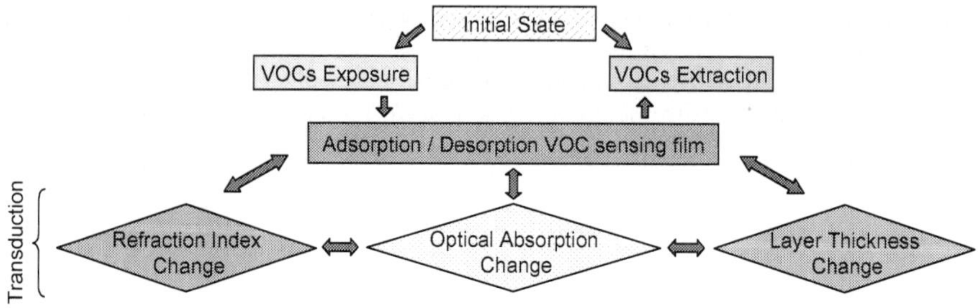

Fig. 6. Scheme with the parameters present in the response of a reflection sensor.

relevant application so far based on luminescent sensors is the detection and measurement of oxygen (O'Neal et al., 2004).

2.2.3 Choosing the sensing architecture

All the configurations described so far are summarized in Figure 7; as it can be inferred, choosing one of them is challenging. The best way to select one is thinking about the requirements of the final application and check which one better meets them.

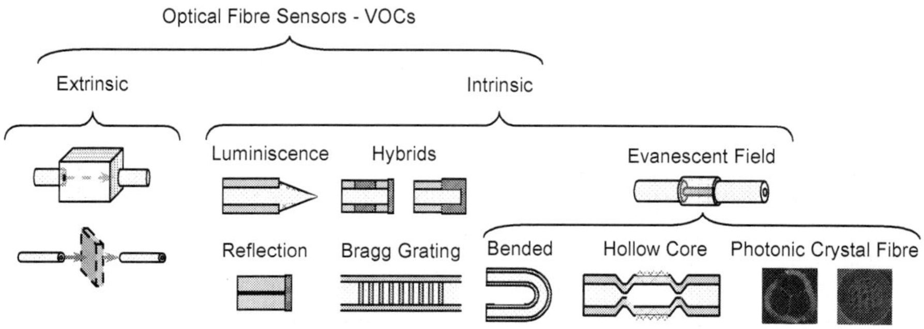

Fig. 7. Classification of sensors described in this section.

3. Construction techniques

The second basic step when preparing OFS is the deposition of the sensing material along or onto a cleaved ended pigtail. This stage is necessary only in intrinsic sensors because with extrinsic sensors, the transduction takes place outside de fiber. The deposition has to be repetitive and as homogeneous as possible. The VOC to be detected has to get diffused into the sensing layer, so its morphology is a critical factor. The construction techniques have been adapted for OFSs, so the final thickness is in the nanometric scale, which ensures a fast response. The most relevant two procedures will be described along this third section.

3.1 Dip coating technique

The simplest method consists of dissolving the compound of interest in a solvent that does not alter its sensing properties and then, dipping the fiber in this mixture. This technique is called *dip coating* (Bariain et al., 2000). The morphology and thickness of the final deposition are affected mainly by the times that the fiber is immersed into the solution of the sensing material and the velocity at which it is dipped (Arregui et al., 2002). An alternative option consists of leaving the fiber dipped in the solution for a certain time (Scorsone, et al., 2004).

This procedure is employed when preparing sensors in transmission configuration. The fiber is dipped perpendicularly into the mixture, so a thin layer is fixed along it. In the case of reflection sensors, the layers deposited onto the end of the pigtail do not conform a regular shape but one similar to a match head (Figure 8). Although simple, when preparing reflection sensors, this method is limited by its poor reproducibility.

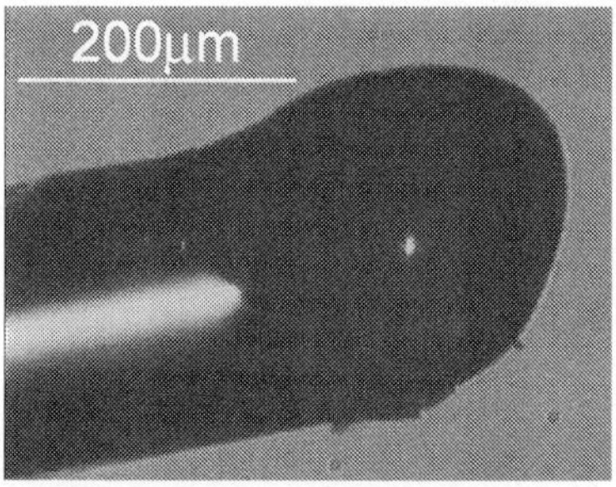

Fig. 8. Image obtained from an optical microscope of a reflection sensor prepared with Dip Coating technique.

3.2 Layer by layer method

The small dimensions of optical fiber can take advantage of deposition techniques that do not depend on the size and morphology of the substrate. One of them is the Layer by Layer method (LbL): it belongs to a series of procedures whose aim consists of creating matrixes or supports on different substrates at a molecular scale. To achieve this, the electrostatic attraction of molecules with opposite electric charge can be used, so that they get assembled one with each other. Long polymer chains are used in this way: they exhibit ionic groups when they are dissolved. Polymers that exhibit this property are called polyelectrolytes. Multilayer structures based on this idea have been studied since the 90s (Decher, 1997). One important feature is that the LbL is carried out at room temperature; moreover, it allows the thickness of the layers deposited to be controlled (Choi&Rubner, 2005).

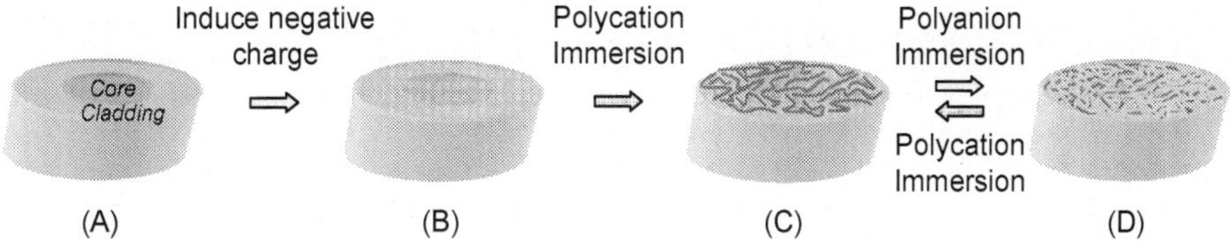

Fig. 9. Main steps of LbL method.

The LbL does not depend on the geometry of the substrate, so it can be used to develop either transmission or reflection sensors. In case of the last ones, the construction process can be registered thanks to the interferometric nature of that configuration. Every time a monolayer gets fixed, the thickness increases, so that the reflected power changes, recording a response similar to the one of an interferometer as the layers are deposited. The thickness of each individual layer is at nanometric scale, which allows the growth of the nanocavity to be studied by the Ray Theory (Arregui et al., 1999). Employing the experimental set up shown in Figure 10, the construction process can be controlled: plotting the reflected power in function of the immersions, an interferometric sinusoidal response is obtained.

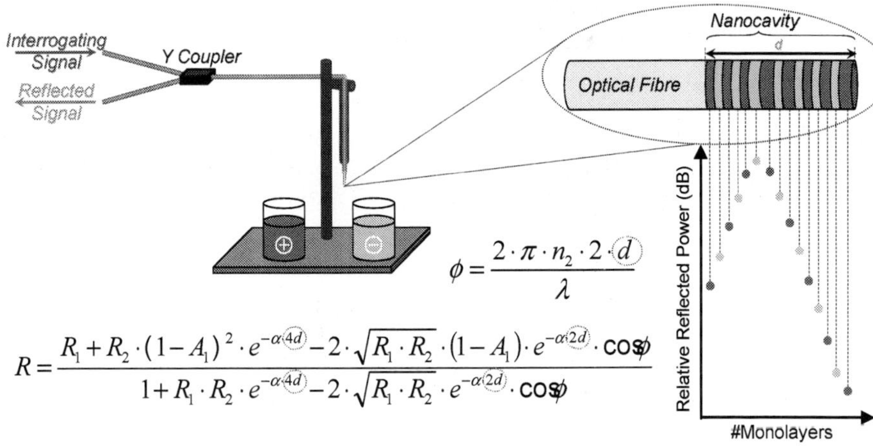

Fig. 10. The nanostructure is supposed to be homogeneous, its optical absorption α, scattering losses A_1 and refraction index n_2 are constant; therefore, R_1 and R_2 too (reflectivities of the fiber/nanocavity and nanocavity/media interfaces respectively). In this manner, the reflected signal depends on the thickness d of the structure.

4. Experimental set ups and multiplexing networks

This section will be about the photonic equipment used to study the sensors individually and multiplexed in optical networks. It is assumed that all the devices are intensity modulated: with other modulations, the optical source and detector would show different devices, but the topology would be similar. The multiplexing networks described are supposed to work with reflection sensors, although they can be generalized to any type.

4.1 Individual sensor experimental set ups

Extrinsic sensors are easy to interconnect: the light path is "interrupted" by the cell where the gas to detect interacts with the ray and is coupled again to the fiber. Therefore, the cell is connected between the light source and the receiver. For transmission extrinsic sensors, it is just the same situation: the OFS is connected between the emitter and the receiver (Figure 11). The signal level is much higher than noise, so that it is not necessary to use high sensitivity detectors or high power optical sources.

Regarding to reflection sensors, the interrogating signal must reach the sensor head from the optical source; thereafter, the response signal has to be guided back to the receiver. The

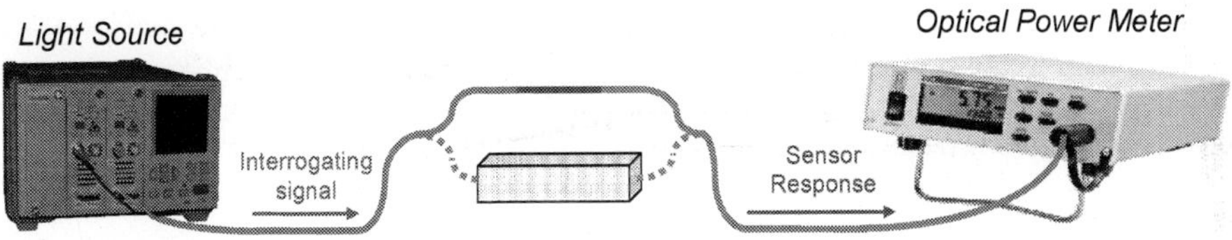

Fig. 11. Experimental set up for an extrinsic sensor (doted line) and an intrinsic transmission one (solid line).

photonic device most used to achieve it is the optical coupler: typically, it has 4 ports, but for sensing applications, one of them is removed, being called "Y couplers" (Figure 12). These devices are widely used, so it is easy to find one for almost any kind of fiber. On the other hand, there is a device with three ports that can be use to guide back the sensor response introducing only insertion losses: it is the optical circulator. It is able to guide the signal from its port 1 to its port 2 and from that port to its port 3. In this manner, the sensing signal is coupled entirely to the receiver.

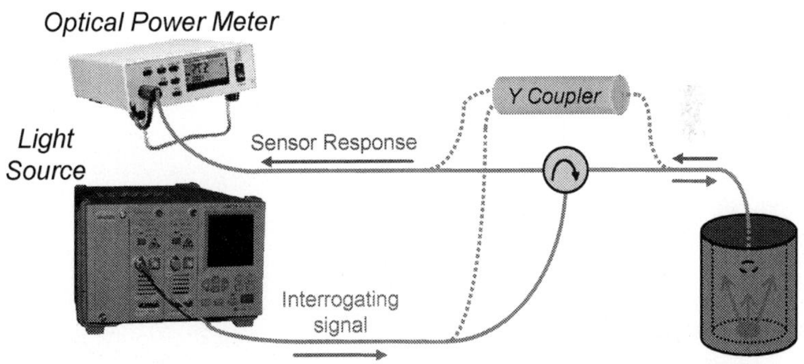

Fig. 12. Reflection experimental set up with an optical coupler (doted line) and with a circulator (solid line). A thermal jacket is used to ensure isothermal measurements.

4.2 Multiplexing networks

One of the most relevant features of optical fiber is the multiplexing ability. Actually, if several sensors have to be used at the same time, it is critical to share resources such as the optical source, the receiver and the light path. It is important to keep in mind that it has to be done with devices used in telecommunication networks: therefore, the sensors have to be implemented with standard monomode fiber. The configurations to be described are based on the optical wavelength division multiplexer: it is able to multiplex in one direction (join several channels into a single fiber) and demultiplex in the other one (separate several channels into several fibers) (Orazi et al., 1996). Thereby, the interrogating signal can be guided to each sensor through it; meanwhile the responses can be directed back to the receiver. Sensors connected to the multiplexor in this manner follow the *star topology*.

The active components of the system are grouped in a modular way, making the system more robust in case of failures. As an example, the optical source and the receiver can be place at the same unit, so that they can be repaired or replaced easily. This unit is called *Optical Header*: this a concept widely used in remote sensing applications (Vallejo, 2009).

4.2.1 Wavelength division combined with time division multiplexing

Intuitively, each sensor should be interrogated with a wavelength that matches each channel of the multiplexor. Using one laser per sensor would increase significantly the cost, and thereby, the scalability of the system would be poor. Following the idea of illuminating the sensors with signals at different wavelengths, a tunable laser can be employed. In order to interrogate the sensors, the laser is tuned at the wavelength of one of the multiplexer optical channels, then the signal from the sensor is registered, and finally, the laser is tuned at another channel wavelength. There is no moment when signals at two different wavelengths travel together through the fiber: therefore, an optical power meter can be used instead of an Optical Spectrum Analyzer (Figure 13).

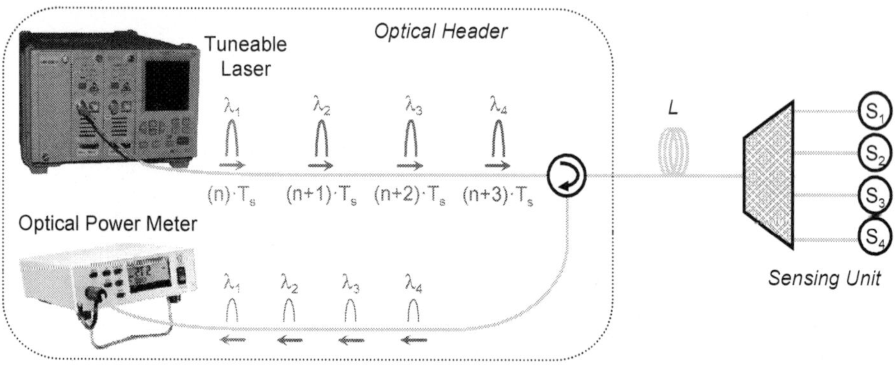

Fig. 13. Experimental set up that combines wavelength and time division multiplexing. Every λ is emitted at multiples of sampling period Ts, as it is shown in the figure for the nth sampling cycle.

The experimental set up combines Wavelength Division Multiplexing (WDM) in the sensing unit with the Time Division Multiplexing (TDM) of the tunable laser. This configuration is limited by the sampling rate: more specifically, by the tuning speed of the laser. A low sampling rate would produce an information loss. It would not be a problem in sensors whose recovery or response times were slow or when they could be characterized only with the power level; nevertheless, it is a great drawback in real time applications.

4.2.2 Multiplexing based on a white light source

The WDM technique allows all the sensors to be interrogated and sampled at the same time; moreover, all the signals share the optical path. It is better to use only one optical source than one per sensor to keep the scalability of the system. Taking advantage of the filtering capability of the multiplexer, a white light source can be used to illuminate the sensors: the multiplexer will filter the broadband signal towards the sensors and it will guide back the narrow band responses of the sensors centered at each optical channel. The number of sensors could be increased using narrower channels and/or wider spectral sources.

Since all the signals share the optical path, they have to be separated at the receiver. The most economical solution consists of using a demultiplexer in the optical header whose channels match the ones from the multiplexer in the sensing unit. This configuration sets out two problems: first, as many receivers as the number of sensors are needed, and second, the demultiplexer has to match its optical channels with the multiplexer ones. One alternative to overcome these problems consists of coupling the multiplexed signals into an OSA. The responses from the sensors can be obtained sampling the spectra registered by the OSA at the wavelengths where the channels of the multiplexer are centered (Figure 14). The drawback of using an OSA is its high cost compared to optical power meters.

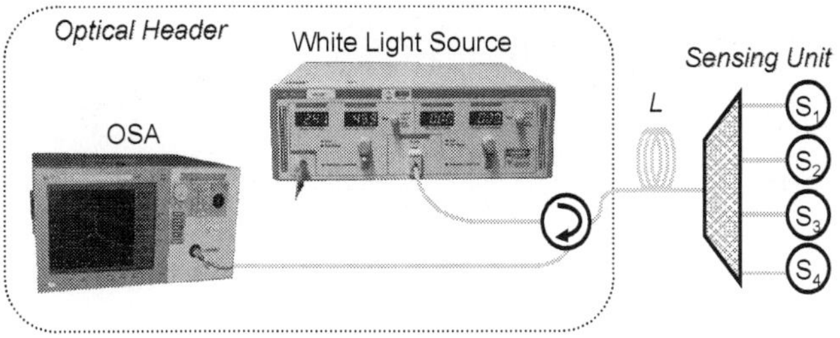

Fig. 14. Experimental set ups used for sensors multiplexing with a white light source.

4.2.3 Multiplexing network architectures

The number of sensors in the array based on the star topology is limited by the number of its optical channels of the multiplexer. The configuration drawn in Figure 14 can handle several sensors but just one sensing unit. There are some applications where it might be necessary to work with several sensing units at the same time: as an example, sensors arrays might be placed in the fermentation tanks of a cellar. In this situation, the optical signal has to be distributed among all the arrays, which can be done only using optical couplers. The sensing units can be multiplexed with the different topologies (star, tree, single bus, ladder): the optimum one depends on the characteristics of the topology (number of arrays, types of sensors and multiplexing technique) and on the requirements of the final application (physical distribution of the points where the arrays are to be placed).

Let us suppose that the arrays are connected in a single bus, which is drawn in Figure 15. The optical header contains the optical source, the receiver and the circulator, as before: now, the broadband signal has to be distributed to all the sensing units by the couplers. Thereby, each one of them couples part of the signal to the array and the rest forward to the bus in order to interrogate the other units. The optimum ratio of the optical couplers (50:50, 90:10, 95:5, 99:1) depends mainly on the number of sensing units (Abad Valtierra, 2002).

5. Signal processing and data mining techniques

The response of any sensor when expose to the VOCs sample or mixture is sampled in almost every case: in this way, it can be stored and processed. Although this section will handle with intensity modulated sensors, it can be applied to any other modulation. If the concentration of a certain VOC has to be detected, a parameter that shows a linear relationship with it is typically used (intensity, wavelength shift, just to mention some).

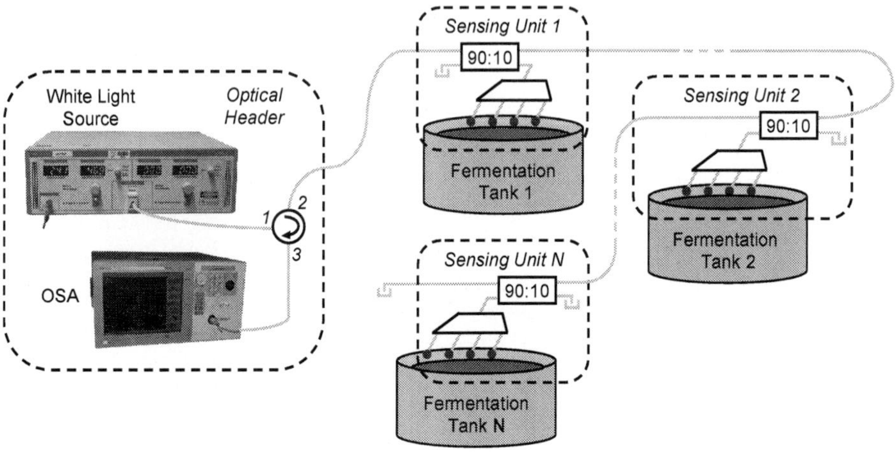

Fig. 15. Example of a network in which the arrays are connected in a single bus topology and the sensors of each unit follow a star configuration.

Nevertheless, if the sample is to be identified and not quantified, it is not necessary a linear response: either linear or not linear data mining techniques can be used to extract features from the sensor signal and so, to identify the sample.

5.1 Spectral analysis and temporal response

The signal detected from the sensor can be registered in terms of optical power along the time (with an optical power meter) or converted into the frequency domain (by an OSA). The last option is employed with broadband signals or when several sensors are wavelength multiplexed. The shape of this spectrum can be used to identify different VOCs or mixtures, and its amplitude can determine the concentration of a certain volatile compound. Figure 16 illustrates three examples of signals converted to the frequency domain of some sensors exposed to organic vapors. In these cases, the spectra are compared before and after the exposition to the organic vapors. This kind of measurements is known as *Absorbance Spectra*, and it is expressed in dB.

Another manner to obtain information from a spectrum consists of monitoring the amplitude at a certain wavelength: moreover, it can be achieved by using an optical source centered at that wavelength combined with a power meter as receiver. The optical power variation or response/recovery times of these signals are used to characterize the response of the sensors. As an example, in Figure 17 are plotted the signal fluctuations of a sensor in presence of different concentrations of organic vapors individually: there is a linear relationship between the power variation (in dB) and each vapor concentration.

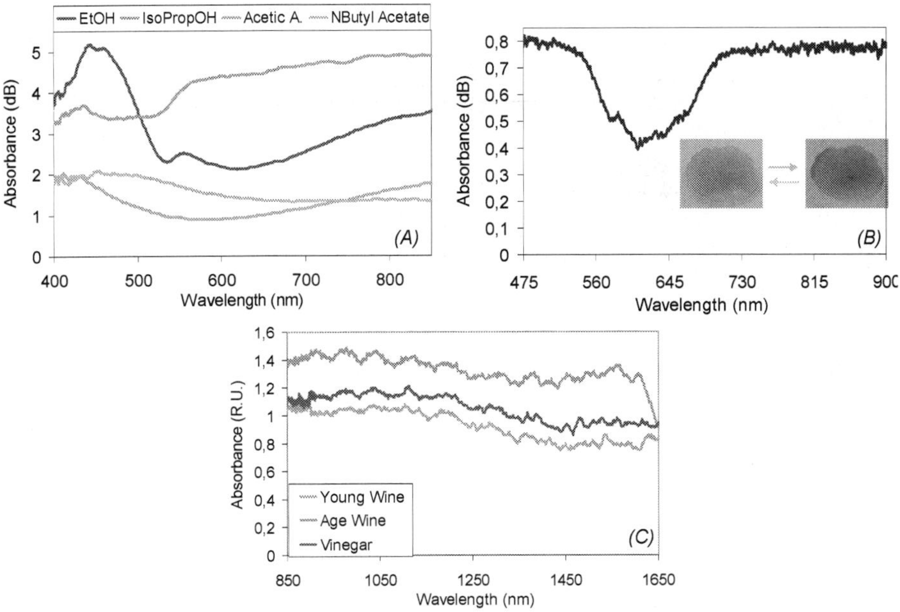

Fig. 16. Different spectra from distinct sensors: (A) signals registered for a sensor when exposed to different VOCs in the VIS; (B) color change from a sensing material shown in its absorbance spectrum; (C) absorbance spectrum in the NIR for different wine related drinks.

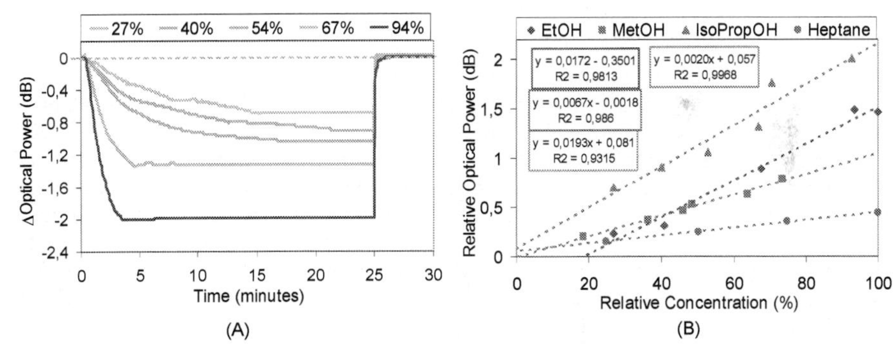

Fig. 17. (A) Signals from the sensor in presence of different isopropanol concentrations and (B) linear approximations for that sensor for distinct vapors.

5.2 Principal component analysis

There is abundant information available in the spectral or in the temporal response of the sensor: it can produce a high computational cost. It exits a linear procedure to compress information that reduces the number of parameters to use, which is the Principal Component Analysis (PCA) (Penza&Cassano, 2003). This transformation is used either with discrete temporal (Guadarrama et al., 2004) and spectral signals (Perez-Hernandez et al., 2009). PCA is based on the idea of removing redundant information: it can be achieved assigning the whole covariance or the cross correlation of the input space to a few components. Thereby, each transformed principal component storages an amount of the initial information. The principal components chosen contain thus a percentage of the total covariance, but not 100%: this difference is the resulting error of the dimensional reduction. Another important factor is how the information is distributed along the components, or better said, the weight that they have: typically, the principal components are ordered from the highest to the lowest weight. Once the PCA components are placed in the transformed space, they can be clustered to identify groups that show similar characteristics. Therefore, PCA is used in applications related with classification of VOC samples (Boholt et al., 2005).

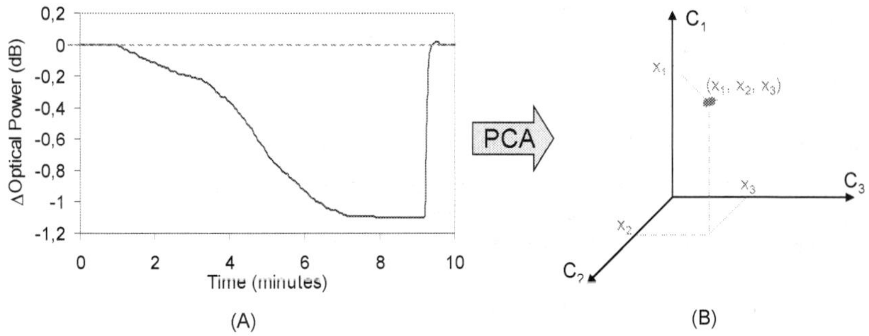

Fig. 18. Conversion of a temporal response (A) into a three dimension space (B) by PCA.

5.3 Artificial neural networks

Whereas PCA is a linear transformation, Artificial Neural Networks (ANN) are non linear algorithms that process information in a manner inspired by brain behavior (Shaham et al., 2005). In its most general form, an ANN is a machine designed to model the way in which the brain performs a particular task or function. Neural networks use a massive interconnection of simple computing cells referred to as neurons. An ANN is similar to the brain with regard to acquiring knowledge from its environment through a learning process, and using interneuron connection strengths (synaptic weights) to store this knowledge. There are several types of ANN, and even nowadays, new models are been optimized. Among all of them, one of the most used so far is the Feed Forward Network (FFN) to process the input information. The network has also to be trained: a popular learning algorithm employed to train the ANN is Back Propagation.

The basic unit of an ANN is the neuron: it is an information-processing unit that is fundamental for neural network operation. In Figure 19, it is shown the model of a neuron which can be used in a large family of neural networks. The basic elements are: a set of synapses, each of which is characterized by a weight of its own; an adder for summing the input signals weighted by the respective synaptic strengths and an Activation Function (non linear but differentiable) for limiting the output amplitude of a neuron.

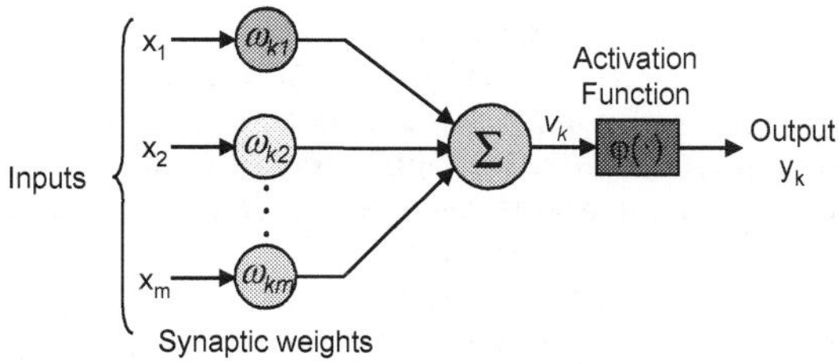

Fig. 19. Nonlinear model of a neuron labeled.

The training stage is performed in an iterative way. Firstly, the synaptic weights of the ANN are randomly initialized; then, the input data feeds the network and the resulting output is compared with the desired one. This error (back propagated) is used to correct the synaptic weights, repeating the process with another sample of the training subset. There are advanced ANNs with no training stage, but this FFN and the Back Propagation Algorithm offer a general idea about how neural networks compute data. Thanks to their learning ability, ANNs have been widely used with VOCs identification tasks (Luo et al., 2004)

6. An application related to VOCs identification: Wine fermentation process

After showing the most important factors related to OFSs that handle with VOCs, a specific application will be described taking into account all the topics exposed so far. Artificial systems able to identify odors can take advantage of the optical fiber features. In these applications, several sensors are grouped in arrays: this amount of information can be transmitted through multiplexing configurations. The data processing has to be optimized because there is much redundant information available. In this section, a system able to distinguish between grape juice, wine and vinegar will be described. Although this task might seem trivial, it is remarkable that the system would allow the fermentation process to be monitored on line and in real time, which is hard to do with traditional methods.

Fig. 20. Some VOCs present along wine fermentation. They characterize each stage: thereby they can be used to identify each one of them.

6.1 Preparation and multiplexation of the sensors

Although reflection sensors show a low signal level, the high mechanical robustness makes them a good choice for this type of application. It is compulsory to use singlemode fibers if the sensors have to be multiplexed, so it could lower the sensitivity of the devices: the core of this fiber is much smaller than the multimode ones. This inconvenience can be overcome by the LbL method: the morphology of the depositions can be altered, so that the roughness is increased, improving the interaction between the VOCs and the sensing materials. Organometallic compounds that suffer changes on its optical properties in presence of organic vapors were chosen as sensing materials. These products belong to a family of products this chemical structure but each one show a distinct molecular marker: it ensures a different selectivity for different VOCs (Luquin et al., 2005). Furthermore, these variations are reversible, which guarantees real time monitoring. Up to 4 different sensing materials are used to develop 4 devices: Sensor A, Sensor B, Sensor C and Sensor D (Figure 21).

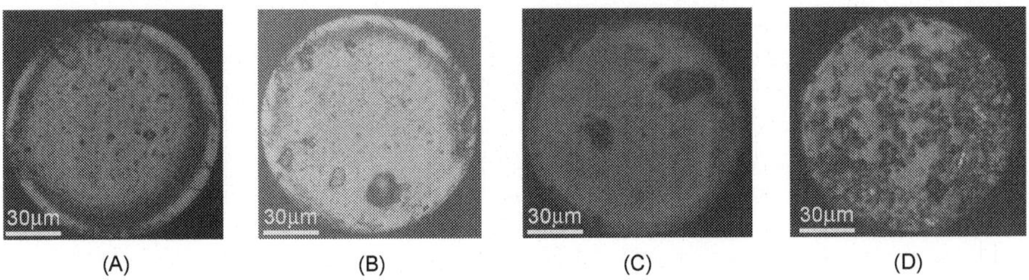

Fig. 21. Plan images from the sensors prepared following LbL to identify the drinks.

The sensors developed are exposed to 1 mL samples of grape juice, wine and vinegar individually. The initial experimental set up uses a laser at 1550 nm, an optical circulator and a power meter. The responses for each drink are plotted in Figure 22.

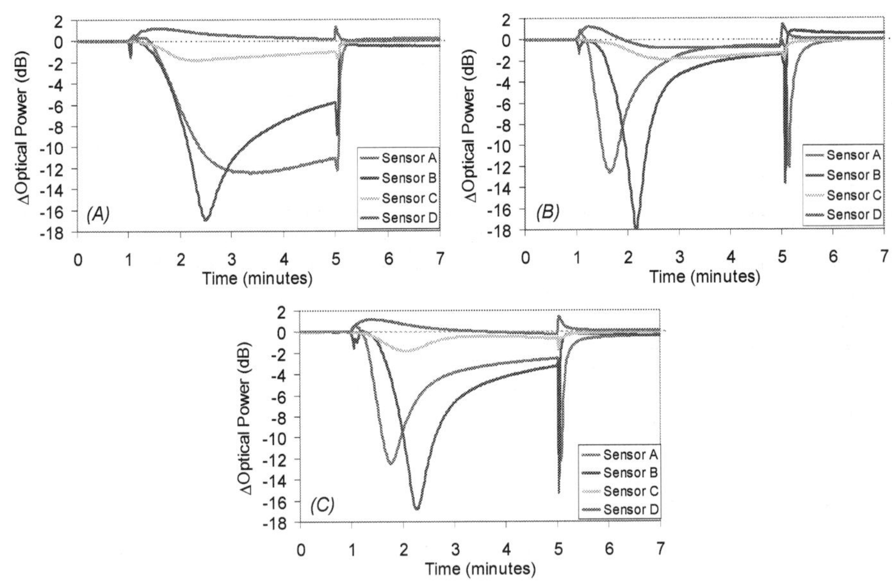

Fig. 22. Temporal responses in terms of optical power from the sensors for each beverage: (A) grape juice, (B) wine and (C) vinegar.

The four devices are connected following a star configuration with an optical multiplexer. As was described in section 4, the array is linked to an optical header through a circulator; inside the header are located a white light source and an OSA (as in Figure 14).

6.2 Processing the array information

Looking the responses from the sensors shown in Figure 22, parameters such as the recovery time or the signal variation could be used to identify the different beverages. Nevertheless, as intensity modulated devices, power fluctuations due to the optical source or to the aging of the sensors, might alter these factors. In order to check these inconveniences, a total of 100 samples per drink along 1 month were recorder from the sensor array (Figure 23). It can be observed that for all the sensors there are severe optical power fluctuations that make impossible the identification of the drinks.

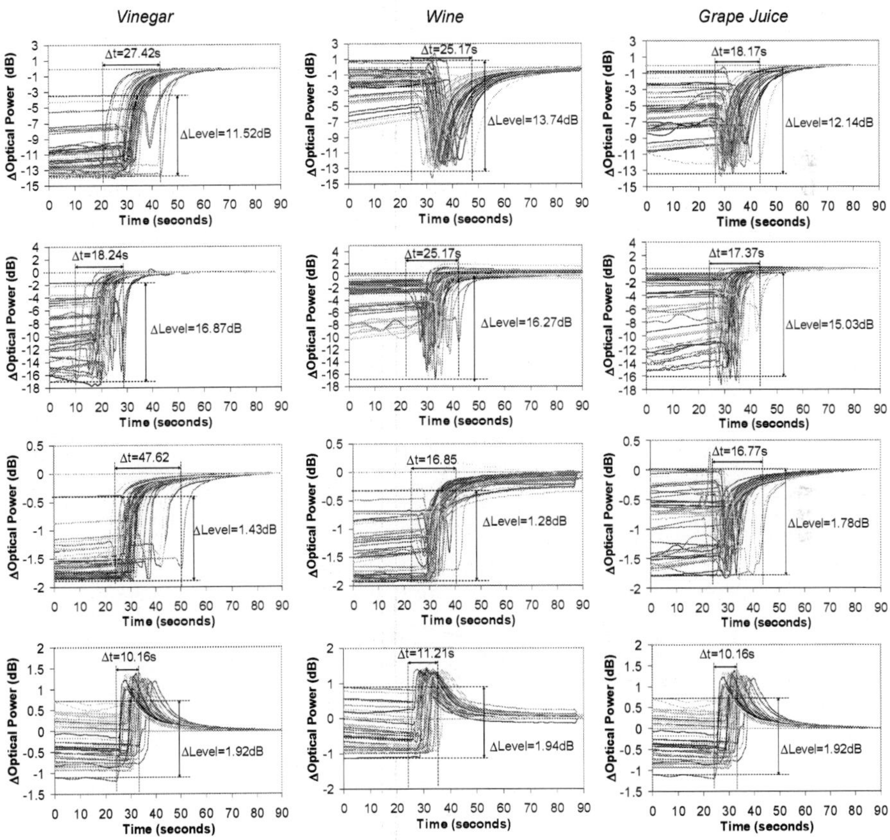

Fig. 23. Recovery responses from 50 samples per each drink and sensor. Each row corresponds to a different sensor: from up to down, Sensor A, Sensor B, Sensor C and D.

It is evident that a more robust data mining is needed to overcome the fluctuations observed in Figure 23. Processing the transitories by an ANN or by PCA is still low efficient, in terms of computational cost. Actually, the transitory is a quick change in the signal level, so that the most relevant information can be obtained by applying the Fast Fourier Transform. This transformation is also a linear operator: once computed, the spectrum of the signal can be shortened eliminating redundant coefficients. The most relevant information is in the low frequencies: therefore, the sequences can be truncated after the second lobule. In any case, the reduced spectrum sequences are still too heavy to be computed: PCA can be applied to compress this information into a three dimension space. The combination of the sensors responses makes the system more robust, but as there might be several fluctuations in the signals, these two linear transformations might not be able to handle with them. The non linear nature of ANN is a good option to deal with this kind of signals. In this way, the whole data set transformed into the frequency domain and then compressed by PCA, can be used to train and test an ANN. Moreover, the network can be designed to have 3 outputs, one per drink to identify, which will behave as binary indicators.

To sum up the complete data mining computation, the recovery response of the sensors is transformed and processed along the data mining process. Initially, the sensor array returns 320 samples from the signals of the 4 sensors, and eventually, they become a 3x1 binary vector that indicates the beverage the array has been exposed to. The process is summarized in Figure 25 for a wine vapors test.

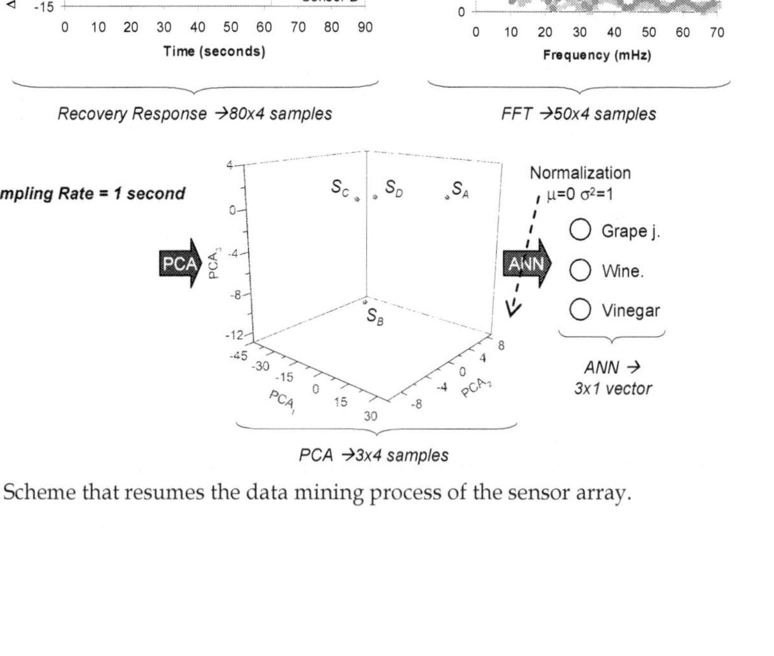

Fig. 24. Scheme that resumes the data mining process of the sensor array.

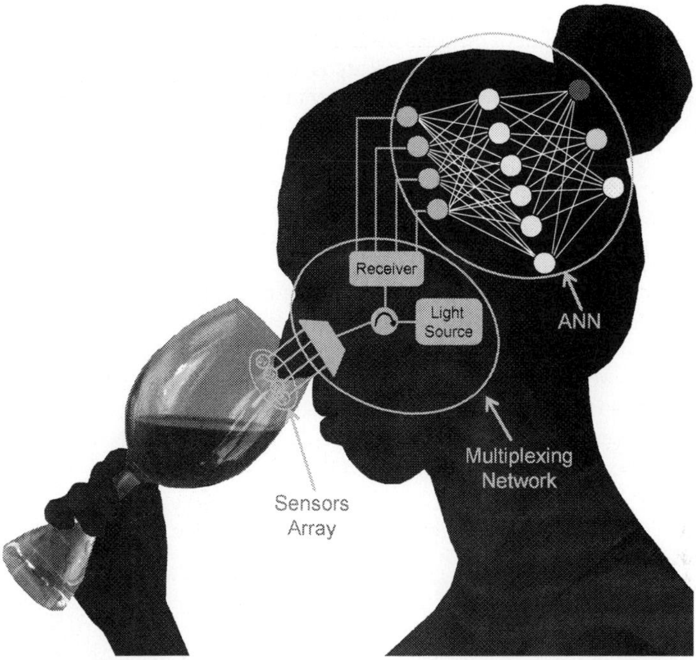

Fig. 25. Analogy between the sense of smell and the optoelectronic nose, together with the blocks involved in the odor identification.

6.3 Features of the whole system

The system described in this section is able to identify the most important stages during the wine fermentation process. It is just one of the potential applications where this technology can be used. In this specific case, the device allows the state of the beverage fermentation to be followed in situ with a maximum delay of 4 minutes. In contrast, traditional analyses require the samples to be taken from the fermentation tanks and brought them to a laboratory, where the alcoholic grade and volatile components are analyzed separately. Nowadays this second option is still preferred, mainly in small installations, and it is also more precise from an analytical point of view. The photonic equipment would be lower cost by multiplexing several optoelectronic noses in one single network. Then, other important magnitudes, such as temperature, humidity, CO_2 or O_2 could be controlled too, taking advantage of the transparency of this type of networks. The VOCs detection based on OFSs can be improved and extended in the future to other applications in the food industry: in fact, some researchers are already focused on it (Garcıa, et al., 2003).

The final system can be divided into three blocks: the sensors array, data transportation and data mining (Brezmes et al., 2000). Making an analogy with mammalian sense of smell, the first one would correspond to the olfactory cells placed in the pituitary; the second one acts as the nervous system that transfers the information from these receptors and the third is the brain, where the processing occurs. It is illustrated in Figure 25.

The block configuration makes it scalable, modular and versatile. Thereby, should one constituent be damaged, only the affected part need to be repaired or replaced, while the others can keep on working properly. Even in the case of the array, if a sensor fails, it could be replaced without compromising the behavior the other devices. Regarding to the versatility, other types of sensing configurations can be integrated in the same network, whether they are discrete or distributed and it even could transmit others types of data.

7. Conclusions

There are several solutions in the VOCs detection field based on electronic sensors: some of them are even commercially available. Even so, there are certain environments where the optical fiber intrinsic features offer important advantages. In the VOCs market niche, optical fiber sensors show a great potential. Electromagnetical immunity, no biasing signals (and no explosion risk) or remote sensing are just some the most relevant features. There is not an optimal configuration to handle with VOCs; actually, there are several different sensing architectures depending on the requirements of the final application. Extrinsic sensors are optimal to detect gases individually with a relatively low interference from other volatile compounds. On the contrary, intrinsic sensors need a sensing material that provides them selectivity to certain VOCs. There also other factors to be taken into account such as the kind of modulation or the type of fiber: all these design parameters have to be chosen depending on the measuring conditions and requirements.

Remote sensing and multiplexing are very useful in VOCs applications. The block design that distinguishes between optical header and sensing unit allows measurements to be done in dangerous places; moreover, the scalability also permits several sensing units to be connected to the same network. Transparency is another important factor: sensors measuring different VOCs or any other parameters such as temperature and humidity can share the same light path, optical source and receiver. Among all the potential market niches, wine fermentation monitoring can take advantages of all the properties exposed in this chapter. The samples have to be analyzed qualitatively, so intrinsic sensors are the best option to handle with complex mixtures of VOCs. Up to 20 sensors could be multiplexed in a single array with the current technology, and of course, several arrays could be connected also to the same bus. Furthermore, thanks to the optical fiber low attenuation, the network might be implemented in medium size cellar, for instance.

Nowadays, it is evident that photonic technology is essential in communications networks. There also some sensing applications that have shown a great utility and are used in many environments. The success of optical fiber sensors to detect and identify VOCs will be determined by the requirements of the market. The best strategies have to find the environments and applications where these sensors offer features that other technologies do not, making them an attractive and affordable option.

8. Acknowledgments

The authors would like to acknowledge the financial support from the Spanish Ministerio de Educación y Ciencia through projects TEC2010-17805 and TEC2010-20224-C02-01. The collaboration of Professor Lopez-Amo and Dr. Perez-Herrera is acknowledged as well.

9. References

Ampuero, S. & Bosset, J. O. (2003). The electronic nose applied to dairy products: a review, *Sensors and Actuators, B: Chemical*, Vol. 94, pp. 1-12, ISSN 0925-4005.

Arregui, F. J.; Matias, I. R.; Liu, Y.; Lenahan, K. M. & Claus, R. O. (1999). Optical fiber nanometer-scale Fabry-Perot interferometer formed by the ionic self-assembly monolayer process, *Optics Letters*, Vol. 24, No.9, pp. 596-598, ISSN 0146-9592

Arregui, F. J.; Otano, M.; Fernandez-Valdivielso, C. & Matias, I. R. (2002). An experimental study about the utilization of Liquicoat solutions for the fabrication of pH optical fiber sensors, *Sensors and Actuators, B: Chemical*, Vol. 87, pp. 289-295, ISSN 0925-4005.

Barbosa, C.; Costa, N.; Ferreira, L. A.; Arajo, F. M.; Varum, H.; Costa, A.; Fernandes, C. & Rodrigues, H. (2008). Weldable fibre Bragg grating sensors for steel bridge monitoring, *Measurement Science and Technology*, Vol. 19, No.12, pp. 125305-10, ISSN 0957-0233.

Bariain, C.; Matıas, I. R.; Arregui, F. J. & Lopez-Amo, M. (2000). Optical fiber humidity sensor based on a tapered fiber coated with agarose gel, *Sensors and Actuators, B: Chemical*, Vol. 69, pp. 127-131, ISSN 0925-4005.

Boholt, K.; Andreasen, K.; Berg, F. d. & Hansen, T. (2005). A new method for measuring emission of odour from a rendering plant using the Danish Odour Sensor System (DOSS) artificial nose, *Sensors and Actuators, B: Chemical*, Vol. 106, pp. 170-176,

Brezmes, J.; Llobet, E.; Vilanova, X.; Saiz, G. & Correig, X. (2000). Fruit ripeness monitoring using an Electronic Nose, *Sensors and Actuators, B: Chemical*, Vol. 69, pp. 223-229, ISSN 0925-4005.

Brudzewski, K.; Osowski, S. & Markiewicz, T. (2004). Classification of milk by means of an electronic nose and SVM neural network, *Sensors and Actuators, B: Chemical*, Vol. 98, pp. 291-298, ISSN 0925-4005.

Burl, M. C.; Doleman, B. J.; Schaffer, A. & Lewis, N. S. (2001). Assessing the ability to predict human percepts of odor quality from the detector responses of a conducting polymer composite-based electronic nose, *Sensors and Actuators, B: Chemical*, Vol. 72, No.2, pp. 149-159,

Consales, M.; Crescitelli, A.; Penza, M.; Aversa, P.; Veneri, P. D.; Giordano, M. & Cusano, A. (2009). SWCNT nano-composite optical sensors for VOC and gas trace detection, *Sensors and Actuators, B: Chemical*, Vol. 138, pp. 351-361, ISSN 0925-4005.

Culshaw, B. (2000). Fiber optics in sensing and measurement, *IEEE Journal on Selected Topics in Quantum Electronics*, Vol. 6, No.6, pp. 1014-1021, ISSN 1077-260X.

Culshaw, B. (2004). Optical Fiber Sensor Technologies: Opportunities and—Perhaps—Pitfalls, *Journal of Ligthwave Technology*, Vol. 22, No.1, pp. 39-50, ISSN 0733-8724.

Culshaw, B. (2005). Research to reality: bringing fibre optic sensors into applications, *Proc. of SPIE*, Vol. 5952, No.01, pp. 01-15, ISSN 1996-756X.

Culshaw, B.; Stewart, G.; Dong, F.; Tandy, C. & Moodie, D. (1998). Fibre optic techniques for remote spectroscopic methane detection - From concept to system realisation, *Sensors and Actuators, B: Chemical*, Vol. 51, No.1-3, pp. 25-37, ISSN 0925-4005.

Chambers, P.; Austin, E. A. D. & Dakin, J. P. (2004). Theoretical analysis of a methane gas detection system, using the complementary source modulation method of correlation spectroscopy, *Measurement Science and Technology*, Vol. 15, No.8, pp. 1629-1636, ISSN 0957-0233.

Choi, J. & Rubner, M. F. (2005). Influence of the Degree of Ionization on Weak Polyelectrolyte Multilayer Assembly, *Macromolecules*, Vol. 38, pp. 116-124, ISSN 0024-9297.

Decher, G. (1997). Fuzzy nanoassemblies: Toward layered polymeric multicomposites, *Science*, Vol. 277, No.5330, pp. 1232-1237, ISSN 0036-8075.

Diaz, S.; Mafang, S. F.; Lopez-Amo, M. & Thevenaz, L. (2008). A high-performance optical time-domain brillouin distributed fiber sensor, *IEEE Sensors Journal*, Vol. 8, No.7, pp. 1268-1272, ISSN 1530-437X.

Elosua, C.; Bariain, C. & Matias, I. R. (2009). Optical Fiber Sensors to Detect Volatile Organic Compunds in Sick Building Syndrome Applications, *The Open Construction and Building Technology Journal*, Vol. pp. ISSN 1874-8368.

Elosua, C.; Matias, I. R.; Bariain, C. & Arregui, F. J. (2006). Volatile organic compound optical fiber sensors: A review, *Sensors*, Vol. 6, No.11, pp. 1440-1465, ISSN 1424-8220.

Frazao, O.; Santos, J. L.; Araujo, F. M. & Ferreira, L. A. (2008). Optical sensing with photonic crystal fibers, *Laser & Photonics Reviews*, Vol. pp. 1-11, ISSN 1530-437X.

Garcıa, M.; Horrillo, M. C.; Santos, J. P.; Aleixandre, M.; Sayago, I.; Fernández, M. J.; Arés, L. & Gutiérrez, J. (2003). Artificial olfactory system for the classification of Iberian hams, *Sensors and Actuators, B: Chemical*, Vol. 96, pp. 621-629, ISSN 0925-4005.

Garriguesa, S.; Taloua, T.; Nesab, D. & Gaset, A. (2001). Modified GC/MS system versus dedicated MS device: comparative application of electronic nose for QC in automotive industry, *Sensors and Actuators, B: Chemical*, Vol. 78, pp. 337-344, ISSN 0925-4005.

Giordano, M.; Russo, M.; Cusano, A. & Mensitieri, G. (2005). An high sensitivity optical sensor for chloroform vapours detection based on nanometric film of δ-form syndiotactic polystyrene, *Sensors and Actuators, B: Chemical*, Vol. 107, pp. 140-147, ISSN 0925-4005.

Goschnick, J.; Koronczi, I.; Frietsch, M. & Kiselev, I. (2005). Water pollution recognition with the electronic nose KAMINA, *Sensors and Actuators, B: Chemical*, Vol. 106, pp. 182-186, ISSN 0925-4005.

Guadarrama, A.; Rodríguez-Méndez, M. L. & Saja, J. A. d. (2004). Influence of electrochemical deposition parameters on the performance of poly-3-methyl thiophene and polyaniline sensors for virgin olive oils, *Sensors and Actuators, B: Chemical*, Vol. 100, pp. 60-64, ISSN 0925-4005.

Hudon, G.; Guy, C. & Hermia, J. (2000). Measurement of odor intensity by an electronic nose, *Journal of the Air and Waste Management Association*, Vol. 50, No.10, pp. 1750-1758, ISSN 1047-3289.

Kawamura, K.; Vestergaard, M.; Ishiyama, M.; Nagatani, N.; Hashiba, T. & Tamiya, E. (2006). Development of a novel hand-held toluene gas sensor: Possible use in the prevention and control of sick building syndrome, *Measurement: Journal of the International Measurement Confederation*, Vol. 39, No.6, pp. 490-496, ISSN 0263-2241.

Khijwania, S. K. & Gupta, B. D. (1999). Fiber optic evanescent field absorption sensor: effect of fiber parameters and geometry of the probe, *Optical and Quantum Electronics*, Vol. 31, No.8, pp. 625-636, ISSN 0306-8919.

Lee, B. (2003). Review of the present status of optical fiber sensors, *Optical Fiber Technology*, Vol. 9, pp. 57-79, ISSN 1068-5200.

Lee, C. E.; Gribler, W. N.; Atknis, R. A. & Taylor, H. F. (1992). In-Line Fiber Farby-Perot Interferometer with High-Reflectance Internal Mirrors, *IEEE Journal of Lightwave Technology*, Vol. 10, pp. 1376-1379, ISSN 0733-8724.

Leung, C. K. Y. (2001). Fiber optic sensors in concrete: the future?, *NDT&E International*, Vol. 34, pp. 85-94, ISSN 0963-8695.

Lopez-Higuera, J. M. (1998). *Optical sensors*, Santander.

Luo, D.; Hosseini, H. G. & Stewart, J. R. (2004). Application of ANN with extracted parameters from an electronic nose in cigarette brand identification, *Sensors and Actuators, B: Chemical*, Vol. 99, pp. 253-257,

Luquin, A.; Bariain, C.; Vergara, E.; Cerrada, E.; Garrido, J.; Matias, I. R. & Laguna, M. (2005). New preparation of gold-silver complexes and optical fibre environmental sensors based on vapochromic [Au2Ag2(C6F 5)4(phen)2]n, *Applied Organometallic Chemistry*, Vol. 19, No.12, pp. 1232-1238,

Mitsubayashi, K.; Kon, T. & Hashimoto, Y. (2003). Optical bio-sniffer for ethanol vapor using an oxygen-sensitive optical fiber, *Biosensors & Bioelectronics*, Vol. 19, pp. 193-198, ISSN 0956-5663.

O'Neal, D. P.; Meledeo, M. A.; Davis, J. R.; Ibey, B. L.; Gant, V. A.; Pishko, M. V. & Coté, G. L. (2004). Oxygen Sensor Based on the Fluorescence Quenching of a Ruthenium Complex Immobilized in a Biocompatible Poly(Ethylene Glycol) Hydrogel, *IEEE Sensors Journal*, Vol. 4, No.6, pp. 728-734, ISSN 1530-437X.

Orazi, R. J.; Vu, T. T.; McLandrich, M. N.; Hewett, C. A. & Poirier, P. M. (1996). Cascaded narrow channel fused fiber wavelength division multiplexers, *IEE Electronics Letters*, Vol. 32, No.4, pp. 368-370, ISSN 1350-911X.

Penza, M. & Cassano, G. (2003). Application of principal component analysis and artificial neural networks to recognize the individual VOCs of methanol/2-propanol in a binary mixture by SAW multi-sensor array, *Sensors and Actuators, B: Chemical*, Vol. 89, No.3, pp. 269-284, ISSN 0925-4005.

Perez-Hernandez, J.; Albero, J.; Correig, X.; Llobet, E. & Palomares, E. (2009). Multivariate calibration analysis of colorimetric mercury sensing using a molecular probe, *Analytica Chimica Acta*, Vol. 633, No.2, pp. 173-180, ISSN 0003-2670.

Rao, Y.; Zhou, C. & Zhu, T. (2006). A Novel Chirped-FBG-Based F-P Sensor Network Using SFDM/WDM, *Procedings of 18th OFS*,ISBN 1-55752-817-1 ISBN 1-55752-817-1, 2006.

Santos, J. P.; Arroyo, T.; Aleixandre, M.; Lozano, J.; Sayago, I.; Garcıa, M.; Fernández, M. J.; Arés, L.; Gutiérrez, J.; Cabellos, J. M.; Gilb, M. & Horrillo, M. C. (2004). A comparative study of sensor array and GC–MS: application to Madrid wines characterization, *Sensors and Actuators, B: Chemical*, Vol. 102, pp. 299-307, ISSN 0925-4005.

Scorsone, E.; Christie, S.; Persaud, K. C. & Kvasnik, F. (2004). Evanescent sensing of alkaline and acidic vapours using a plastic clad silica fibre doped with poly(o-methoxyaniline), *Sensors and Actuators, B: Chemical*, Vol. 97, pp. 174-181, ISSN 0925-4005.

Shaham, O.; Carmel, L. & Harel, D. (2005). On mappings between electronic noses, *Sensors and Actuators, B: Chemical*, Vol. 106, pp. 76-82, ISSN 0925-4005.

Smith, C. M.; Venkataraman, N.; Gallagher, M. T.; MÃ¼ller, D.; West, J. A.; Borrelli, N. F.; Allan, D. C. & Koch, K. W. (2003). Low-loss hollow-core silica/air photonic bandgap fibre, *Nature*, Vol. 424, No.6949, pp. 657-659, ISSN 0028-0836.

Stella, R.; Barisci, J. N.; Serra, G.; Wallace, G. G. & Rossi, D. D. (2000). Characterisation of olive oil by an electronic nose based on conducting polymer sensors, *Sensors and Actuators, B: Chemical*, Vol. 63, pp. 1-9, ISSN 0925-4005.

Tao, S.; Winstead, C. B.; Jindal, R. & Singh, J. P. (2004). Optical-Fiber Sensor Using Tailored Porous Sol-Gel Fiber Core, *IEEE Sensors Journal*, Vol. 4, No.3, pp. 322-328, ISSN 1530-437X.

Tao, S.; Xu, L. & Fanguy, J. C. (2006). Optical fiber ammonia sensing probes using reagent immobilized porous silica coating as transducers, *Sensors and Actuators, B: Chemical*, Vol. 115, No.1, pp. 158-163, ISSN 0925-4005.

Whitenett, G.; Stewart, G.; Yu, H. & Culshaw, B. (2004). Investigation of a tuneable mode-locked fiber laser for application to multipoint gas spectroscopy, *Journal of Lightwave Technology*, Vol. 22, No.3, pp. 813-819, ISSN 0733-8724.

Wolfbeis, O. S.; Kovdcs, B.; Goswami, K. & Klainer, S. M. (1998). Fiber-Optic Fluorescence Carbon Dioxide Sensor for Environmental Monitoring, *Mikrochim Acta*, Vol. 129, pp. 181-188, ISSN 0026-3672.

Zamarreño, C. R.; Hernaez, M.; Del Villar, I.; Matias, I. R. & Arregui, F. J. (2010). Sensing properties of ITO coated optical fibers to diverse VOCs, *Procedings of* 2010.

3

Intrinsic Optical Fiber Sensor

Sylvain Lecler and Patrick Meyrueis
Strasbourg University
France

1. Introduction

The use of more and more complex systems in transportation, bio medical devices, defence, production, systems, etc., that are more and more automated, having to comply with many norms related to low energy consumption, miniaturization, insensivity to modifications of the surrounding conditions, security, etc. induce the need of new sensors that have to be reliable, low cost, easy to produce, if possible without clean rooms, avoiding too demanding calibration process and using no poisonous components or materials.

Optical fiber sensors were initially elaborated to validate some light propagation modelling in optical fiber. Initially, they were fragile, only performing in laboratory conditions and, most of the time, they were potentially more expensive to realize than other technology sensors, and they had a reliability factor that was not excellent.

But with the improvement of knowledge in guided optic modelling, lowering of tolerance of manufacturing in guided optics components, improvement of optical fiber connection technology, new way to extract physics modulation parameters from light modulation in polarization, phase, frequency, intensity have appeared.

It was thus discovered that optical fiber sensors have a very wide range of effective applications and that the packaging of optical fiber sensors for industry fit more and more the need for quality and performance that is the one of modern products.

In fact, two kinds of optical fiber sensors arised in the market: the extrinsic and the intrinsic optical fiber sensors. In the first case, the transducer, that is not an optical fiber, is modulating the light, out of a physics phenomena: for instance a vibration, and this modulated light is then sent inside the fiber where it will propagate up to a device converting, at the output of the fiber, the modulated light into a modulated electronic signal. Since only light intensity modulation can be read by these devices it is necessary to convert a light modulation that is not intensity into light intensity modulations. Many optical fiber designs allow this task. The one that will be selected will fit the application needs. Extrinsic sensors are generally not dedicated to high level measurements, but they are convenient for applications where low resolution and low cost are requested.

Intrinsic optical fiber sensors are sensors that have a totally different design. The conversion of the modulation of a physic parameter is done by a portion of an optical fiber. These sensors are much more difficult to elaborate, but they have the best performances. The first ones to be developed for mass production were the Sagnac gyro-sensors: they are based on the speed limit of light in a media with a relativity effect and a close optical loop in which the photons are circulating in opposite guided directions. Their interferences give, with simple relation the output signal that is a rotational acceleration. A gyroscope is thus constituted. Such gyros are intensively used now for transportation navigation, including satellite.

We can also classify the optical fiber sensors by the kind of optical fiber that constitutes them. So we will have multimode optical fiber sensor, and monomode optical fiber sensor. The difference is lying on the performances of the fiber used: multimode fibers have a low band pass but they are easy to assemble, it is the contrary for the monomode device. The effect of perturbation of their surroundings to the light propagating in the fiber is easy to model with a monomode device and more difficult with the multimode fiber. So the initial trend was to use monomode fibers for sensing, but more and more smart uses of multimode fibers are implemented in sensing devices because of their advantages.

The development of multimode fiber sensors induces the use of polymer fiber, much cheaper, and that can be cut and assembled in an economic way, but they cannot assume any kind of measurements. The most rewarding use of polymer fiber in sensing was done for measuring shock and vibration.

The more recent evolution of optical fiber sensing is toward integrated optics. The optical circuit allowing the physics phenomena detection through guided wave is implemented in a chip that will incorporate sensitive optical waveguide. This allows the realization for instance of biochips for sensing biological molecule by constituting some kind of a "lab on a chip".

The very last technical contribution of innovation to optical fiber sensors are plasmonic effects and photonic crystal fiber. They allow to do measurement that were out of reach before like hydrogen concentration, by using new effects connected to the specific way light is propagating in this case.

So it can be said that optical fibers sensing has matured but is still in a development process: even if effective products are more located in the defence, ecology or biomedical domain plenty of other commercial products are now proposed for planes and automotives. These devices are also named because they are a mixing of micro optics, mechanics, electronics, micro optomecatronic devices.

As it is scheduled in the Photonics 21 prepared for the European Union forecast, the future will see many new technologies of this kind. Photonics being an enabling technology acknowledged by the European Union as one of the 5 KET (Key enabling technology). This enabling work will be as relevant in the field of optical fiber sensing than in other areas of photonics.

2. Intrinsic optical fiber sensors: A matured technology

2.1 Extrinsic and intrinsic optical fiber sensors

Initially in the seventies, the aim of optical fibers and components was to transmit light without perturbation. The fiber was used as a connexion to transport information. The possible light perturbations were only studied in order to be minimized. In contrary, an optical fiber sensor is a system where information is created in the optical path way due to an interaction with a physical phenomenon to be measured and before its transmission to a detector using one or several optical fibers. If the optical fiber is only used to transmit the light, we talk about an extrinsic optical fiber sensor. If the modulation of one of the properties of light takes place in the optical fiber due to the physical phenomenon to measure, we talk about an intrinsic optical fiber sensor [Ferretti, 1996].

Sensing principle

A change in pressure, temperature, curvature and so on will modify such properties of the optical fiber propagating light as its phase, polarization, amplitude or spectre.

For each of these physical parameters corresponds a family of intrinsic optical fiber sensors that will be described.

2.2 Optical fiber sensors tools improvements

Any optical fiber sensor is composed of source(s), optical fiber(s), transducer(s), detector(s) and a signal analyzing system. The recent advances in photonics, nanotechnology and telecommunications have increased the number of available technologies of all this components, allowing new innovative sensors.

Available sources

There is always more available sources:

- The incoherent Light Emitted Diode (LED): low cost and available in always more spectral range from UV to infrared using new semiconductor.
- The super-luminescent diode (SLD): more powerful than a classical LED. More generally, a lot of progresses have allowed increasing the LED output power.
- The coherent Laser Diode (LD): available in always more spectral range from UV to infrared. Their cost has decreased. They can be pigtailed (Fig.1) and power controlled.
- The Vertical-Cavity Surface-Emitting Laser (VCSEL), more powerful than classical LD but less coherent. They are often multimode.

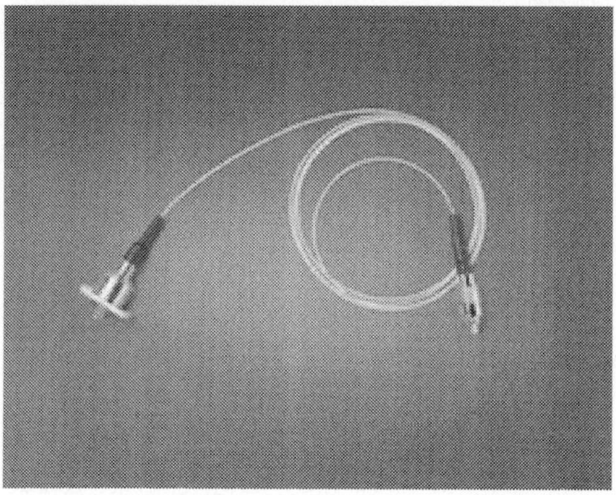

Fig. 1. Example of pigtailed Laser Diode.

Always more optical fibers

Due to telecom applications, the classical step index fibers in silica are very cheap (~0.10$/m). However beyond these specific applications, a large number of new fibers is nowadays available with different materials and geometries. Some of them can be customer-made. Among the available optical fibers, we can distinguish:

- The monomode and multimodes optical fibers, with core diameter from 5 microns to the millimetre.
- The polarization maintaining fibers (generally with an elliptical core).
- The step index fibers and the graded index ones with several index profiles.
- The fibers whose core and cladding are in silicate or polymer (plastic). A lot of improvements have been achieved concerning plastic fiber [Peters, 2011]: loss decrease, temperature resistance, homogeneity, etc.
- The silica fibers with several possible doping (germanium, fluorine, alumina, erbium, etc.).
- The micro-structured fiber in the radial direction such as the photonic crystal fibers.
- The Bragg grating fiber (with a periodic modulation of the refractive index in the longitudinal direction).

Several coatings can be used. Depending on applications, they can allow the sensitization of a fiber to a physical phenomenon to be measured or on the contrary its neutralization. Some possible coatings:

- Plastic (polymer, PPMA, etc.);
- Metallic (Silver, gold, palladium, copper, etc.);
- Magnetostrictive ;
- Electrostrictive ;
- Etc.

Transducers

The transducer is the main part of the sensor. It's the sensitized part, the location where the system has been designed in order to allow an external physical phenomenon to interact with the light propagating in the fiber. The transducer has to be selective, that is to be sensitive to phenomena that have to be measured without being sensitive to others.

In an intrinsic optical fiber sensor, the transducer converts the physical value to be measured into an optical information. It can be a thermo-optic, elasto-optic, magneto-optic, electo-optic, acousto-optic one, etc.

The optical fiber is generally weakly sensitive to external physical parameters (temperature, pressure, vibration, etc.). Generally, a perturbation has to be applied to increase its sensitivity. Example of perturbations:

- Micro-bending;
- Thermo-sensitive coating;
- Mechanical strengths;
- Curvatures.

Available detectors

Depending of the optical parameter that has been modified by the physical phenomenon to be measured, several detectors can be used. The detector is not only a light sensitive area but often a smart micro system:

- Photodiode for amplitude (intensity) monitoring; in silica for visible light detection (Fig.2) and in GaAs for infrared radiation. It is a cheap component. An amplification and transimpedance circuit can be integrated in the component.
- Polarizer, wave-plates and photodiode for polarized state determination. These components can be pigtailed or achieved in micro-optic. Components using liquid crystal or electro-optic components can be dynamically controlled. Alignment of these components is an issue. Their prices decrease but are still expensive.
- Spectrometers for spectral analysis from 0.1 to 5 nm resolution, from small to large range, with punctual, linear or two-dimensional detectors. In some case the use of a monochromater can be replaced by a wavelength-scanning laser. Large improvements have been achieved in the packaging of spectrometer allowing industrial applications. These devices are still quite expensive.

Fig. 2. A photodiode.

2.3 Known advantages of optical fiber sensors

Compared to electric or mechanical sensors, the optical fiber ones have several advantages:

- Their high electromagnetic compatibility, because of the superposition principle in the Maxwell equation. [Grattan et al., 2000 | Lee, 2003].
- The possibility to achieve extended and distributed sensors allowing in situ measurements.
- The compatibility to explosive environments. Generally, no current is needed in the sensitized part.
- The low cost components due to Telecom applications.
- The high integration possibility.
- The long time duration of optical fiber and components.
- The large available dynamic.
- The high sensitivity.
- No moving mechanical part. Useful when the sensor is submitted to high accelerations.

2.4 Methods to sensitize an optical fiber

Historically, optical fibers were achieved in order to carry out long distance propagation and some of their main issues were to minimize losses and possible disturbances: regularity of the geometry, purity of the material, protection of the external jacket. Today 0.2dB/km is achieved for long distance telecom connexions (monomode fibers).

In contrast to optical fiber sensor applications, we want to improve the optical fiber sensibility to an external physical parameter. This can be done mainly in three ways:

- Apply geometric perturbations: curvatures, micro-bending, non axis-symmetric deformations, etc.
- Choose specific sensitive materials: materials with magneto-optic, electro-optic, thermo-optic, elasto-optic properties; materials able to adsorb other ones;
- Remove the external jacket or replace it by a system able to enhance the external physical parameter effect on the fiber.

3. Monomode optical fiber intrinsic sensing: The principles

An optical fiber is monomode when only the fundamental mode is propagative. It is the case when the normalized frequency $V<2.405$. The normalized frequency V of an optical fiber of radius a, numerical aperture NA, with a source having a free space wavelength λ_o is:

$$V = \frac{2\pi}{\lambda_0} a\text{NA} \tag{1}$$

If the refractive index of the core and respectively the optical cladding is n_c and respectively n_g, the numerical aperture of the fiber is:

$$\text{NA} = \sqrt{n_c^2 - n_g^2} \tag{2}$$

Example: the monomode silica/silica fibers are the optical fibers used for long distance communication applications with near infrared sources (λ_o=1.3µm, λ_o=1.5µm). The core is generally doped with germanium n_c=1.465, n_g=1.455, their radius is smaller than 10 microns.

In fact a monomode fiber has 4 degenerate modes, that means 4 different modes with the same propagation constant: a TE and TM fundamental modes in two propagating directions. Perturbations can break this degeneration.

Three main families of optical sensors can be achieved using monomode optical fibers: the sensors where polarization or phase or reflected wavelength is modulated by an external physical parameter. For these three families, a single initial possible propagation constant is needed.

3.1 Polarization modulation

In this case, the physical phenomenon to be measured modifies the polarization of light propagating in a monomode optical fiber. The fiber has to be monomode because of the depolarization process taking place in the multimode optical fiber.

Polarization

The polarization describes the evolution of the electric field direction during time and especially its regularity.

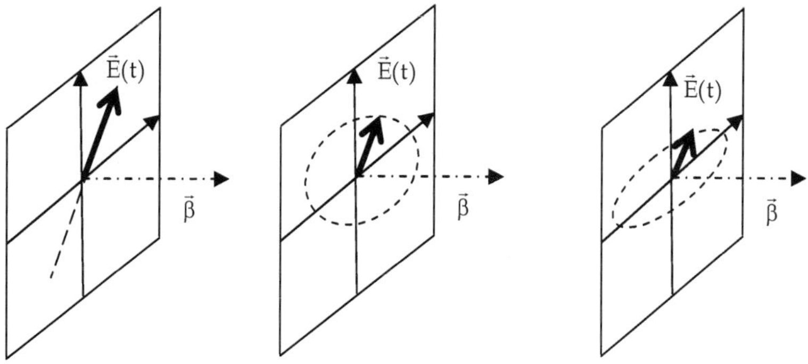

Fig. 3. Linear, circular and elliptical state of polarization.

The polarization is defined only for transverse waves, that is when the electric field or the magnetic one is orthogonal to the propagation vector **β**. The classical defined states of polarization are the linear, circular or elliptical polarization following the shape described in the transverse plane by the electric field vector during time (see figure 3).

More generally, a state of polarization can be described using the Stokes elements. If *x* and *y* are two orthogonal directions in the transverse plane, the Stokes elements are defined as:

$$S = \begin{pmatrix} S_0 \\ S_1 \\ S_2 \\ S_3 \end{pmatrix} = \begin{pmatrix} < E_x E_x^* > + < E_y E_y^* > \\ < E_x E_x^* > - < E_y E_y^* > \\ < E_x E_y^* > + < E_y E_x^* > \\ i(< E_y E_x^* > - < E_x E_y^* >) \end{pmatrix} \qquad (3)$$

Where <*A*> represents the time mean value of *A* and *A** the conjugate one. Generally, the Stokes elements are normalized by S_0, the total radiated power. The Stokes elements are intensity terms which have the advantage to be deduced by means of measurements; however, they do not constitute a vector space, because the sum of two Stokes elements does not necessary correspond to a physical state of polarization. The degree of polarization is given by:

$$p = \frac{1}{S_0}\sqrt{S_1^2 + S_2^2 + S_3^2} \qquad (4)$$

p cannot be larger than 1 and smaller than 0. *p=1* corresponds to a perfectly polarized wave and *p=0* to a non polarized wave. The classical states of polarization are:

S=(1 1 0 0) linear polarization parallel to x ;

S=(1 -1 0 0) linear polarization parallel to y ;
S=(1 0 1 0) linear polarization at 45° of direction x and y ;
S=(1 0 -1 0) linear polarization at -45° of direction x and at -135° of y ;
S=(1 0 0 1) circular right polarization ;
S=(1 0 0 -1) circular left polarization.

All the physical Stokes elements can be described in the Poincare sphere or bead (Fig.4).

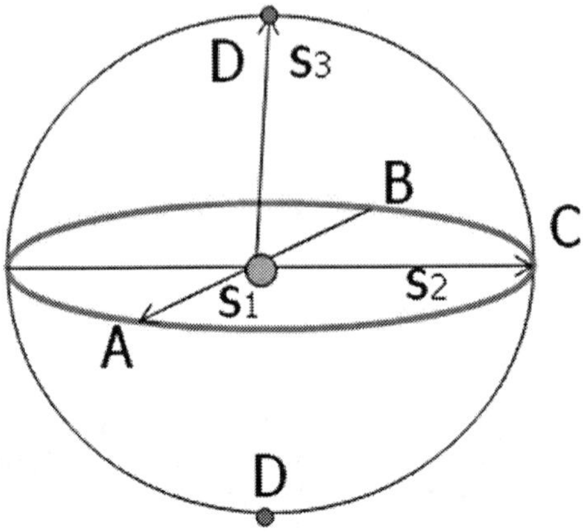

Fig. 4. Poincare sphere describes all the physical possible states of polarization

The perfectly polarized waves are on the sphere surface, the non polarized ones are in the centre, in between we find the partially polarized wave ($0<p<1$). The poles correspond to the two circular polarization states and the equator describes the linear polarization states. The others points on the surface correspond to elliptical polarization states.

Mueller matrix

If the initial polarization state is described by a Stokes element, its evolution can be described by a 4x4 matrix called the Mueller matrix.

Sensor structure and principle

In a polarization modulated optical fiber sensor, the source has to be monochromatic, polarized but not necessary coherent. The change of polarization can be explained by a change of birefringence in the optical fiber due to the physical phenomenon to be measured. The polarization can rotate, change of state or both.

The detector is generally constituted of a polarizer and a photodiode. The change in the polarization state produces an intensity modulation measured by the photodetector after the polarizer. The calibration curve links the modulation intensity ranging from *Imin* to *Imax* and the physical phenomenon to be measured.

Birefringence

The birefringence is an anisotropy of the optical properties. In the main case, this anisotropy will bring about a change in the polarization state.

Depending on the polarization eigen mode, the birefringence can be linear or circular. The eigen polarization modes are the modes whose polarization is conserved after propagation in the medium. In the case of the circular birefringence the eigen modes are the circular left and right polarizations. In the case of the linear birefringence, the eigen modes are two orthogonal linear polarizations parallel to two axes called the extraordinary and ordinary axes. These polarizations have different refractive indexes written n_e and n_o. This birefringence can be uniaxial if the refractive index is n_e or n_o in the third direction or else biaxial.

The birefringence breaks the degeneration of the two modes TE and TM and two propagation constants have to be considered:

$$\Delta\beta = |\beta_e - \beta_o| = \frac{2\pi}{\lambda_0}|n_e - n_o| = \frac{2\pi}{\lambda_0}\Delta n \tag{5}$$

After propagation in a birefringent optical fiber of length l, a phase shift $\Delta\Phi$ appears between the two modes that changes the polarization states.

$$\Delta\Phi = \frac{2\pi}{\lambda_0} l\Delta n = \Delta\beta.l \tag{6}$$

In a monomode optical fiber the birefringence is due to an asymmetrical geometry, if the core has an elliptical shape for example, or an anisotropy of the medium itself. Two kinds of birefringence can be distinguished: the intrinsic birefringence due to the process of fabrication and the natural defaults of the fiber and the extrinsic one due to external perturbations. The intrinsic birefringence is characterized by the polarisation beating length, which is the smallest length of propagation in order to find again the same polarization as the source. In a classical silica monomode optical fiber used for telecom application the polarization beating length is around 1.2m.

An optical fiber sensor by polarization modulation uses the extrinsic birefringence. Several external perturbations can create an extrinsic birefringence. We will present several of them.

Curvature

In an optical fiber curvature, two different phenomena may result in a birefringence, the geometry modification and the elasto-optic effect. The elasto-optic effect is a change in refractive index due to a mechanical stress in the medium. An anisotropic stress may achieve optical anisotropy.

The birefringence due to an optical fiber curvature as described in figure 5 is given by:

$$\Delta n = \frac{1}{4} n_c^3 (p_{11} - p_{12})(1 + \nu)\frac{r^2}{R^2} \tag{7}$$

Where :

R is the curvature radius,
r is the optical fiber core radius,
n_c is the core refractive index,
P11-P12 are the elasto-optic coefficients (*P11=0,121* and *P12=0,27* for silica),
ν is the Poisson coefficient (silica: ν=0,17).

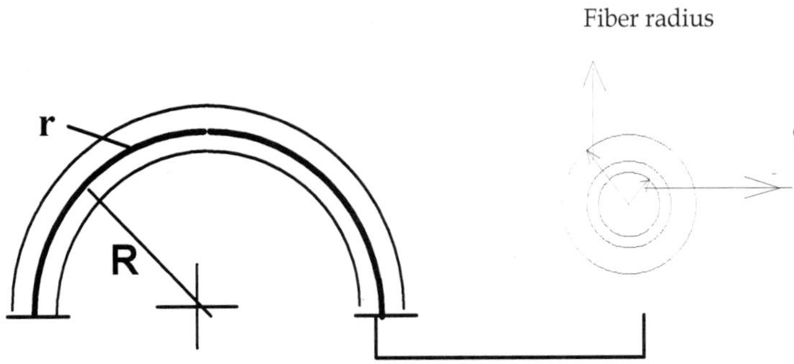

Fig. 5. Optical fiber curved with a radius R, without additional stress

The fast birefringence axis (corresponding to the lowest refractive index) is in the curvature plane and the slow one (corresponding to the highest refractive index) is parallel to the curvature axis.

One of the applications is the Lefèvre loop. In this case, an optical fiber loop is made in order to achieve the equivalent of a half or quart wavelength plate.

Compression

A directional force applied on the optical fiber as described in figure 6 brings about stresses in the optical fiber and, therefore, a linear optical birefringence:

$$\Delta n(F) = 2n_c^3(P_{11} - P_{12})(1+\nu)F/(\pi r E) \tag{8}$$

Where:

-F is the force by fiber length unit L,
-E is the Young modulus of the fiber.

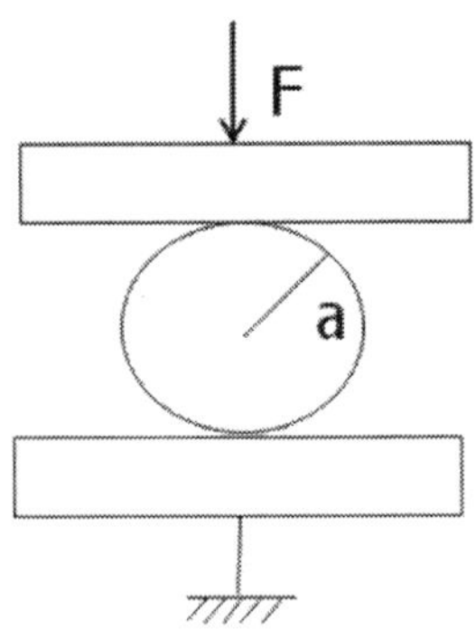

Fig. 6. Directional force applied on an optical fiber.

The fast polarization axis of birefringence is orthogonal to the force F.

Therefore, the phase shift between the two linear polarized modes is:

$$\Phi_{x-y}(F) = 4n_c^3(P_{11} - P_{12})(1+\nu)FL/(\lambda rE) \quad (9)$$

Torsion

When an optical fiber is twisted, a circular birefringence appears. The difference of propagation constant between the two circular polarized modes left and right $\Delta\beta$ is proportional to the twist ratio τ expressed in rad/m:

$$\Delta\beta = g\tau \quad (10)$$

g is a adimensional constant depending on the medium:

$$g = n^2(P_{11} - P_{12})/2 \quad (11)$$

This birefringence makes a linear polarization to rotate at a ratio $g\tau/2$ (rad/m).

Electric and magnetic field

A magnetic field (static or low frequency) may also be able to achieve a circular birefringence because of the Faraday effect. The rotation angle of an incident linear polarization α will be:

$$\alpha = vLB \quad (12)$$

Where:

v is the Verdet constant.
L is the length of fiber where the magnetic field is applied.
B is the magnetic field component parallel to the optical fiber axis.

The electric field due to the Pockels in a crystal can also make change the polarization by a linear birefringence and therefore be detectable. The same method can be used to measure voltage: a voltage V between two conductive plates separated of a distance d achieve an electric field V/d orthogonal to the plates.

3.2 Phase modulation

In this case, the external physical phenomenon to be measured modifies the phase shift between two coherent propagating beams having different paths.

Mach-Zehnder

Most phase modulated optical fiber sensors are interferometers. Using a coupler, the light source is guided through two optical fibers: one is the reference arm, the other one is the measurement arm. Using a second coupler the two beams are again put together (Fig.7). [Yuan et al., 2000 | Ghorai et al., 2005].

The fiber has to be monomode, otherwise each mode would have a different phase shift linked to its eigen propagation constant. The coherence length of the light source has to be larger than the maximum possible difference of the optical path between the two arms.

The phase shift will be linked to a refractive index change in the measurement arm. This refractive index modulation can take place in the optical fiber core or cladding due to a

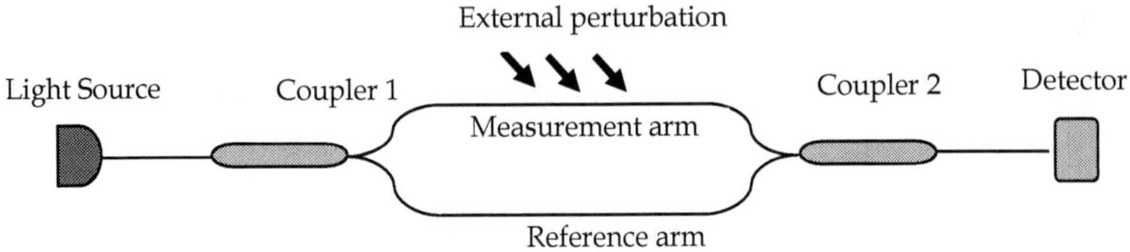

Fig. 7. Mach-Zehnder optical fiber interferometer

large range of possible phenomena (thermal, mechanical, chemical, electromagnetical, etc.).

Sagnac effect

When two counterpropagative waves propagate in an optical fiber loop in rotation (figure 8), a phase shift appears between the two counterpropagative waves which is proportional to the rotation rate Ω (rad/s). This phenomenon is called the Sagnac effect [Sagnac, 1913].

The phase shift is:

$$\Delta\phi_S = \frac{8\pi RL}{\lambda_0 c}\Omega \tag{13}$$

Where :

L is the length of optical fiber in the loop,
Ω is the rotation rate (rad/s).

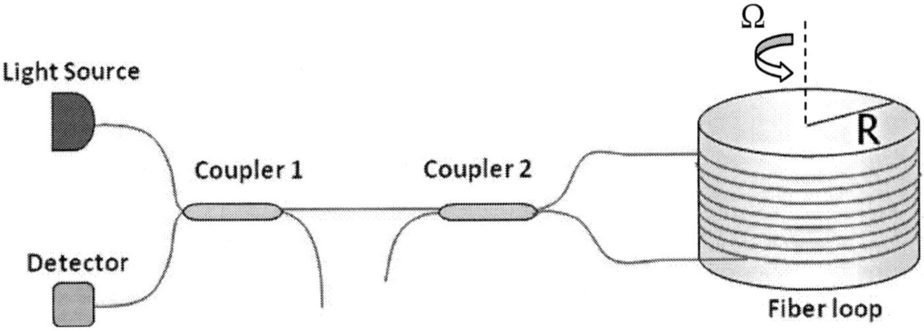

Fig. 8. Sagnac optical fiber interferometer

Interferometers as described in figure 8 allow to measure this phase shift and, therefore, to deduce the rotation rate [Burns et al., 1983][Zarinetchi et al., 1991].

Generally, this kind of gyroscope is made with monomode polarization maintaining fibers. Moreover, in order to allow unambiguous measurements, the maximum difference of the optical pathway cannot be larger than λ_o. Therefore, in the visible range, a LED whose coherence length is around 3μm is enough to create the needed interferences.

3.3 Spectral modulation

In this case, the external physical phenomenon to be measured modifies the source spectre in its transmission or reflexion.

Bragg grating

A fiber Bragg grating is an optical fiber where a periodic modulation of the refractive index has been written in its core. Thus refractive index modulation is often photo-written using UV Eximer laser. Because of this index periodicity, several given wavelengths λ_p will be reflected (Fig.9):

$$\lambda_p = 2n_e\Lambda / p \tag{14}$$

where Λ is the grating period, p an integer and n_e the effective refractive index of the grating.

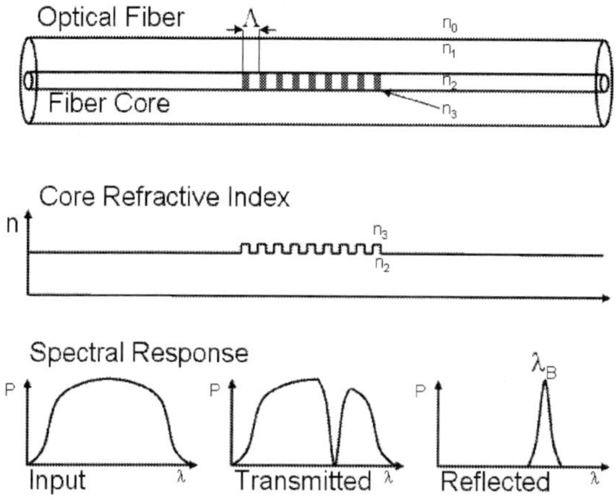

Fig. 9. Fiber Bragg grating principle

Stretching, temperature or pressure may change the reflected wavelength due to a change in the refractive index. To measure this change in the spectrum, a monochromater or a scanning wavelength laser has to be used. These two devices are quite expensive. However the fiber Bragg grating sensors, because of their accuracy, are the more used optical fiber sensors in the industry. They allow the measurement of several physical parameters:

- Stretching and force [Kang et al., 2007 | Ferraro et al., 2002] ;
- Temperature [Zhao et al., 2007 | Zhan et al., 2007 | Hirayama et al., 2000] ;
- Pressure [Takahashi et al., 2000 | Ni et al., 2007] ;
- Two physical parameters together [Hathaway et al., 1999].

The typical sensitivities of fiber Bragg gratings are summarized in table 1 [Lee, 2003].

Parameters	stretching	temperature	pressure
Sensibility $\Delta\lambda$	1pm/µm	12pm/°C	-0.5pm/bar

Table 1. The typical sensitivity of fiber Bragg gratings [Lee, 2003].

Several Bragg gratings with different periods can be used in the same fiber what allows to achieve a distributed sensor with several measurement localizations.

Fabry-Perot

Some spliced fibers are used as a Fabry-Perot cavity whose resonance is shifted as a function of the external physical perturbation to measure. The shift can be due to the elasto-optic effect, the thermo-optic effect, the Sagnac effect, etc. It can take place inside the fiber or at the fiber ends.

4. Multimode optical fiber intrinsic sensor: The principles

With the multimode optical fiber sensor, the physical phenomenon to be measured modifies the transmitted intensity. An optical fiber is multimode when its normalized frequency is larger than 2.405. Multimode fibers often have a larger core radius (compared to the wavelength) or a higher numerical aperture than a monomode fiber. In this case, several modes defined by different transverse intensity distributions and propagation constants can be guided inside the optical fiber. The transmitted intensity modulation is due to mode couplings between the guided modes and the radiated ones. The mode couplings are the exchanges of energy between modes and the radiated modes are the propagative modes which are not guided inside the fiber but go out in the optical cladding and are generally absorbed by the mechanical jacket. The mode couplings will be due to perturbations of the optical fiber.

4.1 Micro bending and mode coupling

The micro bending is a classical perturbation of a multimode fiber. In this case, the fiber is periodically curved (Fig.10). If z is the initial axis of the fiber, the new one is a periodic function $f(z)$. The micro bending may change the local numerical aperture creating losses [Remouche et al., 2007].

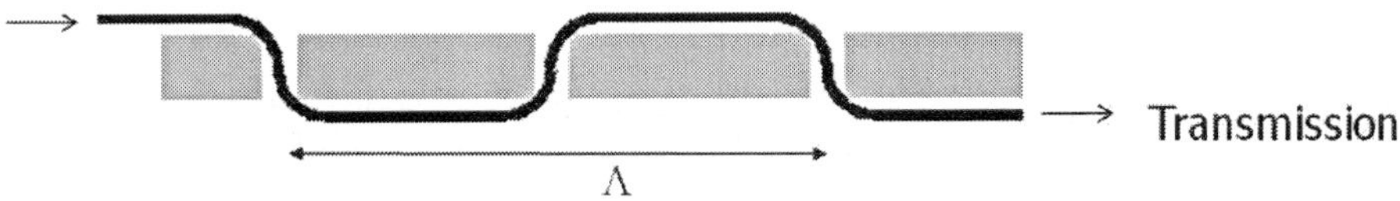

Fig. 10. Micro-bending of a multimode optical fiber

In an optical fiber of refractive index n_c in the core and n_g in the optical cladding, the guided modes have a propagation constant β_g between $2\pi\ n_g\ /\lambda_0$ and $2\pi\ n_c\ /\lambda_0$ whereas the propagation constants of the radiated modes β_r range between 0 and $2\pi\ n_g\ /\lambda_0$ (Fig.11). A guided mode of propagation constant β_g can be coupled with a radiated mode of propagation constant β_r if the geometric perturbation has a period Λ such as:

$$\beta_g - \beta_r = \frac{2\pi}{\Lambda} \quad (15)$$

For a small bending period, the guided modes will be coupled together prior to being coupled with the radiated modes. This method can be used to distribute the source energy

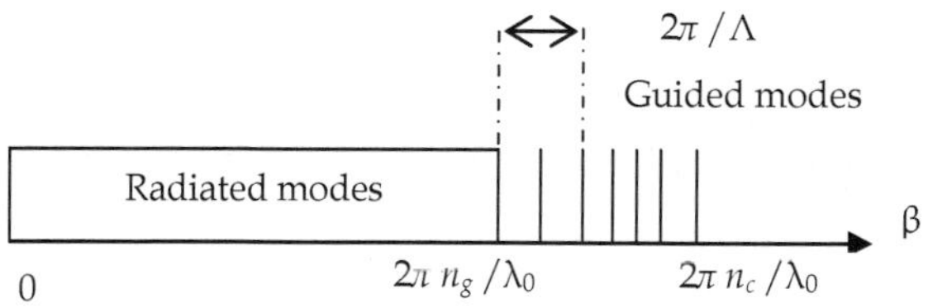

Fig. 11. Propagation constant distribution in an optical fiber: example of two coupled modes

on the entire guided mode of the fiber. In contrary, when the bending is too high, the optical fiber can have inelastic modifications and can be broken. The maximum curvature radius of a classical silica/silica fiber is around 1.5mm.

The losses can be also explained and modelled by using ray tracing (Fig.12).

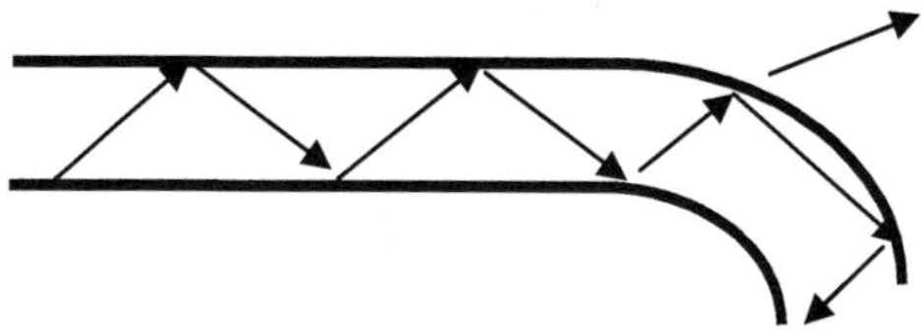

Fig. 12. Ray tracing in a micro bend optical fiber

The physical parameters which can be measured using micro-bending optical fiber sensors are pressure, force [Lagakos et al., 1987], acceleration, temperature [Yadav et al., 2007], Stretching [Luo et al., 1999] and the electromagnetic field [Lagakos et al., 1987].

It is also possible to periodically perturb the optical fiber diameter. The coupling model has been developed by Marcuse [Marcuse, 1991].

4.2 Spatial mode filtering

When a multimode fiber is welded to a monomode one a spatial mode filtering is achieved (Fig.13). This principle has been used to realize a stretching sensor [Ribeiro et al., 2004]: the physical parameter to be measured creates mode couplings in the multimode fiber and the monomode fiber allows to filter only the fundamental mode.

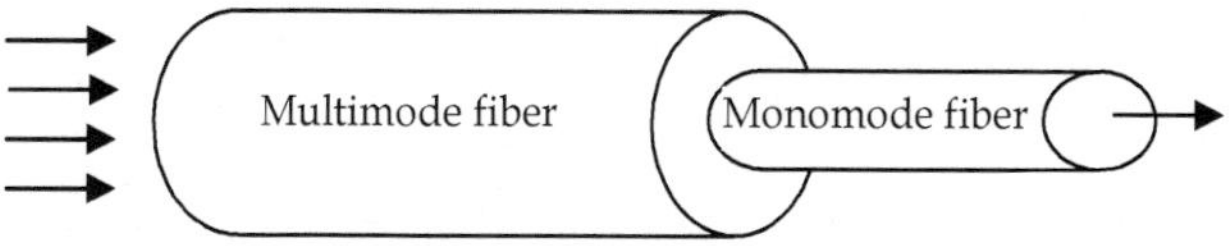

Fig. 13. Spatial mode filtering using welded multimode and monomode fibers

4.3 Scattering and fluorescence

The multimode optical fiber sensors are mainly amplitude modulation sensors. Independently of the geometry several others phenomena can achieve an amplitude modulation linked to a physical parameter to be measured:

- **Absorption**. It is linked to the imaginary part of the refractive index.
- **Rayleigh scattering**. It is an elastic scattering: the wavelength is conserved. The light is spatially dispersed because of optical fiber defects smaller than the wavelength, what makes decrease the transmitted intensity. The Rayleigh scattering can be measured with a spatial localization using an OTDR (Optical Time-Domain Reflectometer).
- **Raman scattering**. It is a non linear inelastic scattering where electromagnetic and acoustic energies are exchange. Stokes rays with frequency smaller than the source and anti-Stokes rays with frequency higher than the source appear. Only the anti-Stokes ray depends of the temperature, what has interest for metrology.
- **Fluorescence**. Molecules are exited for example by absorbing a radiation or a particle and then emit a photon when they come back to their fundamental state. Fluorescence can be linear or not, elastic or not. The typical time between absorption and emission is an important parameter. UV and gamma optical fiber sensors are achieved using this principle, biological sensors also.
- **Brillouin scattering**. It is a non linear scattering: a frequency shift takes place due to interaction between the light wave and acoustic waves (Stokes ray). The frequency shift is in the range of 10 GHz and typically depends of temperature and stress in the fiber. A large range of optical sensors have been achieved using this phenomenon measuring spectrum or absorption. Dynamic and distributed sensors can be done using the time modulation of the acoustic wave.

4.4 Transimpedance circuit

In an amplitude modulation optical fiber sensor the transmission can be measured using a simple photodiode. In this case, a transimpedance circuit (figure 14) allows minimizing the noise level. The photo current out of the diode is converted into a stable output voltage. Since there is no voltage across the diode, there is no black current and the diode response is linear. The resistance R_f allows having an adjustable amplification:

$$V_{out} = R_f I \tag{16}$$

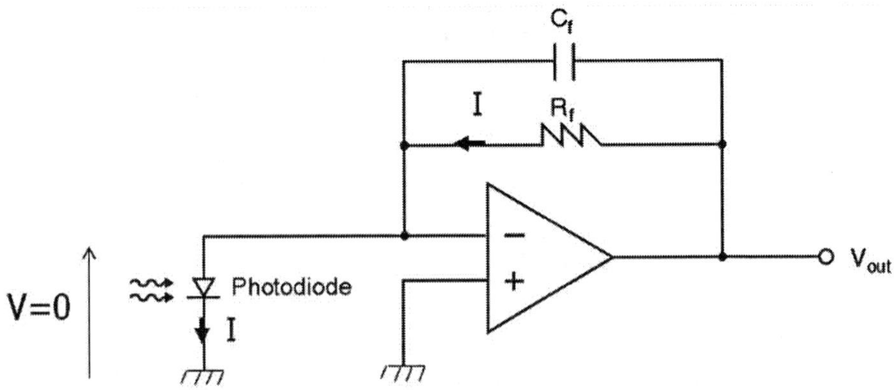

Fig. 14. Transimpedance circuit

5. Example of intrinsic optical fiber sensing engineering and examples

Figure 15 represents the ratio of scientific papers concerning the various physical parameters to be measured with optical fiber sensors [Lee, 2003]. The more studied parameters are stretching, temperature and pressure. The used technologies for measurements using optical fiber sensors are shown in figure 16.

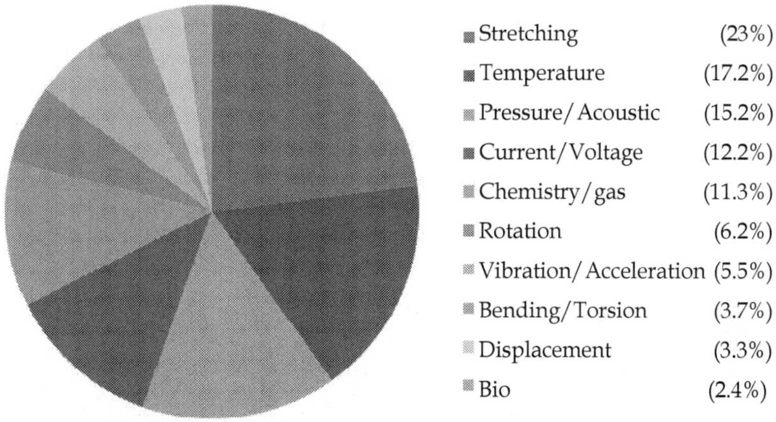

Fig. 15. Ratio of scientific papers concerning the different physical parameters to measure [Lee, 2003]

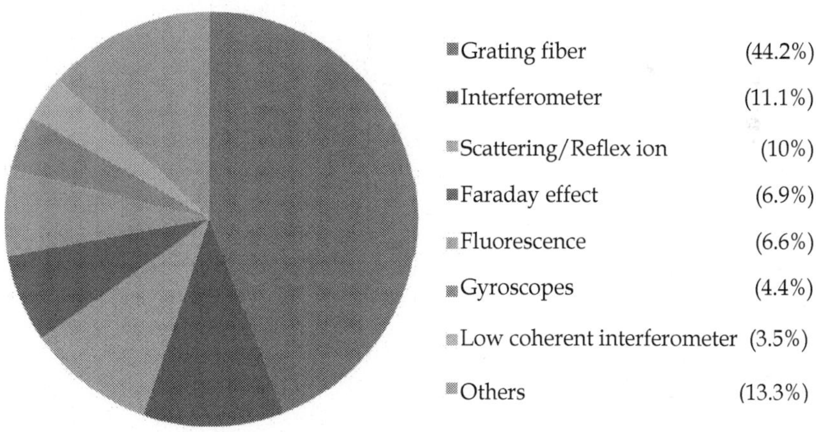

Fig. 16. used technologies for measurements using optical fiber sensors [Lee, 2003]

Bragg grating fiber

The Bragg grating fiber is the more studied technology [Lee, 2003].

Bragg grating optical fiber sensors are used to measure temperature, pressure, stretching, force, acceleration, etc. [Ferdinand et al., 1997]. Inclinometers have also been achieved.

Force, vibration measurements and shock detection

Force, vibration and shock detection measurements can be carried out using micro bends in multimode fibers or using polarization modulation out of monomode fibers. Microphones, hydrophones (US Navy), seismometers, vibrometers have been achieved. Large dynamic can be considered. These technologies allow achieving long lifetime and low cost sensors that could be applied in the automotive applications.

Such measurements can also be done using distributed Bragg grating fibers or Brillouin sensor. For example, Bragg grating fibers have been used for railway monitoring.

Torque measurement

Torque measurement can be achieved using polarization study out of a twist monomode optical fiber. The measurement is indirect; the applied torque will be linked to the measured twist.

Rotation rate measurement

The optical fiber gyroscopes are one of the more accurate gyroscopes. Rotation rates smaller than 0.1°/h could be detected using maintaining polarization fibres. These quite expansive gyroscopes (several 10k$) are used for example for military navigation applications. These sensors have no moving mechanical parts and are resistant to electromagnetic field. Several configurations have been proposed: open loop interferometer, resonant ring, Brillouin gyroscope, etc. Cheaper multimode optical fiber gyroscopes have also been proposed for rate-grade applications [Medjadba et al., 2011] when classical micromechanical systems cannot be used.

Temperature measurement

The main effect that will be used is the thermo-optic effect, which is the change of refractive index of a material due to the temperature. This phenomenon is characterized by the thermo-optic coefficient α:

$$n = n_{T_0} + \alpha(T - T_0) \tag{17}$$

The thermo-optic coefficient of silica is around $1.45.10^{-5}K^{-1}$. This change of refractive index may change the reflected wavelength in a Bragg grating fiber sensor, the transmission in an amplitude modulation sensor, the phase shift in a Mach-Zhender sensor, etc.

In the amplitude modulation sensor, to increase the change of numerical aperture due to temperature, different materials are used in core and cladding. Typically the core will be in silica and a polymer with a negative thermo-optic coefficient will be used. The fiber with polymer can generally not be used for temperature larger than 50°C, whereas silica fiber can be used until 800°C.

The dependence of the Brillouin scattering on the temperature is also used to achieve temperature sensors. Another principle that could be used is the change of Young modulus as a function of the temperature in a perturbed plastic optical fiber.

Gas and pollution detection

Gas and pollution detection are possible using absorption or fluorescence in amplitude modulation sensor. Refractive index change can also be obtained in some materials. Hydrogen is absorbed by palladium coating for example, or Rayleigh scattering due to pollution can increase absorption in hollow photonic crystal fiber.

Shape and stretching measurement

Shape and stretching can be measured by observing bending change or Bragg grating period elongation. Optical stress gauge can be achieved. Distributed sensors have been used in geni-civil applications as strain highway bridge monitoring and mine security control.

Magneto optic current sensor

Polarization modulation sensors or Sagnac interferometer could be used to measure currents due to the induced magnetic field and the Faraday effect. This measure can be carried out without contact and electromagnetic perturbation of the studied system.

Defence applications

The possibility to use optical fiber sensors in explosive environments or in an area where high electric or magnetic field take place, are often researched advantages for military applications. No moving mechanical part is also useful when a sensor is submitted to high accelerations.

One of the more used optical fiber sensor in the defence application is the optical gyroscope for navigation mainly because is high accuracy.

Ecology and biomedical applications

The optical fibers are used for biological studies: grow of cells, kinetic of binding reactions, chemical evolution monitoring, etc. using refractometry, interferometer, amplitude modulation, etc. The biological samples can be located at the fiber end, as an optical cladding or in the hollow core of a photonic crystal fiber. Integrated biochip can be imagined. Gas detection and concentration measurements can be done. The distributed in-situ temperature and strain measurements have also been studied for medical applications, inside the body.

6. Prospectives: Sensing with photonic crystal fibers and plasmonic sensors

New optical fiber structures and materials, but also the improvements of the micro and nano fabrication processes allow the development of new families of optical fiber sensors. These new sensors are often still in development and need additional time before to be mature for industrial applications. However, nowadays a many patents are deposed in order to protect these future technologies.

6.1 Micro structured fibers

The classical optical fibers are made of bulk materials, but now it is possible to micro structure the core or the optical cladding of the fiber. These new fibers can be decomposed into two families: the photonic crystal fiber and effective index optical fiber. Because of their guiding principles these fibers are highly sensitive to external physical parameters (temperature, pressure, bending, etc.). They have been used for example to achieve biochemical sensors [Rindorf et al., 2006], gas detection [Hoo et al., 2003], etc.

The photonic crystal fiber

These fibers (figure 17b) have a hollow core and the guidance properties are due to a photonic band gap in the periodically structured optical cladding [Bjarklev, 2003]. A photonic band gap is a spectral domain for which the light cannot propagate in the medium: it appears in structures where the optical properties are periodic with a period that can be compared to the wavelength and where the refractive index contrast is higher than 2. The periodic structure is generally done by using air and another material. The hollow core allows not only a low dispersive propagation but also the insertion of gas to study inside the fiber. Liquids can also insert in the hollow core, as for example for the photonic liquid crystal fiber.

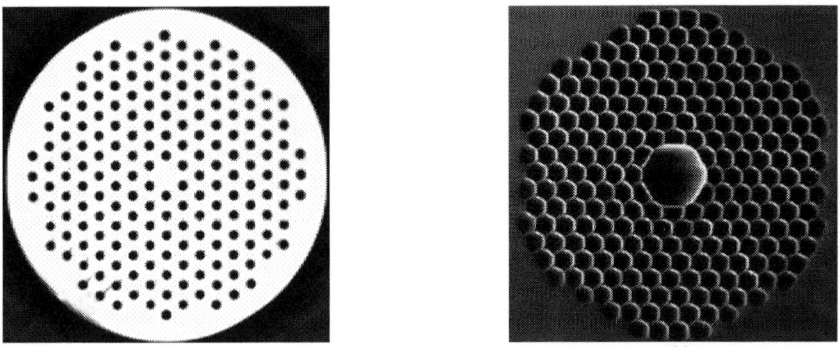

(a) (b)

Fig. 17. Micro structured optical fiber: (a) Effective index fiber, (b) photonic crystal fiber

Effective index optical fiber

The core of these optical fibers is full and the cladding is structured at a scale smaller than the wavelength in order to create an effective medium with an effective refractive index smaller than the one of the core (figure 17a).

Temperature and strain sensors, gas detectors and refractometers have yet been experimented using these new fibers. The polarization maintaining photonic crystal fiber and the large mode area photonic crystal fiber, both monomode, are specifically studied.

6.2 Plasmonic sensors

A plasmon is a surface electromagnetic wave linked to a surface current. These waves propagate along a metallic surface and are evanescent in other directions. Because of the metallic complex refractive index, the plasmons are generally highly absorbed allowing amplitude modulation sensors. Plasmon resonances can occur in thin metal layers (several ten nanometers thickness for visible range resonances). These resonances are very sensitive to the environment (refractive index), therefore sensors have been achieved. In this case we talk about Surface Plasmon Resonance optical fiber sensor.

Figure 18 is an example of plasmonic optical fiber sensor for hydrogen detection [Perrotton et al., 2011]. Locally, the optical cladding has been replaced by a thin 10 nm layer of palladium. The adsorption of hydrogen by the metallic palladium layer changes its complex refractive index and therefore its absorption and resonance spectrum. Ratio of Hydrogen smaller than 4% has to be detected. It's an example of optical sensors used in an explosive environment in order to minimize risks. The plasmon excitation can also be done on a thin layer at the end of the fiber. In this case, the detection takes place in reflexion.

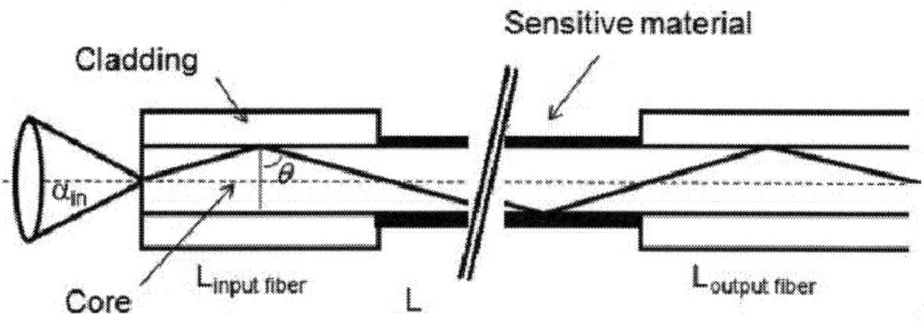

Fig. 18. Plasmonic optical fiber sensor for hydrogen detection. In black: palladium layer

This family of new sensors are sometime classified as evanescent field sensors because a plasmon is a particular case of evanescent waves. This wave is used as a very sensitive and localized probe to sense a system.

7. Conclusion

We have seen that many simple physics principles can be efficiently used for intrinsic optical fiber sensing through an appropriate engineering. With the new potential open by photonics crystal fibers and integrated optics probably many others applications in optical fiber sensing of these physics principles are going to be discovered in the years to come.

The main inconvenient of intrinsic optical fiber sensing is to sense "everything" at the same time. This inconvenient was progressively overcome.

It is possible to say, now, that optical fiber sensing is a mature technology able to solve many measurements and control problems, and not only in the laboratory.

The optical fiber sensing industry is just emerging. Its principal burden is to deal with the lack of trained engineers and technician in this field. What make that even if the basic knowledge on optical fiber sensing is now abundantly available in literature, know-how to proceed is scarcely diffused. An educational effort is necessary, but it will provide positive results only with long delay.

Anyway the engineers and scientists from many fields have to be aware of the existing capacities of optical fiber, to be able to consider them for optimizing their process or devices by trying to find an assistance among the small number of specialized laboratories.

Since innovation is going to be the main motor of the future economy, optical fiber sensing can bring many progress for innovative products, even if it is difficult presently to associate experts in optical fiber sensing to projects because these experts are not enough numerous.

So optical fiber sensing has the potential, in the industrial world, to transform its statute by evolution from a niche technology to a basic enabling industrial growth opportunity that can become strategic in : transportation, biomedical, defense, ecology, energy, etc.

Optical fiber sensing offers a tool to fabricate products more secure, less polluting or poisonous, saving energy and rare materials.

Optical fiber sensing belong at the same time to the past by being related for instance to the Nobel Prize of M. KAO for optical fiber and to the future because it allows to imagine and to realize creative products implementing new functions with many advantages.

In this chapter we have tried to summarize the state of the art in optical fiber sensing. We hope that this chapter will rise the interest of engineers and scientists not aware of the strong opening to innovations of optical fiber sensing, and that they will follow by trying to learn more specifically from optical fiber sensing laboratories that deserve assistances of any kind to contribute to the rewarding accomplishments that, we hope, are going to take place in the future through classical and disruptive optical fiber sensing technologies.

8. Acknowledgment

The authors thank Ayoub Ckakari, Pierre Pfeiffer, Nicolas Javahiraly, Sylvain Fischer, Bruno Serio, Francis Georges, Anthony Bichler, Cédric Perrotton, Hocine Medjadba and Mustafa Remouche that work with us on optical fiber sensors and share their experience.

9. References

A. Bjarklev, A. S. Bjarklev, J. Broeng, *Photonic crystal fibers*, Springer 2003.

W. K. Burns, R. P. Moeller, C. A. Villaruel, and M. Abebe, *Fiber-optic gyroscope with polarization-holding fiber*, Opt. Lett., Vol. 8, pp. 540–542, 1983.

P. Ferdinand, S. Magne, V. Dewynter-Marty, C. Martinez, S. Rougeault, and M. Bugaud, *Applications of Bragg grating sensors in Europe*, in *Optical Fiber Sensors*, OSA Technical Digest Series, 1997

P. Ferraro, G. De Natale, *On the possible use of optical fiber Bragg gratings as strain sensors for geo dynamical monitoring*, Optics and Lasers in Engineering, 37, 2002, pp. 115-130

M. Ferretti, *Capteurs à fibres optiques*, Techniques de l'ingénieur, traité Mesures et Contrôles, R415, 1996

S.K. Ghorai, D. Kumar, B.K. Hura, *Strain measurement in a Mach-Zehnder fiber interferometer using genetic algorithm, Sensors and Actuators* A, 122, 2005, pp. 215-221

K.T.V Grattan, T. Sun, *Fiber optic sensor technology: an overview*, Sensors and Actuators, 82, pp. 40-61, 2000

M.W. Hathaway, N.E. Fisher, D.J. Webb, C.N. Pannell, D.A. Jackson, L.R. Gavrilov, J.W.

N. Hirayama, Y. Sano, *Fiber Bragg grating temperature sensor for practical use*, ISA Transactions, 39, pp. 169-173, 2000

Y. L. Hoo, W. Jin, C. Shi, H. L. Ho, D. N. Wang, and S. C. Ruan, *Design and Modeling of a Photonic Crystal Fiber Gas Sensor*, Appl. Opt. 42, 3509-3515, 2003

L. Kang, D. Kim, J. Han, *Estimation of dynamic structural displacements using fiber Bragg grating strain sensors*, Journal of Sound and Vibration, 305, 2007, pp. 534-542

N. Lagakos, J.H. Cole, J.A. Bucaro, *Microbend fiber-optic sensor*, Applied Optics, Vol. 26, N°11, pp. 2171-2180, 1987

B. Lee, *Review of the present status of optical fiber sensors*, Optical Fiber technology, 9, pp. 57-79, 2003

F. Luo, J. Liu, T.F. Morse, *A fiber optic microbend sensor for distributed sensing application in the structural strain monitoring*, Sensors & Actuators, 75, pp. 41-44, 1999

D. Marcuse, *Theory of dielectric optical waveguides*, Edition : Academic Press, Second Edition, 1991

H. Medjadba, S. Lecler, L. M. Simohamed, J. Fontaine, R. Kiefer , *An optimal open-loop multimode fiber gyroscope for rate-grade performance applications*, Optical Fiber Technology, 2011

X. Ni, Y. Zhao, J. Yang, *Research of a novel fiber Bragg grating underwater acoustic sensor*, Sensors and Actuators A, 138, 2007, pp. 76-80

C. Perrotton, M. Slaman, N. Javahiraly, H. Schreuders, B Dam, *Wavelength response of a surface plasmon resonance palladium-coated optical fiber sensor for hydrogen detection*, Optical Engineering 50(1), 2011

K. Peters , *Polymer optical fiber sensors – a review*, Smart Mater. Struct. 20 013002, 2011.

M. Remouche, R. Mokdad, M. Lahrashe, A. Chakari and P. Meyrueis, *Intrinsic optical fiber temperature sensor operating by modulation of the local numerical aperture*, Opt. Eng. 46, 024401, 2007

R.M. Ribeiro, M.M. Werneck, *An intrinsic graded-index multimode optical fibre strain-gauge*, Sensors and Actuators A, 111, pp. 210-215, 2004

L. Rindorf, J. B. Jensen, M. Dufva, L. H. Pedersen, P. E. Høiby, and O. Bang, *Photonic crystal fiber long-period gratings for biochemical sensing*, Opt. Express 14, 8224-8231, 2006

G. Sagnac, *L'éther lumineux démontré par l'effet du vent relatif d'éther dans un interféromètre en rotation uniforme*, Compte-rendu de l'Académie de Sciences, Vol. 157, pp. 708-710, 1913.

N. Takahashi, K. Yoshimura, S. Takahashi, K. Imamura, *Development of an optical fiber hydrophone with fiber Bragg grating*, Ultrasonics, 38, 2000, pp. 581-585

D. Yadav, A.K. Nadir, *Transformer temperature measurement using optical fiber based microbend sensor*, Sensors & Transducers Journal, Vol. 84, N°10, pp. 1651-1659, 2007

L-B Yuan, L-M Zhou, J-S Wu, *Fiber optic temperature sensor with duplex Michelson interferometric technique, Sensors and Actuators*, 86, pp. 2-7, 2000

F. Zarinetchi, S. P. Smith, S. Ezekiel, *Stimulated Brillouin fiber-optic laser gyroscope*, Opt. Lett., Vol. 16, pp. 229-231, 1991

Y. Zhan, S. Xue, Q. Yang, S. Xiang, H. He, R. Zhu, A novel fiber Bragg grating high-temperature sensor, Opt. Int. J. Light Electron. Opt, 2007

H., L. Zhang, I. Bennion, Combined ultrasound and temperature sensor using a fibre Bragg grating, Optics Communications, 171, 1999, pp. 225-231

C. Zhao, M.S. Demokan, W. Jin, L. Xiao, *A cheap and practical FBG temperature sensor utilizing a long-period grating in a photonic crystal fiber*, Optics Communications, 276, pp. 242-245, 2007

4

Life-Cycle Monitoring and Safety Evaluation of Critical Energy Infrastructure Using Full-Scale Distributed Optical Fiber Sensors

Zhi Zhou, Jianping He and Jinping Ou
School of Civil Engineering, Dalian University of Technology, Dalian
P.R. China

1. Introduction

Critical energy infrastructure, such as huge oil depot and offshore platform, are inclined to be seriously damaged by the coupling effects of environmental erosion and loading in long-term service. The failure of these structures will cause large economic damage, huge personnel loss and environmental pollution. Therefore, it is important to monitor and assess the safety condition of these structures, so that they can be damage alarming and repaired in time. The optical fiber sensing techniques, especially fiber Bragg grating (FBG) sensors and distributed Brillouin fiber sensors, have the advantages of absolute measurement, anti-electromagnetic interfere, high-durability over the conventional electrical sensors, and have been widely applied in structural health monitoring (SHM)(OU and ZHOU,2003-2010; NI,2009; CHAN,2006; INAUDI,2005; WU,2006;BAO,1994; FILIPPO, 2005;JIANG, 2006). However, the development of applicable engineering optical fiber sensors and self-repairing optical fiber sensing network to meet the requirements of the life-cycle monitoring and safety evaluation for these infrastructure is not yet fully studied. In this chapter, some recent advances are introduced for the full-scale distributed optical fiber based structural health monitoring techniques for critical energy infrastructure in Dalian University of Technology (DUT) of mainland China.

2. The engineering full-scale distributed optical fiber sensing technique

Critical infrastructure usually has a long service life, for example, the design service life of a oil reservoir is around 15 years and that of a large-scale bridge can be hundreds of years. In addition, sometimes, these critical structures may be suffering from harsh environments including high temperature, high mechanical strength, or high pressure. However, the present packaging methods for the OF sensors may not be able to meet the requirements of the life-cycle monitoring for these structures from the durability aspect. What's more, the damages of the critical infrastructure could be substantial, deeply hidden and random. In this case, neither FBG sensors nor Brillouin sensors alone can provide effective monitoring to these structures. FBG sensors can only provide local information, and Brillouin sensors show low spatial resolution, low test accuracy, and high cost for strain and temperature measurements. In an effort of solving these problems, in this chapter, the fabrication of engineering optical fiber sensors, the optimization of measuring signals, and a novel multi-signal engineering collinear optical fiber sensing system are introduced. Fig. 1 shows the schematic of the SHM networks based on the newly developed optical sensing techniques including the sensor configuration, signal optimization and processing, and finally its applications.

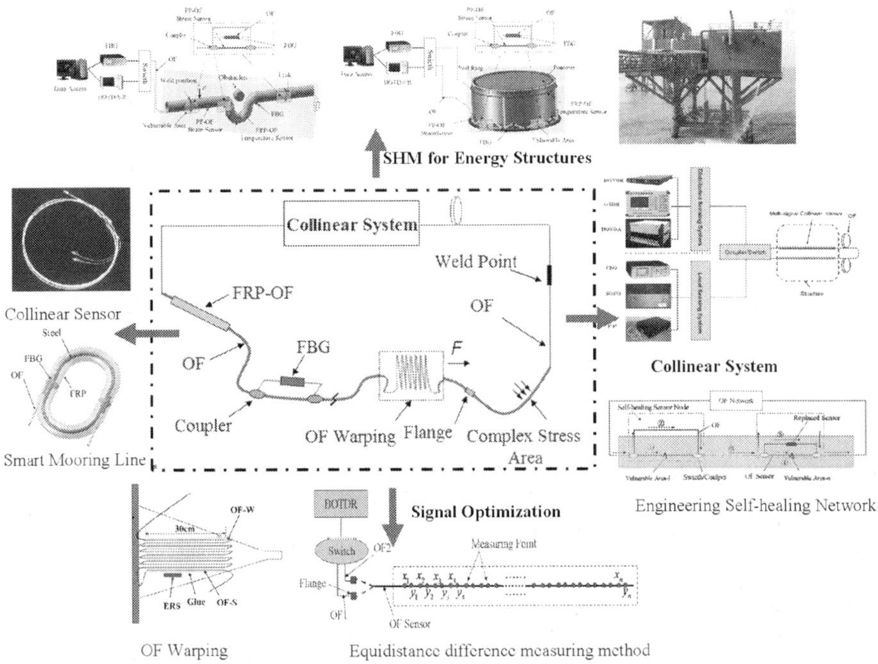

Fig. 1. The sketch of the SHM techniques for major energy infrastructures

2.1 Multi-signal engineering collinear optical fiber sensing system

Combining the advantages of both distributed optical fiber sensors and local high-accuracy optical fiber sensors, a multi-signal engineering collinear optical fiber sensing system is proposed to provide local high-precision and full-scale distributed measurement simultaneously, shown as in Figure 2. The collinear system consists of distributed optical fiber sensing system and local optical fiber sensing system, light switcher or coupler and collinear optical fiber sensors. The distributed optical fiber sensing system includes Brillouin optical time domain analysis/refectometry (BOTDA/R) and Roman optical time domain refectometry (ROTDR). The local optical fiber sensing system, on the other hand, includes fiber Bragg grating and P-T sensing system. The collinear system can provide multi-signal by one sensor, which significantly reduces the cost of the sensor layout, and simplifies the layout process.

As such, an integrated global and local structural health monitoring technology with a single OF is proposed by combining the BOTDA/R for distributed strain measurement and the FBG for local high-precision strain measurement, as illustrated in Figure 3. Light comes out from the light source, comes up to the light switcher and travels through the single-

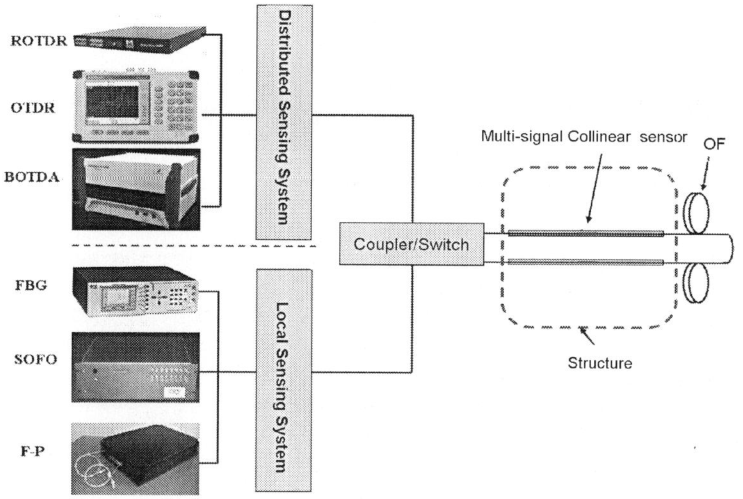

Fig. 2. Multi-signal Engineering Collinear Sensing System

mode optical fiber. When light researches the FBGs, they will be reflected back and the reflected signals are recorded on the FBG and BOTDA/R interrogators. With the light switcher, the light signals from BOTDA/R and FBGs can be recorded in time division independently. In this system, a SI-720 instrument from Micron Optics Inc., has been used as the FBG interrogator and a DiTeSt STA 202 from Omnisens has been used as the BOTDA/R interrogator. On the single-mode optical fiber, several FBGs have been written to form a collinear sensor named OF-FBGs. The OF sensor or the Brillouin sensor provides the distributed strain or temperature information along the optical fiber, and the FBG provides the local strain information at the hot spot. The combined BOTDA/R-FBG sensor can be manufactured in a long length (more than 5 km). This measuring technique has been validated in Lab and field applications (ZHOU and HE, 2009-2010).

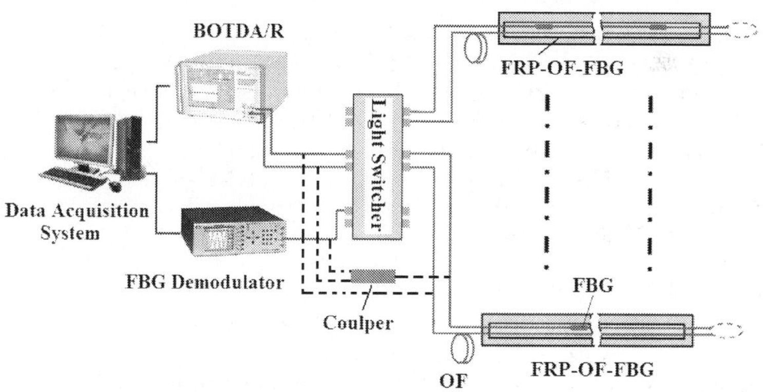

Fig. 3. Configuration of the integrated BOTDA/R-FBG system

2.2 Optimization of distributed Brillouin measuring signals

FBGs use the broadband light source, but BOTDA/R uses a light source with a certain wavelength of 1550nm. Thus the potential signal interactions may exist between the BOTDA/R and FBGs signal, including interaction of the reflected light from different optical light sources. In addition, FBGs are somehow regarded as damages on the single-mode optical fiber. Their structures, typically the Bragg period, change with various mechanical and thermal loading, resulting in a complex Brillouin gain spectrum at the locations of FBGs. Moreover, local complex area or damage at small range may lead to the inaccuracy of measurement due to the limitation of spatial resolution of present commercial Brillouin demodulators such as BOTDR made in Japan and BOTDA made in Switzerland with the smallest spatial resolution of 0.5m. Figure 1 shows the sketch about complex sensing configuration of optical fiber sensing system. There are packaged OF sensor (FRP-OF), bare OF, FBG, local complex stress area, damage point (or weld point), coupler and flange and so on, so the sensing signal at the sensing road is inevitably influent. In follow, we will introduce the optimization techniques of measuring signals.

2.2.1 Interactions between BOTDA/R and FBG signals

Two collinear OF-FBG sensors packaged by GFRP, named as GFRP-OF-FBG12 and GFRP-OF-FBG34, had been tested for strain sensing under tension loads. Each OF-FBG sensor has two FBGs. The initial center wavelengths of GFRP-OF-FBG12 are 1525nm and 1527nm, and that of GFRP-OF-FBG34 are 1540nm and 1545nm, respectively. Figure 4 shows the test results. It can be seen that the strain measured by OF around of the position of FBG agree well with the strain measured by FBG_1 and FBG_2. While the obtained distributed strain measured by OF has a huge intensity loss at the locations closed to FBG_3 and FBG_4, shown as in Figure 12(b). The more the wavelength of FBG is close to 1550nm, the more the intensity loss of the Brillouin signal will be in the collinear OF-FBG sensor. Therefore, the FBGs in a collinear sensor must meet the following two requirements: ① The center wavelength of FBG must be far away from 1550nm. So we should consider the initial center wavelength and the wavelength under some loads in field applications. In short, the wavelength of FBG sensor is not closed to 1550nm in its working time. ② The wavelength

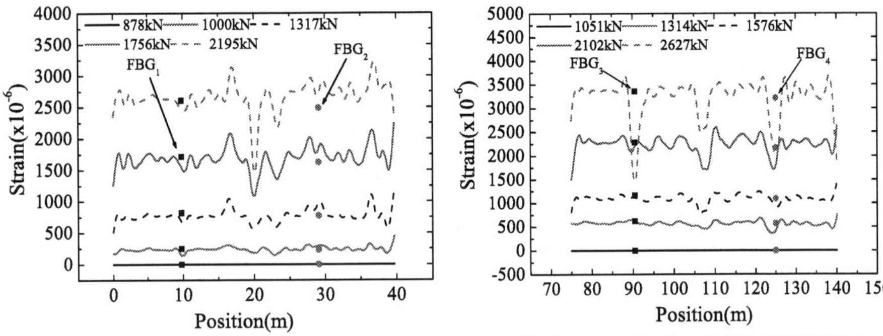

(a) Test results for GFRP-OF-FBG12 (b) Test results for GFRP-OF-FBG34

Fig. 4. Strain measured by GFRP packaged OF-FBGs

of FBGs must be separated from 2-3nm in the collinear sensor for the wavelength division multiplexing (WDM).

2.2.2 Brillouin gain spectrum curve fitting algorithm

Brillouin distributed optical fiber sensing technology is based on the assumption that Brillouin frequency shift (BFS) obtained from Brillouin gain spectrum linearly depends on the environmental temperature or applied strain on optical fiber. The Brillouin gain spectrum of OF applied axial tensile load has a Lorentz profile, while it becomes complex for complex stress condition or damage within the spatial resolution. Figure 5 shows the comparison of the 3-D mapping of one GFRP packaged OF-FBGs' Brillouin gain spectrum to that of an unpackaged OF at certain temperature. Brillouin gain spectrum for the GFRP packaged OF-FBGs sensor turns to be more complex than that of the unpackaged OF, especially at the locations of FBGs. This complexity is caused by the complex initial stress induced by the packaging process and the interaction between the FBGs and the Brillouin sensor. Typically, the algorithm to calculate the Brillouin frequency shift provided by present commercial instruments is not flexible, which cannot fully meet the required measurement accuracy of temperature or strain. Thus, local Lorentz single-peak and localized least square fitting algorithms are proposed to fit the Brillouin gain spectrum. By applying optimization, a selected bandwidth of 50MHz has been employed to fit the measured Brillouin gain spectrum with the proposed algorithms based on the assumption that the peak of Brillouin gain spectrum is single. Experiments had been set up to verify the two proposed spectrum fitting algorithm, as shown in Figure 6. One GFRP packaged OF-FBGs sensor had been placed into a water tank with various temperature measurements, and the center wavelength of FBG is about 1527nm.

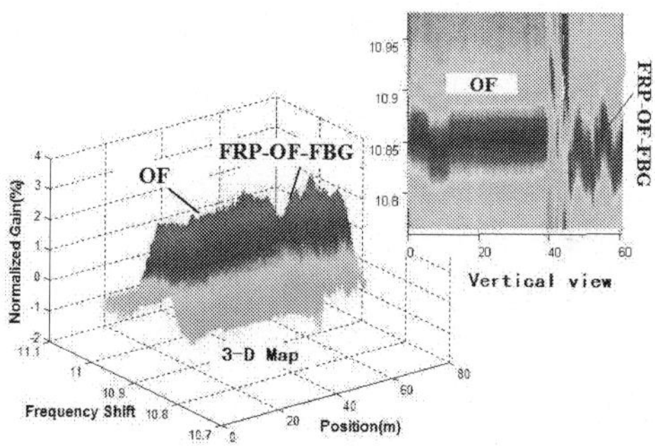

Fig. 5. Mapping of Brillouin gain spectrum

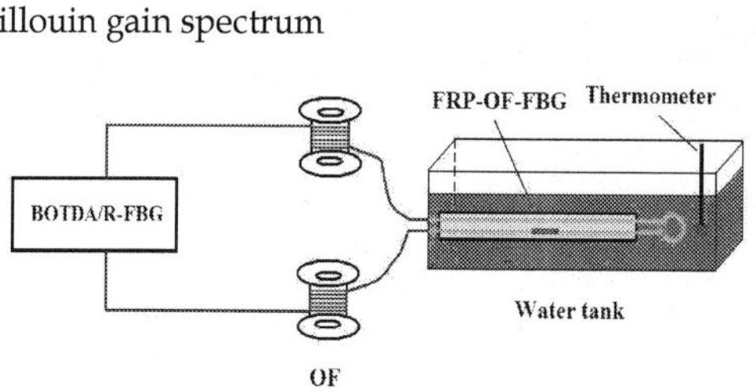

Fig. 6. Setup of temperature test

Figure 7 (a, b) show the Brillouin gain spectrum at FBG location at various temperatures and the corresponding fitted curve by using the local Lorentz single-peak fitting algorithm and the localized least square fitting algorithm. Comparison with the actual temperature, the maximum errors from local single-peak Lorentz, localized least square fitting algorithms and BOTDA/R measurement are 3.7%, 6.9% and 7.5%, respectively. The test results indicated that with the proposed curve fitting algorithms, the interaction of the FBGs to the Brillouin gain spectrum of OF-FBGs sensors can be eliminated.

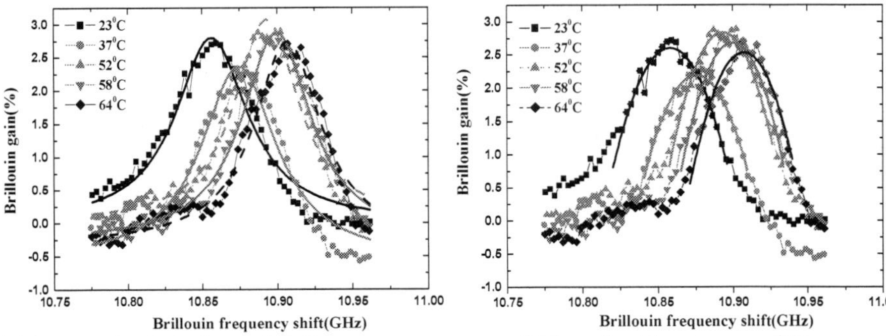

(a) Lorentz local single-peak fitting (b) Localized least square fitting

Fig. 7. Initial spectrum of GFRP-OF-FBGs' BFS at various temperatures

2.2.3 Wavelet denoising for the OF-FBGs sensor

Brillouin sensing technique is one of OF intensity based sensing techniques. Its measuring signal is prone to be interfered by the OF's damage, little bending radius and the FBG written in it. The wavelet denosing technique has been used to eliminate the noise induced in the measuring signals. Figure 8(a) shows the strain distribution of the collinear OF-FBG sensor under various loads. The measured strain spectrum shows a number of burrs. To eliminate the effect from these interactions, a db5 wavelet was selected to denoise the measured signal, and the results are depicted as in Figure 8(b).

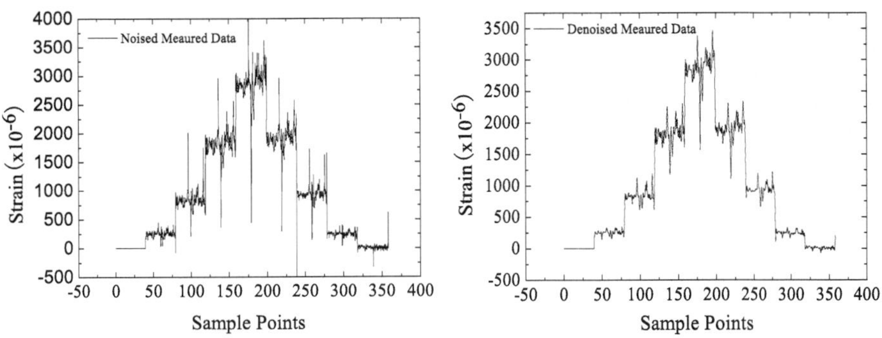

a) Measured strain b) Strain after denoising

Fig. 8. Measured Brillouin signals from OF-FBG sensor.

Figure 9(a) illustrates the strain measurement comparison of the GFRP-OF-FBGs sensor before and after db5 wavelet denoising and the strain gauge. Figure 9(b) shows the corresponding relative error. With the wavelet denoising, both the strain measured by the BOTDA/R and that by the FBGs agree well with that from the strain gauge. The maximum relative error of the GFRP-OF-FBGs sensor is 18% before denoising and 4% after denoising. The results indicate the feasibility of the wavelet denoising method for the data processing of OF-FBGs sensors (HE and ZHOU, 2009).

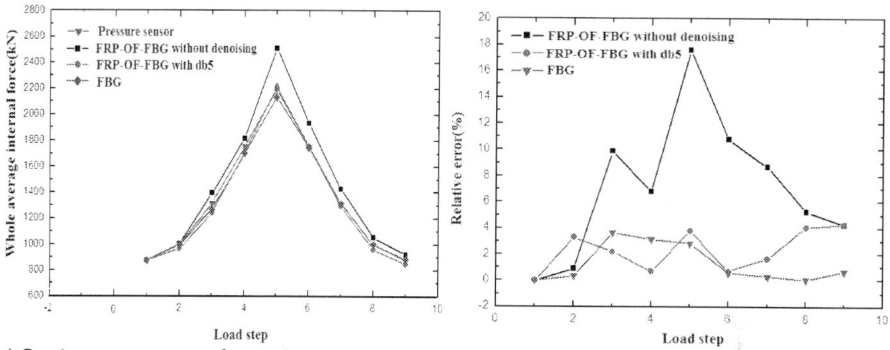

a) Strain measurement by various sensors b) Relative measurement error

Fig. 9. Strain measurement comparison before and after denoising

2.2.4 The optimization of distributed Brillouin signal

In common application, both the spatial resolution and distance resolution of commercial Brillouin sensing system are low. The low spatial resolution will lead to the measurement inaccuracy for local damage and stress of small-scale structures. And improving the distance resolution can greatly add the number of measuring point and enrich the information of measured elements. The equidistance difference measuring method and sensing OF warping method are proposed here to optimize the test signals' spatial and distance resolution.

1) Equidistance difference measuring method

Figure 10 shows the principle of the equidistance difference measuring method. With the transmission length of OF changing followed by certain rules, by using the equidistance difference measuring method, the measured zone of Brillouin sensor can be changed. These rules can be described as follow:

1. The transmission length of OF1 is defined as *L1*, and that of OF2 is defined as *L2*. With *L0* being the difference between *L1* and *L2*, we have $0<|L0|<d0$,where *d0* is the length of the sampling space.
2. When the Brillouin sensor is connected to OF1 by flange, the measuring points X=[*x(1),x(2),…,x(n)*] are obtained. While Brillouin sensor is connected to OF2, the measuring points Y=[*y(1),y(2),…,y(n)*] are gotten. The difference of $|y(i)-x(i)|$ equals to *L0* or $|x(i+1)-y(i)|$ equals to *d0-L0*, where *i=1,2,…n*.
3. Numbers of measuring points Y(i) can be obtained by using OF3, OF4,…, OFn to connect the Brillouin sensor, only if the length satisfies the rule (1).

4. Based on the series of difference set D, the strain information about the measured structure can be refined, where D= | Y(i)-X |.

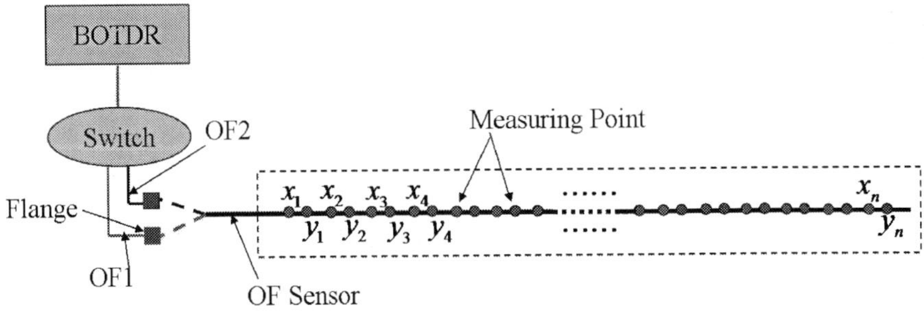

Fig. 10. Equidistance difference measuring method

Figure 11 shows the measuring results of one OF sensor using the equidistance difference measuring method. Here the length of the Brillouin sensor is 7.26m, the length of the transmission optical fiber (denoted as OF1) is 2.0m. Under different load, the strain applied on Brillouin sensor is shown as in Figure 11(a). When the OF1 is shorten with 0.05m, the length of transmission optical fiber (denoted as OF2) is 1.95m, and the strain measured by Brillouin sensor is shown as in Figure 11(b). Obviously, there is large discrepancy of the distributed strain between these two measurements, denoted as Strain-OF1 and Strain-OF2. Figure 12 shows the difference between the Strain-OF1 and Strain-OF2. It can be seen that the strain difference at some locations are much larger than that of others, which may attribute to the local damages or complex stress conditions. In practice, this method is very useful for multiple stress information to be obtained by one sensor.

2) Sensing OF warping method

The spatial resolution of the developed Brillouin measuring system is too low to accurately measure small-scale structures. A novel OF warping method is proposed for the local fine measurement of small-scale structures. The OF is warped back and forth many times at the key position shown as in Figure 13. The warped OF is much longer than the spatial

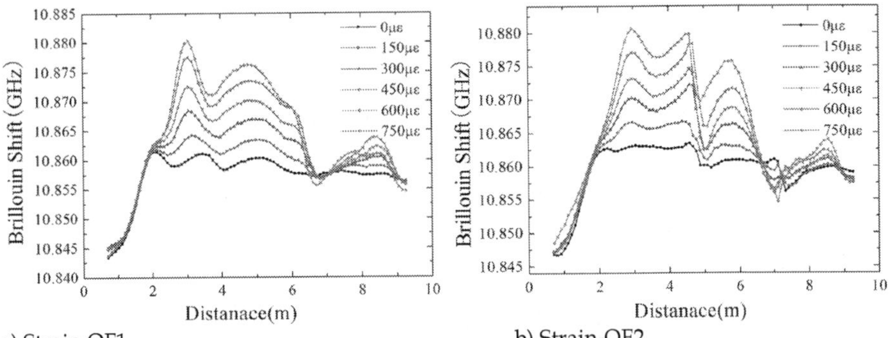

a) Strain-OF1 b) Strain-OF2

Fig. 11. Results of the equidistance difference measuring method

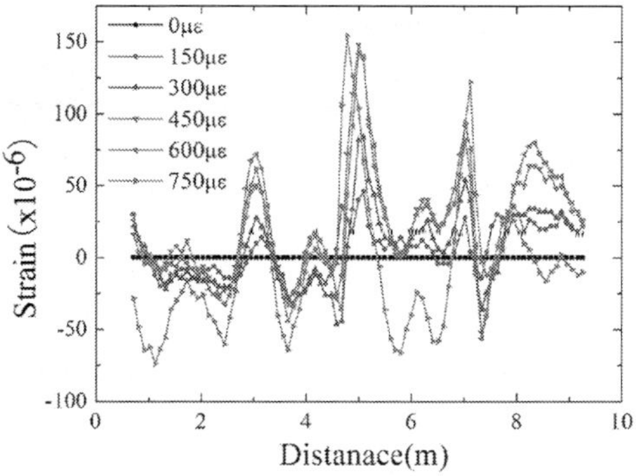

Fig. 12. The different between the Strain-OF1 and Strain-OF2

resolution, which can be characterized as type I and type-II, shown as in Figure 13. The strains at straight segments, denoted as $O_1,\ldots,O_n$, are much larger than that at curve segments. The average strain measured at the straight segments by Brillouin system are closed to the actual strains applied on the OF. With the capability of the warped OF to survive large strain, the OF sensor can detect the occurrence of the whole crack propagation process.

Experiment had been set up to verify the proposed sensing technique, as shown in Figure 14. A common optical fiber is bonded on the surface of uniform strength beam with a length of 40cm. The length of warped OF is 4.5m (denoted as OF-W), and the curve segments are free, the numbers of the straight segments is 15. For comparison, one electronic-resistance strain (ERS) is bonded on the surface of the beam, and another OF with the length of 30cm (denoted as OF-S) is bonded on the beam parallel to the warped OF, shown as in Figure 15. Figure 16 shows the measured distributed strains under various loads. Although the strains on the surface of beam should be equally distributed everywhere, the strain measured by OF is not flat as it should. Two reasons may attribute to this phenomenon: One is that the smallest spatial resolution is larger than the size of beam, and the other one is that the sensing location has optical fiber with free motion. Meanwhile, it can be seen that the strain measured by OF-S has small changes.

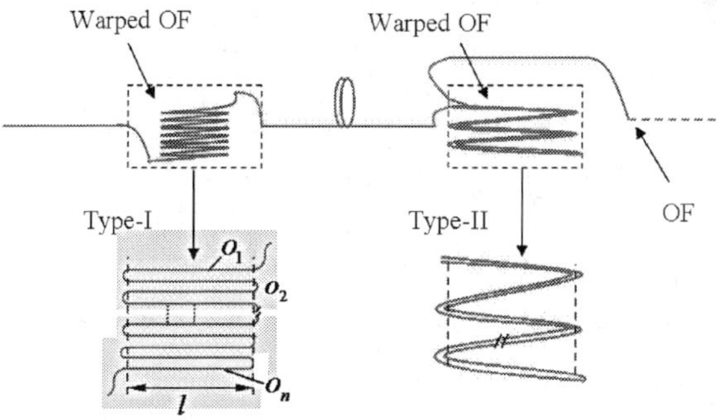

Fig. 13. Sensing OF warping method

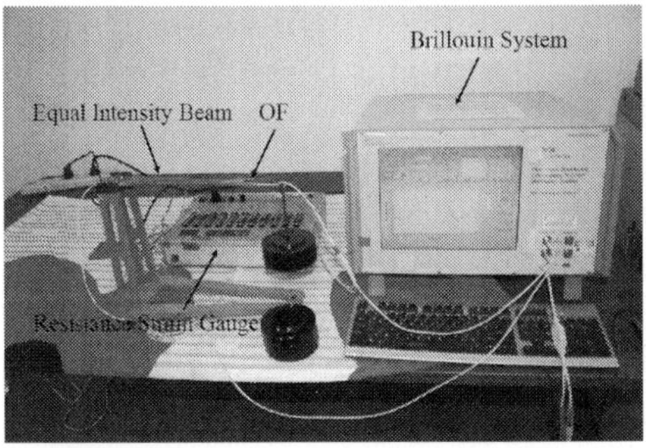

Fig. 14. Test setup of OF warping method

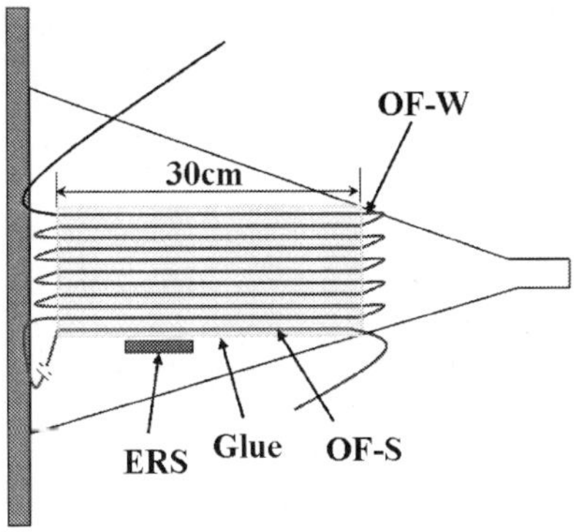

Fig. 15. Sensor layout on the beam

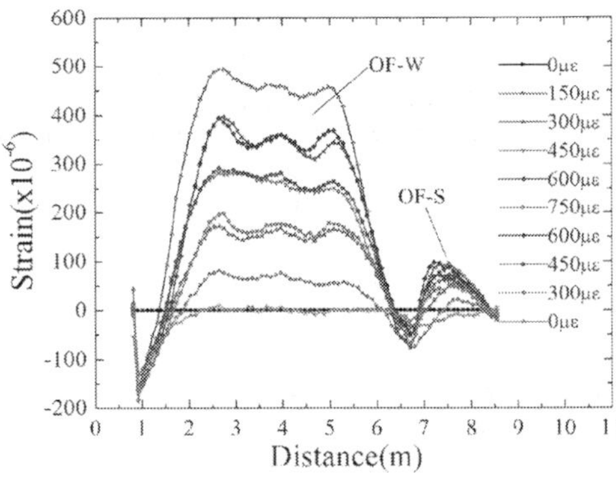

Fig. 16. Distributed strain at various load

Figure 17 gives the comparison of the strains measured by the OF-W, OF-S and the ERS The strain measured by OF-W shows good linearity with a slope of 0.6064. Here, the strain measured by local ERS is acted as the reference value. Using the slope coefficient, the modified value of OF-W can be obtained, which agrees well with the ERS with the maximum absolute error of 40με. In practice, OF-based warping sensor can be fabricated by FRP packaging technique with a pre-calibration correction coefficient.

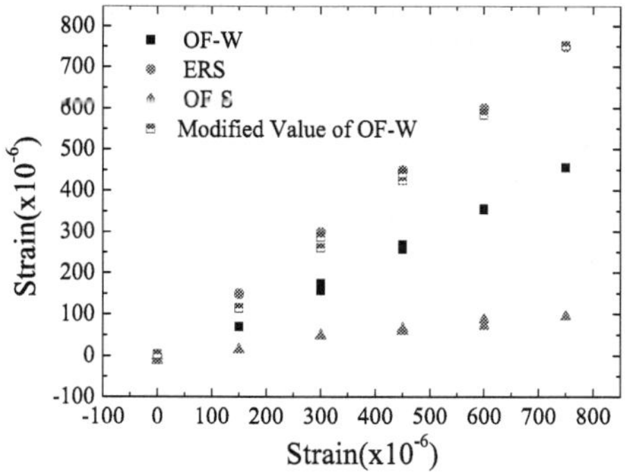

Fig. 17. Strain comparison by different sensors

2.2.5 Temperature self-compensation for BOTDA/R-FBG sensing system

For BOTDA/R-FBG collinear system, another prominent advantage is temperature compensation for strain measurement. Based on the collinear technique, we proposed a novel temperature self-compensation for strain measurement without additional temperature compensation sensor. Without considering the cross sensitivity of strain and temperature, both the Bragg wavelength shift of FBG and the Brillouin frequency shift of BOTDA/R show excellent linearity to the strain and temperature as expressed below in binary matrix:

$$\begin{pmatrix} \Delta\lambda_B \\ \Delta\upsilon_B \end{pmatrix} = \begin{pmatrix} \alpha_\varepsilon & \alpha_T \\ C_\varepsilon & C_T \end{pmatrix} \begin{pmatrix} \Delta\varepsilon \\ \Delta T \end{pmatrix} \tag{1}$$

where $\Delta\lambda_B$ and $\Delta\nu_B$ are the changes of the Bragg wavelength and the Brillouin frequency; $\Delta\varepsilon$ and ΔT are the changes of strain and temperature; α_ε and α_T are the strain and temperature sensitivity of FBGs, which equal to 1.2 $pm/\mu\varepsilon$ and 10.8$pm/°C$, respectively; C_ε and C_T are the strain and temperature sensitivity of BOTDA/R, which have values of 0.05$MHz/\mu\varepsilon$ and1.0$MHz/°C$, respectively. So the Eq. (1) can also be expressed as follows:

$$\begin{pmatrix} \Delta\varepsilon \\ \Delta T \end{pmatrix} = \begin{pmatrix} \alpha_\varepsilon & \alpha_T \\ C_\varepsilon & C_T \end{pmatrix}^{-1} \begin{pmatrix} \Delta\lambda_B \\ \Delta\upsilon_B \end{pmatrix} \tag{2}$$

With measured changes of Bragg wavelength ($\Delta\lambda_B$) and Brillouin frequency shift ($\Delta\nu_B$), the changes of strain ($\Delta\varepsilon$) and temperature (ΔT) then can be predicted accordingly. Thus, the temperature effect to the BOTDA/R-FBG system can be eliminated or self-compensated. Figure 18 (a, b) show the test results of self temperature compensation. The strain measured by the OF-FBG sensor with self temperature compensation agrees well with the reference strain calculated from absolute temperature compensation by OF or FBG, and the maximum relative error is about 9%. Some of the measured strains or temperatures have large deviation with the references because of the temperature instability of the water tank.

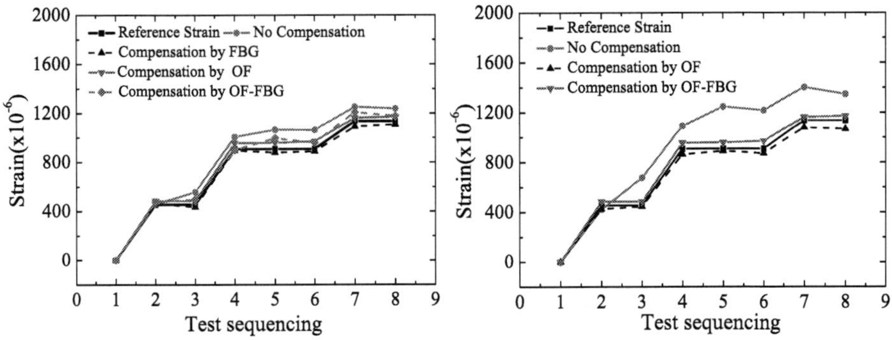

(a) FBG (b) OF

Fig. 18. Test results for self temperature compensation verification of OF-FBG

Figure 19 compares the variance of strain measured with and without self temperature compensation at two temperature states at out door. It can be seen that compared to the strain measured by the OF-FBGs sensor without the self temperature compensation, the corresponding strain with the self temperature compensation varies much smaller from the actual strain.

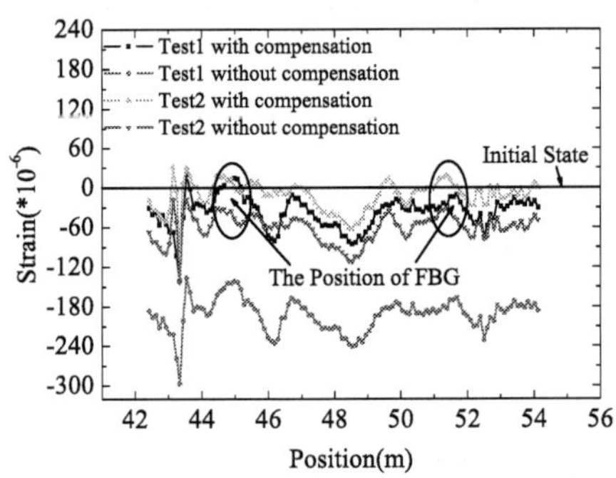

Fig. 19. Strain measurement with temperature inter-compensation

The temperature self-compensation method shows great advantage in practice. As be known, the temperature measured by sensors for temperature compensation may be different from that applied on the strain sensor due to the discrepancy of the sensor layout location, or there are no place for the sensor layout around the strain sensor in some cases. Using the novel temperature self-compensation method, special temperature sensors are not necessary, which can be a promising and cost-efficient sensing technique for field practices (HE and ZHOU, 2010).

2.2.6 Collinear OF sensor based shape monitoring technique

The shape reconstruction is an important way to evaluate the safety of structures such as oil pipeline, flexible pipe of platform and large-scale pillars. Some researchers have used local sensors (ERS or FBG) to perform shape reconstruction (ZHU, 2007; QIAN, 2004). However, the shape reconstruction for the large-scale structures has not been studied fully. Local sensor used for shape reconstruction could not be able to provide the structural health condition of the whole structure. In addition, these local sensors cannot bear the large deformation due to earthquake and other disasters. Based on the advantage of newly developed collinear sensing technique in this chapter, such as large dynamic range and high accuracy, a novel method for shape reconstruction has been proposed by using the FRP or PP packaged collinear sensor. Figure 20 shows the sketch of deformation and sensor layout of oil tank and flexible pipe.

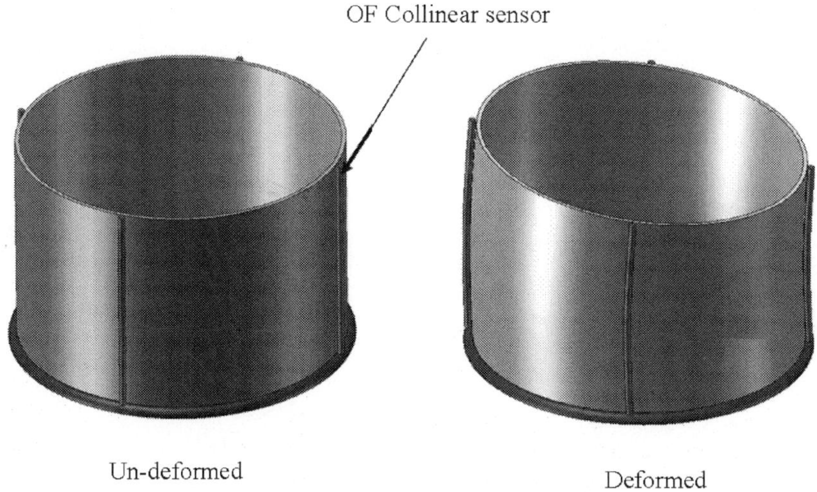

a) Oil tank

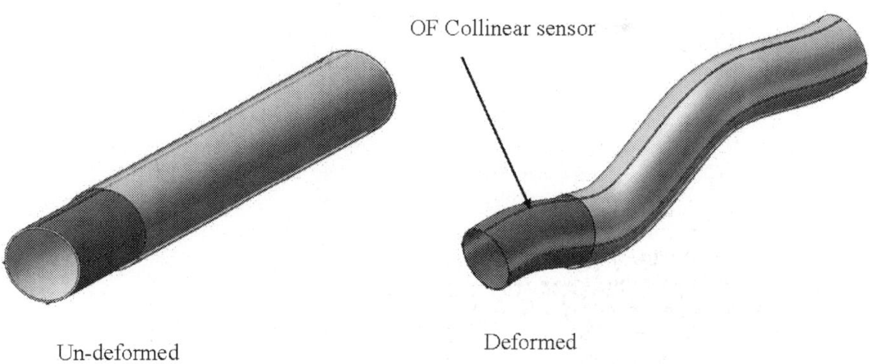

b) Flexible pipe

Fig. 20. Sketch of deformation and sensor layout of oil tank and flexible pipe

Using the distributed strain information, the shape reconstruction can be done through the following procedures:

1. Calculating the curvature components ($k(i)_1$, $k(i)_2$ and $k(i)_3$) through the bending strain measured by different sensor, and synthesizing the curvature vector $K(i)$, where the subscript "*i*" denotes the measuring location.
2. Using the genetic algorithm to optimize the minimum interpolating number between the two neighboring curvatures.
3. Based on the differential geometric method, calculating coordinate position of each measuring location in the osculating plane.
4. 3D visualization based on visual parallel programming platform such as VC++, VB program.

Figure 21 (a, b) shows the simulation results of the shape reconstruction of one plane curve using various numbers of calculated curvatures. For shape reconstruction-I, nine curvatures are used to reconstruct the shape of curve. While for shape reconstruction-II, ninety curvatures are used. The accuracy of reconstruction is highly dependent on the number of curvatures used in the reconstruction. The more the curvatures are used, the higher accuracy of reconstruction can be obtained. Therefore, with a larger number of curvatures obtained by the Brillouin sensor compared to local sensor, the developed sensing system can provide a promising shape reconstruction. In addition, by using the collinear technique, the deformation of key locations can also be checked from the composed FBG sensors.

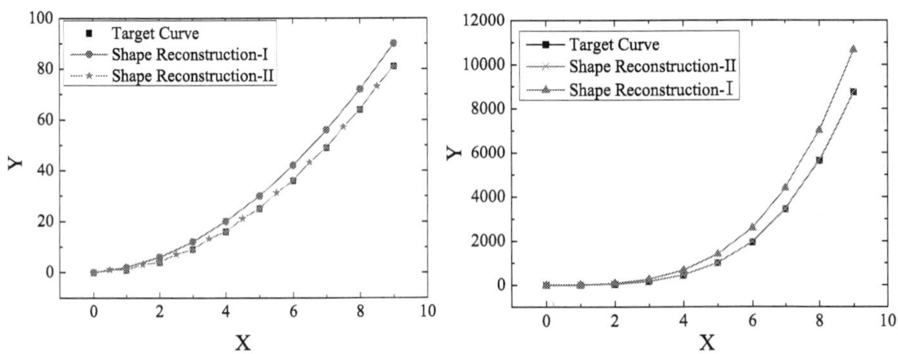

a) $y = x^2$ b) $y = x^4 + 3x^3 + 2$

Fig. 21. Simulation of shape reconstruction for curve

3. Engineering-suitable high-durability optical fiber sensors

The protection methods of OF sensors, for example, by using the capillary packaging method and the glue as coating, showed the disadvantages of low-durability and low environmental resistance, which disable its potential for the life-cycle monitoring of critical infrastructure. Fiber Reinforced Polymer (FRP), with the advantages of corrosion resistance and high strength, has the natural compatibility with optical fibers. In our group, various FRPs including glass fiber reinforced polymer (GFRP), carbon fiber reinforced polymer (CFRP), and basalt fiber-reinforced polymer (BFRP) etc, are investigated to be used as packaging and protecting the OF sensors. Figure 22 shows the fabrication process of a FRP-OF smart bar, by which the optical fiber is incorporated into the FRP to form a optical fiber sensor. One or two optical sensors are placed into the FRP bar, with the consideration of the practice convenience. Moreover, Polypropylene can be chosen to develop surface install strain sensor, named as PP-OF belt strain sensor.

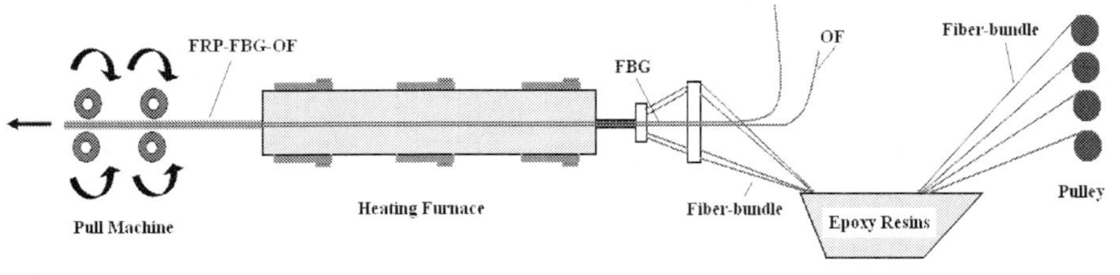

Fig. 22. Sketch of the manufacture process of FRP smart bar

Figure 23 shows the products of FRP packaged optical fiber smart bar and PP packaged optical fiber smart belt. In an effort to meet various layout requirements for different applications, FRP-OF sensor can be produced into different types of sensors. Figure 24(a) and 24(b) show the fabricated prototypes of these two types of FRP-OF sensors. The anchors on both ends of the sensor are designed to improve the bonding between the host matrix and the sensors. The length of these anchors can be designed based on the requirements of the dynamic range of the strain measurements. In addition, with the GFRP packaging layer, several OF-FBGs sensors can be packaged into one GFRP layer for multiple parameters sensing. The PP-OF sensor can be directly bonded on the surface of structures. The FRP-OF bar has good sensitivities to strain and temperature. For the FBG system, the strain and temperature sensitivity coefficients are 1.2 pm/με and 16.8pm/°C, and for the Brillouin system, the strain and temperature sensitivity coefficients are 0.05MHz/με and 1.2MHz/°C. Comparing with the OF sensors without package, the temperature sensitivity coefficient of the OF (or FBG) sensors with FRP package turn to be larger (ZHOU and OU, 2009).

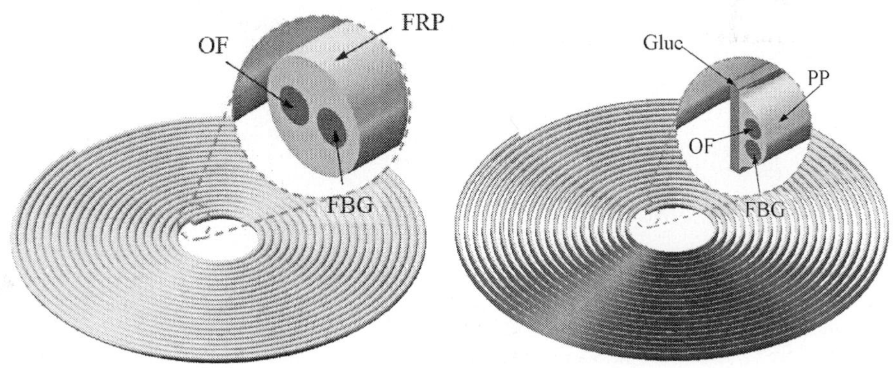

a) FRP-OF rib b) PP-OF belt

Fig. 23. High-durability FRP/PP-OF sensor

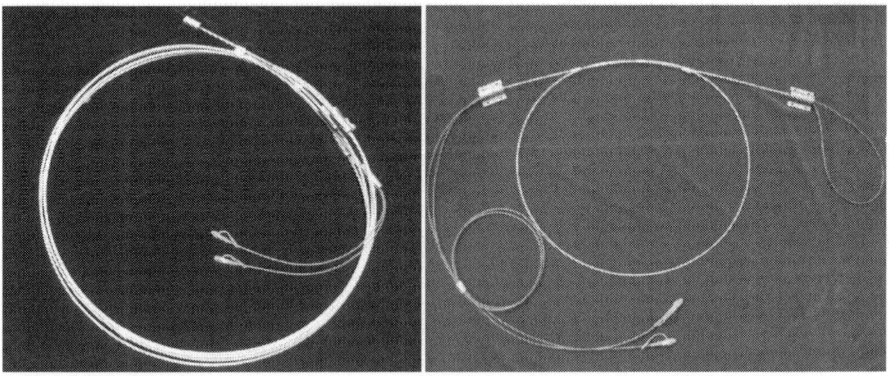

a) The GFRP-OF-FBGs based rod b) The surface installed GFRP-OF-FBGs

Fig. 24. Sensor prototypes of the GFRP-OF-FBGs based sensors

Furthermore, for the distributed strain measurement and shape monitoring of mooring lines and flexible pipes, OF based smart mooring and smart flexible pipe for offshore platform have been investigated.

1) Smart mooring line. In an effort to increase the corrosion resistance of the conventional mooring line, FRP mooring line is developed, shown as Figure 25. Two FBGs are incorporated into the FRP mooring line for the strain measurement. Furthermore, small steel strand is placed into the FRP mooring line to strengthen the elastic module and shear property of FRP depicted as Figure 25(b). This novel smart mooring line acts as either the tension structural member or sensing element. At present, the study on its sensing and mechanical properties is under going.

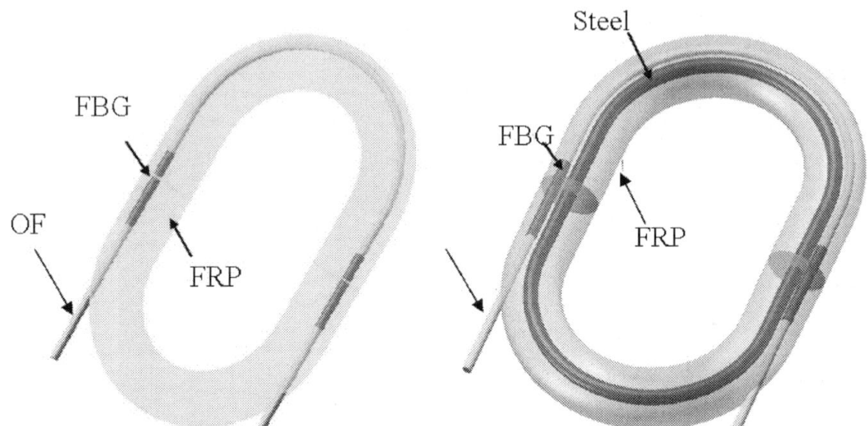

a) Pure FRP smart mooring line b) FRP smart mooring line with steel

Fig. 25. Smart FRP mooring line

2) OF based smart flexible pipe. Figure 26 shows two types of smart flexible pipes. In type-I, four OF sensors are placed into the outside layer of a flexible pipe at the interval of 90 degree: one OF temperature sensor and three OF strain sensors. The temperature sensor is used for the temperature compensation for strain measurement. In type-II, FRP packaged OF sensors are bonded on the surface of a pipe by clamps. The distributed OF sensor can be either optical fiber sensor or collinear optical fiber sensor. Using the collinear sensor, the distributed strain and local strain information can be obtained. Meanwhile, the point curvature can be gotten from the distributed strain, and reconstructed the shape of flexible pipe.

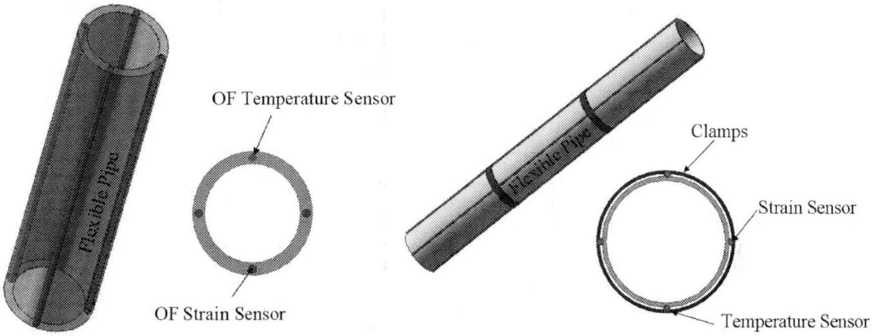

a) Type-I b) Type-II

Fig. 26. Smart flexible pipe

4. The engineering self-healing optical fiber sensing network

Based on the wavelength-division multiplexing (WDM) technique (spatial-division multiplexing or time-division multiplexing), the FBG sensors can be arranged simply in series or in parallel. The full-scale distributed Brillouin sensing technique uses one common single mode optical fiber to sense the continuous strain or temperature. A sensing network system for the health monitoring of structures can be constructed by using several FBG sensors and the Brillouin sensor, the sensing demodulation system, and the data acquisition system. However, The optical fiber sensing network, especially full scale distributed optical fiber Brillouin sensor network, is weak for disaster resistance. Local failure or transmission breakdown of the sensing network can easily lead to the failure of the whole optical fiber sensing network. The vulnerability of the local sensors and the transmission in the optical fiber sensing network will reduce the robustness of the sensing network. Therefore, the reliability and robustness of sensing network becomes a challenge. Some researchers proposed the self-healing networks. Peng and Tseng developed a novel self-healing network with star topology and ring topology by adding light switchers into the optical fiber Bragg grating network (Peng and Tseng, 2003). This network is prone to failure once more than three nodes are failed. Wei developed a FBG fan-shaped network and invented four remote light switch nodes, which greatly improved its robustness (Wei and Sun, 2007). However, only the connection of the presented self-healing network has been studied, and research on the sensing continuity and exchangeability of damaged sensors has not been discussed. In this chapter, a novel self-healing OF sensing network is proposed. The self-healing sensing network has the advantages of high robustness, self-healing, sensing continuity, and exchangeability of damaged sensors.

4.1 Construction of the self-healing network

Based on the vulnerability analysis, the classification of sensor nodes and sensing routes is figured out. The light switches or couplers are used to develop robust sensing nodes or sensing routes. Figure 27 shows a sketch of the developed self-healing full-scale distributed Brillouin sensing network. The sensing network includes two types of robust nodes: link robust node and resume robust node. At the vulnerable area-I, if route① or sensor is broken, the connectivity of the sensing network can be kept by the optical fiber outside the structure②. At the vulnerable area-n, if road④or sensor is damaged, the damaged sensor can be replaced by a local FBG sensor⑤ to continue monitor the damage expansion.

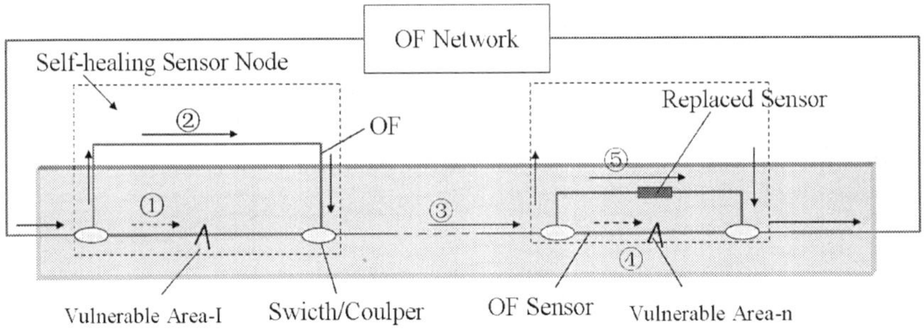

Fig. 27. The proposed self-healing optical fiber sensing network

4.2 Experimental investigation

One steel sheet, bonded with OF sensor on surface, as shown in Figure 28, was used to verify the proposed self-healing optical fiber sensing network. The high strength epoxy was used for sensor bonding. The length of the specimen is 3.7m, and the length of sensing part is approximately 2.5m. In this study, the middle portion of the steel sheet (approximately 0.8m in length) was chosen as the damage position by cutting a minor crack as shown in Figure 28(b). Axial load was applied to the specimen, and the optical fiber was connected by couplers at the vulnerable element. When the OF was broken due to the crack propagation, the broken sensing portion was replaced by local FBG sensor, and forming a collinear system. Furthermore, the FBG sensor can perform crack monitoring continuously. Here the sensing gauge of the FBG sensor is 6cm, which is too small to act as a Brillouin strain sensor.

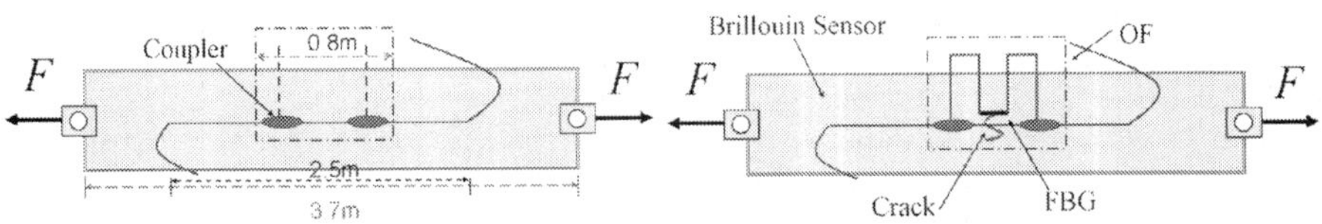

a) Network in good condition
b) Self-healing network

Fig. 28. Test for sheet steel bonded with OF

4.3 Results and discussions

(I) Undamaged steel sheet. Six-stage loading was applied on the specimen with the strain step of 500με. Figure 29 depicts the distributed strain, local strain information at different conditions of the steel sheet. Figure 29 shows that the strain curve of the undamaged steel sheet is flat for each axial load stage. The minor fluctuation in the curve is attributed to the system error and strain inconsistency of the curve segment of the OF.

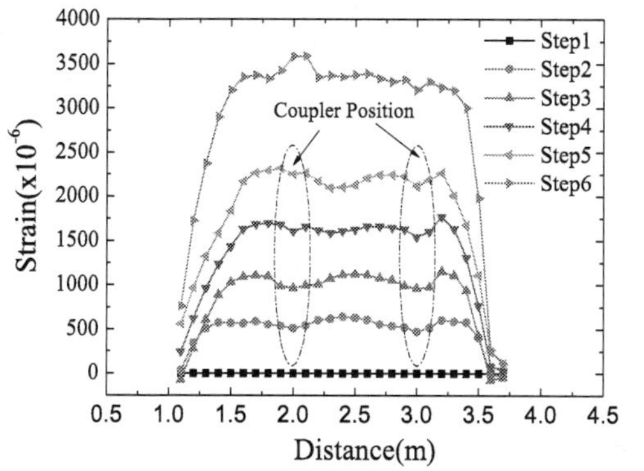

Fig. 29. Distributed strain of undamaged steel

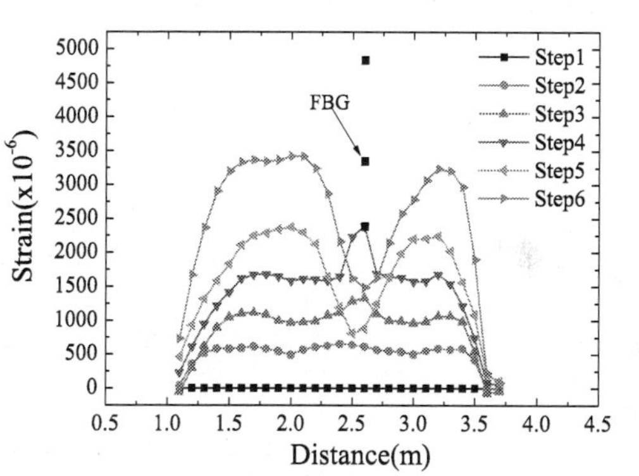

Fig. 30. Strain of damage steel with crack

(II) Pre-damaged steel sheet. In Figure 30, the specimen was pre-damaged by cutting a minor crack at the middle part of steel. Under the second loading step (Step3), there was stress mutation at the strain curve due to the present of the local crack, and other parts of the OF far away from crack keep stable. Under the third loading step (Step4), the maximum strain reached 2384με at the cracking location, and that at other places was about 1500με. At the stage of load maintenance, the OF was broken at the cracking location. The broken sensing path was then reconnected. Under the loading steps of step5 and step6, the strain curves had a large depression at the free part or the FBG location. And the crack can be cautiously monitored by the FBG sensor. The FBG sensor's initial center wavelength was corresponding to the strain of 2384με, the strain at cracking location reached 4900με at step6 shown in Figure 30. The self-healing network performed excellently throughout the period of monitoring despite of some multi-signal interactions, which can be solved by the optimization of distributed Brillouin measuring signals mentioned above.

5. OF based life-cycle monitoring and safety evaluation for huge oil deport

The safety evaluation of critical energy infrastructure, such as huge oil depots, offshore platforms, long oil pipelines and so on, is one of the most important concerns worldwide. The tragedy of the Deepwater Horizon alarmed the industry and government that the structural life-cycle monitoring and safety evaluation needs be taken. Our research group has focused on the OF based life-cycle monitoring of offshore platforms, huge oil depots in recent years and the developed SHM system has been successfully implemented to applications such as in the offshore platform and Casing Pipe in mainland China (ZHOU and OU, 2010). In this section, the safety evaluation based on the developed SHM system of a huge oil deport in Daqing, mainland China has been introduced.

5.1 Sensor layout

A oil depot includes the oil tank, the oil pipeline and some other accessory equipments. It has a huge domino effect, which is a local failure (such as the explosion) prone to induce large-scale structural failure due to the distributed density of oil pipes and storage tanks. The potential damages for oil depot include crack (large strain) and corrosion, which may result in oil leak, tank capsize, and even fire. The proposed full-scale distributed optical fiber sensors have been used to monitor and evaluate the safety of the huge oil depot in long term. Figure 31(a, b) shows the sensor layout for strain and temperature sensing on the oil pipeline, respectively. PP-OF sensor is used to measure the distributed strain of oil pipeline, and have been directly bonded on the surface of the oil pipe. The FRP-OF sensor is bonded on the surface of oil pipe by steel clamps, acted as a distributed temperature sensor to monitor the oil leak.

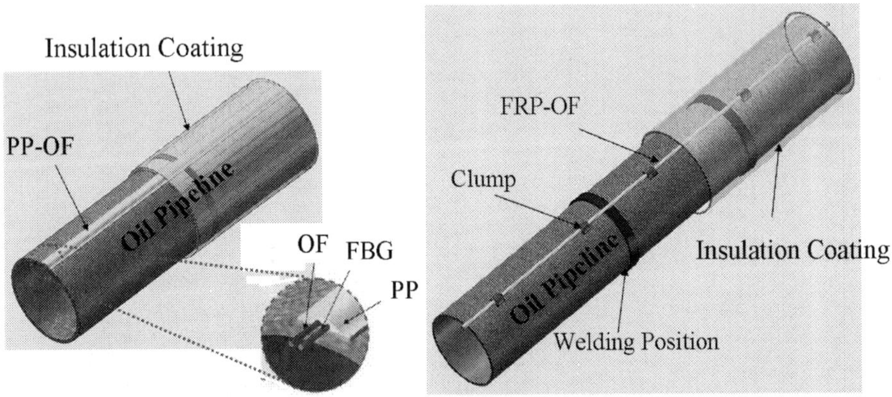

a) PP-OF strain sensing b) FRP-OF temperature sensing

Fig. 31. Sensor layout on the oil pipeline

Figure 32 (a, b) shows the configuration of the SHM system including the signal interrogation for oil pipeline and storage tank, respectively. The FRP-OF rib is installed on the surface of the oil pipe and the storage tanks for oil leak monitoring, and the PP-OF belt strain sensor is placed on the surface of the oil pipe and the storage tanks for large strain and crack monitoring. At the locations prone to damages, robust sensing nodes are designed to form self-healing optical fiber network. In addition to the strain and leak monitoring sensors, for the oil tank, four PP-OF sensors are vertically installed on the oil tank to reconstruct the shape of the oil tank under different loading conditions. Based on the shape reconstruction, the warning of the oil spilling can be provided and the safety of oil tank can be accurately evaluated.

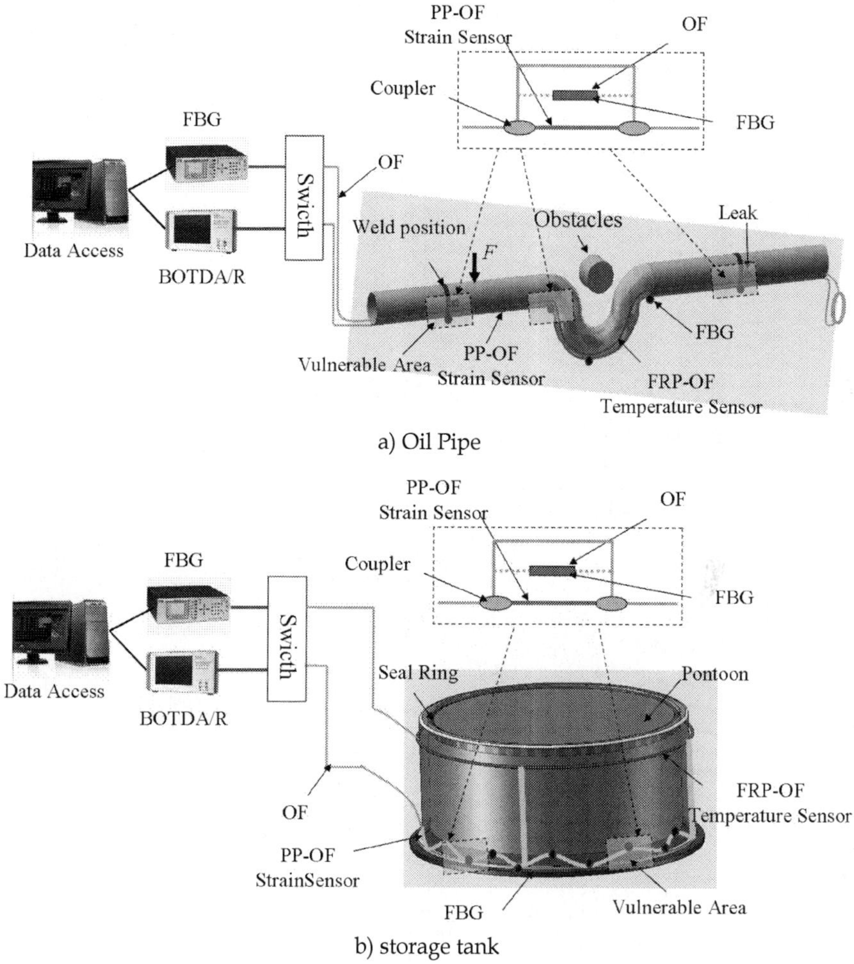

Fig. 32. OF based huge oil depot life-cycle monitoring

5.2 The laboratory test for OF sensor for safety monitoring on oil pipelines

In some cases, the oil transportation pipes are extremely long. Various service environments result in a complex service conditions for the distributed oil pipe. For example, the petroleum transportation pipeline from MoHe, China to Daqing, China is a section of oil transportation pipeline between Russia to China. It has a length of 960km, passing through large mountains such as the Daxing'an and Xiaoxing'an Mountains. The complex service environments of the pipelines involve high water permafrost at the north and seasonally frozen areas at the south. The oil temperature distribution along the oil pipeline is within the range from -6°Cto 10°C. Temperature changes, human activity, and natural disasters will significantly influence the foundation of the oil pipeline. Therefore, monitoring the safety of the oil pipeline is urgently needed. In this study, the proposed collinear BOTDA/R-FBG technique has been suggested to construct the real-time SHM system for the oil pipeline monitoring, as shown in Figure 34. In order to validate the proposed technique, laboratory test has been employed. The polyvinyl chloride (PVC) pipeline is used as the oil pipeline, and the OF based monitoring method is applied on the PVC pipeline to study its behavior under frost-heaving condition in cold areas. This model test is used to investigate the novel monitoring method for the actual pipeline.

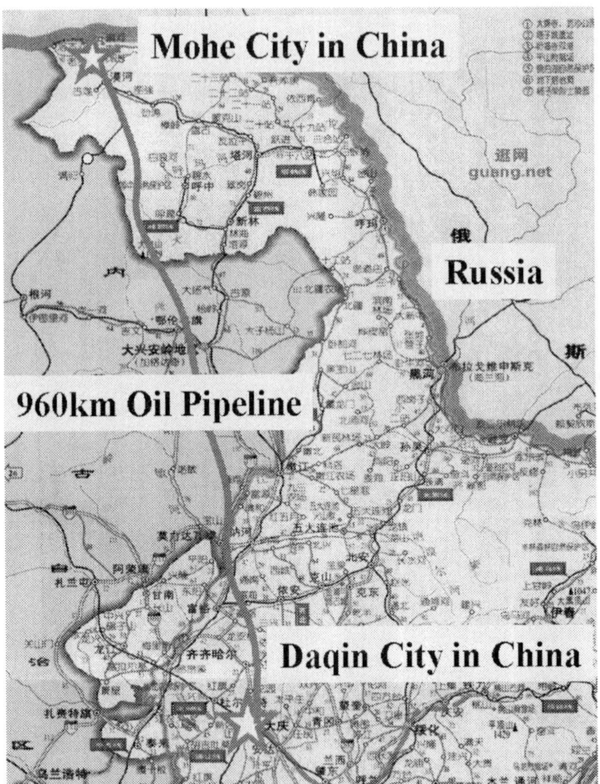

Fig. 33. MoHe-Daqing oil pipeline

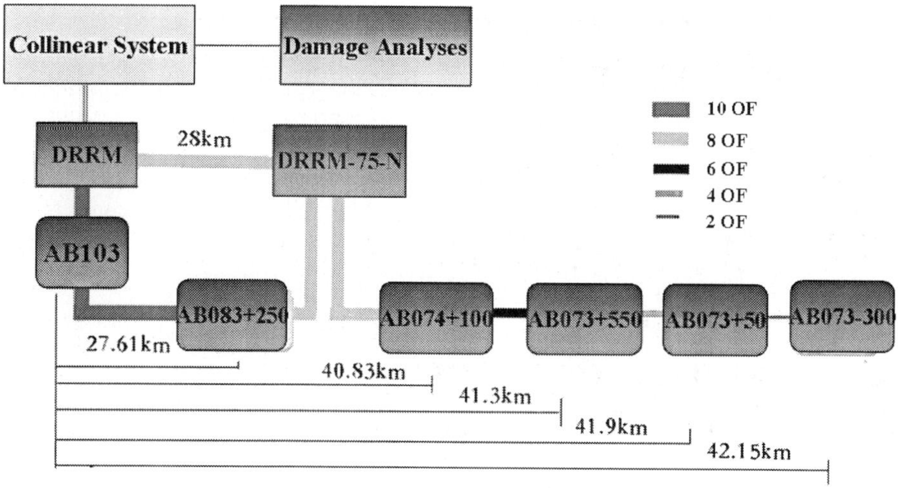

Fig. 34. SHM system of Mohe-Daqing oil pipeline

1) Indoor Test. A concentrated load was employed on one 4m PVC. To monitor the service behavior of the pipeline, the sensors including the ERS and OF sensors were bonded on the surface of PVC, as shown in Figure 35.

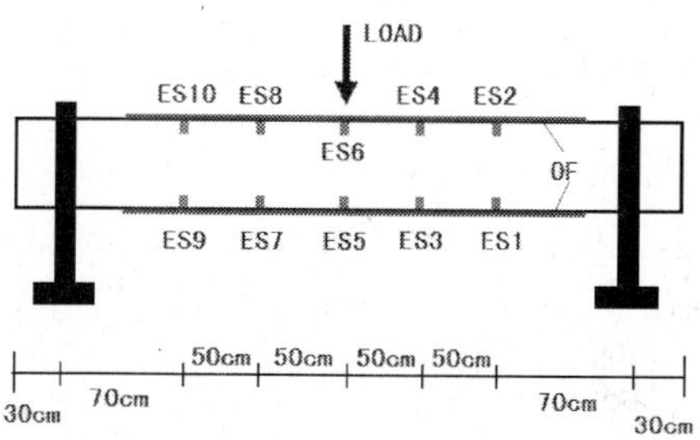

Fig. 35. Sensor installation

The PVC pipe was preloaded three cycles before the data acquisition, and the load was progressively applied on the pipe with a step of 40N. Load had been maintained for two minutes to take data acquisition. The test continued until the PVC pipe or the optical fiber was broken. Figure 36 shows the comparison of the measured strain distribution by the OF sensor and the ERS under various load. It can be seen that the strain measured by the OF agrees well with that measured by the ERS.

2) Outdoor Test. To validate feasibility of the proposed sensing technique in real application environments, tow PVC pipelines were placed into an outdoor water tank with the environmental temperature changing from -10°C to -20°C in winter, as shown in Figure 37. The FBG sensors were symmetrically placed on the pipeline, and the OF sensors were installed on the bottom and top surfaces of the PVC pipeline, shown as in Figure 38.

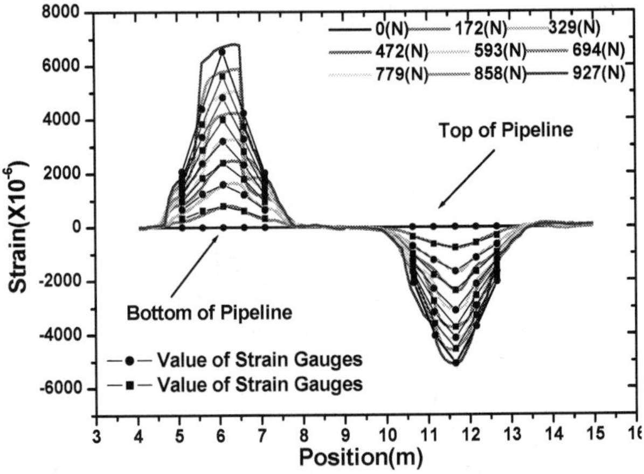

Fig. 36. Distributed strain by OF and ERS

Fig. 37. Instrumented PVC pipe in water tan

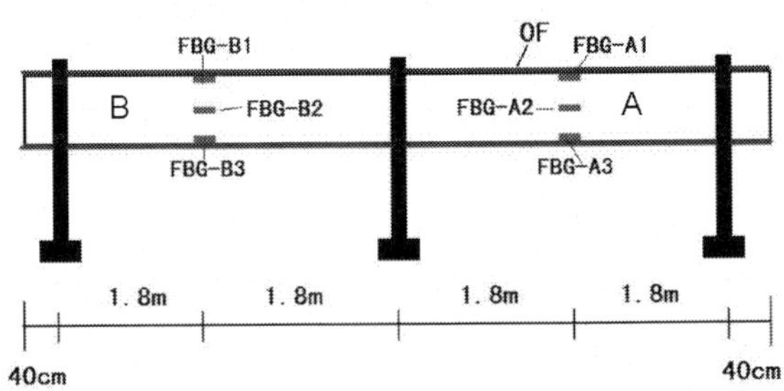

Fig. 38. Sensor installation

In this test, the thawing-settlement was controlled by using the heating rod. Although with the same environments, the frost-heaving was not evenly distributed along the PVC pipeline. Pipelines in "A" area frosted significantly, while that in "B" area almost did not frost, as can be seen in Figure 39. Figure 40 shows the distributed strain of the PVC pipeline from frost-heaving to thawing-settlement. The measured stress by the OF sensor on the top of the PVC pipe was in tensile, and that at the bottom was in compression during the process of frost-heaving. While the stress measured by the OF sensors during the process of thawing-settlement was opposite to that during the process of frost-heaving. Figure 41 shows the strain comparison between the FBG and the OF sensors. It can be seen that the strain trend measured by the OF agreed well with that by the FBG during the process of frost and thaw. Small variance between may be caused by the measuring accuracy difference or irregular frost or thaw. At "A" area of the PVC pipeline, the strain of FBG-A1 decreased from 800με to -480με, because the PVC pipeline broke around FBG-A2.

Fig. 39. Profile of frost-heaving

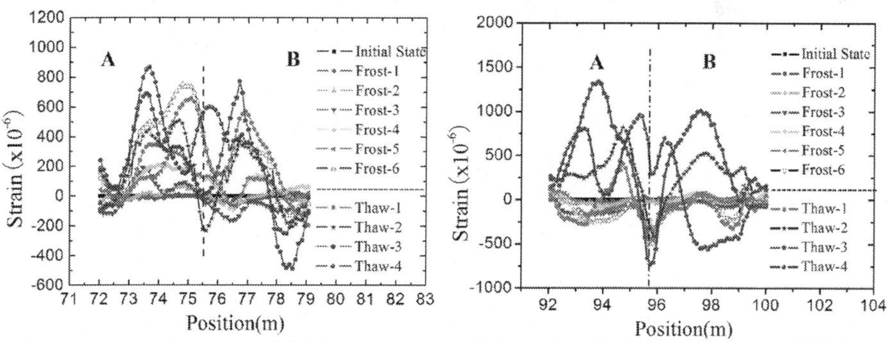

a) Distributed strain at top of PVC

b) Distributed strain at bottom of PVC

Fig. 40. Distributed strain measured by OF at frost-thaw process

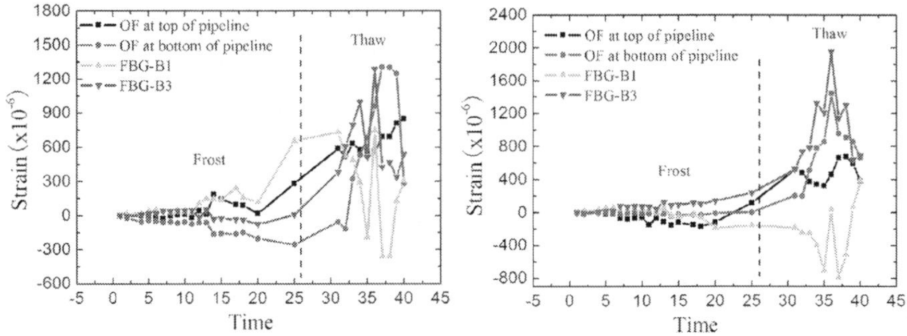

a) A-part b) B-part

Fig. 41. Strain comparison between FBG and OF

6. Applications of full-scale distributed optical fiber sensors in SHM for critical energy infrastructure

6.1 Life-cycle health monitoring of the casing pipe using the OF sensors

Casing pipes in oil well constructions may suddenly buckle as the hydrostatic pressure difference increases between the inside and outside of the pipes. For the safety of workers and the followed oil transportation process, the construction and service of the casing pipes is critical to be monitored. To meet this need, field test of structural health monitoring of casing pipes had been operated by employing both local FBG and Brillouin sensors on casing pipe for long-term monitoring. Figure 42 illustrates a sketch of the employed casing pipe monitoring system on a real application of the proposed casing pipe monitoring in Daqing Oilfield, China. Brillouin sensors or FRP-OF bars were installed on the surface of the casing pipe above non-oil reservoirs and FBG sensors were installed on the hot spots where casing damages often take place.

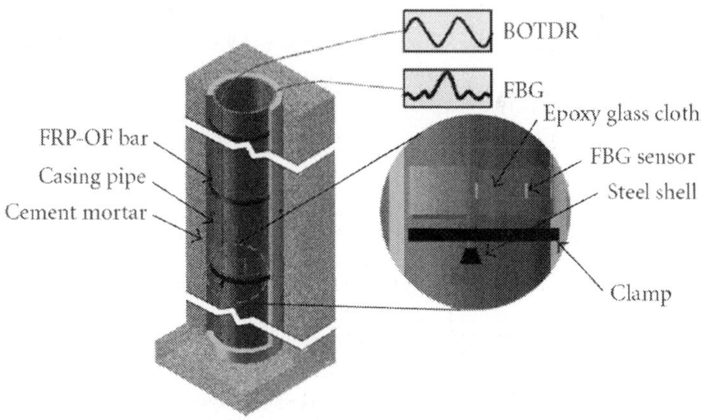

Fig. 42. Sketch of a structural monitoring system for the casing pipe

6.1.1 Sensor installation

1) Brillouin sensors. A FRP-OF bar was installed on the surface of the casing pipe over its entire length. To ensure that the FRP-OF bar deformed together with the casing pipe, several stainless steel encircling hoops were used to fix the FRP-OF bar on the casing pipe at 2m intervals. After placing the pipe in the down-hole well, the casing pipe and the FRP-OF sensor were wrapped with cement mortar to enhance their deformation compatibility. In addition, the two OF sensors were connected in series to form a measurement loop for the BOTDA measurement system. The bottom end of the FRP-OF bar and the splice joints of the OF sensors must be well protected to prevent damage during the construction. For this purpose, a stainless steel shell was used to cover the OF splice joints, and the gap in the shell was sealed with an epoxy iron mixture, as illustrated in Figure 43(a). Figure 43(b) shows the field installation of an FRP-OF bar and a casing pipe in the Daqing Oilfield, China.

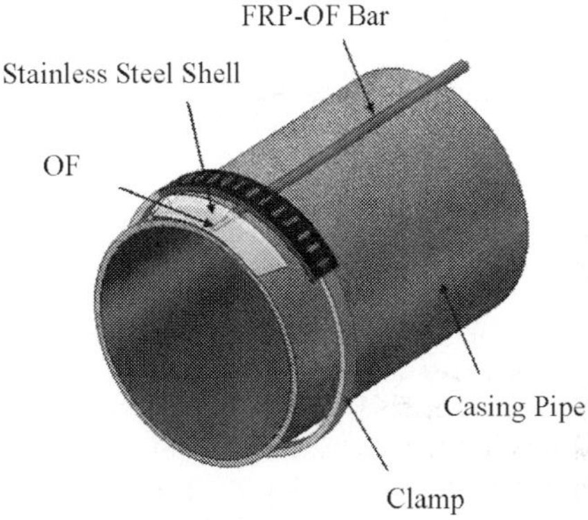

a) Protective steel shell at a solder joint

b) Sketch of the FRP-OF bar installation

Fig. 43. Brillouin sensor installation

2) FBG sensors. Four FBG strain sensors were installed at stress concentration areas of the casing pipe. Figure 44 illustrates their field installation on a casing pipe in the Daqing Oilfield, China. They were distributed around the circumference of the pipe at a 90° interval.

Fig. 44. FBG sensor installation

To perfectly bond to the casing pipe, the installation surface of each FBG strain sensor must be polished and cleaned. The FBG strain sensors were then covered by a piece of epoxy glass cloth for increasing the corrosion resistance and durability. For the purpose of long-term strain/stress measurements, one additional FBG temperature sensor was installed near the FBG strain sensors for temperature compensation.

To separate the strain and temperature information, the FBG temperature sensor was inserted into a small steel tube which was filled with silicone fluid to enhance the thermal conductivity. The initial center wavelengths of the five FBG sensors must be at least 3nm apart for reliable measurements. A special fiber optical cable with 8mm in diameter, protected by a steel strand at free state, was used to transmit the measured optical signals from the OF and FBG sensors.

6.1.2 Field test and discussion

To validate the installation techniques and measurement effectiveness of both FBG and BOTDA optical fiber sensor, the proposed SHM system was successfully employed to three oil wells in the Daqing Oilfield, including Xing10-5Bing3112, Xing10-5Bing3022, and Xing13-4-PB335.

1) Strain measurements at Xing10-5Bing3112 and Xing10-5Bing 3022.

Two FRP-OF bars installed on the casing pipes of these two oil wells had a length of 818m for Xing10-5Bing3112 and of 838m for Xing10-5Bing3022, respectively. Four FBG strain sensors and one FBG temperature sensor were installed in the significant locations of Xing10-5Bing3022 for high accuracy behavior monitoring. Considering the limitations of the oil pipe casing techniques and time required by the BOTDA/R measurements, limited locations with stress concentration on the casing pipe were monitored by the FBG sensors during the construction. The strains measured by the FBG sensors were recorded by an MOI si720 interrogator with a sample frequency of 5Hz and measurement accuracy of 1 $\mu\varepsilon$. Figure 45 and 46 present the strain measured by the FBG sensors during the processes of oil well casing and grouting . The stress states of the casing pipe during the whole construction process had been correctly reflected by the installed SHM system. Figure 45(a) and Figure 46(a) show the raw measured data without eliminating the environmental noise. Figure 45(b) and Figure 46(b), on the other hand, show the corresponding denoised data by Db5 wavelet. It shows that the denoising process significantly improve the quality of the measured signal for real applications.

The two oil wells were opened for operation after the complete of its construction. The strain obtained from the FRP-OF bar was recorded by the BOTDR demodulator produced by NTT in Japan, with the measuring accuracy of ±50$\mu\varepsilon$ and minimum spatial resolution of 1m. Figure 47 presents the strainsmeasured by the FRP-OF sensors on from December 15, 2007 to December 17, 2007. The measured strains remained stable, and all ranged from -120$\mu\varepsilon$ to +120$\mu\varepsilon$. The strains measured over three days by the FRP-OF sensors and its nearby FBG sensors at the stress concentration areas are compared in Figure 48. The strains measured by the FRP-OF sensors fluctuated around those measured by the FBG sensors, which changed little over the three days. This FBG sensor shows more accurancy than the BOTDR/A based FRP-OF sensor. The maximum relative error between these two techniques in strain

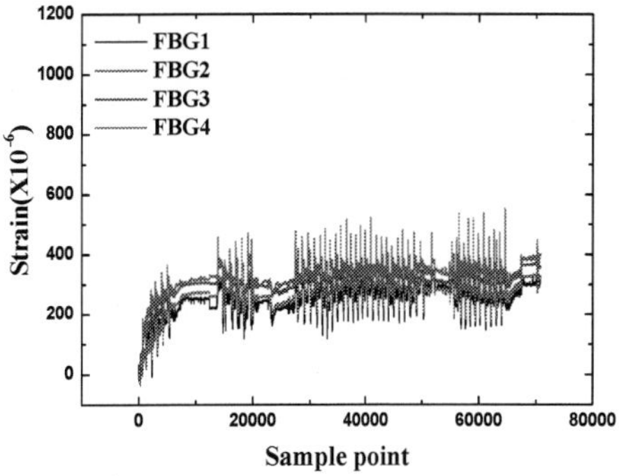

a) Raw data

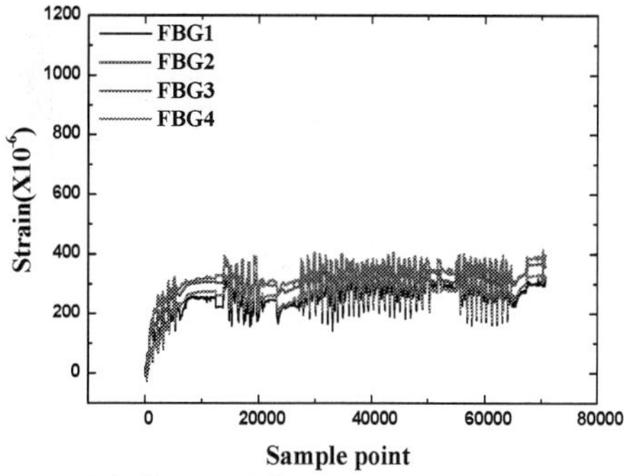

b) Cleansed data

Fig. 45. Strain measurements during casing of the oil well

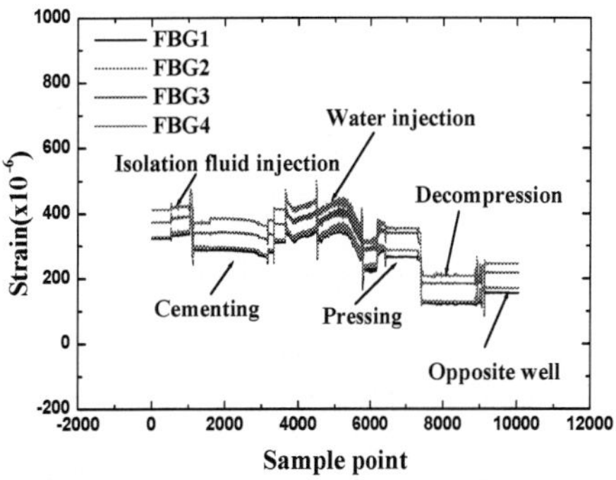

a) Raw data

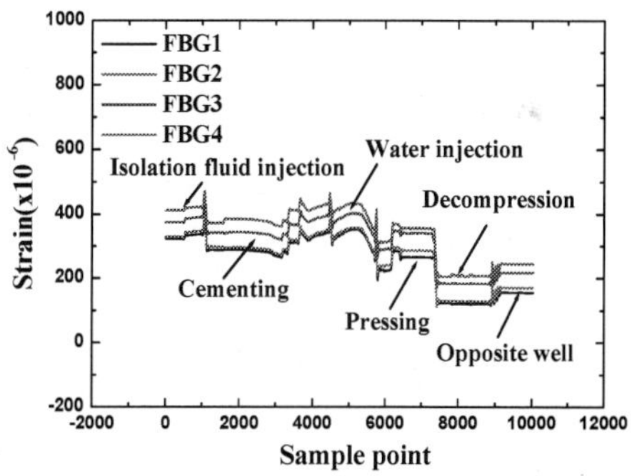

b) Cleansed data

Fig. 46. Strain measurements during grouting of the oil well

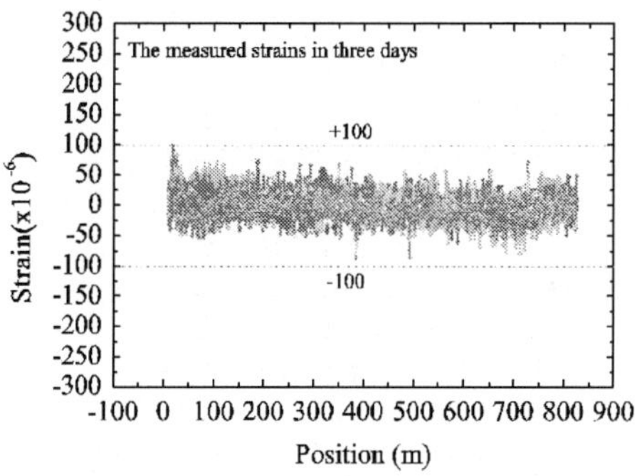

Fig. 47. Strain distribution of the casing pipe

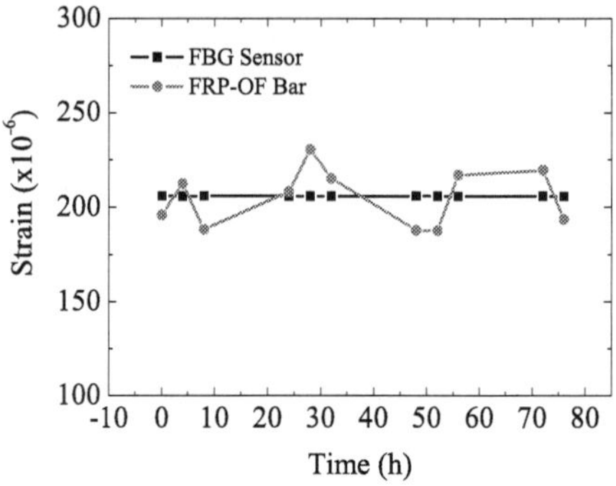

Fig. 48. Strain comparison

measurement is approximately 12%. Throughout the monitoring period, the strain state of the in-service casing pipes was normal, showing no sign of buckling.

2) Geostress measurements and Perforation Experiments for Xing13-4-PB335.

Perforation completion and positioning of the geostress orientation are two key factors in oil and gas explorations.. To measure the geostress or circumferential stress of the casing pipe, a total of 12 FBG strain sensors were installed around the circumference of the casing pipe of Xing13-4-PB335 oil well at a interval of 30° interval. Two FRP-OF bars of 1,133m long were installed along the casing pipe for the stress distribution measurement. The orientations of the 12 FBGs were identified by a multi-arm caliper logging tool. The orientation of the FBG6 corresponding to the marks on the casing pipe as shown in Figure 49 is approximately 311°. The orientations of the perforation are 41° and 221° near the FBG9 and FBG3, respectively. Figure 50 shows in-situ measurements of the strains of the in-service casing pipes. As shown in Figures 51, the orientation of the maximum horizontal stress determined by the cross multipole array acoustic instrument (XMAC) and FBG sensors are approximately 94.67° and 101°, respectively. The relative variance between the two measuring techniques is approximately 6.7%. Figure 52 shows the distribution of circumferential stresses of the casing pipe before and after perforation completion measured by the FBG sensors. The

Fig. 49. Orientation mark of casing pipe

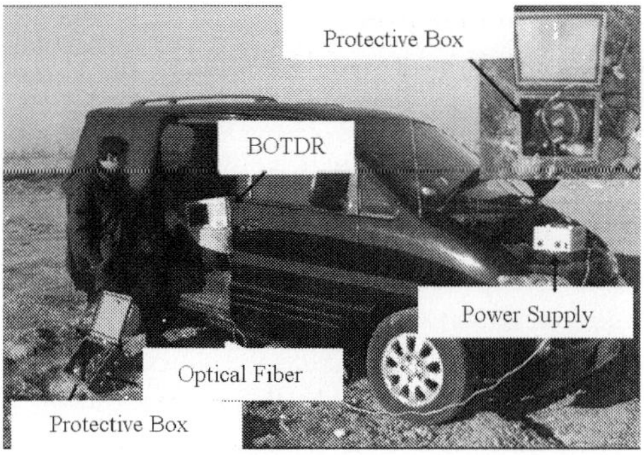

Fig. 50. In-situ measurements of the strains

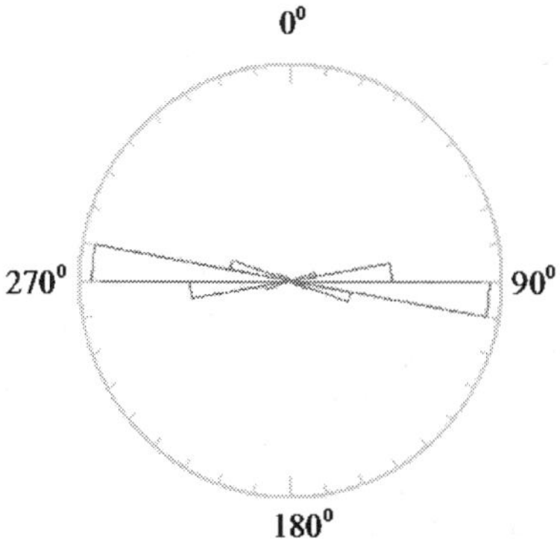

Fig. 51. The orientation of the maximum horizontal stress by XMAC

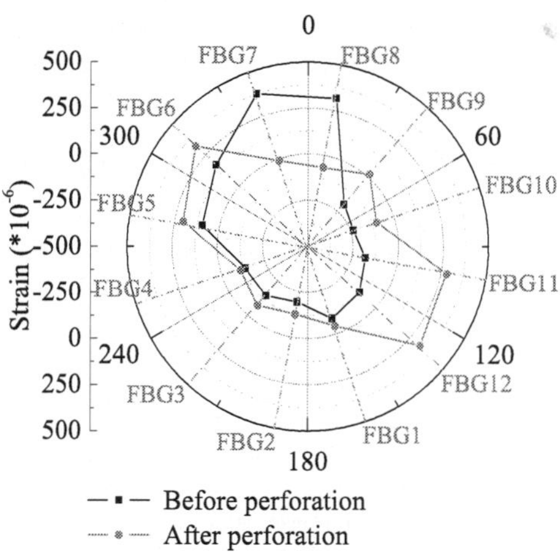

Fig. 52. The distribution of circumferential stress measured by FBG sensors

orientation of the maximum circumferential stress is vertical to that of perforation, which agrees well with the actual stress state applied on the casing pipe.

Figure 53 shows the strain distribution along the casing pipe before and after perforation, which was measured by two FRP-OF sensors. The readings from these two sensors were consistent. The strain difference at any particular position ranges from -200$\mu\varepsilon$ to +200$\mu\varepsilon$, indicating that the axial strain of the casing pipe remains stable before and after the perforation process. The average strain appears to be proportional to the distance to the ground surface due to the effects of gravity.

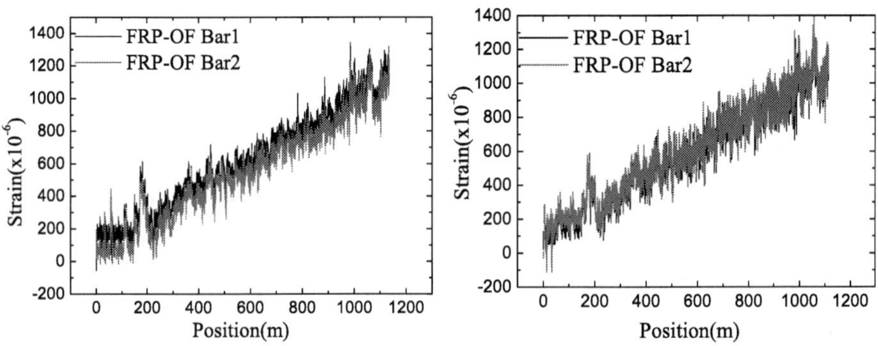

a) Before perforation b) After perforation

Fig. 53. Strain distribution from the FRP-OF

The field application shows that both the FRP-OF sensors and the FBG sensors wrapped in epoxy glass cloth are stable and durable. They have survived the casing and grouting processes of oil wells in a harsh construction environment and it indicates that the sensor installation procedure is feasible.

6.2 The life-cycle monitoring of offshore platform by using OF sensors

Offshore platform plays an important role for oil exploitation and accommodation. The offshore platform, especially deep offshore platform, serviced in a extremely harsh environment. It often involves the coupling effects of wind, tide, current, and corrosion, which could cause the breakdown of the mooring line, the large deformation of the flexible pipe, and even the collapse of the offshore platform. Stress and state-of-the-art are two key parameters correspond to the safety of the offshore platform. The existing sensors applied on offshore platform are off-line sensors, such as self-contained tilt sensors and self-contained strain sensors. The cost of the safety assessment based on these techniques is high, and no in-time safety monitoring and pre-warning capability is provided. In an effort of monitoring the offshore platform in long term, our research group has proposed several different types of sensors to successfully provide the SHM system on offshore platform application in mainland China.

6.2.1 Laboratory model test of T shaped tubular joint

A laboratory model of a T shaped tubular joint of #CB32A offshore platform of China in Bohai Bay was designed based on the similarity theory. Figure 54 shows the geometry and characteristic of the test model. The materials used for the joint are consistent with the actual #CB32A offshore platform, with 20Mn steel for tubes and E4315 for welding. Magnetism particle test and ultrasonic damage detection test were conducted to ensure the welding quality (OU, 2003).

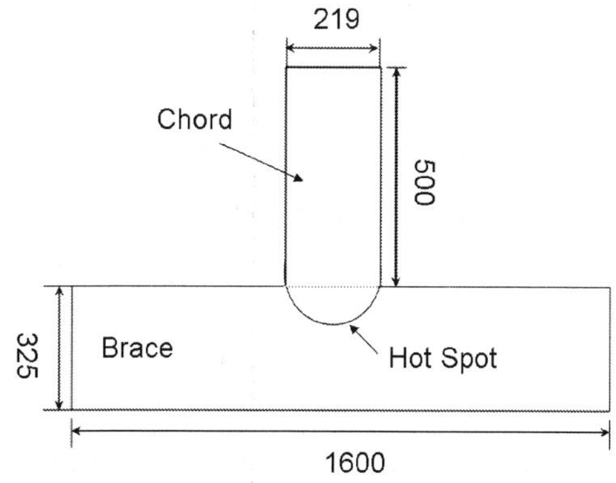

Fig. 54. Geometry dimensions of T shaped joint(mm)

FBG strain sensors were used to measure the nominal stress of the chord and the stress of the hot spots in the bracing tube. Axis force of the chord can then be calculated using nominal stress. Figure 55 shows the sensor layout in plane. Twelve FBG sensors in total were mounted on the joint. Using the Wavelength Multiplex Division (WMD) technique, FBG sensors with different Bragg wavelength can be connected together, and be interrogated at the same time. In this labotorary test, one dynamic interrogator with four channels was used to acquire the twelve FBG sensors' signal with a sampling rate of 250 Hz. Figure 56 shows the setup of the labotorary T shaped joint model test. Two ends of the

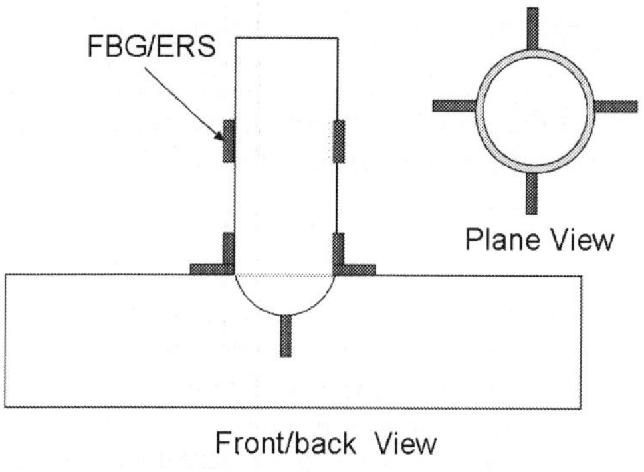

Fig. 55. Sensor installation

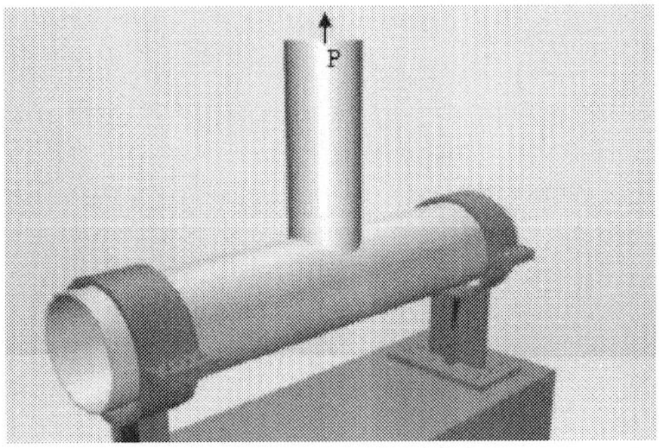

Fig. 56. Setup of T shaped tubular joint

bracing tube were fixed to the foundation by brace racks. Tensile forces were applied on the chord tube in axis direction. Electrical fluid servo actuator was connected to the computer to control the loading process . The maximum force output is ±630KN, and the maximum displacement output is ±250mm.

For the T shaped tubular joint under tensile load, the hot spots were located in the joint conjunction between the chord and the brace tube. Figure 57 shows the finite element analysis results for the maximum principle stress contour of the tested model simulated by ANSYS 9.0. A stress concentration along the welding in the joint conjunction was formed, and the maximum stress is observed at the saddle spot.

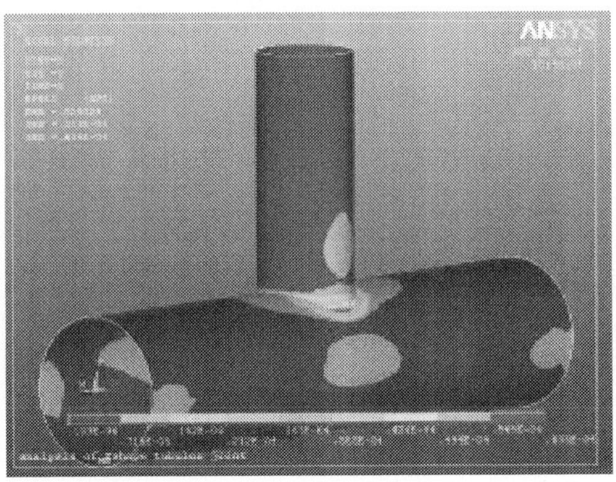

Fig. 57. Principal strain contour result

1. **Measured plane strain of the hot spots.** One particular location on the hot spot areas was selected for strain monitoring as shown in Figure 58. Three FBG strain sensors and three ERSs were installed on the required location. The angle between the three sensors was 45 degree, denoted as S1, S2 and S3, and the three sensors formed as strain rosette. Figure 59 shows the principle strain and strain in X direction measured by the two kinds of rosettes. The FBG sensors show great agreement to the ERSs, verifying their feasibility of principle strain measurements.

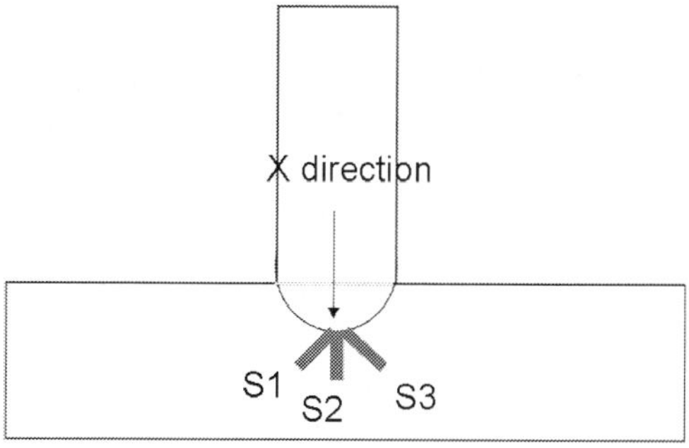

Fig. 58. Sensor installation of plane strain monitoring test

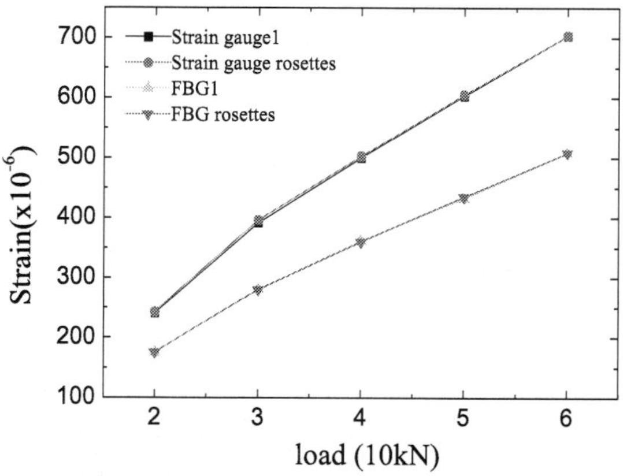

Fig. 59. Comparison between the principle strain and x direction strain

2. **Static test for local strain monitoring** For the local strain monitoring, a static test had been operated on the labotorary T shaped tubular joint model. The load was placed on the top of the chord tube, with an increment of 10KN, from 20KN to 60KN. Three loading cycles were conducted. Figure 60 shows the measured strain comparison between the two kinds of sensors, FBG and ERSs. Linear relationship between loading and strain indicates that strains in hot spot were much larger than that in the chord tube.

6.2.2 Field applications of the proposed SHM system on #CB32A Offshore Platform, Bohai Bay, China

#CB32A Offshore Platform represents an important type of platform structures in the oil industry. During the long term service, the platform will be subjected to extremely harsh environmental loads, such as wind, wave tide, water erosion and current loads. To monitor its performance, approximately 259 FBG sensors, 24 acceleration sensors, 178 PVDF sensors, and various other sensors were installed on the offshore platform's key locations, such as the T-shape tubular joints. As shown in Figure 61, the SHM system included a data

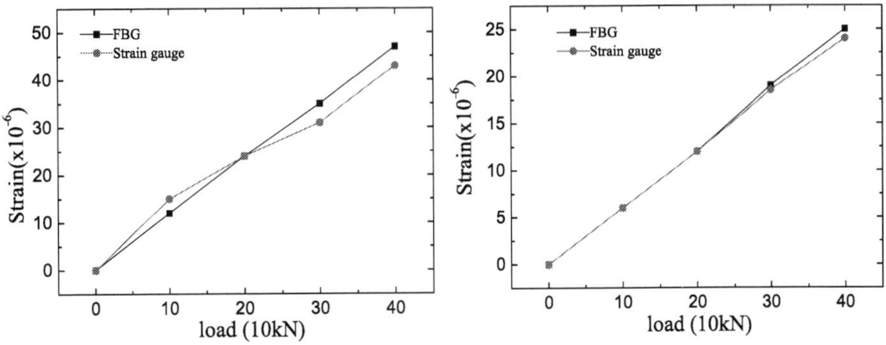

Fig. 60. Strain comparison of two kinds of sensors, FBGs and ERSs

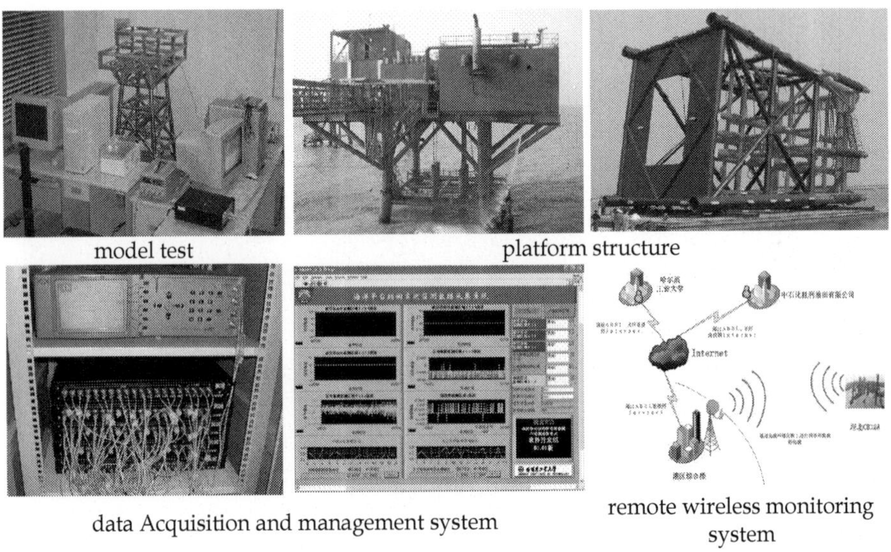

Fig. 61. SHM for CB32A offshore platform

acquisition and management module, a safety assessment module, an early warning module, and a remote wireless real-time monitoring module. The system has been running since 2003, and its stability, monitoring performance, and real-time data collection all exceed the original expectations. Figure 62 shows parts of the monitored data, indicating the applicability and feasibility of the proposed SHM system in the practical applications.

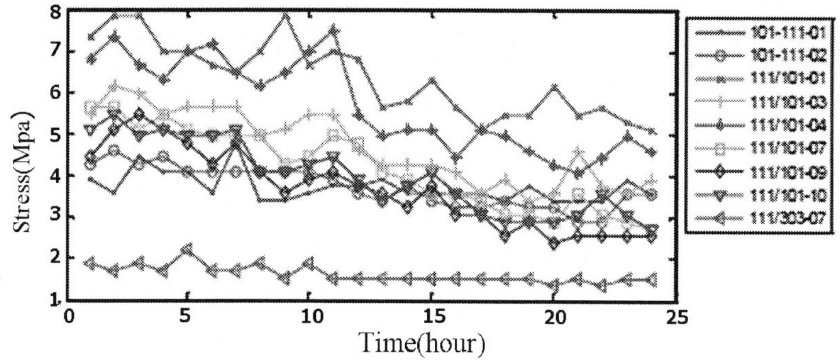

Fig. 62. Parts of the monitored data

7. References

[1] AM Gillooly, L Zhang, I Bennion. High survivability fiber sensor network for smart structures. SPIE, 2004, 5579:99-106.

[2] D Inaudi and B Glisic. Development of Distributed Strain and Temperature Sensing Cables. 17th International Conference on Optical Fibre Sensors. SPIE, 2005,5855:222-225

[3] DS Jiang, YC Hao, SC Liu, Application of FBG load cell on bowstring arch bridge. Instrument Technique and Sensor, 2006,2: 43-44.

[4] Filippo B, Andrea R, Nestore G, et al. Discontinuous Brillouin strain monitoring of small concrete bridges: comparison between near-to-surface and "smart" FRP fiber installation techniques. Proceedings of SPIE 2005; 5765:612-623.

[5] H Li, JP Ou, Z Zhou, Applications of optical fibre Bragg gratings sensing technology-based smart stay cables optics and lasers in engineering, 2009, 47(10):1077-1084.

[6] JP He, Z Zhou, YH Wu, JP Ou. Temperature inter-compensation for single-line optical fiber sensors based on Briilouin and fiber Bragg grating technique. LASER TECHNOLOGY, 2010, 1(34):13-16.

[7] JP He, Z Zhou, GD Chen, JP Ou. Measurement Accuracy Improvement of Brillouin Signal Using Wavelet Denoising Method. Proceedings of SPIE, 2009, 7293(0B):1-7.

[8] JP Ou, H Li. Structural health monitoring in mainland china: review and future trends. Structural Health Monitoring 2010; 9(3):219-231.

[9] JW Qian, QH Zheng, YN Zhang, et al. Deformation sensing and incremental shape reconstruction for intelligent colonoscopy. Optics and Precision Engineering, 2004, 12(5):518-524.

[10] PC Peng, HY seng, S ChiA. A novel fiber-laser-based sensor network with self-healing function. IEEE PHOTONICS TECHNOLOGY LETTERS, 2003, 2(15):275-277.

[11] PC Peng, HY Seng, S ChiA. A hybrid star-ring architecture for fiber Bragg grating sensor system. IEEE PHOTONICS TECHNOLOGY LETTERS, 2003, 9(15):1270-1272.

[12] P Wei, XH Sun. Smart sensor network is enhanced by new models. SPIE-newsroom, 2007, 10.1117/2.1200707.0523.

[13] THT Chan, L Yu, HY Tam, etc. Fiber Bragg grating sensors for structural health monitoring of Tsing Ma Bridge: Background and experimental observation. Engineering Structures 2006; 28(5):648–659.

[14] XJ Zhu, MY Lu, H Cao, et al. 3D Reconstruction Method of Space Manipulator vibration shape based on structural curvatures. The Eighth International Conference on Electronic Measurement and Instruments, 2007, 2:772-775.

[15] XF Zhao, ZD Duan，JP Ou. Local Strain Monitoring Study of Offshore Platform T Shaped Tubular Joint Using Fiber Bragg Grating Sensors. Proceedings of SPIE, 2008,6932(0E):1-11.

[16] XY Bao, DJ Webb, et al. Combined Distributed Temperature and Strain sensor Based on Brillouin Loss in an Optical Fiber.Optics Letters, 1994,19:141-143

[17] YQ Ni, Y Xia, WY Liao, JM Ko. Technology innovation in developing the structural health monitoring system for Guangzhou New TV Tower. Journal of Structural Control and Health Monitoring 2009; 16:73-98.

[18] Z Zhou, JP Ou. Various Types of Optical FBG Sensors and Their Applications SHM. Proceeding of the 2nd International Conference on Structural Monitoring and Intelligent Infrastructures, Shenzhen China, 2005:489-498.

[19] Z.S. Wu, B. Xu, K.J. Hayashi, et al. Distributed Optic Fiber Sensing for A Full-scale PC Girder Strengthened with Prestressed PBO Sheets. Engineering Structures, 2006, 28: 1049~1059.

[20] Z Zhou, JP He, JP Ou. Measurement technique based on the FBG-BOTDA(R) for infrastructures. China Civil Engineering Journal, 2010,3(43):111-118.

[21] Z Zhou, JP He, JP Ou. Fiber-Reinforced Polymer-Packaged Optical Fiber Sensors based on Brillouin Optical Time-Domain Analysis. Optical Engineering, 2008, 47(014401): 1-10

[22] Z Zhou, JP. He, GD Chen, et al. A new kind of smart cable with functionality of full scale monitoring using BOTDR technique. Proceedings of SPIE, 2009, 7293(0L):1-7.

[23] Z Zhou, JP He, JP Ou. Casing Pipe Damage Detection with Optical Fiber Sensors: a Case Study in Oil Well Constructions. Advances in Civil Engineering, 2010, 638967:1-9.

5

Characterization of Brillouin Gratings in Optical Fibers and Their Applications

Yongkang Dong, Hongying Zhang,
Dapeng Zhou, Xiaoyi Bao and Liang Chen
University of Ottawa
Canada

1. Introduction

There is an increasing interest over recent years in Brillouin gratings in optical fibers due to their applications in optical storage (Cao et al., 2008; Kalosha et al., 2008; Zhu et al., 2007), distributed sensing (Dong et al., 2009; Song et al., 2009, 2010; Zou et al., 2009a, 2009b, 2010), optical delay line (Song et al., 2009), and birefringence characterization of polarization-maintaining fibers (Dong et al., 2010). Generally, a Brillouin grating is generated by using two counter-propagating pump waves through stimulated Brillouin scattering (Song et al., 2008). A Brillouin grating is formed by a moving periodically modulated refractive index associated with an acoustic wave, which is created by two pump waves through electrostriction effect. Compared with conventional fiber Bragg gratings, the Brillouin gratings have two unique features: one is that it is a moving grating, which can produce a Brillouin frequency shift on the reflected light with respect to the probe wave, and the other is that it has a lifetime (~10 ns in a silica fiber) for its existence after removing the two pump waves.

For most applications, a high-birefringence polarization-maintaining fiber is used, where two pump waves are launched into one axis of the polarization-maintaining fiber to create a Brillouin grating and a probe wave is launched into the other axis to read the grating; when the frequency difference between the probe and the pump waves satisfies the phase matching condition, a maximum reflection on the Brillouin grating can be observed. So the process involves four light waves, i. e., two pump waves, a probe wave and a reflected wave, and in fact it is a Brillouin enhanced four-wave mixing process. The interaction between the probe wave and the Brillouin grating involves two cases: one case is that the probe wave propagates in the same direction as the Brillouin grating, so coherent Stokes Brillouin scattering could occur and the reflected wave frequency is lower than that of the probe wave by a Ω_B; the other case is for the counter-propagating between the probe wave and the Brillouin grating, so coherent anti-Stokes Brillouin scattering could occur and the reflected wave is frequency up-converted by a Ω_B with respective to the probe wave. The phase matching condition of the coherent Stokes Brillouin scattering or coherent anti-Stokes Brillouin scattering in a polarization-maintaining fiber is $\Delta\nu_{Bire} = \Delta n \nu / n_g$, where Δn is the phase birefringence of the polarization-maintaining fiber, n_g is the group refractive index, and ν is the frequency of probe wave. $\Delta\nu_{Bire}$ is the birefringence-induced frequency shift between the probe and pump waves and is proportional to the local birefringence of the polarization-maintaining fiber.

A Brillouin grating can be generated in frequency correlation domain (Zou et al., 2009a, 2009b, 2010) and time domain (Dong et al., 2009, 2010a, 2010b; Song et al., 2008, 2009a, 2009b, 2010), respectively. With the correlation-based technique, a Brillouin grating is generated using two synchronized and frequency-modulated continuous-wave (CW) pump waves. In time domain, a Brillouin grating can be more conveniently generated through two frequency-locked pump waves, which includes three schemes in terms of pump type. The first scheme uses two CW pumps, and the Brillouin grating can exist in the whole fiber for a uniform Brillouin frequency shift over the fiber (Song et al., 2008; Zou et al., 2009), where the length of Brillouin grating would be limited by the coherent length of the pump lasers. The second scheme includes a CW pump and a pump pulse, where the Brillouin grating is generated following the pump pulse from one end to the other end (Song et al., 2009, 2010). In this case, due to the decay of the acoustic wave, the length of the existing Brillouin grating is limited to $L_B = (t_P + \tau_B)c/n$, where t_P is the width of pump pulse, τ_B is the phonon lifetime. The third scheme includes two short pump pulses (< 10 ns), where the Brillouin grating can be generated at a specific location controlled by the delay between the two pump pulses (Dong et al., 2009, 2010). The interaction length of the two pulses is $L = (t_{P1} + t_{P2})c/2n$, where t_{P1} and t_{P2} are the pulse widths of pump 1 and pump 2, respectively. However, due to different phonon excitation time, the effective length of Brillouin grating is only half of the interaction length, i. e. $L_B = (t_{P1} + t_{P2})c/4n$, for example, 20 cm for two 2 ns pump pulses.

This chapter is structured as follows: in the second section, we will give the coupled wave equations of Brillouin enhanced four-wave mixing associated with a Brillouin grating, from which the analytic solution of the steady uniform Brillouin grating spectrum is deduced without pump depletion; in the third section, we will numerically study the Brillouin grating spectrum with a strong probe wave for both coherent Stokes Brillouin scattering and coherent anti-Stokes Brillouin scattering cases; in the fourth section, we will experimentally study the transient Brillouin grating spectrum by using two pump pulses and a probe pulse; in the fifth section, we will introduce two applications of Brillouin gratings, i. e., distributed birefringence measurement of polarization-maintaining fibers and distributed simultaneous temperature and strain measurement; at the end, a conclusion will be given in the sixth section.

2. Coupled wave equations of Brillouin gratings

In this section, the coupled wave equations of Brillouin enhanced four-wave mixing associated with a Brillouin grating are given first. Then the analytic solution of the steady uniform Brillouin grating spectrum is deduced for the case of a weak probe. And last, the phase matching condition for the case of a polarization-maintaining fiber as well as the birefringence-induced frequency shift is discussed.

2.1 Coupled wave equations

The scheme of interaction between the four optical waves and the acoustic wave within a polarization-maintaining fiber is shown in Fig. 1, where the fiber length is along the z direction, while its two axes are along x and y directions, respectively. In this analysis an acoustic wave of amplitude ρ, frequency Ω, and wave vector **q** is generated by two pump waves, pump1 and pump2, of amplitudes A_1 and A_2, frequencies ω_1 and ω_2, and wave vectors $\mathbf{k}_1$ and $\mathbf{k}_2$ respectively that counter propagate in the x axis of the polarization-maintaining fiber, and ω_1 is larger than ω_2 by a Brillouin frequency shift. Here we discuss the coherent anti-Stokes Brillouin scattering case in which the probe wave of amplitude A_3, frequency ω_3, and wave vector $\mathbf{k}_3$ counter propagates with and scatters off the acoustic wave to form a scattered wave of amplitude A_4, frequency ω_4, and wave vector $\mathbf{k}_4$ in the y axis of the polarization-maintaining fiber.

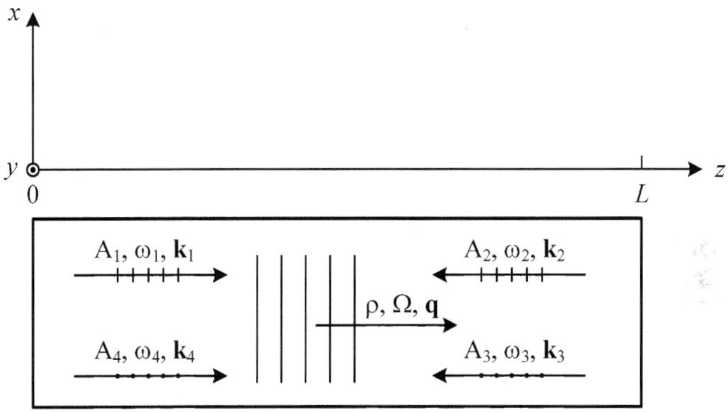

Fig. 1. Scheme of interaction between the optical and acoustic waves in a Brillouin enhanced four-wave mixing process associated with a Brillouin grating.

For the interacting system shown in Fig. 1, the total optical field can be represented as

$$\tilde{E}(z,t) = \tilde{E}_1(z,t) + \tilde{E}_2(z,t) + \tilde{E}_3(z,t) + \tilde{E}_4(z,t) \tag{1}$$

where $\tilde{E}_1(z,t)$, $\tilde{E}_2(z,t)$, $\tilde{E}_3(z,t)$, $\tilde{E}_4(z,t)$ represent pump 1, pump 2, probe and reflection waves, respectively, and can be expressed as

$$\tilde{E}_1(z,t) = A_1(z,t)\exp[i(k_1 z - \omega_1 t)] + c.c. \tag{2a}$$

$$\tilde{E}_2(z,t) = A_2(z,t)\exp[i(-k_2 z - \omega_2 t)] + c.c. \tag{2b}$$

$$\tilde{E}_3(z,t) = A_3(z,t)\exp[i(-k_3 z - \omega_3 t)] + c.c. \tag{2c}$$

$$\tilde{E}_4(z,t) = A_4(z,t)\exp[i(k_4 z - \omega_4 t)] + c.c. \tag{2d}$$

Similarly, the acoustic field can be described in terms of the material density distribution as

$$\tilde{\rho}(z,t) = \rho_0 + \{\rho(z,t)\exp[i(qz - \Omega t)] + c.c.\} \tag{3}$$

where $\Omega = \omega_1 - \omega_2$, $q = k_1 + k_2$, and ρ_0 denotes the mean density of the medium.

Taking the electrostriction effect into account, the equation for the material density can be expressed by

$$\frac{\partial^2 \tilde{\rho}}{\partial t^2} - \Gamma' \nabla^2 \frac{\partial \tilde{\rho}}{\partial t} - \upsilon^2 \nabla^2 \tilde{\rho} = \nabla \cdot \mathbf{f} \tag{4}$$

where υ is the velocity of the acoustic wave and $\Gamma'=(4\eta_s/3+\eta_b)/\rho$ is a damping parameter (Boyd, 2008), with η_s and η_b being the shear and the bulk viscosity coefficients, respectively.

The source term on the right-hand side of Eq. (4) consists of the divergence of the force per unit volume **f**, which is given by

$$\mathbf{f} = -\frac{1}{2}\varepsilon_0 \gamma_e \nabla \left\langle \tilde{E}^2 \right\rangle \tag{5}$$

with the term in the bracket denoting the electrostriction term.

Introducing Eqs. (3) and (5) into the acoustic wave equation (4) and making the slowly varying amplitude approximation, one can obtain the result

$$-2i\Omega \frac{\partial \rho}{\partial t} - i\Omega \Gamma_B \rho - 2iq\upsilon^2 \frac{\partial \rho}{\partial z} = \varepsilon_0 \gamma_e q^2 (A_1 A_2^* + A_3^* A_4 e^{i\Delta k z}) \tag{6}$$

where $\Gamma_B=q^2\Gamma'$ is the Brillouin linewidth whose reciprocal $\tau_p=1/\Gamma_B$ gives the phonon lifetime, and $\Delta k=(k_4+k_3)-(k_1+k_2)$ is the phase mismatch of the four optical waves. As we have mentioned above, the two pumps are polarized in x direction, while the probe and the reflected waves are polarized in y direction, therefore there are only two terms left in the drive term of the acoustic field.

The spatial evolution of the optical waves is described by the wave equation

$$\frac{\partial^2 \tilde{E}_j}{\partial z^2} - \left(\frac{n}{c}\right)^2 \frac{\partial^2 \tilde{E}_j}{\partial t^2} = \frac{1}{\varepsilon_0 c^2} \frac{\partial^2 \tilde{P}_j}{\partial t^2} \tag{7}$$

where j=1, 2, 3 and 4, and the total nonlinear polarization which gives rise to the source term is given by

$$\tilde{P} = \varepsilon_0 \Delta \chi \tilde{E} = \varepsilon_0 \Delta \varepsilon \tilde{E} = \varepsilon_0 \rho_0^{-1} \gamma_e \tilde{\rho} \tilde{E} \tag{8}$$

Introducing Eq. (1) into the wave equation (7) and making the slowly-varying amplitude approximation, one can obtain the coupled equations for the optical waves. Combining the acoustic wave equation (6), the coupled-wave equations for the optical fields and the acoustic wave can be obtained as

$$\left(\frac{\partial}{\partial z} + \frac{n_x}{c}\frac{\partial}{\partial t}\right) A_1 = i g_o \rho A_2 \tag{9a}$$

$$\left(-\frac{\partial}{\partial z} + \frac{n_x}{c}\frac{\partial}{\partial t}\right) A_2 = i g_o \rho^* A_1 \tag{9b}$$

$$\left(-\frac{\partial}{\partial z}+\frac{n_y}{c}\frac{\partial}{\partial t}\right)A_3 = ig_o\rho^* A_4 e^{i\Delta kz} \tag{9c}$$

$$\left(\frac{\partial}{\partial z}+\frac{n_y}{c}\frac{\partial}{\partial t}\right)A_4 = ig_o A_3 \rho e^{-i\Delta kz} \tag{9d}$$

$$\left(\frac{\partial}{\partial t}+\frac{\Gamma_B}{2}\right)\rho = ig_a\left(A_1 A_2^* + A_3^* A_4 e^{i\Delta kz}\right) \tag{9e}$$

The above Eqs. (9) are just the coupled-wave equations for the optical fields and the acoustic wave in the Brillouin enhanced four-wave mixing process, where n_x and n_y are refractive indexes for the two axes of the polarization-maintaining fiber; g_o and g_a are coupling coefficients of optical wave and acoustic wave, respectively, and $g_B=4g_og_a/\Gamma_B$ is the Brillouin gain factor. Note that when deducing Eq. (9e) the term describing the propagation of the acoustic wave in Eq. (6) has been omitted because the velocity of the acoustic wave is far smaller than that of light.

2.2 Phase matching condition

Next we will discuss the phase matching condition for this interaction process. Fig. 2 shows the wave vector mismatch of the four optical waves, according to which one can obtain the relationship between the four optical wave vectors

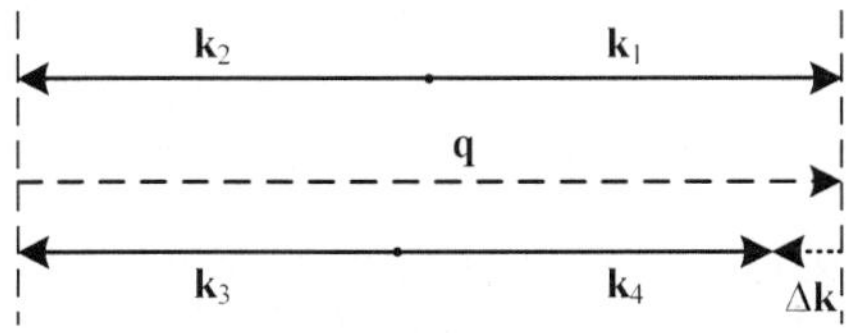

Fig. 2. Wave vector mismatch of the four optical waves.

$$\Delta \mathbf{k} = (\mathbf{k}_4 - \mathbf{k}_3) - (\mathbf{k}_1 - \mathbf{k}_2) \tag{10}$$

Therefore the magnitude of the phase mismatch is given by

$$\begin{aligned}\Delta k &= (k_4 + k_3) - (k_1 + k_2)\\ &= \left(\frac{\omega_4}{\upsilon_y}+\frac{\omega_3}{\upsilon_y}\right) - \left(\frac{\omega_1}{\upsilon_x}+\frac{\omega_2}{\upsilon_x}\right)\\ &\approx \frac{2n_y\omega_3}{c} - \frac{2n_x\omega_2}{c}\\ &= \frac{2n_y}{c}\cdot\Delta\omega\end{aligned} \tag{11}$$

where

$$\Delta\omega = \omega_3 - \frac{n_x}{n_y}\omega_2 \tag{12}$$

and we have made the assumption that $\omega_4 \approx \omega_3$ and $\omega_1 \approx \omega_2$, respectively.

If the frequency of the probe wave ω_3 is chosen to ensure that the condition of perfect phase matching, i.e., $\Delta k=0$ is satisfied, then ω_3 will take the value

$$\omega_3 = \frac{n_x}{n_y}\omega_2 \tag{13}$$

and the birefringence-induced frequency shift can be obtained as

$$\Delta\omega_{Bire} = \omega_3 - \omega_2 = \frac{\Delta n}{n}\omega_2 \tag{14}$$

or

$$\Delta\nu_{Bire} = \frac{\Delta n}{n}\nu \tag{15}$$

where $\Delta n=n_x-n_y$ is the birefringence of the polarization-maintaining fiber.

2.3 Analytic solution for steady uniform Brillouin grating

In the case of steady Brillouin grating generated by two continuous-wave (CW) pumps, all the time derivatives in the coupled-wave equations (9) are considered to be zero. We treat the simple case of uniform Brillouin grating without pump depletion, which indicates that the intensities of the two pump waves are unaffected during the interaction; while the probe wave and the reflected waves are both weak so that the acoustic wave generated by the two pumps does not affected by the probe wave. Therefore the driving term of the acoustic wave caused by the probe and the reflected waves in Eq. (9e) can be dropped and the amplitudes A_1 and A_2 of the pump waves can be considered as a constant in the coupled-wave equations, which then reduce to the simpler set for the probe and reflected waves and the acoustic wave:

$$\frac{\mathrm{d}A_3}{\mathrm{d}z} = -ig_o A_4 \rho^* e^{i\Delta kz} \tag{16a}$$

$$\frac{\mathrm{d}A_4}{\mathrm{d}z} = ig_o A_3 \rho e^{-i\Delta kz} \tag{16b}$$

$$\frac{\Gamma_B}{2}\cdot\rho = ig_a A_1 A_2^* \tag{16c}$$

By introducing Eq. (16c) into Eqs. (16a) and (16b), we obtain

$$\frac{\mathrm{d}A_3}{\mathrm{d}z} = K_1 A_4 e^{i\Delta kz} \tag{17a}$$

$$\frac{dA_4}{dz} = K_2 A_3 e^{-i\Delta kz} \tag{17b}$$

where

$$K_1 = -\frac{1}{2} g_B A_1^* A_2 \tag{18a}$$

$$K_2 = -\frac{1}{2} g_B A_1 A_2^* \tag{18b}$$

where $g_B=4g_o g_a/\Gamma_B$ is the Brillouin gain factor.

The general solution for Eqs. (17) can be written as follows:

$$A_3(z) = (C_1 e^{gz} + C_2 e^{-gz}) e^{i\Delta kz/2} \tag{19a}$$

$$A_4(z) = (D_1 e^{gz} + D_2 e^{-gz}) e^{-i\Delta kz/2} \tag{19b}$$

Where g, which is determined by $g^2=K_1K_2-(\Delta k)^2/4$, gives the rate of spatial variation of the fields and C_1, C_2, D_1, and D_2 are constants whose values depend on the boundary conditions.

Substituting the solutions (19a) and (19b) into Eqs. (17), since the resulting equations must hold for all values of z, the terms that vary as z and must each maintain the equality separately; the coefficients of these terms thus must be related by the following independent equations:

$$C_1(g + \frac{1}{2} i\Delta k) = K_1 D_1 \tag{20a}$$

$$C_2(g - \frac{1}{2} i\Delta k) = -K_1 D_2 \tag{20b}$$

Two additional relations among the quantities C_1, C_2, D_1, and D_2 come from the appropriate boundary conditions:

$$A_3(L) = (C_1 e^{gL} + C_2 e^{-gL}) e^{i\Delta kL/2} \tag{21a}$$

$$A_4(0) = D_1 + D_2 = 0 \tag{21b}$$

The values of C_1, C_2, D_1, and D_2 can be easily evaluated by solving the four independent linear Eqs. (21), and the solutions for Eqs. (17a) and (17b) thus can be obtained as

$$A_3(z) = A_3(L) \cdot \frac{2g\cosh(gz) - i\Delta k \sinh(gz)}{2g\cosh(gL) - i\Delta k \sinh(gL)} \cdot e^{-i\Delta k(L-z)/2} \tag{22a}$$

$$A_4(z) = A_3(L) \cdot \frac{2K_2 \sinh(gz)}{2g\cosh(gL) - i\Delta k \sinh(gL)} \cdot e^{-i\Delta k(L+z)/2} \tag{22b}$$

Therefore for a steady Brillouin grating, the analytic solution for the reflectivity R which is defined by

$$R = \frac{I_4(L)}{I_3(L)} \tag{23}$$

can be finally obtained as

$$R = \frac{\sinh^2(gL)}{\cosh^2(gL) - \Delta k^2 / (4K_1K_2)} \tag{24}$$

The Eq. (24) is just the analytic expression for Brillouin grating spectrum under weak probe condition, from which the reflectivity of a Brillouin grating in a polarization-maintaining fiber can be easily obtained. It should be noted that the Brillouin grating spectrum under weak probe condition, which is described by Eq. (24), is exactly the same as that of a fiber Bragg grating (Erdogan, 1997).

3. Simulations of steady Brillouin grating spectra with a strong probe wave

As we have mentioned above, the Brillouin grating spectrum is the same as the fiber Bragg grating under the condition of a weak probe wave and can be easily obtained through Eq. (24); however, it is not the case for a strong probe wave that can considerably modify the Brillouin grating, therefore numerical simulations by solving the coupled-wave equations (9) must be carried out in order to obtain the Brillouin grating spectrum. In this section, we will focus on numerical studies of the steady Brillouin grating spectrum which is generated by two counter-propagating CW pumps for the strong probe regime that can't be reached by the analytic solution. For all of these simulations some parameters like the Brillouin linewidth, the fiber length and powers of the two pump waves take their values listed in Table 1 otherwise they are stated in particular.

Parameter	$\Gamma_B/2\pi$	L	P_1	P_2	g_B
Value	30MHz	1m	0.1W	0.1W	2.5×10^{-11} m/W

Table 1. Some common parameters used for simulations.

3.1 Time traces of reflectivity

Fig. 3 shows the time trace of reflectivity under phase matching condition for coherent anti-Stokes Brillouin scattering case, where the four figures (a), (b), (c) and (d) correspond to a probe power of 1, 10, 30 and 50W, respectively. It can be seen from these figures that at a lower probe power, the reflection intensity increases monotonously with time and then tends to a fixed value; however, it oscillates several times before turning into a constant value as the probe power increases, and it will oscillate more times at a higher probe power. The oscillation of the reflection in time domain at high probe power can be attributed to the relaxation of the acoustic field.

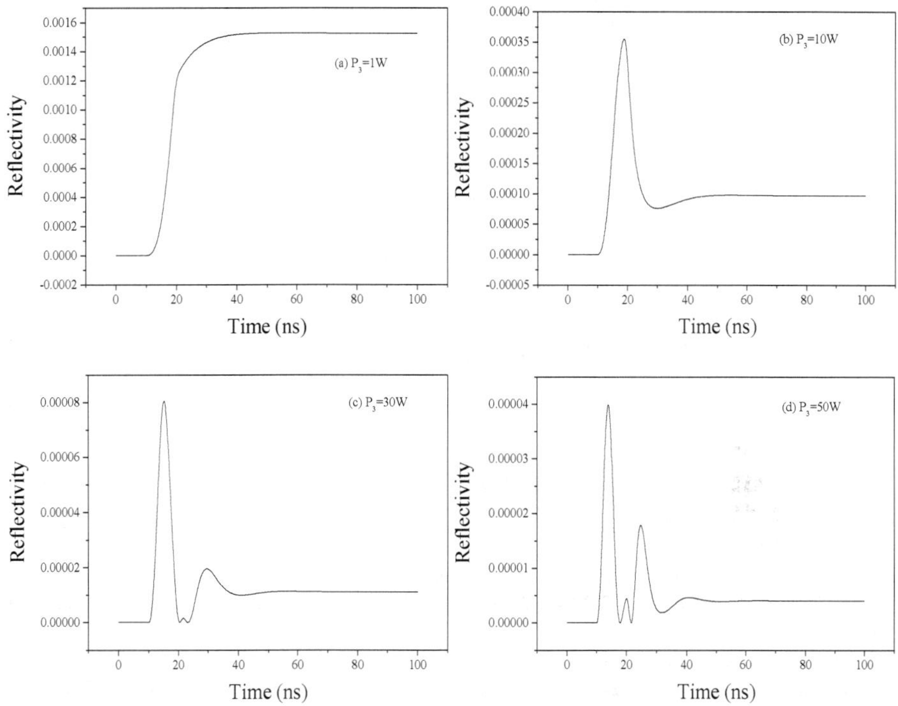

Fig. 3. Time traces of reflectivity under phase matching condition for coherent anti-Stokes Brillouin scattering case.

The time trace of reflectivity under phase matching condition for coherent Stokes Brillouin scattering case is shown in Fig. 4, where the four figures (a), (b), (c) and (d) correspond to a probe power of 1, 10, 20 and 28W, respectively. For the coherent Stokes Brillouin scattering case, the time evolution behaves similarly as that for the coherent anti-Stokes Brillouin scattering case, however, it will take a much longer time for the reflection to reach steady than that for the coherent anti-Stokes Brillouin scattering case, especially with a high probe power. And the higher the probe power is, the longer time it will oscillate before reaching a steady value.

3.2 Brillouin grating spectra

First we do simulations in weak probe regime for both coherent Stokes and anti-Stokes Brillouin scattering cases and compare the results with the analytic solution in Fig. 5. In this simulation a probe power of 0.01W was used. It is seen that for both coherent Stokes and anti-Stokes Brillouin scattering cases, the simulation results agree very well with the analytic solution.

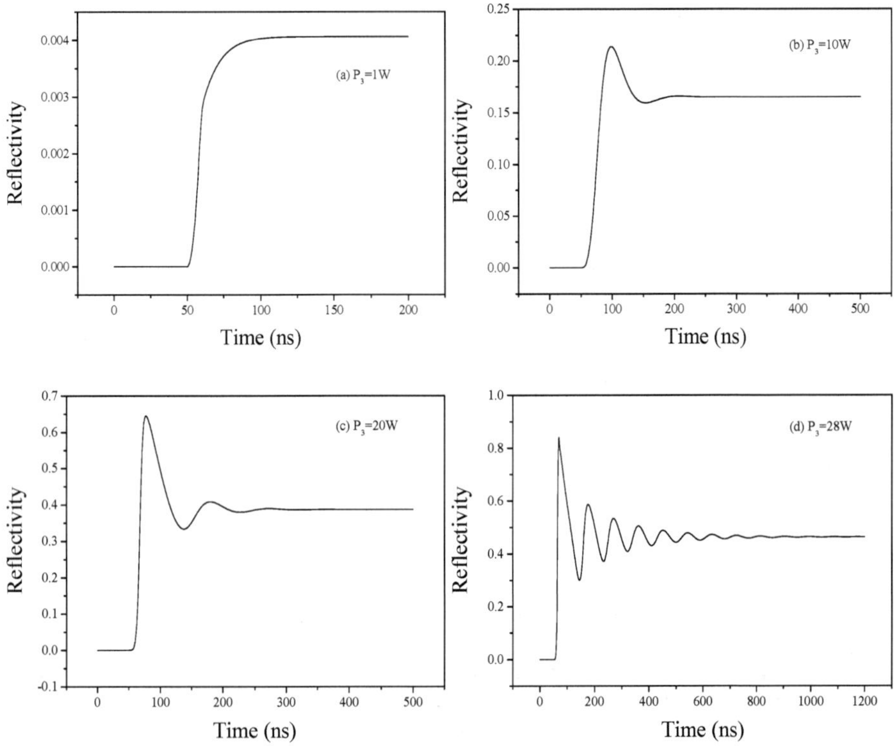

Fig. 4. Time traces of reflectivity under phase matching condition for coherent Stokes Brillouin scattering case.

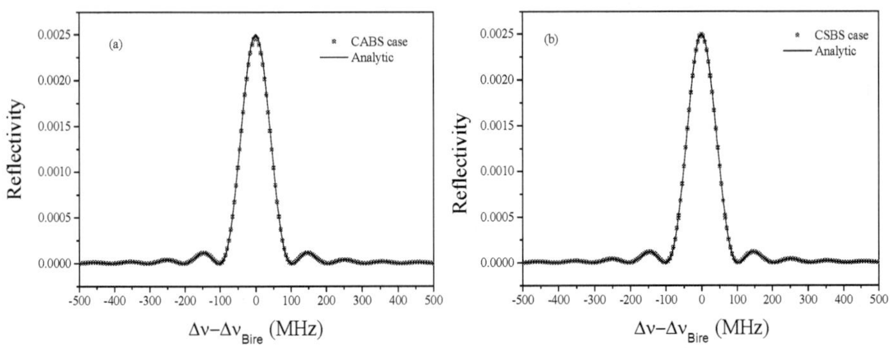

Fig. 5. Comparison between simulation and analytic results for (a) coherent anti-Stokes Brillouin scattering (CABS) and (b) coherent Stokes Brillouin scattering (CSBS).

Fig. 6 shows the normalized reflected spectrum at different probe powers for coherent anti-Stokes Brillouin scattering case. In this figure different curves correspond to a probe power of 0.1, 5, 10 and 20W, respectively. It can be seen from the figure that for all of these cases, the spectra are symmetrical and maximized at line center. The side-lobes vanish gradually as the probe power increases, while the spectrum broadens evidently and tends to a standard Lorentzian shape.

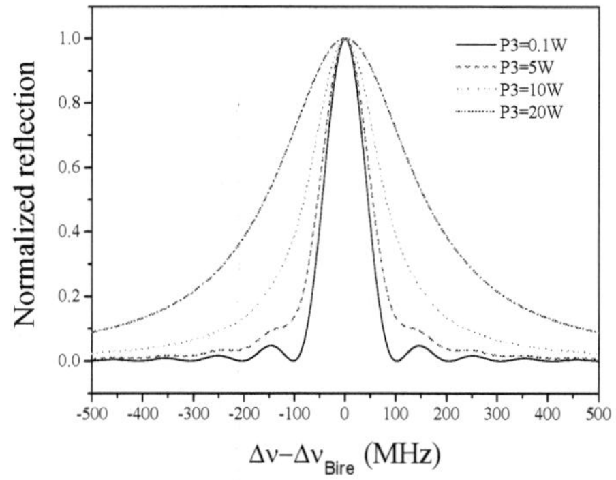

Fig. 6. Reflected spectrum at different probe powers for coherent anti-Stokes Brillouin scattering case.

As for the coherent Stokes Brillouin scattering case, we obtain some results that are very different. The normalized reflected spectrum at different probe powers for coherent Stokes Brillouin scattering case are shown in Fig. 7, where different curves correspond to a probe power of 0.1, 5, 10 and 20W, respectively. From Fig. 7 it can be seen that the spectra are also symmetrical and behave similarly as that for coherent anti-Stokes Brillouin scattering case at a low probe power. However, as the probe power becomes higher, the side-lobes grow remarkably and even exceed the line center, so that the maximum reflectivity shifts from line center to the first side-lobe.

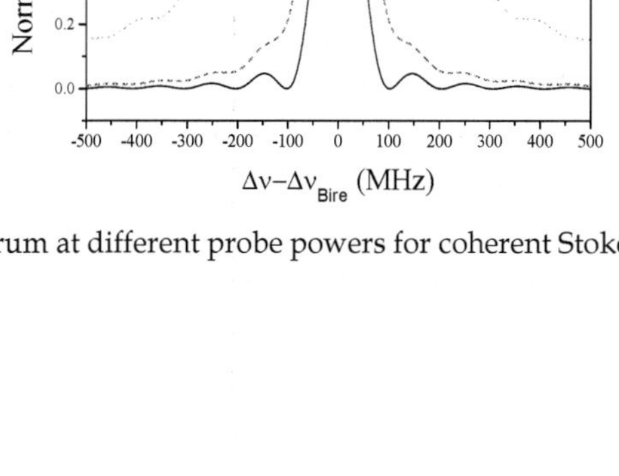

Fig. 7. Reflected spectrum at different probe powers for coherent Stokes Brillouin scattering case.

Fig. 8 shows the reflectivity as a function of probe power under phase matching condition. It is clearly seen that for the coherent anti-Stokes Brillouin scattering case, in which the acoustic wave will be depleted by the probe wave, the reflectivity decreases as the probe power becomes higher, and the maximum reflectivity is not larger than 0.25%. While for the coherent Stokes Brillouin scattering case, in which the acoustic wave will be enhanced by the probe wave, the reflectivity is much higher and increases with the probe power.

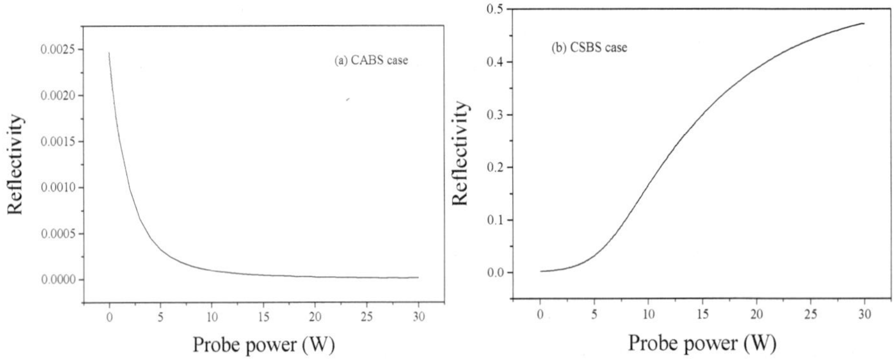

Fig. 8. Reflectivity as a function of probe power under phase matching condition for (a) coherent anti-Stokes Brillouin scattering (CABS) and (b) coherent Stokes Brillouin scattering (CSBS).

Fig. 9 shows the spectrum width as a function of probe power, where (a) is the result of coherent anti-Stokes Brillouin scattering case, while (b) is that of coherent Stokes Brillouin scattering case. It is seen that for both cases the spectrum width remains unchanged in the weak probe regime and becomes larger rapidly as the probe power increases in the strong probe regime.

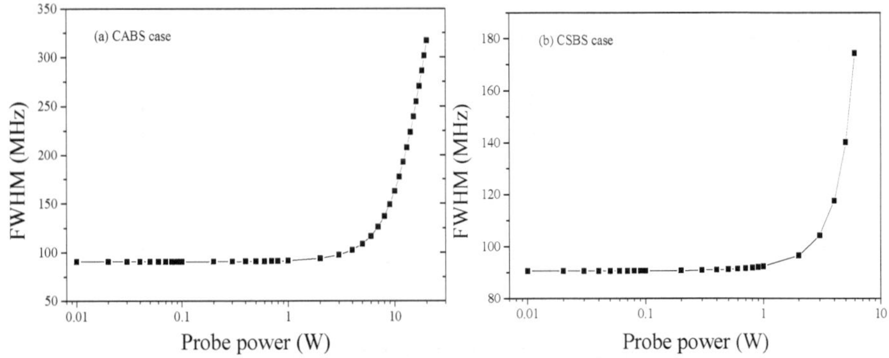

Fig. 9. Spectrum width as a function of probe power for (a) coherent anti-Stokes Brillouin scattering (CABS) and (b) coherent Stokes Brillouin scattering (CSBS).

4. Experimental study of transient Brillouin grating spectrum

In this section, we will experimentally investigate the reflected spectrum of transient Brillouin grating that is generated by two pump pulses. As shown in Fig. 10, the frequency of pump 1 is higher than that of pump 2 by a Ω_B of optical fiber, and the Brillouin grating has the same motion direction as pump 1. Two pump pulses are launched into the slow axis of a polarization-maintaining fiber, and the probe pulse immediately following pump 2 pulse (the leading edge of probe pulse overlapping with trailing edge of pump 2 pulse) is launched into the fast axis to read the Brillouin grating. Because the probe pulse counter-propagates with the Brillouin grating, coherent anti-Stokes Brillouin scattering could occur when the probe pulse interacts with the Brillouin grating, and consequently the scattered light frequency is higher than that of probe pulse by a Ω_B .

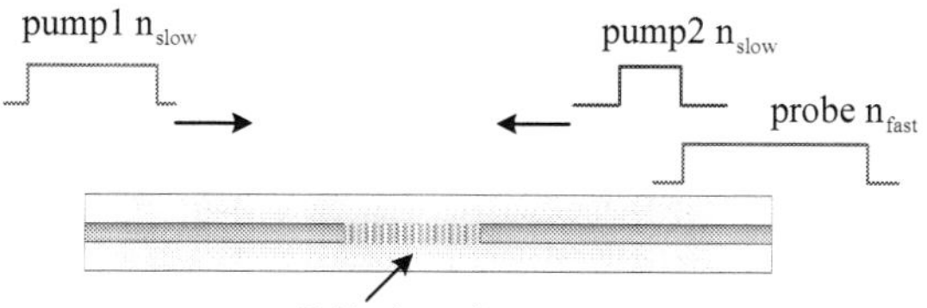

Fig. 10. Schematic diagram of transient Brillouin grating generation and reading using two pump pulses and a probe pulse.

The experimental setup is shown in Fig. 11. Two narrow linewidth (3 kHz) fiber lasers operating at 1550 nm are used to provide the pump1 and pump2, respectively, and their frequency difference is locked by a microwave frequency counter. A tunable laser with a wavelength resolution of 0.1 pm is used as probe wave. The frequency difference between pump1 and probe is monitored and recorded by a 45-GHz bandwidth high-speed detector and a 44-GHz electrical spectrum analyzer. Three high extinction-ratio (ER) electro-optic modulators (> 45 dB) are used to generate pump 1, pump 2 and probe pulses. The power of

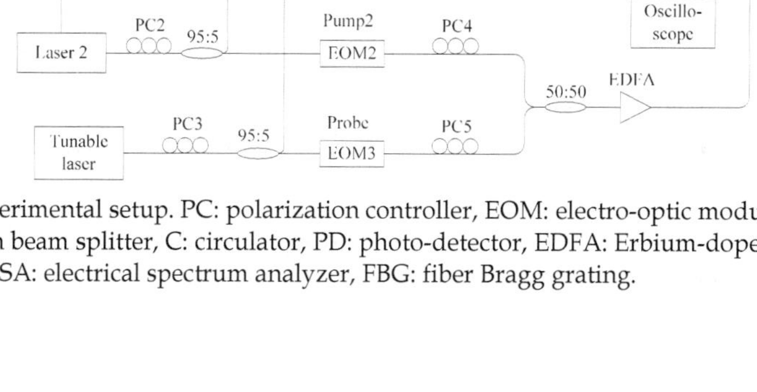

Fig. 11. Experimental setup. PC: polarization controller, EOM: electro-optic modulator, PBS: polarization beam splitter, C: circulator, PD: photo-detector, EDFA: Erbium-doped fiber amplifier, ESA: electrical spectrum analyzer, FBG: fiber Bragg grating.

pump 1 pulse, pump 2 pulse and probe pulse in the PMF are about 0.4W, 30 W and 30 W, respectively. A tunable fiber Bragg grating with a bandwidth of 0.2 nm is used to filter out the transmitted pump1 pulse. A Panda PMF is used in experiment, whose nominal beat-length is smaller than 5 mm at 1550 nm with a Brillouin frequency shift of 10.871 GHz at room temperature.

Then we investigated the Brillouin grating spectra with different grating lengths. The effective length of the transient Brillouin grating is $L_B = (t_{P1} + t_{P2})c/4n$, where t_{P1} and t_{P2} are the pulse widths of pump 1 and pump 2, respectively. In experiment, t_{P2} was fixed at 2 ns and t_{P1} was increased from 2 ns to 6 ns to increase the effective length of the transient Brillouin grating from 0.2 m to 0.4 m. The probe pulse width of 6 ns and 8 ns were chosen to read the Brillouin grating, and the measured reflected spectra of the transient Brillouin grating are shown in Fig. 12. For the case of a 2-ns pump 1 pulse and a 2-ns pump 2 pulse, the effective length of the transient Brillouin grating is 0.2 m and the measured spectra are shown in Fig. 12 (a), which fit well with a Gaussian profile. The intensity of the acoustic wave is determined by the interaction time of the pump and the Stokes waves, and thus the

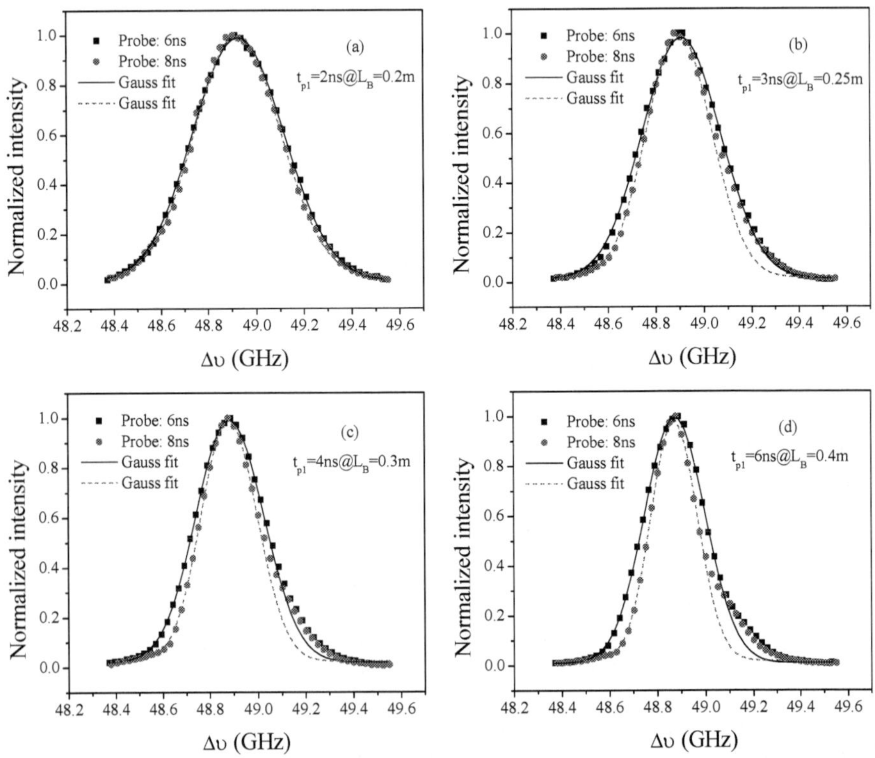

Fig. 12. Measured Brillouin grating spectra with 6 ns and 8 ns probe pulse with different pump 1 pulse-widths of (a) 2 ns, (b) 3 ns, (c) 4 ns, (d) 6 ns. Pump 2 pulse width is constant at 2 ns. (Dong et al., 2010b, with permission from OSA)

generated transient Brillouin grating is a nonuniform-grating with Δn maximizing at the center and decreasing toward two sides. This nonuniform apodization of transient Brillouin grating results in a Gaussian spectrum without side lobes, which are expected in a uniform-grating.

The measured spectrum is the convolution of the probe pulse spectrum and the intrinsic Brillouin grating spectrum. The intrinsic Brillouin grating spectrum can be obtained from deconvoluting the probe pulse spectrum out of the measured spectra, and the spectral widths (FWHM) are plotted in Fig. 13. The black curve shows the FWHM spectrum width of weak fiber Bragg grating described in (Erdogan, 1997). For a 8-ns probe pulse, the measured spectral widths agree very well with the theory of weak fiber Bragg grating; for a 6-ns probe pulse, the experimental results also agree with the theoretical curve when the grating length is shorter than 0.3 m, while the discrepancy from the theory at 0.4 m originates from the effective reading-length of the 6-ns probe pulses. These results directly prove that the moving Brillouin grating and the static fiber Bragg grating follow the same basic theory.

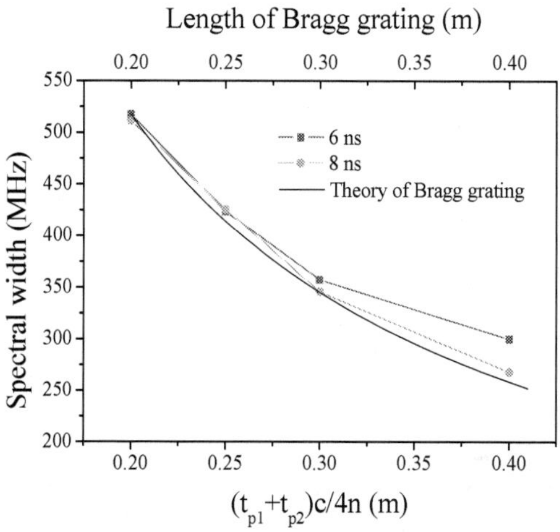

Fig. 13. The intrinsic Brillouin grating spectral width as a function of length. (Dong et al., 2010b, with permission from OSA)

The other important characteristic of the transient Brillouin grating is the acoustic wave relaxation or grating decay. The Brillouin grating is created through electrostriction effect and can only be sustained by keeping the two pump waves. After removing the pump waves, the Brillouin grating will exponentially decay, which characterizes an intrinsic Lorentzian Brillouin gain spectrum. In our scheme as shown in Fig. 10, pump 2 pulse is fixed at 2 ns and pump 1 pulse width is increased to prolong the grating length. The Brillouin grating is generated following pump 2 pulse from the right end to the left end, and the intensity of the whole Brillouin grating exhibit a slope due to the acoustic wave decay. For the probe pulse immediately following pump 2 pulse, the front of the probe pulse always read the strongest part of the grating, while the rear of the probe pulse always read the decayed part, which causes the asymmetry of the spectra especially for long gratings as shown in Fig. 12. It is clearly seen that for long gratings the low-frequency side still agrees well with Gaussian profile, while the high-frequency side has a discrepancy and shows a Lorentzian-like wing, which comes from the apodization of the Brillouin grating induced by acoustic wave decay.

To further demonstrate the effect of the apodization of the Brillouin grating induced by the acoustic wave decay, we switched the pulse width of pump 1 and pump 2 pulses with probe pulse still immediately following pump 2 pulse. Figure 14 shows the measured Brillouin grating spectrum with a 2-ns pump 1 pulse, a 10-ns pump 2 pulse and a 6-ns probe pulse. In this case, the Brillouin grating is created following pump 1 pulse from the left end to the right end, and both of the front and the rear of the probe pulse read the decayed grating. We see that the measured spectrum agrees with a Lorentzian profile, which indicates that the acoustic wave decay play an important role in this case. For the case of a 2-ns pump 1 and a 2-ns pump 2, the interaction duration is much smaller than the phonon lifetime, and consequently the grating decay can be neglected, so that the spectra have a Gaussian profile as shown in Fig. 12 (a).

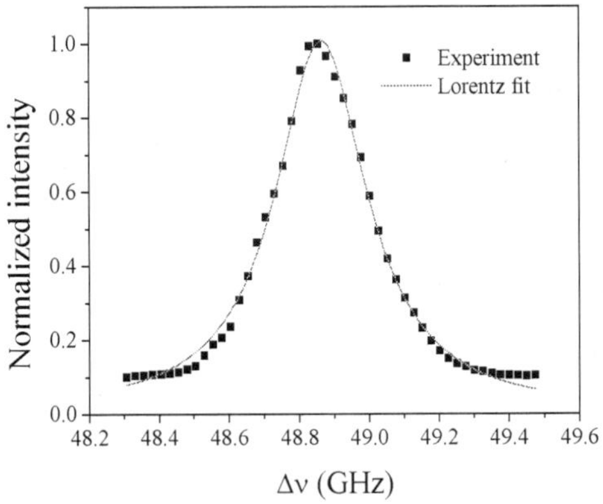

Fig. 14. Measured Brillouin grating spectrum with a 2-ns pump 1 pulse, a 10-ns pump 2 pulse and a 6-ns probe pulse. (Dong et al., 2010b, with permission from OSA)

5. Applications of Brillouin gratings

5.1 Distributed birefringence measurement of polarization-maintaining fibers

Polarization-maintaining fibers have attracted considerable interest in telecommunication and optical sensing community. The phase modal birefringence of a polarization-maintaining fiber is an important parameter for its applications. During the fabrication of the polarization-maintaining fibers, due to the non-uniformity of the materials and environmental condition change in fiber drawing process, it is inevitable to introduce birefringence variation along the fiber. Many methods have been proposed to measure the birefringence of polarization-maintaining fibers (Flavin et al., 2002; Hlubina et al., 2007; Kim et al., 2003; Olsson et al., 1998). However, these methods can only give the average birefringence of the test fibers, and can not characterize the birefringence variation along the fiber. In this section, we study a truly distributed birefringence measurement of polarization-maintaining fibers based on a transient Brillouin grating.

According to the phase matching condition $\Delta\nu_{Bire} = \Delta n\nu/n_g$, the birefringence of the polarization-maintaining fibers can be calculated from the birefringence-induced frequency shift $\Delta\nu_{Bire}$, which can be obtained through the reflected spectrum of the Brillouin grating. In our birefringence measurement scheme, two short counter-propagating pump1 and pump2 pulses, where the frequency of pump1 is larger than that of pump2 by a Brillouin frequency shift ω_B, are launched into one axis of a polarization-maintaining fiber to excite a Brillouin grating through stimulated Brillouin scattering process. Following pump2 pulse, a long probe pulse is launched into the other axis, and energy from the probe pulse could be partly reflected at the expense of Brillouin grating. A maximum reflected probe signal could be obtained when the frequency difference between the pump2 and probe satisfies the phase matching condition.

There are two schemes to perform the birefringence measurement: one is that the Brillouin grating is excited in the fast axis and probed in the slow axis; and the second is that the Brillouin grating is excited in the slow axis and probed in the fast axis, shown as Fig. 15 (a) and (b), respectively. For both cases, the birefringence-induced frequency shift between the pump2 and probe are the same, which is proportional to the local birefringence. As can be seen in the following context, the birefringence induced frequency shifts are about 43 GHz and 49 GHz for Bow-tie and Panda fibers, respectively, which are difficulty to be measured directly. In order to precisely measure the birefringence frequency shift and thus accurately calculate the birefringence, we choose the second scheme. In this scheme, the frequency difference between the pump1 and probe is smaller than the birefringence-induced frequency shift by a Brillouin frequency shift, and can be measured by a high-speed photo-detector and an electrical spectrum analyzer, as shown in Fig. 11. So the actual birefringence-induced frequency shift equals the measured frequency difference between pump1 and probe plus a Brillouin frequency shift.

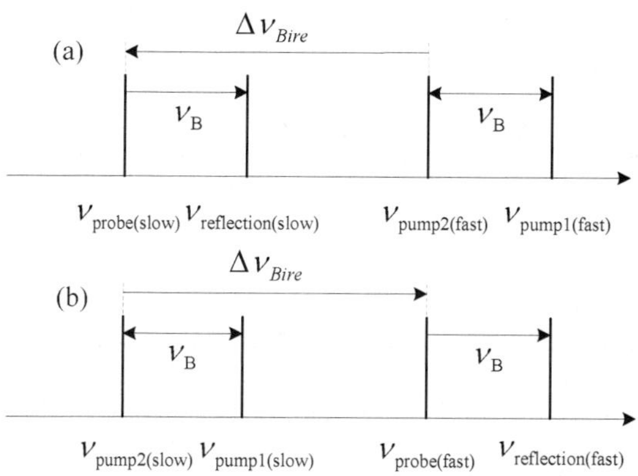

Fig. 15. Frequency relationship of pump1, pump2, probe and reflection: (a) the Brillouin grating is excited in the fast axis and probed in the slow axis, and (b) the Brillouin grating is excited in the slow axis and probed in the fast axis.

An 8-m Bow-tie fiber (HB1500T) and an 8-m Panda fiber (PM-1550-HP) are used as fibers under test. The Bow-tie fiber has a mode field diameter of 10.5 μm and its nominal beat-length is smaller than 2 mm at 633 nm, whose Brillouin frequency shift is 10.815 GHz at room temperature and Brillouin spectrum width is 61 MHz. The Panda fiber has a mode field diameter of 10.5 μm and its nominal beat-length is smaller than 5 mm at 1550 nm, with a Brillouin frequency shift of 10.871 GHz at room temperature and a Brillouin spectrum width of 60 MHz. Both fibers were in loose state to avoid strain-induced birefringence. In experiment, pump 1 and pump 2 pulses are both set at 2 ns, which give a 20-cm spatial resolution. The measured Brillouin grating spectrum is the convolution of the probe pulse spectrum and the intrinsic Brillouin grating spectrum, and thus a longer probe pulse can give narrower spectrum and better frequency and birefringence accuracy. However, the probe pulse width is limited by the phonon lifetime, which is ~10 ns in the optical fiber, so an 8-ns probe pulse is used to obtain the Brillouin grating spectrum.

Figure 16 show the measured results of Bow-tie and Panda fibers, where the delay step between the two pump pulses is set at 1 ns corresponding to 10 cm/point. For the Bow-tie fiber, the birefringence has a characteristic periodic variation with a period of ~3.5m, and exhibits an increasing tendency along the fiber, which indicates that there could be another long period. Different periods could correspond to different disturbance factors during the fiber drawing process; over the 8-m length the birefringence variation range is ~2.4×10^{-6}. For the Panda fiber, there is no obvious periodic characteristic within the measured length except a dip at the location of ~3 m; over the 8-m length the birefringence variation range is ~1.3×10^{-6}. The measured central frequency uncertainty of the Brillouin spectrum is ~3 MHz, which corresponds to a measured birefringence accuracy of ~2×10^{-8}.

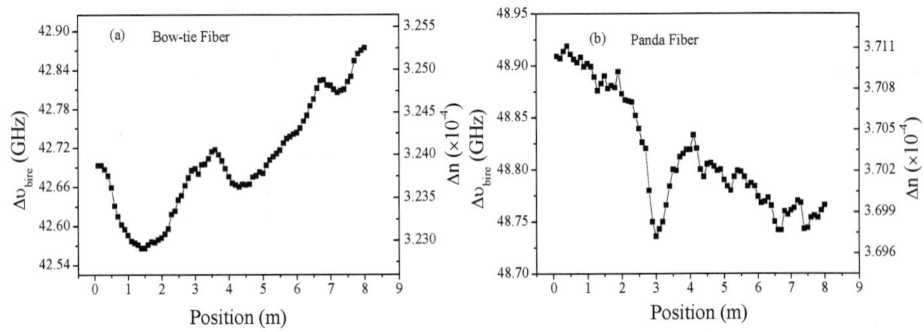

Fig. 16. Measured distributed birefringence of (a) a 8-m Bow-tie fiber and (b) a 8-m Panda fiber. (Dong et al., 2010a, with permission from OSA)

5.2 Distributed simultaneous temperature and strain measurement

Brillouin scattering based distributed fiber sensors have shown the capability of distributed strain and temperature measurement. However, since both strain and temperature can produce a Brillouin frequency shift change, the Brillouin based sensor suffers from the cross sensitivity. Normally it can not be determined that a Brillouin frequency shift change is induced by a change in temperature or strain. In this section, we study simultaneous strain and temperature measurement using Brillouin scattering and birefringence in a polarization-maintaining fiber.

Brillouin scattering is an inelastic light scattering with acoustic phonons and undergoes a frequency shift after collision. Brillouin frequency shift is given by

$$\nu_B = 2nV_a/\lambda \tag{25}$$

where n is the refractive index, V_a is the velocity of acoustic wave, and λ is the light wavelength in vacuum. It was found that the Brillouin frequency shift increases with strain and temperature.

In a polarization-maintaining fiber, temperature and strain both can alter its birefringence, and it was found that the birefringence-induced frequency shift increases with the strain and decreases with temperature, which provides another uncorrelated mechanism to measure the temperature and strain change.

The dependences of the changes of Brillouin frequency shift and birefringence-induced frequency shift, $\Delta\nu_B$ and $\Delta\nu_{Bire}$, on strain and temperature can be expressed by

$$\begin{aligned}\Delta\nu_B &= C_B^{\varepsilon}\Delta\varepsilon + C_B^{T}\Delta T \\ \Delta\nu_{Bire} &= C_{Bire}^{\varepsilon}\Delta\varepsilon + C_{Bire}^{T}\Delta T\end{aligned} \tag{26}$$

where $\Delta\varepsilon$ and ΔT are changes in strain and temperature, respectively; C_B^{ε} and C_B^{T} are coefficients to strain and temperature related to Brillouin frequency shift, respectively; C_{Bire}^{ε} and C_{Bire}^{T} are coefficients to strain and temperature related to birefringence-induced frequency shift, respectively. The strain and the temperature can then be uniquely determined by the measured Brillouin frequency shift and birefringence-induced frequency shift, and can be given by

$$\begin{bmatrix}\Delta\varepsilon \\ \Delta T\end{bmatrix} = \frac{1}{C_B^{\varepsilon}C_{Bire}^{T} - C_B^{T}C_{Bire}^{\varepsilon}}\begin{bmatrix}C_{Bire}^{T} & -C_B^{T} \\ -C_{Bire}^{\varepsilon} & C_B^{\varepsilon}\end{bmatrix}\begin{bmatrix}\Delta\nu_B \\ \Delta\nu_{Bire}\end{bmatrix} \tag{27}$$

Here C_B^{ε}, C_B^{T} and C_{Bire}^{ε} have positive sign, while C_{Bire}^{T} has a negative sign, so $C_B^{\varepsilon}C_{Bire}^{T} - C_B^{T}C_{Bire}^{\varepsilon}$ is of a large value, which ensures a high discrimination accuracy.

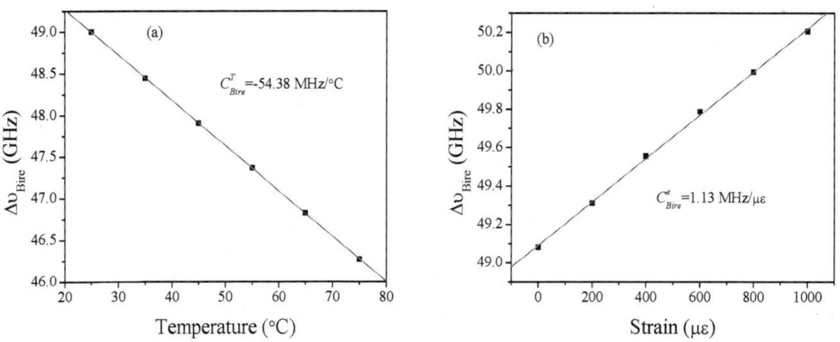

Fig. 17. The dependence of birefringence-induced frequency shift on (a) temperature and (b) strain. (Dong et al., 2010c, with permission from IEEE)

First, we measured the dependences of Brillouin frequency shift on temperature and strain, and the temperature and strain coefficients are $C_B^T = 1.12\,\mathrm{MHz/°C}$ and $C_B^{\varepsilon} = 0.0482\,\mathrm{MHz/\mu\varepsilon}$, respectively. Then we measured the dependences of birefringence-induced frequency shift on temperature and strain, and the results are shown in Fig. 17, where the fitting temperature and strain coefficients are $C_{Bire}^T = -54.38\,\mathrm{MHz/°C}$ and $C_{Bire}^{\varepsilon} = 1.13\,\mathrm{MHz/\mu\varepsilon}$, respectively.

Simultaneous measurement of strain and temperature is performed using a 6-m Panda fiber with 1-m stressed segment, 1-m heated segment and 1-m segment at room temperature and loose state in between, which is shown in Fig. 18. The applied strain is about 670 με and the heated temperature is higher than room temperature by 30 °C.

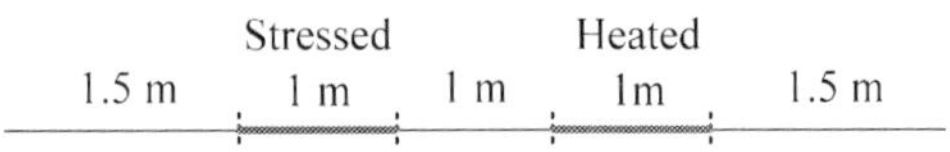

Fig. 18. Layout of the sensing fiber.

A 30/28 ns pulse pair is used to perform Brillouin frequency shift measurement and the spatial resolution is 20 cm, which is determined by the 2-ns pulse-width difference. The measured Brillouin frequency shift difference (subtracting the values at room temperature and loose state) is shown in Fig. 19 (a). The fitting uncertainty of the central frequency is 0.4 MHz, corresponding to a δT of 0.36 °C and a $\delta\varepsilon$ of 8.3 με. For birefringence-induced frequency shift measurement, the spatial resolution is determined by the effective length of the Brillouin grating, which is 20 cm for 2-ns pump pulses. A longer 6-ns probe pulse is used to provide narrower measured Brillouin grating spectrum, and thus give better frequency accuracy. The measured birefringence-induced frequency shift difference (subtracting the values at room temperature and loose state) is shown in Fig. 19 (b). The fitting uncertainty of the central frequency is 3 MHz, corresponding to a δT of 0.06 °C and a $\delta\varepsilon$ of 2.7 με. Note that the birefringence-induced frequency shift provides higher precision with respect to Brillouin frequency shift, so that the discrimination accuracy is limited by the uncertainty of Brillouin frequency shift.

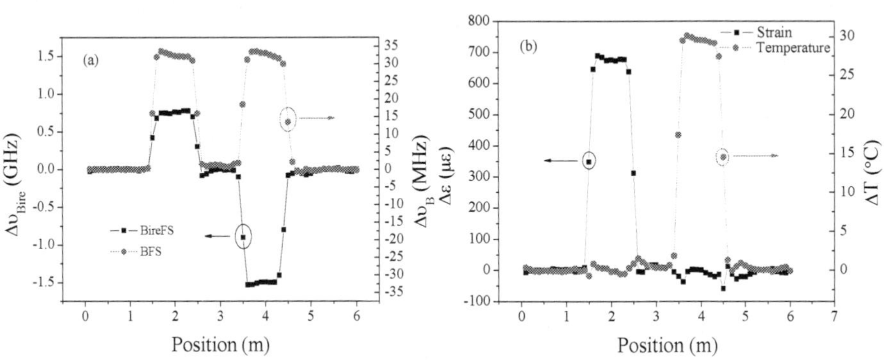

Fig. 19. (a) Measured birefringence-induced frequency shift (BireFS) and Brillouin frequency shift (BFS) and (b) calculated temperature and strain. (Dong et al., 2010c, with permission from IEEE)

6. Conclusion

First, the coupled wave equations of Brillouin enhanced four-wave mixing associated with a Brillouin grating are deduced, and the birefringence-induced frequency shift in a polarization-maintaining fiber is given through phase matching condition; in the case of weak probe, the steady uniform Brillouin grating spectrum is exactly the same as that of a fiber Bragg grating. Second, in the case of strong probe, the Brillouin grating can be considerably modified by the probe wave, resulting in that for the coherent anti-Stokes Brillouin scattering case, as the probe power increases, the reflectivity decreases and the spectrum width increases, and the spectrum profile tends to a Lorentzian shape; for the coherent Stokes Brillouin scattering case, as the probe power increases, both the reflectivity and the spectrum width increases, and the spectrum profile exhibits oscillations. Third, the transient Brillouin gratings generated by two short pump pulses have a Gaussian profile, and the acoustic wave decay can impose an apodization on the gratings. Fourth, two applications of Brillouin grating are given: for the distributed birefringence measurement of a polarization-maintaining fiber, the spatial resolution of 20 cm is obtained by using 2-ns pump pulses and the measurement accuracy can reach to 2×10^{-8}; the birefringence-induced frequency shift in a polarization-maintaining fiber shows dependences on temperature and strain, and simultaneous measurement of temperature and strain is realized by combining a conventional Brillouin fiber sensor based on Brillouin frequency shift.

7. Acknowledgment

The authors would like to thank Natural Science and Engineering Research Council of Canada (NSERC) through discovery and strategic grants and Canada Research Chair program for their financial support of this research.

8. References

Boyd, R. W. (2008). *Nonlinear Optics,* Academic Press, ISBN 978-0-12-369470-6, San Diego, USA

Cao, Y.; Lu P.; Yang Z. & Chen W. (2008). An Efficient Method of All-optical Buffering with Ultra-small Core Photonics Crystal fibers. *Opt. Express,* Vol.16, No.18, (September 2008), pp. 14142-14150, ISSN 1094-4087

Dong Y.; Bao X. & Chen L. (2009). Distributed Temperature Sensing Based on Birefringence Effect on Transient Brillouin Grating in a Polarization-maintaining Photonics Crystal Fiber. *Opt. Lett.,* Vol.34, No.17, (September 2009), pp. 2590-2592, ISSN 0146-9592

Dong Y.; Chen L. & Bao X. (2010). Truly Distributed Birefringence Measurement of Polarization-maintaining Fibers Based on Transient Brillouin Grating. *Opt. Lett.,* Vol.35, No.2, (January 2010), pp. 193-195, ISSN 0146-9592

Dong Y.; Chen L. & Bao X. (2010). Characterization of the Brillouin Grating Spectra in a Polarization-maintaining fiber. *Opt. Express,* Vol.18, No.18, (August 2010), pp. 18960-18967, ISSN 1094-4087

Dong Y.; Chen L. & Bao X. (2010). High-Spatial-Resolution Time-Domain Simultaneous Strain and Temperature Sensor Using Brillouin Scattering and Birefringence in a

Polarization-Maintaining Fiber. *IEEE Photon. Technol. Lett.,* Vol.22, No.18, (September 2010), pp. 1364-1366, ISSN 1041-1135

Flavin D. A.; McBride R. & Jones J. D. C. (2002). Dispersion of Birefringence and Differential Group Delay in Polarization-maintaining Fiber. *Opt. Lett.,* Vol.27, No.12, (June 2002), pp. 1010-1012, ISSN 0146-9592

Erdogan T. (1997). Fiber Grating Spectra. *J. Lightwave Technol.,* Vol.15, No.8, (August 1997), pp. 1277-1294, ISSN 0733-8724

Hlubina P. & Ciprian D. (2007). Spectral-domain Measurement of Phase Modal Birefringence in Polarization-maintaining Fiber. *Opt. Express,* Vol.15, No.25, (December 2007), pp. 17019-17024, ISSN 1094-4087

Kim C. S. et al. (2003). Optical Fiber Modal Birefringence Measurement Based on Lyot-Sagnac Interferometer. *IEEE Photon. Technol. Lett.,* Vol.15, No.2, (February 2010), pp. 269-272, ISSN 1041-1135

Olsson B. E.; Karlsson M. & Andrekson P. A. (1998). Polarization Mode Dispersion Measurement Using a Sagnac Interferometer and a Comparison with the Fixed Analyzer Method. *IEEE Photon. Technol. Lett.,* Vol.10, No.7, (July 1998), pp. 997-999, ISSN 1041-1135

Song K. Y.; Zou W.; He Z. & Hotate K. (2008). All-optical Dynamic Grating Generation Based on Brillouin Scattering in a Polarization-maintaining Fiber. *Opt. Lett.,* Vol.33, No.9, (May 2008), pp. 926-928, ISSN 0146-9592

Song K. Y.; Zou W.; He Z. & Hotate K. (2009). Optical Time-domain Measurement of Brillouin Dynamic Grating Spectrum in a Polarization-maintaining Fiber. *Opt. Lett.,* Vol.34, No.9, (May 2009), pp. 1381-1383, ISSN 0146-9592

Song K. Y.; Lee K. & Lee S. B. (2009). Tunable Optical Delays Based on Brillouin Dynamic Grating in Optical Fibers. *Opt. Express,* Vol.17, No.12, (June 2009), pp. 10344-10349, ISSN 1094-4087

Song K. Y. & Yoon H. J. (2010). High-resolution Brillouin Optical Time Domain Analysis Based on Brillouin Dynamic Grating. *Opt. Lett.,* Vol.35, No.17, (January 2010), pp. 52-54, ISSN 0146-9592

Zhu, Z.; Gauthier D. G. & Boyd R. W. (2007). Stored Light in an Optical Fiber via Stimulated Brillouin Scattering. *Science,* Vol.318, No.5857, (December 2007), pp. 1748-1750, ISSN 0036-8075

Zou W.; He Z. & Hotate K. (2009). Complete Discrimination of Strain and Temperature Using Brillouin Frequency Shift and Birefringence in a Polarization-maintaining Fiber. *Opt. Express,* Vol.17, No.3, (February 2009), pp. 1248-1255, ISSN 1094-4087

Zou W.; He Z.; Song K. Y. & Hotate K. (2009). Correlation-based Distributed Measurement of a Dynamic Grating Spectrum Generated in Stimulated Brillouin Scattering in a Polarization-maintaining Optical Fiber. *Opt. Lett.,* Vol.34, No.7, (April 2009), pp. 1126-1128, ISSN 0146-9592

Zou W.; He Z. & Hotate K. (2010). Demonstration of Brillouin Distributed Discrimination of Strain and Temperature Using a Polarization-maintaining Optical Fiber. *IEEE Photon. Technol. Lett.,* Vol.22, No.8, (April 2010), pp. 526-528, ISSN 1041-1135

Kalosha V.P.; Li W.; Wang F.; Chen L. & Bao X. (2008). Frequency-shifted Light Storage via Stimulated Brillouin Scattering in Optical Fibers. *Opt. Lett.,* Vol.33, No.23, (December 2008), pp. 2848-2850, ISSN 0146-9592

6

Synthesis of Two-Frequency Symmetrical Radiation and Its Application in Fiber Optical Structures Monitoring

Oleg Morozov, German Il'in, Gennady Morozov and Tagir Sadeev
Tupolev Kazan National Research Technical University,
Institute of Radio Electronics and Telecommunications
Russia

1. Introduction

Photonics plays the leading role in the decision of some major social, scientific and technical problems, such as, development, exploitation safety and QoS increasing of telecom lines. Efficiency of photometric methods is defined, first of all, by high informative effects of direct interaction of optical radiation with optical fiber.

Method of optical reflectometry (MOR), as one of the most effective photometric methods, is the powerful tool for creation of optical fiber structures monitoring systems (Barnoski & Jensen, 1976). The present stage of reflectometric systems development is directed on the further improvement of their characteristics, research of new principles of probing and registration of backscattering information, development of precision measuring converters and is based on use of systems with continuous wave radiation. Such accent is explained, firstly, by power equivalence of pulse probing with high peak power and small duration of a pulse and continuous probing with low power of radiation and big time of supervision, secondly, by the fulfilled technique of spatially - resolved measurements based on the linear frequency modulation method, and thirdly, by the significant progress in the field of creation of hi-tech and inexpensive element base (sources of radiation with the big coherence length, broadband devices of radiation parameters control and high-speed photoreception devices).

Overwhelming majority of open and built-in continuous wave reflectometric systems represent homodyne systems in which carrying frequencies of referent and measuring channels coincide. Such systems possess a simple design and an opportunity of direct allocation and registration of information signal. However during photo-electric transformation the low-frequency noise characteristic of structural units, noise characteristics of radiation sources and photo detectors play essential role that considerably worsens metrological characteristics, and also functionalities of the specified systems. The decision of homodyne reflectometric system problems is based on the use of two-frequency methods. In this case, systems will be transformed in heterodyne type where frequencies of basic and measuring channels do not coincide, and displacement of frequencies is achieved due to use of devices forming two-frequency laser radiation.

We have proposed and demonstrated in recent years an unique technique to synthesize the two-frequency radiation by manipulating the amplitude and the phase of initial one frequency lightwave (Il'in & O.G. Morozov, 1983, Il'in et al., 1996, O.G. Morozov & Pol'ski, 1996). The output frequencies are symmetrical relatively to suppressed initial one and have equal amplitudes and alternative phases.

We consider the set of the analysis and synthesis problems which should be solved by development of the theory and technical equipment of optical reflectometers for fiber optical structures (FOS) monitoring in optical two-frequency domain (OTFDR) (Pol'ski & O.G. Morozov, 1996, 1998, Zalyalov et al., 1998, O.G. Morozov et al., 1999). Based upon this technique, the synthesis of two-frequency symmetrical probing radiation (TFSPR), various photonic sensing of fiber optic structures and optical information-processing applications have been developed (Natanson et al., 2005a, 2005b, O.G. Morozov et al., 2006, O.G. Morozov & Aybatov, 2007, O.G. Morozov et al., 2008a, O.G. Morozov et al., 2010).

In this chapter, the principle of TFSPR synthesis is summarized. A series of functional optical-sensing OTFDR systems, including fiber-optic reflectometres, multiplexed fiber Bragg grating (FBG) or fiber Fabry-Perault interferometers (FFPI) sensors are introduced. Fully distributed fiber-optic strain sensing system based on Brillouin frequency shift is highlighted (Natanson et al., 2005a, G.A. Morozov et al., 2011), and four wave mixing compensating system (O.G. Morozov et al., 2008b) based on optimized TFSPR is also presented. Miscellaneous OTFDR applications in biosensors (O.A. Stepustchenko et al., 2011) and two-frequency signal forming for photonic microwave filters synthesis (O.G. Morozov et al., 2009, O.G. Morozov & Sadeev, 2011) are also reviewed.

2. Synthesis of two-frequency symmetrical probing radiation

Theoretical motivation of original method of one-frequency radiation transformation into two-frequency ones (named as Il'in-Morozov method) is considered in this part (Il'in & O.G. Morozov, 1983).

Realization of this method allows getting an output two-frequency oscillation with high degree of spectrum purity. Herewith, there are ensured high transformation factor and stability of output spectrum parameters at the deflection of transformation parameters from optimum, as well as possibility of differential frequency tuning on the given law and with the given velocity.

2.1 Theoretical basis

Initial single-frequency oscillation (fig. 1) is described as

$$e(t) = E_0 \sin\varphi(t)\,, \tag{1}$$

where $\varphi(t) = \omega_0 t + \varphi_0$ is phase change, and E_0, ω_0, φ_0 are its permanent amplitude, frequency, and initial phase correspondingly.

From theory of modulated signals it's known that phase modulation with certain indexes could decrease (eliminate) amplitude of initial (carrier) frequency and produce two symmetric side lobes (Kharckevich, 1962). Let's consider an oscillation with following $\varphi(t)$ commutation

$$e(t) = E_0 \begin{cases} \sin\omega_0 t & \text{when} \quad T(2p-1)/2 < t \le Tp \\ \sin(\omega_0 t + \theta) & \text{when} \quad Tp < t \le T(2p+1)/2 \end{cases}, \tag{2}$$

where $T=2\pi/\Omega$, $\Omega=\omega_0/k$, θ are period, frequency, and phase change, correspondingly, $k>>1$ is integer, and $p = 0, 1, 2,...$ When $\theta=\pi$ Fourier expansion of equation (2) is given by

$$e(t) = \frac{2E_0}{\pi}\sum_n \frac{1}{n}\{\cos(\omega_0 + n\Omega)t - \cos(\omega_0 - n\Omega)t\}, \tag{3}$$

where $n = 1, 3, 5$ is harmonic number of phase change frequency expansion.

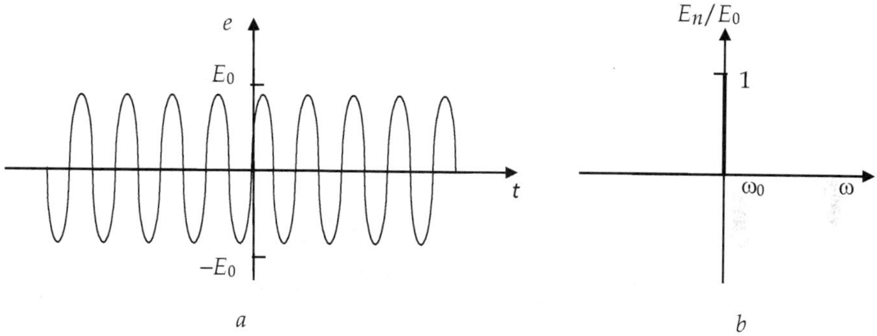

Fig. 1. Timing chart (*a*) and spectrum of Fourier series coefficients (*b*) of (1)

Thus, oscillation (2) is two-band multi frequency oscillation with intercarrier frequency between spectral components equals 2Ω, and suppressed carrier (fig. 2). Initial phases of spectral components inside bands are equal, and difference between initial phases of lower and upper bands is π.

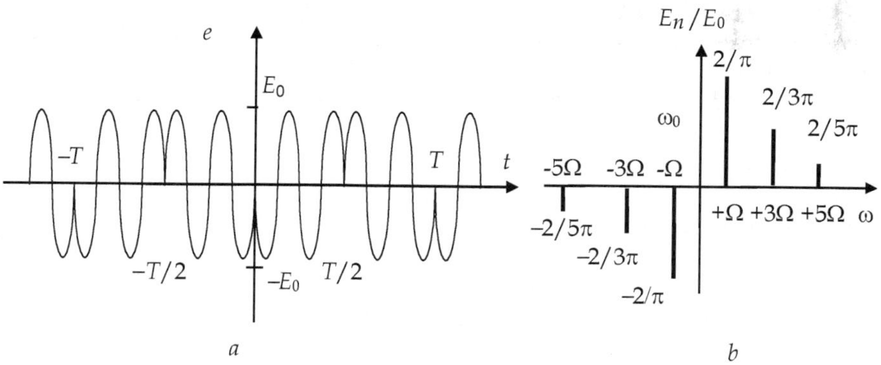

Fig. 2. Timing chart (*a*) and spectrum of Fourier series coefficients (*b*) of (2)

The task of following investigation consists in searching methods of forming two-frequency oscillation from two-band oscillation (3).

2.2 Methods of two-frequency oscillation forming

According to the theory of modulated signals (Kharckevich, 1962) and analysis of spectrum characteristics (3) we can suppose that suppression of its spurious products for forming two frequency oscillation is possible after amplitude modulation of oscillation (2) by some signal S(t), which satisfies the following demands.

During amplitude modulation of single-frequency oscillation by S(t) is produced amplitude modulated signal with frequency spacing between carrier and the nearest side components and between side components at each band equals 2Ω. The difference between carrier initial phase and side band components initial phase is π. Initial phases of side components inside bands are equal.

The simplest oscillation, which satisfy that demands, is $S_1(t) = S_1 \cos(2\Omega\, t + \pi)$, where S_1 is its permanent amplitude, π is initial phase. In case of oscillation (2) amplitude modulation by signal $S_1(t)$, we will get the resulting oscillation (fig. 3)with the following spectrum

$$e\ (t) = \frac{2E_0}{\pi} \sum_n \left\{ \left[\frac{1}{n} - \frac{m}{2} \left(\frac{1}{n-2} + \frac{1}{n+2} \right) \right] \left[\cos(\omega_0 + n\Omega)\, t - \cos(\omega_0 - n\Omega)\, t \right] \right\}, \qquad (4)$$

where m is the amplitude modulation index.

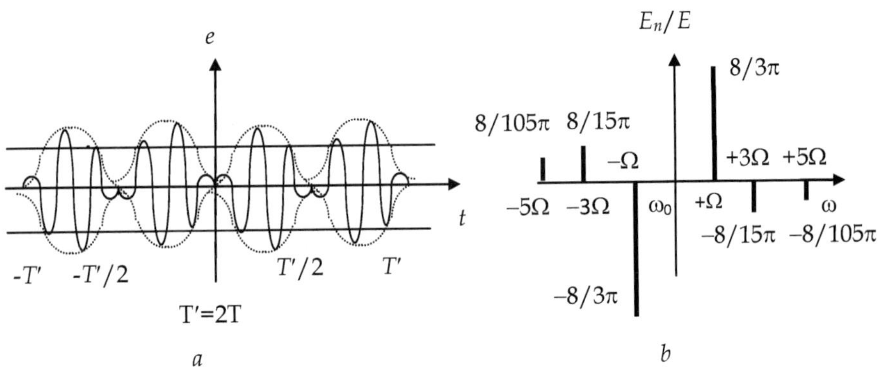

Fig. 3. Timing chart (*a*) and spectrum of Fourier series coefficients (*b*) of (4)

From (4) follows that the first item in square brackets defines spectrum of Fourier series indexes (3), and the second and the third items describes suppression influence at its components. The level of suppression depends on modulation index m. If we take $E_3 = 0$, we will receive the optimal modulation index $m_{opt} = 5/9$, and the resulting oscillation will be almost two-frequency (E_1=0,76E_0), because the amplitude of spectral components $E_n \leq E_1/15$ for $n \geq 5$. Varying the modulation index within (0,85 - 1,15)m_{opt} the output oscillation nonlinear-distortions coefficient would not exceed one percent.

Total suppression of side components with numbers $n \geq 3$ we could achieve with modulating signal $S_2(t) = S_0 |\sin\Omega t|$.

In that case the resulting oscillation will have the following spectrum

$$e(t) = \frac{2E_0}{\pi}(1-b)\sum_n \frac{1}{n}\left[\cos(\omega_0 + n\Omega)t - \cos(\omega_0 - n\Omega)t\right] + \frac{\pi E_0 b}{4}\left[\cos(\omega_0 + \Omega)t - \cos(\omega_0 - \Omega)t\right], \quad (5)$$

where b is amplitude modulation index.

Spectral components amplitudes will be defined by Fourier series indexes and for n=1 $E_1=[2E_0/\pi][1-b]+[\pi E_0 b/4]$, for $n\geq 3$ $E_n=[2E_0/\pi n][1-b]$. When $b_{opt}=1$ the spectrum contains two useful components with frequencies $\omega_0+\Omega$ and $\omega_0-\Omega$, and spurious components are suppressed (fig. 4). Varying the modulation index within $(0{,}7–1)b_{opt}$ the output oscillation nonlinear-distortions coefficient would not exceed one percent.

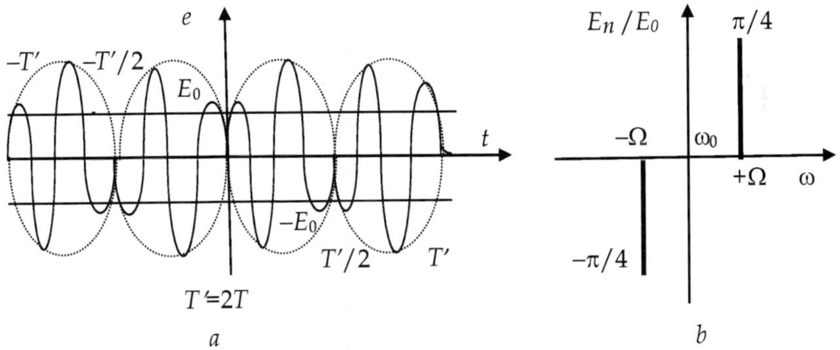

Fig. 4. Timing chart (*a*) and spectrum of Fourier series coefficients (*b*) of (5)

2.3 Discussion of the results

Thus, amplitude-phase method of one-frequency oscillation conversion into two-frequency ones, or Il'in-Morozov method, can be formulated as: conversion of a single-frequency coherent radiation in the two-frequency, based on an external electro-optic modulation, which performs the amplitude modulation of single-frequency coherent radiation by a factor equal to 1 and the phase commutation of the received amplitude modulated radiation at π for each envelope passage through its minimum.

Intercarrier frequency of two-frequency oscillation (5) when $b_{opt}=1$ is defined by frequency Ω of phase commutation θ. Its stability unambiguously concerned with frequency stability of driving voltages and instabilities of commutations devices. Really achievable value of intercarrier frequency's instability with simple thermo stating of driving generators is 10^{-6}. Intercarrier frequency's retuning, which is required in some types of measurements, is rather simple realized with the use of amplitude-phase conversion, minimum frequency shift is determined by modulators gain slope, and maximum frequency shift is determined by higher cutoff frequency of modulator and correlation between modulatory and modulated frequencies.

Energy equality of side lobes and the effectiveness of their conversion are of great importance in multi frequency systems. Using the derived equations for the spectrums of output oscillations and taking into consideration properties of amplitude-phase conversion, i.e. using the additional power of amplitude modulation and phase commutation for forming the side lobes, we could determine, that power of the last one is nearly 60 percent of initial single-frequency oscillation, and conversion index equals unit value without taking into account the loss in real modulators.

3. Information structure of symmetric two-frequency oscillation

Symmetric two-frequency oscillation is complex signal with amplitude and phase modulation. Main feature of this signal is that all its components (envelope A(t), instantaneous phase θ(t) and frequency ω(t)) are periodic functions with period $T = (1/2\pi)\Omega$ and depend on its components amplitude ratio A_1/A_2. Let's define the parameters, which could be used for registration in measuring system and carry information about interaction between two-frequency oscillation and structure under investigation. As such parameters could be used amplitude modulation factor, gradients of phase and frequency changes.

1. Let's evaluate modulation rate of two-frequency signal. Thereto we will use the modulation coefficient (Gonorovski, 1977) $m = 2(A_{max} - A_{min}) / 2(A_{max} + A_{min})$. Calculation results of $m(A_1/A_2)$ are shown on fig. 5. Analysis of the dependence $m(A_1/A_2)$ allows determination of components amplitude ratio with high precision during the measurement of modulation index, and effective realization the feedback process by controlling the instantaneous amplitude of two-frequency signal using its components amplitude ratio.

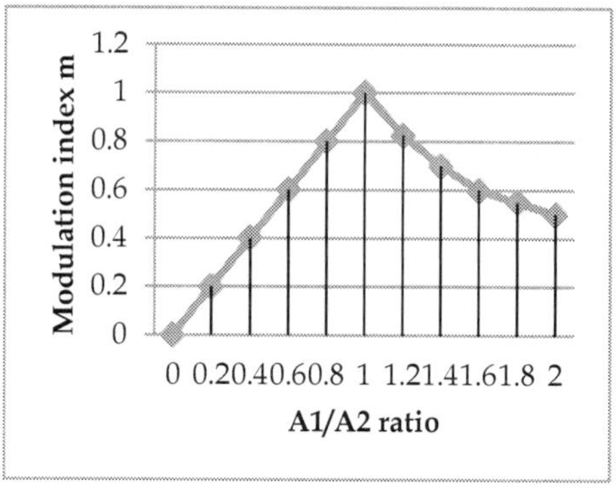

Fig. 5. Dependence $m(A_1/A_2)$

2. Let's define phase gradient of two-frequency signal under its amplitude changing. Calculation results of $\Delta\theta(A_1/A_2)$ are shown at fig. 6. Analysis of the dependence $\Delta\theta(A_1,A_2)$ allows determination of components amplitude ratio with high precision during the measurement of modulation phase, and effective realization the feedback process by controlling the instantaneous phase of two-frequency signal using its components amplitude ratio.

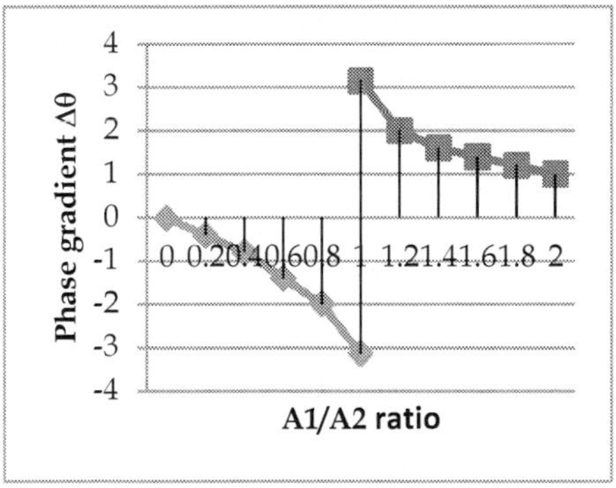

Fig. 6. Dependence $\Delta\theta(A_1/A_2)$

3. Let consider gradient of two-frequency signal instantaneous frequency under its amplitude changing. Calculation results of instantaneous frequency overriding $|\Delta\omega|$ dependence on amplitude components ratio A_1/A_2, $\Delta\omega(A_1, A_2)$ are shown at fig. 7.

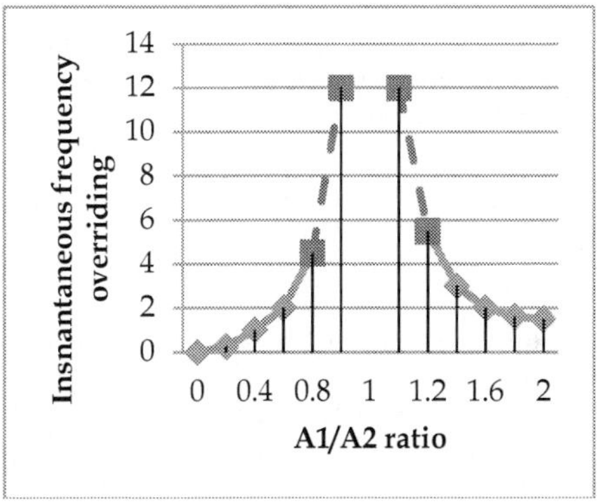

Fig. 7. Dependence $|\Delta\omega|(A_1/A_2)$

Unlike previous cases analysis of the dependence $|\Delta\omega|(A_1/A_2)$ allows determination of components amplitude ratio during the changes of modulation frequency, and effective realization the feedback process by controlling the instantaneous frequency of two frequency signal using its components amplitude ratio in some limits (overriding).

Dependence of instantaneous frequency normalized to frequency distortion $\omega/\Delta\omega$ $(\Delta\omega = \omega_2 - \omega_1)$ on amplitude components ratio A_1/A_2 is more informative. Calculation results of $(\omega/\Delta\omega)$ as function from (A_1/A_2) are shown at fig. 8. Average frequency of two-frequency signal fluctuates with changes of components amplitude ratio in range limited by frequencies of two-frequency signal. When components amplitudes are equal, the average frequency is situated in the middle of that range and has the value $\omega_a = (\omega_1 + \omega_2)/2$.

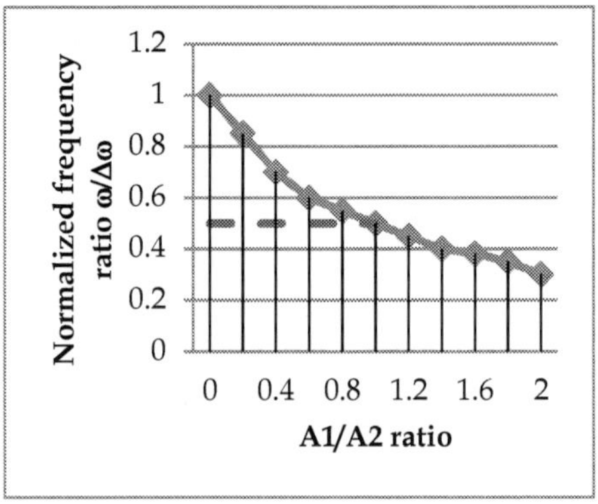

Fig. 8. Dependence $|\omega/\Delta\omega|(A_1/A_2)$

On the basis of two frequencies signal information structure investigation with purpose of finding out main principles of combined interaction of instantaneous values of amplitude, phase and frequency has been defined.

1. Instantaneous phase of two-frequency signal has a sawtooth dependence. Speed of instantaneous phase changes is defined by components amplitude ratio. If $A_1 / A_2 = 1$ the maximal speed of phase changing is observed. When two-frequency envelope has its minimum value, instantaneous phase has a shift, which depends on harmonic components amplitude ratio. If the amplitudes of two-frequency signal components are equal, phase shift is equal to π .
2. It has been shown, that instantaneous frequency of two-frequency signal changes with deviation of it components amplitude ratio in range limited by signal frequencies. In case of components amplitude equality, instantaneous frequency coincides with average frequency of two-frequency signal $\omega_a = (\omega_1 + \omega_2)/2$. When the envelope of two-frequency signal has its minimum value, frequency overriding occurs, which depends on harmonic components amplitudes ratio. When amplitudes of two frequency components are equal, value of frequency overriding is infinity.
3. Modulation coefficient has a linear dependence on amplitudes ratio of two-frequency components. When amplitudes of frequency components are equal, modulation coefficient is maximal and equals unit value.

All obtained dependences could be used in symmetric reflectometric systems as informative and directive functions and in order to get information from selective optical fiber structures, like FBG, IFP, embedded multilayer thin film filter and so on.

4. Questions of method realization in optical range

One of the main purposes of our investigations is to determine the most rational methods of two-frequency laser radiation obtaining and devices for their realization in different regions of IR band. Comparative analysis of radiators characteristics was done in order to achieve this purpose. The main parameters for analysis were following: reproduction of waves length and their difference frequency, difference frequency range, stability and equality of amplitudes on both radiating frequencies, re-tuning possibility of radiating frequencies by desirable law with desirable speed, linearity of drive characteristic, peculiarities of utilizing in FOS monitoring system composition and cost of technical realization. Some results of analysis are presented in table 1.

Two-frequency radiators and their characteristics	Difference frequency range, MHz	difference frequency instability	Equality and stability of amplitude	Possibility of re-tuning	Purity of output spectrum
Diode laser	1000	10^{-2}	–	NSM (*)	+
Two-mode laser	100	10^{-8}	+	NSM	+
Zeeman's laser	10^{-3}...100	10^{-9}	+	NSM	+
Acoustooptic	10^{-2}...100	10^{-6}	–	+(**)	± (***)
Electrooptic:					
polarize	0...20	10^{-6}	–	+	± (***)
phase	0...5	10^{-6}	–	+	± (***)
amplitude-phase	0...1000	10^{-6}	+	+	± (***)

(*) NSM - in these radiators it is necessary to use special means for organizing of re-tuning; (**) + - the organizing of re-tuning is very simple; (***) – it is necessary to use stabilization measures to decrease parasitic components caused by non-stability of modulating parameters and their deviation from optimal.

Table 1. Results of comparative analysis

We can affirm with some care that diode lasers, acoustooptic and electrooptic converters will be widely used. However complexity of control canal realization in first and some uncomfortability of optical schemes in second pull out on first plan the using of electrooptical devices (if the process of their characteristics improvement is rapid and qualitative). It's necessary for this to raise the stability of spectral characteristics and the degree of output radiation spectral purity when the parameters of transformation are deviated from optimal.

So, electro optical devices, possessing the best features for the transformation of coherent radiation parameters, particularly, under the modulating effects, belonging to the microwave range of electromagnetic waves, are choose as the base elements for realization of method in the optical range. Let's examine the problem of output spectrum purity and transformation parameters effect on it, as characteristics which directly determine the accuracy of monitoring on the whole.

Our converter is based on classic amplitude electro-optic modulator on 3m class metaniobat lithium crystal. The output spectrums of converter radiation in different cases are shown on fig. 9. We are the first, who shown (Il'in et al., 1996, 1997), that when converter operated in zero point of modulation characteristic described above Il'in-Morozov method was realized (fig. 9, c) and output spectrum was described as

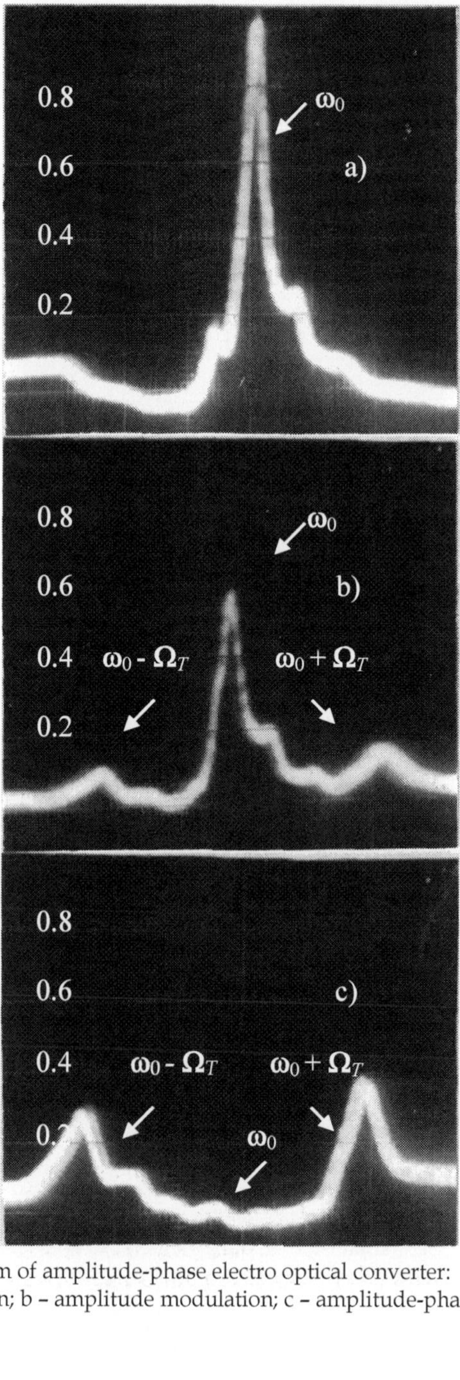

Fig. 9. Output spectrum of amplitude-phase electro optical converter:
a – without modulation; b – amplitude modulation; c – amplitude-phase modulation

$$e(t) = -jE_0 \exp(j\omega_0 t) \times \left[2\sum_{k=0}^{\infty} J_{2k+1}(z)\sin(2k+1)\Omega_T t \right], \tag{6}$$

where j is the index, which shows us, that output radiation is orthogonal to initial one; z is parameter, which is determined by the crystal and transforming field characteristics; $J_{2k+1}(z)$ is Bessel's function of (2k+1)-order.

When the value of controlling voltage U_m is equal to half wave voltage $U_{\lambda/2}$ of modulator, then $z= (\pi/2)(U_m/U_{\lambda/2}) = \pi/2$ and $J_1(z)=0{,}64$, $J_3(z)=0{,}06$.

The main causes which make spectral purity worse are temperature non-stability, deviation of transformation parameters from optimal and desalignment of converter. All of them cause the deviation of operating point from zero ($z \neq \pi/2$) and fast growth of parasitic spectral components amplitudes

$$e(t) = -j\frac{\sqrt{2}}{2}\exp(j\omega_0 t) \times \left[\begin{array}{c} \sum_{k=0}^{\infty} J_0(z) + \\ +2\sum_{k=0}^{\infty} J_{2k+1}(z)\sin(2k+1)\Omega_T t + \\ +2\sum_{k=1}^{\infty} J_{2k}(z)\cos 2k\Omega_T t \end{array} \right]. \tag{7}$$

Simple special countermeasures, using in converter, make their effect on spectrum purity very inappreciable. The maximum transformation coefficient for (6) is equal to 0,64, but when we choose transformation coefficient equal to 0,58 the coefficient of nonlinear distortions is not exceed 1%. Two more important things we must underline are equality of amplitudes of both spectrum components without dependence from location of working point and simplicity of frequency retuning, which is explained by using of one modulating signal.

Great results are obtained in electro optical integral Mach-Zehnder modulators (MZM) which are broadband, have dimensions 40-60 micrometers with an output of 20-40 mW when integrated micro-and nanotechnologies are used. In most papers devoted to the synthesis of two-frequency MZM formers, the modulation curve of MZM is generally considered in intensity domain as shown in Fig. 10,a. In these circumstances the phase is not taken into consideration and may be determined indirectly.

Consider the spectrum of the MZM output signal for different positions of the bias point on modulation characteristic with respect to the electric field E, as shown in Fig. 10,b. The modulation characteristic of electric field has four distinct regions. We consider the spectrum of the output oscillations under constant bias on behalf of two-frequency former.

The most specific is null bias point. This regime corresponds to the minimum output voltage $\Phi_{bias}=\pi$. The output signal is described by (8-9), where E_0, ω - amplitude and angular frequency of the electric field input CW optical signal; b - modulation index; Ω, V_{mod} - angular frequency and amplitude of modulating signal; V_π - half-wave voltage of the modulator.

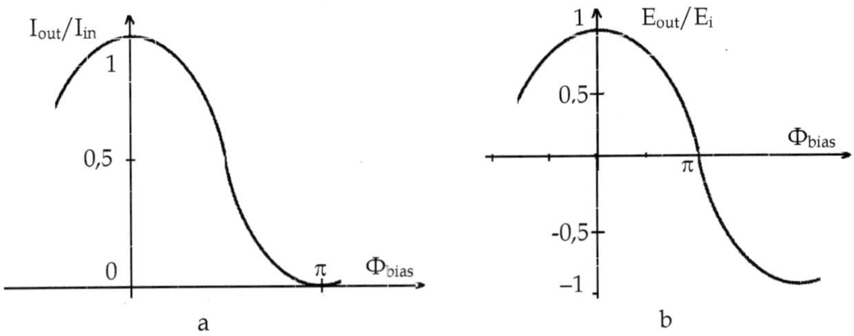

Fig. 10. MZM intensity (a) and electric field (b) modulation curve

This regime will also be available when $\Phi_{bias}=(2k-1)\pi$, however, the choice of k>1 does not satisfy the condition that the bias voltage should be of low amplitude otherwise the modulator's parameters undergo serious deviation, which will affect the spectrum conversion efficiency.

$$E_{out}(t)=\left(E_0/\sqrt{2}\right)e^{j\omega t}\times\left[e^{jb\sin(\Omega t+\pi)}+e^{jb\sin(\Omega t)+j\Phi_{bias}}\right]= \\ =\left(E_0/\sqrt{2}\right)e^{j\omega t}\times\left[2\sum_{n=1}^{\infty}J_{2n-1}(b)\left[e^{j(2n-1)\Omega t}-e^{-j(2n-1)\Omega t}\right]\right], \quad (8)$$

Analysis of (8) shows that the first term n=1 detects the presence in the spectrum of harmonics at frequency of $\omega_{+1}=\omega_0+\Omega$ and $\omega_{-1}=\omega_0-\Omega$, the second n=2 – the presence of harmonics at a frequency of $\omega_{+3}=\omega_0+3\,\Omega$ and $\omega_{-3}=\omega_0-\Omega$, etc. So, the Il'in-Morozov method is realized.

The spectrum of the MZM output signal, when $\Phi_{bias}=\pi$, is multi-frequency and consists of two bands. They are located symmetrically around the suppressed carrier, the spectral components are at frequencies $\omega=\omega_0\pm n\Omega$, where n = 1, 3, 5, ... Even harmonics are suppressed. The initial phases of the harmonics that make up the lower side band are π - radians different from the initial phases of the harmonics of the upper band.

The modulation index b has major influence on the spectral distribution of the output signal. This effect is examined further. Spectral distribution according to (8) is shown on fig. 11.

To obtain the two-frequency signal one should set the level of total harmonic distortion (THD) that is determined by the expression:

$$THD=\sqrt{\left(E_3^2+E_5^2\right)/E_1^2}\,. \quad (9)$$

Due to the smallness of the 7, 9, ...-orders harmonics' relative amplitudes their consideration is neglected. According to specified level of THD <1%, when the level of unwanted harmonics is negligible comparing to the amplitudes of the useful components, and the plot

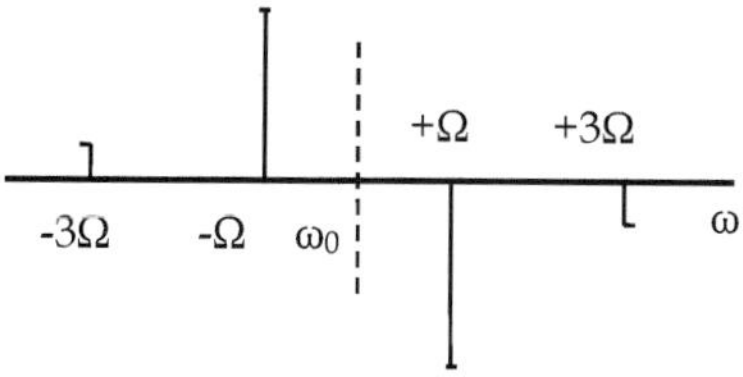

Fig. 11. Spectrum of (8)

of $J_n(b)$, we find that the fundamental condition for the two-frequency signal is b<0,49. Further increase of modulation index will lead to the fact that the relative amplitudes of 3rd and 5th orders harmonics will amount to the same order as the amplitude of spectral components of the 1st order. In particular the increase in the relative amplitudes of harmonics of third order leads to the fact, that for b = 3 the output signal will comprise four spectral lines.

5. Two-frequency reflectometry for FOS monitoring systems

Questions of OTFDR application efficiency for FOS monitoring on the examples of OTDR with two frequencies filling of probing pulse, non coherent OTFDR with LFM modulated components, and non coherent OTFDR for selective FOS probing are examined in this part.

5.1 OTDR with two frequencies filling of probing pulse

Structure of OTDR with two frequencies filling of probing pulse is synthesized on the basis of CO-OTDR (Jasenek, 2000). The main advantage of CO-OTDR is the opportunity of the maximal photo detector sensitivity maintenance which in an ideal case is determined by its quantum-mechanical limit $P_S / P_N = \eta P_0 / \hbar\omega\Delta F$. The basic destabilizing factors of CO-OTDR are instability of heterodyne frequency and default of polarizing conditions and spatial overlapping of measuring and basic beams at photo mixture. Undoubtedly, we can remove the mentioned above factors in fiber systems much easier, than in open, however methods used at it not always lead to desirable result.

In two-frequency OTDR system the frequency mode is realized inside a probing pulse with frequency components diversity on Ω. Ω is equal to tens or hundreds MHz and determines the central frequency of selective intermediate amplifier. Thus, measurements with carry of backscattering signal spectrum to area with a minimum level of photodetector noise are carried out.

Structure of OTDR with two frequencies filling of probing pulse is shown on fig. 12.

Optical radiation with angular frequency ω_0 divides on two parts in optical splitter OS1. Its one part is shifted on frequencies $\omega_0 \pm \Omega$ in TFLR forming electro optical modulator ((Il'in et al., 1997, O.G. Morozov & Sadeev, 2011) and received probing pulse with two frequencies filling incomes through OS2 into fiber under test. Backscattering radiation develops in OS3 with laser signal. Laser signal with power P_{LO} and backscattering signal with power $P_{RBS}(\omega_0 \pm \Omega)$ income on photodetector.

Application of methods and devices of symmetric two-frequency reflectometry in TFSC-OTDR allows to remove instability of frequency by application of two-frequency heterodyne

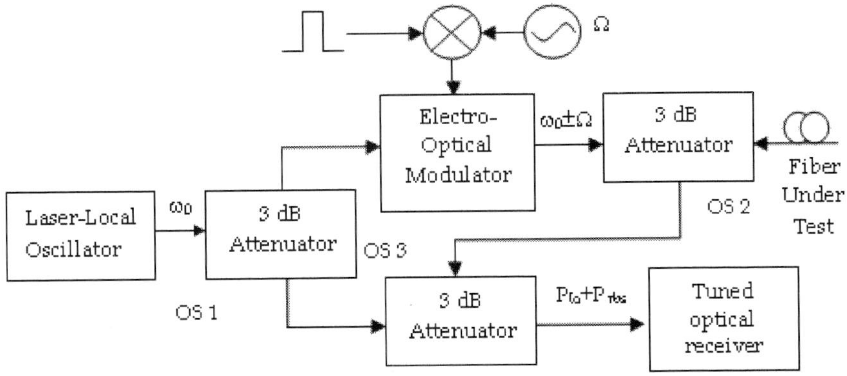

Fig. 12. Structure of OTDR with two frequencies filling of probing pulse

and also to provide performance of photo heterodyning conditions at direct two-frequency probing. In the second case for the signal/ noise ratio we shall receive

$$\frac{P_S}{P_N} = \frac{2P_1P_2(e\eta/\hbar\omega)^2}{2e\Delta F\left[i_T + P_{RBS}(e\eta/\hbar\omega)\right]} \approx \frac{P_1P_2(\eta/\hbar\omega)}{P_{RBS}\Delta F}. \tag{10}$$

The analysis of (10) at $P_{1,2} > P_{RBS}$ allows to approve, that, despite of reduction in sensitivity of the receiver in comparison with coherent reception, the bonus is received due to carry of an information signal spectrum to area with a minimum level of photo detector noise in comparison with direct detecting.

5.2 Non coherent OTFDR with LFM modulated components

Structure of non coherent OTFDR with LFM modulated TFSPR is synthesized on the basis of NC-OFDR (Abe et al., 1989, Froggatt, 1998, Pierce, 2000). We shall consider a piece of a fiber which is submitted on fig. 13. On a piece $0 < x < x_1$ there are the intrinsic losses α_1 caused by Raleigh scattering. On a piece $x_1 < x < x_2$ there are external losses caused by fiber microbends (defect, fig. 14). As at the appendix of mechanical tension ξ we receive decrease in a level of backscattering signal in comparison with a condition of rest, so losses in defect zone are equal $\alpha_2 = \alpha_i(1 - \Delta\alpha\xi)$, where α_i - initial losses caused by twisting. On a piece $x_2 < x < L$ with locked ends, where L - general length of a fiber, we come back to intrinsic losses α_1. Fibers remain permanently bent along their length after twisting. When the applied axial mechanical impact (e.g., stretching), the level of losses in the sensor becomes lower, as the curvature decreases. The relationship between optical loss and mechanical stress was obtained theoretically in the form shown below. Fiber radius of curvature R after two optical fibers twisting is given by

$$R = 1/(2\pi)^2 l; \; l = [p^2 + (2\pi r)^2]^{1/2} \tag{11}$$

where p_t – a twist step, r - half of the distance between the centers of two fibers, as shown in fig. 14, and l - length of the fiber, attributable to step twist.

When mechanical stress is applied uniformly over the entire length of the sensor, step twist changes and takes the value

$$p = p_i(1+\xi), \tag{12}$$

where p_i - the initial step of twisting, when the force is not applied. On the other hand, the distance r decreases as optical fibers, stretching, compressed, and their main shell deformed.

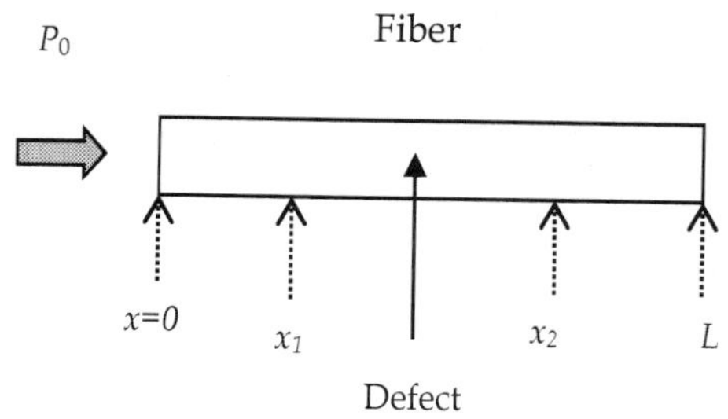

Fig. 13. Structure of fiber under the test

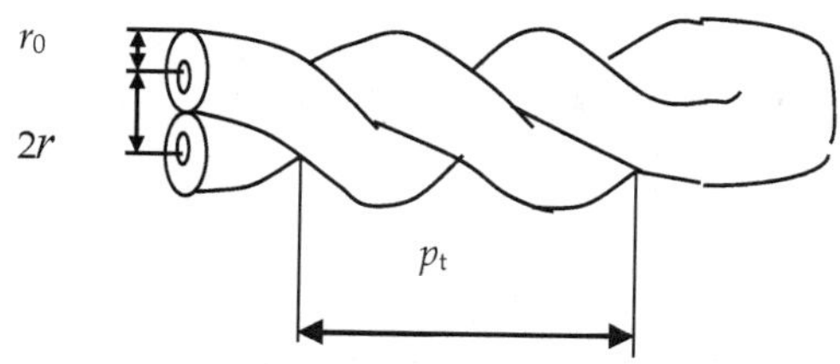

Fig. 14. Structure of sensor part

We take into account closure of all FOS on the two twisted OF, under the assumption that the untwisted fiber, connecting the ends of the sensor has its own losses $\alpha_1 << \alpha_i$, then

$$\alpha(\xi) = 2\alpha_i(1-2\Delta\alpha\xi). \tag{13}$$

In fact, (13) shows that ceteris paribus, $\xi << 1/\Delta\alpha$, $\alpha_1 << \alpha_i$, FOS on two twisted fiber with locked ends provides two fold the gradient changes of losses under the same applied load than the FOS on a twisted fiber with unlocked ends. Fig. 15 shows the calculated characteristics of FOS for various tensions and cross-elasticity *k*.

5.3 Non coherent OTFDR for resonant FOS monitoring

One of the main TFSPR applications is defined by an opportunity of testing resonant FOS, for example FBG or IFP. To analyze interaction of two-frequency signal we use some abstract circuit with normalized amplitude and phase dependences from frequency corresponding to FBG. Generalized amplitude-frequency characteristic of that abstract circuit we can define as follows

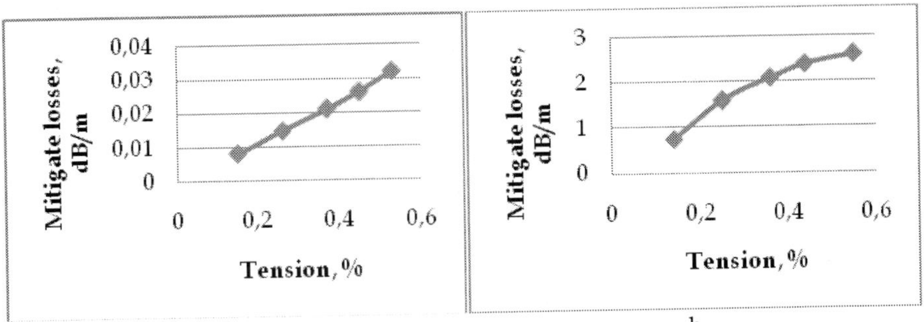

a b

(k = 38 kg/mm² (a) and k = 1,2 kg/mm² (b))

Fig. 15. Characteristics of FOS on two twisted fiber with closed ends

$$Y(\varepsilon)=1/\sqrt{1+\varepsilon_{0k}^{2}}\ . \tag{14}$$

Here $\varepsilon_{0k}=Q\left[(\omega/\omega_0)-(\omega_0/\omega)\right]$ is generalized circuit distortion.

Amplitude-frequency characteristic of circuit with two-frequency input signal is presented on fig. 16.

According to the average generalized signal distortion ε_0 the amplitude-frequency characteristic of abstract circuit could be divided on three parts:

1. if average generalized distortion lies in the interval $-\infty<\varepsilon_0<-\varepsilon_0/2$ (I area), then; $\varepsilon_0>\varepsilon_{02}$, $A_1/A_2>1$;
2. average generalized distortion lies in the interval $-\varepsilon_0/2<\varepsilon_0<\varepsilon_0/2$ (II area);
3. if average generalized distortion lies in the interval $-\varepsilon_0/2<\varepsilon_0<\infty$ (III area), then $\varepsilon_0>\varepsilon_{02}$, $A_1/A_2<1$.

Taking into account that $\varepsilon_{01}=\varepsilon_0+\Delta\varepsilon/2$ and $\varepsilon_{02}=\varepsilon_0-\Delta\varepsilon/2$, we will get the equations for phase dependences of output two-frequency signal components on generalized distortion $\varphi_{1out}(\varepsilon_0)$ and $\varphi_{2out}(\varepsilon_0)$:

$$\varphi_{1out}(\varepsilon_0)=-\mathrm{arctg}(\varepsilon_0+\Delta\varepsilon/2), \tag{15}$$

$$\varphi_{2out}(\varepsilon_0)=-\mathrm{arctg}(\varepsilon_0-\Delta\varepsilon/2), \tag{16}$$

Amplitude and phase of two-frequency signal, as it follows from fig.16, depend on values of average generalized distortion ε_0 and frequency distortion $\Delta\varepsilon$.

Now we consider the changes of amplitude and phase of output two-frequency signal envelope $U_{output}(t)$ while it passes through the abstract circuit. Output two-frequency signal envelope is described as follows:

$$U_{output}(t)=E_{output}(t)\cos(\omega_M t+\varphi_{output}(t)), \tag{17}$$

where $E_{output}(t)$ - resulting value of output two-frequency signal amplitude, $\varphi_{output}(t)$ - instantaneous phase, ω_M - instantaneous frequency of two-frequency signal.

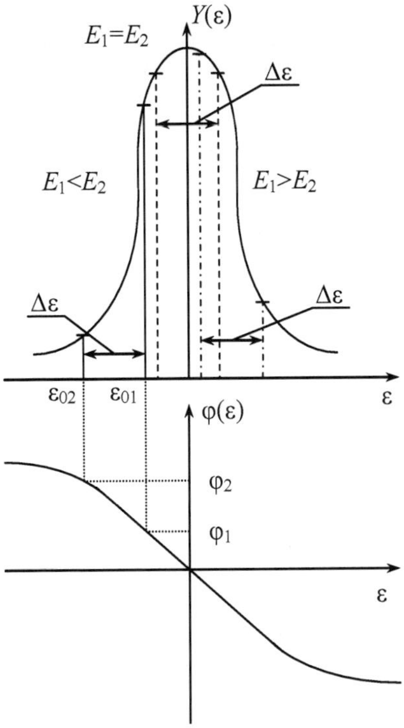

Fig. 16. Circuit amplitude-frequency characteristic with two-frequency input signal

Define the resulting value of two-frequency signal envelope amplitude E_{output} and then find the dependence of modulation index:

$$m = \sqrt{1+(\varepsilon_0+\Delta\varepsilon/2)^2} \,/\, \sqrt{1+(\varepsilon_0-\Delta\varepsilon/2)^2} \,. \quad (18)$$

From this equation we see that modulation index depends on average generalized distortion on two-frequency signal ε_0 and on distortion between two-frequency signal components $\Delta\varepsilon$. Dependence of modulation index on average generalized distortion of two-frequency signal $m(\varepsilon_0)$ with different distortions between frequency components $\Delta\varepsilon$ is presented on fig. 17.

Fig. 17 shows that curve $m(\varepsilon_0)$ is continuous function, which achieve its maximum at $\varepsilon_0=0$. Function decreases to its minimum, then increases to unit value, and decreases again. Steepness of curve $m(\varepsilon_0)$ depends on frequency distortion $\Delta\varepsilon$. Let estimate that dependence.

Dependence of curve steepness $m(\varepsilon_0)$ on frequency distortion, $S(\Delta\varepsilon)$, is presented on fig. 18. Fig. 18 shows that this curve has maximum at $\Delta\varepsilon=2$, i.e. maximal curve steepness corresponds the case, when frequency distortion $\Delta\varepsilon$ is equal abstract circuit pass band. Consequently, when $\Delta\varepsilon=2$, estimation error of output signal envelope amplitude will be minor.

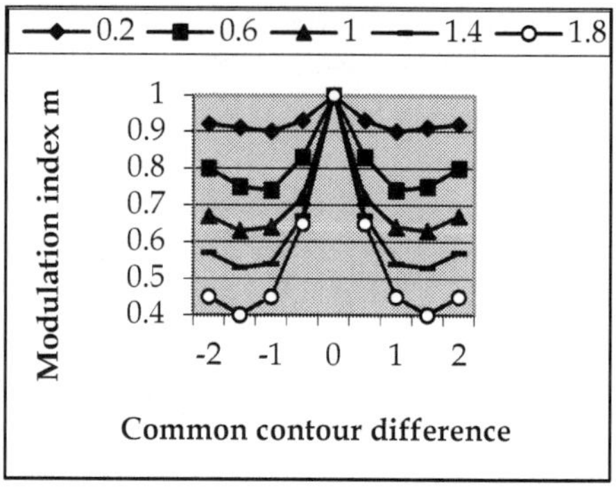

Fig. 17. Function $m(\varepsilon_0)$

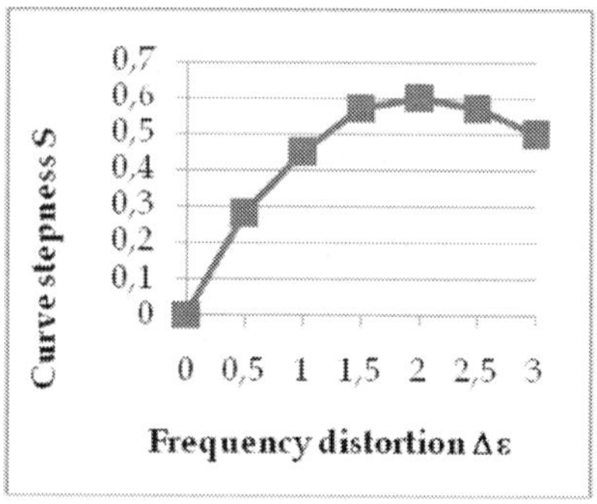

Fig. 18. Function $S(\Delta\varepsilon)$

Now consider dependence of phase shift of output signal components on average generalized signal distortion ε_0 with different values of distortion between frequencies $\Delta\varepsilon$. Calculation results are presented on fig. 19.

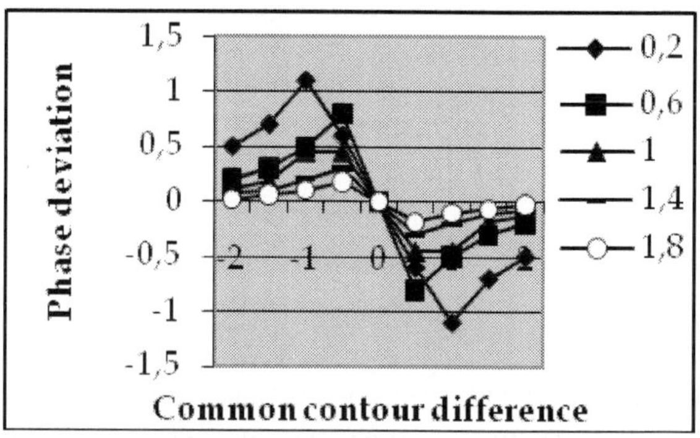

Fig. 19. Function $\Delta\varphi(\varepsilon_0)$

Fig. 19 shows that plot $\Delta\varphi(\varepsilon_0)$ is the third order curve which pass through zero when $\varepsilon_0=0$. Curve $\Delta\varphi(\varepsilon_0)$ is a continuous function which increases to maximum value, then decreases to minimum, and increases again.

Now we define the phase shift φ_{com} of output two-frequency signal envelope relatively input signal phase.

Phase shift of output signal envelope relatively input signal phase will consist:

$$\varphi_{com} = \varphi + \varphi_{1out}\left(-\infty < \varepsilon_0 < -\varepsilon_0 / 2 : \varepsilon_{01} > \varepsilon_{02}, E_1/E_2 > 1\right), \tag{19}$$

$$\varphi_{com} = \varphi + \varphi_{2out}\left(+\infty < \varepsilon_0 < +\varepsilon_0 / 2 : \varepsilon_{01} > \varepsilon_{02}, E_1/E_2 < 1\right), \tag{20}$$

where φ_{1out}, φ_{2out} – phase shifts of the first and the second output two-frequency signal components, φ – phase shift of resulting amplitude of output signal E_{output}.

After a series of complex transformations with notation of ε^+ and ε^- as $\varepsilon_0+\Delta\varepsilon/2$ and $\varepsilon_0-\Delta\varepsilon/2$ correspondingly we can rewrite equations (19) and (20) in a new way (upper index for ε^+ and lower – for ε^-):

$$\varphi_{com} = -\mathrm{arctg}\left(\varepsilon^{\pm}\right) + \mathrm{arctg}\left[\frac{\sin\left[-\mathrm{arctg}\left(\varepsilon^{\mp}\right) + \mathrm{arctg}\left(\varepsilon^{\pm}\right) + \Omega t\right]}{\left\{\left[1/\sqrt{1+(\varepsilon^+)^2}\right] / \left[1/\sqrt{1-(\varepsilon^-)^2}\right]\right\} + \cos\left[-\mathrm{arctg}\left(\varepsilon^{\mp}\right) + \mathrm{arctg}\left(\varepsilon^{\pm}\right) + \Omega t\right]}\right]. \tag{21}$$

From equation (21) we see that phase shift φ_{com} of output two-frequency signal envelope relatively input signal phase depends on value of average generalized signal distortion ε_0 and on value of frequency distortion $\Delta\varepsilon$. Calculation results of $\varphi_{com}(\varepsilon_0)$ for the case of $\varepsilon_{01} > \varepsilon_{02}$, $(E_{1out} / E_{2out})>1$ and different $\Delta\varepsilon$ are presented on fig. 20. Fig. 20 shows, that phase shift of output two-frequency signal envelope relatively phase of input signal will be equal to zero, when value of average generalized distortion $\varepsilon_0=0$.

Maximal sensitivity to average frequency of two-frequency signal change will be observed in case of equality of distortion between signal components $\Delta\varepsilon$ and pass band of abstract circuit.

Common equation for output two-frequency signal of abstract circuit for $-\infty< \varepsilon_0 < -\varepsilon_0/2$: $\varepsilon_{01} > \varepsilon_{02}$, $E_1/E_2>1$ is written as follows:

$$U_{output}(t) = \sqrt{E_{1out}^{\;2} + E_{2out}^{\;2} + 2E_{1out}E_{2out}\cos\left[\left(\varphi_{2out} - \varphi_{1out}\right) + \Omega t\right]}\cdot$$
$$\cdot\cos(\omega_M t + \left\{\varphi_1 + \arctan\left[\frac{\sin\left[\left(\varphi_{2out} - \varphi_{1out}\right) + \Omega t\right]}{E_{1out} / E_{2out} + \cos\left[\left(\varphi_{2out} - \varphi_{1out}\right) + \Omega t\right]}\right]\right\}) \tag{22}$$

Calculation results of output signal envelope are shown on fig. 20-21. From fig. 20-21 we see, that at the same moment, when the average frequency of two-frequency signal achieve resonance frequency of abstract circuit, output signal envelope has the same phase as input two-frequency signal envelope. In this case modulation index of output two-frequency signal will be equal a unit value.

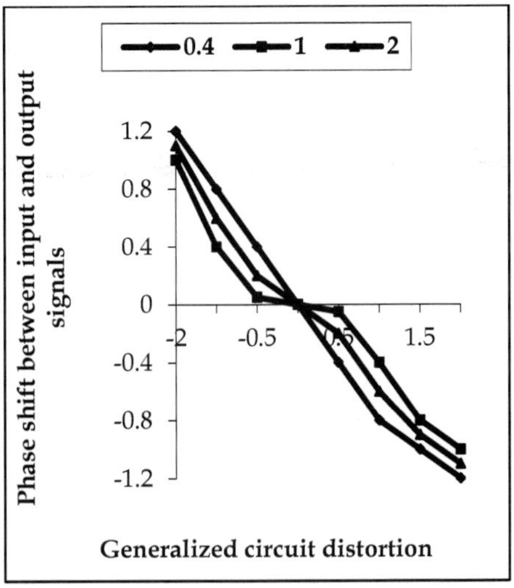

Fig. 20. Dependence of $\Delta\ \varphi(\varepsilon_0)$ for different $\Delta\ \varepsilon$

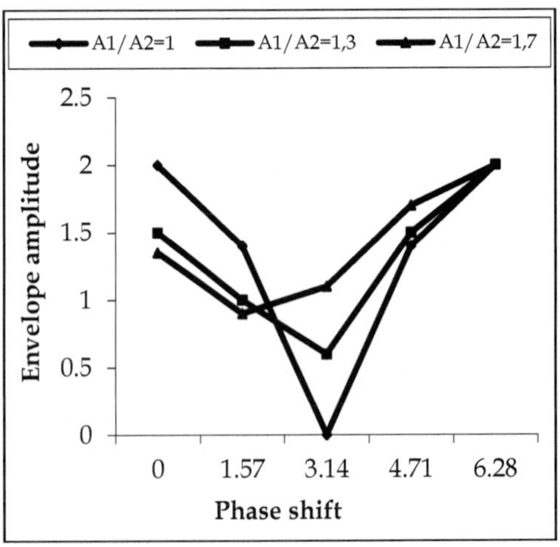

Fig. 21. Two-frequency output signal envelope $U_{output}(\Omega t)$

In the research was considered the propagation of symmetrical two-frequency laser radiation through the widespread types of FBG. The following FBG types were considered: homogeneous FBG with high reflection index (the reflection spectrum contains intensive side lobes), FBG with Gaussian envelope of index profile (side lobes on the long-wave part of spectrum from the main peak are suppressed), FBG with phase π-shift (allow to form narrow band transmission areas), and chirped FBG (the resonance wavelength is changing along the grating in a specified way). This method also allows registering the FBG reflection spectrum displacement due the influence of temperature or strain.

6. OTFDR application for nonlinear effects measuring and compensation

6.1 Two-frequency detector of Brillouin scattering

Brillouin scattering carries vast amount of information about fiber conditions but has low energy level. That's why it's necessary to detect these types of scattering and determine their properties. Applying photomixing allows significantly increase the reflectometer systems sensitivity under the condition of weak signals and receives information from frequency pushing of backscattered signal spectrum. We offer to use two-frequency heterodyne (TFHet) and the second nonlinear receiver in the structure of Brillouin reflectometer (fig. 22).

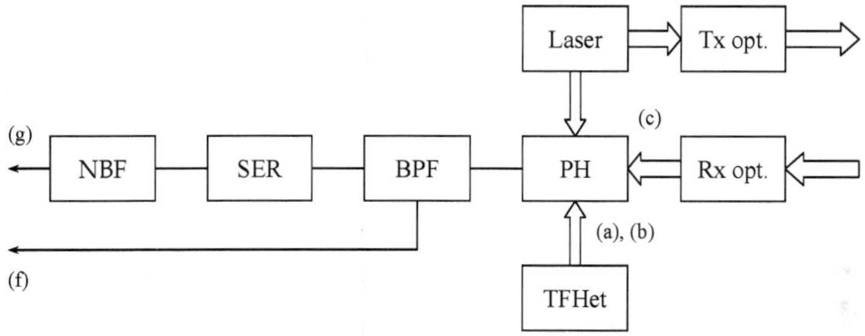

Fig. 22. Structure of two-frequency heterodyne system for OFS monitoring: NBF - narrow band filter, SER - square electronic receiver, BPF- band pass filter, PH - photo heterodyne, Rx opt - receiving optics, Tx opt - transmitting optics, TFHet - two-frequency heterodyne

Now we consider the signal passing through such system. Assuming that within photo detector aperture is provided the first order spatial error, amplitudes E_1 and E_2 of two-frequency signal components have the same polarization we can write the follow equation for electrical field intensity on receivers input:

$$E(t) = E_1 \cos(\omega_1 t + \varphi_1) + E_2 \cos(\omega_2 t + \varphi_2) + E_s \cos(\omega_s + \varphi_s), \tag{23}$$

where ω_1, ω_2 and ω_s - angular frequencies of two-frequency signal components and receiving signal, φ_1, φ_2 and φ_s - its phases.

Taking into consideration Stoletov's rule, square characteristics of optical receivers and significant exceeding of heterodyne signal intensity above input signal we can say that output receiver signal consists on constant components and components with difference frequencies $(\omega_1-\omega_s)$, $(\omega_1-\omega_s)$ and $(\omega_1-\omega_2)$. Believing, that $A_1=A_2=A_h$, and the signal from photo detector output through blocking filter on frequency ω_1- ω_2 acts on the second square-law electronic receiver, we shall receive

$$I^2 \approx 4k^2 A_{bs}^2 A_h^2 \left\{ \begin{array}{c} 1 + \frac{1}{2}\cos 2(\omega_1 - \omega_{bs})t + \frac{1}{2}\cos 2(\omega_2 - \omega_{bs})t + \\ + \cos(\omega_1 + \omega_2 - 2\omega_{bs})t + \cos(\omega_1 - \omega_2)t \end{array} \right\}. \tag{24}$$

If we install the narrow-band passing filter (NBF) on frequency ω_1-ω_2 on output of the second square-law electronic receiver (SER), its target signal will represent only change of amplitude of a backscattering Brillouin signal without taking into account frequencies instability of the transmitter $I^2 \approx 4\beta^2 A_{bs}^2 A_h^2 \cos\Delta\omega_{bs} t$. The component with frequency $\omega_1+\omega_2-2\omega_{bs}$ characterizes displacement of Brillouin signal frequency concerning frequencies ω_1 and ω_2, equal $\Delta\omega_{bs}= [(\omega_1+\omega_2)/2]-\omega_{bs}$.

Thus, using of two-frequency heterodyne (TFHet) and second nonlinear electronic receiver (SER) allows to divide amplitude and frequency information channels, and also to generate the basic channel for elimination of frequency instability influence of the transmitter on the data of measurements. As a result of such modernization heterodyning system for OFS monitoring can be submitted as follows (fig. 22).

For confirmation of system serviceability its mathematical modeling (results put on fig. 23) and researches at the experimental stand has been carried out.

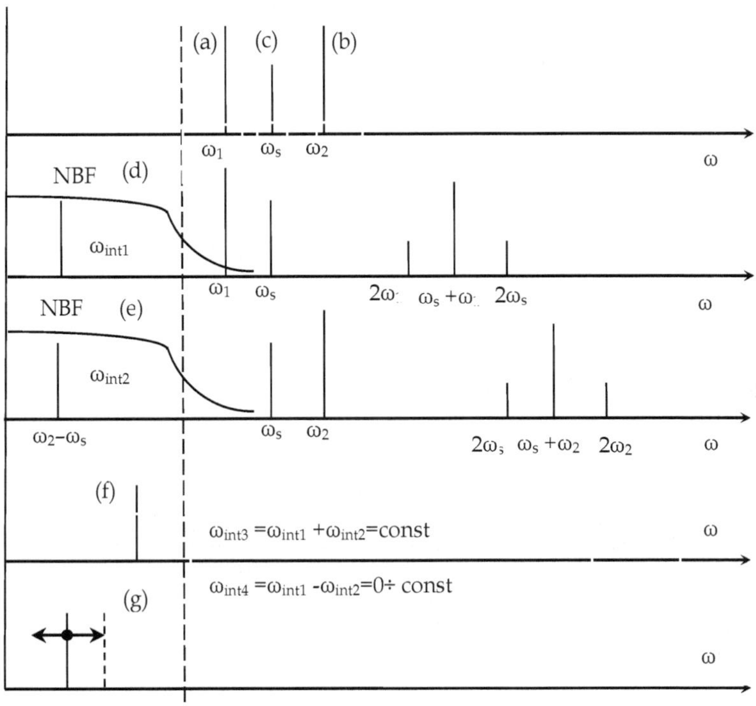

Fig. 23. Spectrograms in control points

The purpose of both researches was confirmation of opportunity to receive stable component on difference frequency and allocation of signal $\Delta\omega_{bs}$. Letters on fig. 22 designate control points, spectrograms in which are submitted on fig. 23. Experimentally received signals in the specified points are coincide with theoretic ones.

Thus, the opportunity of two-frequency heterodyne synthesis for OTFDR has been experimentally shown. It allows separately registering of useful signals of Brillouin scattering, as frequency information is separated from amplitude information. Using of two-frequency heterodyne and second nonlinear electronic receiver allow to divide amplitude and frequency information channels, and also to generate the basic channel for elimination of frequency instability influence of the transmitter on the data of measurements. Applying OTFDR method for Brillouin scattering detecting doesn't need a series of measurements with retuning scanning frequency, as all information is receiving through one pass that significantly simplifies the measurement process in comparison with homodyne detection systems.

6.2 Double mode system for FWM reducing

Let's examine block of compensating signal source. It consists of light source and double mode radiation source (DMRS) as shown on Fig. 24a. In [2] detailed explanation of conversion method is given. According to this method we can get double mode radiation from single mode with known distribution of phases and amplitudes. Amplitudes of $\omega_0+\Omega=\omega_3$, $\omega_0-\Omega=\omega_1$ harmonics according to method are $E_n=\pi E_0/4$, where E_0 is amplitude of single mode radiation.

For polarization alignment of interacting signals quarter-wave plate can be used. According to this approach scheme shown on Fig. 24a undergoes some changes that are presented on Fig. 24b.

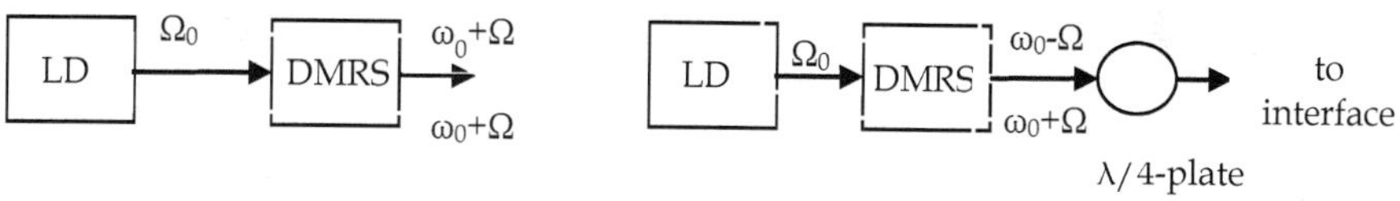

a b

Fig. 24. Block-scheme of compensating signal source without (a) and with (b) polarization being taken in account

Selecting ω_0, Ω according to following well-known expressions with accordance to Fig.25a it's possible to suppress FWM - signal. Taking $\omega_0=\omega_2$, and Ω equal to interchannel interval for we reduce FWM harmonic that is located on ω_3, the same mechanism is applied for FWM - product reduction on ω_4, furthermore communication signals which are located on ω_1, ω_2 are amplified. Mechanism is briefly illustrated on Fig. 25,a and Fig. 25,b. As a variation of above described method another can be used in this case compensating signal comprises an amplitude modulated signal that is modulated according to expression below, such type signal's spectrum is illustrated on Fig. 26.

$$e(t)=E_0\sin\omega_0 t[1+m\cos(\Omega t+\pi)]=E_0\sin\omega_0 t-\frac{mE_0}{2}\sin(\omega_0 t+\Omega+\pi)-\frac{mE_0}{2}\sin(\omega_0 t-\Omega-\pi) \quad (25)$$

As a result FWM signals on ω_3 and ω_4 are reduced from $A_{fwm.init}$ to A'_{fwm} and communication signals on ω_1 and ω_2 are amplified from $A_{sig.init}$ to A'_{sig} according to expression (26), (27), where *m* is modulation coefficient.

$$A'_{fwm}=A_{fwm.init}-mE_0/2, \quad (26)$$

$$A'_{sig}=A_{sig.init}-(mE_0/2)+E_0. \quad (27)$$

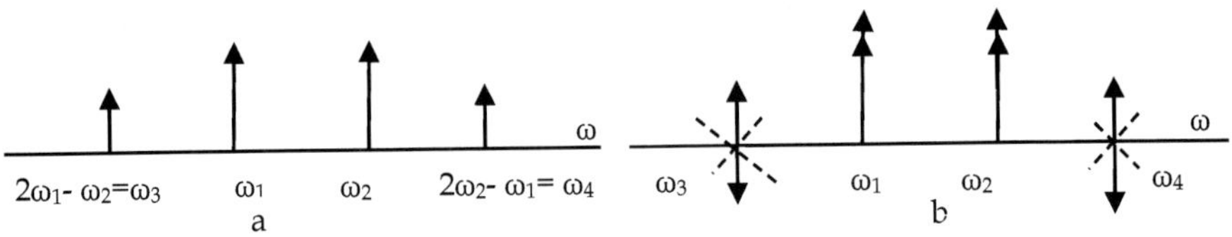

Fig. 25. FWM (a) and interaction (b) between FWM, communication signal and double mode compensating signal

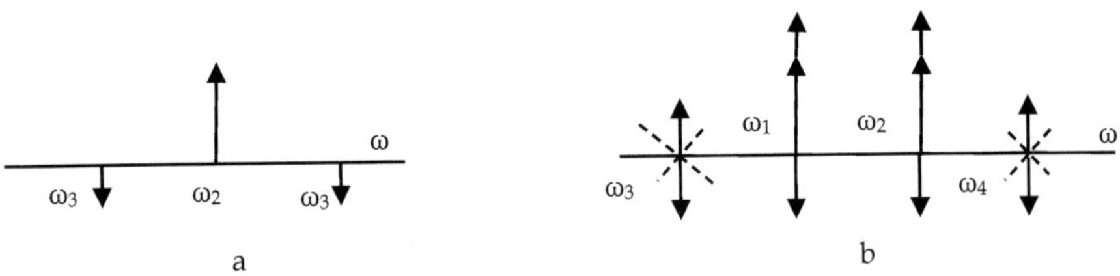

Fig. 26. AM-compensating signal (a) and interaction (b) between FWM, communication signal and AM-compensating signal.

For FWM products suppression it's claimed to use method based on conversion of single mode radiation into double that meets demands of amplitude, phase, and polarization alignment. This method is active i.e. communication channel that undergoes fading due to FWM is amplified, FWM products are suppressed to minimum levels also there is no need to use complex registration and measurement devices and components. In AM utilizing method it's necessary to select amplitudes of interacting signals carefully in order to prevent extra communication signal fading.

7. Miscellaneous applications

The miscellaneous application of OTFDR in bio sensing and TFSPR utilization in photonic microwave filter synthesis are mainly presented in (O.A. Stepustchenko et al., 2011) and (O.G. Morozov & Sadeev, 2011).

7.1 OTFDR in biosensors

Utilization of FBG with phase π-shift ROB assumes for measurement the central wavelength shift value of transmission Lorentz's contour with a half-width equal to pm units. For such values their estimation in terms of frequency is standard. Thus, the width of investigated contour spectrum is units of GHz. It is obvious, that application of broadband sources and optical spectrum analyzers for the decision of given problem is inefficient. High accuracy and resolution of measurements can reach at use of scanning Mickelson interferometer and the frequency-synchronized narrow-band laser sources. However maintenance of phase measurements stability in the first case and impossibility of high-speed feedback maintenance in the second do the decision very problematic as in case of static and dynamic measurements. The offered decision can be found at use of OTFDR. As we show in 5.3

OTFDR allows raising considerably power parities for modulation frequencies and to spend detecting on amplitude, phase and sign on a phase of differential frequency envelope with high stability of metrological characteristics.

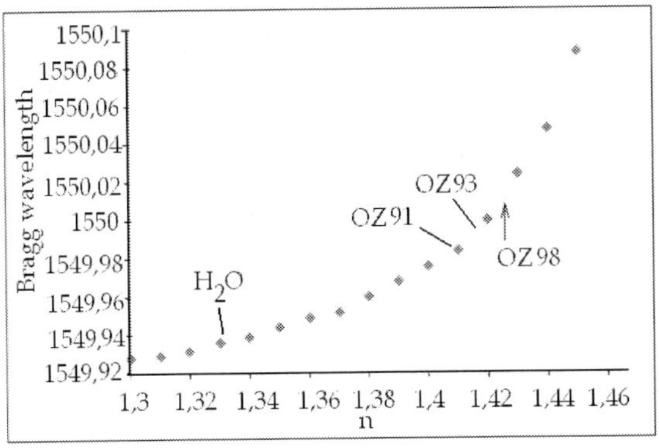

Fig. 27. The Bragg center wavelength as a function of the refractive factor of ambient material

Modeling results of biosensor in Optiwave Graiting software for red shifts of different materials concerning to water (n=1,33) are presented (fig. 27): gasoline with 91, 93, 98 octane numbers accordingly.

As comparison with results of another authors the dynamic range of wavelength shift is smaller, but enough to get sufficient resolution. From the other side experimental probing of this type ROB was realized by narrow band laser (kHz or MHz) with differential frequency in MHz range and its forming in Mach-Zehnder modulator. So SNR increase of measurements in 10-50 times were predicted. Experimental results was got on the base of SMF-28 Corning fibers.

7.2 All-optical microwave photonic filter based on two-frequency optical source

In this part we presented two tap photonic filter. Tunability is achieved by time delay adjusting which is performed by subcarriers separation. Proposed filter has the potential to implement several positive and negative components through bias voltage adjusting. First order notch located at 9 GHz might be tuned either by ω_0 or Ω adjusting.

To implement filter with two alternative coefficients we propose to utilize two-frequency radiation sources that was obtained by phase-amplitude modulation of initial single wavelength laser. The spectrum is shown on fig. 28,a, and spectrum after RF-modulation on fig. 28,b.

Overall filter frequency response H(f) is given on fig. 29.

To verify our mathematically based suggestions we have conducted simulation of filter in OptiSystem 7.0 simulation software tool and experimental base of Fiber Optic R&D Centre of our university. Firstly we obtained two separated wavelength by F GHz with the initial F/2 GHz signal. Then this signal was modulated in quadrature point in MZM by ω_{RF} – radio

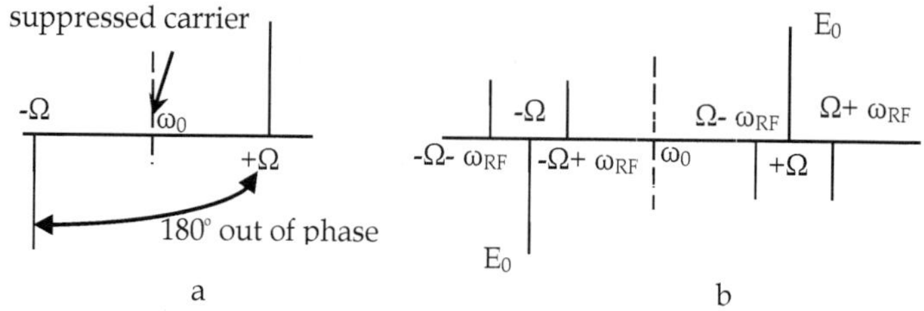

Fig. 28. Filter coefficients (a) and final spectrum of RF modulated signal

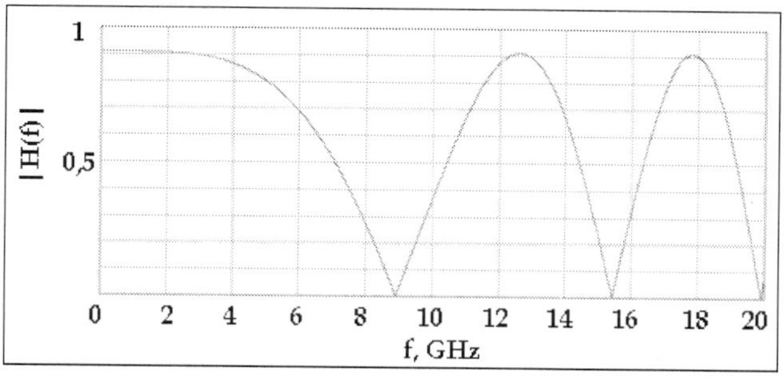

Fig. 29. Two tap filter frequency response *H(f)*

signal to be filtered. We used z=50 km of SMF fiber with D=16,5 ps/nm km dispersion thus achieving delay time T=ΔλLD. Frequency response was observed through RF spectrum analyzer tool and exactly matched FSR=1/T.

8. Conclusion

We reviewed the principle and the applications of the TFSPR technique. A variety of multiplexed sensing functions can be provided by the technique. As the examples, we introduced the systems of optical reflectometry, distributed lateral stress location sensors, multiplexed FBG and FFPI sensors. These systems possess the advantages of the continuous-wave operation, high resolution, high accuracy, and high speed. We have also developed a system to measure the strain distribution along an optical fiber through the Brillouin scattering. This system has already achieved 5-cm spatial resolution and 100 times higher dynamic range, respectively, than the conventional pulse-based time-domain techniques. Applications in the optical biosensing and microwave signal processing have also been demonstrated. System performance is being improved further with the perfection of two-frequency sources and application of four, six and etc. frequencies based on application of MZM comb-generators and elements of WDM technologies.

9. Acknowledgements

This work was supported by Federal Program «Tupolev Kazan State Technical University (Kazan Aircraft Institute) - National Research University».

10. References

Abe, T., Mitsunaga, Y., Koga, H. (1989). A strain sensor using twisted optical fibers, *Journal of Lightwave Tech.*, Vol. 7, № 3, pp. 525-531.

Barnoski, M.K. & Jensen, S.M. (1976). Fiber waveguides a novel technique for investigating attenuation characteristics, *Applied Optics*, Vol. 15, N9, pp.212-216.

Gonorovskii, I. S. (1977). *Radio electronics circuits and signals*, Moscow, Russia.

Froggatt, M. & Moore, J. (1998). Distributed measurement of static strain in an optical fiber with multiple Bragg gratings at nominally equal wavelengths, *Applied Optics*, Vol. 37, pp.1741-1746.

Il'in, G.I. & Morozov, O.G. (1983). Method of one frequency coherent radiation convertion into two frequency, USSR patent N1338647, 20.04.2004.

Il'in, G.I. et al. (1995). LFM lidar with frequency conversion, *Atmospheric and Ocean Optics*, Vol. 8, N12, pp. 1871-1874.

Il'in, G.I. et al. (1996). Two-frequency oscillator for interferometers with polarization dived channels, In: *Optical Inspection and Micromeasurements*, edited by Christophe Gorecki, Proceedings of SPIE Vol. 2782 (SPIE, Bellingham, WA) pp. 814-819.

Jasenek, J. (2000). The theory and application of the fiber optic sensors with spread parameters, Slovak University of Technology in Bratislava, 2000, Available from http://www.eaeeie.org/theiere_bratislava/index.html.

Kharckevich, A.A. (1962). *Spectrums and signals*, Moscow, Russia.

Morozov, O.G. et al. (1999). Metrological features of laser interferometers with frequency conversion in supporting channel, In: *Optical Measurement Systems for Industrial Inspection*, edited by Malgorzata Kujawinska, Wolfgang Osten, Proceedings of SPIE Vol. 3824 (SPIE, Bellingham, WA) pp. 162-168.

Morozov, O.G. et al. (2006). Two-frequency analysis of fiber-optic structures, In: *Optical Technologies for Telecommunications 2005*, edited by V.A. Andreev, V.A. Burdin, A.H. Sultanov, Proceedings of SPIE Vol. 6277 (SPIE, Bellingham, WA) pp. 62770E.

Morozov, O.G. et al. (2008). Metrological aspects of symmetric double frequency and multi frequency reflectometry for fiber Bragg structures, In: *Optical Technologies for Telecommunications 2007*, edited by Vladimir A. Andreev, Vladimir A. Burdin, Oleg G. Morozov, Albert H. Sultanov, Proceedings of SPIE Vol. 7026 (SPIE, Bellingham, WA)pp. 70260J.

Morozov, O.G. et al. (2008). Methodology of symmetric double frequency reflectometry for selective fiber optic structures, In: *Optical Technologies for Telecommunications 2007*, edited by Vladimir A. Andreev, Vladimir A. Burdin, Oleg G. Morozov, Albert H. Sultanov, Proceedings of SPIE Vol. 7026 (SPIE, Bellingham, WA) pp. 70260I.

Morozov, O.G. et al. (2008). Double mode system for FWM reducing, In: *Optical Technologies for Telecommunications 2007*, edited by Vladimir A. Andreev, Vladimir A. Burdin, Oleg G. Morozov, Albert H. Sultanov, Proceedings of SPIE Vol. 7026 (SPIE, Bellingham, WA) pp. 702603.

Morozov, O.G. et al. (2009). All-optical microwave filter for ROF WDM systems based on double mode method, In: *Optical Technologies for Telecommunications 2008*, edited by Vladimir A. Andreev, Vladimir A. Burdin, Oleg G. Morozov, Albert H. Sultanov, Proceedings of SPIE Vol. 7374 (SPIE, Bellingham, WA) pp. 73740A.

Morozov, O.G. et al. (2010) Sensor on the twisted fibers for building construction monitoring, In: *Optical Technologies for Telecommunications 2009*, edited by Vladimir

A. Andreev, Vladimir A. Burdin, Oleg G. Morozov, Albert H. Sultanov, Proceedings of SPIE Vol. 7523 (SPIE, Bellingham, WA) pp. 75230J.

Morozov, G.A. et al. (2011). Structural minimization of fiber optic sensor nets for monitoring of dangerous materials storage, In: *Optical Technologies for Telecommunications 2010*, edited by Vladimir A. Andreev, Vladimir A. Burdin, Albert H. Sultanov, Oleg G. Morozov, Proceedings of SPIE Vol. 7992 (SPIE, Bellingham, WA) pp. 79920E.

Morozov, O.G. & Aibatov, D. L. (2007). Two-frequency scanning of FBG with arbitrary reflection spectrum, In: *Optical Technologies for Telecommunications 2006*, edited by Vladimir A. Andreev, Vladimir A. Burdin, Albert H. Sultanov, Proceedings of SPIE Vol. 6605 (SPIE, Bellingham, WA) pp. 660506.

Morozov, O.G. & Pol'ski, Y.E. (1996). Perspectives of fiber sensors based on optical reflectometry for nondestructive evaluation, In: *Nondestructive Evaluation of Materials and Composites*, edited by Steven R. Doctor, Carol A. Nove, George Y. Baaklini, Proceedings of SPIE Vol. 2944 (SPIE, Bellingham, WA) pp. 178-183.

Morozov, O.G. & Sadeev, T.S. (2011). All-optical microwave photonic filter based on two-frequency optical source, In: *Optical Technologies for Telecommunications 2010*, edited by Vladimir A. Andreev, Vladimir A. Burdin, Albert H. Sultanov, Oleg G. Morozov, Proceedings of SPIE Vol. 7992 (SPIE, Bellingham, WA) pp. 79920C.

Natanson, O.G. et al. (2005a). Development problems of frequency reflectometry for monitoring systems of optical fiber structures, In *Optical Technologies for Telecommunications*, edited by Vladimir A. Andreev, Vladimir A. Burdin, Albert H. Sultanov, Proceedings of SPIE Vol. 5854 (SPIE, Bellingham, WA) pp. 215-223.

Natanson, O.G. et al. (2005b). Reflectometry in open and fiber mediums: technology transfer, In: *Optical Technologies for Telecommunications*, edited by Vladimir A. Andreev, Vladimir A. Burdin, Albert H. Sultanov, Proceedings of SPIE Vol. 5854 (SPIE, Bellingham, WA) pp. 205-214.

Pierce, S.G. et al. Optical frequency domain reflectometry for microbend sensor demodulation", *Applied Optics*, Vol. 39, N25, pp.4569-4573.

Pol'ski, Y.E. & Morozov, O.G. (1998). Joint field of integrated fiber optic sensors for aircraft and spacecrafts safety parameters monitoring, In: *Nondestructive Evaluation of Aging Aircraft, Airports, and Aerospace Hardware II*, edited by Glenn A. Geithman, Gary E. Georgeson, Proceedings of SPIE Vol. 3397 (SPIE, Bellingham, WA) pp. 217-223.

Stepustchenko, O.A. et al. (2011). Optical refractometric FBG biosensors: problems of development and decision courses, In: *Optical Technologies for Telecommunications 2010*, edited by Vladimir A. Andreev, Vladimir A. Burdin, Albert H. Sultanov, Oleg G. Morozov, Proceedings of SPIE Vol. 7992 (SPIE, Bellingham, WA) pp. 79920D.

Zalyalov, R.G. et al. (1999). Optical frequency domain reflectometer for fiber structural testing, In: *Optical Measurement Systems for Industrial Inspection*, edited by Malgorzata Kujawinska, Wolfgang Osten, Proceedings of SPIE Vol. 3824 (SPIE, Bellingham, WA) pp. 274-279.

7

A Novel Approach to Evaluate the Sensitivities of the Optical Fiber Evanescent Field Sensors

Xuye Zhuang[1], Pinghua Li[2] and Jun Yao[1]
[1]*State Key Laboratory of Optical Technologies for Microfabrication, Institute of Optics and Electronics, Chinese Academy of Sciences*
[2]*Department of Electronic Information Technology, Sichuan Modern Vocational College*
China

1. Introduction

Optical fiber evanescent field sensors(OFEFS) are playing more and more important roles in detecting chemical, bacilli, toxin, and environmental pollutants for their high efficiency, good accuracy, low cost, and convenience (Maria et al., 2007; Wolfbeis, 2006; Angela et al., 2007). In general, these sensors exploit the interactions between evanescent field of the sensing fibers and the surrounding analyte under investigation. Concentrations of analyte can be related to the power attenuation caused by these interactions, and the more energy interacted with the analyte, the higher the sensitivities of the sensors. In order to expose the evanescent filed of the fiber to the analyte, the fiber cladding is often removed and the fiber core is surrounded by the detecting material. To make clear the operation principle of the evanescent field interacting with the analyte and to estimate the sensitivity according to this principle are extremely important to the sensor's designers. This field has attracted a lot of research groups extraordinarily and been widely discussed both based on geometric optics and optical waveguide theory (Yu et al., 2006; Messica et al.,1996; Gupta & Singh, 1994; Guo & Albin, 2006). However a big difference between the theoretical and experimental values is often observed especially when the sensor's absorbency is low (Payne & Hale, 1993; Wu et al., 2007; Deng, et al., 2006).

In this chapter, a thoroughly theoretical study of the OFEFS is presented. The reason why the sensor's absorbency, A, shows nonlinear increase with the sensing fiber's length, l, is given and analyzed. Also, a new method to estimate the sensitivity of the sensor is proposed and confirmed with experimental results, which shows a good agreement with the experimental results and is more accurate than the methods used before. The fabrication technologies of the sensing fibers are introduced. The practical importance of this study lies in helping for understanding, design and construction of optical fiber evanescent field sensors.

This chapter is organized as follows. In section 2, the theory basis of the optical fiber evanescent field sensor is introduced and analyzed. The nonlinear relationship between A and l is discussed theoretically. The new method to estimate the sensitivity of the sensor is proposed in section 2.3. The fabrication method of the sensing fiber with acicular encapsulation is described in section 3. Also, the experiments used to confirm the new method and the nonlinear relationship between A and l are implemented and the results are analyzed. Finally, the contents of the chapter are summarized briefly in section 4.

2. Theory analysis of optical fiber evanescent field sensors

2.1 Evanescent field of the optical fiber

Optical fiber consists of a cylindrical core and a surrounding cladding, both made of silica, as illustrated in Fig. 1. The core is generally doped with germanium to make its refractive index n_{co} slightly higher than the refractive index of the cladding n_{cl}.

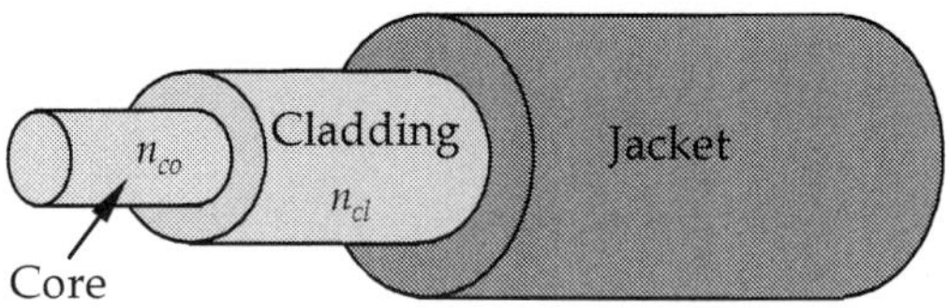

Fig. 1. The structure of step-index fiber

If a ray is launched into the fiber and its propagation angle θ_2 is greater than the critical angle θ_c, which is defined by the Eq.(1), the light will propagate along the fiber by total internal reflection (TIR).

$$\theta_c = \arcsin(n_{cl} / n_{co}) \tag{1}$$

Fig. 2 shows a hypothetical ray of light propagating along the fiber. n_0 is the refractive index of the air. θ_0 is the incident angle of the light at the core-air interface, which must obey the relationship described in Eq.(2) in order that the incident light can propagate along the fiber by TIR.

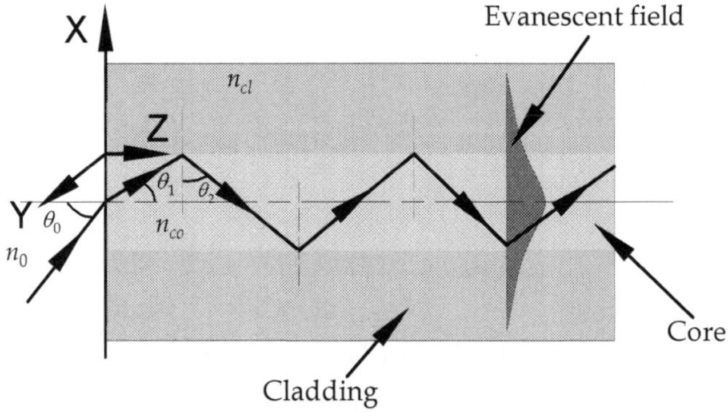

Fig. 2. The schematic of the light rays transmitting in the evanescent field sensing fiber

$$\sin\theta_0 < \frac{1}{n_0}(n_{co}^2 - n_{cl}^2)^{\frac{1}{2}} \tag{2}$$

θ_m is the maximum value of θ_0 and defined by Eq.(3).

$$n_0 \sin\theta_m = (n_1^2 - n_2^2)^{\frac{1}{2}} = NA \tag{3}$$

If the incident angle of a ray is smaller than θ_m, that ray of light will propagate along the fiber by TIR. *NA* is the numerical aperture of the fiber, which is a dimensionless number that characterizes the range of the incident angles over which the light can transmit by TIR in the fiber.

Though the light propagates along the fiber by TIR, there is energy penetrating into the cladding. The electric field amplitude of this field decays exponentially from the core-cladding interface and is described as Eq.(4).

$$E_x = E_0 e^{-x/d_p} \tag{4}$$

Where x is the distance from the core-cladding interface, E_0 is the electric field magnitude of the field at the interface, and d_p represents the penetration depth, which is the distance where the evanescent field decreases to $1/e$ of its value at the core-cladding interface and is mathematically given by (Zhuang et al., 2009a)

$$d_p = \frac{\lambda_0}{2\pi(n_{co}{}^2 \sin^2\theta_2 - n_{cl}{}^2)^{1/2}} \tag{5}$$

where λ_0 is the wavelength of the light source in the vacuum. Because of the non-absorbing cladding, no energy loss is measured in the fiber. When an absorbing media substitutes for the cladding, the intensity of the evanescent field is attenuated and the total transmitted energy in the fiber is reduced. These losses are caused by the interactions between the analyte and the evanescent field. A bigger d_p indicates a wider scope covered by the field, more chances of the energy interacts with the analyte, a bigger loss of the energy and a higher sensitivity of the sensor.

2.2 Mathematical mode of the optical fiber evanescent field sensor

The structure of the optical fiber evanescent field sensor is shown in Fig.3. One portion of the fiber's cladding is removed, and its core is exposed to the analyte directly. So the energy of the evanescent field can interact with the analyte, and the concentrations of the analyte can be detected by measuring the loss of the output power of the sensing fiber. The extent of the energy loss can be marked by the absorbency of the sensor, *A*, which is described as follows.

$$A = -\log_{10}\left(\frac{P_{out}}{P_{in}}\right) \tag{6}$$

where P_{out} and P_{in} are the output power of the optical fiber evanescent field sensors with and without an absorptive analyte for sensing. When detecting the same sample with the same concentrations, the bigger the value, the more sensitive the sensor. It shows clearly in Eq.6 that the sensitivity of OFEFS can also be evaluated by the value of P_{out} / P_{in}. When detecting the same sample, the smaller the value, the higher the sensor's sensitivity.

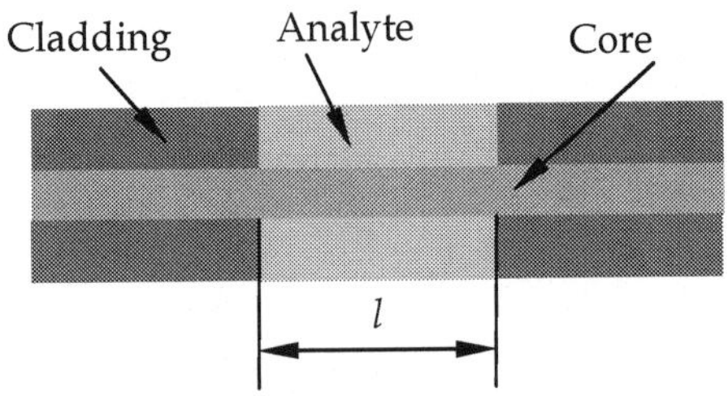

Fig. 3. Schematic diagram of the optical fiber evanescent field sensor

According to Beer's law, P_{out} is expressed as (Wu et al., 2007)

$$P_{out} = \sum_{j=1}^{N} r_j P_{in} \exp(-\alpha \eta_j l), \tag{7}$$

Subscript j stands for the jth mode optical waveguide transmitting in the sensing fiber. Hence, r_j is the fraction of the optical power carried by the jth mode in the sensor. When using incoherent light source, the energy carried by each mode transmitting in the sensing fiber is almost the same (Gloge, (1971). Also, after a long length of transmitting, the energy of each mode is almost equal through energy couples between different modes in the fiber (Zhuang et al., 2009a). So Eq.(8) can be modified as (Payne & Hale, 1993),

$$P_{out} = \frac{P_{in}}{N} \sum_{j=1}^{N} \exp(-\alpha \eta_j l) \tag{8}$$

where l is the length of the sensing fiber whose cladding is removed, α is the absorption coefficient of the analyte and is expressed as

$$\alpha = \frac{\varepsilon c}{\log_{10} e} \tag{9}$$

where ε and c are molar absorption coefficient and concentration of the analyte, respectively.

η is the fraction of power transmitting in the fiber cladding that carried by evanescent field of the sensor and is depicted as,

$$\eta = \frac{P_{cl}}{P_{cl} + P_{co}} \tag{10}$$

where P_{co} is the energy transmitting in the core; P_{cl} is the energy transmitting in the cladding in the form of evanescent field, which is the only energy interacted with the analyte. η_j is the energy fraction of the jth mode transmitting in the cladding. η can be calculated using Eq.(11).

$$\eta = 1 - \frac{P_{co}}{P_{cl} + P_{co}} = 1 - \frac{\int_0^a \int_0^{2\pi} R_e(\vec{e} \times \vec{h}^*)\vec{l}_z r dr d\theta}{\int_0^\infty \int_0^{2\pi} R_e(\vec{e} \times \vec{h}^*)\vec{l}_z r dr d\theta} \tag{11}$$

The integral part of Eq.(11) is the Poyning Vector Integral in cylindrical coordinates. The vector $\vec{l}_z$ indicates that the Z axis of the coordinate is along the fiber's axis. The particular explanations of each parameter used in Eq.(11) are listed in the reference (Gloge, 1971).

For multimode sensing fibers, η_j can be calculated using Eq.(12).

$$\eta_j = j \Big/ N(2N - 2j)^{0.5} \tag{12}$$

where N is the number of the modes supported by the fiber and defined by

$$N = 4 \times V^2 \Big/ \pi^2 \tag{13}$$

V is the normalized frequency of the fiber and expressed as

$$V = \frac{2\pi a}{\lambda}(n_{co}^2 - n_{cl}^2)^{1/2} \tag{14}$$

where a is the radius of the core, λ is the wavelength of the light. As to less-mode or single mode optical fibers, η of different modes should be calculated carefully based on optical waveguide theory (Zhuang, 2009b).

2.3 Nonlinear relationship between absorbency and the length of the sensing fiber

From Eq.(6) and Eq.(8), we can obtain Eq.(15) which describes the absorbency of the sensing fibers.

$$A = -\log_{10}\left(\frac{\alpha l}{N}\sum_{j=1}^{N}\exp(-\eta_j)\right) \tag{15}$$

The sensitivity of the optical fiber evanescent field sensors, M, is defined as the instantaneous change of the sensors' absorbency relative to the analyte.

$$M = dA / dC \tag{16}$$

dA is the sensitivity of the photoelectric detector, which is the least variance of the absorbency the equipment can distinguish. According to Eq.(15) and dA, the lowest concentrations of the analyte to be detected can be approximately estimated.

In experiments, the actual sensitivity of the sensor can be obtained by Eq.(17).

$$M' = \Delta A \Big/ \Delta C \tag{17}$$

where ΔA is the difference of the sensors' absorbency obtained from two different detections, ΔC is the difference of the analytes' concentrations.

From Eq.(15) and Eq.(16), it is found the absorbency A and the sensitivity M of the sensor linearly increase with l. However, non-linear dependence of A and M on the l is often observed in experimental tests (Zhuang, 2009a, 2009b). So the equations listed above should be modified to obtain a more accurate simulation results.

It shows clearly in Eq.(5) that d_p is decided by the parameters λ, n_{co}, n_{cl} and the propagation angle θ_2 of the ray. As the evanescent field energy transmitting along the fiber, none of those parameters change. So the value of d_p keeps unchanged during the process. Because of the interactions between the evanescent field and the analyte, the amplitude of the field decreases quickly which is shown clearly in Fig.4. The portion of the energy transmitting in the cladding is defined by Eq.(11), which indicates the amplitude of the evanescent field of the sensing fiber. For multimode fibers, η_j is given approximately by Eq.(12), which is the function of N. N is decided by the normalized frequency V of the sensing fiber. As expressed in Eq.(14), V is determined by the characters of the sensing fiber, the wavelength of the light source and the analyte's refractive index. After an experiment system is set up, V and N are determined, hence η_j is determined.

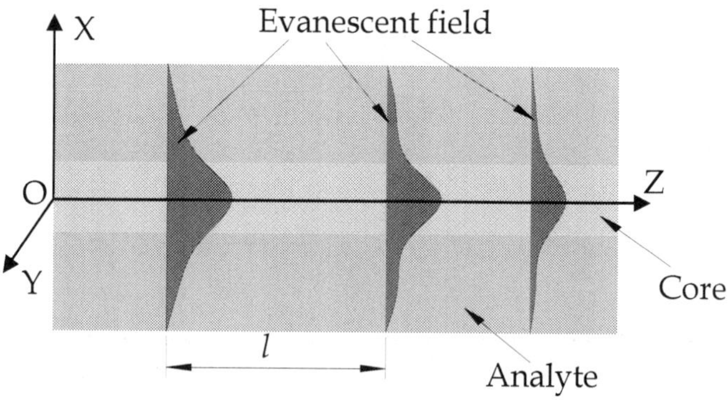

Fig. 4. Schematic of the evanescent field energy distribution along the sensing fiber

So η_j keeps constant along the sensing fiber's axis. As l grows, the evanescent field energy carried by the sensing fiber of unit length, which interacts with the analyte, decreases. That is to say the contribution per unit length of the sensing fiber to A decreases as l increased. So the linear relationship between A, M and L does not exist.

Consider a sensing fiber whose refractive index of the core is 1.4682 and the refractive index of the detecting material is 1.3334. The wavelength of the light source is 632.8nm, the sensing fiber is 14mm in length, 8μm in diameter, and its η_{120} and η_{200} are 0.0643 and 0.1104, respectively. Assume the energy carried by these two modes is both 1W, and the energy carried by the evanescent field is totally absorbed by the analyte. The calculated results of P_{cl} on different locations of the sensing fiber are shown in Fig.5. For η_{200}, the ratio of P_{cl} on the end of the sensing fiber to that on the beginning is just 19.47%, and it is 39.50%

for η_{120}. As l grows, the value of P_{cl} tends to be small. P_{cl} decreases much faster, when the order of the waveguide is higher. It indicates that the contribution per unit length of the sensing fiber to A becomes small as l grows, especially the ones close to the end of the sensing fiber. This is the essential reason why A does not increase linearly with l.

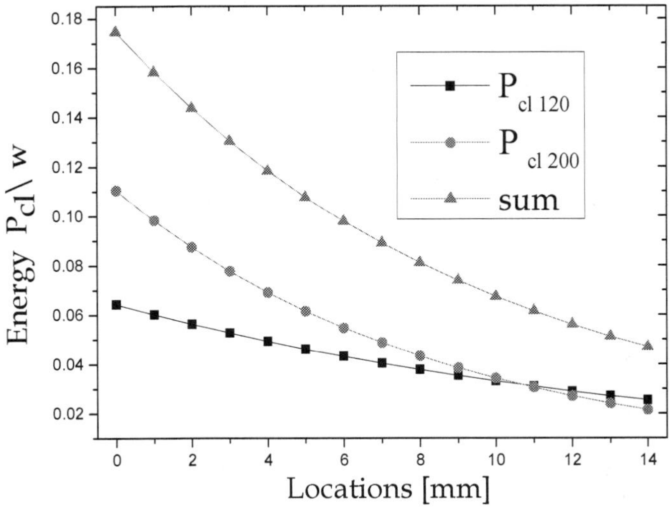

Fig. 5. P_{cl} on different locations of the sensing fiber

2.4 New method to estimate the sensitivity of the OFEFS

Considered that only P_{cl} interacts with the analyte and η is constant along the fiber axis, a new method to estimate the output power of OFEFS is proposed. In this method, the sensing fiber is evenly divided into n fractions with each part to be Δl in length as shown in Fig.6. As P_{cl} transporting along the sensing fiber, it is attenuating following Beer's law. Because η is constant along the whole sensing fiber, there is power replenishment to P_{cl} from the fiber core simultaneously. For each fraction, Δl is very small, so it can be assumed that the energy replenishment occurred at the end of Δl. The total energy left will be redistributed according to η at the beginning of the next Δl. The process repeats until the calculations on all the fractions have finished. As n trends to be infinite, the calculated result trends to be the real value. In this way the two most important factors, that is only the cladding power interacts with the analyte and η is constant along the fiber, are embodied significantly.

Now consider the power transmitting in the jth mode between the kth and the $(k+1)th$ parts of the sensing fiber. $P_{clj(k)}$ is the initial power transmitting in the cladding of the fiber of the kth Δl, and $P_{coj(k)}$ is the initial power in the core. $P'_{clj(k)}$ is the residual energy of $P_{clj(k)}$ after passing Δl. According to Beer's law

$$P'_{clj(k)} = P_{clj(k)} \exp(-\alpha\Delta l) = (r_j P_{in} \eta_j)_k \exp(-\alpha\Delta l) \tag{18}$$

Where $r_j P_{in} \eta_j$ means the power flowing in the cladding i.e. the power interacts with the analyte, and the subscript k means the kth Δl, hence $P_{clj(k)}$ stands for the cladding power in the kth Δl. Here η does not work as the index of the EXP function as in Eq.(8) but it is multiplied by $r_j P_{in}$ to form the power that interacts with the analyte. At the next Δl, the cladding power $P_{clj(k+1)} = \eta_j (P'_{clj(k)} + P_{coj(k)})$ is redistributed according to Eq. (10). Then the next calculation cycle begins.

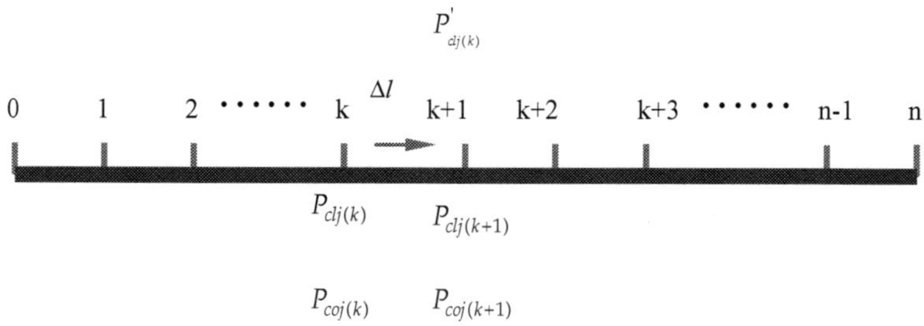

Fig. 6. Schematic diagram of the sensing fiber divided into n parts and the energy distribution

The absorbed energy $\Delta P_{clj(k)}$ caused by the analyte can be written by

$$\Delta P_{clj(k)} = P'_{clj(k)} - P_{clj(k)} = (\exp(-\alpha \Delta l) - 1) P_{clj(k)} \tag{19}$$

When $n \to \infty$, $\Delta P_{clj(k)} \to dP_{clj}$ and $\Delta l \to dl$, Eq. (19) becomes

$$\frac{dP_{clj}}{P_{clj}} = \exp(-\alpha dl) - 1 \tag{20}$$

Together with the initial condition, Eq.(20) and Eq.(10) form the mathematical model of OFEFS.

$$\begin{cases} \dfrac{dP_{clj}}{P_{clj}} = \exp(-\alpha dl) - 1 \\ \dfrac{P_{clj}}{P_{clj} + P_{coj}} = \eta_j \\ P_{clj(0)} + P_{coj(0)} = r_j P_{in} \end{cases} \tag{21}$$

Where P_{in} is the initial input power of the OFEFS. $P_{clj(0)}$ and $P_{coj(0)}$ are the initial values of P_{clj} and P_{coj}, respectively.

It is difficult to get an analytic solution of Eq.(21), but more convenient to solve it by numerical integration. Here we prefer to solve this question using recursion. The output power of the sensor after passing the whole sensing fiber can be expressed as

$$P_{out} = \frac{\sum_{j=1}^{N} P_{totj} - \sum_{j=1}^{N}\sum_{k=0}^{n} \Delta P_{clj(k)}}{\sum_{j=1}^{N} P_{totj}} P_{in} \tag{22}$$

where $P_{totj} = r_j P_{in}$ is the initial energy of the jth mode.

In order to facilitate calculation and obtain an estimated value of good accuracy, the convergence condition of Eq.(22) is set as

$$\begin{cases} \dfrac{(P_{out}/P_{in})_{n1} - (P_{out}/P_{in})_{n2}}{(P_{out}/P_{in})_{n1}} < 0.1\% \\ n_2 - n_1 > 1000 \end{cases} \tag{23}$$

where n_1 and n_2 are the iterative times of the recursion.

3. Experiment

3.1 Preparation of the sensing fiber

Optical fiber evanescent field sensors with acicular encapsulation are fabricated using MEMS silicon photolithography technology and silica wet-etching technology. This type of OFEFS is small, consuming little reagent, suffering from less deformation during the detection process and protecting the sensors from pollution. The sensors' sample cell consists of a cap and a bottom. The characters of the cap and the bottom are illustrated in Fig. 7.

The sizes of the cap and the bottom are both 1mm×20mm. There are two holes fabricated in the cap, which are the channels for the sample flow in and the waster drain out. The fiber slit in the bottom is used to fix the sensing fiber, which will facilitate the process of encapsulation greatly. The slit is 2.5mm in length, 250μm in width and 150μm in depth. The work of the cone that manufactured in the bottom is to guide the sample/de-ionized water into or out of the sample pool. The island stand in the sample pool is used to support the sensing fiber. The function of the slit decorated in the island is to fix the sensing fiber, which can avoid the sensing fiber from shaking. The slit is 140μm in width and 100μm in depth. The slit is 50μm shallower than the fiber slit. The fiber with cladding is thicker than the fiber whose cladding is removed. If these slits are made with the same depth, the sensing fiber will warp, the repeatability of the sensor will be poor and the results will be distorted. So using slits with different depth to support different parts of the fiber can help the sensing fiber keep straight in the sample pool. Sample pool is the place where the analyte interacts with the evanescent field. The width of the sample pool is 200μm, and 150μm in depth. The length of the pool can be chosen discretionary according to experimental requirements. When the length of the pool is 7mm, the total reagent the sensor consumed is only 0.21μl. Both the cap and the bottom are fabricated based on MEMS technologies.

In order to obtain high sensitivities, the radius of the sensing fiber is usually very small, such as 4μm. So the strength of the sensing fiber is weak, and the sensing fiber is fragile and

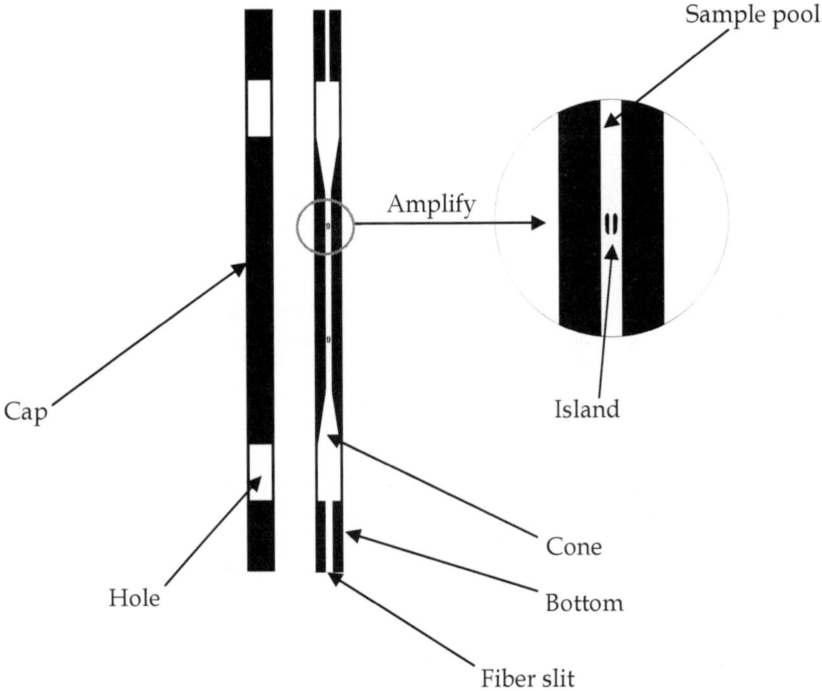

Fig. 7. The structure of the cap and the bottom

easy to break. In order to solve this problem, the fiber is fixed to the bottom first. Then, the cap is bound to the bottom to form the sensor's cell using glue. The construction of the cell is illustrated in Fig.8.

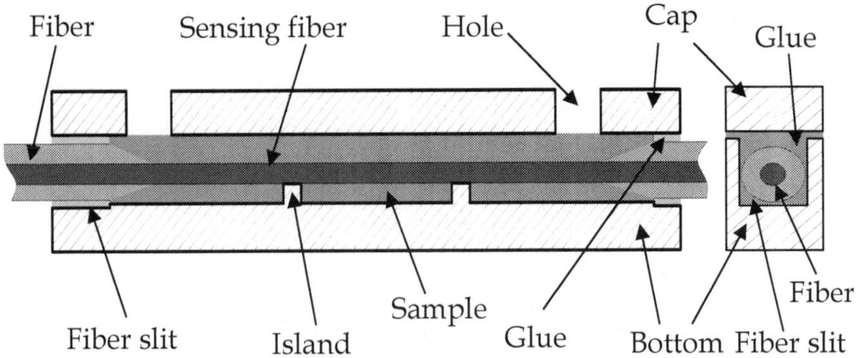

Fig. 8. The schematic structure of the sample cell

The fiber cladding is removed using wet etching, the corrosion solution consists of HF, de-ionized water and CH_3COOH. The function of HF is to etch silica with the well-known reactions:

$$4HF+SiO_2 \rightarrow SiF_4+2H_2O$$
$$6HF+SiO_2 \rightarrow H_2SiF_6+2H_2O \quad (24)$$

By adding CH_3COOH as a buffer, the etching process becomes much more gently and the quality of the core surface would be improved greatly.

The velocity of etching is mainly influenced by the parameters list as follows:

1. the concentration of HF. The higher the concentration of the HF in the corrosion solution, the more F^- reacts with silica, the higher the speed of the etching is;
2. the microstructure of silica. Because the silica used in cladding and core is doped with different elements, the microstructures of them are inconsistent, thus the etching speeds of them are different. A compact structure of silica indicates a low etching speed.
3. the temperature. The chemical reaction equations shown in Eq.(24) are endothermic reactions. When the temperature is high, the reaction between HF and the silica is exquisite and the etching speed is high.

As the reaction goes on, the concentration of F^- in the corrosion solution decreases gradually, and the etching speed slows down. However, the radius of the fiber becomes small, which indicates the area-volume ratio of the fiber becomes bigger. In this case, the speed of the etching will increase. So it's really difficult to control the speed of the etching speed carefully. Here, two methods to monitor the fiber's radius are proposed.

1. by monitoring the output power of the fibers. As shown in Fig.9, a powermeter is used to monitor the transmitted power which is a function of the core diameter. Fig.10 shows the relationship between the output power and the etching time. The output power decreases slightly as the cladding is etched by the HF. When the cladding is eaten up and the core is etched by the etching solution, the strength of the output power decreases dramatically. When the core is eaten up, nothing is left in the sample cell, the relative strength of the output power keeps constant, which is the strength of the background noise. The volume ration among HF, de-ionized water, CH_3COOH in the corrosion solution used here is 1.3 : 1 : 1. According to the relative strength of the output power the radii of the fibers can be estimated accurately.
2. based on the experimental experience. According to the experimental experience, the etching speeds of corrosion solutions of different proportions can be approximately estimated. First, using corrosion solutions with plentiful HF to etch the cladding. When cladding is to be eaten up, solutions with little HF should be chosen to reduce the etching speed and improve the quality of the sensing fibers' surface. During this process, the radius of the fiber should be checked using microscope frequently. When appropriate radius is obtained, the etching process is terminated. In room temperature, as the volume ratio among HF, de-ionized water, CH_3COOH in the solution is 1 : 1 : 1, the etching speed of the cladding is 20μm per hour, and 30μm per hour of the core.

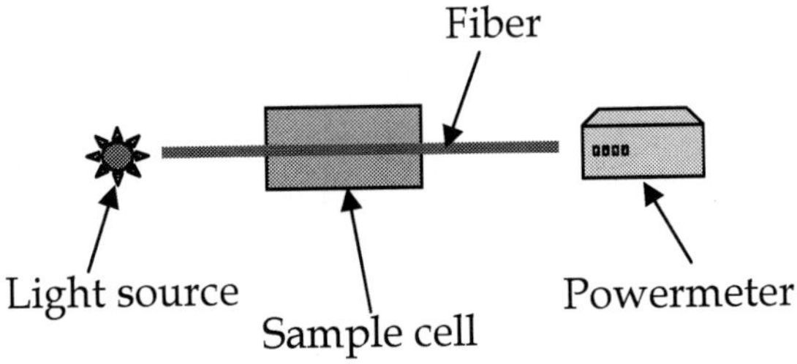

Fig. 9. Experimental set-up of the etching process

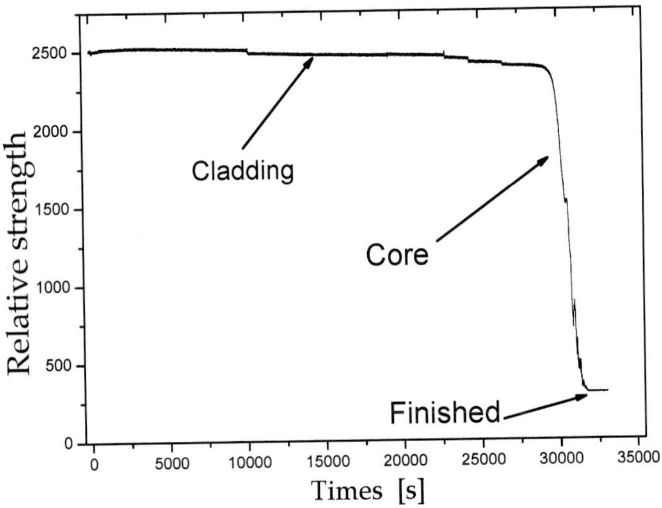

Fig. 10. The relationship between the output power and the etching time

Fig. 11 is the sensing fiber fabricated using the method described above. A cone is obviously observed in the end of the sensing fiber. As shown in Fig.12, when HF erodes the fiber towards the core along the radial direction, cauterization occurs in the fibers' axial direction simultaneously. This is the main reason why the sensing fibers' end is tapered.

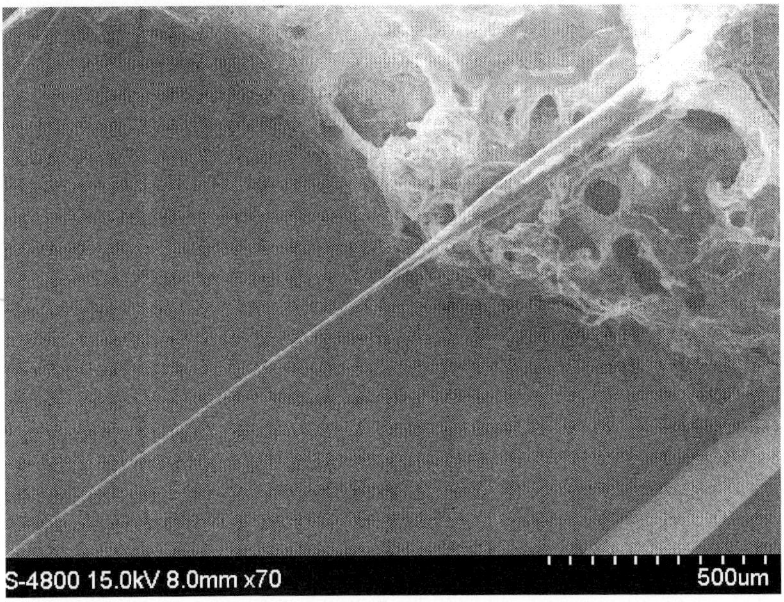

Fig. 11. SEM of the sensing fiber

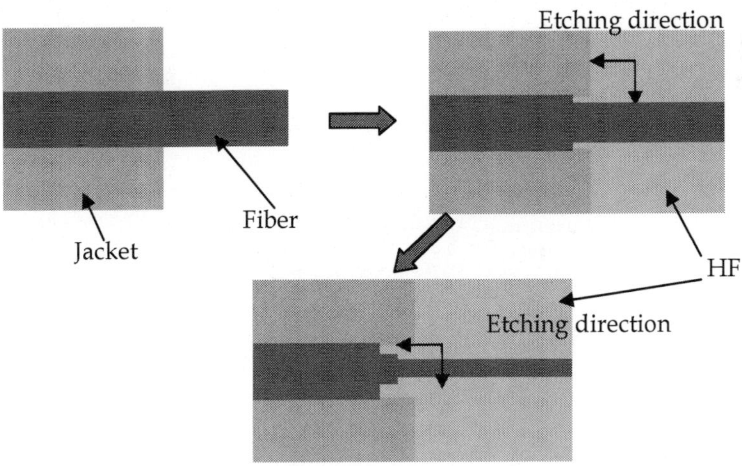

Fig. 12. Schematic of the etching directions of the HF

When the etching terminated, the core of the fiber is washed by de-ionized water and baked to release the remaining liquid. Then the sample cell with the sensing fiber can be used as an optical fiber evanescent field sensor directly, as shown in Fig. 13.

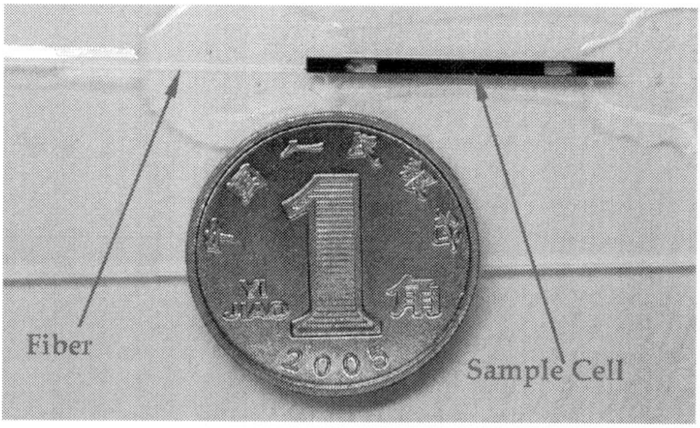

Fig. 13. The optical fiber evanescent field sensor used in the experiments

3.2 Experimental set-up

The experimental set-up is shown in Fig.14. The optics source used here is He-Ne laser (Type 260A，632.8nm). Its minimum output power is 1.8 mW, the maximum drift of it is ±3%, the beam diameter is 2mm and the beam divergence is 1.5 mrad. As shown in Fig.14, the laser is focused and collimated using a couple of lens, then the light is injected into the fiber and the output power is measured by a spectrometer. After the spectrum is analyzed using a computer, the information of the analytes' concentrations is obtained. After the detection, the sensing fiber is refreshed by de-ionized water bath, and the sensing fiber can be used repeatedly.

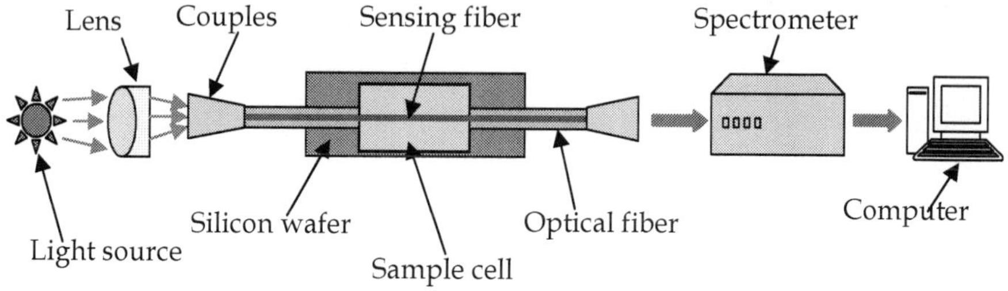

Fig. 14. The experimental set-up

The absorption medium used for experiments is methylene blue (MB) dissolved in de-ionized water and the absorbing peak is 664 nm. Fig.15 is the absorbency spectrum of the methylene blue in visible region. The concentration of the methylene blue used here is 1×10^{-6} mol/L.

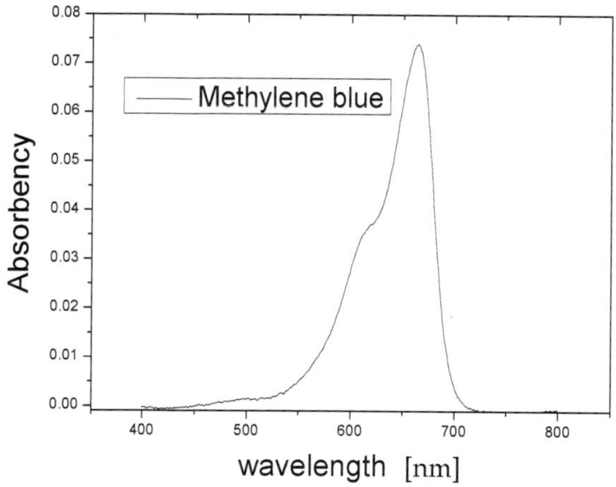

Fig. 15. Absorption spectrum of methylene blue solution in de-ionized water

3.3 Nonlinear relationship between absorbency and the length of the sensing fiber

Fig.16 shows the experimental results of A. In the experiments two sensors with different L, 7mm and 14mm, are used. The radii of the sensing fibers are both 4μm, and the refractive indexes are both 1.4682. As shown in Fig.16, the ratio of the sensor's absorbency with $L=14mm$ to that with $L=7mm$ is smaller than two. It shows clearly the sensing efficiency declines as L increases. The contribution per unit length of the sensing fiber to A drops as L increased. When the concentration of MB grows, the ratio increases. This phenomenon is mainly blamed on the sensing fibers' blemish. When the concentration of the analyte is high, the effects of the sensing fibers' disfigurations on A become obvious. The most common defects of the sensing fibers are displayed in Fig.17. When constructing optical fiber evanescent field sensors, it is not advised to use long sensing fibers because the sensing efficiency of the sensing fiber per unit length decreases as the sensing fiber becomes long. Also, long sensing fibers are difficult to fabricate and easy to disfigure and break.

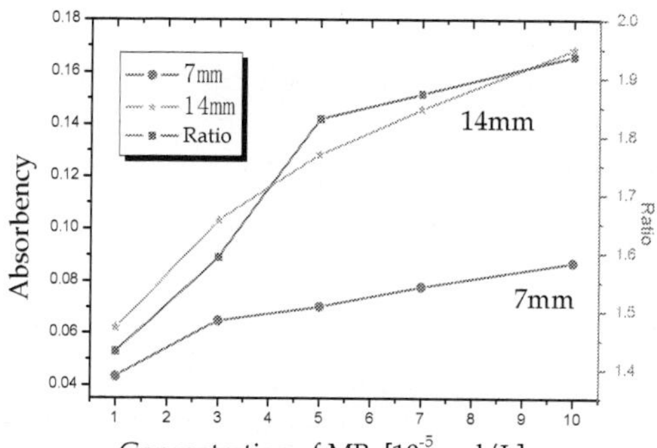

Fig. 16. The experimental absorbency of the sensors with different L

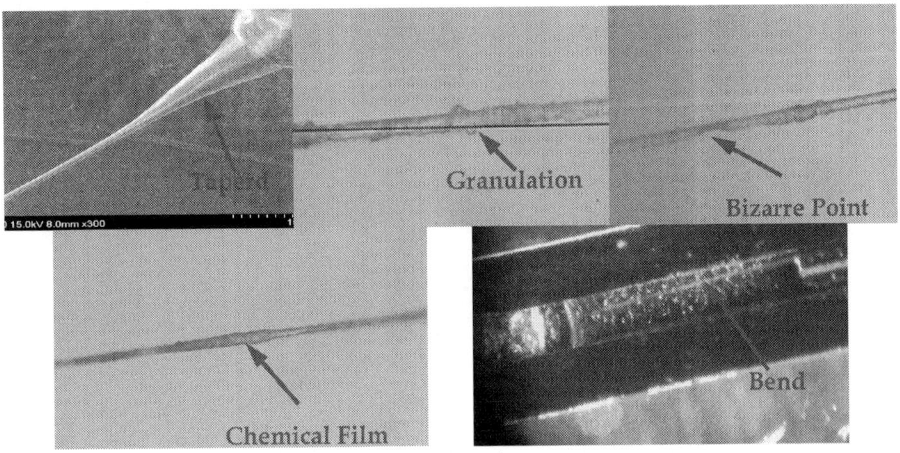

Fig. 17. Defects of the sensing fibers

3.4 The new method to estimate the sensitivity of the sensor

Fig.18a is the picture of the sensor used in the experiments to confirm the accuracy of the novel method proposed here. Fig.18b is the amplificatory image of the sensing fiber. In order to obtain a sensing fiber of high quality and eliminate the deformation caused in the process of encapsulation, the diameter of the sensing fiber used here is 100μm. Sensing fibers with small radii are easy to disfigure and disturb the experimental results. The error of the experimental results can be reduced by using sensing fibers with large radii.

The length of the sensing fiber is 14mm, the refractive index of the fiber core is 1.4577. The wavelength of optic source is 632.8nm. In experiment, the analyte is methylene blue, and its refractive index is adjusted to 1.4550 using glycerol with molar absorption coefficient $27200L / mol \bullet cm$. The diameter of the sensing fiber is etched to 100μm using hydrofluoric acid.

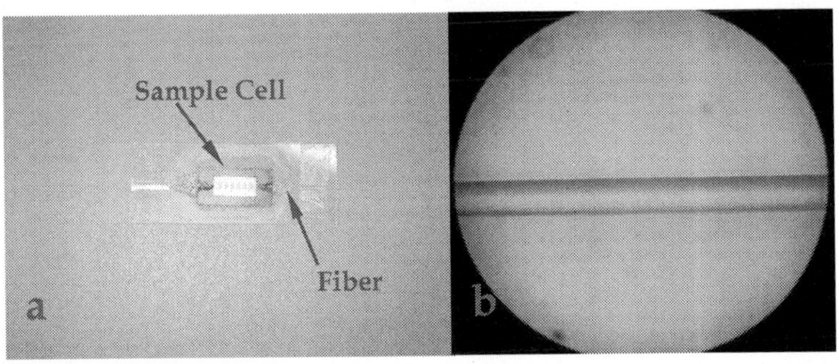

Fig. 18. Sensing fiber used in the experiments

P_{out} / P_{in} against the concentrations of MB obtained from experiments and different calculation schemes are shown in Fig. 19. The curve *P* represents the method used before. The curves $n = 140$ and $n = 1400$ are the results which obtained using the method proposed here with iterative times 140 and 1400. As shown in Fig.19, the values calculated by our method are much closer to the experimental results, especially when P_{out} / P_{in} is high.

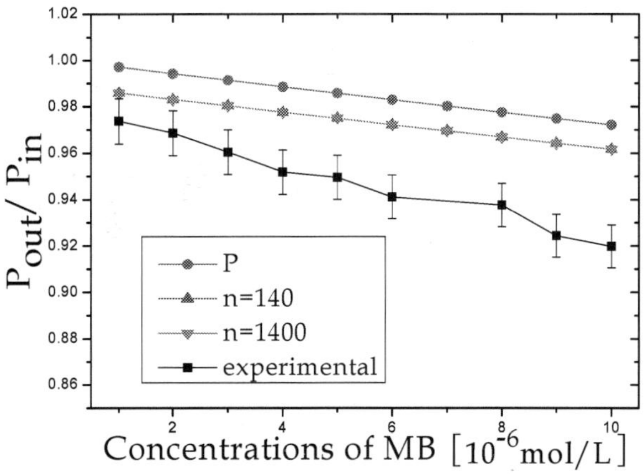

Fig. 19. P_{out} / P_{in} versus the concentrations of MB

From Fig. 20 it is found when $P_{out} / P_{in} > 0.7$ the results obtained using the new method are in a good agreement with the experimental results and more accurate than the results obtained from the methods used before. When the concentration of MB is high, the evanescent field energy absorbed by the analyte is huge. So a small *dl* should be used to get a more accurate emulational result, which indicates a big n should be chosen, as shown in Fig. 20.

However, when $P_{out} / P_{in} \leq 0.7$, a difference between experimental and theoretical values is found. Various reasons have been assigned to this discrepancy. It is mainly caused by the deposition of the methylene blue molecules on the sensing fiber's surface and the disfigurement of the sensing fiber (Ruddy & Maccrait, 1990 ;Potyrailo & Hobbs, 1998; Zhuang et al., 2009b).The effects of the sensing fibers' disfigurements on the results are listed as follow: (1) the surface of the etched core is not perfect and scattering occurs on the interface between the analyte and the sensing fiber, which will decrease the output power of the sensor in experiment; (2) the fiber diameter is unsymmetrical and there are errors in the measurement, which will introduce mistakes to the simulation; (3) the bending loss increases the energy loss in the sensing region, and the values of P_{out} / P_{in} is depressed; (4) when the unclad optical fiber etched, taper will occur in the two ends of the sensing region as described in section 3.1. When light passes through the taper, energy loss happened because of the discontinuity of the modes supported by these two different parts of the fiber and the experimental results will be enhanced (Ahmad & Hench, 2005). As the concentration of MB is high, more molecules will be deposited on the sensing fibers' surface, the absorbency of the sensor will be improved and the discrepancy of the theoretical results from the experimental ones will be enlarged as shown in Fig.20.

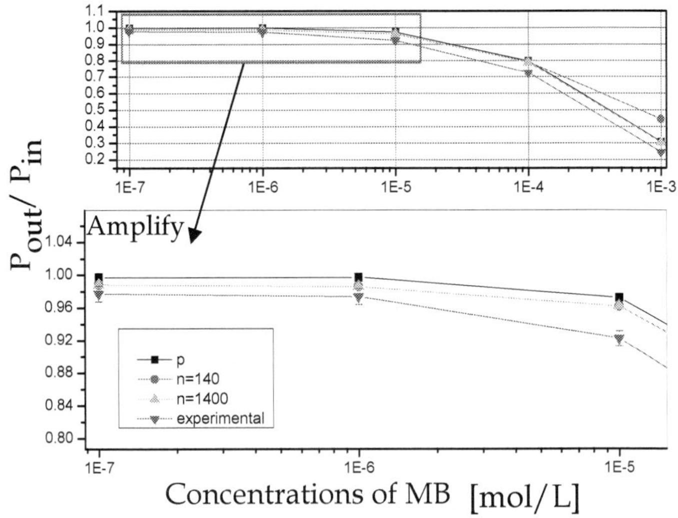

Fig. 20. P_{out} / P_{in} versus the concentrations of MB

4. Conclusion

In this chapter, we have proposed and demonstrated a novel method to estimate the sensitivity of optical fiber evanescent field sensors. In section 1, the importance of constructing mathematical method to estimate the sensitivity of the sensor is described. And the brief introduction of this chapter is also presented in section 1. In section 2, the operation principle of OFEFS is theoretically studied. The effects of the sensing fibers' length, l, on the sensors' absorbency, A, are analyzed. The nonlinear relationship between A and l is also explained in section 2.2. It is found that the contribution per unit length of the sensing fiber to A decreases as l grows. So it does not suggest using long sensing fibers to construct OFEFS. The new method to estimate the sensitivity of OFEFS is proposed in section 2.3. The operation principle of OFEFS is embodied significantly by the mathematical model of the method.

In section 3, the experimental set-up of the sensor is built. The fabrication of the sensor with acicular encapsulation is introduced, and the factors which effect the preparation of the sensing fibers are discussed. The nonlinear relationship between A and l is confirmed with experiments. The new method is prove to be in a good agreement with the experimental results and is more accurate than the methods used before. The divergence of the theoretically results from the experimental ones is explained.

The practical importance of this study lies in helping for understanding, design, and construction of optical fiber evanescent field sensors with high sensitivity.

5. Acknowledgements

The authors acknowledge the supports from West Light Foundation of the Chinese Academy of Sciences (A11K011) and National Natural Science Foundation of China

(60978051). The experiments were done in the national key lab of applied optics of China. The authors would like to thank Professor Yihui Wu for her benefited discussions and helps.

6. References

Ahmad M., Hench L. L., (2005). Effect of Taper Geometries and Launch Angle on Evanescent Wave Penetration Depth in Optical Fibers, *Biosensors and Bioelectronics*, Vol.(20), (2005), pp 1312-1319, ISSN0956-5663

Angela, L., Shankar P. M., Mutharasan R., (2007). A Review of Fiber-optic Biosensors. *Sensors and Actuators B: Chemical*, Vol.125, No.2,(2007), pp 688-703,ISSN 0925-4005

Deng X. H., Wu Y. H., Zhou L. Q., Xu M., (2006). Parameters Analysis of Few-modes Optical Fiber Evanescent Absorption Sensor, *SPIE*, Vol.6032, (2006), pp 603208, ISSN 0277-786X

Gloge D., (1971). Weakly Guiding Fibers, *Appied Optics*, Vol.10, No.10, (1971), pp 2252-2258, ISSN 0003-6935

Guo S., Albin S., (2006). Transmission Property and Evanescent Wave Absorption of Cladded Multimode Fiber Tapers, *Opt. Express*, Vol.11, No.3, (2006), pp 215-223, ISSN 2156-7085

Gupta B. D., Singh C. D., (1994). Fiber-optic Evanescent Field Absorption Sensor: A Theoretical Evaluation, *Fiber and Integrated Optics*, Vol.3, No.4, (1994), pp 433-443, ISSN 0146-8030

Maria E. B., Antonio J. R. S., Fuensanta S. R.,Catalina B. O., (2007). Recent Development in Optical Fiber Biosensors. *Sensors*, Vol.6, No.7,(June 2007), pp 797-859, ISSN 1424-8220

Messica A., Greenstein A., Katzir A., (1996). Theory of Fiber-optic, Evanescent-wave Spectroscopy and Sensors, *Appied Optics*, Vol.35, No.13, (1996), pp 2274-2284, ISBN 0003-6935

Payne F. P., Hale Z. M., (1993). Deviation from Beer's Law in Multimode Optical Fibre Evanescent Field Sensors, *International Journal of Optoelectronics*, Vol.8, (1993), pp 743-748, ISSN 0952-5432

Potyrailo R. A., Hobbs S. E., (1998). Near-ultraviolet Evanescent-wave Absorption Sensor Based on A Multimode Optical Fiber, *Anal. Chem.*, Vol.70, No.8, (1998), pp 1639-1645, ISSN 0003-2700

Ruddy V., Maccrait B. D., (1990). Evanescent Wave Absorption Spectroscopy Using Multimode Fibers, *J.Appl.Phys.* Vol.67, No.10, (1990),pp 6070-6074, ISSN 0021-8979

Wolfbeis O. S., (2006). Fiber-optic Chemical Sensors and Biosensor, *Anal. Chem.*, Vol.78, No.12, (2006), pp 3859-3874, ISSN 0003-2700

Wu Y. H., Deng X. H., Li F., Zhuang X. Y., (2007). Less-mode Optic Fiber Evanescent Wave Absorbing Sensor: Parameter Design for High Sensitivity Liquid Detection, *Sens. Actuators B*, Vol.122, (2007), pp 127-133,ISSN 0925-4005

Yu X., Cottendena A., Jones N. B., (2006). A Theoretical Evaluation of Fibre-optic Evanescent Wave Absorption in Spectroscopy and Sensors. *Optics and Lasers in Engineering*, Vol. 44, (2006), pp 93-101, ISBN 0143-8166

Zhuang X. Y., Wu Y. H., Wang S. R., Zhang P., Liu S. Y., (2009a). Research on the Fiber-optic Evanescent Field Sensor Based on Microfabrication and the Effect of Fiber Length on Its Properties, *Acta Physica Sinca*, Vol.58, No.4, (May 2009), pp 2501-2506, ISSN 1000-3290

Zhuang X. Y., (2009b), Study on the Art of Fiber-optic Evanescent Field Sensor with High Sensitivity, (July 2009), PhD paper of Chinese Academy of Sciences, Beijing, China.

8

Tapered Optical Fibers – An Investigative Approach to the Helical and Liquid Crystal Types

P. K. Choudhury
Institute of Microengineering & Nanoelectronics (IMEN)
Universiti Kebangsaan Malaysia, Bangi, Selangor
Malaysia

1. Introduction

Optical waveguides find multifarious applications in integrated optical devices as well as various forms of communication systems. In order to get a good understanding of the optical features of waveguides, the basic requirement is to analyze the propagation characteristics of electromagnetic (EM) waves through the guide. In this context, the study of EM wave propagation through various types of optical fibers and waveguides has been greatly dealt with, which can be witnessed through the vast research reports by the relevant R&D community, as appeared in the literature. In this context, several forms of optical waveguides implementing different composite materials [1-6] and/or new forms of geometrical cross-sections [7–13] have been developed most of which are now widely recognized for their use in the fabrication of integrated optical circuits and in the laser beam technology. These include unconventional types of waveguide geometries too [3,13].

The study of EM wave propagation through conventional optical fibers does not involve much difficulty. Among the various types of fiber structures, fibers with tapers and flares have received considerable attention over the years because of their promising applicability in the area of fused fiber couplers [14–30]. These tapered optical fibers (TOFs) are of practical interest for the fabrication of *all-optical* components as these miniaturized optical components find enormous potential applications in the areas related to optical sensors as well as other in-line integrated optic applications where directional couplers and beam expanders are being used [31–34].

It is noteworthy that dielectric TOFs can be used with much efficiency in evanescent wave fiber-optic absorption sensors. This way, TOFs become one of the most promising means in high-speed transmission systems. Optically, the taper transitions transform the local fundamental mode from a core mode in the untapered fiber to a cladding mode in the taper waist - the basis of many of its *all-optical* applications including multiplexing and modulation. It is desirable for the transition to be as short as possible, allowing the resulting component to be compact and insensitive to the degradation of the environment.

Treatment of TOFs presents formidable mathematical difficulties, and therefore, a recourse has to be taken to a variety of approximate methods. In the present article, we make a review of the different types of reported TOFs. Starting with the very fundamental form of TOF, which is the dielectric TOF, we venture into the investigations of their other complicated forms, viz. helical clad TOFs (or TOFs with twists) and the liquid crystal TOFs.

Fundamental studies on tapered guides include emphasis on optical fiber up-tapers for single-mode hardware applications and laser transmission evaluation. Among the various techniques, the implementation of split-step method would be one of the potential ways for the analytical treatment of a TOF, wherein the guide itself can be viewed in the analogy with a stack of large number of fibers of increasing (or decreasing) cross-sectional dimension with an end-to-end contact. The present article discusses the formulation of TOFs with regard to the various possible modes as it falls in the region of interest where one would require a suitable fiber design. In this context, the study of dispersion characteristics of fibers/waveguides has importance in the determination of their performance. Apart from that, the propagation of power also plays the determining role in selecting the suitability of the guide for specific applications.

Helical structures find useful applications in all low- and medium-power traveling wave tubes, the analyses of which generally include waveguides under slow-wave structures with conducting sheath and tape helixes [35]. Such a concept can be implemented in the case of optical fibers taking into consideration the case of sheath helix windings [36]. A closer look at the dispersion behavior of such TOFs would be rather interesting as the helix pitch angle itself plays a major role to control the dispersion behavior of the guide. The present article covers all these issues corresponding to particular values of the helix pitch angle - windings perpendicular and parallel to the direction of EM wave propagation. Under the assumption of a small variation of the core radius in the longitudinal direction, it is demonstrated that the helix pitch greatly affects the propagation through the guide, making thereby the angle of pitch as one of the dominant controlling parameters, so far as the fiber design is considered. This is further verified through the comparison of results with those obtained for dielectric TOFs without helical windings.

Liquid crystal fibers (LCFs) exhibit polarization anisotropy, which make them of much technological interest and useful for many optical applications [5,37–39]. It is interesting to note that the macroscopic optical properties of liquid crystals can be manipulated by suitably applying the external electrical fields [40]. This feature essentially remains of promising use in optical sensing. Radial and azimuthal would be two different types of anisotropy, which LCFs may possess. The present article also incorporates a section that reports the study of an amalgamation of the taper structure and the clad anisotropy; it is demonstrated that the LCF structure supports enhanced power in the outer clad region - a much desirable characteristics for optical sensing and field coupling devices.

The arrangement of the present article is made in this way - the entire volume is divided into five different sections. Apart from the ongoing Section 1, we described the dielectric TOFs in Section 2, where the attempts are made to discuss the dispersion relations of dielectric TOFs along with the sustained transmission of power. In Section 3, twisted TOFs are described where the twists are introduced through the helical wraps at the core/clad interface. In this section, the effects on the dispersion relations due to the angle of twists are emphasized. Section 4 incorporates the study of LCFs where the effects of introducing a radially anisotropic layer at the outermost TOF region are investigated. The transmission of power through such LCFs is also touched upon. A concluding remark is presented in Section 5 that wraps the themes of the results obtained corresponding to the different types of TOFs.

2. Dielectric tapered fibers

The analytical treatment of TOFs remains much complicated. In order to deal with the problem, among the available techniques, the split-step method is the one which can be utilized with much accuracy. In this method, a TOF can be viewed in the analogy with a stack of large number of tiny sections of fibers with increasing (or decreasing) cross-sectional dimension with an end-to-end arrangement.

Fundamental studies of TOFs have been appeared in the literature [31–34]. However, the dispersion characteristics of TOFs are less discussed. In this section, we present the formulation of TOFs in respect of possible sustained modes as it remains important to work on the fiber design based on the requirements. Further, the study of dispersion characteristics of fibers/waveguides is much important in the determination of their performance. In the following sub-sections, using Maxwell's field equations, we make an attempt to present the dispersion characteristics and cutoff situations of a dielectric TOF with the special mention of the dispersion curves corresponding to different modes. The features of such TOFs in respect of the transmission of power are also described emphasizing the power confinement factor of the guide.

2.1 Theoretical development

To deduce the exact field solutions for dielectric TOFs, the full set of Maxwell's equations resulting in the vector-wave equation for the system must be implemented. Fig. 1 illustrates the transverse view of a step-index dielectric TOF where the core/clad regions have refractive index (RI) values as n_1 and n_2, respectively. For simplicity, we consider the clad section as infinitely extended. We use the cylindrical polar coordinate system (ρ, ϕ, z) for the analysis with the z-axis as the direction of propagation. We consider fiber with a linearly tapered core having varying taper radius ρ represented by the function

$$\rho = \begin{cases} a_0, & z \le z_1 \\ a_0\left[1+(\alpha-1)\sin^2\left\{\dfrac{\pi}{2}\left(\dfrac{z-z_1}{z_2-z_1}\right)\right\}\right], & z_1 < z < z_2 \\ \alpha a_0, & z \ge z_2 \end{cases} \tag{1}$$

In eq. (1) α is the overall geometrical magnification factor, a_0 is the fiber core radius at the input end and (z_2 - z_1) is the length of the taper [41]. Eq. (1) represents the function to determine the size of the varying radius along the tapered core. The function ρ and its derivative remain continuous at $z = z_1$ and $z = z_2$. The region $z < z_1$ corresponds to the fiber pigtail whereas $z > z_2$ to the expanded cylindrical section (straight tip) following the taper. For the analytical treatment, we follow the method reported earlier by Amitay *et al.* [41], i.e. $z_1 = 0$.

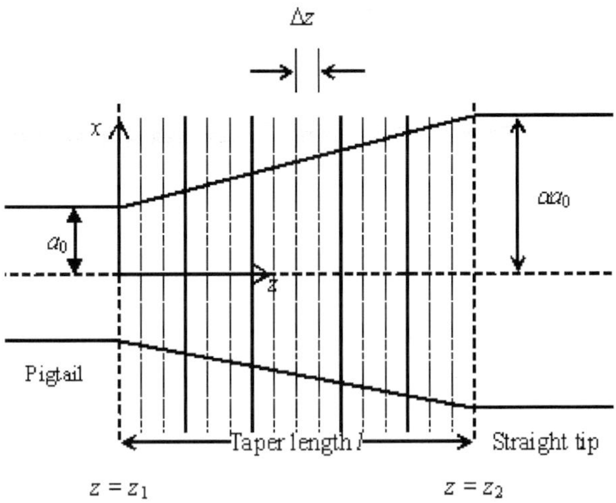

Fig. 1. Transverse view of dielectric TOF.

In another way, the linearly tapered geometrical profile of the fiber core may also be defined by

$$\rho(z) = \rho_i - \frac{z}{l}(\rho_i - \rho_o), \tag{2}$$

where $\rho = \rho_i$ and $\rho = \rho_0$ represent the radius of the input and the output ends, respectively, of the tapered fiber, and l is the length of the taper.

We assume that the core/clad regions are made of linear, homogeneous, isotropic and non-magnetic (i.e. $\mu_1 = \mu_2 \cong \mu_0$, the free-space permeability) materials, and they have the permittivity as ε_1 and ε_2, respectively. The time t-harmonic and the axis z-harmonic electric and magnetic fields will have their functional dependences of the form

$$\mathbf{E} = \mathbf{E}_0\,(\rho)\, e^{j\nu\phi}\, e^{j(\omega t - \beta z)}, \tag{3a}$$

$$\mathbf{H} = \mathbf{H}_0\,(\rho)\, e^{j\nu\phi}\, e^{j(\omega t - \beta z)} \tag{3b}$$

with ω as the angular frequency, β as the propagation constant in the axial direction and ν as the azimuthal periodicity. In the present case, β is no longer a constant owing to the gradual variation of the cross-sectional dimension with increasing distance z. Instead, it becomes a function of z. Thus, assuming a small variation of the core radius, we can use Taylor series expansion [13] for β so that

$$\beta = \beta_0 + \left(\frac{\partial \beta}{\partial z}\right) z, \tag{4}$$

where the higher order terms are suppressed. In eq. (4) β_0 is the axial component of the propagation vector at the origin $z = 0$.

It can be shown that, using the transverse field components, the system of tapered guide will be represented by the equation [13]

$$\frac{\partial^2\psi}{\partial\rho^2}+\frac{1}{\rho}\frac{\partial\psi}{\partial\rho}+\frac{1}{\rho^2}\frac{\partial^2\psi}{\partial\phi^2}+\eta^2\psi=0\,, \tag{5}$$

where ψ is the field (i.e. E_z or H_z, as the case may be) and $\eta^2 = \omega^2\mu\varepsilon - \beta^2$ with ε and μ ($\cong \mu_0$) as the permittivity and the permeability of the medium, respectively. It can also be shown that eq. (5) will have Bessel functions [42] as the solutions, and therefore, fields in the core/clad sections will be described by Bessel and the modified Bessel functions, and/or their linear combinations. The core region will have the expressions for E_z and H_z as

$$E_z(\rho<\rho_a)=C_1J_\nu(u\rho)e^{j\nu\phi}e^{j(\omega t-\beta z)} \tag{6a}$$

and

$$H_z(\rho<\rho_a)=C_2J_\nu(u\rho)e^{j\nu\phi}e^{j(\omega t-\beta z)}\,, \tag{6b}$$

respectively, whereas the clad section will have those as

$$E_z(\rho>\rho_a)=C_3K_\nu(w\rho)e^{j\nu\phi}e^{j(\omega t-\beta z)}\,, \tag{7a}$$

$$H_z(\rho>\rho_a)=C_4K_\nu(w\rho)e^{j\nu\phi}e^{j(\omega t-\beta z)}\,. \tag{7b}$$

In eqs. (6) and (7), C_1, C_2, C_3 and C_4 are arbitrary constants, the values of which can be determined by using the boundary conditions, and the parameters u and w are defined as

$$u^2=\omega^2\mu_0\varepsilon_1-\beta^2=k_1^2-\beta^2 \tag{8a}$$

and

$$w^2=\beta^2-\omega^2\mu_0\varepsilon_2=\beta^2-k_2^2\,, \tag{8b}$$

respectively. Implementing the longitudinal field components, one can explicitly obtain the transverse field components at an arbitrary radial point $\rho = \rho_a$ as

$$E_{\phi1}=-\frac{j}{u^2}\left\{\frac{j\nu\beta}{\rho_a}C_1J_\nu(u\rho_a)-\mu_0\omega C_2uJ'_\nu(u\rho_a)\right\}e^{j\nu\iota}\,, \tag{9a}$$

$$H_{\phi1}=-\frac{j}{u^2}\left\{\varepsilon_1\omega C_1uJ'_\nu(u\rho_a)+\frac{j\nu\beta}{\rho_a}C_2J_\nu(u\rho_a)\right\}e^{j\nu\iota}\,, \tag{9b}$$

$$E_{\phi2}=\frac{j}{w^2}\left\{\frac{j\nu\beta}{\rho_a}C_3K_\nu(w\rho_a)-\mu_0\omega C_4wK'_\nu(w\rho_a)\right\}e^{j\nu\iota}\,, \tag{10a}$$

$$H_{\phi 2} = \frac{j}{w^2}\left\{ \varepsilon_2 \omega C_3 w K_v'(w\rho_a) + \frac{jv\beta}{\rho_a} C_4 K_v(u\rho_a) \right\} e^{jvt}. \tag{10b}$$

In eqs. (9) and (10), the suffixes 1 and 2, respectively, refer to the situations in the core and the clad sections, and the prime represents the differentiation with respect to the argument. Further, the propagation constant β is essentially governed by eq. (4).

2.2 Dispersion characteristics

According to the continuity conditions, the transverse field components must have a smooth match at the core/clad boundary, which will ultimately yield four equations, the solutions to which would exists only if the determinant formed by the coefficients is zero. This will finally provide the eigenvalue equation in β as [43]

$$(\aleph_v + \wp_v)(k_1^2 \aleph_v + k_2^2 \wp_v) - \left\{ \frac{v}{\rho_a}\left(\frac{1}{u^2} + \frac{1}{w^2} \right) \right\}^2 \left\{ \beta_0 + z\left(\frac{\partial \beta}{\partial z} \right) \right\}^2 = 0 \tag{11}$$

with

$$\aleph_v = \frac{J_v'(u\rho_a)}{uJ_v(u\rho_a)},$$

$$\wp_v = \frac{K_v'(w\rho_a)}{wK_v(w\rho_a)}.$$

Under the approximation of very small RI difference, one would have $k_1^2 \approx k_2^2 \approx \beta^2$, and eq. (11) will acquire the form

$$\aleph_v + \wp_v = \pm \frac{v}{\rho_a}\left(\frac{1}{u^2} + \frac{1}{w^2} \right). \tag{12}$$

Using the recurrence relations for $J_v'(\cdot)$ and $K_v'(\cdot)$, there are two sets of equations for eq. (12) corresponding to the positive and the negative signs. The positive sign will yield

$$\frac{J_{v+1}(u\rho_a)}{uJ_v(u\rho_a)} + \frac{K_{v+1}(w\rho_a)}{wK_v(w\rho_a)} = 0, \tag{13}$$

the solution to which will yield a set of EH modes. Corresponding to the negative sign, one would have

$$\frac{J_{v-1}(u\rho_a)}{uJ_v(u\rho_a)} - \frac{K_{v-1}(w\rho_a)}{wK_v(w\rho_a)} = 0 \tag{14}$$

giving thereby a set of HE modes. Once again, using the recurrence relations, one can match all the tangential field components at the layer interface to obtain

$$\frac{uJ_{m-2}(u\rho_a)}{J_m(u\rho_a)} = -\frac{wK_{m-2}(w\rho_a)}{K_m(w\rho_a)}. \tag{15}$$

Eq. (15) represents the characteristic equation for the linearly polarized (LP) modes, where m is a new parameter defined as

$$m = \begin{cases} 1 & \text{for TE and TM modes} \\ \nu+1 & \text{for EH modes} \\ \nu-1 & \text{for HE modes} \end{cases} \tag{16}$$

However, under the limit $w \to 0$, one can obtain

$$\lim_{w\to 0} \frac{wK_{m-1}(w\rho_a)}{K_m(w\rho_a)} \to 0 \quad \text{for} \quad m = 0, 1, 2, \ldots \tag{17}$$

Under the circumstance, eq. (15) will assume the form

$$\frac{uJ_{m-2}(u\rho_a)}{J_m(u\rho_a)} = 0. \tag{18}$$

The solution to eq. (17) provides the knowledge of the dependence of the normalized propagation constant b on the normalized frequency parameter V [44]. The number of sustained modes in a guide (as a function of V) may be represented in terms of a normalized propagation constant

$$b = \frac{\rho_a^2 w^2}{V^2} = \frac{(\beta / k)^2 - n_2^2}{n_1^2 - n_2^2}. \tag{19}$$

In eq. (19), k is the free-space propagation constant. For a small index difference, eq. (19) reduces to the form

$$b \approx \frac{\left(\frac{\beta}{k}\right) - n_2}{n_1 - n_2}. \tag{20}$$

Under the limiting condition $w^2 \to 0$, the cutoff characteristic equation for the lowest mode ($\nu = 0$) of the fiber can be obtained as

$$J_0(u\rho_a) = 0. \tag{21}$$

If S_n represents the solutions to eq. (21) (with $n = 1, 2, 3$ etc. correspond to first, second, third etc. solutions), we would then have

$$u\rho_a = S_n. \tag{22}$$

One would finally obtain by implementing eqs. (4), (8a) and (21) that

$$z^2\left(\frac{\partial\beta}{\partial z}\right)^2 + 2\beta_0 z\left(\frac{\partial\beta}{\partial z}\right) + \left\{\beta_0^2 - k_1^2 + \left(\frac{S_n}{\rho_a}\right)^2\right\} = 0\,. \tag{23}$$

Eq. (23) is essentially a quadratic equation in $\partial\beta/\partial z$, and it can be shown that, under the weak guidance approximation, the factor S_n in eq. (23) will have its value as 2.405 for the fundamental LP_{11} mode [44]. Further, eq. (23) can be solved for $\partial\beta/\partial z$ numerically for different values of z. For a conventional fiber, the z-component of the propagation constant decreases with the decrease in fiber radius, and therefore, the propagation vector β should also decrease, making thereby a negative $\partial\beta/\partial z$. On the other hand, for increasing fiber radius (with z), $\partial\beta/\partial z$ must be positive. We thus consider the roots of eq. (23) so that those meet the physical situation. Fig. 2 depicts a plot of the variation of $\partial\beta/\partial z$ with z under the assumption of decreasing taper cross-section with the increase in z. We notice that the quantity $\partial\beta/\partial z$ increases rapidly with z near the origin; with the increase in the taper length, it shows a very small increase.

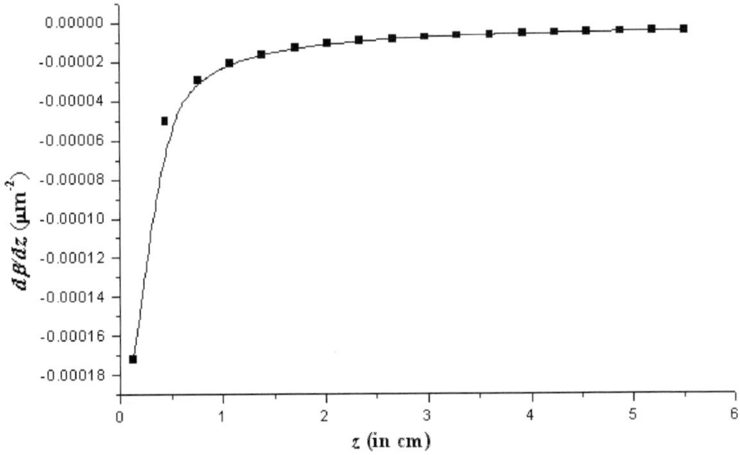

Fig. 2. Variation of $d\beta/dz$ with the longitudinal distance z.

We now make an attempt to analyze the dispersion relation for the TOF. We consider the fiber core as made of polysterene material (with RI n_1 = 1.6) and the clad of methyl methacrylate (RI n_2 = 1.48). Step-index guides made of such materials are good candidates for low cost links. The operating wavelength is kept fixed at 1.55 μm because the attenuation of polysterene materials is relatively low at this wavelength.

Fig. 3 illustrates the variation of the left hand side of eq. (11), as abbreviated as $f(u,w)$, with the dimensionless quantity $u\rho$. We observe the existence of the first intersection of the curve with the $f(u,w)$ = 0 axis near $u\rho$ = 1. Fig. 4 depicts the variation of the left hand side of eq. (17), as abbreviated as $f(u)$, with $u\rho$, which shows the cutoff value to be near $u\rho$ = 0.7 – a value less than that obtained corresponding to fig. 3, as discussed above. As such, the results of figs. 3 and 4 remain consistent.

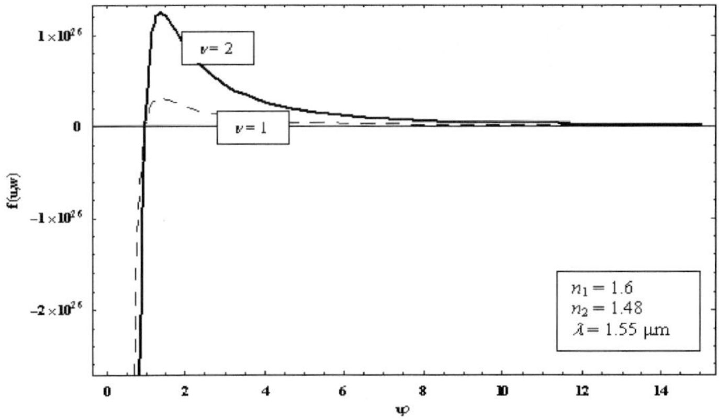

Fig. 3. Variation of the function $f(u, w)$ with $u\rho$.

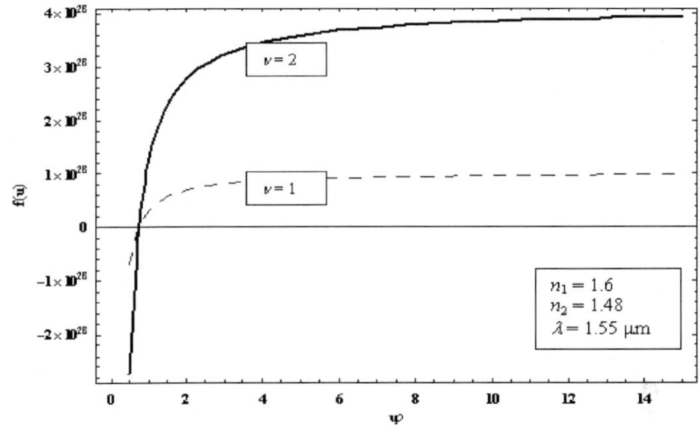

Fig. 4. Variation of the function $f(u)$ with the dimensionless quantity $u\rho$.

Fig. 5 represents the dispersion pattern of the TOF – the plot of b (in terms of β/k) against V – for a few low order modes. We observe each mode to exist only for values of $V < 2$. More explicitly, the cutoff values are crowded together in the region $V = 0.15 - 1.50$. The curves exhibit a tendency towards saturation near $V > 2.5$. Further, there exists another crowding for V-values exceeding 4.5. The modes are cutoff when $\beta/k = n_2$, and in this case, the modes exhibit cut off when $\beta/k = 1.48$. Only five modes (fig. 5) – HE_{11}, TE_{01}, TM_{01}, EH_{11} and HE_{12} – are found to exist in the tapered region of the TOF.

Fig. 6 depicts the variation of the normalized propagation constant b against the normalized frequency parameter V in terms of LP modes; five such modes are found to exist. We notice that, for a TOF with $0 < V < 0.5$, there will be only one guided mode, namely, LP_{01} mode. Similarly, for $0 < V < 1$, only LP_{01} and LP_{11}; for $1 < V < 1.5$, only LP_{01}, LP_{21} and LP_{11}; for $1.5 < V < 2$, only LP_{02} and LP_{03} guided modes exist. The number of modes decreases rapidly as V is reduced.

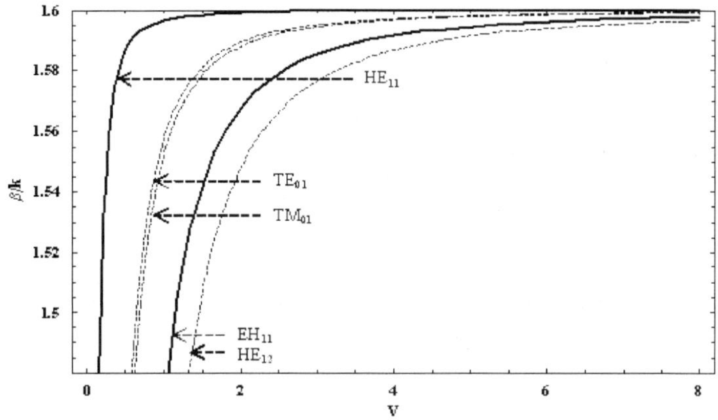

Fig. 5. Variation of β/k with the normalized frequency parameter V.

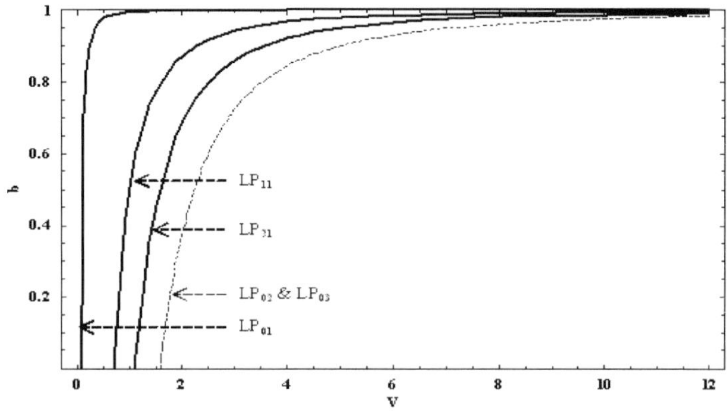

Fig. 6. Variation of the normalized propagation constant b with the normalized frequency parameter V.

A conventional circular core fiber becomes single-mode when $V < 2.405$ for which only the HE_{11} (or the LP_{01}) mode exists [44]. Comparing it with the present case of dielectric TOF, we observe that it becomes single-mode corresponding to the value of V approximately less than 0.7 (HE_{11} or LP_{01} mode exists in the fiber) – the feature attributed to the tapered geometrical nature of the guide. Thus, it can be inferred that the cutoff value is observed to greatly reduce in the case of dielectric TOFs.

2.3 Power propagation characteristics

The transmission of power [4,45] through the dielectric TOF remains important in the applications point of view. In the present subsection, we attempt to report the analysis of the power confinement pattern in a dielectric TOF made of polysterene core and cladded with methyl methacrylate. Such plastic clad fibers (PCF) exhibit considerably greater attenuation than the conventional silica fibers. But, high numerical apertures and large acceptance angles of these PCFs essentially result in an easy launching of optical signals. The technique of split-step method, as discussed earlier, is implemented to deduce the formulations. The analysis becomes much rigorous through the implementation of Maxwell's field equations, and the results are ultimately obtained with much greater accuracy. The results presented in this subsection is aimed to provide the variation of the power confinement factor in the dielectric TOF core and the clad sections corresponding to some low order modes and different taper lengths.

As stated above, the continuity conditions require that the transverse EM field components should have a smooth match at the core/cladding interface. Based on this, and the above field eqs. (9) and (10), the amount of power [46] transmitted through the core and the clad sections can be explicitly given by the following expressions [47]:

$$\begin{aligned} P_{core} = \frac{1}{2}\Re e\int_0^{2\pi}\int_0^{\rho_a} C_1^2 \Xi \Bigg[\Bigg\{ & \omega\varepsilon_1 u^2 \beta J_v'^2(u\rho) - \frac{\omega\varepsilon_1 u v^2 \beta}{\rho\rho_a}\Psi J_v(u\rho)J_v'(u\rho) \\ & + \frac{v^4\beta^3}{\rho^2\rho_a}\psi^2 J_v^2(u\rho) - \frac{uv^2\beta^3}{\omega\mu_0\rho_a\rho}\Psi J_v(u\rho)J_v'(u\rho)\Bigg\} \\ & - \Bigg\{\frac{uv^2\beta^3}{\omega\mu_0\rho_a\rho}\Psi J_v(u\rho)J_v'(u\rho) - \frac{\omega\varepsilon_1 v^2\beta}{\rho^2}J_v^2(u\rho) - \frac{u^2v^2\beta^3}{\rho_a}\Psi^2 J_v'^2(u\rho) \\ & + \frac{\omega\varepsilon_1 u v^2\beta}{\rho\rho_a}\Psi J_v(u\rho)J_v'(u\rho)\Bigg\}\Bigg]\rho d\rho d\phi \end{aligned} \tag{24}$$

and

$$\begin{aligned} P_{clad} = \frac{1}{2}\Re e\int_0^{2\pi}\int_{\rho_a}^{\infty} C_1^2 \Phi \Bigg[\Bigg\{ & -\frac{\omega\varepsilon_2 w v^2\beta}{\rho\rho_a}\Omega\frac{J_v(u\rho_a)}{K_v(w\rho_a)}K_v(w\rho)K_v'(w\rho) + \frac{v^4\beta^3}{\rho^2\rho_a}\Omega^2 K_v^2(w\rho) \\ & + \frac{\omega\varepsilon_2 w^2\beta J_v^2(u\rho_a)K_v'^2(w\rho)}{K_v^2(w\rho_a)} - \frac{v^2 w\beta^3}{\omega\mu_0\rho_a\rho}\Omega\frac{J_v(u\rho_a)}{K_v(w\rho_a)}K_v(w\rho)K_v'(w\rho)\Bigg\} \\ & - \Bigg\{\frac{v^2 w\beta^3}{\omega\mu_0\rho_a\rho}\Omega\frac{J_v(u\rho_a)}{K_v(w\rho_a)}K_v(w\rho)K_v'(w\rho) - \frac{\omega\varepsilon_2 v^2\beta}{\rho^2}\frac{J_v^2(u\rho_a)}{K_v^2(w\rho_a)}K_v^2(w\rho) \\ & - \frac{v^2w^2\beta^3}{\rho_a}\Omega^2 K_v'^2(w\rho) + \frac{\omega\varepsilon_2 v^2 w\beta}{\rho\rho_a}\Omega\frac{J_v(u\rho_a)}{K_v(w\rho_a)}K_v(w\rho)K_v'(w\rho)\Bigg\}\Bigg]\rho d\rho d\phi \end{aligned} \tag{25}$$

In eqs. (24) and (25), we used the symbols with their meanings as

$$\Xi = \frac{1}{\left(\omega^2 \mu_0 \varepsilon_1 - \beta^2\right)^2}, \quad \Psi = \frac{J_\nu(u\rho_a)\left(\frac{1}{u^2}+\frac{1}{w^2}\right)}{\frac{J'_\nu(u\rho_a)}{u} + \frac{J_\nu(u\rho_a)K'_\nu(w\rho_a)}{wK_\nu(w\rho_a)}},$$

$$\Omega = \frac{J_\nu^2(u\rho_a)\left(\frac{1}{u^2}+\frac{1}{w^2}\right)}{\frac{J'_\nu(u\rho_a)K_\nu(w\rho_a)}{u} + \frac{J_\nu(u\rho_a)K'_\nu(w\rho_a)}{w}}.$$

Further, in eqs. (24) and (25), the used constant C_1 is involved with the remaining constants C_2, C_3 and C_4 as follows:

$$C_2 = \frac{j\nu\beta\Psi}{\omega\mu_0\rho_a} C_1, \tag{26a}$$

$$C_3 = \frac{J_\nu(u\rho_a)}{K_\nu(w\rho_a)} C_1, \tag{26b}$$

$$C_4 = \frac{j\nu\beta\Omega}{\omega\mu_0\rho_a} C_1. \tag{26c}$$

The constant C_1 in eqs. (26) can be determined by a normalization condition considering the input power.

Now, in the context of investigating the distribution of power in different TOF sections, figs. 7–9 illustrate the dependence of the power confinement factor along the tapered length. For the analysis, we keep the radius of the taper input ρ_i fixed to 50 μm, and the output taper cross-sectional size is varied. As stated earlier, the TOF core and the clad regions are, respectively, made of polysterene (with RI n_1 = 1.6) and methyl methacrylate (RI n_2 = 1.48). The operating wavelength λ_0 is kept fixed at 1.55 μm. We present the plots of power for meridional (ν = 0) and the lowest skew (ν = 1) modes.

Figs. 7 and 8 illustrate the power confinement in the tapered core (against the taper length l) corresponding to the azimuthal mode indices ν = 0 and 1, respectively. We take into account three different values of the output core radius, namely, 100 μm, 150 μm and 200 μm. We observe that the power confinement increases with the increase in output cross-sectional dimension, and also, the taper length. Corresponding to the meridional mode, the increase in power is fairly linear throughout the taper length; doubling of the output cross-section makes the power to increase by 20%. This increase in power is very much expected because fibers with larger dimensions transmit more amount of power.

Fig. 8 corresponds to the case of lowest skew mode with ν = 1. We observe that, for smaller taper lengths, the power patterns show initially a sharp and fairly linear increase with the increase in taper length l. However, with the further increase in l, power increases, and ultimately reaches almost a saturation stage. Moreover, the nature of saturation becomes

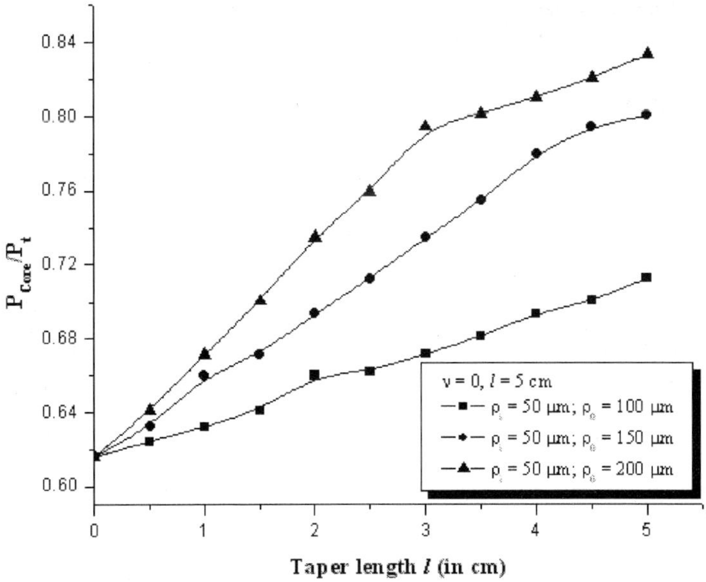

Fig. 7. Variation of core confinement factor corresponding to $\nu = 0$.

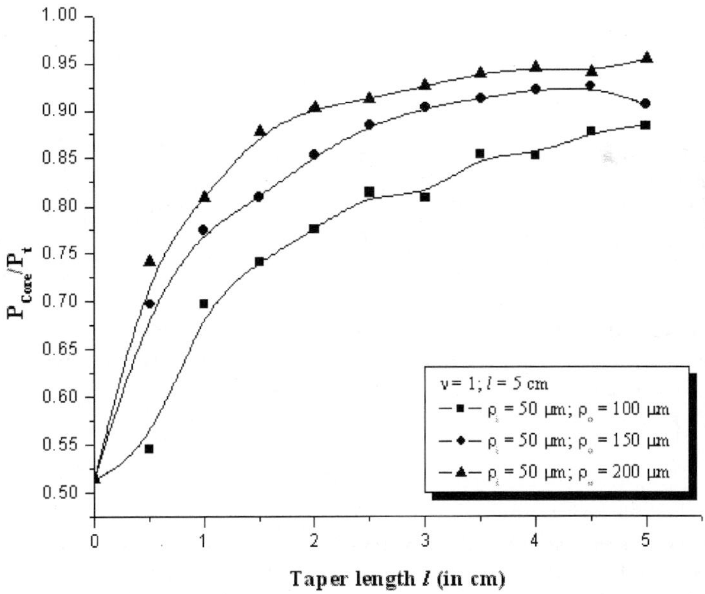

Fig. 8. Variation of core confinement factor corresponding to $\nu = 1$.

more pronounced corresponding to larger cross-sectional size. Fig. 8 also demonstrates that, for ρ_i = 50 μm and ρ_o = 200 μm, and for larger taper lengths, more than 95% of power is confined in the TOF core, which determines that the skew modes transmit more amount of power as compared to the meridional modes. Further, skew modes transmit power better uniformly with increasing value of l - the rise in power is about 45% with the increase in output cross-section, and with the increase in taper length. As such, for skew modes, a fairly uniform distribution of power takes place among the sustained modes in this case - a desirable aspect for optical communications.

Fig. 9 depicts the power confinement with taper length corresponding to the meridional mode (ν = 0). We observe that the confinement almost linearly decreases with increasing l, with a pronounced decrease for larger output cross-sections. Corresponding to the skew mode (ν = 1; fig. 10), we observe that the confinement factor exhibits an initial sharp decrease in the clad section with increasing l, followed by relatively slower decrease with larger values of l. Further, the decrease in power is more corresponding to larger taper output cross-section (e.g. 200 μm). Thus, the noticeable fact remains that, by increasing the output cross-sectional size, a markedly higher power is transmitted through the TOF core, and a pronounced decrease in power is observed in the TOF clad.

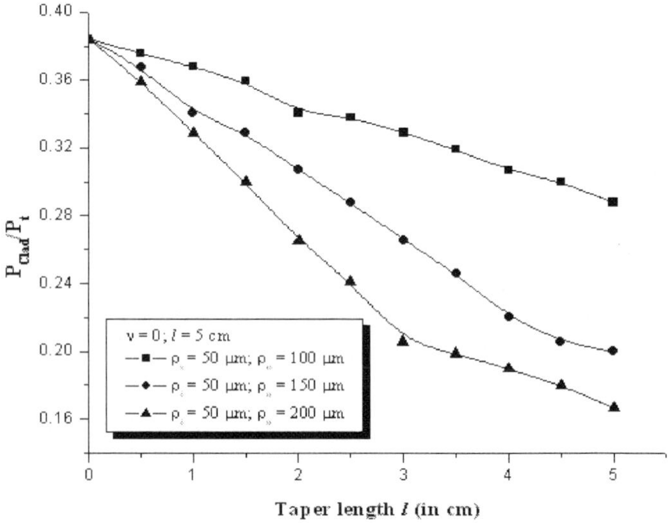

Fig. 9. Variation of clad confinement factor corresponding to ν = 0.

Comparison of figs. 7–10 yields that a larger taper output dimension causes a fairly good uniformity in the distribution of power. However, it is noteworthy that the power distribution is also affected by the cross-sectional size at the taper input end. But, in our analysis, we fixed the taper input radius to 50 μm. A physical interpretation of the uniformity in power distribution can be given in a way that, when the radii ρ_i and/or ρ_o of the tapered sections are small, it behaves like a few-mode component, making thereby the

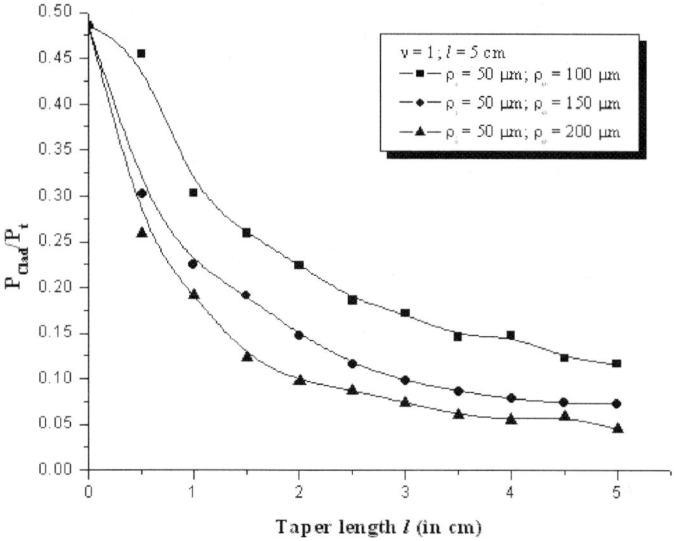

Fig. 10. Variation of clad confinement factor corresponding to $\nu = 1$.

power distribution patterns to bear some sort of *noisy* characteristics. The component starts behaving like a multimode structure corresponding to larger radii values - the feature more prominent in the case of skew modes.

3. Helical clad tapered dielectric optical fibers

Helical structures find enormous applications in low- and medium-power traveling wave tubes (TWTs) [35], the analyses of which generally include waveguides under slow-wave structures with conducting sheath and tape helixes. The investigators have reported the implementation of this concept in the case of optical fibers [36]. As such, helical clad fibers, particularly with sheath helix patterns, fall among the category of new fiber types [48–54].

The analytical treatment of helical clad tapered fibers remains much complicated owing to the boundary conditions. The tapered structure may be handled by implementing the split-step method, similar to as discussed in the previous sections. An investigation of the dispersion characteristics of such tapered fibers with helical clads would be rather interesting as the helix pitch angle essentially plays a determining role to alter the dispersion behavior [50–54] of the guide.

In following subsections, we will deal with the preliminary ground work of helical clad dielectric TOFs, implementing a sheath helical pattern within the core-clad interface, because the purposely introduced helix pitch angle essentially brings in modifications in the dispersion profile. The use of Maxwell's equations can ultimately provide the dispersion relations for the structure, and the cutoff situations can be obtained under the chosen values of helix pitch.

3.1 Analytical approach

We consider a step-index linearly tapered helical clad optical fiber, as shown in fig. 11, the clad section of which is considered to be infinitely extended. A helical winding is introduced at the core/clad interface with the pitch angle as ψ. With the use of cylindrical polar coordinated (ρ,ϕ,z), the wave propagation takes place along the direction of z, and the linear tapered function for structure is defined by eq. (2). The region $z < z_1$ in fig. 11 corresponds to the fiber pigtail, and that with $z > z_2$ to the expanded cylindrical section following the tapered section. The values of helix pitch angle $\psi = 0°$ and $\psi = 90°$ correspond to the two extreme situations - the former one indicates the winding to be perpendicular to the longitudinal direction whereas the latter one states those to be parallel to the z-axis. We assume that the core/clad sections are made of linear and isotropic non-magnetic materials, and therefore, $\mu_1 = \mu_2 \cong \mu_0$, the free-space permeability. Further, ε_1 and ε_2 are, respectively, the permittivity values of the core/clad sections.

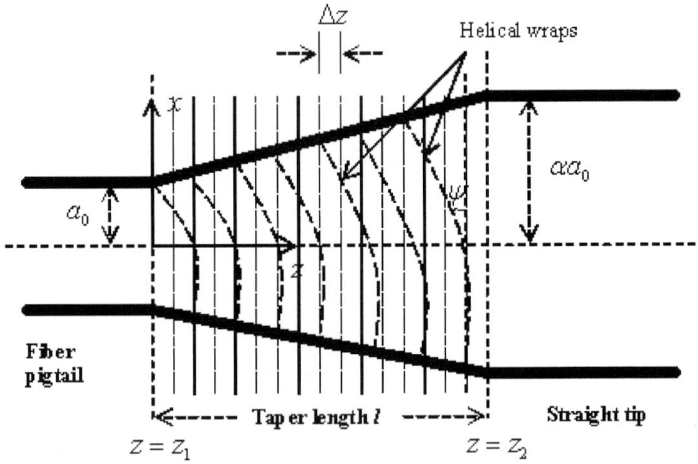

Fig. 11. Transverse view of TOF with helical wraps.

Considering the time t-harmonic and the axis z-harmonic EM fields, β becomes a function of the longitudinal distance z. Assuming a small variation of the core radius, the dependence of β may be expressed in the form of Taylor series expansion, as given in eq. (4). The longitudinal electric/magnetic fields in the core/clad sections will be described by Bessel and the modified Bessel functions, as shown in eqs. (6) and (7). Further, the transverse field components at an arbitrary radial point $\rho = \rho_a$ will be as given in eqs. (9) and (10).

3.2 Dispersion characteristics

The continuity conditions require that the transverse components of electric/magnetic fields at the core/clad boundary should get a smooth match. However, in the present case of helical clad fibers, the boundary conditions are essentially modified [35,55], which follows from the fact that the components of the electric field vanish in the direction of helix. Following that, a 4×4 matrix may be formed from the coefficients of the four unknowns in four different equations, and the matrix should be vanishing in order to have a non-trivial solution to those four equations. In this way, the eigenvalue equation in β may be finally deduced as [56]

$$u\frac{J_\nu(u\rho_a)}{J'_\nu(u\rho_a)}\left[\left(\sin\psi+\frac{\{\beta_0+(\partial\beta/\partial z)z\}\nu}{u^2\rho_a}\cos\psi\right)^2\right]$$

$$-w\frac{K_\nu(w\rho_a)}{K'_\nu(w\rho_a)}\left[\left(\sin\psi+\frac{\{\beta_0+(\partial\beta/\partial z)z\}\nu}{w^2\rho_a}\cos\psi\right)^2\right]$$

$$-\left[\frac{(k_1n_1)^2}{u}\frac{J'_\nu(u\rho_a)}{J_\nu(u\rho_a)}-\frac{(k_2n_2)^2}{w}\frac{K'_\nu(w\rho_a)}{K_\nu(w\rho_a)}\right]\cos^2\psi=0\,. \tag{27}$$

Eq. (27) is the characteristic dispersion relation for helical clad tapered fibers. We may now consider two extreme values of the helix pitch angle ψ, namely ψ = 0° and ψ = 90°, to be used to observe the effect of helix introduced at the core/clad interface. In this stream, when eq. (27) is solved for ψ = 0° (i.e. when the helical turns are perpendicular to the optical axis of the TOF), we finally obtain the characteristic dispersion relation as [56]

$$u\frac{J_\nu(u\rho_a)}{J'_\nu(u\rho_a)}\left[\left(\frac{\{\beta_0+(\partial\beta/\partial z)z\}\nu}{u^2\rho_a}\right)^2\right]-w\frac{K_\nu(w\rho_a)}{K'_\nu(w\rho_a)}\left[\left(\frac{\{\beta_0+(\partial\beta/\partial z)z\}\nu}{w^2\rho_a}\right)^2\right]$$

$$-\left[\frac{(k_1n_1)^2}{u}\frac{J'_\nu(u\rho_a)}{J_\nu(u\rho_a)}-\frac{(k_2n_2)^2}{w}\frac{K'_\nu(w\rho_a)}{K_\nu(w\rho_a)}\right]=f_1(u,w)(say)=0\,. \tag{28}$$

Corresponding to the case of helical turns being parallel to the optical axis (i.e. ψ = 90°) of the TOF, eq. (27) provides [56]

$$u\frac{J_\nu(u\rho_a)}{J'_\nu(u\rho_a)}-w\frac{K_\nu(w\rho_a)}{K'_\nu(w\rho_a)}=f_2(u,w)(say)=0\,. \tag{29}$$

Much changes in the shape of eqs. (28) and (29) essentially indicate the dominant effect of the helical clad in controlling the dispersion behavior of the guide. A deeper look at the dispersion relations under the helix pitch angles 0° and 90° (by solving eqs. (27) and (28), respectively) will numerically provide the values of the modal propagation constant β. In general, the dispersion equations may have multiple solutions for each value of azimuthal mode index ν. These solutions may be represented as $\beta_{\nu m}$; m being the order of the solution corresponding to a given value of ν. Each value of $\beta_{\nu m}$ corresponds to one possible mode. The limiting condition $w^2 \to 0$ determines the cutoff characteristics of the guide. Under the cutoff situations, corresponding to the helix pitch angles ψ = 0° and ψ = 90°, we write the obtained equations in symbolic forms as $f_1(u)=0$ and $f_2(u)=0$, respectively.

In our computations, we consider the helical TOF as made of polystyrene core (n_1 = 1.6) and methyl methacrylate clad (n_2 = 1.48). Also, similar to the previous cass, λ_0 = 1.55 μm is the operating wavelength.

Fig. 12 illustrates the plot of $f_1(u,w)$ (based on eq. (28)) against the dimensionless quantity $u\rho$ corresponding to the perpendicular (to the axis) helical turns, i.e. ψ = 0°. We chose two different illustrative values of the azimuthal mode index, viz. ν = 1 and ν = 2, and observe that the curves intersect the horizontal axis at around $u\rho$ = 0.65. As such, there exist modes at this intersection point, and the exact values of the modal propagation constants may be evaluated.

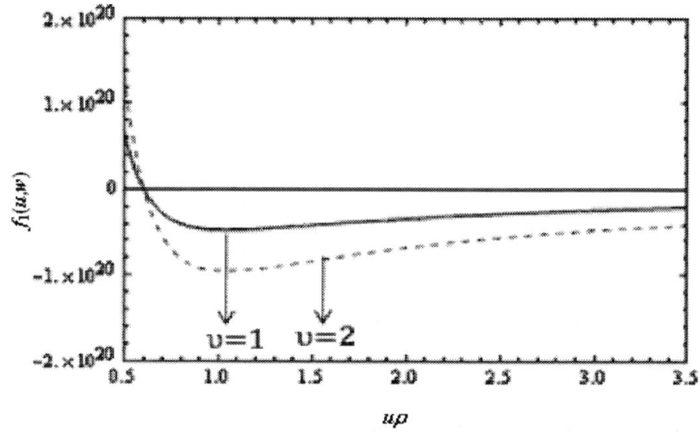

Fig. 12. Variation of the function $f_1(u,w)$ with $u\rho$ corresponding to ψ = 0°.

Fig. 13 depicts the variation of $f_1(u)$ with the dimensionless quantity $u\rho$; $f_1(u)$ being the form of eq. (28), when solved under the limit $w^2 \rightarrow 0$ (not explicitly shown here). Thus, the cutoff variation of $f_1(u)$ corresponding to two values of the azimuthal index ν (namely, 1 and 2)

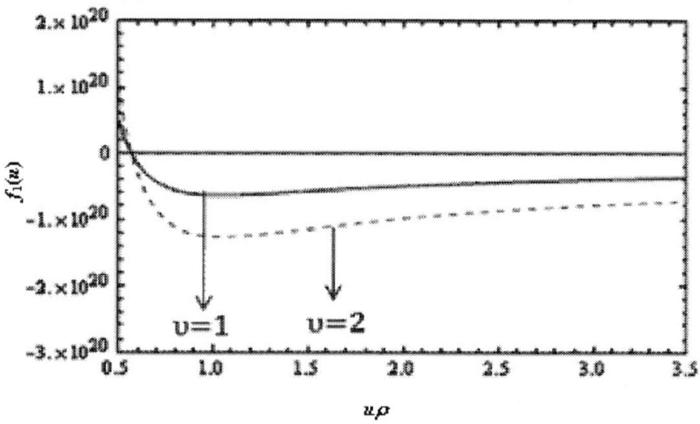

Fig. 13. Variation of the function $f_1(u)$ with $u\rho$ corresponding to ψ = 0°.

shows the intersection with the horizontal axis at around $u\rho$ = 0.60, which is less than the value of $u\rho$ (= 0.65), as observed in fig. 12. This shows the consistency of the results obtained in figs. 12 and 13. However, comparing the results with those reported earlier corresponding to the case of conventional dielectric TOFs without helical windings (Section 2; figs. 3 and 4), we observe that the existence of a helical wrap has the effect to increase the modal propagation constants.

Corresponding to the case of helical windings being parallel to the optical axis (i.e. ψ = 90°), fig. 14 presents the plot of the left hand side $f_2(u,w)$ of eq. (29) against the dimensionless quantity $u\rho$ corresponding to the azimuthal mode index value as ν = 1. We notice that the plot has the form of almost a straight line with the intersection at about $u\rho$ = 0.70. Thus, as compared to the case of ψ = 0° (fig. 12), the $u\rho$-crossing value becomes larger in this case, and therefore, the value of modal propagation constant decreases.

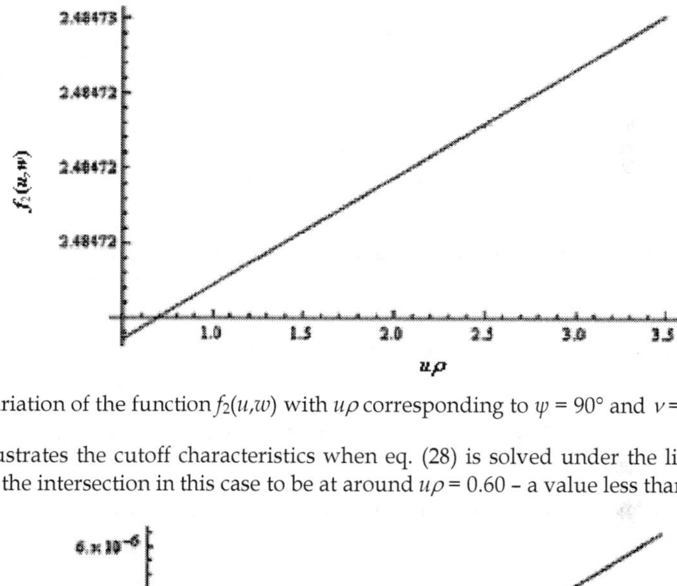

Fig. 14. Variation of the function $f_2(u,w)$ with $u\rho$ corresponding to ψ = 90° and ν = 1.

Fig. 15 illustrates the cutoff characteristics when eq. (28) is solved under the limit $w^2 \rightarrow 0$. We notice the intersection in this case to be at around $u\rho$ = 0.60 – a value less than that found

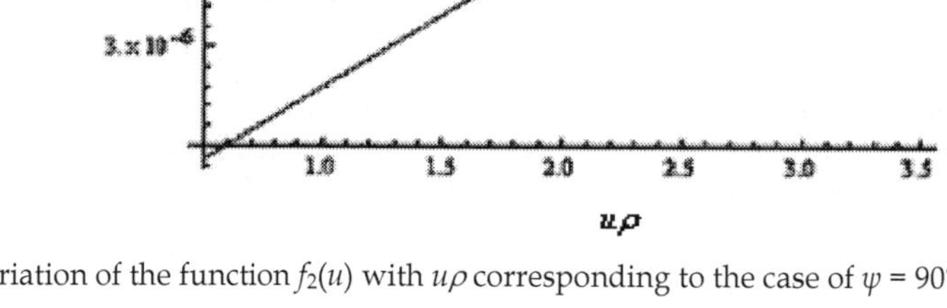

Fig. 15. Variation of the function $f_2(u)$ with $u\rho$ corresponding to the case of ψ = 90° and ν = 1.

in fig. 12, indicating thereby the results of figs. 12 and 15 to be consistent. However, upon comparing the results with the case of dielectric TOFs without helical turns, we observe the general tendency that the helical clad TOFs increase the modal propagation constants. This essentially determines the controlling feature of the helical turns at the core/clad interface.

It can thus be inferred that, in helical clad TOFs, modes travel with higher propagation constants under the situation when the helical wraps are perpendicular to the optical axis of the guide than the case of parallel ones. However, compared with the situation of conventional dielectric TOFs, we observe that the existence of helix has the tendency to increase the modal propagation constants - the important role of the helix to control the propagation characteristics of dielectric TOFs.

4. Tapered fibers with liquid crystal clad

Liquid crystal optical fibers [37–39] possess several technological applications due to the fairly large polarization anisotropy [40] of liquid crystals. Owing to the large electro-optic coefficient of liquid crystals, the macroscopic optical properties of them can be manipulated by suitably applying external electrical fields.

In this section, we deal with the case of a three-layer liquid crystal tapered optical fiber (LCTOF) structure for which Maxwell's equations are implemented for a rigorous analysis. We consider the LCTOF having the outermost clad made of radially anisotropic liquid crystal material; the fiber core and the inner clad regions being made of isotropic dielectrics. It is to be pointed out here that the radial anisotropy of liquid crystals may be obtained by the capillary action after inserting the liquid crystal section into a capillary tube coated with N, N-dimethyl-N-octadecyl-3-aminopropyltrimethoxysilyl chloride.

A tapered structure of the fiber core is considered along the longitudinal direction because, as stated before, TOFs are of immense use in optical sensors and other in-line integrated optic applications. As such, an amalgamation of features in respect of fiber geometry and the material, i.e. a tapered core fiber with radially anisotropic outermost liquid crystal clad, would provide enhanced usefulness of the guide. With the advancement of the discussions, we concentrate on the dispersion profile corresponding to the low order TE and TM modes. Though the studies related to radially anisotopic LCTOFs have been reported before by Choudhury *et al.* [5], the present study provides a blend of liquid crystal material and tapered structure. We treat the LCTOF structure by implementing the split-step method, as discussed before.

4.1 Theory

Fig. 16(a) illustrates the cross-sectional view of LCTOF, which is made of homogeneous, isotropic and non-magnetic core and the inner clad. The outermost clad of the LCTOF is infinitely extended, and consists of radially anisotropic liquid crystal material. A comparison of the cases of radial and azimuthal anisotropies of the liquid crystal section makes the situation much clear; fig. 16(b) illustrates the case when the liquid crystal molecules assume azimuthal anisotropy; in figs. 16(a) and 16(b), the elongation of liquid crystal molecules is presented by dashed lines.

Fig. 17 represents the transverse view of the tapered nature of the fiber cross-section, where the core/inner clad sections have the RI values as n_1 and n_2, respectively, with $n_1 > n_2$. Also,

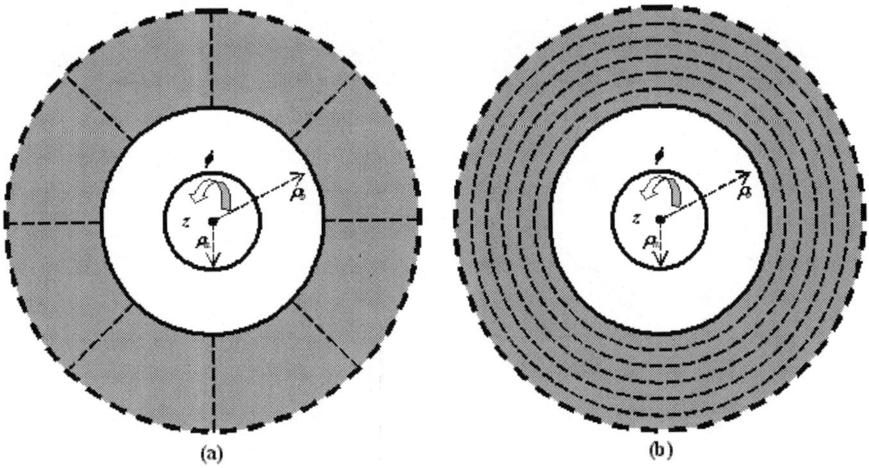

Fig. 16. LCTOF cross-section with nematic liquid crystal outermost clad (a) radial anisotropy, and (b) azimuthal anisotropy.

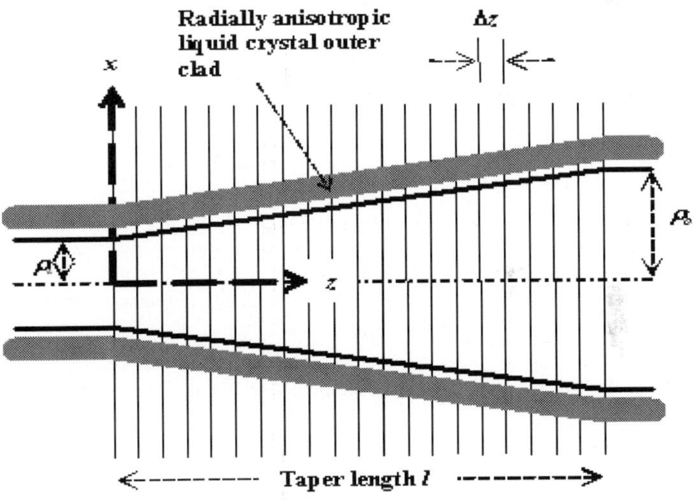

Fig. 17. Transverse view of the LCTOF of taper length l.

we consider the presence of radially anisotropic liquid crystal molecules in the infinitely extended outer clad with the ordinary and the extraordinary RI values as n_o and n_e, respectively. The linear taper nature of the fiber is defined by eq. (2).

The fiber geometry essentially needs the use of cylindrical polar coordinate system (ρ, ϕ, z) wherein the z-axis remains along the direction of wave propagation, and coincides with the principal axes of the outer clad. The extraordinary principal axis has a radial orientation, and therefore, the infinitely extended outer clad possesses the RI distribution given as

$n_\rho = n_e$ and $n_\phi = n_z$ with $n_e > n_1 > n_2 > n_o$,

as depicted figs. 18(a) and 18(b). Here n_ρ, n_ϕ and n_z are, respectively, the RI values along the ρ-, ϕ- and z-directions.

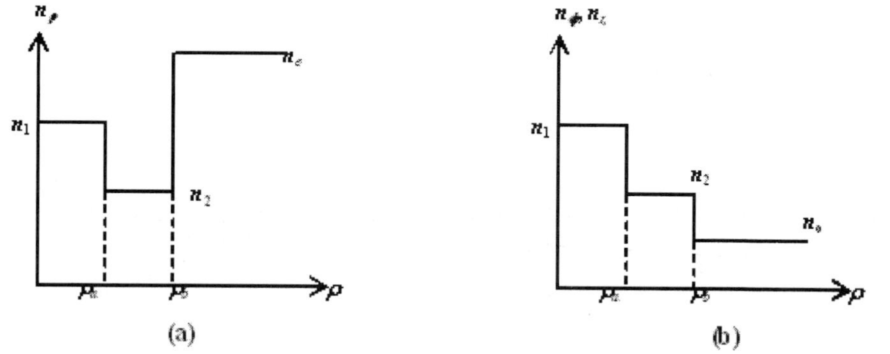

Fig. 18. The RI distribution pattern of the LCTOF.

Considering the time t-harmonic and the axis z-harmonic EM fields, coupled wave propagation equations for the transverse field components can be written as [5]

$$\left(\nabla_t^2 + k_0^2 n_\rho^2 - \beta^2 - \frac{1}{\rho^2}\right) e_\rho = -\frac{2}{\rho^2}\frac{\partial e_\phi}{\partial \phi} + \left(1 - \frac{n_\rho^2}{n_z^2}\right)\frac{\partial}{\partial \rho}\left\{\frac{1}{\rho}\frac{\partial}{\partial \rho}\left(\rho e_\rho\right)\right\}$$

$$+\left(1 - \frac{n_\phi^2}{n_z^2}\right)\frac{\partial}{\partial \rho}\left(\frac{1}{\rho}\frac{\partial e_\phi}{\partial \phi}\right), \tag{30a}$$

$$\left(\nabla_t^2 + k_0^2 n_\phi^2 - \beta^2 - \frac{1}{\rho^2}\right) e_\phi = \frac{2}{\rho^2}\frac{\partial e_\rho}{\partial \phi} + \frac{1}{\rho^2}\left\{\left(1 - \frac{n_\rho^2}{n_z^2}\right)\frac{\partial}{\partial \rho}\left(\rho\frac{\partial e_\rho}{\partial \phi}\right) + \left(1 - \frac{n_\phi^2}{n_z^2}\right)\frac{\partial^2 e_\phi}{\partial \phi^2}\right\}. \tag{30b}$$

In eqs. (30), ∇_t^2 is the Laplacian operator in the cylindrical coordinate system, k_0 is the free-space propagation constant. Owing to the variation of the propagation constant β due to the varying cross-sectional dimension with distance z, β-values in eqs. (30) are governed by Taylor series expansion, as presented in eq. (4).

It is to be stated at this point that, in anisotropic guides, TE_{mn}, TM_{mn} and hybrid modes contain all the three electric field components E_ρ, E_ϕ and E_z. However, there are some special modes which do not contain all the three electric field components for which the index profiles in the zero electric field directions are irrelevant to the mode cutoff conditions. In our computations, we consider the lower order TE and TM modes, viz. TE_{01} and TM_{01}. For the TE_{01} mode, there is only one transverse electrical field component e_ϕ, which is independent of the coordinate ϕ. Thus, corresponding to this mode, we have $e_\rho = 0$ and $\partial e_\phi / \partial \phi = 0$. On the other hand, for TM_{01} mode, there is only one non-zero component e_ρ, which is independent of the coordinate ϕ. Therefore, corresponding to the

TM_{01} mode, we have $e_\phi = 0$ and $\partial e_\rho / \partial\phi = 0$. We will now treat the cases of TE and TM modes individually.

4.1.1 Transverse electric (TE) mode

As stated earlier, in the case of TE_{01} mode we have $e_\rho = 0$ and $\partial e_\phi / \partial\phi = 0$. Using these in eq. (30b), the wave equation corresponding to TE modes in LCTOFs can be deduced as [5]

$$\frac{\partial^2 e_\phi}{\partial\rho^2} + \frac{1}{\rho}\frac{\partial e_\phi}{\partial\rho} + \left(k_0^2 n_\phi^2 - \beta^2 - \frac{1}{\rho^2}\right) e_\phi = 0 . \quad (31)$$

In eq. (31), ρ and β are defined by eqs. (2) and (4), respectively. It can be shown that the solutions to eq. (31) will be in the form of combinations of Bessel and the modified Bessel functions. Using those solutions, harmony of the EM fields and Maxwell's field equations, the field components in the case of TE_{01} mode may finally be obtained as follows [5]:

$$H_\rho = -\frac{\beta}{\omega\mu_0} e_\phi \exp\{j(\omega t - \beta z)\} \quad \text{with} \quad H_\rho = h_\rho \exp\{j(\omega t - \beta z)\} , \quad (32a)$$

$$H_z = \frac{j}{\omega\mu_0}\left(\frac{\partial e_\phi}{\partial\rho} + \frac{e_\phi}{\rho}\right)\exp\{j(\omega t - \beta z)\} \quad \text{with} \quad H_z = h_z \exp\{j(\omega t - \beta z)\} . \quad (32b)$$

In eqs. (32), μ_0 is the free-space permeability as the mediums are considered to be non-magnetic in nature. The use of eqs. (32) and the solutions to eq. (31) will ultimately provide the field components in the different fiber sections as follows [5]:

Core region:

$$\left.H_\rho\right)_I = -A_{\phi 1}\frac{\beta}{\omega\mu_0} J_1(u\rho)\exp\{j(\omega t - \beta z)\} , \quad (33a)$$

$$\left.H_z\right)_I = A_{\phi 1}\frac{j}{\omega\mu_0}\left\{uJ_1'(u\rho) + \frac{1}{\rho}J_1(u\rho)\right\}\exp\{j(\omega t - \beta z)\} . \quad (33b)$$

Inner clad region:

$$\left.H_\rho\right)_{II} = -\frac{\beta}{\omega\mu_0}\left\{A_{\phi 2}K_1(w\rho) + A_{\phi 3}I_1(w\rho)\right\}\exp\{j(\omega t - \beta z)\} , \quad (34a)$$

$$\left.H_z\right)_{II} = \frac{j}{\omega\mu_0}\left[A_{\phi 2}\left\{wK_1'(w\rho) + \frac{1}{\rho}K_1(w\rho)\right\}\right.$$

$$\left. + A_{\phi 3}\left\{wI_1'(w\rho) + \frac{1}{\rho}I_1(w\rho)\right\}\right]\exp\{j(\omega t - \beta z)\} . \quad (34b)$$

Outer clad region:

$$H_{\rho})_{III} = -A_{\phi 4}\frac{\beta}{\omega\mu_0}K_1(v\rho)\exp\{j(\omega t-\beta z)\}\,, \tag{35a}$$

$$H_z)_{III} = A_{\phi 4}\frac{j}{\omega\mu_0}\left\{vK_1'(v\rho)+\frac{1}{\rho}K_1(v\rho)\right\}\exp\{j(\omega t-\beta z)\}\,. \tag{35b}$$

In eqs. (33), (34) and (35), $J(\bullet)$, $K(\bullet)$ and $I((\bullet)$ represent Bessel and the modified Bessel functions, and the prime stands for the differentiation with respect to the argument of the function. Also, $A_{\phi1}$, $A_{\phi2}$, $A_{\phi3}$ and $A_{\phi4}$ are the arbitrary constants to be determined by the boundary conditions.

4.1.2 Transverse magnetic (TM) mode

In the case of TM_{01} mode, since $e_\phi = 0$ and $\partial e_\rho/\partial\phi = 0$, eq. (30a) yields the wave equation for LCTOF as [5]

$$\frac{\partial^2 e_\rho}{\partial\rho^2}+\frac{1}{\rho}\frac{\partial e_\rho}{\partial\rho}+\left\{\left(k_0^2 n_\rho^2-\beta^2\right)\left(\frac{n_z}{n_\rho}\right)^2-\frac{1}{\rho^2}\right\}e_\rho = 0\,. \tag{36}$$

In eq. (36) too, ρ is defined by eq. (2). Further, it can be shown that the non-zero field components for LCTOF in this case will assume the form [5]

$$E_z = -j\frac{n_e^2}{\beta n_o^2}\left(\frac{\partial e_\rho}{\partial\rho}+\frac{e_\rho}{\rho}\right)\exp\{j(\omega t-\beta z)\}\,, \tag{37a}$$

$$H_\phi = n_\rho^2 e_\rho\frac{\omega\varepsilon_0}{\beta}\exp\{j(\omega t-\beta z)\}\,. \tag{37b}$$

Once again, in eqs. (37) too, ρ and β are considered to be according to as defined by eqs. (2) and (4), respectively. By using eqs. (37) and the solutions to eq. (36), the EM field components will be represented in this case as follows [5]:

Core region:

$$E_z)_I = -B_{\rho1}\frac{j}{\beta\eta}\left\{uJ_1'(u\rho)+\frac{1}{\rho}J_1(u\rho)\right\}\exp\{j(\omega t-\beta z)\}\,, \tag{38a}$$

$$H_\phi)_I = B_{\rho1}\frac{\omega\varepsilon_0}{\beta}n_\rho^2 J_1(u\rho)\exp\{j(\omega t-\beta z)\}\,. \tag{38b}$$

Inner clad region:

$$E_z)_{II} = -\frac{j}{\beta\eta}\left[B_{\rho2}\left\{wK_1'(w\rho)+\frac{1}{\rho}K_1(w\rho)\right\}\right.$$

$$+B_{\rho 3}\left\{wI_1'(w\rho)+\frac{1}{\rho}I_1(w\rho)\right\}\Bigg]\exp\{j(\omega t-\beta z)\}\,, \tag{39a}$$

$$H_\phi\big)_{II}=\frac{\omega\varepsilon_0}{\beta}n_\rho^2\{B_{\rho 2}K_1(w\rho)+B_{\rho 3}I_1(w\rho)\}\exp\{j(\omega t-\beta z)\}\,. \tag{39b}$$

Outer clad region:

$$E_z\big)_{III}=-B_{\rho 4}\frac{j}{\beta\eta}\left\{v'K_1'(v'\rho)+\frac{1}{\rho}K_1(v'\rho)\right\}\exp\{j(\omega t-\beta z)\}\,, \tag{40a}$$

$$H_\phi\big)_{III}=B_{\rho 4}\frac{\omega\varepsilon_0}{\beta}n_\rho^2K_1(v'\rho)\exp\{j(\omega t-\beta z)\}\,. \tag{40b}$$

In eqs. (38), (39) and (40), $B_{\rho 1}$, $B_{\rho 2}$, $B_{\rho 3}$ and $B_{\rho 4}$ are arbitrary constants to be defined by the boundary conditions, and the new quantities η and v' have the meanings as

$$\eta=(n_0/n_e)^2 \tag{41}$$

and

$$v'=\frac{n_o}{n_e}\sqrt{n_e^2k_0^2-\beta^2}\,, \tag{42}$$

respectively.

4.2 Dispersion characteristics

4.2.1 Dispersion relation for TE mode

In order to obtain the dispersion relation, the continuity conditions of fields are to be imposed at the layer interface. As such, we consider the localized values of radial parameters (of LCTOF) to define the interface, e.g. we state $\rho = \rho_a$ as the core-inner clad interface and $\rho = \rho_b$ as the inner clad-outer clad interface; ρ_a and ρ_b, respectively, being the local parametric values of the LCTOF core and the inner clad radii. We now match the fields at the defined layer interfaces, which provide four equations altogether. Now, collecting the coefficients of the unknown arbitrary constants from those four equations, after a few lengthy steps, a 4×4 matrix can be deduced, as follows [57]:

$$\begin{pmatrix}\xi_{11} & -\xi_{12} & -\xi_{13} & 0\\ 0 & \xi_{22} & \xi_{23} & -\xi_{24}\\ \xi_{31} & -\xi_{32} & -\xi_{33} & 0\\ 0 & \xi_{42} & \xi_{43} & -\xi_{44}\end{pmatrix}=\Delta_{TE}\quad(\text{say})=0. \tag{43}$$

In eq. (43), various symbols have their meanings as follows:

$$\xi_{11}=(\beta/\omega\mu_0)J_\nu(u\rho_a)\,,\qquad \xi_{12}=(\beta/\omega\mu_0)K_\nu(w\rho_a)\,,$$

$$\xi_{13} = (\beta / \omega\mu_0) I_v(w\rho_a), \quad \xi_{12} = (\beta / \omega\mu_0) K_v(w\rho_b),$$

$$\xi_{23} = (\beta / \omega\mu_0) I_v(w\rho_b), \quad \xi_{24} = (\beta / \omega\mu_0) K_v(v\rho_b),$$

$$\xi_{31} = (j / \omega\mu_0)\{uJ'_v(u\rho_a) + (1/\rho_a) J_v(u\rho_a)\},$$

$$\xi_{32} = (j / \omega\mu_0)\{wK'_v(w\rho_a) + (1/\rho_a) K_v(w\rho_a)\},$$

$$\xi_{33} = (j / \omega\mu_0)\{wI'_v(w\rho_a) + (1/\rho_a) I_v(w\rho_a)\},$$

$$\xi_{42} = (j / \omega\mu_0)\{wK'_v(w\rho_b) + (1/\rho_b) K_v(w\rho_b)\},$$

$$\xi_{43} = (j / \omega\mu_0)\{wI'_v(w\rho_b) + (1/\rho_b) I_v(w\rho_b)\},$$

$$\xi_{44} = (j / \omega\mu_0)\{vK'_v(v\rho_b) + (1/\rho_b) K_v(v\rho_b)\}.$$

Eq. (43) represents the dispersion relation for the LCTOF in the case when the TE modes are excited, and the solutions to this equation will provide different TE modes existing in the fiber under consideration. Further, it must be remembered that eq. (43), when solved under the limit $w^2 \to 0$, will provide the cutoff characteristics of the LCTOF.

4.2.2 Dispersion relation for TM mode

Following the procedure as discussed above for the case of TE modes, matching the field components at the localized radial parameters $\rho = \rho_a$ (core-inner clad interface) and $\rho = \rho_b$ (inner clad-outer clad interface), we finally obtain (after collecting the coefficients of the unknown constants) the dispersion relation (in the form of a 4×4 matrix) corresponding to the case of TM mode excitation of the LCTOF, as follows [57]:

$$\begin{pmatrix} \zeta_{11} & -\zeta_{12} & -\zeta_{13} & 0 \\ 0 & \zeta_{22} & \zeta_{23} & -\zeta_{24} \\ \zeta_{31} & -\zeta_{32} & -\zeta_{33} & 0 \\ 0 & \zeta_{42} & \zeta_{43} & -\zeta_{44} \end{pmatrix} = \Delta_{TM} \quad \text{(say)} = 0 \tag{44}$$

where the symbols have their meanings as given in the following:

$$\zeta_{11} = (j / \beta\eta^2)\{uJ'_v(u\rho_a) + (1/\rho_a) J_v(u\rho_a)\},$$

$$\zeta_{12} = (j / \beta\eta^2)\{wK'_v(w\rho_a) + (1/\rho_a) K_v(w\rho_a)\},$$

$$\zeta_{13} = (j / \beta\eta^2)\{wI'_v(w\rho_a) + (1/\rho_a) I_v(w\rho_a)\},$$

$$\zeta_{22} = (j / \beta\eta^2)\{wK'_v(w\rho_b) + (1/\rho_b) K_v(w\rho_b)\},$$

$$\zeta_{23} = \left(j / \beta\eta^2\right)\left\{wI_v'\left(w\rho_b\right) + \left(1/\rho_b\right)I_v\left(w\rho_b\right)\right\},$$

$$\zeta_{24} = \left(j / \beta\eta^2\right)\left\{vK_v'\left(v\rho_b\right) + \left(1/\rho_b\right)K_v\left(v\rho_b\right)\right\},$$

$$\zeta_{31} = \left(\omega\varepsilon_0 / \beta\right)n_\rho^2 J_v\left(u\rho_a\right), \qquad \zeta_{32} = \left(\omega\varepsilon_0 / \beta\right)n_\rho^2 K_v\left(w\rho_a\right),$$

$$\zeta_{33} = \left(\omega\varepsilon_0 / \beta\right)n_\rho^2 I_v\left(w\rho_a\right), \qquad \zeta_{42} = \left(\omega\varepsilon_0 / \beta\right)n_\rho^2 K_v\left(u\rho_a\right),$$

$$\zeta_{43} = \left(\omega\varepsilon_0 / \beta\right)n_\rho^2 I_v\left(w\rho_b\right), \qquad \zeta_{44} = \left(\omega\varepsilon_0 / \beta\right)n_\rho^2 K_v\left(v\rho_b\right).$$

The solutions to eq. (44) will provide different TM modes existing in the LCTOF. Also, solving eq. (44) under the condition $w^2 \rightarrow 0$ will provide the LCTOF cutoff features.

4.2.3 Comparison of dispersion characteristics for TE and TM modes

Eqs. (43) and (44), corresponding to the TE and the TM mode excitations, provide the eigenvalue equations for the LCTOF. In the case of isotropic guides, viz. liquid crystal waveguides, TE and TM modes are difficult to separate owing to the reason that the direction independent RI values yield identical propagation constants and field cutoffs. However, the TE and the TM modes undergo different polarizations in such guides because of the direction dependent RIs, and therefore, these modes possess different values of propagation constants and field cutoffs.

We will now analyze these to determine the LCTOF characteristics in respect of their dispersion behavior and field cutoffs. The LCTOF in our consideration has three different sections with radially anisotropic liquid crystal outer clad. We take the RI values of the core and the inner clad as n_1 = 1.5 and n_2 = 1.46, respectively, and the outermost section has nematic liquid crystal as BDH mixture 14616 having the respective ordinary and the extraordinary RI values as n_o = 1.457 and n_e = 1.5037. We consider the excitation of low-order modes (with the azimuthal index v as 1, 2 and 3), and the taper length l is taken to be 5 cm. Further, the localized values of the core and the inner clad radii are taken as ρ_a = 60 μm and ρ_a = 120 μm, respectively, and the operating wavelength is kept as 1.55 μm.

Fig. 19(a) presents the plots of the left hand side of eq. (43) corresponding to three different values of v. The crossings of the curves with the horizontal axis represent the existence of modes with a particular value of the propagation constant β. We observe in fig. 19(a) that the zero crossings show a little increase with the increase in v. We notice that the propagation constant β of the first mode corresponding to v = 1 is close to 5.935×10^6 m^{-1}; β-values corresponding to the other existing modes can also be estimated from such intersections of the curves.

Eq. (43), when solved under the limit $w^2 \rightarrow 0$, gives the cutoff features under the TE mode excitation. Fig. 19(b) illustrates the cutoff situation of the TE modes. We observe that, when compared with the plots of fig. 19(a), the zero crossings of the curves indicate lesser cutoff β-values (for v = 1, the cutoff β-value is close to 5.918×10^6 m^{-1}) than the β-values of the first existing mode. This essentially indicates that the results of figs. 19(a) and 19(b) are consistent.

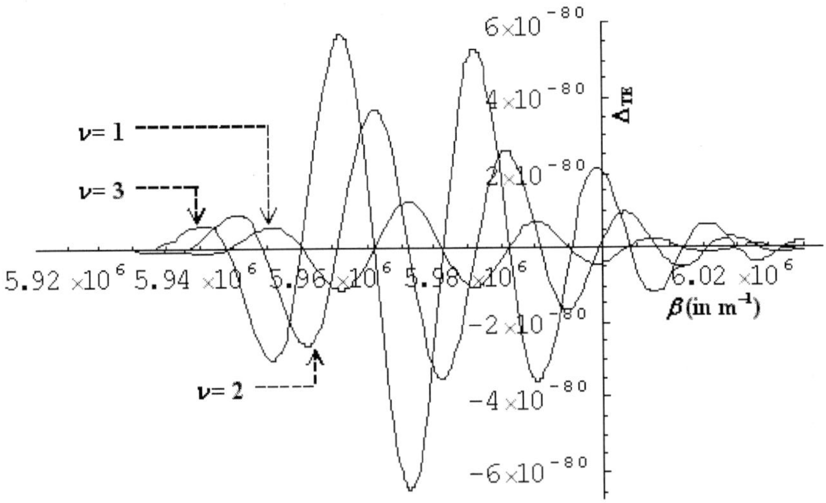

Fig. 19a. Plot of the dispersion relation for the TE modes.

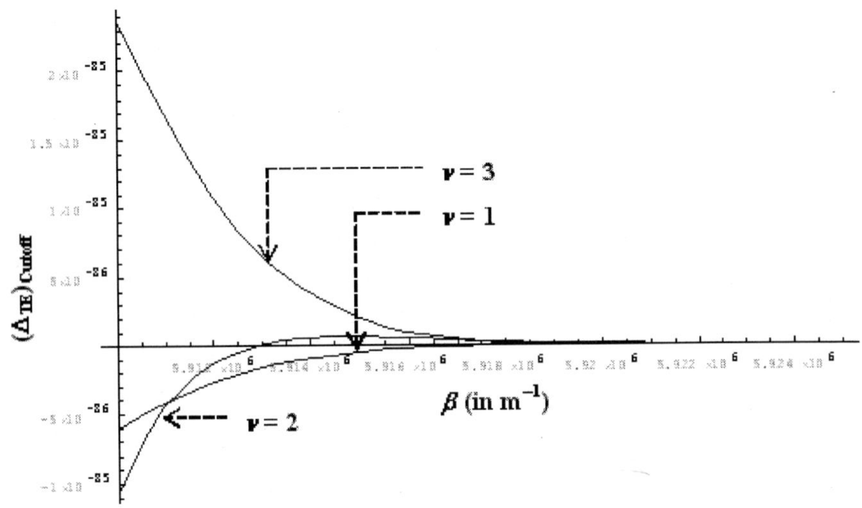

Fig. 19b. Plot of the cutoff characteristics for the TE modes.

Figs. 20 present the results corresponding to the TM modes for three different values of ν, namely 1, 2 and 3. Fig. 20(a) illustrates the plots of the left hand side of eq. (44), and fig. 20(b) corresponds to the cutoff plots when eq. (44) is solved under the limit $w^2 \rightarrow 0$. From fig. 20(a) we notice that the propagation constant for the first mode corresponding to $\nu = 1$ exists around $\beta = 6.076 \times 10^6$ m^{-1}, which is smaller than the correspondingly observed zero crossing in the case of TE modes. Thus, as compared to the TE modes, the TM modes have the

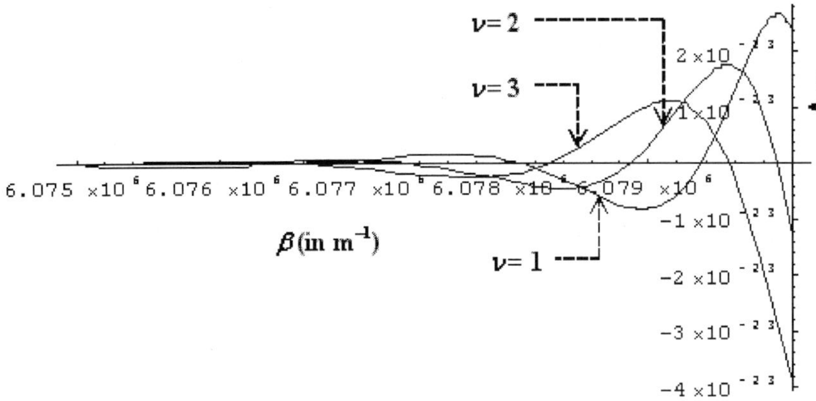

Fig. 20a. Plot of the dispersion relation for the TM modes.

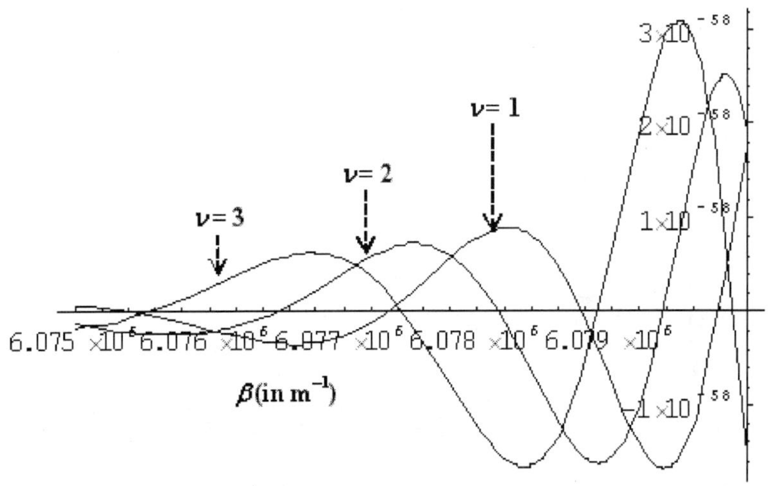

Fig. 20b. Plot of the cutoff characteristics for the TM modes.

tendency the lower the β-values. We also notice that the propagation constants increase with the increase in azimuthal index ν, which is similar to the situation observed in the case of TE modes. As observed from fig. 20(b) corresponding to the cutoff feature in the case of TM modes, the first zero crossings of the curves exist earlier than those seen in the corresponding plots in fig. 20(a), indicating thereby the consistency of the results presented in figs. 20(a) and 20(b). The analysis of the dispersion characteristics reveals the noticeable fact that the TE eigenmodes propagate in LCTOF with larger propagation constants as compared to the TM eigenmodes.

4.3 Features of power transport

As stated before, the individual useful properties of liquid crystal fibers and tapered fibers essentially motivate for the investigation of LCTOFs - the subject matter of the ongoing section. The applications of optical fibers are determined based on the power confinement characteristics. Thus, apart from the propagation behavior of the guide in terms of dispersion features and mode cutoffs, a glimpse of the power confinement factor remains equally important in order to emphasize the usefulness of the guide. As such, the discussion of the present subsection is pivoted to the analytical investigation of the power confinement in LCTOF structure. Using Maxwell's equations, a rigorous analysis is made of the confinement factors of the low order TE and TM modes sustained in the different LCTOF sections. In this context, variations in the core/clad dimensions are considered, and the illustrations are made of the power confinement factors against the length of the tapered section of the LCTOF.

4.3.1 Expressions of power for TE mode

In eqs. (33), (34) and (35), the values of arbitrary constants $A_{\phi 1}$, $A_{\phi 2}$, $A_{\phi 3}$ and $A_{\phi 4}$ can be evaluated by implementing the continuity conditions. Finally, after deducing the values of $A_{\phi 2}$, $A_{\phi 3}$ and $A_{\phi 4}$ in terms of $A_{\phi 1}$, the power [46] transmitted through the different sections of anisotropic LCTOF can be obtained as [58,59]

$$P_I = A_{\phi 1}^2 \frac{\pi}{\omega \mu_0} \left\{ u \int_0^{\rho_a} \rho J_1(u\rho) J_1'(u\rho) d\rho + \int_0^{\rho_a} (J_1(u\rho))^2 \, d\rho \right\}, \tag{45}$$

$$P_{II} = A_{\phi 1}^2 \frac{\pi}{\omega \mu_0} \Bigg[C_1^2 \left\{ w \int_{\rho_a}^{\rho_b} \rho I_1(w\rho) I_1'(w\rho) d\rho + \int_{\rho_a}^{\rho_b} (I_1(w\rho))^2 \, d\rho \right\}$$
$$+ C_2^2 \left\{ w \int_{\rho_a}^{\rho_b} \rho K_1(w\rho) K_1'(w\rho) d\rho + \int_{\rho_a}^{\rho_b} (K_1(w\rho))^2 \, d\rho \right\} + C_1 C_2 \Bigg\{ 2 \int_{\rho_a}^{\rho_b} K_1(w\rho) I_1(w\rho) d\rho$$
$$+ w \int_{\rho_a}^{\rho_b} \rho K_1(w\rho) I_1'(w\rho) d\rho + w \int_{\rho_a}^{\rho_b} \rho I_1(w\rho) K_1'(w\rho) d\rho \Bigg\} \Bigg], \tag{46}$$

$$P_{III} = A_{\phi 1}^2 \frac{\pi}{\omega \mu_0} \left(\frac{C_1 I_1(w\rho_b) + C_2 K_1(w\rho_b)}{K_1(v\rho_b)} \right)^2 \left[\int_{\rho_b}^{\infty} \{K_1(v\rho)\}^2 \, d\rho + v \int_{\rho_b}^{\infty} \rho K_1(v\rho) K_1'(v\rho) d\rho \right] \tag{47}$$

where

$$C_1 = \frac{w K_1'(w\rho_a) J_1(u\rho_a) + K_1(w\rho_a) \left\{ \frac{1}{\rho_a} (J_1(u\rho_a) - J_1(u\rho_a)) - u J_1'(u\rho_a) \right\}}{w K_1'(w\rho_a) I_1(w\rho_a) + K_1(w\rho_a) \left\{ \frac{1}{\rho_a} (I_1(w\rho_a) - K_1(w\rho_a)) - w I_1'(w\rho_a) \right\}}, \tag{48}$$

$$C_2 = \frac{J_1(u\rho_a)\left\{wI_1'(w\rho_a) + \frac{1}{\rho_a}(K_1(w\rho_a) - I_1(w\rho_a))\right\} - uJ_1'(u\rho_a)I_1(w\rho_a)}{K_1(w\rho_a)\left\{wI_1'(w\rho_a) + \frac{1}{\rho_a}(K_1(w\rho_a) - I_1(w\rho_a))\right\} - wK_1'(w\rho_a)I_1(w\rho_a)}. \tag{49}$$

Eqs. (45), (46) and (47), respectively, determine the power propagating through the fiber core, inner dielectric clad and the outer liquid crystal clad of the LCTOF under the situation when the TE modes are excited. In eqs. (45)–(49) ρ_a and ρ_b are the localized values of the LCTOF core and the inner clad radii, respectively; the outermost liquid crystal section is having an infinite extension. We used the split-step technique to analyze the problem, and ρ_a and ρ_b are the values of radii of a particular step. It is also to be remembered that the values of propagation constant β in all the equations are defined by eq. (4).

Further, in eqs. (45), (46) and (47), the constant $A_{\phi 1}$ can be determined by a normalization condition considering the input power. Now, if P_T is the total power transported by the TE_{01} modes of LCTOF, i.e.

$$P_T = P_I + P_{II} + P_{III}, \tag{50}$$

then P_I/P_T, P_{II}/P_T and P_{III}/P_T will, respectively, determine the power confinement factor in the fiber core, inner clad and the outer clad of the LCTOF.

4.3.2 Expressions of power for TM mode

Following the above procedure implemented for TE modes, by the use of eqs. (38), (39) and (40) along with the continuity conditions, the expressions of power in the different sections of LCTOF under the TM mode excitation can finally be derived as [58,59]

$$p_I = B_{\rho 3}^2 \frac{\pi\omega\varepsilon_0 n_1^2}{\beta} \frac{I_1(w\rho_a)}{J_1(u\rho_a)}\left[1 + \left\{\frac{1-\left(\frac{n_1}{n_2}\right)^2}{\left(\frac{n_1}{n_2}\right)^2 - 1}\right\}\right]^2 \left\{\int_0^{\rho_a} \rho\{J_1(u\rho)\}^2 d\rho\right\}, \tag{51}$$

$$p_{II} = B_{\rho 3}^2 \frac{\pi\omega\varepsilon_0 n_2^2}{\beta}\left\{\int_{\rho_a}^{\rho_b} \rho\left\{\frac{I_1(u\rho_a)\left\{1-\left(\frac{n_1}{n_2}\right)^2\right\}}{K_1(w\rho_a)\left\{\left(\frac{n_1}{n_2}\right)^2 - 1\right\}} K_1(w\rho) + I_1(w\rho)\right\}^2 d\rho\right\}, \tag{52}$$

$$p_{III} = B_{\rho 3}^2 \frac{\pi\omega\varepsilon_0 n_e^2}{\beta}\left[\frac{1}{K_1(v'\rho_b)}\left\{I_1(w\rho_b) + \frac{K_1(w\rho_b)I_1(w\rho_a)}{K_1(w\rho_a)}\left(\frac{1-\left(\frac{n_1}{n_2}\right)^2}{\left(\frac{n_1}{n_2}\right)^2 - 1}\right)\right\}\right]^2$$

$$\times \left\{ \int_{\rho_b}^{\infty} \rho \{K_1(v'\rho)\}^2 \, d\rho \right\}. \tag{53}$$

In eqs. (51), (52) and (53), the constant $B_{\rho 3}$ can be determined by the normalization condition considering the input power. Now, if p_T is the total power transported by the TM modes of the LCTOF, i.e.

$$p_T = p_I + p_{II} + p_{III}, \tag{54}$$

then p_I/p_T, p_{II}/p_T and p_{III}/p_T will, respectively, determine the relative power distributions in the LCTOF core, inner clad and the outer clad regions. In eqs. (51), (52) and (53) too, the parameters ρ_a, ρ_b and β will assume the forms as described above in the case of TE modes.

4.3.3 Analysis of power transmission by TE and TM modes

We now analyze the LCTOF characteristics in respect of the power confinement factors (or the relative power distributions) corresponding to the cases of TE and TM modes. In our computations, we consider the core/inner clad RI values as n_1 = 1.462 and n_2 = 1.458, respectively. Further, as stated before, the infinitely extended outermost section is taken to be nematic liquid crystal BDH mixture 14616 having the respective ordinary and extraordinary RI values as n_o = 1.457 and n_e = 1.5037. For simplicity, we considered the modes with the azimuthal index value ν = 1. The taper length l is taken to be 5 cm and the operating wavelength as 1.55 μm.

Figs. 21a, 21b and 21c illustrate the logarithmic plots of the power confinement patterns in the core, the inner clad and the outer clad, respectively, of the LCTOF under the case of TE mode excitation, and the azimuthal mode index ν = 1. While obtaining this, the input end core radius of the LCTOF tapered region is taken to be fixed (as 60 μm) whereas the output end core radius is varied (to be as 80 μm, 100 μm, 120 μm and 140 μm). We observe in these figures that the confinement factor increases with the increase in taper length along the direction of propagation; the lowest value of confinement corresponds to the situation when the outer core radius remains minimum (i.e. 80 μm). This becomes explicit because lesser amount of power is transported by the guides of lower dimensions. Apart from this, a gradual increase in power confinement remains due to a steady increase in the LCTOF dimension as the wave propagates across the tapered section. We also notice that the confinement reaches a kind of saturation in the region near the output end of the tapered section.

Further analysis of figs. 21 reveals that, apart from the trend of increase in confinement factor, a very small amount of power is confined within the tapered core section (fig. 21a), and this feature sustains with all the chosen dimensions of LCTOF. Corresponding to the similar values of fiber dimensions and other operating conditions, the confinement is increased in the inner clad (fig. 21b) along with the trend of its variation remaining almost same, as that noticed in fig. 21a. The confinement shows a pronounced enhancement in the outer clad (fig. 21c), and remains maximum in this region. It is also seen that the increase in confinement in the outermost liquid crystal section simultaneously causes to decrease the same in the LCTOF core or the inner clad, and this may be viewed in a way as if the power

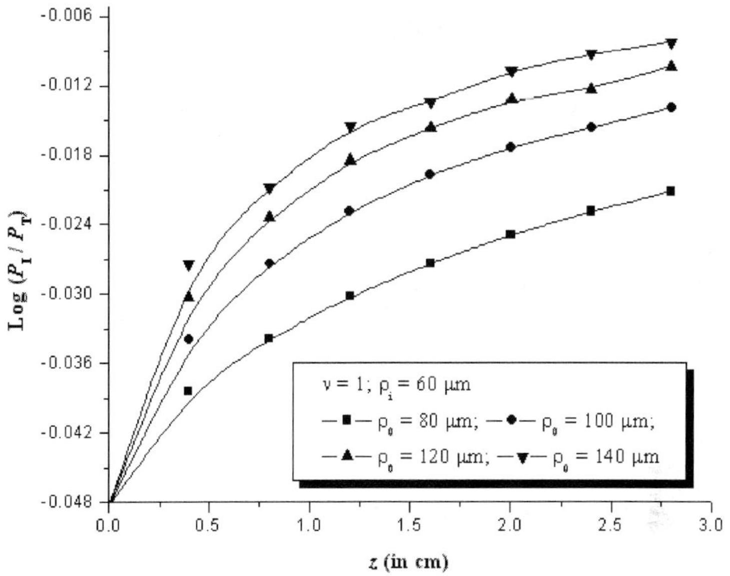

Fig. 21a. TE mode power confinement in LCTOF core.

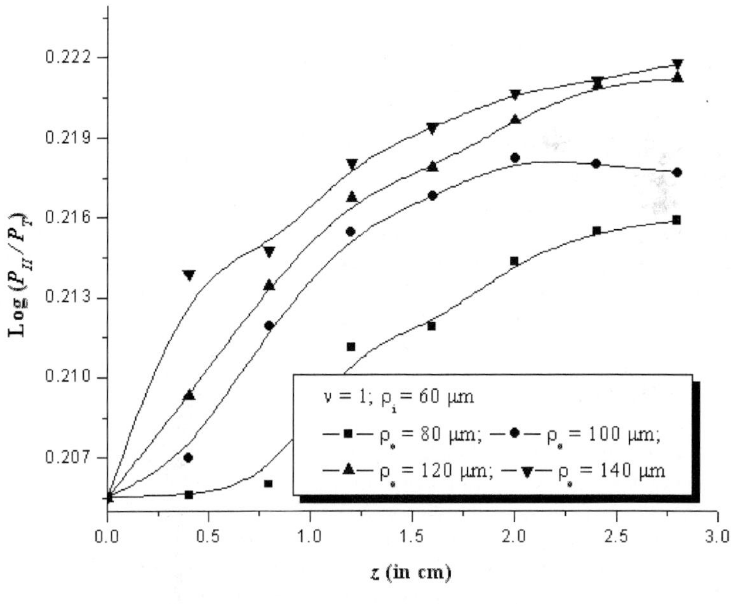

Fig. 21b. TE mode power confinement in the inner clad of LCTOF.

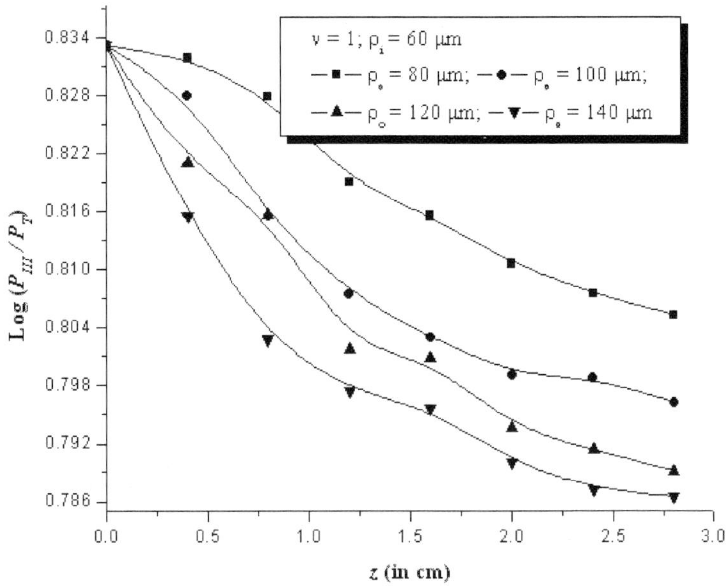

Fig. 21c. TE mode power confinement in the outermost clad of LCTOF.

is *leaking off* the fiber core, and propagating through the clad. This feature is essentially attributed to the presence of liquid crystal in the outermost clad region, and is of much use in sensing applications.

Figs. 22 correspond to the logarithmic variations of the power confinement for TE modes with $\nu = 1$ when the output end radius of the tapered core is fixed as 100 μm, and the input end radius is varied. We observe from the figures that the power confinement remains maximum corresponding to the minimum value of the input core radius (i.e. 10 μm). With the increase in LCTOF core input end radius, the confinement becomes uniform without exhibiting much change along the taper length, and this is very much obvious as the taper section undergoes maximum variation in respect of its dimensional structure along its length with minimum value of radius of the input end.

Fig. 22a shows that the power confinement remains almost uniform when the core radius at the input end is 70 μm, and it presents maximum variation corresponding to the situation when the input end core radius is 10 μm. On comparing figs. 22a, 22b and 22c we observe that the maximum amount of power confinement is attained in fig. 22c. This is just a replica of the situation that we observe in figs. 21c – the maximum amount of power is confined in the outermost section of the LCTOF. This feature of LCTOF is expected to find prominent usefulness in optical sensing, particularly the situations when the evanescent field sensing remains the best option. Apart from sensing, such a characteristic of LCTOF would also be useful in field coupling devices, wherein a relatively high amount of power is required to be present in the outermost section of the fiber.

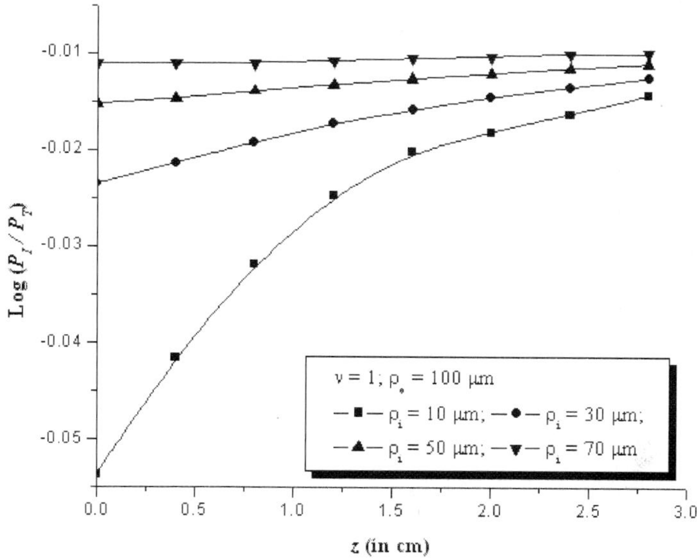

Fig. 22a. TE mode power confinement in LCTOF core.

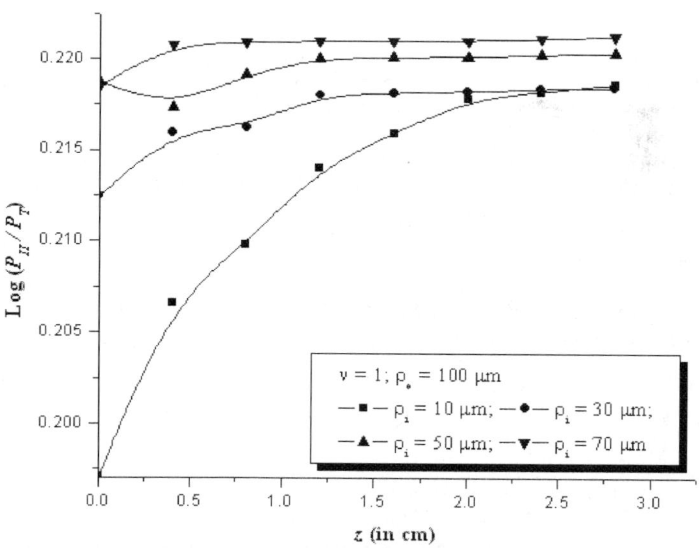

Fig. 22b. TE mode power confinement in the inner clad of LCTOF.

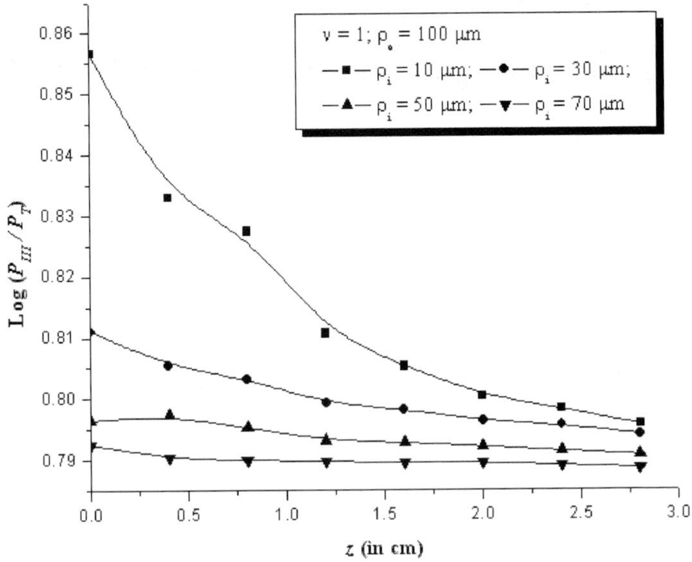

Fig. 22c. TE mode power confinement in the outermost clad of LCTOF.

Figs. 23a, 23b and 23c illustrate the logarithmic plots of the relative power distributions in the LCTOF core, the inner clad and the outermost liquid crystal clad, respectively, along the taper length. These figures present a comparative feature of the cases with different values of the azimuthal index. We consider two illustrative values (viz. 10 μm and 30 μm) of the core radius of the input taper end, and the core output end radius is taken to be 100 μm. We observe from fig. 23a that, for both the values of input core radius, the modes with higher azimuthal index ($\nu = 2$) transport a little higher amount of power as compared to that by the lower azimuthal index value ($\nu = 1$), and the difference increases as we move to look at the situations of power confinements in the inner and the outer clads. We notice from fig. 23c that the confinement remains maximum in the outermost liquid crystal clad, and also, the modes with $\nu = 2$ carry substantially large amount of power than those with $\nu = 2$.

The aforesaid discussions correspond to the situations when the TE modes are excited in an LCTOF. Now, considering the excitation of the TM modes, the logarithmic variations of the power confinement factor are illustrated in figs. 24 and 25 for the azimuthal index $\nu = 1$ and the tapered section length as 5 cm.

Fig. 24a corresponds to the power transmission through the LCTOF core when the different parameters are chosen similar to the case of TE modes, i.e. a fixed input end core radius as 60 μm and varying output end core radius as 80 μm, 100 μm, 120 μm and 140 μm. We observe from fig. 24a that a relatively high amount of power is confined into the core as compared to the situation of TE modes (fig. 21a). Further, fig. 24b shows that the highest amount of power is found to be sustained in the inner clad, which is in contrast to the feature observed corresponding to the case of TE modes which presents the maximum amount of power to be confined in the outermost clad. In the present case, of TM mode

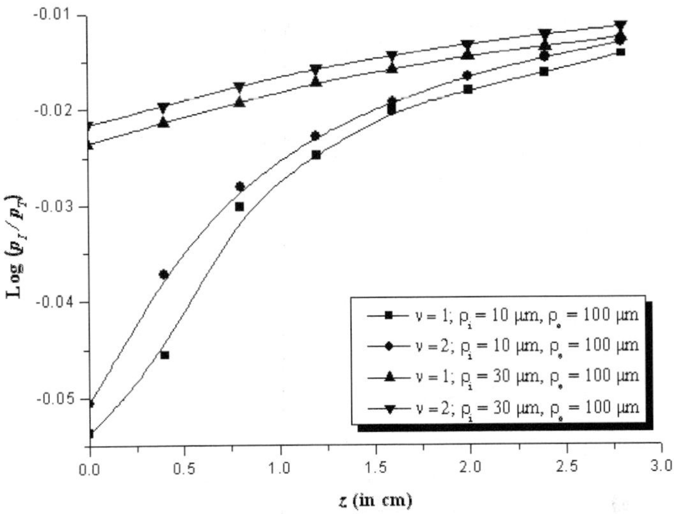

Fig. 23a. TE mode power distribution in the LCTOF core.

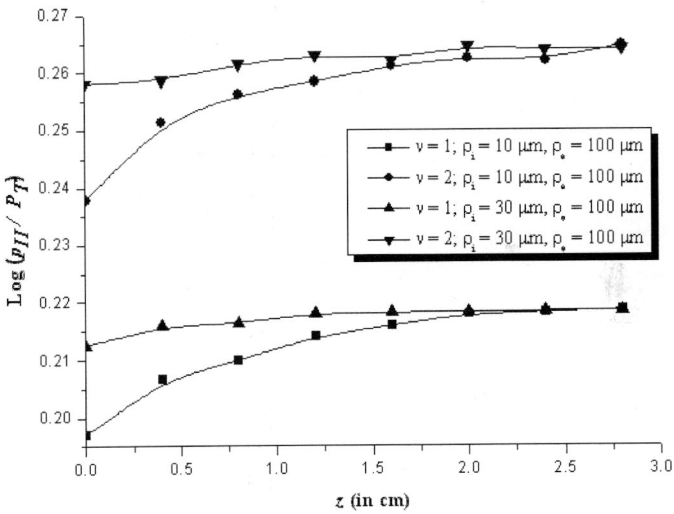

Fig. 23b. TE mode power distribution in the LCTOF inner clad.

excitation, we observe that the power confinement in the outermost liquid crystal clad (fig. 24c) is slightly less than that in the inner clad region.

We also observe that the trend of variation of the confinement factor in TM mode excitation seems to be opposite to that noticed corresponding to the case of TE modes. This is because, in figs. 24a and 24b, confinement falls with the increase in the taper length, and remains

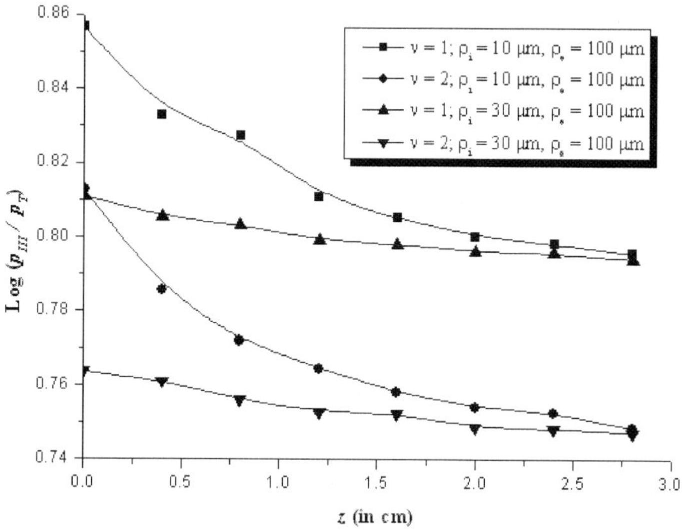

Fig. 23c. TE mode power distribution in the LCTOF outer clad.

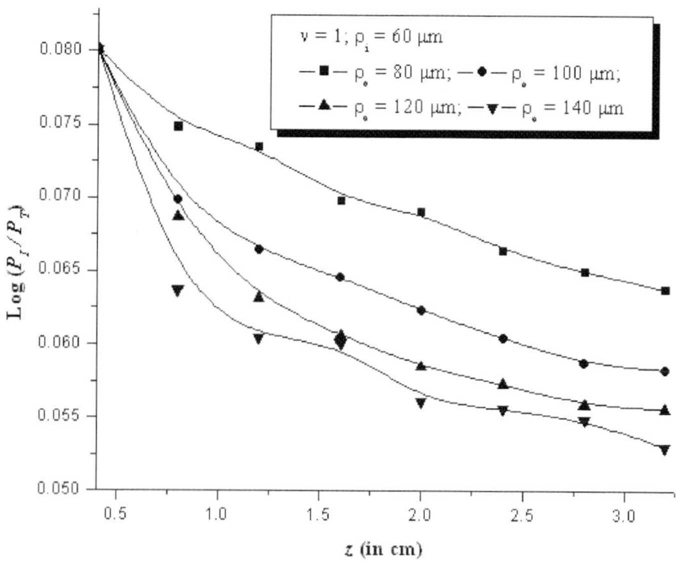

Fig. 24a. TM mode power confinement in LCTOF core.

minimum at the output end of the taper section. This is attributed to the TM mode properties whereby the power is gradually being coupled to the neighboring regions of the guide with increasing fiber dimensions.

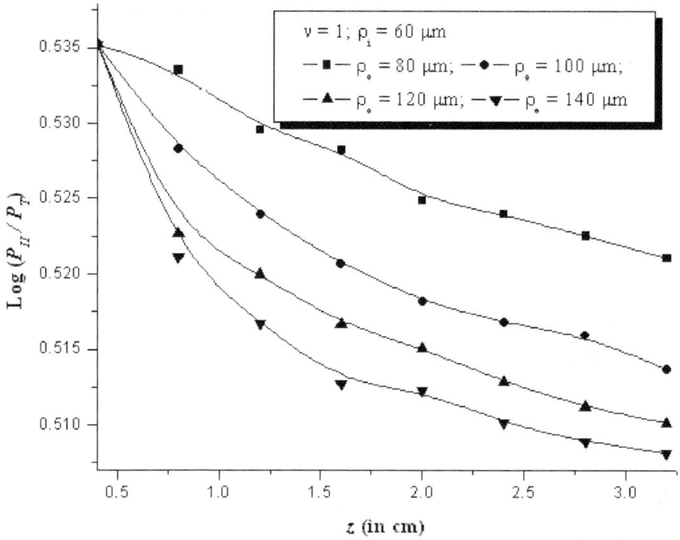

Fig. 24b. TM mode power confinement in the inner clad of LCTOF.

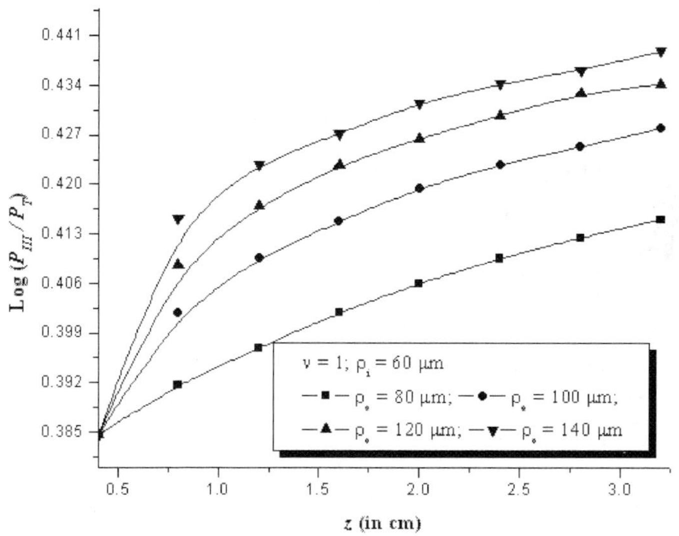

Fig. 24c. TM mode power confinement in the outermost clad of LCTOF.

Figs. 25 illustrate the logarithmic plots of power confinement in different fiber sections with fixed output end dimension of the taper section and varying size of the input end. Confinement in the fiber core is presented in fig. 25a, whereas those in the inner and the outer clads are illustrated in figs. 25b and 25c, respectively. We consider the output end

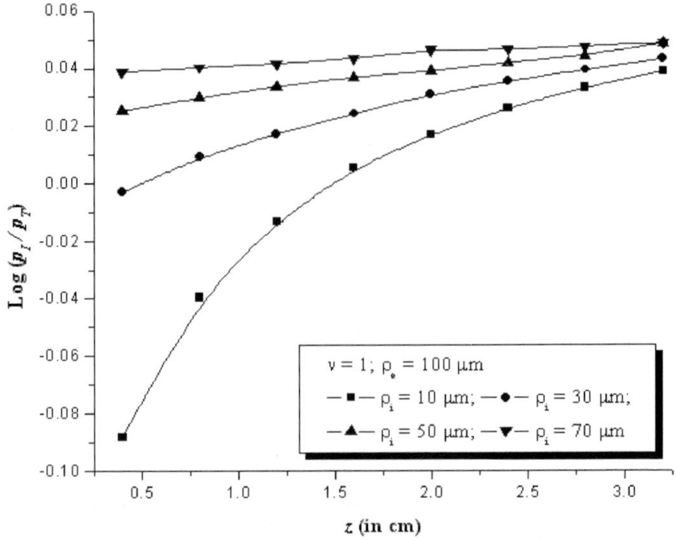

Fig. 25a. TM mode power confinement in LCTOF core.

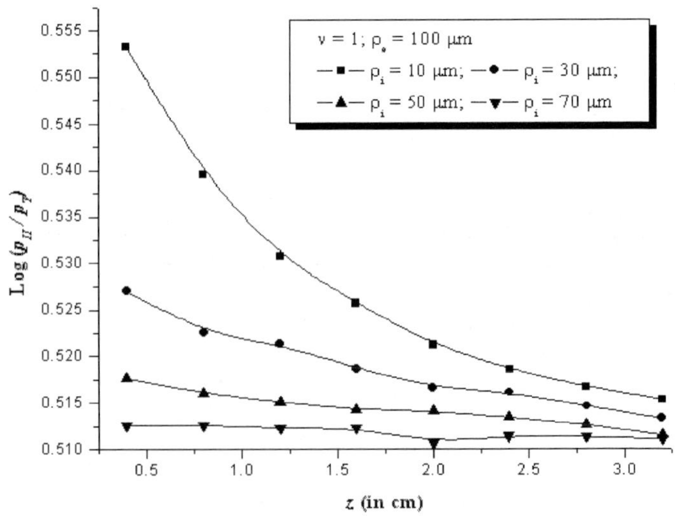

Fig. 25b. TM mode power confinement in the inner clad of LCTOF.

radius to be 100 μm whereas the different input end radii are 10 μm, 30 μm, 50 μm and 70 μm. We observe in these figures that the confinement remains strongest corresponding to the largest input end dimension, i.e. 70 μm radius. Also, with such a fiber structure, the confinement remains most uniform too. This is attributed to the fact that the waves face maximum variations in respect of their boundaries with large difference in the output and

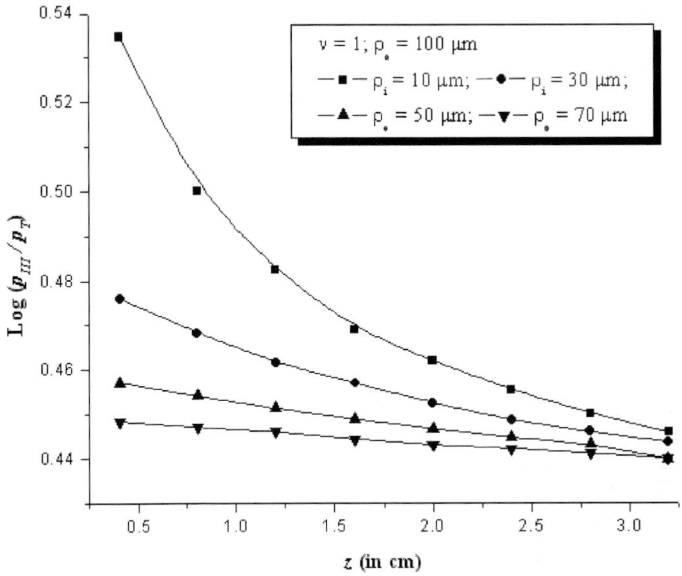

Fig. 25c. TM mode power confinement in the outermost clad of LCTOF.

the input end dimensions. However, the noticeable fact remains that the maximum amount of power is confined in the inner clad section, and the outermost clad sustains slightly less amount of power than the inner clad; the core section confines the minimum amount of power.

In order to have a comparative look at the power confinements by the modes with different azimuthal indices under the TM mode excitation, figs. 26a, 26b and 26c illustrate the logarithmic plots of the power confinement factor in the LCTOF core, the inner clad and the outermost liquid crystal clad, respectively, along the taper length. Once again, we use in our computations two illustrative values (viz. 10 μm and 30 μm) of the core radius of the input taper end, and the core output end radius is taken to be fixed as 100 μm. We observe from these figures that the modes with $\nu = 1$ and $\nu = 2$ transport almost similar amount of power in all the three LCTOF sections, which is unlike the situation observed in the case of the TE mode excitation where the modes with higher azimuthal index transmit substantially large amount of power than those with lower azimuthal index value.

It can thus be inferred that, in LCTOFs with radially anisotropic liquid crystal outermost clad, the TE modes transmit the maximum amount of power in the outermost section of the fiber. This feature is not that pronounced corresponding to the TM mode excitation as the confinement mostly remains in the inner clad section in this case. As stated before, the higher amount of power distribution in the outermost clad section can be interpreted as if the power is *leaking off* the LCTOF core, and transferred to the fiber clad. This feature is strongly observed in the case of TE mode excitation, and the phenomenon is attributed to the presence of radially anisotropic liquid crystal material in the outermost clad.

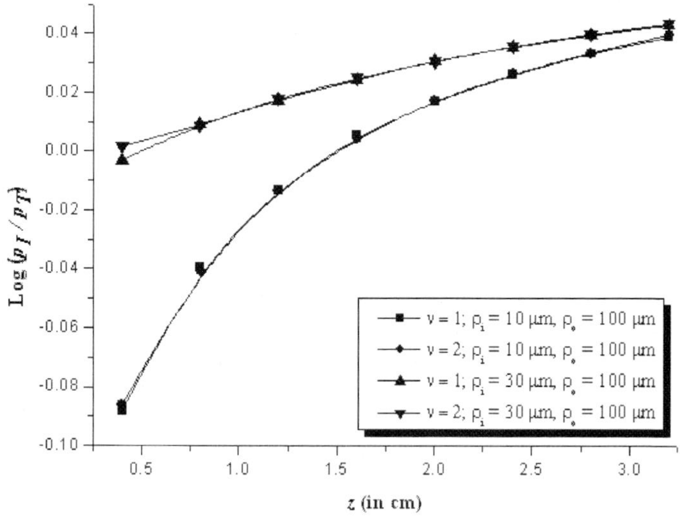

Fig. 26a. TM mode power distribution in LCTOF core.

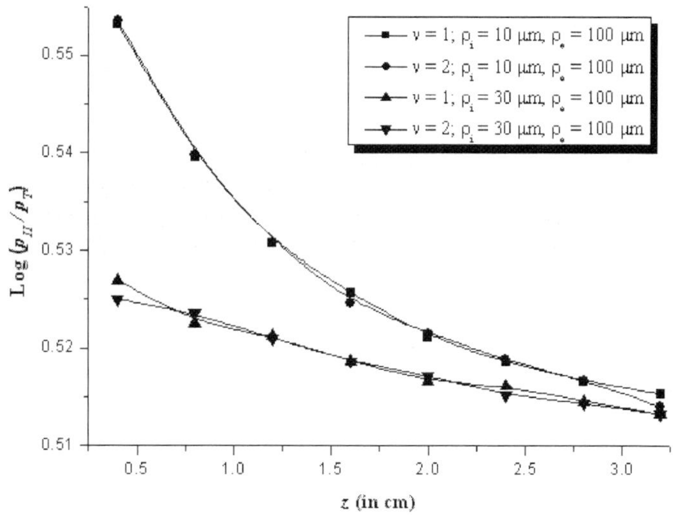

Fig. 26b. TM mode power distribution in LCTOF inner clad.

Apart from the material used in LCTOF fabrication, the taper structure of guide essentially plays a vital role to transfer power to the outermost clad. TOFs are already established in coupling applications, and the inclusion of liquid crystal section makes the structure more

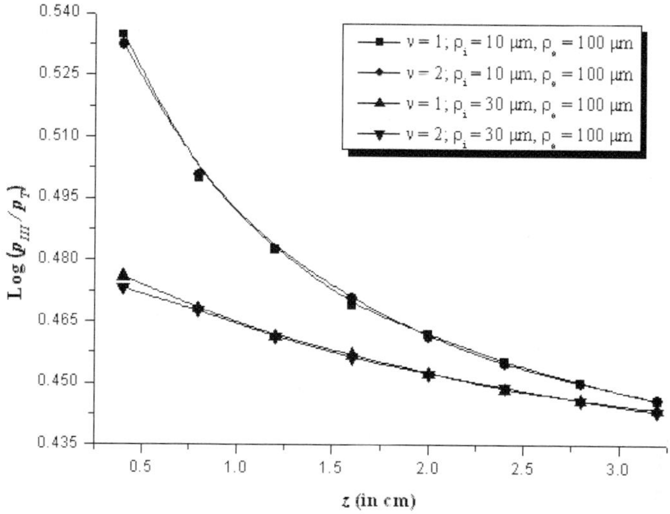

Fig. 26c. TM mode power distribution in LCTOF outer clad.

attractive for promising use. Thus, an amalgamation of the taper structure and the clad anisotropy brings in an enhanced confinement of power in the outermost clad section, and this feature opens up a demanding usefulness of LCTOFs in optical sensing and other coupling applications.

5. Conclusion

The article basically describes the EM wave propagation characteristics of TOFs considering various forms of them, viz. simple dielectric TOFs, dielectric TOFs with twists in the form of helical clad, and TOFs with radially anisotropic liquid crystal clad. The study of dielectric TOFs is presented under the assumption of a small variation of the core radius with the fiber length. The eigenvalue equations for such TOFs are developed, and the dispersion characteristics are described, which show almost similar features as that for conventional dielectric fibers. However, the study reveals that the tapered feature of fibers greatly affects the cutoff value of the normalized frequency parameter, and it is much reduced as compared to that exhibited by conventional fibers. The power transmission characteristics of such dielectric TOFs is also described in the form of the relative power distribution over the taper length considering the cases of meridional and the lowest skew modes. It is noticed that, in the case of meridional mode, the relative power exhibits some sort of linear dependence on the taper length. Corresponding to the skew mode, the power remains relatively uniform over the taper length, and the uniformity increases with the increase in the dimension of the taper output. The feature of uniform distribution of optical power in fibers remains much useful for communication purposes.

While dealing with the case of helical clad TOFs, the dispersion characteristics and the cutoff situations are deduced corresponding to two particular values of the helix pitch angle – helical turns being parallel (i.e. 0°) and perpendicular (i.e. 90°) to the direction of wave propagation. It is found that the modes travel through the fiber with higher propagation constant under the situation when the helical turns are perpendicular to the optical axis than the case of parallel wraps. However, when the propagation features are compared with the situation of fibers without helical wraps, it is noticed that the helical turns possess the tendency to increase the modal propagation constants. This exhibits the importance of the helix pitch angle in controlling the propagation characteristics of such TOFs.

Study of LCTOFs with radially anisotropic liquid crystal clads is also touched upon, and the dispersion characteristics as well as the features of power transmission corresponding to the TE and TM mode excitations are reported. It is found that the TE eigenmodes propagate in the guide with larger propagation constants as compared to the TM ones. The results further reveal that, in the case of TE mode excitation, a large amount of power remains in the outermost liquid crystal region with substantial difference in the power sustained in the fiber core and the inner dielectric clad. The existence of a large amount of power in the outermost clad region is attributed to the presence of radially anisotropic liquid crystal medium used in the LCTOF. Further, the tapered structure of guide also plays the role for the proliferation of *power transfer* to the outermost clad. As such, a large amount of TE mode power in the outermost liquid crystal clad essentially indicates the prominent use of LCTOFs in optical sensing and/or coupling applications. At this point, it is noteworthy that TOFs are proved much promising for field coupling, and the incorporation of liquid crystal section makes them more demanding, as demonstrated through the results.

6. Acknowledgement

This work is partially supported by the Fundamental Research Grant Project (FRGS/1/2011/TK/UKM/01/16) sanctioned by the Ministry of Higher Education, Malaysia. The author is thankful to Prof. Burhanuddin Yeop Majlis for constant encouragement and help. He is also grateful to Prof. S. Shaari for some stimulating discussions and fruitful suggestions.

7. References

[1] N. Engheta and P. Pelet, "Modes in chirowaveguides," *Opt. Lett.*, Vol. 14, 593–595, 1989.

[2] H. Cory and I. Rosenhouse, "Electromagnetic wave propagation along a chiral slab," *IEE Proc. H*, Vol. 138, 51–54, 1991.

[3] Kh.S. Singh, P.K. Choudhury, V. Misra, P. Khastgir, and S.P. Ojha, "Field cutoffs of three-layer parabolically deformed planar chirowaveguides," *J. Phys. Soc. Jpn.*, Vol. 62, 3778–3782, 1993.

[4] P.K. Choudhury and T. Yoshino, "Dependence of optical power confinement on core/cladding chiralities in a simple chirofiber," *Microw. and Opt. Tech. Lett.*, Vol. 32, 359–364, 2002.

[5] P.K. Choudhury and T. Yoshino, "TE and TM modes power transmission through liquid crystal optical fibers," *Optik*, Vol. 115, 49–56, 2004.

[6] A. Nair and P.K. Choudhury, "On the analysis of field patterns in chirofibers," *J. Electromag. Waves and Appl.*, Vol. 21, 2277–2286, 2007.

[7] A. Kumar, K. Thyagarajan, and A.K. Ghatak, "Analysis of rectangular core dielectric waveguides: an accurate perturbation approach," *Opt. Lett.*, Vol. 8, pp. 63–65, 1983.

[8] P.K. Choudhury, P. Khastgir, and S.P. Ojha, "Analysis of the guidance of electromagnetic waves by a deformed planar waveguide with parabolic cylindrical boundaries," *J. Appl. Phys.*, Vol. 71, pp. 5685–5688, 1992.

[9] P.K. Choudhury, P. Khastgir, S.P. Ojha, and K.S. Ramesh, "An exact analytical treatment of parabolically deformed planar waveguides near cutoff," *Optik*, Vol. 95, pp. 147–151, 1994.

[10] P.K. Shukla, P.K. Choudhury, P. Khastgir, and S.P. Ojha, "Comparative aspects of a metal-loaded triangular waveguide with uniform and non-uniform distribution of Goell's matching points," *J. Inst. Electron. Telecommun. Eng.*, Vol. 41, pp. 217–220, 1995.

[11] P.K. Choudhury, "On the preliminary study of a dielectric guide having a Piet Hein geometry," *Ind. J. Phys.*, Vol. 71B, pp. 191–196, 1997.

[12] P. Sharan, S.P. Ojha, P. Khastgir, and P.K. Choudhury, "Modal cutoff of an optical waveguide having a leminiscate of Bernoulli-type core cross-section," *Microw. and Opt. Tech. Lett.*, Vol. 14, pp. 170–172, 1997.

[13] P.K. Choudhury and O.N. Singh, "Some multilayered and other unconventional lightguides," In: *Electromagnetic fields in unconventional structures and materials* (O.N. Singh and A. Lakhtakia, Eds.), John Wiley: New York, pp. 289–357, 2000.

[14] D.T. Cassidy, D.C. Johnson, and K.O. Hill, "Wavelength-dependent transmission of monomode optical fiber taper," *Appl. Opt.*, Vol. 24, pp. 945–950, 1985.

[15] R.P. Payne, C.D. Hussey, and M.S. Yataki, "Modeling fused single-mode-fiber couplers," *Electron. Lett.*, Vol. 21, pp. 461–462, 1985.

[16] A.C. Boucouvalas and G. Georgiou, "Biconical taper coaxial optical fiber couplers," *Electron. Lett.*, Vol. 21, pp. 864–865, 1985.

[17] J.V. Wright, "Variational analysis of fused tapered couplers," *Electron. Lett.*, Vol. 21, pp. 1064–1065, 1985.

[18] S. Lacroix, F. Gonthier, and J. Bures, "All-fiber wavelength filter from successive biconical tapers," *Opt. Lett.*, Vol. 11, pp. 671–673, 1986.

[19] F. Gonthier, J. Lapierre, C. Veilleus, S. Lacroix, and J. Bures, "Investigation of power oscillations along tapered monomode fibers," *Appl. Opt.*, Vol. 26, pp. 444–449, 1987.

[20] S. Lacroix, R. Bourbouriars, F. Gonthier, and J. Bures, "Tapered monomode optical fibers: understanding large power transfer," *Appl. Opt.*, Vol. 26, pp. 4421–4425, 1987.

[21] H.S. Monteiro, R. Arradi, D.C. Dini, S.L.A. Carrara, and E. Conforti, "A spot reflection alignment system for fabrication of polarization-maintaining optical fiber couplers," *Microw. and Opt. Tech. Lett.*, Vol. 19, pp. 75–77, 1988.

[22] R.J. Black, F. Gonthier, S. Lacroix, J. Lapierre, and J. Bures, "Tapered fibers: an overview," *Proc. SPIE*, Vol. 839, pp. 2–19, 1988.

[23] W.-C. Chang, S.-J. Chan, H.-C. Hou, and C.-Y. Sue, "Measurement of coupling coefficients of Ni:LiNbO directional couplers," *Microw. and Opt. Tech. Lett.*, Vol. 19, pp. 24–27, 1988.

[24] M.C. Gabriel and N.O. Whittaker, "Measurement of optical waveguide coupling coefficients using multiple waveguide systems," *J. Light. Tech.*, Vol. 7, pp. 1343–1350, 1989.

[25] A. Bolle and L. Lundgren, "Analytical solution of the field in a fiber up-taper with a parabolic index profile," *IEE Proc. J.: Optoelectron.*, Vol. 137, pp. 301–304, 1990.

[26] K. Ono and H. Osawa, "Excitation characteristics of fundamental mode in tapered slab waveguides with nonlinear cladding," *Electron. Lett.*, Vol. 27, pp. 664–666, 1991.

[27] K. Ono, S. Tsuruta, T. Sakai, H. Osawa, and Y. Okamoto, "Highly efficient coupling between two planar waveguides with different refractive index profiles by using nonlinear cladding," *IEEE Phot. Tech. Lett.*, Vol. 4, pp. 381–384, 1992.

[28] A.K. Singh, P.Khastgir, O.N. Singh, and S.P. Ojha, "Local field configuration modes in a weakly guiding optical fiber with a conically annular core: an analytical study," *Jpn. J. Appl. Phys.*, Vol. 32, pp. L71–L74, 1993.

[29] D.B. Singh, P. Khastgir, U.K. Singh, and O.N. Singh, "Modal analysis of a dielectric waveguide with planar guiding-nonguiding boundaries with a flare in the direction of propagation," *Microw. and Opt. Tech. Lett.*, Vol. 17, pp. 394–398, 1998.

[30] D.B. Singh, P. Khastgir, U.K. Singh, and O.N. Singh, "Unattenuated global modes in a planar dielectric waveguide with a flare in the direction of propagation," *Microw. and Opt. Tech. Lett.*, Vol. 19, pp. 77–80, 1998.

[31] A.R. Nelson, "Coupling in optical waveguides by tapers," *Appl. Opt.*, Vol. 14, pp. 3012–3015, 1975.

[32] M.G.F. Wilson and G.A. The, "Tapered optical directional couplers," *IEEE Trans. Microw. Theory Tech.*, Vol. MTT-23, pp. 85–92, 1975.

[33] T.K. Lin and J.P. Marton, "An analysis of optical waveguide tapers," *J. Appl. Phys.*, Vol. 18, pp. 53–62, 1979.

[34] R.J. Black, F. Gonthier, S. Lacroix, and J.D. Love, "Tapered single-mode fibres and devices: I. adiabaticity criteria," *IEE Proc. -J*, Vol. 138, pp. 343–354, 1991.

[35] J.R. Pierce, *Traveling wave tubes*, D. Van Nostrand, New Jersey, 1950.

[36] U.N. Singh, O.N. Singh II, P. Khastgir, and K.K. Dey, "Dispersion characteristics of a helically cladded step-index optical fiber: an analytical study," *J. Opt. Soc. Am. B*, Vol. 12, pp. 1273–1278, 1995.

[37] C. Veilleux, J. Lapierre, and J. Bures, "Liquid-crystal-clad tapered fibers," *Opt. Lett.*, Vol. 11, pp. 733–735, 1986.

[38] M. Green and S.J. Madden, "Low loss nematic liquid crystal cored fiber waveguides," *Appl. Opt.*, Vol. 28, pp. 5202–5203, 1989.

[39] H. Lin, P.P. Muhoray, and M.A. Lee, "Liquid crystalline cores for optical fibers," *Mol. Cryst. Liq. Cryst.*, Vol. 204, pp. 189–200, 1991.

[40] S.-T. Wu and U. Efron, "Optical properties of thin nematic liquid crystal cells," *Appl. Phys. Lett.*, Vol. 48, pp. 624–636, 1986.

[41] N. Amitay and H.M. Presby, "Optical fiber up-tapers modeling and performance analysis," *J. Light. Tech.*, Vol. 7, pp. 131–137, 1989.

[42] M. Abramowitz and I.A. Stegun, *Handbook of mathematical functions*, New York: Dover; 1965.

[43] M.H. Lim, S.C. Yeow, P.K. Choudhury, and D. Kumar, "Towards the dispersion characteristics of tapered core dielectric optical fibers," *J. Electromag. Waves and Appl.*, Vol. 20, 1597–1609, 2006.

[44] D. Gloge, "Weakly guiding fibers," *Appl. Opt.*, Vol. 10, pp. 2252–2258, 1971.

[45] P.K. Choudhury and T. Yoshino, "A rigorous analysis of the power distribution in plastic clad annular core optical fibers," *Optik*, Vol. 113, pp. 481–488, 2002.

[46] A.H. Cherin, *An introduction to optical fibers*, McGraw-Hill, New York, 1987.

[47] S.C. Yeow, M.H. Lim, and P. K. Choudhury, "A rigorous analysis of the distribution of power in plastic clad linear tapered fibers," *Optik*, Vol. 117, 405–410, 2006.

[48] D. Kumar and O.N. Singh II, "Some special cases of propagation characteristics of an elliptical step-index fiber with a conducting helical winding on the core-cladding boundary - an analytical treatment," *Optik*, Vol. 112, pp. 561-566, 2000.

[49] D. Kumar and O.N. Singh II, "An analytical study of the modal characteristics of annular step-index waveguide of elliptical cross-section with two conducting helical windings on the two boundary surfaces between the guiding and the non-guiding regions," *Optik*, Vol. 113, pp. 193-196, 2002.

[50] D. Kumar and O.N. Singh II, "Modal characteristic equation and dispersion curves for an elliptical step-index fiber with a conducting helical winding on the core-cladding boundary – an analytical study," *J. Light. Tech.*, Vol. 20, pp. 1416 –1424, 2002.

[51] D. Kumar, P.K. Choudhury, and F.A. Rahman, "Towards the characteristic dispersion relation for step-index hyperbolic waveguide with conducting helical winding," *Prog. in Electromagn. Res.*, Vol. PIER 71, pp. 251–275, 2007.

[52] D. Kumar, P.K. Choudhury, and O.N. Singh II, "Towards the dispersion relations for dielectric optical fibers with helical windings under slow- and fast-wave considerations – a comparative analysis," *Prog. in Electromagn. Res.*, Vol. PIER 80, pp. 409–420, 2008.

[53] A.H.B.M. Safie and P.K. Choudhury, "On the field patterns of helical clad dielectric optical fibers," *Prog. in Electromagn. Res.*, Vol. PIER 91, 69–84, 2009.

[54] P. K. Choudhury and D. Kumar, "On the slow-wave helical clad elliptical fibers," *J. Electromagn. Waves and Appl.*, Vol. 24, pp. 1931-1942, 2010.

[55] D.A. Watkins, *Topics in electromagnetic theory*, Wiley, USA, 1958.

[56] C.C. Siong and P.K. Choudhury, "Propagation characteristics of tapered core helical clad dielectric optical fibers," *J. Electromagn. Waves and Appl.*, Vol. 23, pp. 663–674, 2009.

[57] P.K. Choudhury and P.T.S. Ping, "On the dispersion relations of tapered core optical fibers with liquid crystal clad," *Prog. in Electromagn. Res.*, Vol. PIER 118, pp. 117-133, 2011.

[58] P.K. Choudhury and W.K. Soon, "On the tapered optical fibers with radially anisotropic liquid crystal clad," *Prog. in Electromagn. Res.*, Vol. PIER 115, pp. 461–475, 2011.

[59] P.K. Choudhury and W.K. Soon, On the transmission by liquid crystal tapered optical fibers," *Optik*, Vol. 122, pp. 1061–1068, 2011.

9

Robust Fiber-Integrated High-Q Microsphere for Practical Sensing Applications

Ying-Zhan Yan[1,2], Shu-Bin Yan[1], Zhe Ji[3], Da-Gong Jia[3], Chen-Yang Xue[2], Jun Liu[1], Wen-Dong Zhang[1,2] and Ji-Jun Xiong[1,2]

[1]*Key Laboratory of Instrumentation Science and Dynamic Measurement (North University of China), Ministry of Education, Taiyuan,*

[2]*Science and Technology on Electronic Test and Measurement Laboratory, North University of China, Taiyuan,*

[3]*Key Laboratory of Opto-Electronics Information and Technical Science, Tianjin University, Tianjin*

P. R. China

1. Introduction

Whispering gallery modes (WGM) resonators (K. J. Vahala, 2003; A. B. Matsko et al., 2006) have been extensively studied in a large variety of geometrical shapes and in a wide range of promising studies and applications (A. Chiasera et al., 2010), both fundamental like *CQED* and practical like low threshold lasers and sensors. The Silica spherical microresonators, made from commercial optical fibers, could support WGM with ultra-high quality factors (*Q*) (about $10^{7\sim9}$). With the merits of very small model volume and good compatibility with fiber-integrated optics, they are promising for a number of passive and active devices as filters, lasers and modulators (A. Chiasera et al., 2010). In WGMs, the electromagnetic wave is strongly confined within the microcavity in the manner of "totally internally reflection" (TIR). Thus, there is minimal reflection optical losses at the cavity interface. With the negligible silica absorption loss, the resonant modes can reach the ultra-high *Q*. Benefitting from the high-*Q* (M. L. Gorodetsky et al., 1996), the WGMs behave as extreme-narrow linewidth resonant dips. Therefore, a small shift of the resonant dip can be detected with high resolution, meaning that the WGMs are potential in the sensing researches (F. Vollmer et al., 2008; Y. Sun et al., 2008; M. Sumetsk et al., 2007; F. Xu et al., 2008; I. M. White et al., 2008). Intuitively (as shown in Fig. 1), photons can travel around many trips inside the high-*Q* microcavity, and interact with the detected matters around the microcavity many times (Fig. 1-a). By contrast, in the traditional fiber sensors each photon can interact with the detected matters only once (Fig. 1-b). Therefore, the microcavity-based sensors can have more superior performance than the conventional fiber sensors, demonstrating a higher sensitivity or resolution.

However, despite the promising prospect, there are great challenges in the practical microcavity-based sensing devices. The microsphere needs an external fiber taper to excite its WGMs by an external fiber taper (M. Cai et al., 2000). This discrete coupling system can't provide sufficient stability, because the effective coupling can be affected, and even be broken when loading the vibration on the taper or the microsphere. Moreover, the exposure of the

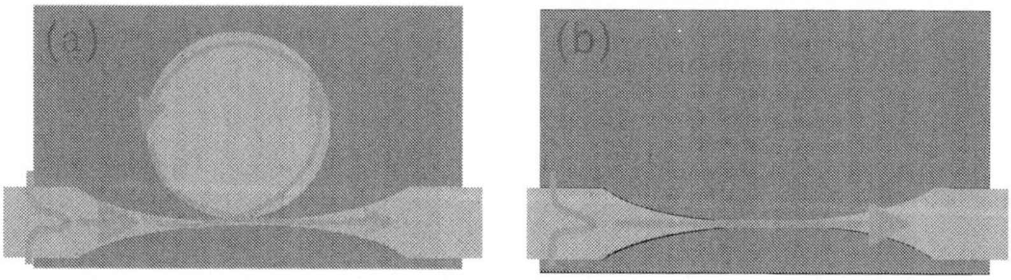

Fig. 1. Comparison of the sensing mechanisms of the microcavity-based sensors and the traditional fiber sensors. (a) The sensing mechanism for the microcavity-based sensor, while (b) for the traditional fiber sensors.

coupling system also makes the Q-maintenance challenging greatly, because the water and the dust in the air could spoil the Q drastically. These above problems are some challenges when promoting the sensing research into practical applications.

In this chapter, we propose one solution to these problems so as to construct a robust structure. Furthermore, we demonstrate the practical thermal sensing application based on the robust fiber-integrated microsphere coupling structure. The flow of this chapter is as follows: First, we introduce the construction of the traditional microcsphere coupling system, and point out the problems which hinder the practical application in the traditional microcavity sensing systems. Then we illustrate the solution to these problems, which for the first time, is proposed by our group. Afterwards, by using our improved robust structure, thermal sensing experiments will be demonstrated to illustrate the practical application. The last section of this chapter is the conclusions and the expectations.

2. Fiber-taper coupled microsphere system

2.1 Fabrication

The excitation of the microsphere WGMs needs an external coupler. Here, the optical tapered fiber is used to excite the microsphere optical modes evanescently. Tapered fibers are first suggested in applications as bidirectional fiber couplers and polarizers in the early 1980s. Mode field evolution in a tapered fiber has been thoroughly studied since then. In order to fulfill a high coupling efficiency between the two parts, phase matching is required where the propagation constant of the fiber taper should be matched to that of the microsphere cavity (J. C. Knight et al., 1997). Consequently, microspheres with different diameters (D) require the tapered fibers with different taper region and different taper angle. To fabricate tapers meeting various demands, a versatile fabrication method is used. Fig. 2-a is the tapered fiber fabrication setup for the heat-drawing manner used in our lab. Firstly, a section of a standard single mode fiber is prepared by removing the plastic coating and then cleaning it with anhydrous alcohol. After that, clamp it on a pair of motorized flats which should be paralleled, and each with a synchronous tunable speed. The thermal resource is a hydrogen microtorch placing underneath the optical fiber prepared. A microscope is installed vertically for monitoring the pulling process. During the procedure, the microtorch is lighted while the two flats move apart in opposite directions at a certain drawing speed. Low loss is essential in this process, and the taper loss is monitored by detecting the power transmission through the tapered fiber with a New-Focus dynamometer. As shown in the insets of Fig. 2-b, a low loss of 2.9 db is achieved through adjusting the two stages to keep them horizontal. This is

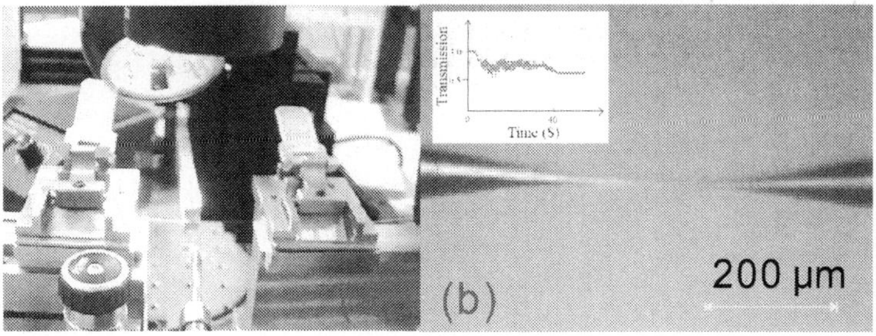

Fig. 2. (a) Tapered fiber fabrication setup in our lab. (b) Typical micrography of a taper, insets: transmission power monitoring during taper fabrication.

the critical point to fabricate the tapered fiber. In addition, the flame position, hydrogen flow and the pulling speed are crucial factors. Through controlling and adjusting these parameters precisely, tapered fibers with desired tapered waist diameter, taper region and taper angle can be prepared, as shown in Fig. 2-b.

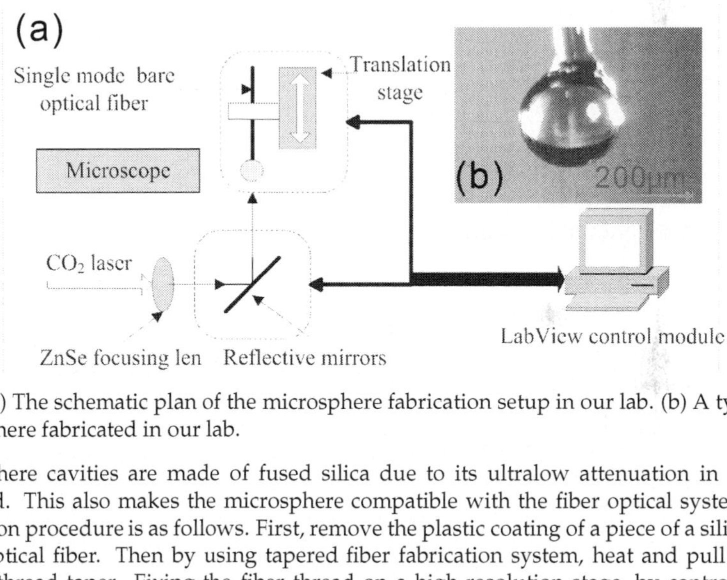

Fig. 3. (a) The schematic plan of the microsphere fabrication setup in our lab. (b) A typical microsphere fabricated in our lab.

Microsphere cavities are made of fused silica due to its ultralow attenuation in the 1550 nm band. This also makes the microsphere compatible with the fiber optical systems. The fabrication procedure is as follows. First, remove the plastic coating of a piece of a silica single mode optical fiber. Then by using tapered fiber fabrication system, heat and pull it into a slender thread taper. Fixing the fiber thread on a high-resolution stage, by controlling the CO_2 laser position, a microsphere grows automatically due to the surface tension. According to the volume conservation, we can control the size of the microsphere by controlling the heating length. The residual fiber attaching to the sphere can help us handle the microcavity conveniently. Fig. 3-a is a schematic plan of the fabrication setup in our lab, in which the coherent CO_2 laser ($10.6\mu m$ in wavelength) is used as the heating source, due to the strong absorption of silica in the infrared band. The heating laser is guided by reflective mirrors. It is then focused towards the control stages where microspheres are made. A 450 $\sim$ 570X microscope is placed vertically to monitor the procedure by using a *CCD* camera. Other heating sources, such as the hydrogen flame is also used by several other groups. However, hydrogen flame heating can lead-in hydroxide ion which can induced extra optical loss in the 1550 nm band.

2.2 Resonant characteristic testing

Fig. 4-a shows the schematic plan of the microsphere testing system. Narrow linewidth ($< 300KHZ$), external-cavity New Focus Velocity Laser at the telecommunications band ($1520\ nm \sim 1570\ nm$) is used to pump the microsphere WGMs. Using a polarization controller, the polarization state is controlled to research the influence on the resonance. High efficient coupling between the tapered fiber and the microsphere is obtained through precisely controlling their gap by using a piezoelectric translation stage. Dual microscopes are used to simultaneously monitor the coupling system, horizontally and vertically, respectively. The output signal is collected by the photoreceptor and displayed on the digital oscillograph. Signal-generator generating a triangular wave with frequency $500HZ$ with the peak value less than $3V$ is used to modulate the laser scanning with a modulation width $30GHZ$. Selecting a microsphere with the diameter (D) about $150\mu m$, a corresponding tapered fiber with waist diameter $2.1\mu m$ is used to excite the microcavity WGMs evanescently. As shown in Fig. 4-b, the microsphere is fixed on the piezoelectric high-resolution translation stage (step resolution about 20 nm), to adjust the appropriate coupling position. Fig. 4-c shows a typical microcavity coupled with a taper. Fig. 4-d shows a typical resonant dip of the microsphere. It's important to note that experimental environment should be shockless to ensure the stable transmission spectra.

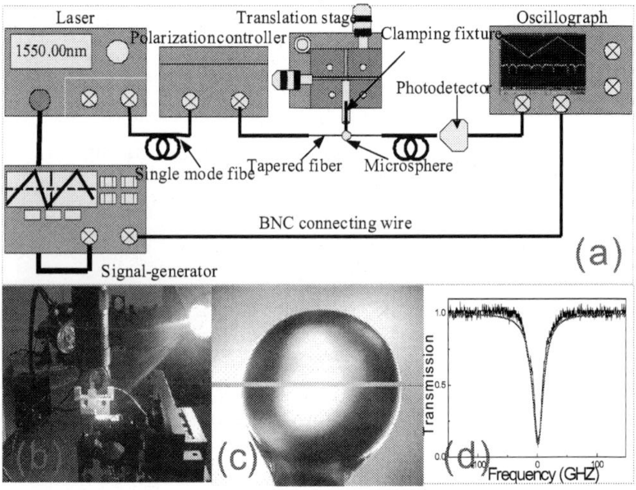

Fig. 4. (a) The schematic plan of the microsphere testing setup in our lab. (b) The experimental setting of the microsphere testing system in our lab. (c) A typical micrography of a microcavity coupled with a taper. (d) A typical resonant dip of the microsphere.

2.3 The challenging for practical application developments

Although there is great potential for the microcavity-based novel devices as illustrated in the "INTRODUCTION", there are some drawbacks of the traditional microsphere-taper system when promoting the above researches into practical applications.

First, the coupling efficiency can be affected, and the effective coupling can even be broken when loading the vibration on the taper or the microsphere, which challenges the robustness greatly. The robustness directly determines if the device is practical and can be used extensively. Thus, the robustness is the precondition for the microcavity-based practical devices.

Second, WGMs are very sensitive to the microcavity surroundings, which is an advantage for sensors, but makes the structure invalid in high RI materials or non-uniform dielectrics. For example, we can't measure the temperature of the Benzene, the Acetone or the Carbon disulfide disulphide et al, because the RI of these substances is higher than that of the silica which is frequently used to fabricate microcavities due to its low attenuation at the telecom band. The higher RI of the detected substance can destroy the resonance and subsequently fail the test. From this point of view, some special treatment are needed to improve this problem.

Third, the exposure of the traditional coupling system to the open environments makes the Q-maintenance challenging greatly, because the water and the dust in the air could spoil the Q drastically. The Q reduction can result in error signals to the functional devices. Therefore, realizing the Q maintenance is another aspect need to be solved.

Finally, high-resolution $3D$ translation stages are necessary in the traditional coupling system to adjust the coupling gap precisely. However, the stages are expensive and bulky, limiting the mobility in applications.

3. The solution to the problems

In this section, we propose one solution to the problems of the traditional microsphere-taper coupling system addressed above, and a package technology is proposed.

3.1 wholly-package manner

For the first time, we propose and realize a novel wholly packaged microsphere-taper coupling structure (PMTCS) experimentally (Y.-Z. Yan et al., 2011). The wholly-packaged structure is sketched by Fig. 5-a. We demonstrate that the PMTCS can avoid the problems addressed above, which is stable without $3D$ translation stages, robust against the vibration and free to move. Moreover, as a protective layer, the package body isolates the microsphere-taper from the surroundings, and maintains the Q above 10^6 for a long time.

To package the microsphere-taper coupling system, ultraviolet (UV) glue is used. The glue is fibre coating material with low RI ($n_g \approx 1.35$), which is made from special silicone acrylate with the product specification $KD - 310$. The packaging process is comprised of five steps, as shown in Fig. 5-b1 to b5. First of all, optimal coupling in the air with desirable resonant dips needs to be achieved, as shown in Fig. 5-b1. And then the UV glue is coated on the microsphere-taper by using a glue spreading machine in a dropping manner, as shown in Fig. 5-b2. Afterward, the package is solidify through 10 minutes exposure under a UV lamp, as shown in Fig. 5-b3. In the forth step, the microsphere stem, mounted on the 3D stages,

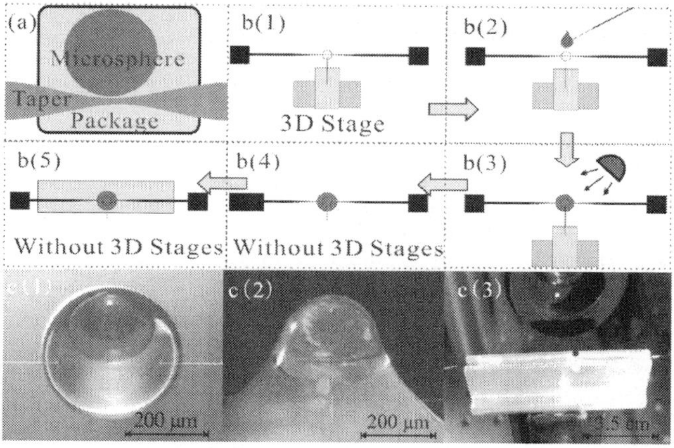

Fig. 5. (a) The schematic diagram of the wholly-packaged structure. b(1)-b(5) Illustration of the package process. c(1) The micrograph of the free microsphere coupling system after the forth step of the package. c(2) The micrograph of the semi-finished products of the packaged microcavity unit. c(3) A typical sealed packaged microsphere-taper coupling system.

is truncated by using a heat burning manner. The PMTCS here is independent of the 3*D* stages and can be moved freely, as shown in Fig. 5-a4 and Fig. 5-c1. Finally, similar to the potted circuit module we further package the structure using a designed slot as the mould to package the fragile taper totally (Fig. 5-b5). Fig. 5-c2 shows the semi-finished products of the wholly-packaged microcavity unit. Fig. 5-c3 shows a typical wholly-packaged module. In this module the microcavity, its coupling system and the fragile taper are all solidified into an entire body. The whole procedure is monitored by two microscopes, horizontally and vertically. It is worth noting that we need to re-adjust the coupling before the solidification, because the initial coupling is affected by the glue dropping which changes the environment and the surface tension around the microsphere as well as the taper. The resonant spectra fluctuate randomly at the first moment of the glue dropping, and gradually stabilize when the glue inosculates with the coupling structure after a few minutes. Particular attention has also to be paid to avoid the taper cracking, especially at its fragile taper waist.

Fig. 5-c1 shows a typical micrograph (side view) of a PMTCS ($D = 340\mu m$), in which the package body is asymmetric due to the gravity. As the backbone, the tapered fiber traverses across the package body to hold the PMTCS. At the package boundary (marked with blue triangles), there are two contact points which cause extra scattering loss (less than 20%). In the package experiments, the scattering loss can be reduced through using a shorter taper and encapsulating it completely, as shown in Fig. 5-c2 and Fig. 5-c3. This is also an effective way to improve the robustness which is determined by the un-stretched single-model fiber after packaging while depending on the fragile taper for an unpackaged system. In addition, the evanescent decay length (d) of WGMs ($d = \lambda/(2\pi\sqrt{n_s^2 - n^2})$, n_s and n are the RI of the microsphere and the surroundings, respectively.) increases from 0.1516λ to 0.3008λ, which causes a larger field overlap between WGMs and the taper. This makes the critical coupling easier to fulfill for a thicker taper. Besides, the coating layer with the minimum thickness about 2500μm ($\gg d$) ensures the complete isolation of WGMs from the outside, as shown in Fig. 5-c3.

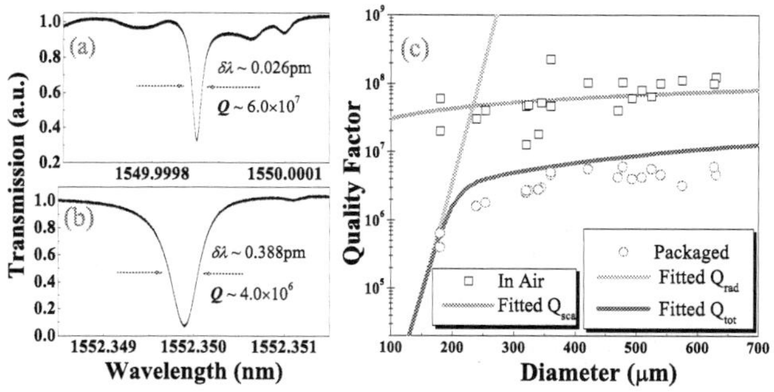

Fig. 6. (a) and (b) are contrast resonant spectra for a microsphere ($D = 421\mu m$) before and after the packaging, respectively. (c) Tested and fitted Q versus microsphere D.

By comparing the spectra before and after the packaging, a red shift to the longer wavelength (S) has been observed, which can be estimated through $S = \frac{\lambda^2}{\pi D}(\frac{1}{\sqrt{n_s^2-n_g^2}} - \frac{1}{\sqrt{n_s^2-n_{air}^2}})$, where n_{air} is the air RI. For a microsphere with $D \approx 421\mu m$, the S we observed is $2.35nm$, which agrees well with theoretical prediction of $2.3nm$, as shown in Fig. 6-a and Fig. 6-b. The package can also eliminate high radial order WGMs and offer much more regular spectra. Because after the package, the relative RI (n_s/n) decreases from 1.44 to 1.07, corresponding to an increase of the total reflection critical angle from 43.62 to 70.92 degree, which results in much larger leakage for high radial order WGMs.

As the most important parameter of the microsphere, the intrinsic Q (Q_{tot}) of WGMs can be expressed as

$$Q_{tot}^{-1} = Q_{abs}^{-1} + Q_{sca}^{-1} + Q_{rad}^{-1}, \quad (1)$$

where Q_{abs}, Q_{sca} and Q_{rad} are related to the absorption, surface scattering and radiation loss, respectively (D. W. Vernooy et al., 1998). In our experiments, the microspheres are made of the commercial fiber, where high purity silica is used. This ensures Q_{abs} greater than 10^{10} in the air. Q_{sca} originates from the surface rayleigh scattering, and is mainly determined by the surface smoothness. For a microsphere induced by the surface-tension, the surface roughness is nanometer-sized ($1nm - 10nm$), which indicates the Q_{sca} above 10^8. The radiation loss of WGMs strongly depends on the size and RI of the cavity. At $1.55\mu m$ waveband, $Q_{rad} > 10^8$ can be achieved in the air for silica microspheres with D larger than $25\mu m$. In Fig. 6(c), we plot the measured Q_{tot} against D for the unpackaged microspheres by using white squares, where the recorded maximum Q_{tot} is around 10^8. Here, D is larger than $100\mu m$, indicating that the Q_{tot} is limited by the surface scattering loss, which agrees well with the fitted Q_{sca}.

However, compared with the unpackaged structure, Q_{tot} decreases apparently in the PMTCS. As shown by red circles in Fig. 6-c, the measured Q_{tot} decreases sharply when the diameter is less than $200\mu m$, and Q_{tot} is always smaller than 10^7 for larger microspheres. The similar phenomenon has been reported in microtoroids embedded in the water by Armani et al (A. M.

Armani, 2005). Former experiments have indicated that a coating on the microcavity surface can greatly reduce the Q_{sca}. Thus, the loss in the PMTCS mainly originates from the radiation loss and the glue absorption. We fit the Q_{tot} (blue line) with $n_g = 1.351 + 5 \times 10^{-6}i$ by analytically solving the WGMs in microsphere at $\lambda = 1550$nm. The results agree with the measurements greatly. Here, the imaginary part of n_g is corresponding to the amount of the absorption loss when the light propagates in the glue. With a real n_g=1.351, we obtain the Q_{rad} (green line) which decreases exponentially with the reduction in the diameter. In addition, particular attention should be paid to prevent contaminating the UV glue, because the contaminants attached to the microcavity or the taper can increase the overall scattering loss (M. L. Gorodetsky et al., 2000). On the contrary, mixing nano-particles at a certain concentration in the glue provides a feasible way to control the backscattering (X.-W. Wu et al., 2009). It is worth noting that the Q of the PMTCS is still much higher than the highest Q (2×10^5) of a packaged microfiber coil resonator (Y. Jung et al., 2010).

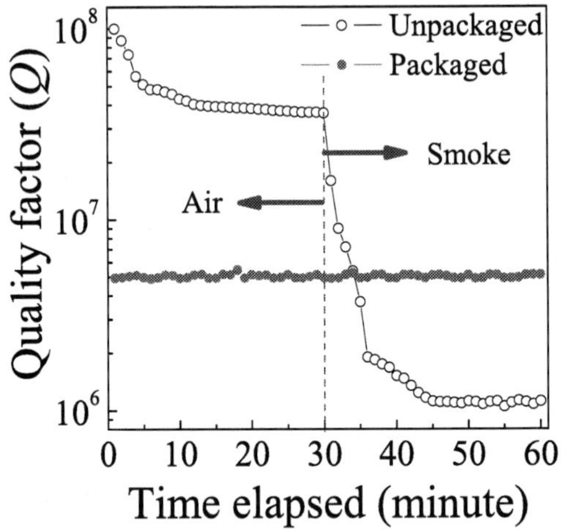

Fig. 7. Tested Q versus elapsed time of a microsphere coupling system in the air and in the smoke, for un-packaged (in black) and packaged (in red) samples, respectively.

It is obvious that the package body isolates the whole coupling system from the surroundings, excluding Q spoiling factors from the dust and water in the air. As shown in Fig. 7, when exposing a microsphere in the air, the Q shows a quick decay due to the water absorption in a few minutes after its fabrication. When putting it in the smoke, the Q has another remarkable decrease due to the scattering by the dust adhering to the surface. By contrast, the Q of PMTCS is much more stable and independent of the surrounding influences. In fact, we have maintained the Q above 10^6 for a few months. Thus, the package provides a feasible way to maintain the Q, paving the way for devices research in practical application.

3.2 Spot-package manner

WGMs of microspheres are attracting more and more attention recently due to their ultra-high quality factor (Q) values, very small volume and good compatibility with fiber-integrated optics, promising for a number of passive and active devices as filters, lasers and modulators. In the WGMs, light are well confined through total internal reflection (TIR) at the boundary, and a small portion of energy exists outside the microsphere in the form of the evanescent field. The evanescent field is very important because it enables the WGMs to interact with the outside matters, such as atoms, nanoparticles, chemical and biological molecules. Thus, the WGMs are very potential for ultrasensitive sensors (J. T. Gohringa et al., 2010; J. Zhu et al., 2010; T. Lu et al., 2011).

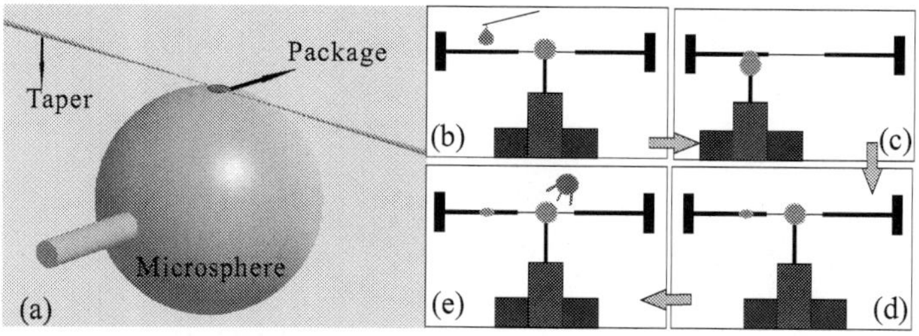

Fig. 8. (a) The schematic diagram of the spot-packaged microsphere-taper coupling system. (b)-(e) Illustration of the spot-package process. Red color represents the package glue.

In most applications, the fiber tapers are utilized to excite and collect the microsphere WGMs with high coupling efficiency. However, there are two principal drawbacks in the traditional microsphere-taper coupling system (MTCS) when exploring practical evanescent sensing devices. On one hand, in the external evanescent sensing applications the coupling region can be inserted or filled by the detected substance. The intrusion of the ectocrine in the coupling region can interfere in or destroy the microsphere-taper coupling, giving rise to fluctuations in the transmission spectra, and finally induce error signal for sensing. Therefore, the coupling region should be protected to avoid the insertion or filling of the detected substance. On the other hand, the MTCS lacks robustness, and a tiny vibration can change the relative positions of the taper and microsphere, subsequently influence the coupling efficiency and the resonant frequency. These instability limits the development of the practical sensors. Recently, we have realized a robust silica microsphere-taper coupling platform by using low refractive index (RI) curable polymer to encapsulate the coupling system wholly. This wholly-package enhances the robustness, isolates the coupling system from the surroundings, and makes the structure get rid of the burdensome fiber taper holders. These advantages are of immense importance for microcavity based gyro and temperature sensors. However, the wholly-package isolates the microsphere completely from external environments, preventing the interaction of the evanescent field with the matters outside.

In this part, we further improve the package technology for the MTCS. A novel spot-packaged structure is proposed and realized experimentally (Fig. 8-a). This structure not only holds the advantage of the robustness and stability, protects the coupling region, but also keeps the feasibility of the evanescent field sensing. This is potential for in-line optical sensors, and multiple microspheres can be integrated with the standard fiber, which can be applied to remote or distributed sensing applications in harsh environment.

The package process is comprised of four steps, as shown in Fig. 8-b to Fig. 8-e. First of all, after obtaining the optimal coupling in the air, the glue is dripped on one end of the un-stretched taper (Fig. 8-b). And then, by using two microscopes (horizontally and vertically, respectively) as the monitoring system, we adjust the microsphere to approach the glue attached to the fiber, as shown in Fig. 8-c. When an appropriate amount of the glue is agglutinated on the microsphere, we adjust the microsphere away from the fiber. Afterward, we move the microsphere close to the taper again, as shown in Fig. 8-d. In this process, the coated spot-surface is controlled to contact with the taper gently. The coupling position is also modulated to achieve the fine resonant dips again. In the last step, the package is solidified through 10 minutes exposure under a UV lamp, as shown in Fig. 8-e. In addition, the stem of the microsphere can be burned out to release the microsphere from the $3D$ stages. Finally, we obtain the spot-packaged MTCS as shown in Fig. 8-a, which is integrated with fiber and can be moved freely and conveniently.

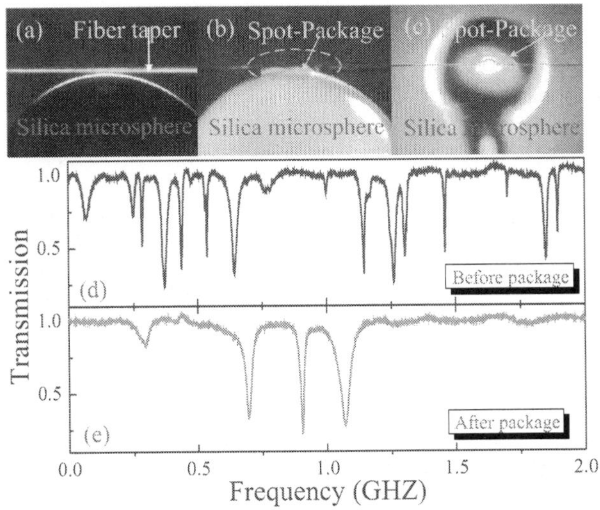

Fig. 9. (a) and (b) are the micrograph of a typical microsphere-taper coupling system before and after the spot-package process, respectively. (c) The side view of the spot-package. (d) and (e) are two typical transmission spectra of a microsphere before and after the spot-package process, respectively.

In Fig. 9-a and Fig. 9-b, we show typical micrographs of the MTCS without and with the spot-package. As shown clearly in Fig. 9-b, the coupling region surrounded by the glue is well protected from the external intrusion and vibration, while the rest of the silica microsphere surface are not affected. Therefore, the evanescent field can still interact with the matters near the microsphere. In addition, resulting from the liquid surface tension the glue surface is very smooth (similar to silica surface), which indicates that the package-induced scattering losses are almost negligible. Generally, the spot-package has an average area of about $30\mu m^2$, meaning that the spot-package occupies a circular arc length about $6\mu m$ at the equator of the silica microsphere (Fig. 9-c). As we will study below, the area of the package is the key factor to both the Q and the mechanic stability of the structure. Simply, the influence on WGMs is proportional to the rate $6/(\pi \times D)$ which describes the relative area of the package. From this perspective, microspheres with larger D are more suitable for this spot-package manner than the smaller ones.

Fig. 9-d and (e) are the transmission spectra before and after the spot-package, respectively. It is obvious that the resonant dips are broadened after the package, corresponding to a reduction of Q. It should be noted that this spot-packaged structure is essentially different from the wholly-packaged microsphere. In the wholly-packaged microsphere, light is confined at the silica-polymer interface. Since the boundary is rotational symmetric, the light ray undergoes continuous TIR at the boundary. The radiation loss and the polymer absorption are dominant, which respectively originates from the tunneling leak out at the low refraction contrast interface and the large polymer absorption. However, in the spot-packaged structure here (Fig. 9-b), the boundary of the spot-packaged microsphere is slightly deformed, which is a kind of asymmetric resonant cavities (ARC) (Y.-F. Xiao et al., 2009). In the ARC the properties are very different, because the rotational symmetric is broken. Thus, the light ray reflection at the boundary can't conserve the incident angle. Consequently, some rays refract out of the cavity after several reflection at the boundary. Hence, the Q is reduced after the spot-package process. Fig. 10-a, b and c show the numerical simulated electric field intensity of resonant modes in the spot-packaged microsphere. The critical total reflect angle at polymer/silica interface is $\chi_C = 68$ degree. Rays with large incidence angle $\chi > \chi_C$ can reflect at the silica boundary, so the incidence angle is well conserved. Therefore, for low radial order (high-Q) modes as shown in Fig. 10-a, the glue only gives small perturbations, and the energy is localized near the boundary, forming the modes as WGMs. For higher radial order (low-Q) modes as shown in Fig. 10-b and Fig. 10-c, rays reflect at the asymmetric boundary, resulting in chaotic ray dynamics and directional emission at the boundary. Therefore, only a few high-Q mode with large χ can survive in this ARC, leading to the much more regular spectra of the spot-packaged microsphere (Fig. 9-e).

To investigate the influence of the package on the Q, the Q before and after the spot-package was tested respectively by using the linewidth manner, in which the laser power was controlled to be lower than $5\mu W$ to avoid the thermal effect. Fig. 10(d) shows the measured highest Q for untreated and spot-packaged microspheres. It reveals that the Q for the spot-packaged microspheres are lower than that of the corresponding untreated sample. Although there is a decreasing of the loaded Q, the spot-packaged structure still has an average Q_{loa} about 1.02×10^7. With a tested average Q for the untreated microspheres about 4.2×10^7, the decreased factor has a average value about 4. The tested Q for the spot-packaged microsphere in our experiments is consistent with the former researches, which have demonstrated that high Q ($> 10^7$) WGMs can exist in a small deformation microsphere.

The spot-packaged structure also offers great mechanical stability, as the microresonator and the taper are integrated together via the solidified UV polymer. It provides feasibility for the microresonator to extricate itself from the bulky translation stages and makes the packaged structure be moved freely, which is of great significance for practical applications. By using a force measurement gauge, we tested the bearable loaded adhesive force for the packaged structure, as shown in Table I. The bearable loaded force grows with the increasing of the spot area. The force is about $50mN$ with the spot area about $30\mu m^2$ as mentioned above, which indicates that in such a regime a microsphere with the D of $300\mu m$ can endure an acceleration about $8.2 \times 10^4 g$. Actually, the robustness of this spot-packaged structure is limited by the strength of the fiber taper. As the maximum stress that a silica fiber taper can withstand is about $\delta_f \approx 10GPa$, a maximum force acting on the taper with the waist diameter $1\mu m$ is about $7.8mN$. Improvement of the stability of the spot-packaged regime by using a much more robust coupler, such as side-polished fiber and channel waveguide coupler, is undergoing.

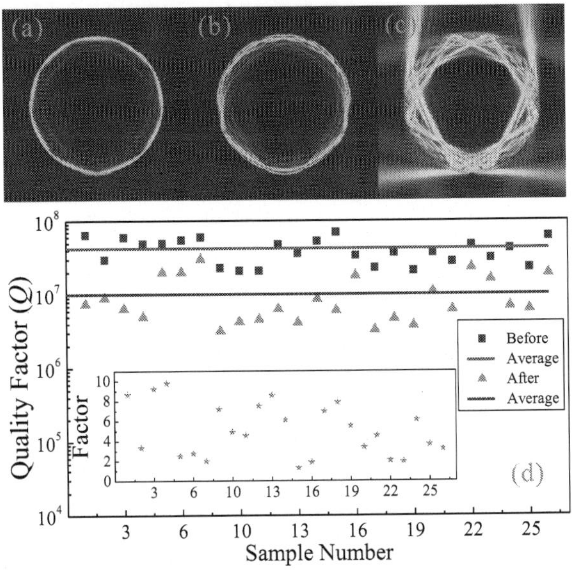

Fig. 10. (a), (b) and (c) The electric intensity distribution of three typical modes in the spot-packaged microsphere. (d) Tested Q of different microspheres before and after the spot-package. Insets: The decreased factor of the Q corresponds to different samples.

Spot Area (μm^2)	22	26	30	32	35	42	50	58	70
Force1 (mN)	35	39	51	54	58	59	65	72	82
Force2 (mN)	37	42	49	53	57	60	67	75	85
Force3 (mN)	32	46	53	55	60	61	69	74	87

Table 1. Force tested for different spot-package area

In this part, we have studied a novel spot-packaged MTCS by only encapsulating the coupling region with low RI polymer. The spot-package not only protects the coupling region, but also provides both mechanical and optical stability. In the packaged structure, Q greater than 10^7 is demonstrated with an average decreasing factor about 4 compared with the unpackaged microspheres. The Q decreasing is mainly due to the asymmetric boundary shape. The packaged MTCS can be integrated with fiber, and are expected to be used in practical external evanescent applications, including chemical and biological sensors. Especially, we believe that these in-line robust sensor can be used in harsh environments such as rockets, missiles and airborne crafts.

4. Practical sensing applications

To verify the practicability of the PTMCS, thermal sensing experiments are carried out in complex dynamic water environment. Generally, the temperature sensing is carried out by detecting the thermo-induced resonant wavelength shift. However, the shift can also be induced by the changing of the microsphere surroundings, which is usually used to research the refractive index sensing. This phenomenon is the main obstacle for practical temperature sensors, because the refractive index of the open air is mutable. Here, in order to improve its sensing selectivity to only response to the temperature, the wholly-packaged microsphere is used as the sensing head. The package isolates the microsphere from the outside surroundings and eliminate the influence of the RI variation.

In sensing experiments, a beaker containing 600*ml* water with a stirrer in it is adopted as the testing environment, as shown in Fig. 11-a. The stirrer, not only helps to equalize the temperature, but also simulates a flowing water environment. A thermocouple and a heater are placed in the vicinity of the PMTCS, to measure and change the temperature, respectively.

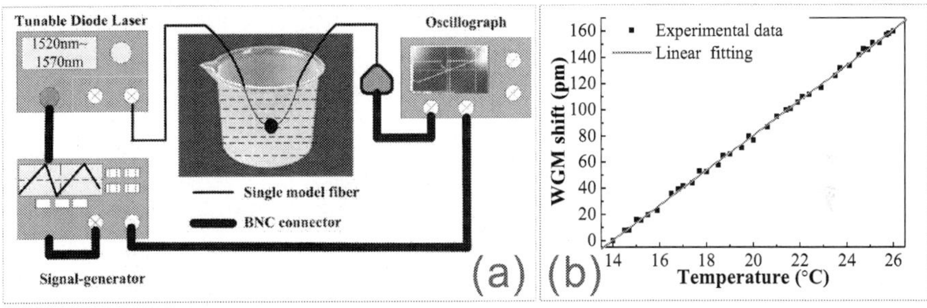

Fig. 11. (a) Temperature testing setup used in our expetiments. (b) WGM wavelength shifts as a function of the surrounding temperature, in NaCl resolution and in pure water, respectively.

The robustness is confirmed by the undisturbed spectra in the water flow. Furthermore, the completeness of the encapsulation is verified through the unshifted spectra when adding Sodium Chloride (NaCl) in the water gradually (keep the same temperature) to change the water RI about 2×10^{-3}, which could cause a WGM wavelength shift about 20*pm* for an unpackaged silica microsphere. Wavelength shifts against the temperature are recorded, in saturated NaCl solution and in pure water, respectively. As shown in Fig. 11-b, wavelength shifts are not affected by RI changes of the water. The temperature variation leads to changes both in the size and the RI of the silica microsphere, which subsequently causes resonance wavelength shifts (B.-B. Li et al., 2010; C.-H. Dong et al., 2009; Q. Ma et al, 2008; Y. Wu et al., 2009). The resonant wavelength shows a red shift about 160.39*pm* when the temperature increases from 14°C to 26°C, which indicates a sensitivity of $13.37 pm/°C$ with a relevant coefficient in linear fitting about 0.998. Taking into account the spectral resolution ($\Delta\lambda_{min}$) of our system, which is 0.015*pm*, we estimate the resolution ($\Delta T_{min} = \Delta\lambda_{min}/(d\lambda/dT)$) of the PMTCS microsphere temperature sensor as 1.1×10^{-3}°C.

This resolution is in the same order of magnitude with the traditional silica microcavity thermal sensor. What's more, the practicability is greatly enhanced in the PMTCS, in which only the temperature change causes the WGM shift, while the change of external RI fails to cause the shift. Besides, the PMTCS thermal sensor shows an excellent anti-jamming ability, and can be used in harsh environments. Furthermore, the PMTCS can provide much more stable performance (for instance, the resolution) due to its superior Q maintenance ability.

Additionally, the bulky translation stages are no longer unnecessary in the PMTCS. The portable structure makes these sensors easy to move and miniature, which is important in practical applications, especially in microsystem technology research.

Further efforts will be focused on exploring new package materials and improving the package technology to enhance the packaged Q_{tot}. The PMTCS could be extended to multi-packaged microsphere-taper system on one fiber to monitor the real-time temperature at different locations, or package multi-microspheres in a certain point to study the classical analog to electromagnetic induced transparency (EIT) (Y.-F. Xiao et al, Y.-F. Xiao; J. Scheuer et al., 2006), even its application on gyroscopic (Y.-Z. Yan et al., 2011).

In addition, the wholly-packaged structure is also beneficial to the microcavity based filters, lasers et al. However, the wholly-package destroys the external evanescent field sensing which bases on the interaction of the WGMs with the detected substance. The spot-packaged structure can cover this shortage, making spot-packaged structure promising in RI, concentration or biochemical molecules.

In a word, when designing practical sensors we should choose an appropriate packaged structure (wholly-package or spot-package) according to the style of the sensing.

5. Conclusion

In summary, this chapter here proposed an novel package technology to solve the problems which hinders the improvements of microcavity based devices. Two package manners are realized, wholly-package and spot-package respectively. The fabrication and the tests of the two packages are illustrated in detail. The results show that, the packaged structures (wholly-package and spot-package) hold the striking merits to promote the developments of practical devices. In addition, the two manners are each with its own distinguishing features, and we should choose proper manner to use according to the devices we designed.

6. Acknowledgment

The authors greatly thank Chang-Ling Zou in University of Science and Technology of China for the helpful discussion. The work was supported by the National Basic Research Program of China under Grant No. 2009CB326206 and the Innovation Project under Grant Nos. 7130907 and 9140C1204040706. Y. Z. Yan was also supported by Innovation Projects under Grant Nos. 20093076 and 100115122.

7. References

A. B. Matsko and V. S. Ilchenko. (2006). Optical Resonators With Whispering-Gallery Modes-Part I: Basics. *IEEE J. Quantum Electron*, Vol. 12, 3–14, ISSN: 0018-9197

A. Chiasera, Y. Dumeige, P. Fl̦eron, M. Ferrari, Y. Jestin, G. Nunzi Conti, S. Pelli, S. Soria, and G.C. Righini. (2010). Spherical whispering-gallery-mode microresonators. *Laser Photonics Rev.*, Vol. 4, 457–482, ISSN: 1863-8880

A. M. Armani, D. K. Armani, B. Min, K. J. Vahala, and S. M. Spillane. (2005). Ultra-high-Q microcavity operation in H_2O and D_2O. *Appl. Phys. Lett.*, Vol. 87, 151118, ISSN: 0003-6951

B.-B. Li, Q.-Y. Wang, Y.-F. Xiao, X.-F. Jiang, Y. Li, L. Xiao, and Q. Gong. (2010). On chip, high-sensitivity thermal sensor based on high-Q polydimethylsiloxane-coated microresonator. *Appl. Phys. Lett.*, Vol. 96, 251109, ISSN: 0003-6951

C.-H. Dong, L. He, Y.-F. Xiao, V. R. Gaddam, S. K. Ozdemir, Z.-F. Han, G.-C. Guo, and L. Yang. (2009). Fabrication of high-Q polydimethylsiloxane optical microspheres for thermal sensing. *Appl. Phys. Lett.*, Vol. 94, 231119, ISSN: 0003-6951

C.-H. Dong, F.-W. Sun, C.-L. Zou, X.-F. Ren, G.-C. Guo, and Z.-F. Han. (2010). High-Q silica microsphere by poly(ethyl methacrylate) coating and modifying. *Appl. Phys. Lett.*, Vol. 96, 061106, ISSN: 0003-6951

D. W. Vernooy, V. S. Ilchenko, H. Mabuchi, E. W. Streed, and H. J. Kimble. (1998). High-Q measurements of fused-silica microspheres in the near infrared. *Opt. Lett.*, Vol. 23, 247–249, ISSN: 0146-9592

F. Vollmer, and S. Arnold. (2008). Whispering-gallery-mode biosensing: labelfree detection down to single molecules. *Nat. Methods*, Vol. 5, 591–596, ISSN: 1548-7091

F. Xu, and G. Brambilla. (2008). Demonstration of a refractometric sensor based on optical microfiber coil resonator. *Appl. Phys. Lett.*, Vol. 92, 101126, ISSN: 0003-6951

F. Xu, V. Pruneri, V. Finazzi, and G. Brambilla. (2008). An embedded optical nanowire loop resonator refractometric sensor. *Opt. Express*, Vol. 16, 1062–1067, ISSN: 1094-4087

I. M. White and X. Fan. (2008). On the performance quantification of resonant refractive index sensors. *Opt. Express*, Vol. 16, 1020–1028, ISSN: 1094-4087

J. C. Knight, G. Cheung, F. Jacques, and T. A. Birks. (1997). Phase-matched excitation of whispering-gallery-mode resonances by a fiber taper. *Opt. Lett.*, Vol. 22, 1129–1131, ISSN: 0146-9592

J. Scheuer, and A. Yariv. (2006). Sagnac Effect in Coupled-Resonator Slow-Light Waveguide Structures. *Phys. Rev. Lett.*, Vol. 96, 053901, ISSN: 0031-9007

J. T. Gohringa, P. S. Daleb, X. Fan. (2010). Detection of HER2 breast cancer biomarker using the opto-fluidic ring resonator biosensor. *Sens. Actuators, B*, Vol. 146, 226-230, ISSN: 0925- 4005

J. Zhu, S. K. Ozdemir, Y. F. Xiao, L. Li, L. He, D. R Chen, L. Yang. (2010). On-chip single nanoparticle detection and sizing by mode splitting in an ultrahigh-Q microresonator. *Nat. Photonics*, Vol. 4, 46-49, ISSN: 1749-4885

K. J. Vahala. (2003). Optical microcavities. *Nature*, Vol. 424, 839–846, ISSN: 0028-0836

M. L. Gorodetsky, A. A. Savchenkov, and V. S. Ilchenko. (1996). Ultimate Q of optical microsphere resonators. *Opt. Lett.*, Vol. 21, 453–455, ISSN: 0146-9592

M. Cai, O. Painter, and K. J. Vahala. (2000). Observation of Critical Coupling in a Fiber Taper to a Silica-Microsphere Whispering-Gallery Mode System. *Phys. Rev. Lett.*, Vol. 85, 74–77, ISSN: 0031-9007

M. L. Gorodetsky, A. D. Pryamikov, and V. S. Ilchenko. (2000). Rayleigh scattering in high- Q microspheres. *J. Opt. Soc. Am. B*, Vol. 17, 1051–1057, ISSN: 0740-3224

M. Hossein-Zadeh, and K. J. Vahala. (2006). Fiber-taper coupling to Whispering-Gallery modes of fluidic resonators embedded in a liquid medium. *Opt. Express*, Vol. 14, 10800–10810, ISSN: 1094-4087

M. Sumetsky, R. S. Windeler, Y. Dulashko and X. Fan. (2007). Optical liquid ring resonator sensor. *Opt. Express*, Vol. 15, 14376-14381, ISSN: 1094-4087

Q. Ma, T. Rossmann and Z. Guo. (2008). Temperature sensitivity of silica micro-resonators. *J. Phys. D: Appl. Phys.*, Vol. 41, 245111, ISSN: 0022-3727

X.-W. Wu, C.-L. Zou, J. M. Cui, Y. Yang, Z.-F. Han and G. C. Guo. (2009). Modal coupling strength in a fibre taper coupled silica microsphere. *J. Phys. B: At. Mol. Opt. Phys.*, Vol. 42, 085401, ISSN: 0953-4075

Y. Sun, and X. Fan. (2008). Analysis of ring resonators for chemical vapor sensor development. *Opt. Express*, Vol. 16, 10254–10268, ISSN: 1094-4087

Y. Jung, G. S. Murugan, G. Brambilla, and D. J. Richardson. (2010). Embedded optical microfiber coil resonator with enhanced high-Q. *Photon. Technol. Lett.*, Vol. 22, 1638–1640, ISSN: 1041-1135

Y. Wu, Y.-J. Rao, Y.-H. Chen and Y. Gong. (2009). Miniature fiber-optic temperature sensors based on silica/polymer microfiber knot resonators. *Opt. Express*, Vol. 17, 18142–18147, ISSN: 1094-4087

Y.-F. Xiao, X.-B. Zou, W. Jiang, Y.-L. Chen, and G.-C. Guo. (2007). Analog to multiple electromagnetically induced transparency in all-optical drop-filter systems. *Phys. Rev. A*, Vol. 75, 063833, ISSN: 1050-2947

Y.-F. Xiao, C.-H. Dong, C.-L. Zou, Z.-F. Han, L. Yang, and G.-C. Guo. (2009).Low-threshold microlaser in a high-Q asymmetrical microcavity. *Opt. Lett.*, Vol. 34, 509-511, ISSN: 0146-9592

Y.-Z. Yan, C.-L. Zou, S.-B. Yan, F.-W. Sun, Z. Ji, J. Liu, Y.-G. Zhang, L. Wang, C.-Y. Xue, W.-D. Zhang, Z.-F. Han, and J.-J. Xiong. (2011). Packaged silica microsphere-taper coupling system for robust thermal sensing application. *Opt. Express*, Vol. 19, 5753-5759, ISSN: 1094-4087

Y.-Z. Yan, Z. Ji, S.-B. Yan, J. Liu, C.-Y. Xue, W.-D. Zhang, J.-J. Xiong. (2011). Enhancing the Robustness of the Microcavity Coupling System. *Chin. Phys. Lett*, Vol. 28, 034208, ISSN: 0256-307X

10

Optical Effects Connected with Coherent Polarized Light Propagation Through a Step-Index Fiber

Maxim Bolshakov, Alexander Ershov and Natalia Kundikova
Joint Nonlinear Optics Laboratory of the Electrophysics Institute and South Ural State University
Russia

1. Introduction

Spin-orbital interaction of a photon manifests itself in two effects. The first one is the influence of the trajectory on the light polarization. The second one is the influence of the polarization on the trajectory of light. The influence of the trajectory on the light polarization has been investigated theoretically (Chiao & Wu, 1986; Rytov, 1938; Vladimirskii, 1941) and the change in the azimuth of linearly polarized light has been observed experimentally under light propagation through a helical (coiled into a spiral) single mode optical fiber (Tomita & Chiao, 1986). The influence of light polarization on trajectory was demonstrated for the first time under total internal reflection. It was shown that reflected rays suffer a longitudinal shift for the linearly polarized light (Goos & Hanchen, 1947; 1949; Picht, 1929) and transverse shift for the circularly polarized light (Fedorov, 1955; Imbert, 1972). The value of the shifts is on the order of the light wavelength and depends on the state of light polarization. All mentioned effects have been investigated independently till 1990. The influence of the trajectory on polarization and the influence of the polarization on the trajectory were considered as mutually inverse effects for the first time by Zel'dovich and Liberman in 1990 (Zel'dovich & Liberman, 1990). The interpretation of these effects in terms of quantum mechanics as an interaction between the orbital momentum of photon and its spin (polarization) has been done by Zeldovich, Dooghin, Kundikova and Liberman in 1991 (Doogin et al., 1991). This interpretation was based on prediction and experimental investigation of optical Magnus effect. The effect manifests itself as speckle pattern rotation of circular polarized light transmitted through a multimode step-index fiber under circular polarization sign change (Doogin et al., 1992; 1991).

The chapter describes the results obtained in our laboratory, namely the results of investigation of optical effects connected with coherent polarized light propagation through a step-index fiber. This chapter is based on published results, but some recent results are added. The chapter is organized in the following way. The first section is devoted to the theoretical description of coherent polarized light wave propagation through a multimode step-index fiber. The approximation of weak guiding modes is used. The results of the next sections are based on the first section. The second section presents the results of the theoretical prediction and the experimental investigation of the isolated wave front dislocation formation (Darsht et al., 1995). The optical Magnus effect (Doogin et al., 1992)and its peculiarity in a few modes fiber are discussed in the third section. It also presents the results of computer simulation and the experimental results. The optical Magnus effect is observed under the switch of the circular polarization sign. The influence of light polarization on the speckle pattern of the polarized light transmitted through a multimode optical fiber will be described in the fourth section (Bolshakov & Kundikova, 2003; Bolshakov et al., 2006). The influence of external magnetic field on the speckle pattern behavior is considered in the fifth section, as well as the results of computer modeling and the experimental results. (Anikeev et al., 2001; Ardasheva et al., 2002; Baranova & Zel'dovich, 1994; Darsht et al., 1994). The sixth section deals with the applications results for fiber sensors (Bolshakov et al., 2008; 2011).

2. Theoretical description of coherent polarized light wave propagation through a multimode step-index fiber

Let's consider propagation of polarized light in an axial symmetric optical fiber with the following refractive index profile:

$$n^2(R) = n_{\mathrm{co}}^2\left[1 - 2\Delta f(R)\right], \tag{1}$$

where

$$\Delta = \frac{1}{2}\left(1 - \frac{n_{\mathrm{cl}}^2}{n_{\mathrm{co}}^2}\right), \tag{2}$$

and in the approximation of weakly directing waveguide

$$\Delta \approx \frac{n_{\mathrm{co}} - n_{\mathrm{cl}}}{n_{\mathrm{co}}}. \tag{3}$$

The function $f(R)$ looks like the following:

$$\begin{aligned} f(R) &= 0, & R = r/\rho < 1, \\ f(R) &= 1, & R = r/\rho > 1, \end{aligned} \tag{4}$$

and

$$\begin{aligned} n(r) &= n_{\mathrm{co}}, & r/\rho < 1, \\ n(r) &= n_{\mathrm{cl}}, & r/\rho > 1. \end{aligned} \tag{5}$$

Here $r = |\mathbf{r}|$, $(x, y) = \mathbf{r}$ are the transverse coordinates, ρ is the radius of the fiber core, n_{co} and n_{cl} are the refractive indexes of the core and the cladding respectively. In the approach of weakly directing waveguide the requirements of symmetry allow to write down four polarization modes in the following form for any orbital angular momentum m ($m \geq 0$) and radial quantum number N (Snyder & Love, 1984):

$$\begin{aligned} \mathbf{e}_{m,N}^{(1)}(r,\varphi) &= \left[\cos(m\varphi)\,\mathbf{e}_x - \sin(m\varphi)\,\mathbf{e}_y\right] \cdot F_{m,N}(r), \\ \mathbf{e}_{m,N}^{(2)}(r,\varphi) &= \left[\cos(m\varphi)\,\mathbf{e}_x + \sin(m\varphi)\,\mathbf{e}_y\right] \cdot F_{m,N}(r), \\ \mathbf{e}_{m,N}^{(3)}(r,\varphi) &= \left[\sin(m\varphi)\,\mathbf{e}_x + \cos(m\varphi)\,\mathbf{e}_y\right] \cdot F_{m,N}(r), \\ \mathbf{e}_{m,N}^{(4)}(r,\varphi) &= \left[\sin(m\varphi)\,\mathbf{e}_x - \cos(m\varphi)\,\mathbf{e}_y\right] \cdot F_{m,N}(r). \end{aligned} \tag{6}$$

Here $\mathbf{e}_x$, $\mathbf{e}_y$ are the eigenvectors, $x = r \cdot \cos\varphi$, $y = r \cdot \sin\varphi$. The radial distribution function $F_{m,N}(R)$ looks like:

$$F_{m,N}(R) = \frac{J_m(U_N R)}{J_m(U_N)}, \qquad R = r/\rho < 1,$$
$$F_{m,N}(R) = \frac{K_m(W_N R)}{K_m(W_N)}, \qquad R = r/\rho > 1, \tag{7}$$

where J_m and K_m are Bessel and McDonald functions accordingly, quantities U_N and W_N for each value m are determined from the equation:

$$U_N \frac{J_{m+1}(U_N)}{J_m(U_N)} = W_N \frac{K_{m+1}(W_N)}{K_m(W_N)}, \tag{8}$$

where $V^2 = W_N^2 + U_N^2$, $V = \rho k \sqrt{2n_{co}(n_{co} - n_{cl})}$, $k = 2\pi/\lambda$, λ is the wavelength of light in the air. In the scalar approach all modes propagate with an equal velocity, determined by the propagation constant $\beta_{m,N}$:

$$\beta_{m,N} = \frac{V}{\rho(2\Delta)^{1/2}} \left\{ 1 - 2\Delta \frac{U_N^2}{V^2} \right\}^{1/2}. \tag{9}$$

The influence of the state of polarization of each of the four modes on its propagation velocity is taken into account by the introduction of the polarization corrections $\delta\beta_{m,N}^{(j)}$ to the propagation constants. The polarization corrections are determined as follows:

$$\begin{aligned}
&\delta\beta_{0,N}^{(1)} = \delta\beta_{0,N}^{(3)} = \delta\beta_{0,N}^{(2)} = \delta\beta_{0,N}^{(4)} = I_1, \\
&\delta\beta_{1,N}^{(1)} = \delta\beta_{1,N}^{(3)} = I_1 - I_2, \\
&\delta\beta_{1,N}^{(2)} = 2\left(I_1 + I_2\right), \\
&\delta\beta_{1,N}^{(4)} = 0, \\
&\delta\beta_{m,N}^{(1)} = \delta\beta_{m,N}^{(3)} = I_1 - I_2, \\
&\delta\beta_{m,N}^{(2)} = \delta\beta_{m,N}^{(4)} = I_1 + I_2.
\end{aligned} \tag{10}$$

Here

$$I_1 = I_1(m,N) = \frac{(2\Delta)^{3/2}}{4\rho V} \frac{\int_0^\infty RF_{m,N}\left(\frac{dF_{m,N}}{dR}\right)\delta(R-1)dR}{\int_0^\infty RF_{m,N}^2 dR},$$
$$I_2 = I_2(m,N) = m\frac{(2\Delta)^{3/2}}{4\rho V} \frac{\int_0^\infty F_{m,N}^2 \delta(R-1)dR}{\int_0^\infty RF_{m,N}^2 dR}. \tag{11}$$

If the fiber has a step like refractive index profile, the expressions will be the following:

$$I_1(0,N) = -\frac{(2\Delta)^{3/2}}{2\rho} \frac{W_N U_N^2}{V^3} \frac{K_0(W_N)}{K_1(W_N)},$$
$$I_1(m,N) = -\frac{(2\Delta)^{3/2}}{4\rho} \frac{W_N U_N^2}{V^3} \frac{K_{m-1}(W_N) + K_{m+1}(W_N)}{K_{m-1}(W_N)K_{m+1}(W_N)} K_m(W_N),$$
$$I_2(m,N) = mI_{20}(m,N) = m\frac{(2\Delta)^{3/2}}{2\rho} \frac{U_N^2}{V^3} \frac{K_m^2(W_N)}{K_{m-1}(W_N)K_{m+1}(W_N)}. \tag{12}$$

Taking into account Eq. (12) let us obtain the following expressions for the polarization corrections to the propagation constants in the case of a fiber with a step like refractive index profile:

$$\begin{aligned}
&\delta\beta_{0,N}^{(1)} = \delta\beta_{0,N}^{(3)} = \delta\beta_{0,N}^{(2)} = \delta\beta_{0,N}^{(4)} = -\frac{(2\Delta)^{3/2}}{2\rho}\frac{W_N U_N^2}{V^3}\frac{K_0(W_N)}{K_1(W_N)},\\
&\delta\beta_{1,N}^{(1)} = \delta\beta_{1,N}^{(3)} = -\frac{(2\Delta)^{3/2}}{2\rho}\frac{W_N U_N^2}{V^3}\frac{K_1(W_N)}{K_0(W_N)},\\
&\delta\beta_{1,N}^{(2)} = -\frac{(2\Delta)^{3/2}}{\rho}\frac{W_N U_N^2}{V^3}\frac{K_1(W_N)}{K_2(W_N)},\\
&\delta\beta_{1,N}^{(4)} = 0,\\
&\delta\beta_{m,N}^{(1)} = \delta\beta_{m,N}^{(3)} = -\frac{(2\Delta)^{3/2}}{2\rho}\frac{W_N U_N^2}{V^3}\frac{K_m(W_N)}{K_{m-1}(W_N)},\\
&\delta\beta_{m,N}^{(2)} = \delta\beta_{m,N}^{(4)} = -\frac{(2\Delta)^{3/2}}{2\rho}\frac{W_N U_N^2}{V^3}\frac{K_m(W_N)}{K_{m+1}(W_N)}.
\end{aligned} \tag{13}$$

As $\delta\beta_{m,N}^{(1)} = \delta\beta_{m,N}^{(3)}$ for all values of m and $\delta\beta_{m,N}^{(2)} = \delta\beta_{m,N}^{(4)}$ for all values $m \neq 1$, any linear combinations of the modes $\mathbf{e}_{m,N}^{(1)}$ and $\mathbf{e}_{m,N}^{(3)}$ for all values m and any linear combinations of the modes $\mathbf{e}_{m,N}^{(2)}$ and $\mathbf{e}_{m,N}^{(4)}$ in the case $m \neq 1$ will be also eigenmodes. It is easy to show that combinations of the modes $\mathbf{e}_{m,N}^{(1)} \pm i\mathbf{e}_{m,N}^{(3)}$ and $\mathbf{e}_{m,N}^{(2)} \pm i\mathbf{e}_{m,N}^{(4)}$ ($i = \sqrt{-1}$) represent the modes with the homogeneous on the section of the fiber circular polarization $\mathbf{e}_x + i\sigma\mathbf{e}_y$. Here $\sigma = +1$ for light with the right circular polarization and $\sigma = -1$ for light with the left circular polarization. These new modes are those:

$$\begin{aligned}
\mathbf{e}_{+,m,N}^{+}(r,\varphi) &= \mathbf{e}_{m,N}^{(1)}(r,\varphi) + i\mathbf{e}_{m,N}^{(3)}(r,\varphi) = (\mathbf{e}_x + i\mathbf{e}_y)\, e^{im\varphi} F_{m,N}(r),\\
\mathbf{e}_{-,m,N}^{-}(r,\varphi) &= \mathbf{e}_{m,N}^{(1)}(r,\varphi) - i\mathbf{e}_{m,N}^{(3)}(r,\varphi) = (\mathbf{e}_x - i\mathbf{e}_y)\, e^{-im\varphi} F_{m,N}(r),\\
\mathbf{e}_{+,m,N}^{-}(r,\varphi) &= \mathbf{e}_{m,N}^{(2)}(r,\varphi) + i\mathbf{e}_{m,N}^{(4)}(r,\varphi) = (\mathbf{e}_x - i\mathbf{e}_y)\, e^{im\varphi} F_{m,N}(r),\\
\mathbf{e}_{-,m,N}^{+}(r,\varphi) &= \mathbf{e}_{m,N}^{(2)}(r,\varphi) - i\mathbf{e}_{m,N}^{(4)}(r,\varphi) = (\mathbf{e}_x + i\mathbf{e}_y)\, e^{-im\varphi} F_{m,N}(r).
\end{aligned} \tag{14}$$

Thus, if polarized radiation with the right or left circular polarization is incident on the input of the fiber, the modes with $m = 0$ and $m > 1$ will retain the state of polarization.

Polarized light with arbitrary state of polarization of light can be presented as superposition of light with the left and right circular polarization. Let us use Jones calculus (Azzam & Bashara, 1977; Gerrard & Burch, 1975) to describe the propagation of polarized quasi-monochromatic plane light wave along z-axis. Normalized Jones vector of the wave with arbitrary state of polarization can be presented in the following reductive form:

$$\mathbf{E} = \begin{pmatrix} A \\ B \cdot e^{i\delta} \end{pmatrix}, \tag{15}$$

here $A = |E_x|/|\mathbf{E}|$, $B = |E_y|/|\mathbf{E}|$, $|\mathbf{E}| = \sqrt{|E_x|^2 + |E_y|^2}$ and $\delta = \delta_x - \delta_y$. Arbitrary Jones vector can be represented as a linear combination:

$$\mathbf{E} = C_L \cdot \frac{1}{\sqrt{2}}\begin{pmatrix} 1 \\ -i \end{pmatrix} + C_R \cdot \frac{1}{\sqrt{2}}\begin{pmatrix} 1 \\ +i \end{pmatrix} \tag{16}$$

The coefficients C_L and C_R determine the contribution of left-handed and right-handed circular polarized light into the state of polarization defined by Jones vector **E**. The coefficients depend on the angle of ellipticity ε, and the angle ε depends on ellipticity e in the following way:

$$e = \mathrm{tg}\varepsilon, \qquad e = \frac{b}{a}, \tag{17}$$

where a is the length of a large semi axis of polarization ellipse and b is the length of a small semi axis of polarization ellipse. Following (Azzam & Bashara, 1977) represent the coefficients C_L and C_R as a function of the angle of ellipticity ε:

$$\begin{aligned} C_L &= \frac{1}{\sqrt{2}}(\cos\varepsilon - \sin\varepsilon), \\ C_R &= \frac{1}{\sqrt{2}}(\cos\varepsilon + \sin\varepsilon). \end{aligned} \tag{18}$$

Let $\mathbf{E}_0(r,\varphi)$ is the field of light wave falling on the input end of an optical fiber with the length z. The arbitrary state of polarization of the wave is defined by a normalized Jones vector Eq. (15). Then

$$\mathbf{E}_0(r,\varphi) = C_L\frac{1}{\sqrt{2}}\begin{pmatrix}1\\-i\end{pmatrix}\cdot E_0^-(r,\varphi) + C_R\frac{1}{\sqrt{2}}\begin{pmatrix}1\\+i\end{pmatrix}\cdot E_0^+(r,\varphi), \tag{19}$$

were $E_0^-(r,\varphi)$ is the amplitude of light wave with left-handed circular state of polarization at the optical fiber input, $E_0^+(r,\varphi)$ is the amplitude of light wave with right-handed circular state of polarization at the optical fiber input. To describe the propagation of linearly polarized light or light with the arbitrary state of polarization through the fiber, it is possible to consider propagation through the fiber of light with circular polarization at the fiber input. According to the results of the paper (Darsht et al., 1995):

$$\begin{aligned} E_0^+ = A_{\text{right}}(r,\varphi,z=0) &= e^{-i\varphi}\sum_N B_{1,N}F_{1,N}(r) + \sum_{m\neq 1}\sum_N B_{m,N}e^{-im\varphi}F_{m,N}(r) \\ &+ \sum_m\sum_N A_{m,N}e^{im\varphi}F_{m,N}(r), \end{aligned} \tag{20}$$

$$\begin{aligned} E_0^- = A_{\text{left}}(r,\varphi,z=0) &= e^{i\varphi}\sum_N A_{1,N}F_{1,N}(r) + \sum_{m\neq 1}\sum_N A_{m,N}e^{im\varphi}F_{m,N}(r) \\ &+ \sum_m\sum_N B_{m,N}e^{-im\varphi}F_{m,N}(r). \end{aligned} \tag{21}$$

Here those modes which do not retain circular polarization under propagation through the fiber are selected. It is necessary to underline that in case of the linear polarization at the fiber input the modes $\mathbf{e}^+_{+,m,N}(r,\varphi)$ and $\mathbf{e}^-_{+,m,N}(r,\varphi)$ give equal contribution to the propagating light, and this contribution is determined by the coefficient $A_{m,N}$. The modes $\mathbf{e}^+_{-,m,N}(r,\varphi)$ and $\mathbf{e}^-_{-,m,N}(r,\varphi)$ give the equal contribution to the propagating light as well, and this contribution is determined by the coefficient $B_{m,N}$. Values of coefficients $A_{m,N}$ and $B_{m,N}$ determine the intensity distribution at the fiber input. They are selected by means of a random-number generator in our investigation, but the optimum condition of the effect observation can depend on their relative value. Following the results of the paper Darsht et al. (1995) we have the following distribution of the amplitude of the light field $\mathbf{E}^-(r,\varphi,z)$ and $\mathbf{E}^+(r,\varphi,z)$ at the fiber output:

$$\mathbf{E}^{+}(r,\varphi,z) = \mathbf{E}^{++}(r,\varphi,z) + \mathbf{E}^{+-}(r,\varphi,z) =$$
$$\frac{1}{\sqrt{2}}\begin{pmatrix}1\\ i\end{pmatrix}\cdot\left\{\sum_{m\neq 1}\sum_{N} B_{m,N}e^{-im\varphi}F_{m,N}(r)e^{iz\left(\beta_{m,N}+\delta\beta^{(2)}_{m,N}\right)}\right.$$
$$+\sum_{m}\sum_{N} A_{m,N}e^{+im\varphi}F_{m,N}(r)e^{iz\left(\beta_{m,N}+\delta\beta^{(1)}_{m,N}\right)}$$
$$\left.+\sum_{N} B_{1,N}e^{-i\varphi}F_{1,N}(r)e^{iz\beta_{1,N}}\left(e^{i2z\delta\beta^{(2)}_{1,N}}+1\right)\right\}$$
$$+\frac{1}{\sqrt{2}}\begin{pmatrix}1\\ -i\end{pmatrix}\cdot\left[e^{+i\varphi}\sum_{N} B_{1,N}F_{1,N}(r)e^{iz\beta_{1,N}}\left(e^{i2z\delta\beta^{(2)}_{1,N}}-1\right)\right]. \quad (22)$$

$$\mathbf{E}^{-}(r,\varphi,z) = \mathbf{E}^{--}(r,\varphi,z) + \mathbf{E}^{-+}(r,\varphi,z) =$$
$$\frac{1}{\sqrt{2}}\begin{pmatrix}1\\ -i\end{pmatrix}\cdot\left\{\sum_{m\neq 1}\sum_{N} A_{m,N}e^{+im\varphi}F_{m,N}(r)e^{iz\left(\beta_{m,N}+\delta\beta^{(2)}_{m,N}\right)}\right.$$
$$+\sum_{m}\sum_{N} B_{m,N}e^{-im\varphi}F_{m,N}(r)e^{iz\left(\beta_{m,N}+\delta\beta^{(1)}_{m,N}\right)}$$
$$\left.+\sum_{N} A_{1,N}e^{+i\varphi}F_{1,N}(r)e^{iz\beta_{1,N}}\left(e^{i2z\delta\beta^{(2)}_{1,N}}+1\right)\right\}$$
$$+\frac{1}{\sqrt{2}}\begin{pmatrix}1\\ i\end{pmatrix}\cdot\left[e^{-i\varphi}\sum_{N} A_{1,N}F_{1,N}(r)e^{iz\beta_{1,N}}\left(e^{i2z\delta\beta^{(2)}_{1,N}}-1\right)\right]. \quad (23)$$

The intensity of the light at the optical fiber output $I^{+}(r,\varphi,z)$ and $I^{-}(r,\varphi,z)$ are the following:

$$I^{+}(r,\varphi,z) = \mathbf{E}^{+}(r,\varphi,z)\cdot(\mathbf{E}^{+})^{*}(r,\varphi,z) = |C_{R}(z)|^{2}\cdot\left|\mathbf{E}^{++}(r,\varphi,z)\right|^{2} + |C_{L}(z)|^{2}\cdot\left|\mathbf{E}^{+-}(r,\varphi,z)\right|^{2}.$$

$$I^{-}(r,\varphi,z) = \mathbf{E}^{-}(r,\varphi,z)\cdot(\mathbf{E}^{-})^{*}(r,\varphi,z) = |C_{R}(z)|^{2}\cdot\left|\mathbf{E}^{-+}(r,\varphi,z)\right|^{2} + |C_{L}(z)|^{2}\cdot\left|\mathbf{E}^{--}(r,\varphi,z)\right|^{2}.$$

One can see from Eqs. (22) and (23) that circular polarization does not conserve under the light propagation through an optical fiber with step index profile. All described above is used in the following sections.

3. Theoretical prediction and the experimental investigation of the isolated wave front dislocation formation

Here the possibility of the formation of a wavefront with an isolated screw dislocation of a given sign during the propagation of circularly polarized light through a multimode optical fiber is described. It follows from Eq.(22) and Eq. (23), that only the modes $\mathbf{e}^{2}_{1,N}$ and $\mathbf{e}^{4}_{1,N}$ contribute to the "foreign" polarization under circularly polarized light propagation through a multimode fiber. In the geometrical optics interpretation the modes $\mathbf{e}^{2}_{1,N}$ and $\mathbf{e}^{4}_{1,N}$ correspond to meridional rays for which circular polarization is not conserved because of symmetry consideration (Baranova & Zel'dovich, 1994; Snyder & Love, 1984). If a "circular analyzer" transmitting circular radiation opposite in sign to σ at the fiber input is placed at the fiber exit, only the corresponding modes with $m = 1$ will be transmitted through that analyzer. Let us consider the field passing through the "circular analyzer" in greater detail. This field is described by the last term in Eq. (22)) and Eq. (23). If light with $\sigma = 1$ enters the fiber the field with $\sigma = -1$ transmitted through the "circular analyzer" will have the following form:

$$\mathbf{E}^{+-} = \exp(i\varphi) \cdot \frac{1}{\sqrt{2}} \begin{pmatrix} 1 \\ -i \end{pmatrix} \cdot \left\{ \sum_N B_{1,N} F_{1,N}(r) \exp(iz\beta_{1,N}) \left[\exp\left(i2z\delta\beta_{1,N}^{(2)} \right) - 1 \right] \right\}. \quad (24)$$

The factor $\exp(i\varphi)$ indicates that the wavefront of the field (Eq. 24) contains one positive screw dislocation. If the entrance of the waveguide is illuminated with circularly polarized radiation with $\sigma = -1$ and circularly polarized radiation with $\sigma = +1$ is extracted at the fiber exit, the field E^{-+} after the "circular analyzer" will have the following form:

$$\mathbf{E}^{-+} = \exp(-i\varphi) \cdot \frac{1}{\sqrt{2}} \begin{pmatrix} 1 \\ +i \end{pmatrix} \cdot \left\{ \sum_N A_{1,N} F_{1,N}(r) \exp(iz\beta_{1,N}) \left[\exp\left(i2z\delta\beta_{1,N}^{(2)} \right) - 1 \right] \right\}. \quad (25)$$

It means that the light field (Eq. 25) contains one negative screw dislocation.

It follows from Eq. (24) and Eq. (25) that the intensity distribution $I^{+-} = E^{+-} \cdot \left(E^{+-}\right)^*$ across the fiber does not depend on the angle φ, i.e., the intensity distribution has axial symmetry. In the special case of an optical fiber with a small number of modes, such as $N_{\max} = 1$ for $m = 1$, the intensity of the "foreign" polarization is a periodic function of the fiber length:

$$I^{+-} = \left(B_{1,1} F_{1,1}(r)\right)^2 \left[1 - \cos\left(2z\delta\beta_{1,1}^{(2)}\right) \right]. \quad (26)$$

$$I^{-+} = \left(A_{1,1} F_{1,1}(r)\right)^2 \left[1 - \cos\left(2z\delta\beta_{1,1}^{(2)}\right) \right]. \quad (27)$$

The intensity I^{+-} and I^{-+} are maximum for fiber lengths

$$z_n = \frac{\pi(2n-1)}{2\delta\beta_{1,1}^{(2)}}. \quad (28)$$

The intensity I^{+-} and I^{-+} are minimum for fiber lengths

$$z_n = \frac{n\pi}{\delta\beta_{1,1}^{(2)}}. \quad (29)$$

Therefore, if circularly polarized radiation with a definite sign of σ is injected into a multimode fiber with a step like index profile and circularly polarized radiation of the opposite sign σ is extracted at the fiber exit, the transmitted radiation will consist of a light wave with an isolated wavefront dislocation, and a sign change of σ at the fiber input together with a corresponding sign change of the "circular analyzer" will result in the change of the screw dislocation sign.

Figure 1à shows the calculated intensity distribution of the interference pattern in the case of a plane wave and a converging wave $\mathbf{E} \sim \mathbf{A}(\mathbf{r}, z, t)e^{i\varphi}$ with a positive dislocation, and Fig. 1b shows calculated intensity distribution of the interference pattern in the case of a plane wave and a converging wave with a negative dislocation. As one can see in Fig. 1, the interference pattern is a spiral which unwinds in the clockwise direction in the case of a

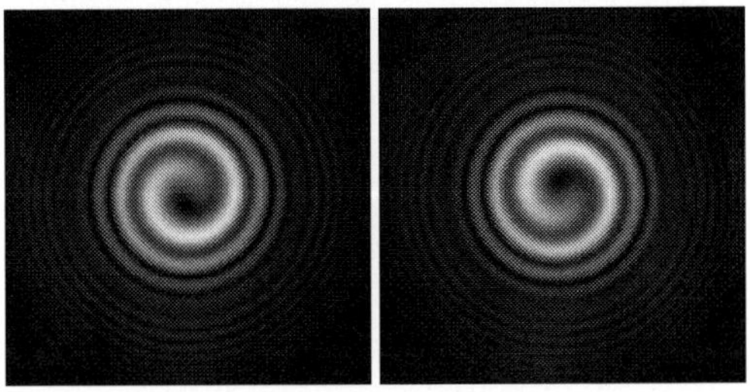

(a) (b)

Fig. 1. Interference patterns in the case of a plane wave and a converging wave with a positive dislocation (a) and calculated intensity distribution of the interference pattern in the case of a plane wave and a converging wave with a negative dislocation (b).

single positive screw dislocation and in the counterclockwise direction in the case of a single negative screw dislocation. It follows from Fig. 1 that the change of the screw dislocation sign results in the change of the spiral direction unwinding. That means, that when the sign of the circular polarization at the fiber input and output changes, the direction of the spiral unwinding should also change.

The experimental generation of a light wave with a wavefront dislocation has been done using an optical fiber with a step like index profile and the following parameters: fiber core diameter $2\rho = 9\ \mu$m, difference between the refraction index of the core and cladding $\delta n = n_{\rm co} - n_{\rm cl} = 0.004$. According to Eq. (8) twelve modes can propagate through the fiber:

$$\begin{array}{llll} M_1 = \mathbf{e}_{0,1}^{(1)}(r,\varphi) & M_2 = \mathbf{e}_{0,1}^{(3)}(r,\varphi) & M_3 = \mathbf{e}_{0,2}^{(1)}(r,\varphi) & M_4 = \mathbf{e}_{0,2}^{(3)}(r,\varphi) \\ M_5 = \mathbf{e}_{1,1}^{(1)}(r,\varphi) & M_6 = \mathbf{e}_{1,1}^{(2)}(r,\varphi) & M_7 = \mathbf{e}_{1,1}^{(3)}(r,\varphi) & M_8 = \mathbf{e}_{1,1}^{(4)}(r,\varphi) \\ M_9 = \mathbf{e}_{2,1}^{(1)}(r,\varphi) & M_{10} = \mathbf{e}_{2,1}^{(2)}(r,\varphi) & M_{11} = \mathbf{e}_{2,1}^{(3)}(r,\varphi) & M_{12} = \mathbf{e}_{2,1}^{(4)}(r,\varphi). \end{array} \quad (30)$$

The optimum fiber length has been calculated and determined in accordance with Eq.(28). The polarization correction $\delta\beta_{1,1}^{(2)}$ has been found to be 0.20526 $^{-1}$. As it was stated above the part of "foreign" polarization in intensity of light at the end of a fiber is periodical, that can also be seen in Fig. 2 for a fiber with parameters described above. The length of the fiber under investigation $z = 114$ cm corresponded to $n = 7$ in Eq. (28). A Mach-Zender interferometer scheme was used to record the wavefront with an isolated dislocation. It should be stressed that an adjustable polarization system with the quarter wave property (Goltser et al., 1993) has been used for "circular polarizer" and "circular analyzer". The device provided up to 98% circular polarization.

Figure 3 shows photos of the interference pattern of a plane wave and the converging wave under investigation. Figure 3a illustrates the situation when circularly polarized light with $\sigma = +1$ is incident upon the fiber input and a circularly polarized light wave with $\sigma = -1$, containing a positive screw wavefront dislocation, is extracted by the "circular analyzer." As one can see in Fig. 3a, the interference pattern is a clockwise spiral. Comparing with Fig. 1a shows that this result agrees completely with theory. Figure 3b displays a photo of the interference pattern of a plane wave and the wave under investigation with a negative wavefront dislocation. The interference pattern unwinds in the counterclockwise direction

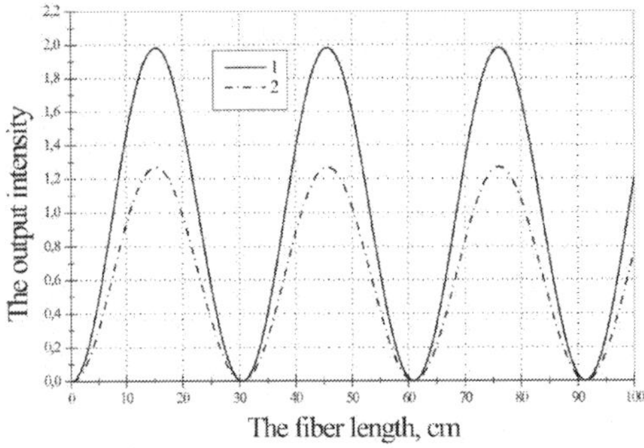

Fig. 2. The dependence of "foreign" polarization part in the output intensity of light on the fiber length at the illumination the fiber with right hand (1) and left hand (2) circularly polarized light.

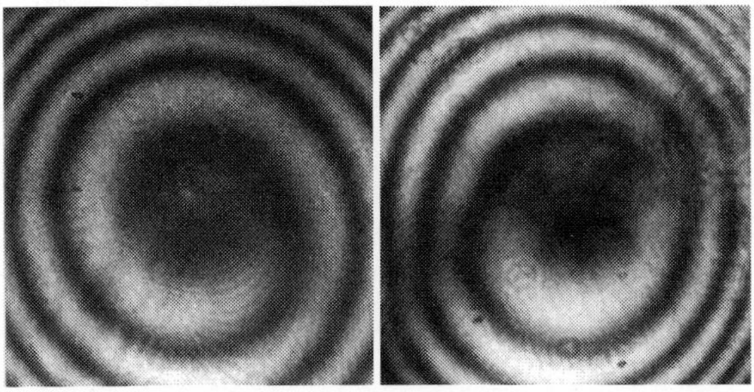

(a) (b)

Fig. 3. Photos of the interference pattern on the screen. a - Circularly polarized radiation with $\sigma = +1$ is incident at fiber input and circularly polarized radiation with $\sigma = -1$ is extracted from the output radiation. b – Circularly polarized radiation with $\sigma = -1$ is incident at fiber input and circularly polarized radiation with $\sigma = +1$ is extracted from the output radiation.

like the spiral in Fig. 1b. Therefore, this result also agrees completely with the expected result. It should be emphasized that only one spiral is observed in the interference pattern in Figs. 3a and 3b; this corresponds to an isolated wavefront dislocation.

As a result of this work, the possibility of the formation of a light wave with an isolated wavefront dislocation having a fixed sign has been proved. Such a wave is formed when polarized radiation passes through a multimode optical fiber placed between crossed "circular polarizers". The sign of the dislocation changes when the sign of the "circular polarizers" at the fiber input and exit changes.

4. The optical Magnus effect and its peculiarity in a few modes fiber

As it has been mentioned above, optical Magnus effect manifests itself as speckle pattern rotation of circular polarized light transmitted through a multimode step-index fiber under circular polarization sign change (Doogin et al., 1992; 1991). Using Eq. (22) and Eq. (23) let us consider a case when we illuminate a multimode optical fiber by right and left polarization in turn, but with strictly the same modal distribution at the input and omit all modes with $m = 1$. Eq. (22) and Eq. (23) take the following form.

$$\mathbf{E}^{+}(r,\varphi,z) = \frac{1}{\sqrt{2}}\begin{pmatrix}1\\ i\end{pmatrix}\cdot\left\{\sum_{m\neq 1}\sum_{N} C_{-,m,N}e^{-im\varphi}F_{m,N}(r)e^{iz\left(\beta_{m,N}+\delta\beta^{(2)}_{m,N}\right)} + \sum_{m}\sum_{N} C_{+,m,N}e^{+im\varphi}F_{m,N}(r)e^{iz\left(\beta_{m,N}+\delta\beta^{(1)}_{m,N}\right)}\right. \tag{31}$$

$$\mathbf{E}^{-}(r,\varphi,z) = \frac{1}{\sqrt{2}}\begin{pmatrix}1\\ -i\end{pmatrix}\cdot\left\{\sum_{m\neq 1}\sum_{N} C_{+,m,N}e^{+im\varphi}F_{m,N}(r)e^{iz\left(\beta_{m,N}+\delta\beta^{(2)}_{m,N}\right)} + \sum_{m}\sum_{N} C_{-,m,N}e^{-im\varphi}F_{m,N}(r)e^{iz\left(\beta_{m,N}+\delta\beta^{(1)}_{m,N}\right)}\right. \tag{32}$$

Let us compare the angular intensity distributions for the left handed ($\sigma = -1$) and right handed ($\sigma = +1$) circular polarizations in some cross-section. Fig. 4 shows the

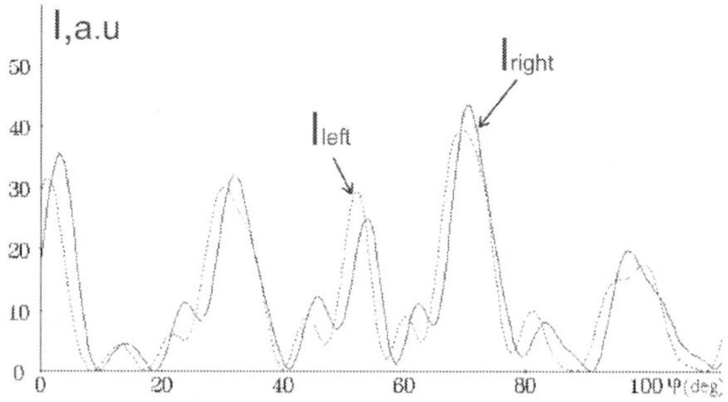

Fig. 4. The angular intensity distributions for the left handed ($\sigma = -1$) and right handed ($\sigma = +1$) circular polarizations in some fiber cross-section

speckle-patterns particular realization for the left and right circular polarizations. It can be seen from Fig. 4 that the whole picture suffers angular shift (rotation) saving main features and being slightly distorted under the change of the circular polarization. To extract the pure rotation from the whole change of the speckle-pattern the correlation function has been calculated.

The first experimental observation of the speckle pattern rotation under the change of the circular polarization has been done using a fiber with the following parameters. The fiber diameter was $2\rho = 200\ \mu$m; the difference between refractive indices of quartz core and polymeric cladding $\Delta n = n_{co} - n_{cl} = 0.006$ was measured via the critical angle of propagating rays; the fiber length was 96 cm.

Changing circular polarization of propagating light from left-handed to right-handed resulted in the speckle pattern flowing clock-wise in accordance with the theoretical prediction. During that flowing some details of speckle-pattern changed, but the main features were conserved under rotation. The fragments of the speckle-patterns for both circular polarizations are shown in Fig. 5 . The arrow points out the bright sport that was rotated slightly changing

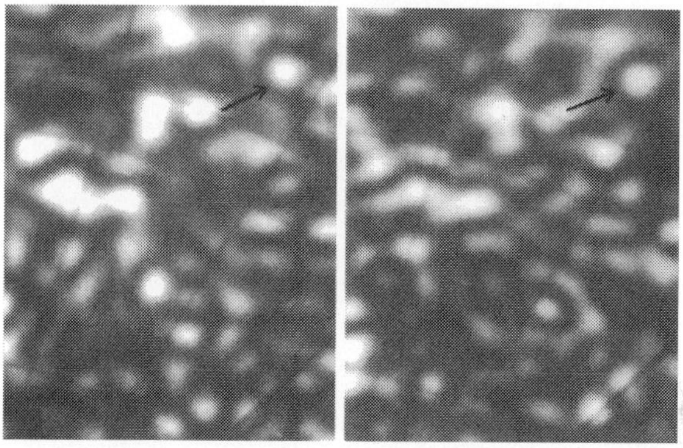

(a) $\sigma = -1$ (b) $\sigma = +1$

Fig. 5. The fragment of the real speckle-patterns of right-hand ($\sigma = +1$) and left-hand ($\sigma = -1$) circularly polarized light at the screen. The arrow points out the bright spot as an example of rotation under circular polarization sign change

its shape (just like it was in the computer experiment, see Fig.4).

In order to determine the "rotation angle" an experimental analog of the correlation function technique was used in the paper (Doogin et al., 1992). The negative slide for the right-handed circular polarization speckle-pattern was projected to the positive photograph of the left-handed one and the least contrast was achieved by their mutual rotation. The angle between the two coordinate nets corresponding to the contrast minimum was regarded as the "rotation angle". It was $(1.4 \pm 0.5)^{\circ}$ for the fiber under investigation.

For the first time optical Magnus effect has been observed in a multi-mode optical fiber (Doogin et al., 1992; 1991), and all calculations of the speckle pattern rotation angle have also been done (Doogin et al., 1992) for a multimode fiber, but the influence of meridional rays has been neglected. Using the wave interpretation we can associate meridional rays with TE and TM modes. Their combination is circular polarized, but due to the difference of their propagation velocity circular polarization does not keep unchanged in contrast to all other skew rays. In the case of multimode fiber contribution of meridional rays can be neglected but the speckle pattern rotation angle in such fibers is too small, of order 0.1 degree per 1 centimeter of fiber's length. In a few mode fibers rotation angle is two orders larger, but the contribution of meridional rays is noticeable.

One can see from Eq. 22 and Eq. (23) that during propagation of light with "original" polarization "foreign" polarization appears due to the presence of modes with $m = 1$ (TE and TM modes). The contribution of these modes has a periodical nature along the fiber length.

To calculate the light intensity distribution across the fiber cross section Eq. 22 and Eq. (23) were used. In Fig. 6 speckle patterns at the end of the fiber with following parameters: $2\rho = 9$

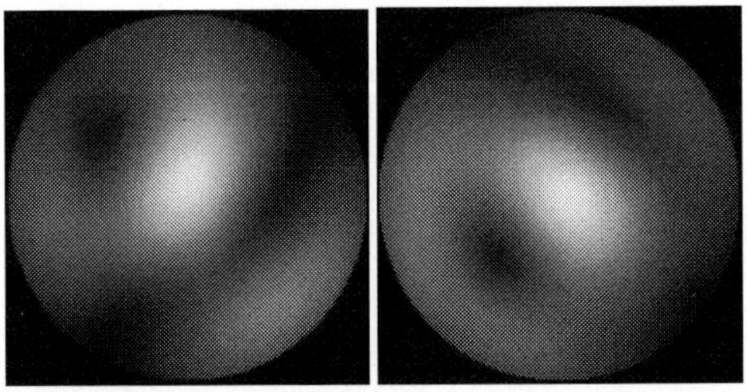

(a) $\sigma = +1$ (b) $\sigma = -1$

Fig. 6. Speckle patterns at the end of the fiber with following parameters: $2\rho = 9$ μm, $n_{co} = 1.47$, $\delta n = 0.004$, $z = 18$ cm, $\lambda = 0.6328$ μm.

μm, $n_{co} = 1.47$, $\delta n = 0.004$, $z = 18$ cm, $\lambda = 0.6328$ μm are presented. It can be seen in Fig. 6 that this fiber is a few mode fiber. A rotation of speckle patterns can be observed. The number of maximal remains the same, but their intensities change due to modes with $m = 1$.

Regardless the modes with $m = 1$, the rotation with no change of maximal intensity will be observed. It is well known that the number of modes propagating in optical fiber depends on the radius of the fiber's core, the difference between refractive indices of the core and the cladding and the wavelength. To determine the fiber parameters for which the contribution of modes with $m = 1$ becomes noticeable the speckle pattern rotation angles for different diameters of fiber core (Fig. 7) were calculated. The calculation was realized in two ways:

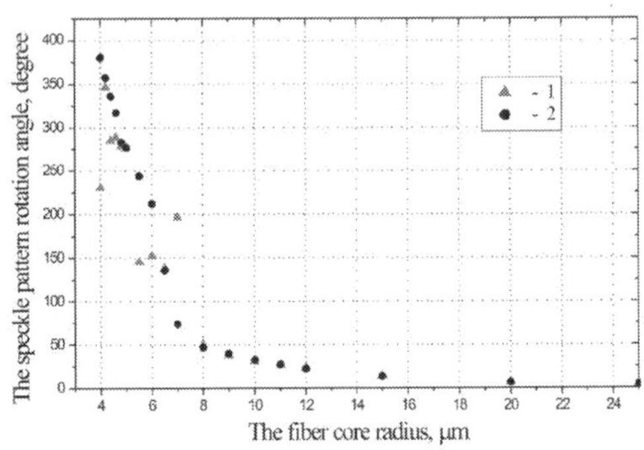

Fig. 7. The dependence of speckle pattern rotation angle on the fiber core radius: taking modes with $m = 1$ into consideration (1) and neglecting those modes (2).

taking modes with $m = 1$ into consideration and neglecting those modes. Figure 7 indicates that at $\rho \geq 8$ μm we can neglect the contribution of modes with $m = 1$ and the rotation angle is $\sim 50^{\circ}$ at length 18 cm.

As it can be seen in Fig. 2 the contribution of modes with $m = 1$ is equal to 0 at length multiple to $\sim$ 30 cm. We should suppose that for this fiber the rotation of speckle pattern can take place only at this length. At the length multiple to $\sim$ 15 cm it will be no rotation due to maximal contribution of modes with $m = 1$. In order to examine this supposition the dependence of speckle pattern rotation on fiber length was computed (Fig. 8). The software calculating the

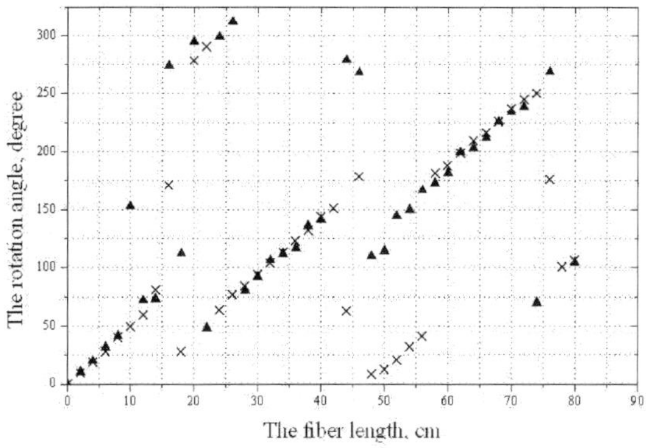

Fig. 8. The speckle pattern rotation angle dependence on the fiber length for two different realization of speckle pattern

rotation angle between two speckle patterns yields the value of rotation angle even if there is no rotation. In this case the obtained value differs from the expected one immensely. This is a cause of the scattering of points in Fig. 8. At the length multiple to $\sim$ 15 cm the scattering is maximal (rotation is absent). At the length multiple to $\sim$ 30 cm there is almost no scattering. In this case the rotation can be observed visually on the computer screen that confirms our supposition. Thus, the optimal length for optical Magnus effect observation in the fiber with present parameters is multiple to $\sim$ 30 cm.

Experimental investigations were carried out at a fiber with step like index profile and $2\rho = 9$ μm, $n_{co} = 1.47$, the numerical aperture $N_A = 0.11$ at the wavelength $\lambda = 0.6326$ μm. Optical Magnus effect was distinctly observed at the small fiber length in the range from 4 till 20 cm. The experimental results are presented in Fig.9. The calculated dependence of the speckle pattern rotation angle on the fiber length is shown in Fig. 9 too. One can see in Fig. 9 that the angle of the speckle pattern rotation depends on the fiber length linearly, but there is difference in the slope of the experimental and theoretical lines. It can be connected with the speckle pattern distortion.

The influence of the meridional modes contribution on the optical Magnus effect in fibers with different numbers of modes has been demonstrated.

5. The influence of light polarization on the speckle pattern of the polarized light transmitted through multimode optical fiber

Let us consider the propagation of arbitrary polarized light through an axially symmetric optical fiber with a step like refractive index profile. The arbitrary state of polarization

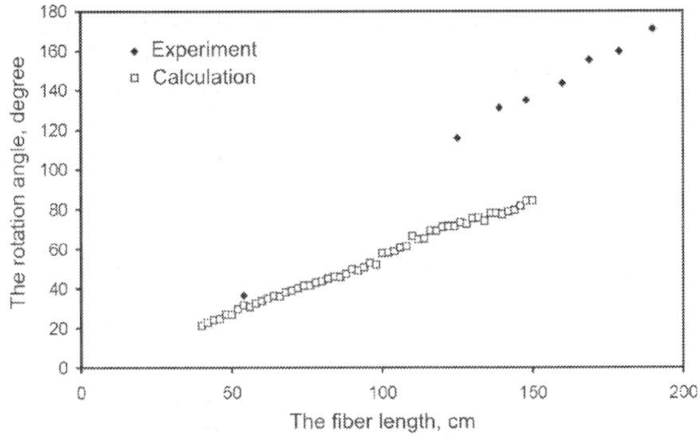

Fig. 9. Experimental and calculated the speckle pattern rotation angle as a function of the fiber length for the fiber with the diameter $2\rho = 9\ \mu$m, the core refractive index $n_{co} = 1.47$, numerical aperture $N_A = 0.11$ at the wavelength $\lambda = 0.6328\ \mu$m.

is defined by the ellipticity angle ε. The contribution of the intensity of the right circular polarized light to the whole intensity β_r is defined by the following way:

$$\beta_r = |C_R|^2 = \frac{1}{2}(\cos\varepsilon + \sin\varepsilon)^2. \tag{33}$$

Here the coefficient C_R is defined by Eq.18. In order to determine the light intensity distribution across the fiber section at the fiber output end (speckle pattern) $I(r, \varphi, z)$ as a function of β_r at the fiber input Eq. (24) has been used.

A multimode optical fiber with a step like index profile and $\rho = 50\ \mu$m, $n_{co} = 1.47$, n_{cl} $n_{co} - n_{cl} = \Delta n = 0.016$ has been chosen for experimental investigation. Optical Magnus effect is distinctly observed in this fiber. Calculation of the speckle patterns according to Eq. (24) has been carried out for a fiber with the same parameters. Figure 10 shows the dependence of the speckle pattern angle rotation ψ on the fiber length z for two various mode structure (different sets of coefficients $A_{m,N}$ and $B_{m,N}$) of light propagating through the optical fiber under investigation. One can see in Fig. 10, that there is some critical length $z_{max} \approx 15$. If the fiber length is less than critical length z_{max}, the dependence $\psi(z)$ will be linear and the rotation angle will increase with the increase of the fiber length. If the fiber length is greater than critical length z_{max} the rotation angle will decrease with the increase of the fiber length. This behavior can be explained by the relative angular size of the single speckle $\varphi_{speckle}$. The speckle pattern rotation will be distinctly observed only if $\psi < \varphi_{speckle}$.

Two pieces of the fiber of different length have been chosen for experimental investigation. The length of the first fiber was almost critical ($z = 15.4$ cm) and the length of the second fiber was greater than the critical length($z = 20.5$ cm). Radiation of He-Ne laser with wave length $\lambda = 0.628\ \mu$m has been used under experimental investigation. The polarizing system consisting of a polarizer and an adjustable quarter wave plate (Goltser et al., 1993) allowed to change the light polarization state from right circular polarization $\beta_r = 1$ to left circular polarization $\beta_r = 0$. Value $\beta_r = 0.5$ corresponds to linear polarized light. The images of speckle patterns of light at the fiber exit have been registered in a digital form. The special software was used for the speckle patterns analysis of the determination and for the rotation angle.

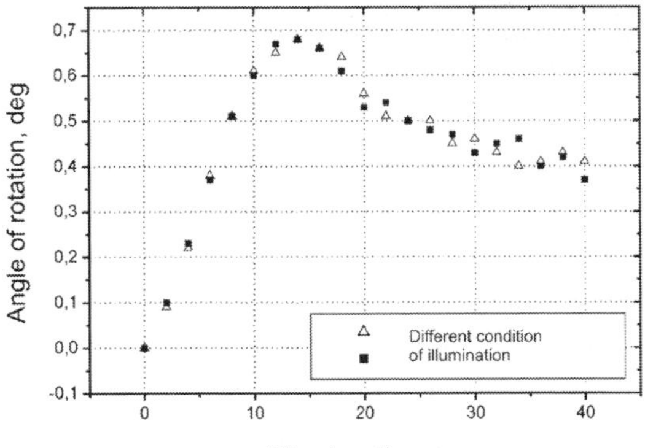

Fig. 10. Dependence of the speckle pattern angle rotation on the fiber length for two various mode structure of light propagating through the optical fiber under investigation.

Figure 11 shows speckle patterns of the light at the output end of the fibers with the length

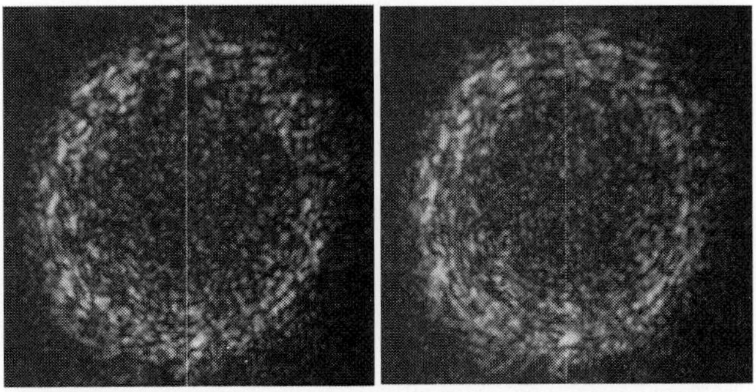

(a) $\beta_r = 0$ (b) $\beta_r = 0.75$

Fig. 11. Images of speckle patterns of radiation with the left circular polarization (a) and elliptic polarization (b) at the fiber output.The fiber length is $z = 15.4$ cm.

$z = 15.4$ cm. Image in Fig. 11a corresponds to the left circular polarization at the fiber input ($\beta_r = 0$), image in Fig. 11b corresponds to the elliptical polarization at the fiber input with $\beta_r = 0.75$. Radiation was entered into the fiber under some angle to the fiber axis, as a result the speckle pattern had the form of a ring (Zel'dovich et al., 1996). It can be seen in Fig. 11, that the main speckle pattern features are kept unchanged under the change of the relative contribution of the right circular polarized light, the speckle rotation was distinctly observed visually. The speckle pattern rotation was also confirmed by the computer processing of the images, the angle of rotation was determined. The speckle pattern rotation was also observed under other states of polarization change. Figure 12 and Fig. 13 show the experimental and calculated angles of the speckle pattern rotation ψ as a function of the contribution of light with the right circular polarization β_r into the elliptical polarization for the fiber length $z =$

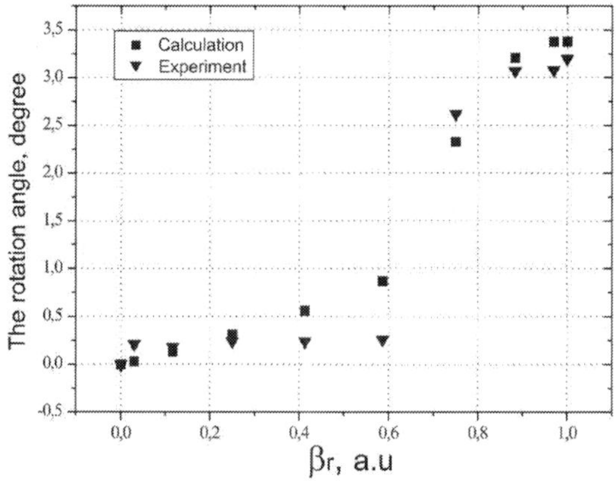

Fig. 12. The experimental and calculated angles of the speckle pattern rotation ψ as a function of the contribution of light with the right circular polarization β_r into the elliptical polarization for the fiber length $z = 20.5$ cm.

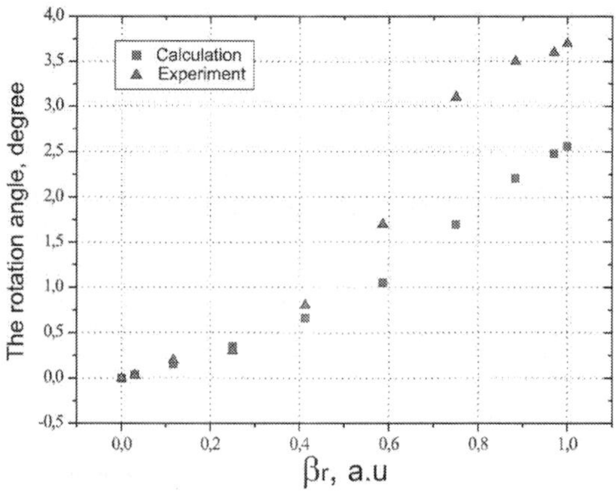

Fig. 13. The experimental and calculated angles of the speckle pattern rotation ψ as a function of the contribution of light with the right circular polarization β_r into the elliptical polarization for the fiber length $z = 15.4$ cm

15.4 cm and $z = 20.5$ cm. Figure 12 and Fig. 13 demonstrate qualitative and quantitative coincidence of the experimental results and the results of calculation.

To sum up, the change of the light ellipticity at the optical fiber input results in speckle pattern rotation at the fiber output and the angle of rotation depends linearly on the contribution of the light intensity with right-handed circular polarization into the intensity with the elliptical state of polarization. The length of the fiber where the linear dependence of $\psi\left(\beta_{\text{r}}\right)$ is observed is determined by optical fiber parameters.

6. The influence of external magnetic field on the speckle pattern behavior

It is well known that under light propagation through an optical medium placed in an external magnetic field a rotation of the polarization plane of linearly polarized light is observed. This rotation is a result of the changes of the refractive indexes of light with the left and right circular polarization under the influence of a magnetic field. However no marked change of the trajectory of light was observed. It has been shown by (Baranova & Zel'dovich, 1994), that the influence of a magnetic field on light propagation in an optical fiber located in a longitudinal magnetic field can result in noticeable rotation of a speckle pattern and the angle of rotation φ will be proportional to the Verdet constant V, value of the magnetic field H and the length of the fiber in the magnetic field z:

$$\varphi \approx VHz.$$

For the observation of this effect it was proposed to illuminate the input of a fiber with linearly polarized light, and to choose the linearly polarized component with the same azimuth in light passed through a fiber. In the paper (Baranova & Zel'dovich, 1994) theoretical analysis of the effect was carried out within the framework of the elementary model in which light propagation in an axial symmetric optical fiber which keeps in scalar approach only three modes was considered.

The experimental results obtained in the paper (Darsht et al., 1994) have shown that such a rotation can really be observed, and the direction of the rotation varies on the change of the external magnetic field direction. The effect of the magnetic rotation of the speckle pattern was observed under light propagation through a quartz fiber. The value of the Verdet constant for quartz at the wavelength $\lambda = 0.633\ \mu\text{m}$ is equal to $V = 1.4 \cdot 10^{-2}\ \text{min}/(\text{Gs} \cdot \text{cm})$ that can provide the rotation of the speckle pattern over the angle $0 \approx 2^{\circ}$ at the value of the magnetic field $H = 500$ Gs and the length of the fiber in the magnetic field 21.5 cm. To enhance the effect (Darsht et al., 1994) the fiber was passed through a solenoid with a magnetic field some times, so that the length of the fiber in the magnetic field increased up to 140 cm. The angle of the rotation of the speckle pattern was determined at the change of the direction of the magnetic field $H = 500$ Gs and it was equal to 15°. The angle of the Faraday rotation would be $\approx 40^{\circ}$ under the same conditions.

Let us consider light propagation through an optical fiber placed into longitudinal magnetic field in detail. Following the paper (Baranova & Zel'dovich, 1994) let us suppose that longitudinal magnetic field influences the propagation constant of only modes which retain their circular polarization and the change of the propagation constant value is $\delta\beta = -\sigma VHz$. Using Eq. (22) and Eq. (23) one can obtain the dependence of the fields $\mathbf{E}^{-}(r, \varphi, z, H)$ and

$\mathbf{E}^+(r,\varphi,z,H)$ on the magnetic field in the following form:

$$\mathbf{E}^+(r,\varphi,z,H) =$$

$$\frac{1}{\sqrt{2}}\cdot\begin{pmatrix}1\\+i\end{pmatrix}\cdot\left\{\sum_{m\neq 1}\sum_{N} B_{m,N}e^{-im(\varphi+VHz/m)}F_{m,N}(r)e^{iz\left(\beta_{m,N}+\delta\beta^{(2)}_{m,N}\right)}\right.$$

$$+\sum_{m}\sum_{N} A_{m,N}e^{+im(\varphi-VHz/m)}F_{m,N}(r)\exp\left[iz\left(\beta_{m,N}+\delta\beta^{(1)}_{m,N}\right)\right]$$

$$\left.+\,e^{-i\varphi}\sum_{N} B_{1,N}F_{1,N}(r)e^{iz\beta_{1,N}}\left(e^{iz\delta\beta^{(2)}_{1,N}}+1\right)\right\}$$

$$+\frac{1}{\sqrt{2}}\cdot\begin{pmatrix}1\\-i\end{pmatrix}\cdot\left[e^{i\varphi}\sum_{N} B_{1,N}F_{1,N}(r)e^{iz\beta_{1,N}}\left(e^{iz\delta\beta^{(2)}_{1,N}}-1\right)\right], \quad (34)$$

$$\mathbf{E}^-(r,\varphi,z,H) =$$

$$\frac{1}{\sqrt{2}}\cdot\begin{pmatrix}1\\-i\end{pmatrix}\cdot\left\{\sum_{m\neq 1}\sum_{N} A_{m,N}e^{+im(\varphi+VHz/m)}F_{m,N}(r)e^{iz\left(\beta_{m,N}+\delta\beta^{(2)}_{m,N}\right)}\right.$$

$$+\sum_{m}\sum_{N} B_{m,N}e^{-im(\varphi-VHz/m)}F_{m,N}(r)\exp\left[iz\left(\beta_{m,N}+\delta\beta^{(1)}_{m,N}\right)\right]$$

$$\left.+\,e^{+i\varphi}\sum_{N} A_{1,N}F_{1,N}(r)e^{iz\beta_{1,N}}\left(e^{iz\delta\beta^{(2)}_{1,N}}+1\right)\right\}$$

$$+\frac{1}{\sqrt{2}}\cdot\begin{pmatrix}1\\+i\end{pmatrix}\cdot\left[e^{-i\varphi}\sum_{N} A_{1,N}F_{1,N}(r)e^{iz\beta_{1,N}}\left(e^{iz\delta\beta^{(2)}_{1,N}}-1\right)\right]. \quad (35)$$

It can be seen from the expressions (34) and (35) that the case, when the circular polarization of the light at the input and output of the fiber is different, special, namely, the magnetic field does not influence the intensity distribution at the fiber output. This situation can be used for experimental verification of the adequacy of the used model. If the circular polarization at the input and output of the fiber has the same sign, the magnetic field will influence a speckle pattern at the fiber output.

The analysis of the expressions (34) and (35) shows that the modal structure of the radiation, propagating through a fiber, influences the behavior of the speckle pattern in the magnetic field. The angle of the speckle pattern rotation described by the expressions (34) and (35) is determined by the factors $\exp\left[+m\left(\varphi - VHz/m\right)\right]$ and $\exp\left[-m\left(\varphi + VHz/m\right)\right]$. These expressions show that the increase of the number of a mode results in the decrease of the angle of the speckle pattern rotation. The different angular contribution of the modes of different orders should result in such change of the speckle pattern which cannot be regarded as rotation. It means, that the rotation of the speckle pattern in a magnetic field should be manifested well in fibers which retain modes of the lowest orders, for such a case we shall consider the situation when the input light is linearly polarized, and the output linearly polarized light has the same azimuth of polarization. Let's consider the influence of a magnetic field on linearly polarized light propagating through a few modes optical fiber retaining only modes with $m = 0$ and $m = 1$. Suppose that $N = 1$ and omit the coefficient N. Let the linearly polarized radiation at the fiber input be polarized along the x-axis, and the linearly polarized radiation at the fiber output have the same azimuth of polarization, then the amplitude of the field of the light wave E_x^x at the fiber output will look like the following:

$$\begin{aligned}
&\mathbf{E}_x^x(r,\varphi,z,H) = \\
&\frac{1}{\sqrt{2}} \cdot F_0(r) e^{iz\left(\beta_0+\delta\beta_0^{(1)}\right)} (A_0 + B_0) \left(e^{-iVHz} + e^{iVHz}\right) \\
&+\frac{1}{\sqrt{2}} \cdot F_1(r) e^{iz\beta_1} \left[e^{iz\delta\beta_1^{(1)}} \left(A_1 e^{+i(\varphi - VHz)} + B_1 e^{-i(\varphi - VHz)}\right) \right. \\
&\qquad \left. + \left(e^{iz\delta\beta_1^{(2)}} + 1\right) \left(A_1 e^{+i\varphi} + B_1 e^{-i\varphi}\right) \right] \\
&+\frac{1}{\sqrt{2}} \cdot F_1(r) e^{iz\beta_1} \left(e^{iz\delta\beta_1^{(2)}} - 1\right) \left(A_1 e^{-i\varphi} + B_1 e^{i\varphi}\right). \qquad (36)
\end{aligned}$$

In the simplest special case, when $A_0 = B_0 = 0$ and $A_1 = B_1 = A$ the expression (36) becomes:

$$\mathbf{E}_x^x(r,\varphi,z,H) = A\sqrt{2} F_1(r) e^{iz\beta_1} \left(e^{iz\delta\beta_1^{(1)}} \cos(\varphi - VHz) + 2e^{iz\delta\beta_1^{(2)}} \cos\varphi \right). \qquad (37)$$

As it follows from the expression (37), even in the simplest case which is difficult to implement experimentally, the rotation of the speckle pattern under the magnetic field change will be observed on some steady background which is determined by the length of the fiber. It means that the best conditions of a speckle pattern observation are determined not only by the mode structure of the radiation propagating through a fiber, but also by the fiber length.

The light intensity at the fiber output is the following:

$$\begin{aligned}
&I_x^x(r,\varphi,z,H) = |E_x^x(r,\varphi,z,H)|^2 = \\
&= |A|^2 F_1^2(r) \left(5 + \cos(2\varphi - 2VHz)\right) + \\
&+ |A|^2 F_1^2(r) \left(2\cos\left(z\left(\delta\beta_1^{(2)} - \delta\beta_1^{(1)}\right)\right)\left(\cos(VHz) + \cos(2\varphi - VHz)\right) + \cos 2\varphi\right). \qquad (38)
\end{aligned}$$

It is possible to group together the terms in the expression (38) on their dependence on the variables r, φ, z and H:

$$I_x^x(r,\varphi,z,H) = I_1^{xx}(r,\varphi,z) + I_2^{xx}(r,z,H) + I_3^{xx}(r,\varphi,z,H) + I_4^{xx}(r). \qquad (39)$$

Only the third term $I_3^{xx}(r,\varphi,z,H)$ in Eq. (39) depends on the magnetic field and influences the angular distribution of the intensity at the fiber output. Let us consider this term in detail:

$$I_3^{xx}(r,\varphi,z,H) = |A|^2 F_1^2(r) I_3^{xx}(\varphi,z,H). \qquad (40)$$

Here

$$I_3^{xx}(\varphi,z,H) = D(z)\cos(2\varphi + \gamma(z,H)), \qquad (41)$$

where

$$D(z,H) = \sqrt{1 + C(z)^2 + 2C(z)\cos(VHz)}, \qquad (42)$$

$$C(z) = 2\cos\left(z\left(\delta\beta_1^{(2)} - \delta\beta_1^{(1)}\right)\right) \qquad (43)$$

and

$$\tan\gamma(z,H) = -\frac{\sin(2VHz) + C(z)\sin(VHz)}{\cos(2VHz) + C(z)\cos(VHz)}. \quad (44)$$

It follows from the expression (41) that the speckle pattern rotation under the magnetic field influence will be determined by the angle $\gamma(z,H)/2$. In the approach $VHz \ll 1$ it is possible to present the expression (44) as follows:

$$\tan\gamma(z,H) \approx -\frac{VHz\,(2 + C(z))}{1 + C(z)}. \quad (45)$$

Thus, it follows from the expression (45) that under the considered approach the angle of rotation linearly depends on the value of the applied field at the given length of the fiber. Dependence of the rotation angle on the length of the fiber at a specified value of a magnetic field has a nonlinear character. The linear dependence is possible in two cases, when $C(z) \ll 1$ or $C(z) \gg 1$. The function $C(z)$ is periodic and varies in the range from -2 up to $+2$, therefore, linear dependence of the rotation angle on the length of the fiber is possible only in a narrow range of lengths of the fiber.

Eqs. (34) and (35) were used for numerical modeling. The following parameters of an optical fiber with a step like refractive index profile were used: $n_{co} = 1.47$, $\Delta n = 0.004$ and $\rho = 4.5$ μm. The Verdet constant V was accepted to be equal to $V = 1.4 \cdot 10^{-2}$ min/(Gs $\cdot$ cm). The coefficients in the expressions (34) and (35) were chosen by means of the random number generator. Calculation was carried out for different states of polarization at the fiber input and output, it was revealed that the state of polarization plays the important role only in one, the above-stated case. In Fig. 14 the calculated dependences of the angle of rotation of the

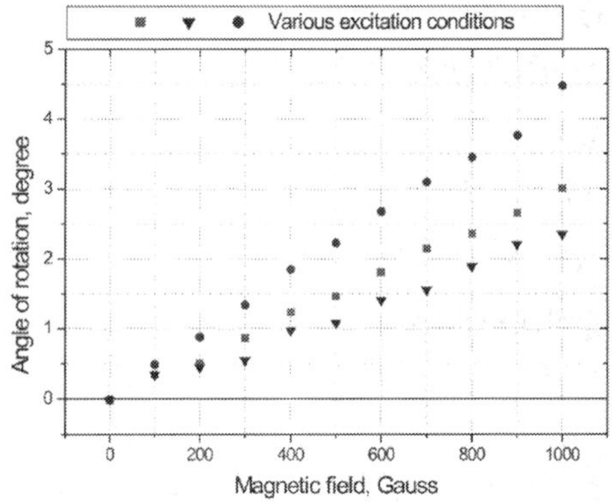

Fig. 14. Dependence of the angle of the speckle rotation on the value of a magnetic field at a different mode structure of radiation, the total length of the fiber $L = 100$ cm, the length of the fiber in the field $L_{mf} = 23$ cm.

speckle pattern of light passed through an optical fiber placed in a longitudinal magnetic field at the change of a direction of a magnetic field on the magnetic field value are presented. Figure 14 shows that the angle of rotation of the speckle pattern of light propagating in a few modes fiber placed in a longitudinal magnetic field linearly depends on the value of the applied magnetic field, however, the angle of rotation at the fixed value of the magnetic field depends on the mode structure of the radiation propagating through the fiber.

The experimental research of the propagation of the polarized radiation in a few modes germanosilicate optical fiber was carried out using a fiber with the following parameters: the difference of the refractive indexes of the core and the cladding $\delta n = 0.004$, the diameter of the core of the fiber $\rho = 4.5\ \mu$m. A He-Ne laser was used as the radiation source, the experimental setup allowed to replace this laser with another one radiating at a required wavelength. For the analysis of obtained images of speckle patterns (for the demonstration of an angle of rotation) the special software for computer image processing was developed which allowed to compare two images, to determine the angle of mutual rotation with high accuracy and also the degree of identity of two images. The special software was used for processing both the experimental results and the results of the computer simulation. Images of speckle patterns of the radiation with the wavelength $\lambda = 633$ nm propagating in the same a few mode germanosilicate optical fiber placed in the longitudinal magnetic field of the value 500 Gs are presented in Fig. 15 . The images were registered under opposite directions of

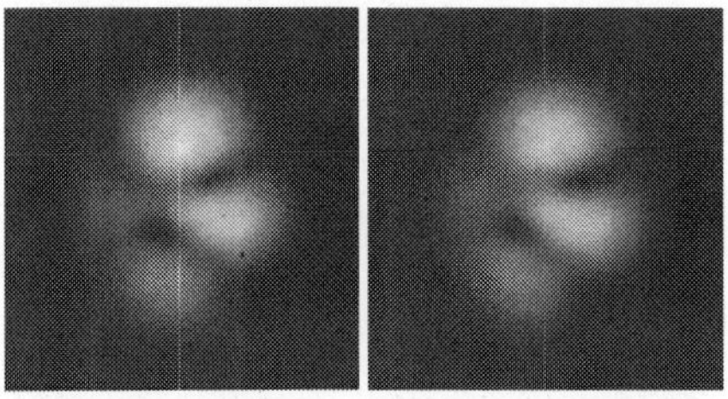

(a) $H = +500$ Gs (b) $H = -500$ Gs

Fig. 15. images of the speckle patterns of the radiation with the wavelength $\lambda = 633$ nm propagating in a few modes germanosilicate optical fiber placed in the longitudinal magnetic field of the value 500 Gs at the opposite directions of the magnetic field.

the magnetic field. The speckle pattern is practically unchanged, and the visual observation confirms the presence of the rotation.

However, the change of the conditions of the radiation input has shown that the change of the mode structure can result in a significant speckle pattern change under the change of the magnetic field direction, but this change is not characterized by rotation. To investigate the influence of the magnetic field on the speckle pattern rotation under different conditions, the conditions for input of the He-Ne laser radiation were chosen such that provided the speckle pattern change of the rotation type. Dependence of the rotation angle of the speckle pattern of the radiation of the He-Ne laser passed through the germanosilicate optical fiber placed in the longitudinal magnetic field on the value of the magnetic field is presented in Fig. 16 . One can see in Fig. 16 that the angle of rotation of the speckle pattern is linearly dependent on the value of the applied magnetic field. The investigation has shown that the change of the mode structure of the radiation propagating through the fiber results in the change of the slope of the linear dependence, however, the angle of rotation always was less than the Faraday angle.

The influence of a state of polarization at the fiber input and output on the behavior of the speckle pattern in the magnetic field was studied. It is revealed that generally the state of polarization does not influence the behavior of the speckle pattern under the change of the direction of the magnetic field. Essentially the other situation is observed in the case when circular polarization at the fiber input and output has opposite sign. According to the

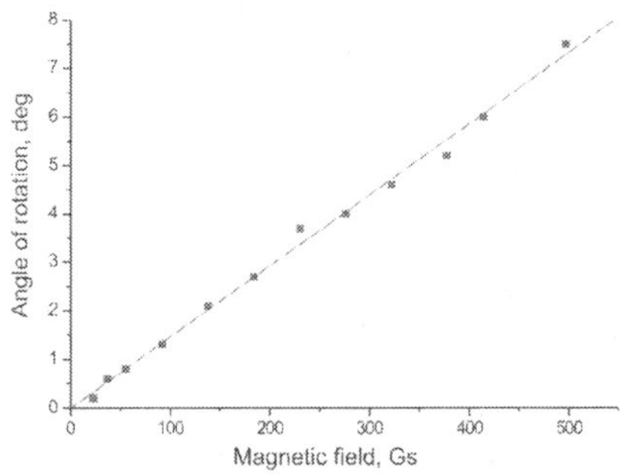

Fig. 16. Dependence of the rotation angle of the speckle pattern of radiation passed through the germanic silicate optical fiber placed in the longitudinal magnetic field on the value of the magnetic field. Length of the fiber in the magnetic field is 90 cm.

theoretical predictions no change of the speckle pattern at the change of the magnetic field direction was observed under any conditions of radiation input into the fiber.

The results of the investigations carried out for germanosilicate fibers, along with the comparison of the theoretical and experimental results, allow to conclude that the magnetic rotation of the speckle pattern of the light transmitted through an optical fiber placed in a longitudinal magnetic field will be best observed in optical fibers which keep only modes of the two lowest orders. The values of the magnetic field H, the Verdet constant V, and the fiber length in the magnetic field z should be such that $VHz \ll 1$.

7. Applications for fiber sensors

The analysis of the Eqs. (34) and (35) shows that modal structure of the radiation, propagating through a fiber, influences the behavior of the speckle pattern in the magnetic field. The angle of rotation of the speckle pattern described by the expressions (34) and (35) is determined by the factors $\exp\left[+m\left(\varphi - VHz/m\right)\right]$ and $\exp\left[-m\left(\varphi + VHz/m\right)\right]$. The speckle pattern rotation angle is determined by superposition of the rotation angles of all the modes.The angular intensity profile in a given crosssection of each mode is determined by the factor

$$\delta\varphi = \pm VHz/m. \quad (46)$$

Equation (46) shows that the increase of the number of a mode results in the decrease of the angle of the speckle pattern rotation. Thus, the different angular contribution of the modes of different orders should result in such change of the speckle pattern which cannot be regarded as rotation.

Figure 17 and Fig.18 show the images of the speckle patterns calculated using Eqs. (34) and (35) for the fiber with $\rho_0 = 4.5\ \mu$m, $N_A = 0.11$, $n_{co} = 1.47$, and the fiber length in the magnetic field $L_{mf} = 20$ cm for the light wavelength $\lambda = 0.532\ \mu$m and the magnetic field strength $H = 1000$ Gs. The calculations were performed for the case when all the modes that can be excited at the given wavelength propagate through the fiber and for the case when only modes with $m = 0$ and $m = 1$ propagate through the fiber. As can be seen in Fig. 17 an Fig.

18, the speckle pattern is distorted when the magnetic field is switched on, but when only the modes with $m = 0$ and $m = 1$ propagate, the speckle pattern is rotated.

It can be expected that the distortion of the speckle pattern will be proportional to the magnetic field strength. To prove this hypothesis, the dependencies of the speckle pattern rotation angle and the rms deviation of two speckle patterns on the magnetic field strength have been calculated for the case when only the modes with $m = 0$ and $m = 1$ are retained in the fiber and for the case when all possible fiber modes propagate. The results of calculations for the fiber with the same parameters ($\rho_0 = 4.5$ μm, $N_A = 0.11$, $n_{co} = 1.47$, $L_{mf} = 20$ cm, $\lambda = 0.532$ μm) are shown in Fig. 19 . It can be seen in Fig. 19 that, for the case when pure rotation of the speckle pattern in the fiber can be observed, the angle of rotation of the speckle pattern and the rms deviations of the speckle pattern obtained at $H = 0$ and $H \neq 0$ vary linearly with the magnetic field strength, though the slopes of these two straight lines are different. When all modes propagate in the fiber and only the distortion of the speckle pattern can be observed, the rms deviations of the speckle patterns obtained at $H = 0$ and $H \neq 0$ also vary linearly with the magnetic field strength; however, the slope of the line is smaller than in the case when only modes with $m = 0$ and $m = 1$ are retained in the fiber. Thus, to detect changes in the magnetic field strength, one can use both the dependence of the speckle pattern rotation

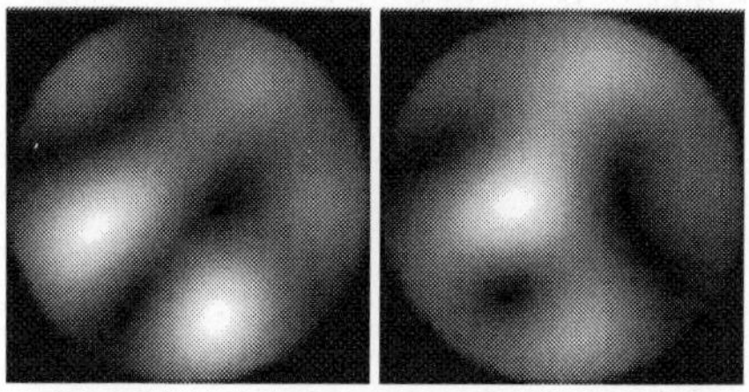

(a) $H = 1000$ Gs (b) $H = 0$ Gs

Fig. 17. Calculated intensity distribution in speckle pattern of light transmitted through optical fiber with $\rho_0 = 4.5$ μm, $N_A = 0.11$, $n_{co} = 1.47$, length of the fiber in magnetic field $L_{mf} = 20$ cm, $\lambda = 0.532$ μm for the case when all the modes that can propagate in the fiber are taken into account at H = 0 and H =1000 Gs.

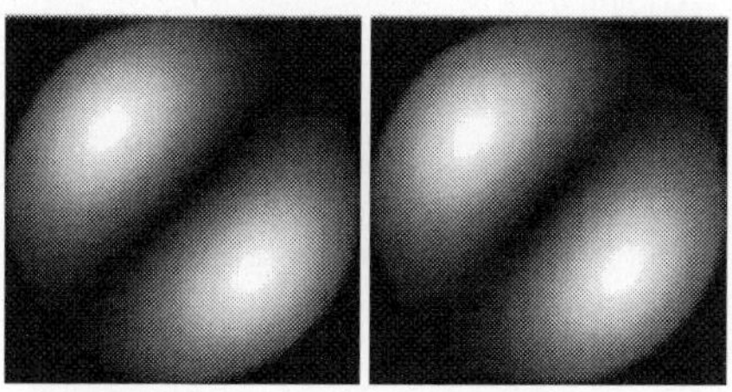

(a) $H = 1000$ Gs (b) $H = 0$ Gs

Fig. 18. Calculated intensity distribution in speckle pattern of light transmitted through optical fiber with $\rho_0 = 4.5$ μm, numerical aperture $N_A = 0.11$, $n_{co} = 1.47$, length of the fiber in magnetic field $L_{mf} = 20$ cm, $\lambda = 0.532$ μm for the case when only the modes with $m = 0$ and $m = 1$ are considered at H = 0 and H =1000 Gs.

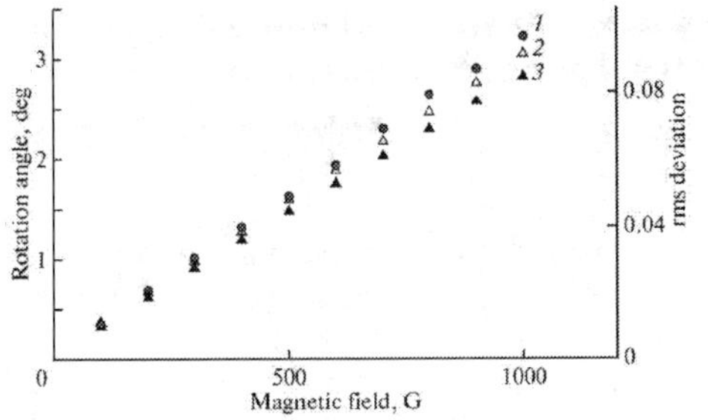

Fig. 19. Dependence of speckle pattern rotation angle (1) and rms deviation (2, 3) of the two speckle patterns obtained at $H = 0$ and $H \neq 0$ on magnetic field strength for cases when only modes with $m = 0$ and $m = 1$ are retained in the fiber (1 and 2) and when all possible modes propagate in the fiber (3).

angle on the magnetic field strength and the dependence of the rms deviation of the speckle pattern on the magnetic field strength.

Changes in the speckle pattern can be easily detected by means of a CCD camera and a computer. However, this method of detection is not suitable for detecting changes in the speckle pattern caused by rapidly varying or pulsed magnetic fields. As a promising method of detecting changes in the speckle pattern, the method of recording dynamic holograms in photorefractive crystals may be considered (Zel'dovich et al., 1995).

Let us consider the records and reconstruction of holograms in a photorefractive crystal by an inhomogeneous light field (by a speckle pattern). The hologram of interference between the signal beam (speckled field) and the reference Gauss beam is written in the photorefractive crystal in the absence of the magnetic field. The signal beam is a speckled light field transmitted through the optical fiber. If the reference beam is blocked after the hologram has been written, then the signal beam illuminating the photorefractive crystal will restore the Gaussian beam. The influence of the magnetic field results in the speckle pattern distortion. Under these conditions, the intensity of the restored Gaussian beam decreases in proportion with the degree of the speckle pattern distortion and, hence, in proportion with the magnetic field strength.

A few mode optical fiber with a step like refractive index profile was used for experimental investigation. The length of the fiber is 36 cm, $\rho_0 = 4.5\ \mu$m, $n_{co} = 1.47$ and $N_A = 0.11$. A photorefractive crystal of barium-sodium niobate $Ba_2NaNb_5O_{15}$ (BNN), $3.5 \times 5.0 \times 7.5\ mm^3$ in size was used as a nonlinear medium for recording dynamic hologram. The pulsed magnetic field was produced by a solenoid. A beam of laser operated at a wavelength $\lambda = 0.532\ \mu$m was used. The intensity of the diffracted beam was detected upon excitation of different modes in the optical fiber, i.e., at different kinds of the speckle pattern at the fiber end. Figure 20 shows the experimentally obtained dependence of the reconstructed Gauss beam intensity on the value of magnetic field on the solenoid axis. It can be seen in Fig. 20 that the reconstructed Gauss beam intensity depends on the value of magnetic field linearly. It means that the form of the pulse magnetic field can be easily determined.

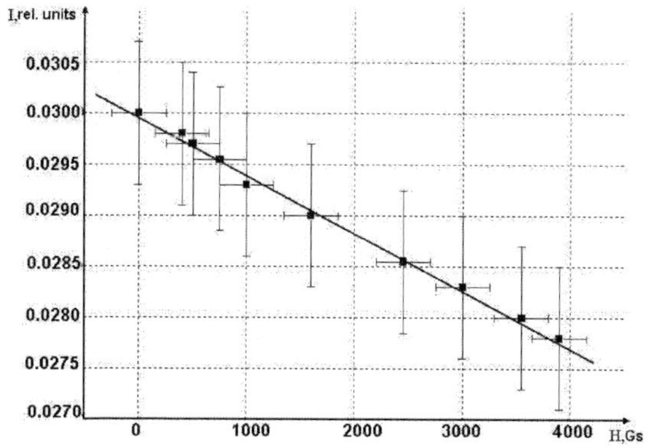

Fig. 20. The dependence of the reconstructed beam intensity on the value of magnetic field on the solenoid axis.

As a result, the usage of the effect of distortion of the speckle pattern of the light transmitted by the optical fiber in an external longitudinal magnetic field in combination with dynamic holography of the speckle fields can be used to detect the magnetic field pulse shape.

8. References

Anikeev, V. V., Bolshakov, M. V., Kundikova, N. D., Valeyev, A. I. & Zinatulin, V. S. (2001). The influence of a magnetic longitudinal field on the behavior of the speckle-pattern of the light, transmitted through optical fiber, *XVII International Conference on Coherent and Nonlinear Optics, "ICONO 2001"*, Technical Digest, Minsk, Belarus.

Ardasheva, L. I., Sadykova, M. O., Sadykov, N. R., Chernyakov, V. E., Anikeev, V. V., Bolshakov, M. V., Valeev, A. I., Zinatulin, V. S. & Kundikova, N. D. (2002). Rotation of the speckle pattern in a low-mode optical fiber in a longitudinal magnetic field, *J. Opt. Technol.* 69: 10–15.

Azzam, R. M. A. & Bashara, N. M. (1977). *Ellipsometry and polarized light*, North - Holland Pablishing Company, New York.

Baranova, N. B. & Zel'dovich, B. Y. (1994). Rotation of a ray by a magnetic field, *JETP Lett.* 59: 681–684.

Bolshakov, M. V., Ershov, A. V. & Kundikova, N. D. (2008). Fiber-optic system for magnetic field change registration, *International Commission for Optics ICO 21 Ű 2008 Congress Optics for the 21 st Century*, Book of Proceeding, Sydney, Australia, p. 286.

Bolshakov, M. V., Ershov, A. V. & Kundikova, N. D. (2011). Optical method for detecting variations of the magnetic field strength, *Optics and Spectroscopy* 110: 624–629.

Bolshakov, M. V. & Kundikova, N. D. (2003). The influence of the light polarization state on the speckle pattern of light transmitted through a fiber, *Proceedings of the Chelyabinsk scientific center* 4: 26–31.

Bolshakov, M. V., Vaganova, N. S. & Kundikova, N. D. (2006). Experimental investigation of the coherent light propagation with the arbitrary polarization state through an optical fiber, *Proceedings of the Chelyabinsk scientific center* 1: 10–15.

Chiao, R. Y. & Wu, Y.-S. (1986). Manifestations of berry's topological phase for the photon, *Phys. Rev. Lett.* 57: 933–936.

Darsht, M. Y., Zel'dovich, B. Y., Kataevskaya, I. V. & Kundikova, N. D. (1995). Formation of single wave front dislocation, *JETP* 107: 1464–1472.

Darsht, M. Y., Zel'dovich, B. Y., Zhirgalova, I. V. & Kundikova, N. D. (1994). Observation of a "magnetic rotation" of the speckle of light passed through an optical fiber, *JETP Lett.* 59: 763–765.

Doogin, A. V., Kundikova, N. D., Liberman, V. S. & Zel'dovich, B. Y. (1992). Optical magnus effect, *Phys. Rev. A* 45: 8204–8208.

Doogin, A. V., Zel'dovich, B. Y., Kundikova, N. D. & Liberman, V. S. (1991). The influence of circular polarization on light propagation in optical fiber, *JETP Lett.* 53: 186–188.

Fedorov, F. I. (1955). On the theory of total internal reflection, *Dokl. Akad. Nauk SSSR* 105(5): 465–469.

Gerrard, A. & Burch, J. M. (1975). *Introduction to matrix methods in optics*, John Wiley and sons, New York.

Goltser, I. V., Darsht, M. Y., Kundikova, N. D. & Zel'dovich, B. Y. (1993). An adjustable quarter-wave plate, *Optics Communications* 97: 291–294.

Goos, F. & Hanchen, H. (1947). Ein neuer und fundamentaler versuch zur totalreflexion, *Ann. Physik.* 436(7-8): 333–346.

Goos, F. & Hanchen, H. (1949). Neumessung des strahlversetzungseffektes bei totalreflexion, *Ann. Physik.* 440(3-5): 251–252.

Imbert, C. (1972). Calculation and experimental proof of the transverse shift induced by total internal reflection of a circularly polarized light beam, *Phys. Rev. D.* 5: 787–796.

Picht, J. (1929). Beitrag zur theorie der totalreflexion, *Ann. Physik.* 395(4): 433–496.

Rytov, S. M. (1938). On transition from wave to geometrical optics, *Dokl. Akad. Nauk SSSR* 18: 263–266.

Snyder, A. W. & Love, J. D. (1984). *Optical Waveguide Theory*, Methuen, London.

Tomita, A. & Chiao, R. Y. (1986). Observation of berryŠs topological phase by use of an optical fiber, *Phys. Rev. Lett.* 57: 937–940.

Vladimirskii, V. V. (1941). The rotation of a polarization plane for curved light ray, *Dokl. Akad. Nauk SSSR* 21: 222–225.

Zel'dovich, B. Y., Kataevskaya, I. V. & Kundikova, N. D. (1996). Inhomogeneity of the optical magnus effect, *Quantum Electron* 26: 87 – 88.

Zel'dovich, B. Y. & Liberman, V. S. (1990). Rotation of the plane of a meridional beam in a graded-index waveguide due to the circular nature of the polarization, *Quantum Electronics* 20: 427–428.

Zel'dovich, B. Y., Mamaev, A. V. & Shkunov, V. V. (1995). *Speckle-Wave Interactions in Application to Holography and Nonlinear Optics*, CRC Press.

11

Long Period Fibre Gratings

Alejandro Martinez-Rios, David Monzon-Hernandez,
Ismael Torres-Gomez and Guillermo Salceda-Delgado
Centro de Investigaciones en Optica, Leon, Guanajuato
Mexico

1. Introduction

In essence, a long period fibre grating (LPFG) is an all-fibre device with wavelength dependent loss. As a band rejection filter, all light in a spectral slice is discarded without affecting the amplitude and phase of neighbouring wavelengths, with the additional advantage of low insertion losses and minimum backreflection (Daxhelet, 2003). LPFGs have been used as band rejection filters, sensors, wavelength selective elements in fibre lasers and amplifiers (Su &Wang, 1999), beam shaping (Mohamed & Gu,2009), etc. In addition, in-series LPFG pairs can be used to form all-fibre Mach-Zender interferometers that are useful as wavelength selective filters for WDM applications; while parallel LPFGs act as wavelength selective couplers or wavelength multiplexers (Chan & Yasumoto, 2007; Shu et. al.,2002). In a common LPFG, the core mode couples to co-propagating cladding modes supported by the cladding-external medium waveguide structure. The specific cladding through which the core mode is coupled is determined by the following resonance condition:

$$\frac{2\pi}{\Lambda} = \frac{2\pi}{\lambda}\left(n_{eco} - n_{ecla}^{i}\right) \tag{1}$$

where n_{eco} is the effective index of the fundamental core mode, n_{ecla}^{i} is the effective index of the ith cladding mode, λ is the resonance wavelength, i.e. the wavelength where the notch band is formed, and Λ is the period of the grating. As can be clearly seen from this relation the resonance wavelength is determined by the effective refractive indices of the core and cladding modes, so that any geometrical, photo-induced, thermal-induced, or mechanically-induced change (periodic) will modify the position of the resonance wavelength. This property is what makes LPFGs so useful for applications in fields as diverse as optical sensing, biological and chemical sensing, as wavelength dependent loss elements in telecommunication systems and laser and many more. In this chapter we will make a review of the relevant aspects of LPFGs. In section 2.2 we will review the fabrication methods used to write LPFGs in optical fibres; in section 3 we will review the theory behind the operation of LPFGs; in section 4 we will talk about the applications of LPFGs; and, in section 5 we give the conclusions of the present chapter.

2. Fabrication methods of long-period fibre gratings

The inscription of long-period gratings on optical fibre basically consists in the generation of a periodical perturbation of the refractive index in the core, the cladding, or both along the optical fibre (Vengsarkar, 1996). The refractive index perturbation requires a length of 3-5 cm, an amplitude in the range of 10^{-4} to 10^{-5} similar to Bragg gratings. In contrast, the perturbation period in LPFGs is usually larger than 100 μm, which is two orders larger than Bragg gratings. In this way, different physical properties of the optical fibre glass or the variation of the optical fibre structure can be used to produce this periodical refractive index perturbation in the fibre, such as; the UV photosensitivity, the residual thermal stress, the photoelasticity and the geometrical modulation of the cladding structure (Hill et. al., 1990; Dianov et. al. 1997; Rao et. al. 2004; Hwang et. al., 1999). In the first case, an ultraviolet light source is used to write the core index modulation in Ge-doped fibres by the photosensitive mechanism, on the other hand, a heating source such as a CO_2 laser (Hwang et. al., 1999), or an electric arc system (Bjarklev, 1986), are employed to produce simultaneously the index modulation in the core and the cladding via the residual thermal stress. In a similar way, a mechanical pressure system temporally produces the index modulation in the core and the cladding by the photoelasticity effect (Savin et. al. 2000). Meanwhile, the geometrical modulation of the cladding structure can be done by chemical etching (Vaziri & Chin-Lin, 1997) and other methods. Except the UV photosensitive mechanism, the others one can be used in any type of optical fibre.

In the last 15 years, alternative writing methods of long-period fibre gratings have been proven after the demonstration of UV radiation method. Some of the most used methods include; the heating, the mechanical stress and chemical etching between others (Fujimaki et. al.,2000; Jeong et. al., 2000). The inscription method and the optical fibre type impact in the physical properties and optical spectral transmission characteristics of the rejection bands of the LPFGs. Each inscription method produces a particular transversal and longitudinal refractive index profile. The transversal index profile determines the mode coupling between the core fundamental mode and the cladding discrete high order modes involved (Anemogiannis et. al., 2003). Meanwhile, the longitudinal index profile determines the location and shape of the rejection loss bands (DeLisa et. al., 2000). The combination of the total index profile and the fibre type define the optical properties of the rejection bands such as; position of the central resonant wavelength, bandwidth, isolation depth, insertion loss, induced birefringence, and polarization dependent loss. In the same way, this combination directly influence in the rejection bands sensitivity to physical variables such as; temperature, tension, torsion, bending, and external refractive index.

The post-processing of LPFGs is actually used to adjust or stabilize the spectral transmission, as well to improve the sensitivity of the rejection bands in the LPFGs. The sensitivity improve or sensitization of the rejection band to a particular external parameter include, the deposition of some materials over the LPFG (Gu et. al., 2006) and the etching process to reduce the cladding diameter to expose the high cladding modes to the external medium (Chiang et. al., 2001). On other hand, the annealing process is used to tune and stabilize the attenuation bands of the LPFGs (Ng & Chiang, 2002).

2.1 UV long-period fibre gratings

There are two common techniques to inscribe a LPFG in Ge-doped optical fibres by using the UV radiation method: the amplitude mask and the point to point technique (Othonos & Kalli, 1999; Vengsarkar, 1996). The key parameters in the UV inscription method of the LPFG over the Ge-doped fibres are: the UV selected wavelength (193-360 nm), the density power of the laser, and the exposition time (Othonos & Kalli, 1999). In the first technique, the optical fibre is scanned by the UV beam through an amplitude mask (AM) as is illustrated by Fig. 1(a). The AM contains an array of transparent windows that forms an illuminated-shadow periodical pattern with a period Λ over the fibre. In case of low UV sensitivity optical fibres, the hydrogenation or the flame brushing process can be used to improve the UV sensitivity in low Ge-doped fibres (Lemaire et. al., 1993; Bilodeau et. al., 1993). Actually, B-Ge codoped optical fibres have been developed to achieve high UV sensitive optical fibre (Williams et. al., 1993). In the point to point technique, the UV beam illuminates a section the optical fibre by a thin slit aperture (S). In this setup, the beam spot is fixed and the optical fibre moves in the z exes with a period Λ. Both techniques are relative simple to implement; however, the first one is more attractive to mass production, although the point to point technique is more flexible and cheaper than the amplitude mask technique. The main limitation of this technique is that only works in UV photosensitivity fibre.

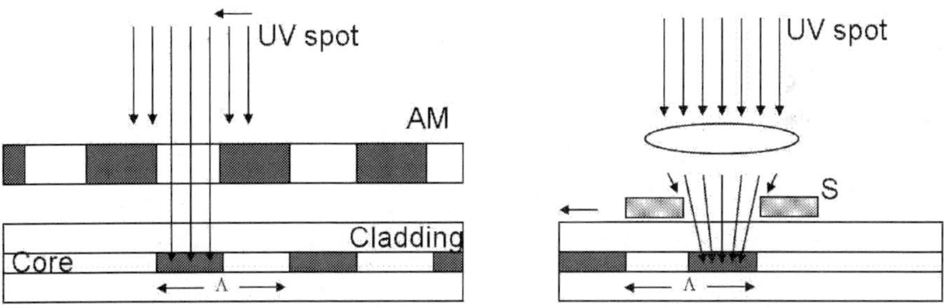

Fig. 1. LPFG inscription by the UV techniques: a) phase amplitude mask, b) point to point.

The expected z index profile with UV techniques is Gaussian and azimuthally uniform in the transversal plane. In this case, this profile supports the mode coupling between the fundamental mode LP_{01} in the core and the discrete LP_{0m} in the cladding. The main spectrum transmission characteristics of the rejection bands of UV LPFGs are: wide range wavelength location from visible to infrared, the lowest loss insertion loss < 0.2 dB, the isolation depth is larger than 25 dB and the lowest induced birefringence group. The disadvantages of UV induced LPFGs are temperature limited operation range <250° C and high cost fabrication. The table 1 displays some properties and the sensitivities to some physical parameters. Because the UV LPFG couples modes with azimuthal symmetry the rejection bands are not sensible to twist or bending direction.

2.2 Residual thermal stress long period fibre gratings

The heating and fast cool down of the glass alter the viscosity, it in turns, slightly modified the refractive index. In this case, the heating process and follow by a fast cool, of the glass can be use to froze a periodical index modulation across the optical fibre structure (core and cladding). There are two used methods to generate the periodical index modulation by CO_2 radiation (heating absorption) and electrical arc. Both methods can be use in any kind of fibre. These methods induce a no uniform refractive index profile in the transversal plane, allowing the coupling to symmetric and asymmetric cladding modes. In the CO_2 radiation method, the optical fibre under tension is radiated though a germanium lenses with a infrared signal at 10 μm point to point. At this wavelength, the silica is not transparent and absorbs the energy from the CO_2 laser, so that the fibre is heated and cooled at room temperature. The required power in optical fibres is 10 W in continuous wave CO_2 lasers. On the other hand, in the electric arc discharge method, the optical fibre is heated by an electric arc generated usually by a fusion splicing machine. For standard optical fibres, the parameters of electric arc discharge are: 200 ms, current 50 mA at tension of few grams. According with the price arc discharge method is cheaper than CO_2 laser.

The spectral characteristics of thermal induced LPFGs are high stability in wide range of temperature, low loss insertion loss. The disadvantages of thermal induced LPFGs are: the loss bands shift and the depth decrease of the bands with the time.

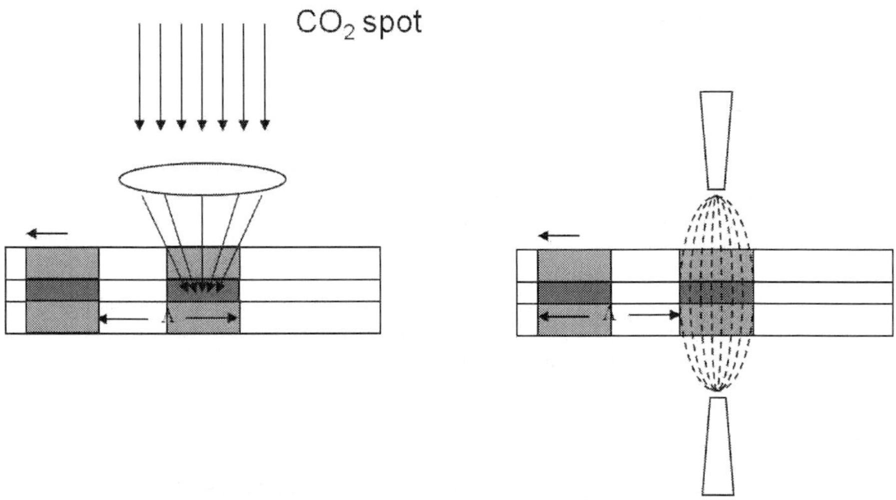

Fig. 2. LPFG inscription by thermal stress techniques; a) CO_2 irradiation, b) electric arc discharge

2.3 Mechanical stress induced long period fibre gratings

The optical refractive index of glass can be modulated when the glass is exposed to stress. This change in optical refractive index is due to the photo-elastic response of glass. In this case, the photo-elasticity can be used to induce a temporal periodical modulation of the refractive index in the core and the cladding to generate long-period fibre gratings. According to this, microbending and periodical pressure points in the fibre can be used to induce long-period fibre gratings. Different techniques have been reported to generate microbending such as: metallic grooved plates, strings, flexure acoustic waves. In the case of periodical pressure points; metallic grooved plates, coils springs and torsion. Figure 3 illustrates a typical example of periodical microbending and pressure points. Because, its flexibility and cost relation plates can be use for microbending or pressure is one of the most used techniques. This technique presents many variants in the corrugated design.

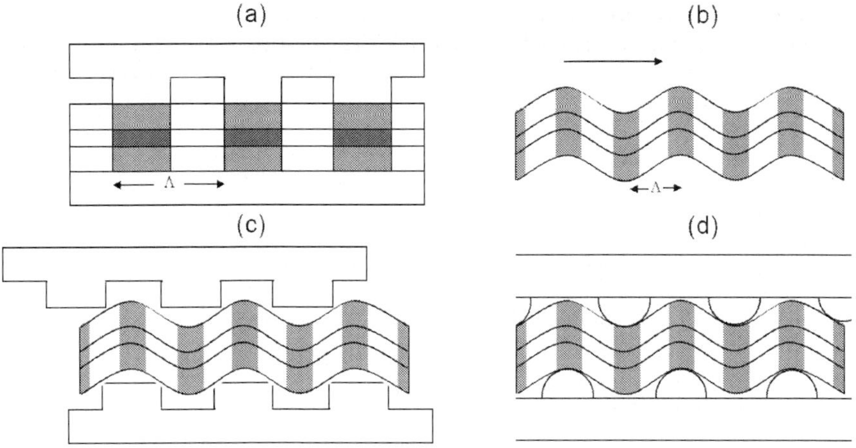

Fig. 3. Stress induced LPFGs, pressure points (a,b) microbending (c,d)

The photoelasticity of the glass allows the inscription of LPFGs practically in any kind of fibre, this include, standard telecommunication fibres, photonic crystal fibres, between others. The amplitude of the refractive index periodical perturbation is the order of 10^{-4} to 10^{-5}, so the pressure requite is less than 500 gr cm^{-1}, in the microbending case, the amplitude of the perturbation needed is in order of 200- 500 μm. The stress induced LPFGs present loss bands with attractive and flexible spectral transmission optical properties. The loss bands are erasable, simple control of the depth (0-20 dB), bandwidth control (10-50 nm) and a large tuning range >250 nm. The drawbacks are high sensitivity to ambient temperature, and insertion loss are in the order of 0.2- 0.3 dB.

2.4 Etching induced long-period fibre gratings

The effective index in the core can be modulated by the geometric modulation of the optical fibre diameter by etching chemical method. The etching induced LPFGs are also known as corrugated LPFGs. This method, consist in locate a photomask over the fibre, and then the optical fibre is immersed in a hydrofluoric acid solution, in this way alternated regions of the fibre are protected of the hydrofluoric solution meanwhile other regions are divested hydrofluoric attack on the silica glass. The diameter in the desvasted regions is controlled by the exposition time, as illustrated in the figure 4. This method is not as popular as UV, thermal or mechanical stress because, their low repeatability and the difficult control of corrugated structured of the fibre. Physically the corrugated LPFGs are fragile to tension or beding stress. The loss bands are high sensitivity to twist because the torsion stress focus in the core diameter.

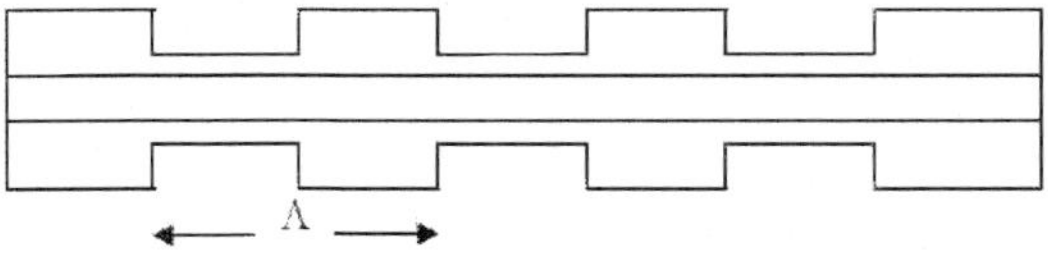

Fig. 4. Corrugated LPFG

Properties	UV	Residual thermal stress	Mechanical stress	Etching
Length [cm]	2 - 4	2 - 5	3 - 5	2 - 3
Isolation depth [dB]	10 – 30	10 – 25	10 – 20	10 – 15
$\Delta\lambda$ [nm]	> 10	> 10	> 15	> 10
Insertion loss [dB]	0.1 – 0.2	0.1 – 1	0.1 - 0.5	0.1 - 0.3
Modal birefringence	2 $(10)^{-7}$	1.7$(10)^{-6}$	7$(10)^{-6}$	--
Λ [μm]	> 100	> 300	> 250	> 500
σ_T [nm /°C]	- (0.04– 0.347)	±(0.05 – 0.204)	0.180 – 200	--
σ_τ [nm /(rad/cm)]	15 -	± (6.7 – 25)	6 – 12	-(70 – 100)
σ_σ [pm/με]	0.204 – 1.34	0.1 – 0.4	0.2 – 0.4	1.5 – 2
σ_b [nm/m]	2.5 - 14	-(1.6 – 12)	--	-(18 – 25)
$\sigma_{ext.\ index}$ [nm/10^{-3}]	-(0.003 – 0.019)	-(0.017 – 0.54)	0.22 – 0.5	--
Range of temperature operation [°C]	< 250	< 1100	< 60	< 1200

Mode order: m = 3rd,4th. Ranges: [Λ=550-650 nm][σ_T:0-200°C], [σ_τ:0-5 rad/cm], [σ_σ:0-1000 με], [σ_b:0-5 m^{-1}], [$\sigma_{ext.\ index}$:1-1.3].

Table 1. Rejection bands properties reported in LPFG with different fabrication methods.

3. Theory of long period gratings

In this section we review the theoretical framework to describe long-period fibre gratings. We consider the most common case of a step-index fibre with three layers: the central core, the cladding, and the external medium. We follow the analysis given by Erdogan (Erdogan, 1997) for the calculation of the core and cladding effective refractive indices.

3.1 Characteristic equation for core and cladding modes

The most important parameter for the modelling of long period-fibre gratings are the effective indices of the core and cladding modes, which determine the coupling wavelength between co-propagating modes according to the following relation:

$$\lambda_{res} = (n_{efco} - n_{efcla})\Lambda \tag{2}$$

where λ_{res} is the resonance wavelength, n_{efco} is the effective index of the core mode, assumed to be single-mode, n_{efcla} is the effective index of a given cladding mode, and L is the period of the refractive index modulation. Here, for the calculation of the effective index, we consider that the core mode is not affected by the external medium that surrounds the cladding, so that the cladding can be considered as an infinitum medium as far as the core modes concerns. In this particular case, the characteristic equation from which the effective index of the core fundamental mode can be written as (Erdogan):

$$V\sqrt{1-b}\frac{J_1(V\sqrt{1-b})}{J_0(V\sqrt{1-b})} = V\sqrt{b}\frac{K_1(V\sqrt{b})}{K_0(V\sqrt{b})} \tag{3}$$

where J_m is the first class Bessel function of mth-order, K_m is the modified Bessel function of second class and m-order, $V(=2\pi a_{core}NA/\lambda)$ is the normalized frequency, b is the normalized propagation constant, a_{core} is the core diameter, NA is the core numerical aperture, and λ is the light wavelength. Figure 5 shows a plot of effective index as a function of wavelength for the fundamental core mode of a fibre with parameters corresponding to standar SMF-28 fibre. This plot was calculated using the Mathematica-like code shown in Table 2.

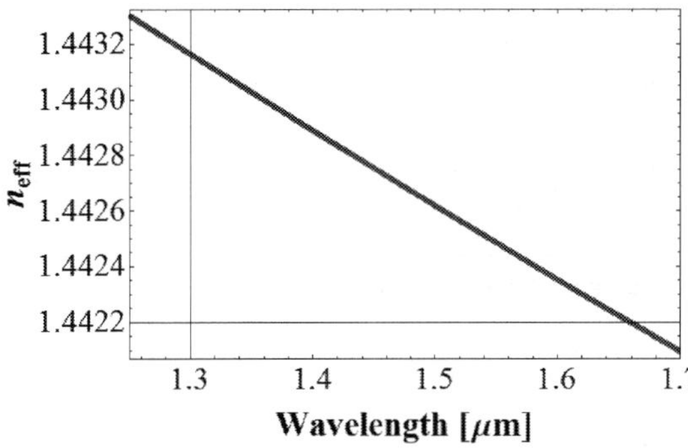

Fig. 5. Effective index of the fundamental core mode as a function of wavelength.

```
ClearAll("Global`*")
NA=0.14; (*Numerical aperture*)
n1=1.4458; (*Core refractive index*)
n2=Sqrt(n1^2-NA^2); (*Cladding refractive index*)
a1=4.1; (*Core radius*)
V[λ_,a_]=(2*Pi/λ)*a*NA; (*Normalized frequency*)
(*Argument of Bessel functions*)
c1[λ_,a_]=V[λ,a]*Sqrt[1-bw];
c2[λ_,a_]=V[λ,a]*Sqrt[bw];
(*Bessel functions*)
Jmn1=BesselJ[m-1,c1(λ,a)];Jmd1=BesselJ[m,c1(λ,a)]; Kmn1=BesselK[m-1,c2(λ,a)]; Kmd1=BesselK[m,c2(λ,a)];
(*Characteristic equation*)
fun1[a_,λ_,bw_,m_]=c1[λ,a]*(Jmn1/Jmd1)+c2[λ,a]*(Kmn1/Kmd1);
(*Plot the function to provide an approximated value for the normalized propagation constant*)
Plot[{fun1(a1,1.0,bw,0)},{bw,0,1}]
(*Calculation of the effective index of the core fundamental mode as a function of wavelength*)
ncore={{0,0}};
λ=1.7; (*Initial wavelength*)

be1=0.45; (*Initial value of the normalized propagation constant*)
(*For cycle to obtain the effective index at several wavelength*)
For[i=1,i<=450,i++,
r1=FindRoot[fun1(a1,\(Lambda),bw1,0)==0,{bw1,be1},WorkingPrecision->10];
nc=Sqrt[NA^2*r1[[1]][[2]]+n2^2];
ncore=Append[ncore,{λ,nc}];
λ=λ-0.001;];
nefco=Drop[ncore,1]; (*List of effective index versus wavelength *)
(*Plotting the effective index as a function of wavelength*)
ListPlot[{nefco},Frame->True,PlotRange->{{1.25,1.7},All},PlotStyle->Thickness(0.01),FrameLabel->{Style("Wavelength
(\(Mu)m)",Bold, 34),Style("Subscript(n, eff)",Bold, 34)},LabelStyle->Directive(Black,"Helvetica",32),Joined->True]
```

Table 2. Mathematica code for the calculation of the fundamental core mode effective index

In the case of the cladding modes the characteristic equation is given as (Erdogan, 1997):

$$\frac{1}{\sigma_2}\frac{u_2\left(J\cdot K+\frac{\sigma_1\sigma_2 u_{21}u_{32}}{n_2^2 a_1 a_2}\right)\cdot p_m(a_2)-K\cdot q_m(a_2)+J\cdot r_m(a_2)-\frac{1}{u_2}s_m(a_2)}{-u_2\cdot\left(\frac{u_{32}}{n_2^2 a_2}J-\frac{u_{21}}{n_1^2 a_1}K\right)\cdot p_m(a_2)+\frac{u_{32}}{n_1^2 a_2}q_m(a_2)+\frac{u_{21}}{n_1^2 a_1}r_m(a_2)}$$
$$=\sigma_1\frac{u_2\left(\frac{u_{32}}{a_2}J-\frac{n_3^2\cdot u_{21}}{n_2^2\cdot a_1}\right)\cdot p_m(a_2)-\frac{u_{32}}{a_2}\cdot q_m(a_2)-\frac{u_{21}}{a_1}\cdot r_m(a_2)}{u_2\cdot\left(\frac{n_3^2}{n_2^2}J\cdot K+\frac{\sigma_1\sigma_2 u_{21}u_{32}}{n_1^2 a_1 a_2}\right)\cdot p_m(a_2)-\frac{n_3^2}{n_1^2}K\cdot q_m(a_2)+J\cdot r_m(a_2)-\frac{n_2^2}{n_1^2\cdot u_2}s_m(a_2)} \tag{4}$$

where the parameters in (3) are defined by the following relations {Erdogan, 1997):

$$\sigma_{1`}=\frac{im}{\eta}\cdot n_{efcla}\ ;\quad \sigma_{2`}=im\cdot\eta\cdot n_{efcla} \tag{5a}$$

$$u_{21}=\frac{1}{u_2^2}-\frac{1}{u_1^2}\ ;\ u_{32}=\frac{1}{w_3^2}-\frac{1}{u_2^2} \tag{5b}$$

$$u_1=\frac{2\pi}{\lambda}\sqrt{n_1^2-n_{efcla}^2}\ ;\ u_2=\frac{2\pi}{\lambda}\sqrt{n_2^2-n_{efcla}^2}\ ;\ w_3=\frac{2\pi}{\lambda}\sqrt{n_{efcla}^2-n_3^2} \tag{5c}$$

$$J=\frac{1}{2}\frac{J_{m-1}(u_1a_1)+J_{m+1}(u_1a_1)}{u_1J_m(u_1a_1)}\ ;\ K=-\frac{1}{2}\frac{K_{m-1}(w_3a_2)-K_{m+1}(w_3a_2)}{w_3K_m(w_3a_2)} \tag{5d}$$

$$\begin{aligned}
p_m(r)&=J_m(u_2r)Y_m(u_2a_1)-J_m(u_2a_1)Y_m(u_2r)\\
q_m(r)&=\frac{1}{2}J_m(u_2r)(Y_{m-1}(u_2a_1)-Y_{m+1}(u_2a_1))-Y_m(u_2r)(J_{m-1}(u_2a_1)-J_{m+1}(u_2a_1))\\
r_m(r)&=\frac{1}{2}Y_m(u_2a_1)(Y_{m-1}(u_2r)-Y_{m+1}(u_2r))\\
s_m(r)&=\frac{1}{4}(J_{m-1}(u_2r)-J_{m+1}(u_2r))(Y_{m-1}(u_2a_1)-Y_{m+1}(u_2a_1))\\
&\quad-(J_{m-1}(u_2a_1)-J_{m+1}(u_2a_1))(Y_{m-1}(u_2r)-Y_{m+1}(u_2r))
\end{aligned} \tag{5e}$$

where Y_m is a Bessel funtion of the the second kind, $\eta=(\mu_0/\varepsilon_0)^{1/2}$ is the vacuum impendance, and m is the azimuthal mode number. Figure 6 shows the calculated effective index difference between the fundamental core mode and the HE_{1m} (left graph) and HE_{2m} (right graph) cladding modes assuming a standard single-mode fibre, where we have used Eqs. (3) and (4) to find the effective refractive index of the core and cladding, respectively. Table 3 shows the Mathematica-like code used to calculate the effective index of the cladding modes.

```
(*Cladding modes*)
Z0=377; (*Free space impedance*)
nco=n1; (*Core refractive index*)
ncl=n2; (*Cladding refractive index*)
acl=62.5; (*Cladding radius*)
next=1; (*External refractive index*)
NAcla=Sqrt(ncl^2-next^2); (*Cladding numerical aperture*)
(*Parameters of the Erdogan Characteristic equation*)
σ1=I*m*nefcla/Z0; σ2=I*m*nefcla*Z0;
u1=Sqrt((2*Pi/λ cla)^2*(nco^2-nefcla^2)); u2=Sqrt((2*Pi/λcla)^2*(ncl^2-nefcla^2));
w3=Sqrt((2*Pi/λcla)^2*(nefcla^2-next^2));
u21=(1/(u2^2))-(1/(u1^2)); u32=(1/(w3^2))+(1/(u2^2));
J=(1/2)*((BesselJ(m-1,u1*aco)+BesselJ(m+1,u1*aco))/(u1*BesselJ(m,u1*aco)));
Kp=(-1/2)*((BesselK(m-1,w3*acl)-BesselK(m+1,w3*acl))/(w3*BesselK(m,w3*acl)));
p(m_,r_)=BesselJ(m,u2*r)*BesselY(m,u2*aco)-BesselJ(m,u2*aco)*BesselY(m,u2*r);
q(m_,r_)=(1/2)*(BesselJ(m,u2*r)*(BesselY(m-1,u2*aco)-BesselY(m+1,u2*aco))
        -BesselY(m,u2*r)*(BesselJ(m-1,u2*aco)-BesselJ(m+1,u2*aco)));
r(m_,r_)=(1/2)*(BesselY(m,u2*aco)*(BesselY(m-1,u2*r)-BesselY(m+1,u2*r)));
s(m_,r_)=(1/4)*((BesselJ(m-1,u2*r)-BesselJ(m+1,u2*r))*(BesselY(m-1,u2*aco)-
          BesselY(m+1,u2*aco))-(BesselJ(m-1,u2*aco)-BesselJ(m+1,u2*aco))*(BesselY(m-1,u2*r)-BesselY(m+1,u2*r)));
num1(m_)=(1/σ2)*((J*Kp+((σ1*\(Sigma)2*u21*u32)/(aco*acl*ncl^2)))*u2*p(m,acl)-Kp*q(m,acl)+J*r(m,acl)-
s(m,acl)/u2);
den1(m_)=-((u32/(acl*ncl^2))*J-
(u21/(aco*nco^2))*Kp)*u2*p(m,acl)+(u32*q(m,acl)/(acl*nco^2))+(u21*r(m,acl)/(aco*nco^2));
num2(m_)=σ 1*(((u32/acl)*J-(u21*next^2/(aco*ncl^2))*Kp)*u2*p(m,acl)-(u32*q(m,acl)/acl)-(u21*r(m,acl)/aco));
den2(m_)=((next^2/(ncl^2))*J*Kp+(σ1*σ2*u21*u32/(aco*acl*nco^2)))*u2*p(m,acl)-
(q(m,acl)*next^2/(nco^2))*Kp+J*r(m,acl)-(s(m,acl)*ncl^2/(u2*nco^2));
Erdoganizq(m_,λcla_,aco_,acl_,nefcla_)=num1(m)/den1(m);
Erdogander(m_,λcla_,aco_,acl_,nefcla_)=num2(m)/den2(m);
(*Modes HE1n (m=1)*)
(*Locate the desired mode visually and find the initial root at the wavelength of 1.7 um*)
h1=Plot({Erdoganizq(1,1.7,a1,acl,nefcla)/I,Erdogander(1,1.7,a1,acl,nefcla)/I},
    {nefcla,1.437,1.43901},PlotRange->{{1.4385,1.43901},{-0.2,0.2}},Frame->True,GridLines->Automatic)
(*Initial value of the effective index*)
nstart=Quiet(FindRoot(Erdoganizq(1,λcla,a1,acl,nefcla1)/I==Erdogander(1,(λcla,a1,acl,nefcla1)/I,
{nefcla1,1.43895},WorkingPrecision->15)((1))((2)));
(*Example: Mode HE11*)
necla={{0,0}};   (*Effectiuve index of the cladding*)
Δneff={{0,0}};(*Effectiuve index difference*)
λcla=1.7;
n11=nstart;
For(i=1,i<=450,i++,
r11cla=FindRoot(Erdoganizq(1,λcla,a1,acl,nefcla1)/I==Erdogander(1,λcla,a1,acl,nefcla1)/I,{nefcla1,n11})((1))((2));
n11=r11cla;
necla=Append(necla,{λcla,r11cla});
(Δneff=Append(Δneff,{λcla,nefco((i))((2))-n11});
λcla=λcla-0.001;);
necla=Drop(necla,{1});
Δneff=Drop(Δneff,{1});
ListPlot(necla);
ListPlot(Δneff);
```

Table 3. Mathematica code for the calculation of the cladding mode effective index

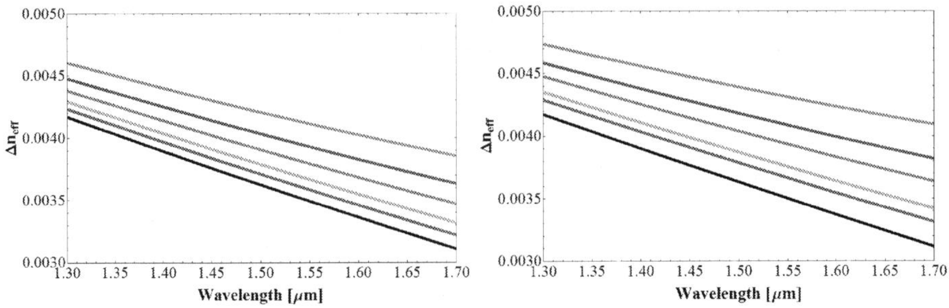

Fig. 6. Effective index difference between the fundamental core mode and HE_{1m} (left graph), and HE_{2m} (right graph) cladding modes.

3.2 Coupled mode equations

In the case of o LPFG, we must consider the coupling between forward propagating modes obeying the coupled mode equations (Bures,2009), which in the case of one forward propagating core mode and forward propagating cladding modes can be written as:

$$\frac{dA_{co}(z)}{dz} = i\sum_{m} B_m(z)C_{co-m}(z) \tag{6a}$$

$$\frac{dB_j(z)}{dz} = iA_{co}(z)C_{j-co} + i\sum_{m\neq j} B_m(z)C_{j-m}(z) \tag{6b}$$

Assuming that the periodic perturbation of the refractive index is much lower that the unperturbed refractive index we can approximate the coupling coefficient by:

$$C_{co-m} = \frac{k}{2\eta_0}\int_{A_\infty} n(r)\delta n(r,z)\,\hat{\mathbf{e}}_{co}^{*}(r)\cdot\hat{\mathbf{e}}_m(r)e^{-i(\beta_{co}-\beta_m)z}dA \tag{7a}$$

where $C_{m-co}=(C_{co-m})^*$, $\hat{\mathbf{e}}_m(r)$ is the normalized electric field, dA is a differential of area, and the integration is over the infinite area.

Here, we consider the uniform refractive index perturbation $\delta n(r,z)$ of a three-layered waveguide written as:

$$\delta n(r,z) = \begin{cases} \delta n_{co} e^{i\left(\frac{2\pi}{\Lambda}z+\phi_{co}\right)} & r \le a_1 \\ \delta n_{cla} e^{i\left(\frac{2\pi}{\Lambda}z+\phi_{cla}\right)} & a_1 \le r \le a_2 \\ 0 & r > a_2 \end{cases} \tag{8}$$

Then the coupling coefficient can be written as:

$$C_{co-m} = \frac{k}{2\eta_0} e^{-i\Delta z + i\phi_{co}} \left\{ n_1 \delta n_{co} I_{1m} + n_2 \delta n_{cla} I_{2m} \right\} \tag{9}$$

where

$$\Delta = \left(\frac{2\pi}{\Lambda} - \frac{2\pi \Delta n_{eff}^{co-m}}{\lambda} \right) \tag{10}$$

and

$$\begin{aligned} I_{1m} &= \int_0^{a_1} \hat{\mathbf{e}}_{co}^*(r) \cdot \hat{\mathbf{e}}_m(r) r dr \\ I_{2m} &= \int_{a_1}^{a_2} \hat{\mathbf{e}}_{co}^*(r) \cdot \hat{\mathbf{e}}_m(r) r dr \end{aligned} \tag{11}$$

In writing (9) and (10) we have used $\beta_{co} - \beta_m = 2\pi(n_{eco} - n_{ecla}) / \lambda = 2\pi \Delta n_{eff}^{co-m}$, i.e., it is expressed in terms of the effective refractive index difference between the core and cladding modes. The integrals in (11) can be seen as overlapping factors between the core and cladding modes, which determine the changes in the transmission spectrum when there are changes in fibre dimensions, the external refractive index, temperature and stress, since in all cases the effect is to change the effective refractive index of the core and cladding modes.

In order to find an analytical solution for the amplitudes of the core and cladding modes we consider only the coupling to one cladding mode, so that, after differentiating Eqs. (6a) and (6b) and solving the resultant equations we can obtain (Bures, 2009):

$$A_{co}(z) = A_1 \cos\left(\sqrt{\left(\frac{\Delta}{2}\right)^2 + |C_{co-m}|^2} z \right) e^{-i\frac{\Delta}{2}z} + B_1 \sin\left(\sqrt{\left(\frac{\Delta}{2}\right)^2 + |C_{co-m}|^2} z \right) e^{-i\frac{\Delta}{2}z} \tag{12a}$$

$$B_m(z) = A_2 \cos\left(\sqrt{\left(\frac{\Delta}{2}\right)^2 + |C_{co-m}|^2} z \right) e^{i\frac{\Delta}{2}z} + B_2 \sin\left(\sqrt{\left(\frac{\Delta}{2}\right)^2 + |C_{co-m}|^2} z \right) e^{i\frac{\Delta}{2}z} \tag{12b}$$

Now we consider specific boundary conditions found in LPFGs, where $A_{co}(0)=1$ and $B_m(z)=0$ which results in (Bures, 2009)

$$A_{co}(z) = \left(\begin{array}{c} \cos\left(\sqrt{\left(\frac{\Delta}{2}\right)^2 + |C_{co-m}|^2} z \right) \\ +i \frac{\Delta}{2\sqrt{(\Delta/2)^2 + |C_{co-m}|^2}} \sin\left(\sqrt{\left(\frac{\Delta}{2}\right)^2 + |C_{co-m}|^2} z \right) \end{array} \right) e^{-i\frac{\Delta}{2}z} \tag{13a}$$

$$B_m(z) = i\frac{C_{co-m}}{\sqrt{(\Delta/2)^2 + |C_{co-m}|^2}}\sin\left(\sqrt{\left(\frac{\Delta}{2}\right)^2 + |C_{co-m}|^2}z\right)e^{i\frac{\Delta}{2}z-\phi} \tag{13b}$$

From here we can now write the transmittance at the core as:

$$T_{co}(z) = \cos^2\left(\sqrt{\left(\frac{\Delta}{2}\right)^2 + |C_{co-m}|^2}z\right) + \frac{(\Delta/2)^2}{(\Delta/2)^2 + |C_{co-m}|^2}\sin^2\left(\sqrt{\left(\frac{\Delta}{2}\right)^2 + |C_{co-m}|^2}z\right) \tag{14}$$

This equation is only valid for uniform LPFGs with a step change. For real LPGs, the index change is much more complicated since, for example, in arc induced LPFGs the zone affected by the electric arc spans >300 μm, and is asymmetric in both, the longitudinal and transverse directions. This fact makes the modeling of such gratings a really challenging task. A qualitative analysis of this and other types of gratings can be done only by calculating their effective indices (core and clad) and matching the observed resonance to the calculated dispersion curves of the effective refractive index.

4. Applications of long-period fibre gratings in optical sensing

One of the more important applications of LPFGs is as sensing elements. Owing to the well-known advantages over conventional sensors, such as high resolution, low cost, light weight, and multiplexing possibilities, several optical fibre sensors have evolved from laboratory experiments into commercial products. Although many sensor schemes have been proposed, probably the most popular ones are based on evanescent field interactions (Villatoro et. al. , 2005), fibre Bragg gratings (FBGs) (Othonos & Kalli, 1999), and fibre long period gratings (LPGs) (Vengsarkar et. al., 1996; Vengsarkar & Bhatia, 1996). Fibre gratings have attracted considerable attention as ideal detectors since the information of the physical phenomena is encoded in wavelength. Fibre gratings are very sensitive to physical perturbations like temperature, strain, and pressure (Othonos & Kalli, 1999; Vengsarkar & Bhatia, 1996; Kersey et. al., 1997), however, in combination with a suitable transducer can monitor other magnitudes such as humidity (Liu et. al., 2007), or electric fields (Abramov et. al., 1999). The sensitivity of LPGs to ambient refractive index is higher than that observed in fibre Bragg gratings, mainly due to the fact that in the case of cladding modes the guiding structure is formed by the cladding and the external medium in such way that any change in the ambient refractive index modifies the spectral response of the LPGs (Vengsarkar et. al., 1996; Vengsarkar & Bhatia, 1996).

The potential applications of LPG are strongly determined by the sensitivity of these devices to the influence of the surrounding environment. In general, the performance of devices used in communication, it is expected to have low sensitivity to changes of the surrounding media. By contrast, the LPGs fabricated for sensing applications the characteristics of the devices are designed to enhance its intrinsic sensitivity. Thus, a complete analysis of the sensitivity of LPGs to external ambient is necessary prior to practical devices design. The coupling of the fundamental core mode to cladding modes is determined by the period, the effective refractive index of the core and the cladding. These three parameters, and in consequence the LPG resonant wavelength, will be affected by changes in temperature and strain of the fibre. Additionally, since the cladding modes interact with the buffer or any other material in close contact with the cladding of the fibre, the wavelength response of the LPG is also affected by the changes in the refractive index of the external medium. Shu et al. (Shu et. al., 2002) have presented a complete investigation of the shift in the center of the wavelength of the resonant peaks of the LPG due to temperature, strain and external refractive index changes, and can be expressed by

$$\Delta\lambda_{res} = \left(\frac{d\lambda_{res}}{dT}\right)\Delta T + \left(\frac{d\lambda_{res}}{dS}\right)\Delta S + \left(\frac{d\lambda_{res}}{dn_{sur}}\right)\Delta n_{sur} \tag{15}$$

where $(d\lambda_{res}/dT), (d\lambda_{res}/dS), and\ (d\lambda_{res}/dn_{sur})$ are the temperature, strain and surrounding refractive index sensitivity of the resonant wavelength, respectively. In the following sections a detailed expression for such sensitivities is going to present and discussed.

4.1 Temperature sensing with long-period fibre gratings

For more than three decades investigation over optical fibre thermometers has been growing. Early fibre-optic temperature sensors were two fibre interferometers in which temperature was measured by fringe displacements (Hocker, 1979). Later, one-fibre interferometers were proposed and demonstrated that had practically no cross sensitivity (Eickhoff, 1981). Recently, a new generation of fibre-optic temperature sensors based on LPGs has been developed in which temperature is measured in the spectral domain by resonant wavelength shifts, rather than fringe displacements. The vast majority of fibre temperature sensors reported so far have been designed to operate in a range from 20 C to 200 C. However, there are a variety of applications in which high-temperature sensing is important, for example, for monitoring furnace operation or volcanic events, or in fire alarm systems, etc. (Brambilla et. al., 2002). There exist a variety of materials and techniques to fabricate optical fibre thermometers for high temperature. Among them, LPG inscribed with a pulsed CO_2 laser (Davis et. al., 1998), or using an electric arc (Kosinski & Vengsarkar, 1998) have been attracting attention due to their simply fabrication process and their wide range of measurement. The expression for the temperature sensitivity of a LPG, whether is used to measure low- or high-temperature, can be expressed as (Shu et. al., 2002),

$$\frac{d\lambda_{res}}{dT} = \lambda_{res}\left(\alpha_{\Lambda} + \frac{\alpha_{co} n_{co}^{eff} - \alpha_{cl} n_{cl,m}^{eff}}{n_{co}^{eff} - n_{cl,m}^{eff}}\right) \tag{16}$$

where α_{Λ} is the thermal expansion coefficient of the fibre, α_{co} and α_{cl} are the thermo-optic of core and cladding materials, respectively and γ is the waveguide dispersion described as (Shu et. al., 2002):

$$\gamma = \frac{d\lambda_{res} / d\Lambda}{n_{co}^{eff} - n_{cl,m}^{eff}} \tag{17}$$

In general for silica-based fibre the effect of thermal expansion coefficient is lower than of the thermo-optic, represented by the second term in the parenthesis. However, its influence in the analysis of temperature sensitivity cannot be ignored. For standard fibre, the second term in the parenthesis could be higher than zero, whereas for B-Ge co-doped fibre this value is lower than zero, because boron doping can decrease significantly the thermo-optic coefficient of the core (Shu et. al., 2001). Thus, it is to be expected that the thermal responses of LPFGs produced in these two fibre types will exhibit opposite trends. It is also evident from eq. (17) that the temperature sensitivity of a LPG is also dependant on the cladding mode excited and in consequence depends on the technique used to fabricate the LPGs.

4.2 Strain sensing with long-period fibre gratings

Grating-based sensors have redefine the concept of "smart structures" since these devices have been successfully integrated or embedded into the materials to allow monitoring of parameters such as load, strain or vibration from which the health of the structure can be evaluated and tracked in a real-time basis. This success is mainly due to the well-known advantages of the optical fibres for sensing, such as EMI immunity, electric passive operation, low weight, small size, and high sensitivity, but also because these grating-based sensors have the inherent self-referencing capability and are easily multiplexed. The strain response of a long-period fibre grating arise due to the physical elongation of the fibre, changing the grating pitch and the effective refractive index of the core and cladding due to the elastooptic effect. The LPG sensitivity to strain can be expressed as follows (Shu et. al., 2002),

$$\frac{d\lambda_{res}}{dS} = \gamma\ \lambda_{res}\left(1 + \frac{\varsigma_{co}\, n_{co}^{eff} - \zeta_{cl} n_{cl,m}^{eff}}{n_{co}^{eff} - n_{cl,m}^{eff}}\right) \tag{18}$$

where ζ_{co} and ζ_{cl} are the elastooptic coefficient of the core and cladding materials, respectively. The strain response of a LPG depends on the cladding mode excited and the elastooptic coefficient of the core and cladding materials. It can be seen form eq. (18) that a LPG is strain-insensitive when the second term in the parenthesis is equal to -1.

4.3 Refractive index sensing with long-period fibre gratings

The refractive index (RI) is a fundamental material property, in a variety of applications, the measurement of the RI is very important since different chemical substances as well as several physical and biological parameters can be detected through measurements of this parameter. In bio-sensing, for example, there is a need to monitor RI changes on the order of 10^{-4} to 10^{-7} which may be caused by molecular binding, chemical or biochemical reactions, or by changes suffered by a thin bio-layer measure very small RI changes in small volumes of fluid. Traditional bulk refractometers are not appropriate for such an application; in such cases optical fibre refractometers (OFRs) constitute an alternative. OFRs have some advantages over their bulk counterparts, such as high resolution, low cost, light weight, and multiplexing possibilities. Several OFRs based on side-polished single-mode fibre, core-exposed or tapered multimode fibres, fibre Bragg gratings (FBGs), or long period gratings (LPGs) have been reported so far (Lopez-Higuera & Ed., 2002). In a LPG the guided light interacts with the external medium so there is no need to modify the fibre. The effective index of the excited cladding modes depends on the refractive index of the core, cladding and external medium materials. This dependence is clearly observed in the expression derived for the sensitivity of the LPG to external refractive index (Shu et. al., 2002; Miyagi & Nishida, 1979)

$$\frac{d\lambda_{res}}{dn_{ext}} = \gamma\,\lambda_{res}\left(\frac{u_m^2\,\lambda_{res}^3 n_{ext}}{8\pi r_{cl}^3 n_{cl}\left(n_{co}^{eff} - n_{cl,m}^{eff}\right)\left(n_{cl}^2 - n_{ext}^2\right)^{3/2}}\right) \quad (19)$$

u_m is the m-th root of the zeroth-order Bessel function of the first kind, r_{cl} and n_{cl} are the radius and refractive index of the fibre cladding, respectively. In this case is more evident that the sensitivity is rather greater for higher order modes and it can be enhanced by reducing the cladding radius (Martínez-Rios, A. et. al. (2010).) since the term in parenthesis is dependent on (r_{cl}^{-3}).

When the refractive index of the external medium adjacent to the LPG augments the wavelength of the resonant peaks shifts to shorter values. The wavelength changes of the resonant peaks of an arc-induced LPG due to the changes in the external refractive index are shown in graph (a) of the figure 7. The calculated sensitivity of the arc-induced LPG to refractive index changes is 166.66 nm/RIU for a refractive index range of 1.40- 1.46, by contrast in the range of 1-1.39 the sensitivity is only 7.692 nm/RIU. This behavior is also expressed in ec. (4.5), where can be observed that an increment in the external refractive index produce an augment of the sensitivity. In the graph (b) of figure 7 we show the shift of the wavelength resonant peak as the refractive index is changed for an arc-induced LPG that was chemically etched, the final diameter of the cladding was ~39 μm. When the cladding radius decreases, the sensitivity of the of the LPG is highly enhanced as is predicted by the ec. (4.5). The sensitivity of the LPG calculated from the experimental data of graph (b) in Fig. 7 is~3,571 nm/RIU in the range of 1.4-1.46. The sensitivity of an etched-LPG is 1 order of magnitude higher than an unetched LPG. In figure 8 is shown the spectra of an etched arc-induced fibre when the external refractive index changes.

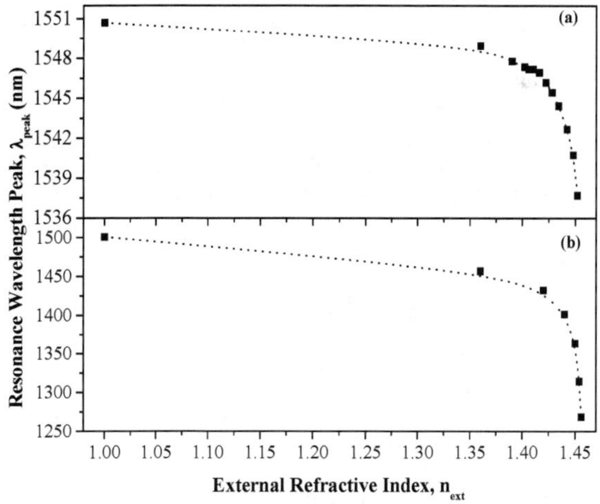

Fig. 7. In these graphs is shown the shift of the resonant wavelength peak when the external refractive index is change. In the upper graph (a) is shown the behavior of an arc-induced LPG when the refractive index of the external medium changes. In the bottom graph (b) the cladding of the LPG fibre was etched, the final diameter was 39 μm.

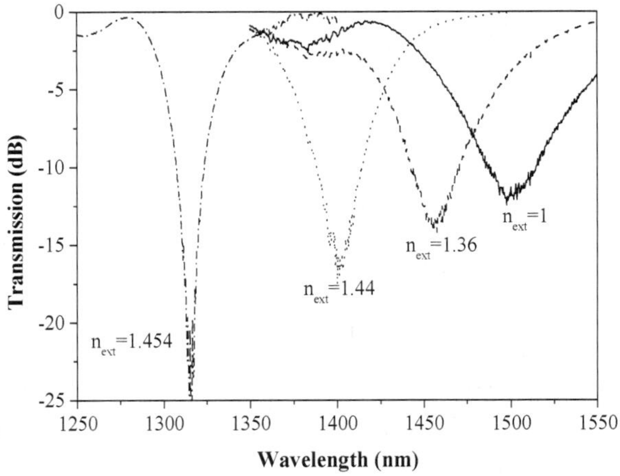

Fig. 8. the spectra transmitted by an arc-induced LPG that was etched to reduce the cladding diameter to 39 μm. As can be seen the resonant peak shifts to shorter wavelengths as the refractive index of the external medium is increased.

5. Conclusion

We have presented a review of the some of the more relevant aspects of LPFGs. Fabrication methods, basic theory and applications have been discussed. As can be clearly seen, LPFGs are sensitive to almost everything, and are used currently in sensor applications and as filters in telecommunication and many other fibre systems. As a sensor their main advantage with respect to fibre Bragg grating based systems is in the lower fabrication cost, since they can be inscribed in almost any type of fibre using a wide variety of forms to change periodically the refractive index in the core, in the cladding, or both. One of the main drawbacks of LPFGs is the bandwidth of notch bands, in the order of 10 nm, which limits the resolution of the measurements. However this can be alleviated by using interferometric arrays consisting of a series of LPFGs which form Mach-Zender interferometers, where the bandwidth may be below 1 nm, similar to fibre Bragg gratings.

6. Acknowledgment

We thank the Centro de Investigaciones en Optica for its support

7. References

Abramov, A.; Eggleton, B.; Rogers, J.; Espindola, R.; Hale, A.; Windeler, R. & Strasser, T. (1999). Electrically Tunable Efficient Broad-band Fibre Filter, *IEEE Photonics Technology Letters*, Vol. 11, No. 4, (April 1999), pp. 445-447, ISSN 1041-1135.

Anemogiannis E., Glytsis E. N. & Gaylord T. K.. (2003). Transmission Characteristics of Long-Period Fibre Gratings Having Arbitrary Azimuthal/Radial Refractive Index Variations. *Journal of Lightwave Technology*, Vol. 21, No. 1, (January 2003), pp. 218-227, ISSN 0733-8724.

Arai, T. & Kragic, D. (1999). Variability of Wind and Wind Power, In: *Wind Power*, S.M. Muyeen, (Ed.), 289-321, Scyio, ISBN 978-953-7619-81-7, Vukovar, Croatia

Bilodeau F., Malo B., Albert J., Johnson D. C., Hill K. O., Hibino Y., Abe M. & Kawachi M. (1993). Photosensitization of optical fibre and silica-on-silicon/silica waveguides. *Optics Letters*, Vol. 18, No. 12, (June 1993), pp. 953-955 , ISSN 0146-9592.

Bjarklev. (1986). Microdeformation losses of single-mode fibres with step- index profiles. *Journal of Lightwave Technology*, Vol. 4, No. 3, (March 1986), pp. 341-346, , ISSN 0733-8724.

Brambilla, G.; Kee, H.; Pruneri, V. & Newson, T. (2002). Optical fibre sensors for earth science: From the basic concepts to optimizing glass composition for high temperature applications, *Optics and Lasers in Engineering*, vol. 37, No. 2–3, (February-March 2002), pp. 215–232, ISSN 0143-8166.

Bures J. (2009). Guided Optics:Optical Fibers and All-Fiber Components. WYLEY-VCH (Weinheim, Germany). ISBN 978-3-527-40796-5.

Chan F. Y. & Yasumoto K. (2007). Design of wavelength tunable long-period grating couplers based on asymmetric nonlinear dual-core fibres. *Optics Letters*, Vol. 32, No. 23, (December 2007), pp. 3376-3378, , ISSN 0146-9592.

Chiang K. S.; Liu Y.; Ng M. N.; Dong X. (2000). Analysis of etched long-period fibre grating and its response to external refractive index. *Electonics Letters*, Vol. 36, No. 11, (May 2000), pp. 966-967, ISSN 0013-5194.

Chunn-Yenn L., Wang L. A., Gia-Wei C. (1997). Corrugated long-period fibre gratings as strain, torsion, and bending sensors. *Journal of Lightwave Technology*. Vol. 19, No. 8, (August 2001), pp. 1159-1168, ISSN 0733-8724.

Davis, D.; Gaylord, T.; Glytis, E. & Mettler, S. (1998). CO_2 laser-induced long-period fibre gratings: spectral characteristics, cladding modes and polarization independence, *Electronics Letters*, Vol. 34, No. 14, (July 1998), pp. 1414-1417, ISSN 00-13-5194.

Daxhelet, X. & Kulishov, M. (2003). Theory and practice of long-period gratings: when a loss become a gain. *Optics Letters*, Vol. 28, No. 9,(May 2003), pp. 686-688, ISSN 0146-9592.

DeLisa M. P., Zhang Z., Shiloach M., Pilevar S., Davis C. C., Sirkis J. S. & Bentley W. E. (2000). Evanescent Wave Long-Period Fibre Bragg Grating as an Immobilized Antibody Biosensor. *Analytical Chemi*stry, Vol. 72, No. 13, (June 2000), pp. 2895–2900, ISSN 0003-2700.

Dianov E. M., Stardubov D. S., Vasiliev S. A., Frolov A. A., & Medvedkov O. I. (1997). Refractive-index gratings written by near-ultraviolet radiation. *Optics Letters* Vol. 22, No. 4 , (February 1997), pp. 221-223, ISSN 0146-9592.

Eickhoff, W. (1981), Temperature sensing by mode-mode interference in birefringent optical fibres, *Optics Letters*, Vol. 6, No. 4, (April 1981), pp. 204-206, ISSN 0146-9592.

Erdogan T. (1997). Cladding-mode resonances in short- and long-period fibre grating filters. *Journal of the Optical Society of America A*, Vol. 14, No.8, (August 1997), pp. 1760-2201, ISSN 0740-3232.

Fujimaki M., Ohki Y., Brebner J. L. & Roorda S. (2000). Fabrication of long-period optical fibre gratings by use of ion implantation. *Optics Letters*. Vol. 25, No. 2, (January 2000), pp. 88-89, ISSN 0146-9592.

Gu Z., Xu Y. & Gao K. (2006).Optical fibre long-period grating with solgel coating for gas sensor. *Optics Letters*, Vol. 31, No. 16, (August 2006), pp. 2405-2407, ISSN 0146-9592.

Hill K. O., Malo B., Vineberg K. A., Bilodeau F., Johnson D. C.& Skinner I. (1990). Efficient mode conversion in telecommunication fibre using externally written gratings. *Electronics Letters*, Vol. 26, No 16, (August 1990), pp. 1270-1272, ISSN 0013-5194.

Hocker, G. (1979). Fibre-optic sensing of pressure and temperature, *Applied Optics*, Vol. 18, No.9, (May 1979), pp. 1445-1448, ISSN 003-6935.

Hwang I. K., Yun S. H. & Kim B. Y.. (1999). Long-period fibre gratings based on periodic microbends. *Optics Letters*, Vol. 24, No. 18, (September 1999), pp. 1263-1265, ISSN 0146-9592.

J. Lopez-Higuera, J. Ed., (2002). *Handbook of Optical Fibre Sensing Technology*, ISBN 978-0-471-82053-6 Wiley, New York, USA

Jeong Y., Yang B., Lee B., Seo H. S., Choi S., Oh K. (2000). Electrically controllable long-period liquid crystal fibre gratings. *Photonics Technology Letters*, Vol. 12, No. 519-521, (May 2000), pp. 519 – 521, ISSN 1041-1135.

Kersey, A.; Davis, M.; Patrick, H.; LeBlanc, M.; Koo, K.; Askins, C.; Putnam, M. & Friebele, E. (1997). Fibre Grating Sensors, *Joruna1 of Lightwave Technology*, Vol. 15, No. 8, pp. 1442-1462, ISSN 0733-8724.

Kosinski, S., & Vengsarkar, A. (1998). Splicer-based long-period fiber gratings, *Optical Fibre Communication Conference*, Vol. 2 of 1998 OSA Technical Digest Series (Optical Society of America, 1998), ISBN 1557525293, paper ThG3..

Lemaire P. J., Atkins R. M., Mizrahi V. & Reed W. A. (1993). High pressure H_2 loading as a technique for achieving ultrahigh UV photosensitivity and thermal sensitivity in GeO_2 doped optical fibres, *Electronics Letters*, Vol. 29, No. 13, (June 1993), pp. 1191-1193, ISSN 0013-5194.

Li, B.; Xu, Y. & Choi, J. (1996). Applying Machine Learning Techniques, *Proceedings of ASME 2010 4th International Conference on Energy Sustainability*, pp. 14-17, ISBN 842-6508-23-3, Phoenix, Arizona, USA, May 17-22, 2010

Lima, P.; Bonarini, A. & Mataric, M. (2004). *Application of Machine Learning*, InTech, ISBN 978-953-7619-34-3, Vienna, Austria

Liu, Y.; Wang, L.; Zhang, M.; Tu, D.; Mao, X. & Lia, L (2007). Long-Period Grating Relative Humidity Sensor with Hydrogel Coating, *IEEE Photonics Technology Letters*, Vol. 19, No. 12,(June 2007), pp.880-882, ISSN 1041-1135.

Martínez-Rios, A.; Monzón-Hernández, D., & Torres-Gomez, I, (2010). Highly sensitive cladding-etched arc-induced long-period fibre gratings for refractive index sensing, *Optics Communications*, Vol. 283, No. 6, (March 2010), pp. 958-962, ISSN 0030-4018.

Miyagi, M. & Nishida, S., (1979). An approximate formula for describing dispersion properties of optical dielectric slab and fibre waveguides, Journal of the Optical Society of America Vol 69, No.2, (February 1997), pp. 291-293, ISSN 0030-3941.

Mohamed W. & Gu X. (2009). Long-period grating and its application in laser beam shaping in the 1.0 μm wavelength region, *Applied Optics*. Vol. 48, No. 12, (April 2009), pp. 2249-2254, ISSN 003-6935.

Ng M. N. & Chiang K. S. (2002). Thermal effects on the transmission spectra of long-period fibre gratings. *Optics Communications*, Vol. 208, No. 4-6, (July 2002), pp. 321-327, ISSN 0030-4018.

Othonos, A & Kalli, K. (1999). *Fibre Bragg Gratings: Fundamentals and Applications in Telecommunications and sensing*, Artech House, ISBN 0-89006-344-3, Norwood, MA, USA

Rao Y.J., Zhu T., Ran Z. L., Wang Y. P., Jiang J. & Hu A. Z. (2004). Novel long-period fibre gratings written by high-frequency CO2 laser pulses and applications in optical fibre communication. *Optics Communications*, Vol. 229, No. 1-6, (January 2004), pp. 209–221, ISSN 0030-4018.

Savin S., Digonnet M. J. F., Kino G. S & H. J. (2000). Tunable mechanically induced long-period fibre gratings. *Optics Letters*, Vol. 25, No. 10, (May 2000), pp. 710 - 712, ISSN 0146-9592.

Shu Y., Lu C., Lacquet B. M., Swart P. L. & Spammer S. J. (2002). Wavelength-tunable add/drop multiplexer for dense wavelength division multiplexing using long-period gratings and fibre stretchers. Optics Communications, Vol. 208, No. 4-6, (July 2002), pp. 337-344, ISSN 0030-4018.

Shu, X.; Allsop, T.; Zhang, L. & Bennion, (2001). High-Temperature Sensitivity of Long-Period Gratings in B–Ge Codoped Fibre, *IEEE Photonics Technology Letters*, Vol. 13, No. 8, (August 2001), pp. 818-820, ISSN 1041-1135.

Shu, X.; Zhang, L. & Bennion, I. (2002). Sensitivity Characteristics of Long-Period Fibre Grating, *Journal of Ligthwave Technology*, Vol. 20, No.2, pp.255-266, ISSN 0733-8724.

Siegwart, R. (2001). Indirect Manipulation of a Sphere on a Flat Disk Using Force Information. *International Journal of Advanced Robotic Systems*, Vol.6, No.4, (December 2009), pp. 12-16, ISSN 1729-8806

Su C. D. & Wang L. A. (1999). Effect of Adding a Long Period Grating In a Double-Pass Backward Er-Doped Superfluorescent Fibre Source. *Journal of Lightwave Technology*. Vol. 17, No. 10, (October 1999), pp. 1896-1903, ISSN 0733-8724.

Van der Linden, S. (June 2010). Integrating Wind Turbine Generators (WTG's) with Energy Storage, In: *Wind Power*, 17.06.2010.

Vaziri M. & Chin-Lin C.. (1997). An etched two-mode fibre modal coupling element. *Journal of Lightwave Technology*. Vol. 15, No. 3, (March 1997), pp. 474-481, ISSN 0733-8724.

Vengsarkar A.N., Lemaire P. J., Judkins J. B., No. Bhatia B., Erdogan T & Sipe J. E. (1996). Long-period fibre gratings as band-rejection filters, *Journal of Lightwave Technology*, Vol. 14, No. 1 , (January 1996), pp. 58-65, ISSN 0733-8724.

Vengsarkar, A. & Bhatia, B. (1996). Optical Fibre Long-Period Fibre Sensors, *Optics Letters*, Vol. 21, No. 9, (May 1996), pp. 692-694, ISSN 0146-9592.

Vengsarkar, A.; Lemaire, P.; Judskin J.; Bhatia, B.; Erdogan, T. & Sipe, J. (1996). Long-Period Fibre Gratings as Band-Rejection Filters, *Journal of Lightwave Technology*, Vol. 14, No. 1, (January 1996) pp. 58-64, ISSN 0733-8724.

Villatoro, J.; Monzón-Hernández, D. & Mejía, E. (200). Fabrication and modeling of uniform-waist single-mode tapered optical fibre sensors, *Applied Optics*, Vol. 42, No. 13, (May 2003), pp. 2278-2283, ISSN 0003-6935.

Williams D. L., Ainslie B. J., Armitage J. R., Kashyap R., Campbell R. (1993). Enhanced UV photosensitivity in boron codoped germanosilicate fibres. *Electronics Letters*, Vol. 29, No. 1, (January 1993), pp. 45-47, ISSN 0013-5194.

Zhang L., Liu Y., Everall L., Williams J. A. R. & Bennion I. (1999). Design and realization of long period grating devices in conventional and high birefringence fibre and their novel applications as fibre-optic load sensor. *IEEE Journal of Selected Topics in Quantum Electronics*, Vol. 5, No. 5, (September/October 1999), pp. 1373-1378, ISSN 1077-260X.

12

Long Period Fiber Grating Produced by Arc Discharges

Julián M. Estudillo-Ayala[1], Ruth I. Mata-Chávez[1],
Juan C. Hernández-García[2] and Roberto Rojas-Laguna[1]
[1]*Universidad de Guanajuato*
[2]*Centro de Investigaciones en Óptica A.C.*
México

1. Introduction

Fiber Gratings are found among the most popular devices widely used in both, optical communications and optical fiber sensing. Within optical communication networks, optical fiber-based devices perform critical operations such as coupling/splitting, wavelength-selective filtering, and optical switching (Han, Y. G. et al., 2004; An H. et al., 2004; Eggleton B. J. et al., 1997). In the field of optical sensing, optical fiber sensors, with unique advantages such as immunity to electromagnetic interference, high sensitivity, resistance to corrosion, and high temperature survivability, are widely used to measure various physical variables such as stress, temperature, pressure, refractive index, etc. and chemical parameters (Udd, E. 1991).

There are two types of fiber gratings that have been developed so far including the Fiber Bragg Grating (FBG) and the Long Period Fiber Grating (LPFG). One important advantage respect to Bragg gratings is that their fabrication process is simpler; their cost is lower. Furthermore, they present low retro-reflection and high sensitivity in sensing applications (Kersey, A. D. et al., 1997).

LPFGs have been traditionally fabricated by exposing photosensitive optical fiber to ultraviolet (UV) light transversely. Either through an amplitude mask or point-by-point to create a periodic refractive index change inside an optical fiber. Unfortunately, the UV-induced refractive index changes inside an optical fiber can only survive at a relatively low temperature. As a result, the UV exposure fabricated-LPFGs are not suitable for high temperature applications. Among the different techniques available for the fabrication of LPFGs, the electric arc technique is one of the few enabling their fabrication, virtually in any type of glass fibers. This technique is simple, inexpensive and harmless when compared to the ones based on laser radiation. Furthermore, arc-induced gratings are suitable for high temperature applications since their formation mechanisms rely on thermal effects. Thus, the temperature reached by the fiber during an electric arc discharge, is a key parameter to understand the properties of gratings (Rego, G. 2010). An LPFG can be considered a special case of a fiber Bragg grating, in which the period of the index modulation satisfies a phase matching condition between the fundamental core mode and the forward-propagating cladding-mode of an optical fiber. Therefore, an LPFG consists of a periodic spatial variation (along the fiber longitudinal axis) in the refractive index of an optical fiber. The periodic refractive index modulation couples light from a forward-propagating core-guided mode, to forward-propagating cladding-guide modes, near certain resonance wavelengths (Vengsarkar, A. M. et al., 1996). The light coupled into the cladding modes, eventually attenuates due to the high loss of the cladding modes. As a result, the transmission spectrum of a LPFG has a series of discrete attenuation bands near the resonance wavelengths. In this chapter we will describe two of the in-fiber LPFGs fabrication methods: optical fiber fattening and optical fiber micro-taper, both methods by means of electric arc (we used a splicer fusion of optical fiber). We will present several configurations of PFGs sensors as well as temperature and curvature sensors. Finally, at the end of this chapter, we will show numerical calculations of the modal behavior of the LPFGs, besides conclusions and references.

2. Fiber grating

One of the most important discoveries, in the field of optical fibers, has undoubtedly been made by Hill et al. in 1978 (Hill, K.O. et al., 1993; Hill, K.O. et al., 1997). Working in their laboratories [The Communications Research Center (CRC) in Ottawa, Canada.], they carried out an experiment to study nonlinear effects in an optical fiber especially designed by Koichi Abe of Bell Northern Research. This special optical fiber had a high germanium concentration, a numerical aperture of 0.22 at 514.5 nm, and the fiber core diameter was of 2.5 mm. They launched intense light of Argon-ion (514.5 nm) laser into the core of a Germania-doped fiber, and several minutes after, they observed the power beam decreasing. Subsequent to this observation, they monitored the light back reflected from the fiber, finding that the power significantly increased with time. The increase in the reflected light was produced by the writing in the core of the fiber by an interference periodic pattern (hologram), corresponding to a periodic modulation of the refraction index in the optical fiber core. This nonlinear effect in optical fiber was called fiber photosensitivity, and the holograms in the core of the fiber are called "Hill gratings" (Hill & Meltz, 1997; Othonos, 1997; Hill, 2000).

After the discovery of the photosensitivity in optical fibers, the effect was forgotten by several years, due to restrictions in the writing of the grating. However, a new interest emerged after the proposal of Meltz et al. 1989, where they described the grating fabrication, producing the overlapping of two coherent beams of ultraviolent light (Meltz, G., et al., 1989). Beams interfered between them, creating a periodic interference pattern with both bright and dark bands, which generated a permanent grating with the same refraction index of the fiber. This technique was called the *transverse holographic technique*. This is possible because of the cladding of the fiber is transparent to the ultraviolet radiation, while the core strongly absorbs this kind of light due to the germanium concentration. Hill et al. 1993 proposed a simple way to fabricate gratings. They did not use an interferometer; instead, they used a beam of ultraviolet light placing a glass phase mask of pure silica, very close to the fiber (Othonos, A. et al. 1997). The grating was formed by exposing the fiber to ultraviolet light through the phase mask. Therefore, it is one of the most efficient methods for inscribing gratings in photosensitive fiber.

This novel method represented the key for simpler and more accelerated fabrication of gratings in special fibers. Nowadays, these gratings are employed in the fabrication of devices for optical communications (fiber lasers, filters, optical switching, and multiplexers) and optical sensing applications.

2.1 Basic concepts

The optical fiber gratings are a special type of filter wavelength selective operating in reflection or transmission with the fundamental principles of reflection, refraction and diffraction of light in periodic structures dimensionless. These basic components are formed when the refractive core index of a fiber section is periodically modulated. According to their operation, there are two types of optical fiber grating: Bragg Gratings (BG) and Long-Period Gratings (LPG). The Bragg gratings are called this way because their operation is based on the Bragg reflection law (Kashyap R., 1999). The Bragg gratings were inscribed by the first time in fiber optics, with the discovery of photosensitivity as it was mentioned previously. The main characteristic of this type of BG is the possibility to reject a narrow band of wavelengths (which meet the Bragg condition). Fig. 1 shows a diagram which illustrates the operation of a Bragg grating. When the signal intensity P_I (broadband light source) is propagated along the BG in the optical fiber, a signal P_B centered at the Bragg wavelength and a narrow band is reflected(< 1nm). On the other hand, the transmitted signal P_T, shows a band rejection to Bragg wavelength.

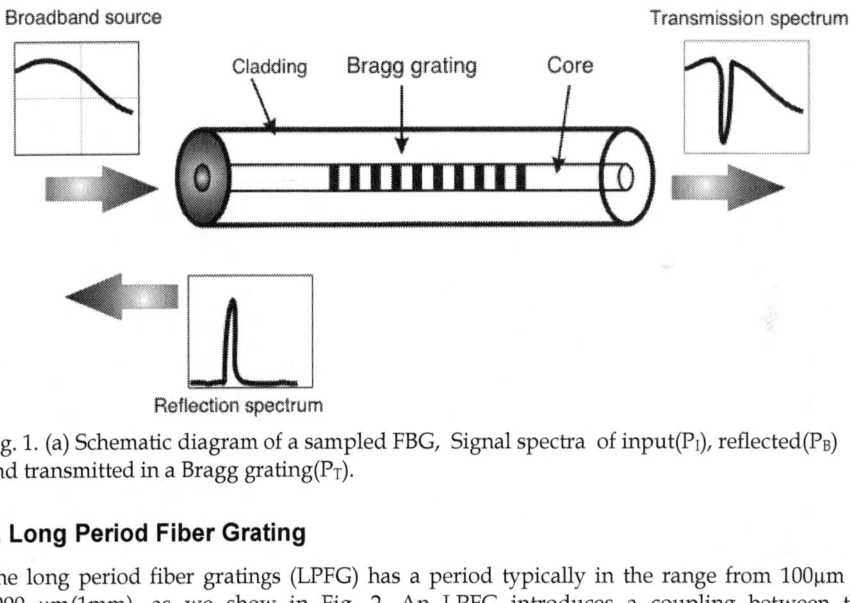

Fig. 1. (a) Schematic diagram of a sampled FBG, Signal spectra of input(P_I), reflected(P_B) and transmitted in a Bragg grating(P_T).

3. Long Period Fiber Grating

The long period fiber gratings (LPFG) has a period typically in the range from 100μm to 1000 μm(1mm), as we show in Fig. 2. An LPFG introduces a coupling between the fundamental guided mode and propagating cladding mode (Riant, I. 2002). In Fig. 3, we show the transmission spectrum of a LPFGs photoimprinted in a standard fiber. The wavelengths of the peaks shown in Fig. 3 are defined as (Erdogan, T. 1997).

$$\lambda^m = (n_{eff}^{01} - n_{eff}^m)\Lambda \tag{3.1}$$

Where n_{eff}^{01} and n_{eff}^{m} represent the effective indices of the fundamental guided mode, the cladding mode of order m coupled to the guide mode by the grating respectively, then Λ is the period of the LPFG.

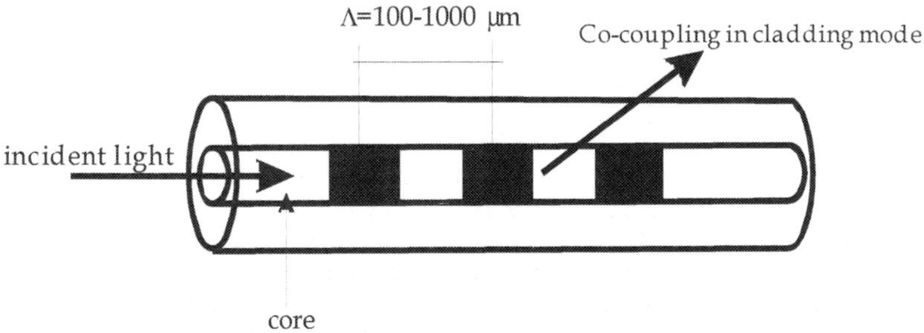

Fig. 2. Schematic diagram of a long-period fiber grating. The period of the grating is Λ.

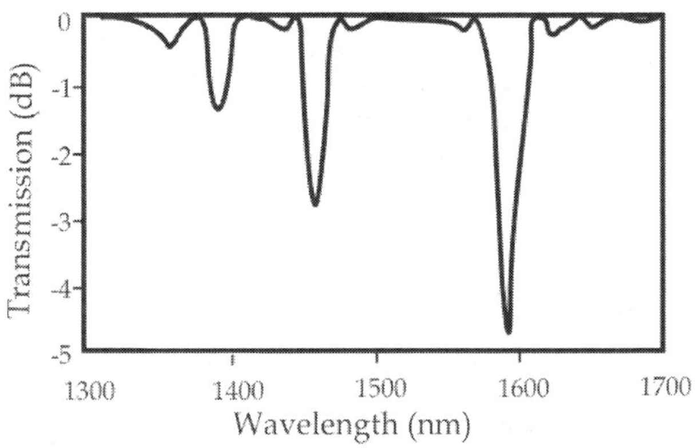

Fig. 3. Depicts transmission spectrum characteristic of a Long Period Grating

Coupling between two propagating modes occurs for periodicity of several hundreds of microns, which are three orders of magnitude larger than for short period grating.

The minimum transmission of the attenuation bands is governed by (Kashyap, R. 1999).

$$T_i = 1 - \sin^2(\kappa_i L) \tag{3.2}$$

Where L is the lenght of the LPFG and κ_i is the coupling coefficient for the ith cladding mode, which is determined by the integral overlap of the core and cladding mode, also by the amplitude of the periodic modulation of the mode propagation constants.

Since the cladding generally has a large radius, it supports a large number of cladding modes. Theoretical analysis has shown that efficient coupling is possible only between core and cladding modes which have a large integral overlap, in other words, modes which have similar electric field profiles (Erdogan, T. 1997). This coupling is observed between the core and circularly symmetric cladding modes of odd order. The effect occurs because the electric field profile of the even-order modes is such that the field amplitude is low within the core (Erdogan, T. 1997). The refractive index of the propagating core mode of the fiber is generally determined using the weakly guided field approximation (Glogle, D. 1971). A simple three-layer slab waveguide model, to determine the cladding mode indices, has been presented, allowing an approximate determination of the coupling wavelengths (Vengsarkar A.M. et al., 1996).

3.1 Fabrication techniques of long-period gratings in optical fiber

The long period grating fabrication in an optical fiber consists of a refractive index periodic modulation of the optical fiber core. This modulation may be permanent or temporary. The permanent modulation can be obtained by irradiation or heating. The temporal modulation index can be achieved by mechanical pressure or acoustic vibrations through the photo-elastic effect. The modulation of the core refractive index has been achieved by being used with different types of optical fibers, for example, standard single mode, photo-sensible, photonic crystal fiber, specialty optical fibers, among others. Techniques like UV irradiation (Ultraviolet) (Vengsarkar A.M. et. al., 1996; Bhatia, V. et. al. 1996; Erdogan, T. 1997), laser CO_2 exposure (Davis, D. D., et. al., 1998; Davis, D. D., et. al., 1999), electric arc discharge, mechanical microbends, etched corrugations, ion beam implantation, femtosecond laser exposure, just to mention a few, have been developed for more than a decade. Different devices based on LPFG have been fabricated for applications in optical communications as well as in sensors.

3.2 Techniques of fabrication by arc discharges

The Long Period Fiber Gratings (LPFG) fabricated with electric arc, have been widely used as filters in telecommunication networks, as well as the sensing element for various physical and chemical parameters. These applications are possible due to their relevant characteristics, such as: very deep attenuation band and narrow bandwidth. In addition, the period is easily adjustable, allowing the wavelength to tune over a wide range (from 1300 to 1700 nm). However, disadvantages of the LPFGs are their low repeatability and the wavelength of the attenuation bands has higher level of losses compared to other technologies (UV-inscribed techniques). Disadvantages are largely due to two issues: first, to the random nature of the arc, and second, the physical deformation that occurs in the optical fiber. In this chapter, we describe the development of LPFG when electric arc discharge is applied. The electric arc can produce geometric deformation, which might increase the coupling of modes, but in many cases it induces losses. There are three main types of modulation: modulation only of the refractive index, the modulation by microtaper, and microbends.

3.3 Modulation only of the refractive index

The best way to inscribe an LPFG is through merely changing the refractive index of the core, without affecting the geometry of the fiber. In this case, the attenuation bands are well defined and out of these ones, there are no losses. Fig. 4 shows the transmission spectrum of a long-period grating inscribed by ultraviolet radiation (LPFG-UV), this kind of gratings meet these conditions. In the LPFG by electric arc (LPFG-EA) this type of modulation can be obtained, but unlike LPFG-UV, in this type of gratings the refractive index of the cladding is

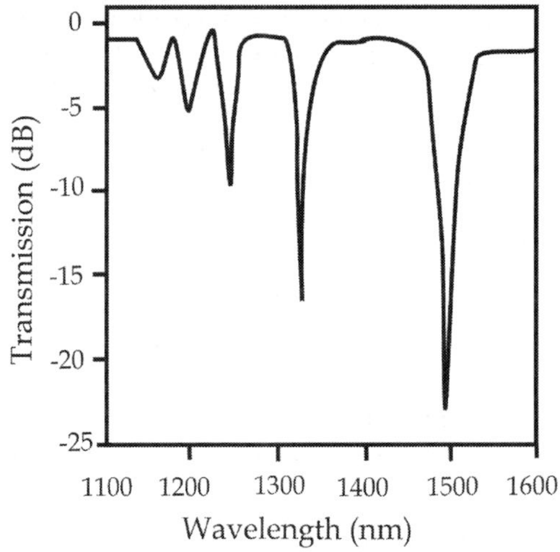

Fig. 4. Transmission spectra of a LPFG modulated only of the refractive index (LPFG-UV) (Stephen, W.J., et. al., 2003).

always modulated and geometric changes occur. The modulation of both refractive indices increases insertion losses, because the single mode condition is different, and moreover, the coupling constant is larger. The latter means that in general, the LPFG-EAs are of shorter length than LPFG-UV and easily, achieve greater levels of attenuation. While LPFG-UV typically require a grating length of 1 inch to achieve a rejection level of 18 dB, the LPFG-EA require approximately 2 cm.

The pure Refractive Index Modulation (RIM), in the electric arc technique, is quite difficult to achieve, as usually, it uses axial stress for greater change in refractive index, producing micro-bending or micro-tapers.

3.4 Periodic modulation by micro-taper

During an electrical arc discharge, an optical fiber can reach temperatures above 1350 ° C (Rego G., 2006). This temperature is sufficient to soften the silica (SiO_2) and make it pliable. This is the principle used for taper manufacture of optical fiber and fiber couplers couplers. This procedure to make a fiber taper is used to inscribe a LPFG. Due to the low weight (less than 200 g) and short electric discharge (hundreds of milliseconds), the tapered length is in the order of micrometers. So, actually, a micro-taper is achieved by every arc discharge, see fig. 5.

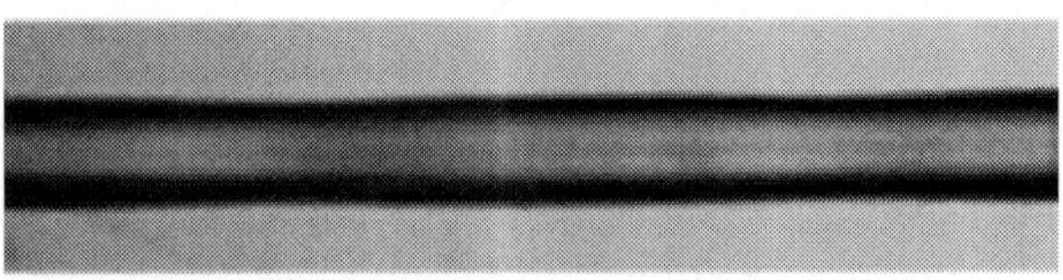

Fig. 5. Micrography of the induced tapers (Myoungwon K., et al., 2002)

In the micro-tapers of fiber, the best option is to reduce the fiber in symetrically shape, as shown in Fig. 6(a). Unfortunately due to factors such as: misalignment, accidental twist or tension, it is not exactly axial, and then an LPFG can be obtained as shown in Fig. 6(b). This mismatch of fiber alignment generates losses as the electromagnetic fields cannot be guided properly. In addition, birefringence is induced because the core is deformed asymmetrically. These features are present in micro-tapers. In the context of the LPFG-EAs, producing micro-tapers with a symmetrical shape is possible because the gratings allow good contrast with shorter length. Also, if handled properly, the procedure of micro-taper can be achieved with minimal levels of losses of RPLA similar to those obtained in other inscription techniques. The asymmetry can be reduced with a good alignment, while for narrow fiber, it is necessary to properly choose the parameters of the technique, in order to achieve an optimal micro-taper.

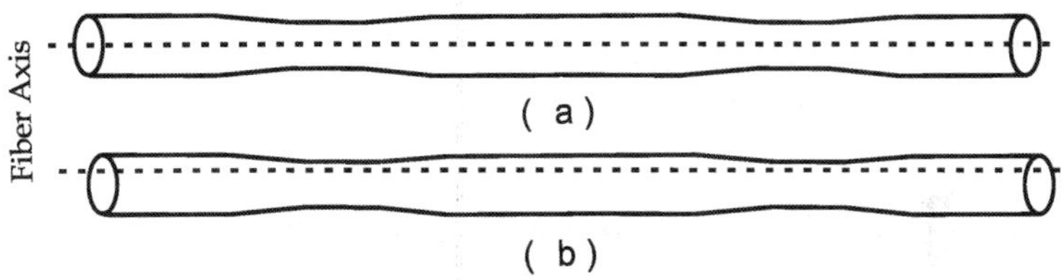

Fig. 6. Two types Microtapers of LPFG: (a) symmetric, (b) asymmetrical

3.5 Periodic modulation by microbends

The Modulation by microbends was presented by In Kag Hwang et al. (1999). In Fig. 7, an unjacketed optical fiber is held straight by two fiber holders, separated by 5.5 cm, which limits the maximum grating length. One of the fiber holders is then displaced (by 100 μm) in the orthogonal direction of the axis fiber so that a lateral stress is induced on the fiber between the two fiber holders. When a local section of the fiber is heated by application of an electric arc, the fiber is slightly deformed owing to the lateral stress, creating a microbend. The amplitude of the microbend is controlled by the duration of the arc and is typically less than 1 μm. Most of the originally applied lateral stress remains in the fiber.

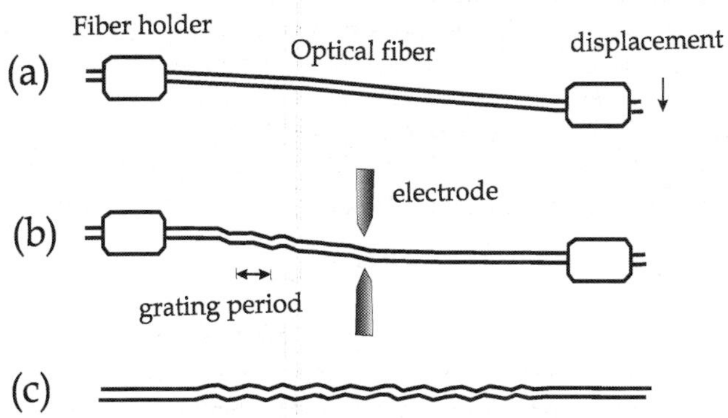

Fig. 7. Fabrication of a LPFG based on periodic microbends by use of an electric arc.(Hwang, I. K., et al. 1999).

Translating the electrodes by the grating period and applying electric arcs, one can create successive microbends without additional displacement of the fiber holder. In this way, a permanent microbend structure is inscribed in an optical fiber, as shown in Fig. 7(c).

Another way to induce microbends is applying many arc discharges at the same point and contrary to the microtapers, when applying the discharge; the glass is compressed axially over the fiber. This will result in a fattening of the area under the electric arc discharge. Fig. 8 shows a long period fiber grating, based in fiber fattening. These LPFGs have very specific characteristics: the number of arc discharges are little, about 8 fattening points per grating and the attenuation bands reach a depth of almost 20 dB. At the same time, the length of the LPFG is very short (about 0.5 cm), but the level of losses that can be produce are at 1 to 2 dB (Mata-Chávez R.I., et al., 2008a).

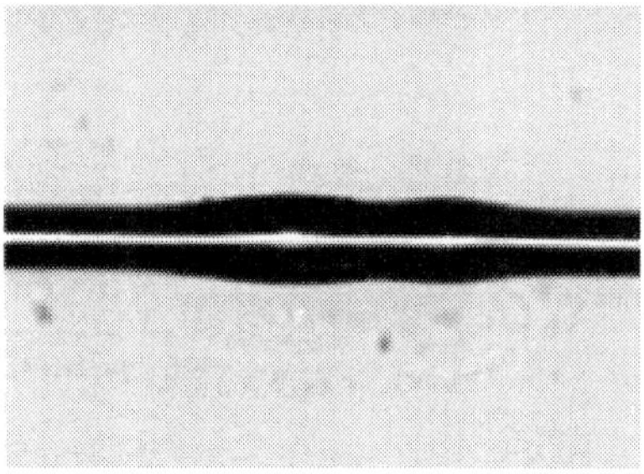

Fig. 8. Photograph of the fattened sections of the optical fiber. (Mata-Chávez R.I., et al., 2008).

3.6 Fabrication of LPFG by the micro-taper method

In these two sections we describe two fabrication methods of LPFGs by electric arc discharges: micro-tapers in which we show the experimental setup as well as data of the parameters used, and the method of fattening.

The gratings were fabricated in commercially avaible standard telecommunication optical fiber (SMF-28) using the point-by-point technique as shown in fig. 9. A bare single mode fiber without its protective coating was placed between the electrodes of a fusion splicer machine (Fitel S-175), while its ends were fixed on two Bare Fiber Adapters. A section of the fiber was fixed on a computer-controlled translation stage (Model NLS4, NEWMARK SYSTEMS INC.) wich a resolution of 1μm.

The fusion splicer is operated in manual mode, and parameters such as arc current, prefusion time, and pushing distance are varied in order to control the grating period. Best results were observed when in the splicer machine the arc current was set up to 11mA, the pre-fusion time was 150 ms, the arc duration time was of 250 ms and the Z-push was set to 1μs, which rise the temperature of the fiber section positioned in between the electrodes to its softening point. The fiber was then moved by the translation stage along a distance equal to the grating period and another arc discharge along with tapering was applied. Care was taken to provide sufficient time between arcs to allow the fiber to get cooled. The periodical tapering continued and the optical transmission of the fiber was monitored during the LPFG

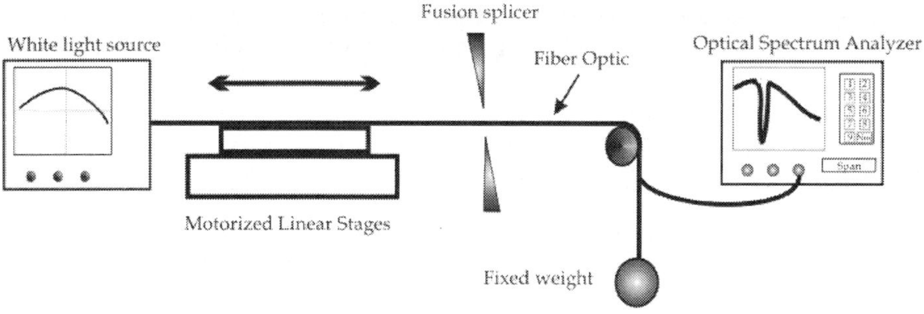

Fig. 9. Experimental setup used for fabrication LPFG by the micro-taper method.

fabrication process in order to obtain the desired spectral attenuation notches. The optical reading setup employed in the experiments were a white light source coupled to one end of the fiber, while the other end was connected to an optical spectrum analyzer, OSA (Yokogawa AQ6370B, Measurement wavelength range 600-1700nm), set to a resolution of 0.1 nm. The transmission spectrum was acquired and processed by a computer to determine the LPFG central wavelength of attenuation and transmittance. The size of the arc limited the minimum grating period Λ, and we got a peak attenuation of 22 dB at 1531 nm after 22 engravings with a grating period of 629μm. The spectral evolution of the LPFG is shown in Fig. 10. illustrating the transmission spectra with various grating period numbers of 10,15,17,20 and 21.

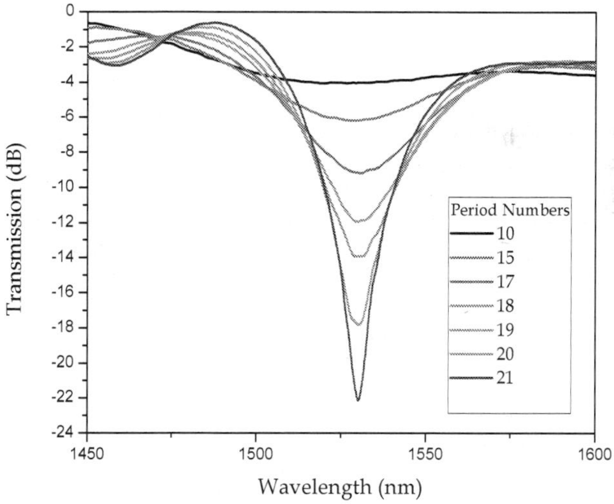

Fig. 10. Evolution of the LPFG spectrum

In fig. 11, it is shown spectra of LPFGs is depicted with different periods, we observed that for shorter periods, the attenuation peak shifts to the left, the LPFGs periods are Λ= 540nm, 560nm, 580nm and 600nm.

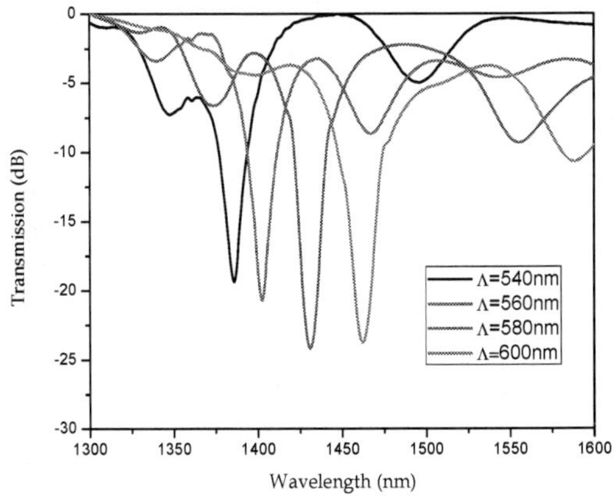

Fig. 11. Normalized transmission spectrum of some LPFGs with different periods.

3.7 Fabrication of LPFG by the fattening method

The fattening method used to fabricate long period fiber gratings consists in gradually enlarging the optical fiber diameter. This is achieved through applying several electrical discharges over an axial region and observing the transmission changes with an optical spectral analyzer. After applying the required discharges by the procedure in one point, the fiber is displaced by the motors of the fusion machine for a distance equal to the grating period (>150 microns) where the process is then repeated to enlarge again the fiber diameter (Mata-Chávez R.I., et. al., 2008a).

This procedure is repeated from two to six times (depending on the grating period) until filtering functions are observed. Fig. 12 depict the transmission loss band achieved by the method in a dispersion shifted fiber.

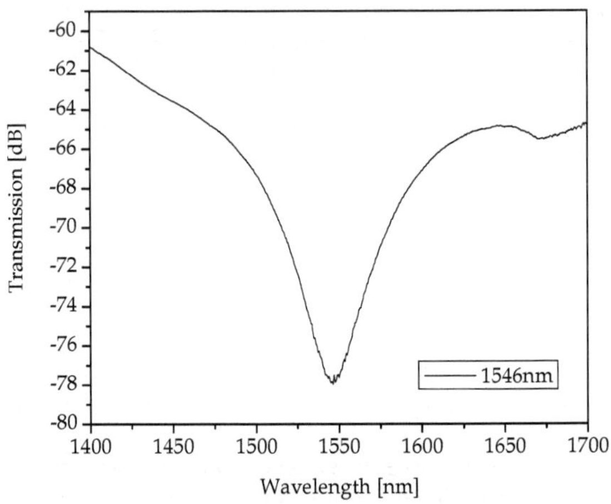

Fig. 12. Loss band of a fattened long period grating near 1550 nm.

The fusion splicer machine has resulted very useful for the fattening method. The standard fusion splice method is used to heat and soften the fiber (known as push delay) followed by a hot push to join the fiber tips (Yablon A. D. 2005). As the procedure is repeated the fiber gets a geometric deformation on its structure for which an energy coupling occurs of the fundamental mode of the fiber core to the radiation or leaky modes in the cladding. Fig. 13 depicts the time delay of the electric discharges applied to enlarge the fiber diameter in a point. Using this approach, wavelength band rejection filters have been obtained, this can be used as optical communication and sensing devices.

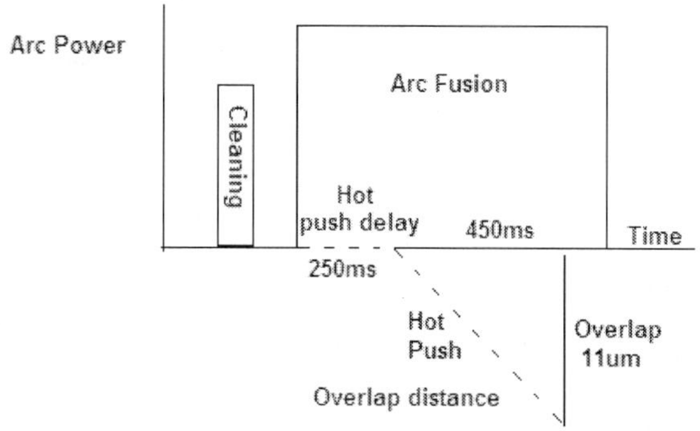

Fig. 13. Time delay of the standard discharged method.

The complete process is done in manual mode after choosing the correspondent optical type program and modifying several variables as arc power, arc duration, pre-fusion time and z-push distance. In fig. 13 can be observed an initial cleaning stage of low power to remove plastic and dust debris. Then, the fusion process continues with an electric arc discharge where the hot push delay heats the fiber to the softening point during 250ms. The hot push stage joins both tips by means of the motors of the machine for a distance of 11 microns. This process is repeated several times until a desired expansion of the diameter is achieved. The following values were initially used to fabricate LPFG in dispersion shifted fiber: Arc power = 85mW, Arc duration = 450ms, hot push delay = 250ms and z push = 11μm.

4. Devices produced by Long Period Fiber Grating

Several LPFG fabrication methods are used with different types of optical fibers like standard single mode, photo-sensible, photonic crystal fiber, specialty optical fibers, etc. Techniques like UV irradiation (Ultraviolet) (Vengsarkar A.M., et. al., 1996a; (Vengsarkar A.M., et al., 1996b; Erdogan, T., 1997), laser CO_2 exposure (Davis D. D., et. al., 1998; Davis D. D., et. al., 1999; Kakarantzas G., et. al., 2001), electric arc discharge (Humbert G., et al. , 2003; Dobb H., et. al., 2004; Rego G. M., et al., 2006b, mechanical microbends (Su L., et al., 2005), etched corrugations (Lin C. Y., et. al., 2001), ion beam implantation (von Bibra M. L., et. al., 2001), femtosecond laser exposure (Grobnic D., et al., 2006; Mihailov S. J., et al., 2006), etc. have been developed for more than a decade. Different devices based on LPFG have been fabricated for applications in optical communications and as sensors.

Sensing applications with LPFG include temperature, strain, bend, torsion, pressure and biochemical sensors because of the sensitivities of resonance wavelength and attenuation amplitude of these measured parameters. Variations of these measurands respond to the fabrication method and optical fiber type that influence on the grating length and optical characteristics of every sensor.

Optical telecommunications find useful applications for LPFG as band rejection filters (Mata-Chávez R.I., et. al., 2008a; Hernández-García J.C., et. al., 2010), gain equalizers for erbium doped fibers (Vengsarkar A.M. et al., 1996a; Vengsarkar A. M., et al., 1996b) , fiber dispersion compensators, core mode blockers, in fiber polarizers (Wang Y.P., et al., 2007), couplers (Liu Y., et. al, 2007), and mode converters (Hill K.O., et al., 1990), because of their bandstop filtering characteristics and selective mode coupling between the fundamental mode and the cladding modes.

5. Sensors of Long Period Fiber Grating

5.1 Curvature sensor

In this section the transmission effects of LPFGs by curves is shown, in Fig. 14 the experimental setup is depicted.

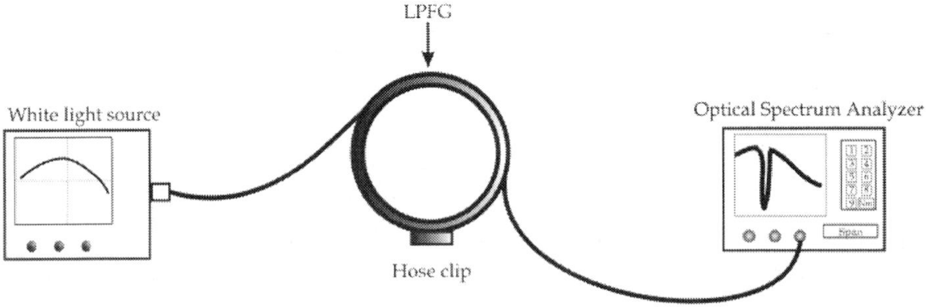

Fig. 14. Experimental setup for sensing diameter of curvature with LPFG by micro-taper modulation

For this experiment, it is necessary to maintain fixed the LPFGs and avoid twisting and LPFGs movements and optical fiber movements, these factors influence the behavior of the attenuation peak of LPFGs. We turned the screw and the hose clip diameter was varied, and then we were measuring the output spectrum of the LPFGs, measurements reduced the diameter of the loop from 13.1 to 12 cm. In Fig. 15, we show the displacement of the attenuation peak to the left as we decrease the diameter of the loop curvature.

In Fig.16, we show the almost linear behavior of the shift in the wavelength of the attenuation peak when the diameter of curvature of the LPFG was decreased. The attenuation peak is shifted by changing the period of the LPFGs. This is due to the change of the diameter of the hose clip. This scheme can serve as a curvature sensor and for tuning a fiber laser.

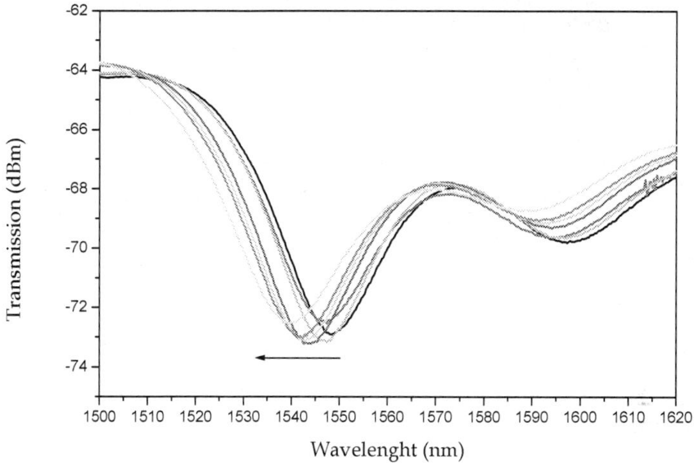

Fig. 15. displacement to left the attenuation peak when decrement the diameter of curvature.

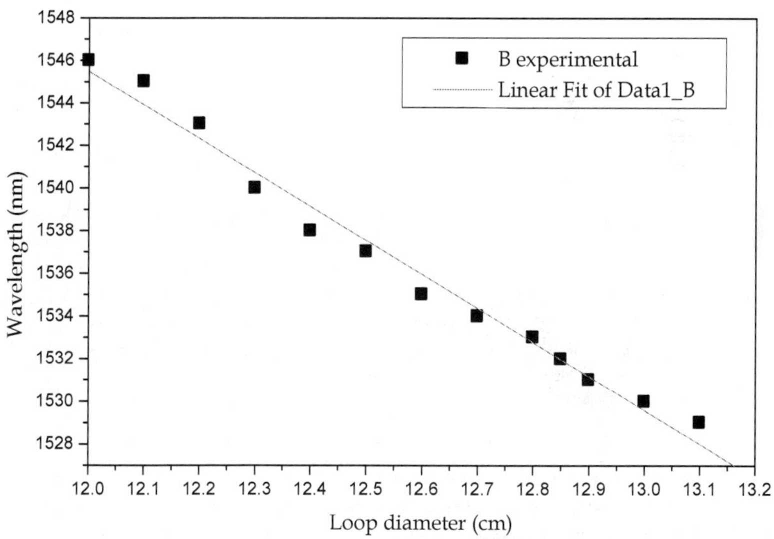

Fig. 16. Linear change of the attenuation peak

5.2 Temperature sensor

Long period gratings are inherently sensitive to changes in temperature of the surrounding environment. This can be observed when a full variation characterizations of the transmission spectra result in shifting of the attenuation band central wavelengths with a possible change in the peak intensity. The chain rule differentiation of the phase matching condition is a common approach to describe the temperature response of the LPFG (Mata-Chávez R.I., et al., 2008b).

$$\frac{d\lambda}{dT} = \frac{d\lambda}{d\left(\delta\left(n_{eff}\right)\right)}\left(\frac{dn_{eff}}{dT} - \frac{dn_{clad}}{dT}\right) + \Lambda\frac{d\lambda}{d\Lambda}\frac{1}{L}\frac{dL}{dT} \tag{5.1}$$

In order to study the temperature effect over the transmission spectrum of the fattened LPG, the experimental setup shown in Fig. 17 was used. The characterization system consists on a fluorescence light source, which is obtained by pumping an Erbium Doped Fiber (EDF) with a laser diode (LD) at 980 nm. A temperature controller in a range of 50-500 °C, and an Optical Spectrum Analyzer (OSA) with a 0.2nm spectral resolution (connected with a dashed line in the drawing).

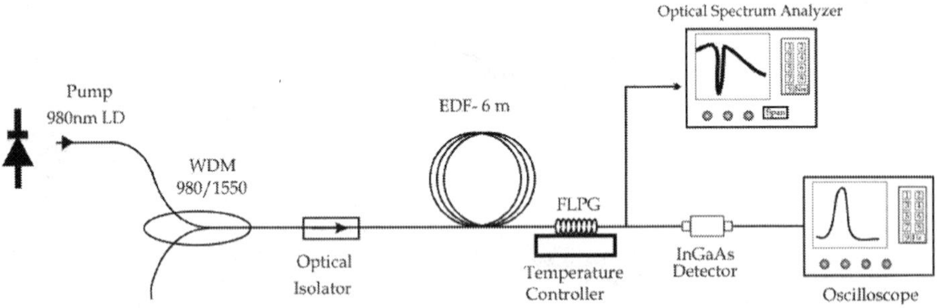

Fig. 17. Experimental setup for LPFG characterizations to study of temperature

When finishing the temperature characterizations, an optical detector and an oscilloscope were placed in the experimental setup, instead of the OSA, to measure the output in volts/100°C for the temperature sensor implementation. The detector is a cheaper and smaller device compared to the OSA that is coupled directly to the optical fiber. It is also a better option to read and process light signals with a smaller electronic acquisition data system.

The experimental setup with the OSA in the log scale shows a wavelength shift response which is linear as temperature increases. Resonance wavelength shifting was observed to longer wavelengths. This behavior is observed almost in all the studies presented in literature for temperature characterizations of LPFG in any type of optical fiber (Rego, G., et. al., 2005). The average sensitivity achieved is about 72 pm/°C. This is a greater value than the obtained in our previous work with DS-LPFG with a resonance wavelength around 1550nm where the fiber was fattened at positions separated by ~25µm (Mata-Chávez R.I., et al., 2008). We think the sensitivity increment is due to the wavelength set inside the loss band and the excited modes within that range as well as the change of the refractive effective index with temperature. As the structure changed, modal group dispersion is also modified, changing the thermo optic coefficients of the core and the ring, contributing to the sensitivity increment.

When switching to the linear scale (mW) , we observed two interesting situations which can be convenient to implement a temperature sensor around 1550 nm. The optical power of the light source was presented as a peak centered at 1560nm and 2.146mW. The LPFG with a resonant wavelength at 1524nm, spliced after the EDF, gradually attenuated the optical power as the loss band of the filter shifted to longer wavelengths with temperature increment. On the other hand, if the LPFG with a resonant wavelength at 1550nm was used, it instantly attenuated the power spectrum. As the loss band shifted to longer wavelengths, the power spectrum increased gradually with temperature (Fig. 18). This measurement approach gives two options to implement a temperature sensor. Attenuating or increasing the output optical power which depends on the position of the central wavelength of the attenuation band and the laser optical power spectrum. Linearity of the photodiode output voltage with temperature is observed in a range of 25-200°C. For temperature over 200-500°C, there is a visible increment of the dV/dT of the photodiode output. The highest sensitivity registered in volts is 0.05mV/°C and the results were repeatable. The insertion losses can be reduced with the improvement of the fabrication method of the DS-LPFG in future works.

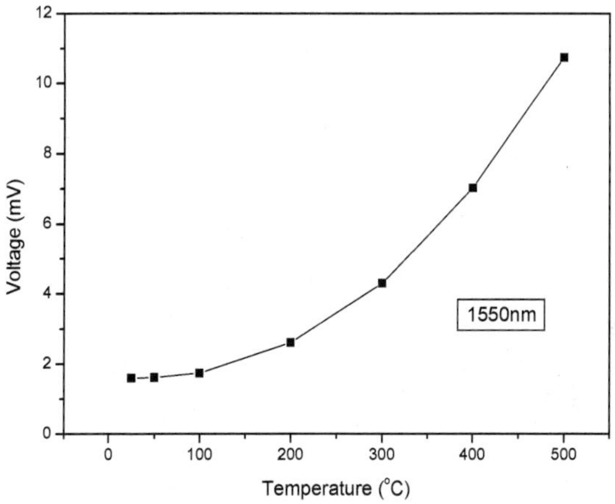

Fig. 18. Power spectrum increment using an LPFG with a central wavelength at 1550nm.

Then, it shows the characterization of an LPFG with micro-taper modulation, for temperature changes measurements. Once the LPFG was fabricated as described in section (3.6), we used a temperature controller (25 °C to 500 °C) for heating the grating. Light from a white light source was introduced to the LPFG.The spectrum for each measurement was obtained at the fiber end (50 °C, 100 °C-500 °C) as shown in the Fig. 19.The attenuation peak is shifted to the right when increasing the temperature from 25 °C to 500 °C. In Fig. 20, it is shown the linear behavior of the attenuation peak against temperature, we obtain those data for each Celsius degree, there was a wavelength variation of 63.15 pm .

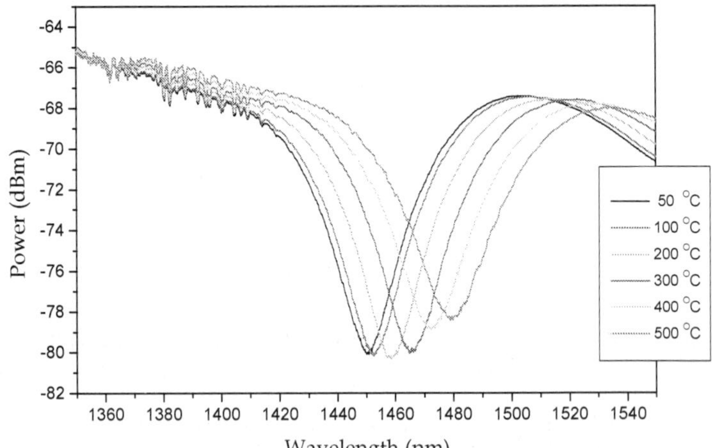

Fig. 19. Displacement of attenuation peak of LPFG to different temperatures.

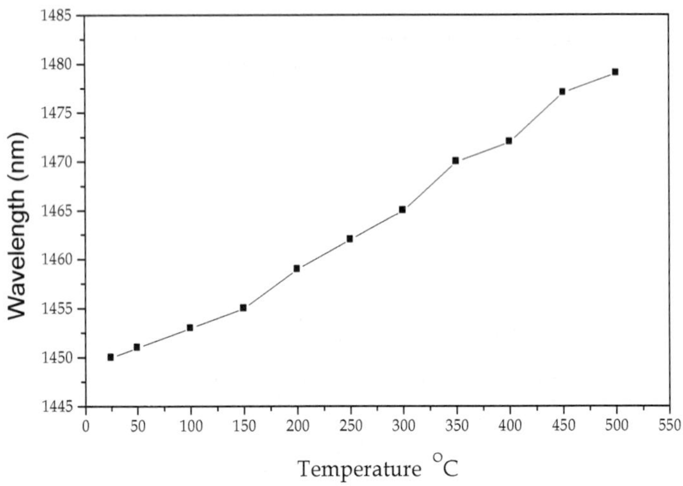

Fig. 20. Linear behavior the attenuation peak against temperature

6. Introduction to modelling individual Long Period Fiber Grating

There are several approaches for LPFG modeling to obtain the spectral properties. Some of them are based on previous studies of LPFG by Vengsarkar (1996), Erdogan, (1997), Kashyap, 1999, Othonos & Kalli, (1999). The coupled-mode theory has been widely applied to model LPFG where an excited mode by the grating can be represented as a linear combination of the eigenmodes of the fiber. The phase matching condition is also considered, where a perturbation of the phase of one mode would match the phase of another at the cladding region, see equation 3.1. A system of two coupled mode equations is given by:

$$\frac{dA_{co}(z)}{dz} = ik_{co-co}A_{co}(z) + i\frac{m_g}{2}k_{co-cl}A_{cl}(z)e^{i\delta c} \quad (6.1)$$

$$\frac{dA_{cl}(z)}{dz} = ik_{cl-co}A_{co}(z)e^{i2\delta c} + i\frac{m_g}{2}k_{cl.cl}A_{cl}(z) \quad (6.2)$$

A_{co} and A_{cl} are the slowly varying amplitudes of the core and cladding modes, k_{co-co}, k_{co-cl} are the coupling coefficients m_g the grating modulation depth and δ is the detuning from the resonant wavelength.

Numerical methods are also applied to the purpose of LPG analysis. It is the case of the Beam Propagation Method (BPM) and transfer-matrix method. These methods are also implemented in commercial software like BPM in RSoft®. With this software we have been able to model individual LPFG with fattening. Grating parameters and structure characteristics are introduced in the software to obtain the excited modes in the grating, the effective index, propagation through the grating, and power losses. A prime model considers an ideal complete structure for the LPFG in a dispersion shifted fiber Fig. 21. This fiber structure has a central core and a ring core which in a second part they are modeled by separate to obtain some characteristics as mode excitation. Different wavelengths were used which correspond to wavelength operation to common laboratory lasers. These are 1.55µm, 1.31µm, and 0.632µm as well as a broad band white light source (400-1700nm).

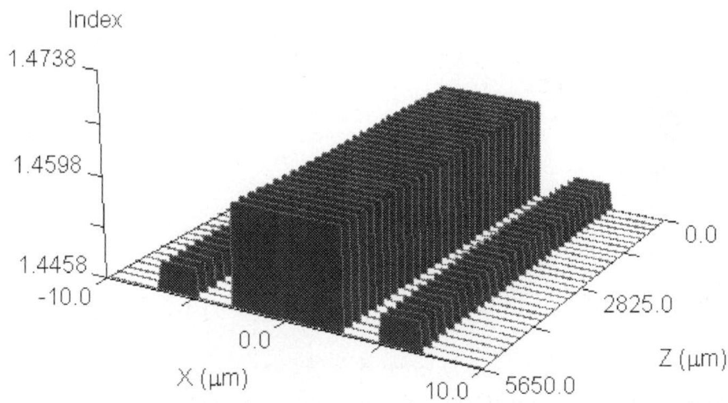

Fig. 21. Idealized step index profile of a dispersion shifted fiber.

The structure radii parameters were: $a = 2.9\mu m$, $b = 5\mu m$, $c = 8\mu m$ and the index refraction values : *n1=1.4598, n2=1.4498, n3=1.4498 and n4=1.4458* (Mata-Chávez R.I., et al., 2008a). To investigate the complete modal characteristics, the fiber structure was divided in two regions as in reference (Mata-Chávez R.I., et. al., 2008b). A central core region and a ring core region, where light is confined and freely propagates along the structure. From this analysis it was encounter that only the LP_{01} and the LP_{11} modes are excited in the optical fiber. A model of the LPFG with fattening is also considered for the modal analysis and such structure is depicted in Fig. 22.

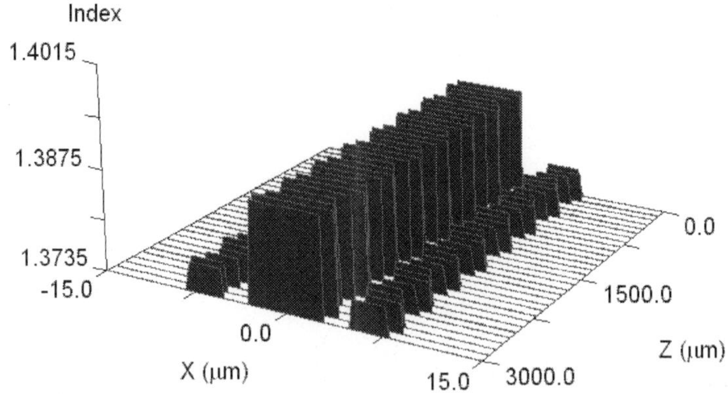

Fig. 22. Idealized step index profile of a dispersion shifted fiber with fattening.

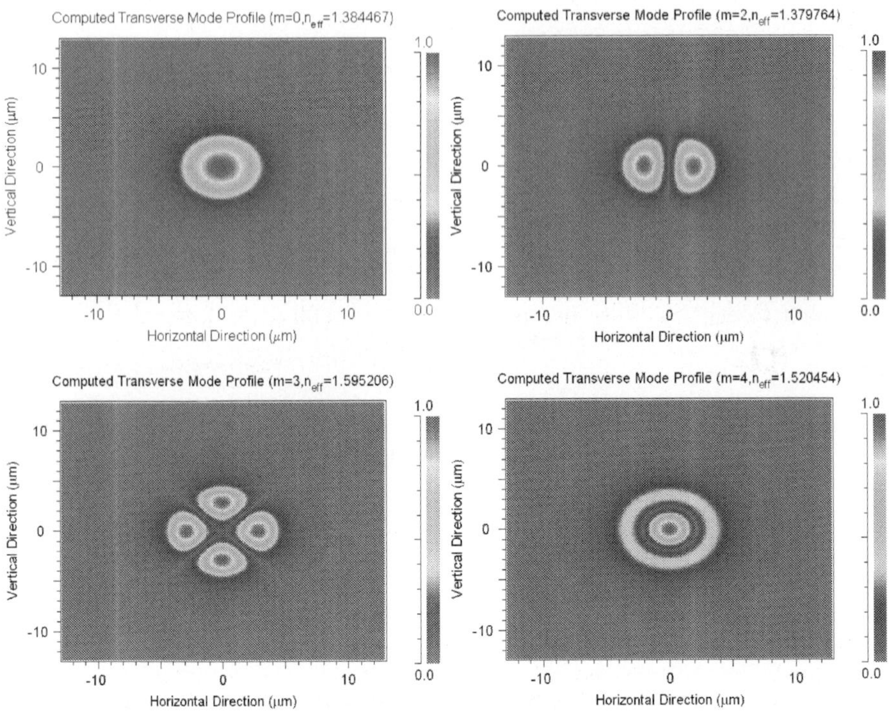

Fig. 23. Excited modes of an LPFG in a dispersion shifted fiber, LP_{01}, LP_{11}, LP_{21}, and LP_{03}.

Propagation of light can also be calculated. Fig. 24 depicts the LPFG with fattening. Several periods were modeled and power losses at the core can be achieved as light propagates. This is because of the energy that couples at the cladding layer due to the phase matching condition. The effective index of refraction can also be calculated as well as the LPFG transmission. This helps to compare the experimental results with the theoretical ones.

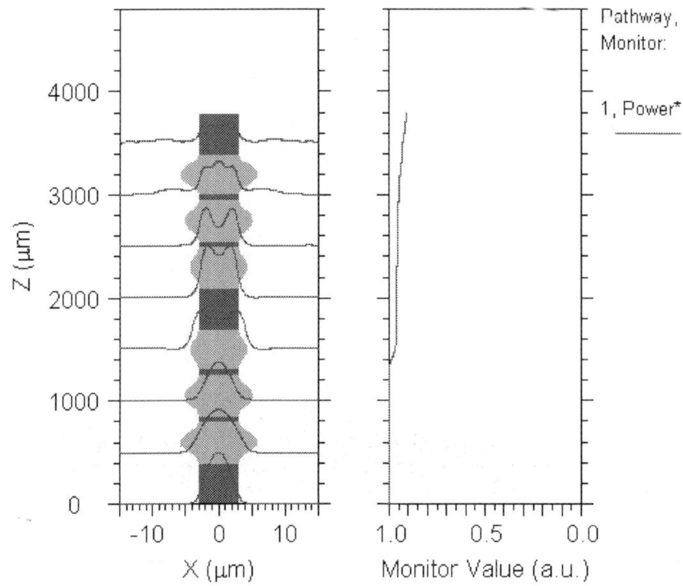

Fig. 24. Power loss at the core in a LPFG with fattening.

7. Conclusion

The main advantage of method of arc dischargers to fabricate long period fiber grating results from the fact that it can be implemented in any commercial fusion splicer, therefore, there are many potentialities of the electric arc technique to produce optical devices in an easy and fast way as sensors, filters and equalizers in any optical fiber. We presented the sensing of curvature and temperature, by means of long period fiber grating that we have fabricated by the method of micro-taper and fattening, and both present good sensitivity as much as 72pm/°C. This type of filter has potential applications in biochemical sensing, environment monitoring, stress, curvature and its principal advantages are that is compact, light and of low cost. One disadvantage of the electric arc method is the unrepeatability of the physical and optical characteristics of the fabricated filters due to the uncontrolled, in most cases, application of the electric arc discharges.

8. Acknowledgment

This work was supported by "Divisi6n de Ingenierías del Campus Irapuato-Salamanca, Universidad de Guanajuato" and DAIP 2010 E20391.

9. References

An H., Ashton B., and Fleming S., (2004), Long-period-grating-assisted optical add-drop filter based on mismatched twin-core photosensitive-cladding fiber, *Opt. Lett.* Vol. 29, pp. 343-345.

Bhatia V. and Vengsarkar A.M., (1996) Optical fiber long-period grating sensors, *Opt. Lett.* Vol 21, pp. 692.

Davis, D. D., Gaylord, T. K., Glytsis, E. N. and Mettler, S. C. (1998) CO2-laser-induced long-period-fibre gratings: Spectral characteristics, cladding modes, and polarisation independence, *Electron. Lett.*, vol. 34, pp. 1416-1417.

Davis, D. D., Gaylord, T. K., Glytsis, E. N. and Mettler, S. C. (1999) Very-high-temperature stable CO2-laser-induced long-period fibre gratings, *Electron. Lett.*, vol. 35, pp. 740-742.

Dobb H., Kalli K., and Webb D. J., (2004) *Electron. Lett.* Vol. 40, pp. 657.

Eggleton B. J., Slusher R. E., Judkins J. B., Stark J. B., and Vengsarkar A. M., (1997) All-optical switching in long-period fiber gratings, *Opt. Lett.* Vol. 22, 883-885.

Erdogan T. (1997) Cladding-mode resonances in short and long-period fiber grating filters, *Journal of the Optical Society of America A,* Vol 14, pp 1760-1773.

Glogle D. 1971 Weakly guiding fiber, *Appl. Opt.* Vol. 10, pp 2252-8.

Grobnic D., Mihailov S. J., Smelser C. W., Becker M., and Rothhardt M. W., (2006) Femtosecond laser fabrication of Bragg gratings in borosilicate ion-exchange waveguides, *IEEE Photonics Technol. Lett.* Vol. 18, pp. 1403-1405

Han Y. -G., Kim S., and Lee S., (2004) Flexibly tunable multichannel filter and bandpass filter based on long-period fiber gratings, *Opt. Express* Vol. 12, pp. 1902-1907.

Hernández-García J.C., Estudillo-Ayala J.M., Rojas-Laguna R., Mata Chávez R.I., Martínez-Ríos A., Gutiérrez J.G., Trejo-Durán M., Vargas-Rodríguez E., Andrade-Lucio J.A. and Alvarado-Méndez E.(2010), Instrumentation and design of a high voltage source to produce optical fiber gratings by the electric arc technique, *Revista Mexicana de Física* Vol 56 pp. 255–2612.

Hill K.O., Malo B., Vineberg K.A., Bilodeau F., Jonson D.C., Skinner I.(1990) Efficient mode conversion in telecommunication fiber using externally written gratings, *Electron. Lett.* Vol. 26, 1270-1272.

Hill, K.O., B. Malo, F. Bilodeau, D.C. Johnson, and J. Albert. (1993). Bragg gratings fabricated in monomode photosensitive optical fiber by UV exposure through a phase mask, *Applied Physics Letters*. Vol. 62, pp. 1035-1037.

Hill, K.O., and G. Meltz. (1997). Fiber Bragg Grating Technology Fundamentals and Overview. *Journal of Lightwave Technology*. Vol. 15, No. 8, pp. 1263-1276.

Hill, K.O. (2000). Photosensitivity in Optical Fiber Waveguides: From Discovery to Commercialization. *IEEE Journal on Selected Topics in Quantum Electronics*. Vol. 6, No. 6, pp. 1186-1189.

Humbert G., Malki A., Fevrier S., Roy P., and Pagnoux D., (2003) *Electron. Lett.* Vol. 39, pp. 349

Hwang, I. K., Yun, S. H. and Kim, B. Y. (1999) Long-period fiber gratings based on periodic microbends, Optics Letters, Vol.24 pp 1263-1265.

Kakarantzas G., Dimmick T. E., Birks T. A., Le Roux R., and Russell P. S. J., (2001), *Opt. Lett.* Vol. 26, pp. 1137.

Kashyap R., (1999) fiber Bragg grating, Academic Press.

Kersey, A. D., Davis M A, Patrick Heather J, LeBlanc M, Koo K P, Askins C G, Putnam M A and Friebele E J (1997) Fiber grating sensors *J. Lightwave Technol. Vol.* 15 pp. 1442-63.

Mata-Chávez R.I. , Martinez-Rios A., Torres-Gomez, I., Alvarez-Chavez, J.A., Selvas-Aguilar, R., Estudillo-Ayala, J. M (2008) Wavelength band-rejection filters based on optical fiber fattening by fusion splicing, *Optics & Laser Technology*, Vol. 40 pp. 671-675.

Mata-Chávez R.I. Martínez- Rios. A. Torres-Gómez I., Selvas-Aguilar R., Estudillo-Ayala J.M. (2008) Mach-Zehnder All- Fiber interferometer Using Two in-Series Fattened fiber grating , *Optical Review*, Vol. 15 pp 1-6.

Meltz, G., W. W. Morey, and W. H. Glenn. (1989). Formation of Bragg Gratings in Optical Fibers by a Transverse Holographic Method. *Optics Letters*, Vol. 14, pp. 823-825.

Mihailov S. J., Grobnic D., Huimin D., Smelser C. W., and Jes B., (2006) Femtosecond IR laser fabrication of Bragg gratings in photonic crystal fibers and tapers, *IEEE Photonics Technol. Lett.* Vol. 18, pp.1837-1839.

Myoungwon K., Dongwook L., Bum I. H. and Haeyang C. (2002). Performance Characteristics of Long-Period Fiber-Gratings Made from Periodic Tapers Induced by Electric-arc Discharge, *Journal of the Korean Physical Society*, Vol. 40 pp. 369-373.

Lin C. Y., Chern G. W., and Wang L. A., (2001), Periodical corrugated structure for forming sampled fibre Bragg grating and long-period fiber grating with tunable coupling strength, *J. Lightwave Technol.*, Vol. 19 pp.1212-1220.

Liu Y., Chiang K.S., Rao Y.J. , Ran Z.L., and Zhu T. (2007) Light coupling between two parallel CO_2-laser written long-period fiber gratings, *Optics Express*, Vol. 15, pp. 17645-17651.

Othonos, A. (1997) Fiber Bragg Gratings. *Reviews of Science Instruments*, Vol. 68, No. 12, pp. 4309-4341.

Rego G., (2006) Chapter 3: Mechanisms of formation of arc-induced long-period fibre gratings" in *Arc-induced long-period fibre gratings: fabrication and their applications in optical communications and sensing*, University of Porto.

Rego G, Marques P.V.S., Santos J.L., Salgado H.M. (2005) Arc-induced long period gratings. *Fiber Integrated Opt*, Vol. 24 pp. 245–59

Rego G. M., Santos J. L., and Salgado H. M., (2006) Opt. Commun. Vol. 262, pp.152

Rego G. (2010) Fiber optic devices produced by arc discharges, *J.Opt.* Vol. 12 pp. 113002.

Riant I. (2002) UV-photoinduce fibre grating for gain equalisation, *Optical Fiber Technology* , Vol. 8, pp. 171-194.

StephenW.J. and Ralph P. T. (2003) Optical fibre long-period grating sensors: characteristics and application Meas. Sci. Technol. Vol 14 pp R49-R61.

Su L., Chiang K. S., and Lu C., (2005), Microbend-induced mode coupling in a graded-index multimode fiber *Appl. Opt.* Vol. 44, pp. 7394-7402.

Udd, E. (1991) Fiber Optic Sensors: An Introduction for Scientists and Engineers. *New York:* ed., *John Wiley and Sons.*

Vengsarkar A. M., Lemaire P. J., Judkins J. B., Bhatia V., Erdogan T., and Sipe J. E., (1996) Long-period fiber gratings as band-rejection filters, *J. Lightwave Technol.*, vol. 14, pp. 58–65.

Vengsarkar A. M., Pedrazzini, J. R., Judkins, J. B., Lemaire, P. J. and Bergamo, N. S. (1996), *Optics Letters.*, Vol 21, pp 336.

Von Bibra M. L., Roberts A., and Canning J., (2001) Fabrication of long-period fiber gratings by use of focused ion-beam irradiation *Opt. Lett. Vol. 26, pp.765*

Wang Y.P., Xiao L.M., Wang D. N., and Jin W., (2007) In-fiber polarizer based on a long-period fiber grating written on photonic crystal fiber *Opt. Lett.,* vol. 32, pp. 1035-1037

Yablon A. D. (2005) Optical Fiber Fusion Splicing. Springer. Chap 8. 226

13

Fibre Sensing System Based on Long-Period Gratings for Monitoring Aqueous Environments

Catarina Silva[1], João M. P. Coelho[1], Paulo Caldas[2,3,4] and Pedro Jorge[3]
[1]*Departamento de Física, Faculdade de Ciências da Universidade de Lisboa*
[2]*Departamento de Física, Faculdade de Ciências da Universidade do Porto*
[3]*Unidade de Optoelectrónica e Sistemas Electrónicos, INESC Porto*
[4]*Escola Superior de Tecnologia e Gestão de Viana do Castelo*
Portugal

1. Introduction

The estuaries and coastal environments are areas of special biodiversity, being extremely sensitive to changes in its boundary conditions, particularly the properties of water, being salinity one of the most important parameters.

Thus, environmental monitoring is a crucial factor in the ecosystems health. Actually, methods for measuring salinity are mainly based on the mobility of ions in water (Zhao et al., 2003) changing the refractive index. However, these methods are inherently electrical and can be affected by any kind of electromagnetic interference. An alternative is the use of sensors based on optical technology, in particular, long-period fiber gratings (LPG) which are intrinsically sensitive to the external refractive index. These gratings devices have the advantage of being immune to electromagnetic interference and having small size, non-intrusive nature, flexible geometry, high speed of response and excellent salt water corrosion resistance.

Long-period gratings can be considered a special class of fiber Bragg gratings in which the period of the index modulation is such that it satisfies a phase matching condition between the fundamental core mode and a forward propagating cladding mode of an optical fiber. The periodic modulation of the refractive index in the fiber core typically has a period in the region from 100 μm to 1000 μm and a length of a few cm (James & Tatam, 2003) which can be induced in the fiber using different process. Irradiation by ultraviolet (UV) source, electric-arc discharges, irradiation with CO_2 source and application of mechanical pressure with adequate periodicity are the most common process.

The first LPG successfully inscribed in an optical fiber was described in 1996 by Vengsarkar (Vengsarkar et al., 1996a, 1996b) and was used as a band-rejection filter. In the same year Bhatia (Bhatia & Vengsarkar, 1996) presented the first LPG device acting as a sensor. Since then, many applications for long-period gratings have been achieved in the last years and have been concentrated on the development of long-period grating-based fiber devices for use in optical communications and fiber optic sensor systems. In optical communication systems, LPGs are applied as gain equalizing filters, wavelength selective devices, bandpass and band rejection filters and wavelength tunable add/drop multiplexers (Vengsarkar et al., 1996a, 1996b; Dong et al., 2009). In the field of sensing systems, they can be applied as structural bend sensors, temperature sensors, axial strain sensors, refractive index sensors and biochemical optical sensors (Patrick et al., 1998a; James & Tatam, 2002; Bhatia, 1999; Falciai et al., 2001).

The high sensitivity of LPGs to the refractive index of the surrounding environment allows them to be used as salinity sensors in aqueous media. However, in their production the removal of the buffer layer makes them a fragile structure when applied in real conditions, usually contaminated with algae and other organic materials. Taking this into consideration, this chapter addresses the potential application of long-period gratings technology in environmental monitoring. It will analyze the response of different gratings to refractive index changes, in order to evaluate and increase their sensitivity, as well as its thermal response. In this chapter, we present the issues related with LPG implementation in a real field application and review the use of LPG based sensing heads for environmental monitoring in field conditions. As a particular example, a prototype of a salinity sensor developed by the authors will be described for in situ research and capable of sustaining harsh operational conditions.

The chapter starts by presenting the main theoretical concepts related with LPGs, focusing on their fundamental aspects and fabrication methods and also its application as sensor elements (section 1). In section 2, the main configurations of refractive index sensors based on LPGs are reported. The problematic of encapsulating the sensing element and make it a robust sensor device will be addressed, in section 3. Mechanical requirements are described and the filtration issues analysed. The easiness of device cleaning is another crucial aspect that will be taken into consideration. Finally, in section 4, a prototype of a refractive index sensor for in situ research will be proposed and characterised and the potential of the sensing system as a remote refractometric device is analysed.

1.1 Fundamentals of long period gratings

As mentioned before, in 1996 (Vengsarkar et al., 1996a) it was reported the principle of operation of an LPG as a device based on the coupling between the fundamental mode propagating in the fiber core with the forward propagating cladding modes. This principle is illustrated in Fig. 1. The cladding modes are quickly attenuated and this results in series of loss bands in the transmission spectrum of the grating. Each of these attenuation bands corresponds to coupling to a discrete cladding mode.

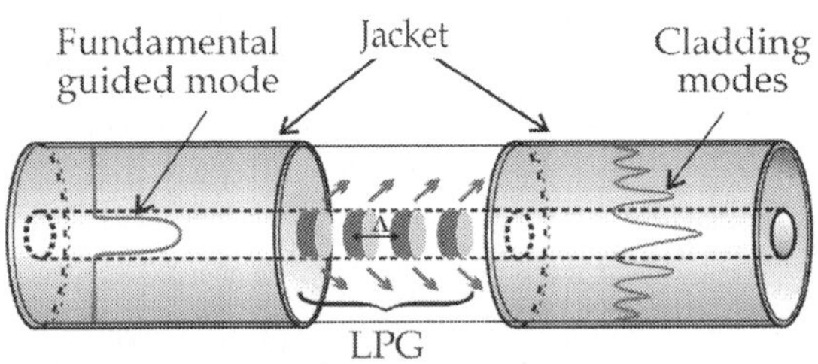

Fig. 1. Coupling of fundamental guided mode to a cladding mode in a long period grating.

The resonant wavelength of light coupling into a particular cladding mode is given by the phase matching condition (Erdogan, 1997a, 1997b):

$$\lambda_{res}^{m} = \left(n_{eff,co} - n_{eff,cl}^{m}\right)\Lambda \tag{1}$$

where Λ is the grating period, $n_{eff,co}$ and $n_{eff,cl}^{m}$ are the effective refractive indexes of the core and mth-cladding modes, respectively.

The light coupled into cladding-guided modes is, most typically, absorbed in the fiber buffer or radiates away from the fiber. Thus, this type of fiber grating acts as a wavelength selective transmission filter. The wavelength transmission response of a typical LPG is shown in Fig. 2, exhibiting the characteristic attenuation bands.

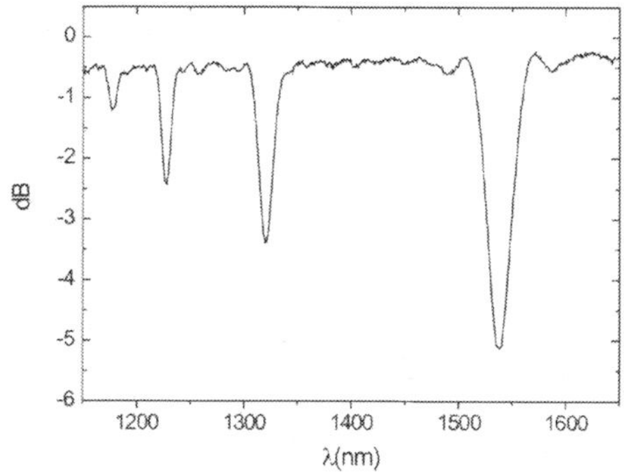

Fig. 2. Transmission response of a typical LPG.

The light transmission through the core follows a sinusoidal function of the core refractive index modulation for the wavelengths in the resonance (Zanlorensi et al., 2009)

$$T = \cos(DL/2) \tag{2}$$

where L is the grating length and D is a coupling coefficient proportional to the core index modulation. The bandwidth of the resonance dips depends on both the coupling coefficient and the difference between the core and cladding indexes:

$$\Delta\lambda_{FWHM} = \frac{\lambda_m^2}{\left(n_{eff,co} - n_{eff,cl}^{m}\right)}\sqrt{\frac{4D}{\pi L}} \tag{3}$$

External changes in parameters like refractive index, temperature or strain can affect the terms in Eqs. (1) and (3), consequently shifting the attenuation dips and altering their bandwidths (Zanlorensi et al., 2009).

1.2 Fabrication techniques

LPGs are created by inducing a periodic refractive index modulation in the core of the optical fiber with period lengths of typically several hundred micrometers. This can be made by permanent modification of the refractive index of the optical fiber core or by the physical deformation of the fibre. Since the first demonstration of LPGs by writing the grating with UV laser light through an amplitude mask in 1996 (Vengsarkar et al., 1996a), several methods have been developed to fabricate theses structures. The most widely used method is the exposition of the photosensitive optical fibres to UV light through an amplitude mask, phase mask or by interferometry (Bhatia & Vengsarkar, 1996; 1998a, James & Tatam, 2002; Erdogan, 1997a). In this case, the mechanism responsible for the refractive index change is related with the formation of Germania-associated defects and the gratings can be written using light with wavelengths between 193 and 355 nm (Bhatia & Vengsarkar, 1996; Zanlorensi et al., 2009). The amplitude mask technique consists of simply imprinting the mask pattern onto the core of a photosensitive optical fiber placed in contact with or in close proximity immediately behind the mask. The fiber exposure to the fringe pattern induces a refractive index modulation in the core with the same periodicity as that of the mask (Nikogosyan, 2006). The mask might be fabricated in chrome-plated silica or metal foil. The disadvantages of this method are that the grating period and length are restricted by the dimensions of the amplitude mask which can also be easily damaged if exposed to UV light whose intensity exceeds its damage threshold. A further useful technique was presented by Hill et al. in 1990 (Hill et al., 1990) in the form of point-by-point periodic grating manufacture. In this technique, gratings are written period after period, and a single pulse of UV light from an excimer laser passes through a mask containing a slit making the refractive index of the core in the irradiated fiber section locally increase. However, all process that depends on UV illumination requires first that the fiber must be receptive to this electromagnetic radiation. In other words, the fiber must be made photosensitive before writing the grating. This can be achieved by doping the fiber silica core with impurity atoms (such as germanium, boron) or by hydrogen loading (Kashyap et al., 2010, Lemaire et al., 1993).

Kondo et al. (Kondo et al., 1999) found that focused irradiation of femtosecond laser pulses causes a permanent refractive index increase in various glasses, and they demonstrated a novel technique for the fabrication of long-period gratings by focused irradiation of femtosecond laser pulses. They induced a refractive index periodic structure in the core of a single-mode fiber by using light that is not absorbed by the core and cladding glasses and a polymer coating. In contrast with the UV-laser induced, LPGs fabricated using infrared (IR) femtosecond lasers have high resistance to thermal decay despite having no special stabilization.

Another possibility to create these structures is by applying the ion implantation technique. This technique induces a refractive index increase in silica-based glass, which is mainly due to the densification of the silica glass induced by atomic collisions (Fujimaki et al., 2000). Disadvantages of LPG manufacture by using this technique includes the necessity of specialized equipment (such as a tandem accelerator), as well as the fact that the cladding material average refractive index is also raised in the process, which leads to large background losses in the transmission spectrum. The main advantage of ion implantation is that it is possible to inscribe LPGs in almost any type of silica-based fiber without prior photo-sensitization.

Gratings inscription was also achieved through the use of thermal effects by exposure to near-IR carbon-dioxide (CO_2) laser light (Davis et al., 1998; Kim et al., 2002) or electric arc discharges (Rego et al., 2001; Falate et al., 2006). Both techniques allow the inscription of the gratings without the need hydrogen-loaded or Ge-doped core fibres and are more flexible than the UV technique since it is not required the fabrication of a specific amplitude mask for each desired resonance wavelength (Humbert & Malki, 2003). The first results of the CO_2 laser application in the fabrication of LPG on fiber optics were published in 1998 (Davis et al., 1998; Akiyama et al., 1998; Liu et al., 2011). In this case, the residual stress relaxation and glass densification are usually considered to be the main mechanisms responsible by the formation of LPGs and have demonstrated very high thermal stability. In this technique, the 10.6 µm wavelength laser radiation is focused to a spot by means of spherical lenses and LPGs are written by periodical local heating. The grating period is determined through the movement of a computer controlled translation stage. During the writing process, a broadband source and an optical spectrum analyzer are used to measure the transmission spectrum.

The other process used to write LPGs by periodical local heating of the fiber is based on electric arc-discharges. In this technique the refractive index modulation results from deformation of fiber due to intense heating caused by an electrical discharge transversal to the fiber. This writing method is very flexible, easy to implement and does not require the use of special fibres or high cost equipment, such as UV or IR lasers. These features make this technique particularly suitable for the low cost fabrication of LPGs. An example of a typical electric arc-discharge setup for LPGs manufacture is schematized in Fig. 3. The fabrication process consists of positioning an uncoated fiber between the electrodes of a commercial splicing machine. A small weight is suspended in one of the fibre's extremities to keep a constant longitudinal tension. The other extremity of the fiber is mounted on a computer controlled translation stage. An electrical arc discharge is applied with the fusion splicer, using an electric current of 8.5–10.0 mA for 0.5–2.0 s. A translation stage moves the fibre, after each discharge, by a distance that represents the grating period (typically 300–700 µm). The refractive index change is created by the fast heating and cooling discharge cycles. Due to the asymmetric nature of electric arcs, the result of each consecutive discharge is

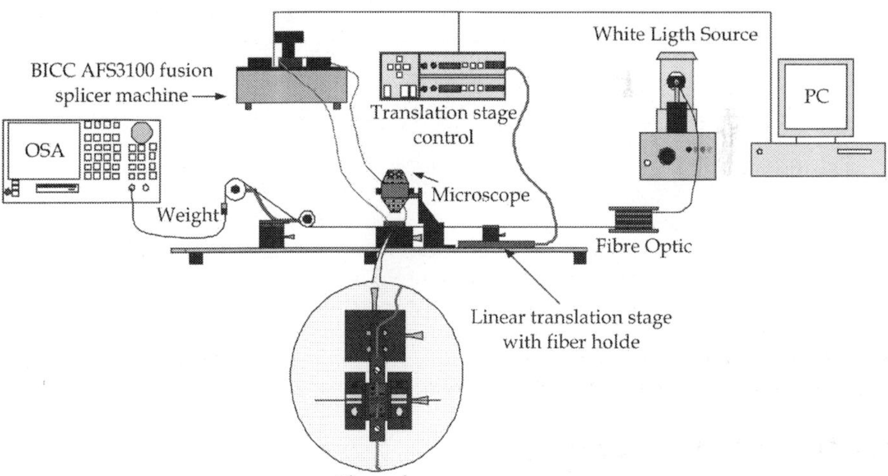

Fig. 3. Schematic diagram of LPFG fabrication system based on electric arc-discharge.

never exactly the same. Therefore, in order to obtain an LPG with a given characteristics, the evolution of the transmitted spectrum is monitored by an optical spectrum analyser and a White Light Source (WLS). When the measured spectrum shows suitable characteristics to the intended application, the process is terminated. Usually 15-100 points are necessary to form the grating.

Since LPGs have periods of the order of hundreds of micrometers they can also be induced mechanically through periodic microbends in optical fiber (Rego et al., 2003; Yokouchi et al., 2005; Cardenas-Sevilla et al., 2009) Several configurations have been developed to produce mechanical gratings but they share the same objective, that is, to induce periodic stress and/or microbending in a single mode fibre (Rego et al., 2003). This technique, as the arc discharge technique, can potentially be applied to any kind of fibre and is also simple, flexible and low-cost. Another important advantage over other techniques is the fact that these gratings can be produced without removing the fibre coating and therefore the fibre keeps its mechanical integrity.

1.3 Sensing characteristics

As mentioned before, LPGs are influenced by alterations that can occur in the environment at the grating location, such as changes in the refractive index or temperature, resulting in measurable changes in the LPG transmission spectrum. This is the basis of the sensing LPGs (Bhatia, 1999). In this section, it is presented an analysis of the implementation of LPGs as an element sensor of physical parameters such as temperature, mechanical deformations and refractive index of the environment.

1.3.1 Temperature sensitivity

The temperature sensitivity of the LPGs may be understood by differentiating the phase matching condition equation (Humbert & Malki, 2002) Since the contribution of the waveguide property $(d\Lambda / dT)$ to the resonance wavelength shift with temperature variation is negligible compared to that of the material property (dn / dT) in LPGs the derivative of equation (1) can be set as

$$\frac{d\lambda_{res}^{m}}{dT} = \frac{d\lambda}{dT}\left(n_{eff,co} - n_{eff,cl}^{m}\right) + \Lambda\left(\frac{n_{eff,co}}{dT} - \frac{n_{eff,cl}^{m}}{dT}\right) \approx \Lambda\left(\frac{n_{eff,co}}{dT} - \frac{n_{eff,cl}^{m}}{dT}\right) \quad (4)$$

Fig. 4 shows the wavelength response and the spectral shift of the LPG (induced in Corning SMF-28 fiber) when subjected to temperature variation from 0 °C (related to a 20 °C ambient temperature). As expected, the resonance peaks shift to longer wavelengths with increasing temperature. The heating causes an increase in the difference between effective refractive indexes of the modes due to the thermo-optic coefficient of the core being larger than that of the cladding, causing a shift in wavelength to higher values. As a coarse approximation, a linear fit to the wavelength shift data yields a slope of ~0.085nm/°C.

1.3.2 Refractive index measurement

An unusual feature of LPGs is their high sensitivity to the refractive index changes in the grating location. This alteration will also change the matching condition expressed by

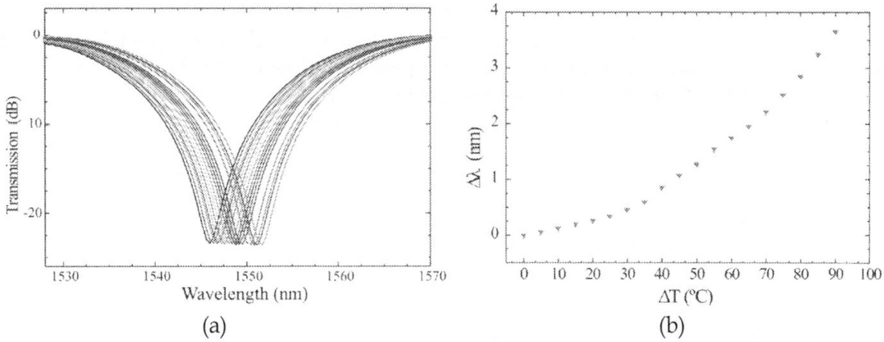

Fig. 4. (a) Transmission spectrum and (b) shift in the wavelengths of the LPG with respect to temperature variation.

equation (1) and will lead to wavelength shifts of the resonance dips in the LPG transmission spectrum. Such displacement occurs due to the dependence of the effective refractive index of the cladding modes on the refractive index of the surrounding medium. The sensibility to refractive index changes increases with the reduction of the difference between the two effective refractive indexes.

Fig.5 shows an example of the evolution of the LPG transmission spectrum when the refractive index of the external medium changes. For calibration purposes, the LPGs were immersed in liquids with different refractive indexes in the range [1.33, 1.43] at room temperature in both situations. It can be seen that as the refractive index of the external medium increases the attenuation peak shifts towards shorter wavelengths. This is an expected displacement because an increase in the effective refractive index of the cladding according to the resonance condition leads to lower resonant wavelengths (Rego et al., 2006).

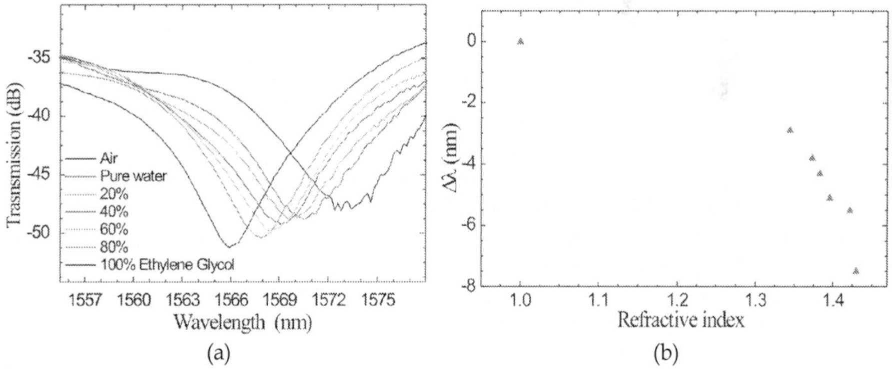

Fig. 5. (a) Transmission spectrum of long period gratings induced in Corning SMF-28, for different refractive index and (b) corresponding variation in the shift of the resonant wavelength.

The sensitivity of LPGs to refractive index and temperature is influenced by the composition of the optical fiber. Rego (Rego et al., 2006) investigated the sensitivity of arc-induced LPGs in pure-silica-core fibres to refractive index changes of the external medium. These results show a two to three times increase of the sensitivity when compared to gratings written in standard fibres. This results from doping the cladding with fluorine which lowers its refractive index and, therefore, reduces the refractive index between the cladding and its surroundings, increasing the grating sensitivity.

As the effective refractive index of the cladding modes is strongly dependent on the cladding diameter, the easiest way to increase the sensitivity of the LPGs is by etching the fiber cladding in the region of the grating to a diameter such that the evanescent field of the mode displays a higher overlap with the immediate surrounding environment. This concept is clearly illustrated in the phase-matching condition (equation 1), where for a fixed period the change in effective refractive index for a higher order m of the cladding mode results in a new resonant wavelength - the effective refractive index of the core remains unaffected. With this configuration, the value of the effective refractive index of the waveguide is directly affected by the refractive index of the medium, and this effect is stronger for higher order modes.

2. Refractometers based on long period gratings

In recent years the development of devices for measuring the refractive index has been the subject of great interest for applications in industrial processes and quality control, biomedical analysis, and environmental monitoring. In this context, the LPGs offer advantages, being extremely sensitive to changes in refractive index of the medium where they are immersed. However, although the use of such devices is advantageous, its use as a sensing element is limited by the relatively large rejection bandwidth (~10 nm) which does not provide a good resolution. In this context, several schemes for refractive index sensing based on LPGs have already been proposed (James et al., 2002; Han et al., 2001).

The simple configuration, consisting of a light source, an LPG immersed in a fluid and an optical spectrum analyzer has been studied by several authors (Falciai et al., 2001; Zhu et al. 2003). Lee et al. (Lee et al., 1997) demonstrated a new analysis method to determine the spectral shift of an LPG as a function of refractive index of the medium. The change of this parameter with the wavelength of the LPG attenuation band was presented by Patrick (Patrick et al., 1998a). As previously mentioned, the sensitivity of the sensor to external refractive index change can be enhanced by decreasing its cladding diameter by chemical etching of the optical fiber (James & Tatam, 2003). The reduction in cladding diameter changes the effective index of the cladding modes, resulting in an increase in the central wavelengths of the attenuation bands.

Gwandu et al. proposed a compact scheme for simultaneous measurements of temperature and refractive index in a medium, using two LPGs recorded in a double cladding hydrogenated fiber (Gwandu et al., 1999). The effect of the thickness of layers deposited on the grating surface regarding its response to refractive index changes has also been investigated (e.g. James et al., 2002). A thin film of organic material was deposited on an LPG through the Langmuir-Blodgett technique. The central wavelength and the minimum transmission band attenuation of the LPG depend on the thickness and refractive index layer of the deposited material, even when the refractive index is higher than the cladding. It was observed that for films with refractive index higher than that of the cladding, the wavelength and amplitude attenuation of the band exhibits a very high sensitivity when the film thickness is of the order of a few hundred nanometers. If the material deposited has a refractive index below that of the cladding, the sensitivity to film thickness is considerably reduced.

LPG pairs have shown high resolution to refractive index measurement compared to the use of a single LPG (Han et al., 2001; Zhang et al., 2005). The advantage of using a LPGP sensor relies upon its principle of operation, where the coupled core and cladding modes from the first LPG combine again at the second matched LPG to form interference fringes. The core and cladding paths constitute the arms of a Mach-Zehnder interferometer (MZI) in the transmission mode (Mishra et al., 2005; Murphy et al., 2007) or a Michelson interferometer in the reflection mode (Swart, 2004). A schematic diagram of these two configurations is presented in Fig. 6.

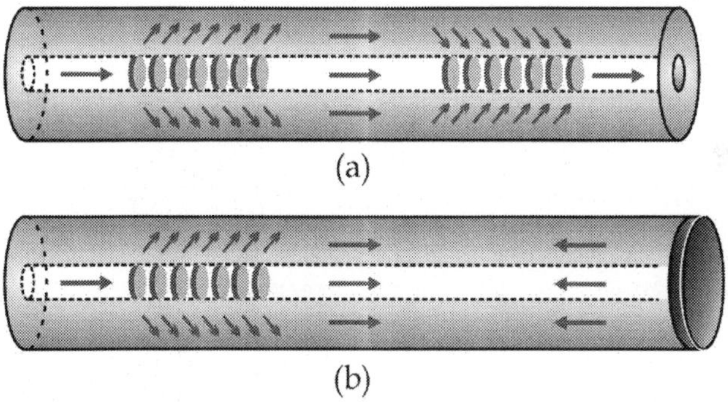

Fig. 6. Schematic diagrams of an LPG-based: (a) Mach-Zehnder interferometer; (b) Michelson interferometer.

The Mach-Zehnder interferometer operate in transmission mode and is obtained by placing in series two identical LPGs inscribed adjacent to one another such that an interferometer cavity of a certain length is formed between the two gratings. When light from a broadband source passes through the first LPG, the part that propagates in the fundamental mode is coupled to the forward propagating cladding modes. At the second LPG, the cladding mode is coupled back into the core. Due the difference between the effective refractive indexes of the core and cladding modes, the light coupled into the core by the second LPG is phase shifted with respect to the light that propagated through the core, giving rise to the interference fringes (Lee & Nishii, 1998). The core and cladding paths constitute the arms of the Mach-Zehnder interferometer. The frequency of the fringes depends on the separation between the two gratings, while the phase is influenced by measured-induced changes in the effective index of the cladding modes (Murphy et al., 2007). The fine interference fringes in the transmission spectrum of the pair of LPGs can enhance the resolution and the sensitivity of measurement. The narrower bandwidth of the fringe facilitates greater resolution in the measurement of the wavelength than is possible with conventional LPGs (Han et al., 2001).

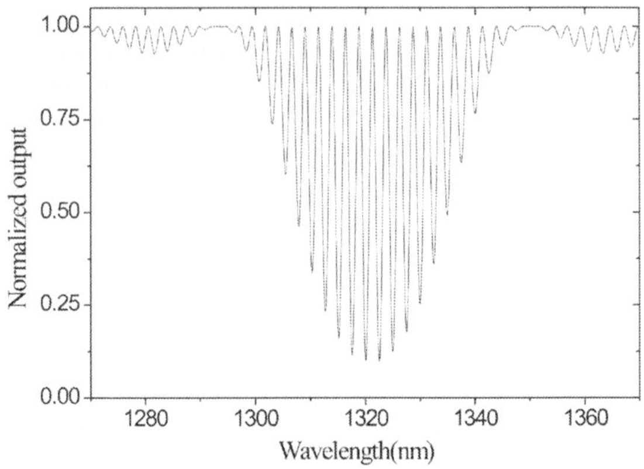

Fig. 7. Transmission spectrum of an optical fiber having two cascaded LPGs as a Mach-Zehnder interferometer.

The in fibre-Michelson interferometer was proposed and studied by Swart (Swart, 2004). This configuration is based on a single LPG and on a fibre end mirror, as show in Fig. 6.b). In this case, the light coupled to cladding modes is reflected by the mirror and finally recoupled back to fibre core. This is a very simple structure and offers several advantages over the Mach-Zehnder configuration such as increased sensitivity, operation in reflection mode and lower manufacture complexity (it avoids the need to fabricate identical LPGs). This reflection mechanism and compact size make this kind of sensing-structure attractive as biomedical probes.

The misaligned splicing point (MSP) formed by a commercial fusion splicer may replace the first LPG in an LPG pair to create an MZI (Dong et al., 2005). In this, the coupling of light from the core to the cladding modes is made trough the MSP, as illustrated in Fig. 8.

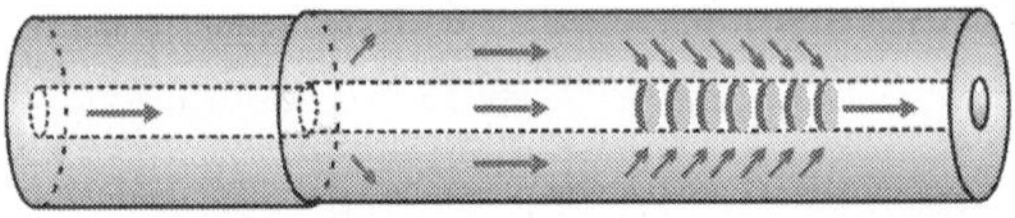

Fig. 8. Misalignment splicing point is formed to interact with the single-mode fibre-based LPG

Ding et al (Ding et al., 2005) reported the fabrication and characterization of a sensor with high sensitivity to refractive index changes based on a couple of LPGs in which it was reduced the fibre section between them (FTS-LPGP-fibre-taper seeded LPG pair). Its principle of operation is schematised in Fig. 9. This configuration allowed an increase by a factor of five in the sensitivity to refractive index change on the external environment when compared to the conventional configuration (Frazão et al., 2006).

Allsop et al. (Allsop et al., 2006), demonstrated that a Tapered-LPG configuration has a high spectral sensitivity to refractive index change in liquids. This configuration is based on the writing of a LPG centred on a taper.

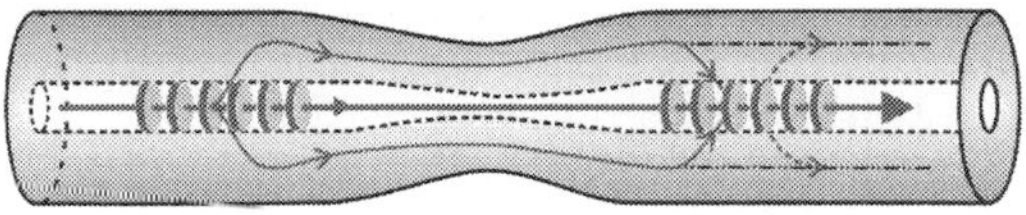

Fig. 9. Schematic diagram of a FTS-LPGP (fiber-taper seeded LPG pair).

Recent advances have been focused in the research of LPGs induced in photonic crystal fibres (PCFs) which consist of a pure-silica core, and a microstructured air-silica cladding (Yu et al., 2006). Because the microstructured cladding consists of air holes, a PCF's cladding index shows strong wavelength dependence. This makes such fibres exhibiting some properties that make them advantageous in face to conventional fibres (e.g. by reducing or even eliminating the cross-section problems).

3. Sensing head

Several authors have experimentally demonstrated different configurations for the application of LPGs as sensor elements. However, when the LPGs are to be deployed in harsh environments, their fragile nature makes the existing technology not satisfactory. One of the harshest environments where fiber-based sensors have a huge potential are natural aqueous media. The existence of biological activity is added to other debris that can not only alter the sensing capability but severely damage the entire sensing head.

This section addresses the several issues that must be considered in the project of a system to host a LPG in order to reduce the potential cross-sensitivity problems and obtain a high performance when applied in real conditions in natural aqueous environments. The issues related to the mechanical requirements in order to achieve a high performance sensor device under the considered conditions will be analysed. Strain constrains, supporting conditions and other operational prescriptions will be addressed. Also, the characteristics of a system developed by the authors will be described in detail.

3.1 Mechanical requirements

The mechanical part of the sensor head should be projected in order not only to bear the sensor element, but also give support to filtering elements (latter addressed) and allow efficient cleaning procedures, part of the unavoidable regular maintenance.

3.1.1 Structure

First considerations should regard the material used to produce the supporting structure. Being in direct contact with the aqueous medium this choice is of prime importance for any project. Resistance to corrosion and structural rigidity are the main characteristics to be taken in consideration, while price and machinability are also usually considered. Two main groups can be considered: metals (including alloys) and polymeric materials.

When metals and alloys interact with an environment, their behaviour is highly dependent on both the thermodynamics and kinetics of the interaction. The chemistry, solubility, structure, thickness, cohesion, adhesion, continuity, and mechanical properties of the reaction compounds contribute to the formation of either passivation or corrosion, two concurrent processes. In fact, passivation can be defined as the spontaneous formation of a hard non-reactive surface film inhibiting further corrosion. This layer is usually an oxide or nitride a few nanometres thick.

Contrary to metals, polymers are not affected by corrosion processes although susceptible to light (mainly, ultra violet) and potentially less structurally stable. Nevertheless, the research in engineering polymers is highly advanced and application-dedicated materials can be developed. So, depending on overall functional requirements, a wide range of materials can be chosen. The structure and components of the sensing head may use carbon steels, low alloy steels, aluminium alloys, copper and copper alloys, stainless steels, superaustenitic stainless steels, nickel-base alloys, cobalt-base alloys, ultra-high strength steels and stainless steels, titanium and titanium alloys, polymers, or composites. Newer materials such as intermetallic compounds, nano-structured alloys, and amorphous alloys are promising materials, but their collective database of respective properties is currently extremely limited (Shifler, 2005).

Choosing the correct material in the subject of this paper depends on very specific data regarding the application. If, as mentioned, the environment is a key point to consider, the kind of measurements to perform, duration of testing, production numbers (mass or small production) and available budget easily superpose. At the end, nowadays there are a sufficient number of materials that can perform well in harsh environments.

3.1.2 Tensing element design

Since the LPGs are intrinsically sensitive to bending curvature, which consequently influences its sensitivity to the various parameters, for field deployment in practical applications, it becomes necessary to design a support system that can avoid these cross sensitivities. A particular solution tested by the authors was a spring-based system designed in order to apply a constant force to the grating regardless of the external handling conditions. In general, the design of a compression spring must consider the following factors:

1. space for housing the spring;
2. rigidity requirements;
3. guidelines for the values of displacement and maximum force regarding the values imposed by the project.

Fig.10 shows a helical spring subjected to a compressive deformation, y, where d is the diameter of wire and p the helix pitch (Shigley, 1986).

The elongation (or contraction) of a spring is determined by the torsion deformation of all the active coils of the spring. Then, the deformation of the spring coils with N_a active coils under the action of a force P is given by

$$y = \frac{8PD_m^3N_a}{d^4G} \quad (5)$$

being G the material's elasticity modulus and D_m the mean diameter or the coils.

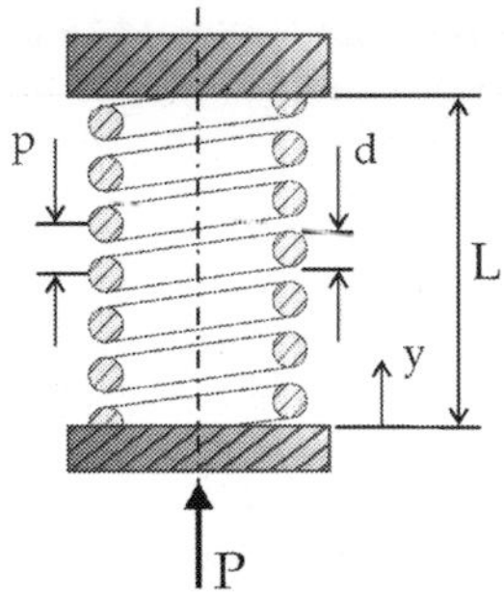

Fig. 10. Helical string subjected to compression.

The helical spring constant or stiffness, κ, which reflects the amount of deformation of a spring, y, when subjected to a load P, can be determined by,

$$k = \frac{P}{y} = \frac{d^4 G}{8 D_m^3 N_a} \tag{6}$$

Representing the length of the unloaded-spring by L_0, the helix step, p, is

$$p = \frac{L_0}{N_a + 1} \tag{7}$$

and considering the spring ends as having close spirals, the total number of coils, N_t, is given by

$$N_t = N_a + 2 \tag{8}$$

3.2 Filtering system

Essential in the sensing head performance is the encapsulation process in view of the specifications of a filtration system. In real conditions, the system hosting the LPGs requires an adequate filtering in order to prevent the deposition of organic materials and the accumulation of particles around the sensor element, hence interfering in the measured value. The filtration process to protect the sensing head is essentially based on the passage of fluid through a porous or permeable material. The encapsulation proposed for the sensor involves two levels of filtration: one capable of retaining larger particles and another for the mud usually existing in natural aqueous environments.

For filtrating larger particles it is common the use of lower pore sizes in order to prevent the deposition of waste or algae. The filtration of mud is achieved through the process of membrane separation. This consists in running a liquid through a membrane, which acts as a selective barrier and when subjected to pressure, forced or natural, leads to the passage of fluid and smaller compounds through their pores.

The membranes can be manufactured with organic or inorganic material. Inorganic membranes exhibit higher mechanical strength, thermal and chemical robustness and have a longer active life. The inorganic membranes can be ceramic (zirconium oxide, aluminium or titanium), metal and graphite. The efficiency of the membrane depends on several factors, such as the properties of membrane and fluid and operating conditions (pressure, temperature, turbulence) (Queiroz et al., 2001).

Their morphology is mainly related with the density and porosity (number and dimension of the pores) of the layers constituting the membrane. The porosity, thickness, pore size and permeability are important factors in the choice of the membrane. These characteristics depend on the material of which it is made off and on the manufacturing technique. The higher the porosity the lower the resistance to the fluid flow through the membrane is. In general, the classification of the membranes is based on the structure and mechanism of separation. Regarding the structure of a membrane, it may be porous or nonporous, and symmetrical or asymmetrical, as is illustrated in Fig. 11. The anisotropic membranes are characterized by presenting a very thin, more packed, layer supported by a porous layer. In the case of porous membranes, the pore size and its distribution will determine which particles or molecules will be retained by the membrane (Queiroz et al., 2001).

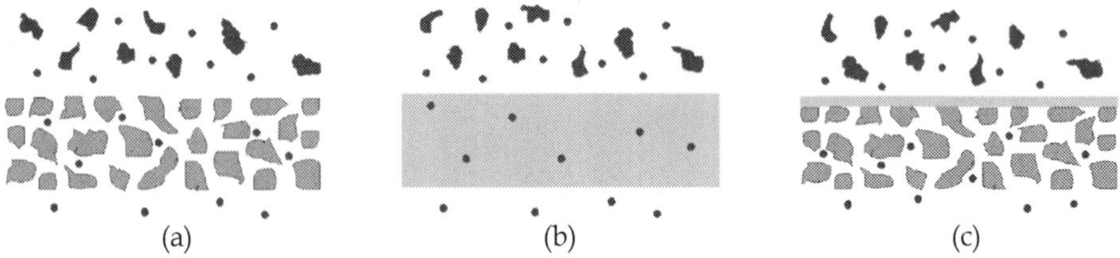

Fig. 11. Schematic representation of (a) symmetric porous and (b) non-porous membranes and (c) asymmetric porous membranes.

Among the existing membrane separation processes, we only address the micro-and ultra-filtration processes. These processes differ among themselves mainly in the pore size and pressure required for operation.

Microfiltration can be defined as a process for solid-liquid separation using membranes with pore diameters ranging from 0.1 µm to 2.0 µm, and operating at pressures between 0.2 bar and 2.0 bar depending on the product to be filtered. The filter is a micro-porous membrane and this technique allows you to retain suspended particles, bacteria and some ions (Queiroz et al., 2001).

The ultrafiltration process operates at pressures between 2.0 bar and 7.0 bar and uses membranes with micro-pores of 1 nm to 100 nm, which allows the passage of small molecules (water, salts) retaining all suspended solids including those causing turbidity and even micro-organisms. The flow through the membrane is controlled by the pressure exerted on the liquid. The main limitation of this process is the membrane blockage, which occurs when particles are deposited on the filter inside leading to a decrease in its filtration capacity. The choice of a membrane with pores smaller than the diameter of the particle filter and subjected to a pressure gradient, prevents this phenomenon. Operating conditions such as flow, velocity and pressure concentrated in the pores are also factors that influence the degree of clogging of the membranes. The application of a pressure gradient in the membrane in contact with the solution promotes a flow of fluid through the membrane.

The main advantages of membrane separation processes are:

1. high selectivity by using one process or the coupling of processes;
2. generally, these processes operate at room temperature, it is not necessary to change the fluid temperature to promote the phase separation;
3. low power consumption: promoting the separation that occurs without phase change;
4. easy to combine with other processes.

The main disadvantage is its high cost. However, costs associated with the application of this technology have been reduced considerably, since the membranes are produced on a larger scale, there is a greater number of firms in the market and, when properly applied, the membranes have a longer service life and flow permeated most stable and high.

3.3 Prototype

The guidelines explained in the previous text were applied in the design of a mechanical device with the functions of supporting and housing a sensor for measuring salinity and temperature, which is based on a LPG inscribed in standard silica optical fibre, with 125μm in diameter. The developed system is intended to operate at half depth in the Ria de Aveiro, inserted in the optical cable, protecting the optical fibre (Silva et al., 2010).

3.3.1 Mechanical device

Being a prototype and not expected to operate for longer periods of time the main driver for the choice of the material was cost an easy of manufacture. For this reason, common aluminium (1060 Alloy) was chosen.

A cylindrical shape was chosen in order to allow the insertion of an outer mud filtrating membrane and a protective mechanical grid. The sensor element is surrounded by enough space to promote the flow of water allowing updating salinity measurements. The optical cable is fixed at both ends by mechanically blocking the fiber at the head's support, leaving about 11 cm of exposed optical fiber. The device allows the adjustment of the tension in the fiber through a spring. Fig. 12 presents a picture of the prototype.

3.3.2 String dimensioning

The coil spring used in the device was scaled by imposing a force (compression) of 20 g over a diameter of 40 mm. The spring, with a free length of 130 mm, was made of stainless steel which has an elasticity modulus of 80 GPa and a wire diameter of 1mm. The number of active coils necessary can be calculated from the equation (5). So, the number of active coils is 14. Assuming squared and ground ends, the total number of the coils is 16 (equation 8).

3.3.3 Filtering system

The selection of the membranes used in this work was done taking into account the environment where they are to be inserted and through well-defined parameters such as its material, thickness and pore size. Three types of membranes were analysed with application in aqueous environments, two with the Nylon and the other with fibreglass mesh. Its specifications are shown in Table 1.

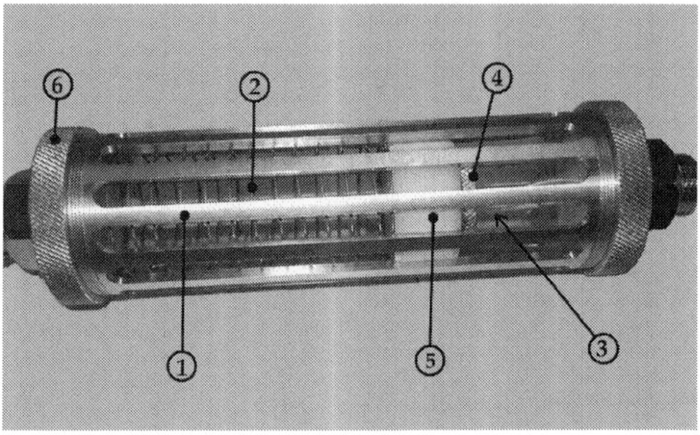

1. Cylinder (membrane support)
2. Spring (control fiber strain)
3. Semi-blocker of optical fibre
4. Plunger
5. Ring small bolt
6. Ring bolt-fixing

Fig. 12. Picture of the mechanical device developed to support the element sensor.

	Mesh thickness [μm]	Pore dimension [μm]	Maximum temperature [K] (dry environment)	Maximum temperature [K] (wet environment)
Fiberglass	0.52	2.7	-	-
Nylon 53	60	53	388	
Nylon 100	78	100	373	

Table 1. Specifications of the analyzed membranes (Cole-Parmer instrument Co).

The analysis to establish the most suitable of these filters was made *in situ*. Supports were projected based on a net applied surrounding two metallic plates, and over which the membrane to test is fixed (Fig. 13). All metallic parts were designed in aluminium. After placing the membranes in the medium, they were immersed in the Ria de Aveiro, Portugal, where they remained for one month. After this period of time, the membranes were analysed regarding their state of blockage using an optical microscope (Meiji). Inside the vessel, another membrane was placed and a similar analysis was made in order to have an idea of the amount of impurities that have passed the outer filter.

In Fig. 14, it can be observed the state presented by the membranes after the specified period under real conditions. It was found that the fiber glass membrane did not supported the conditions of the environment concerned, and may well be conclude that it is not suitable for the application in question. It can also be seen that for the same period, the nylon membrane with pore size of 53 μm shows greater accumulation of dirt compared to the nylon membrane pore with a size of 100 μm, as expected. Fig. 15 shows the microscope images for both nylon membranes. The nylon membrane with larger pore sizes was fully saturated, while the other still had some open pores. However, because of its condition, it appears that the replacement, in real operation, has to be done in less than a month. The nylon membrane with pore size of 53 μm placed inside the support was also relatively clean.

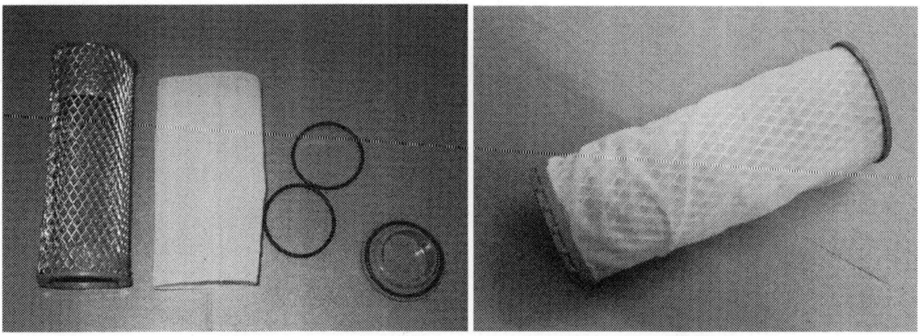

(a) (b)

Fig. 13. (a) Components and (b) membrane mounted on its support for testing.

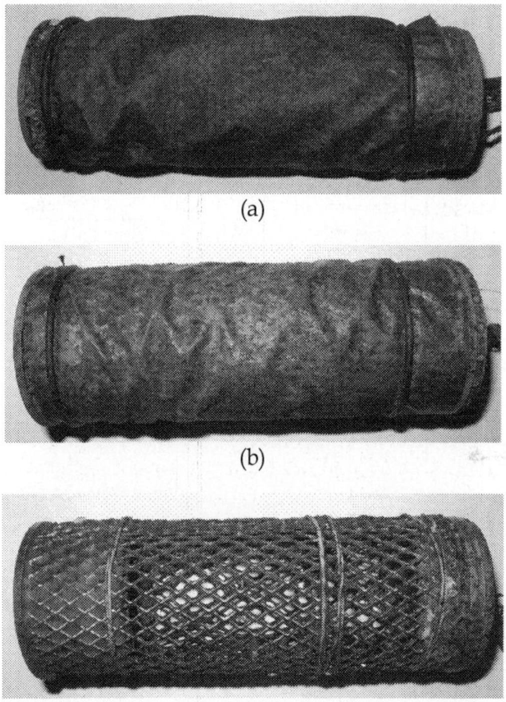

(a)

(b)

(c)

Fig. 14. Condition of the membranes after immersed in the waters of the Ria de Aveiro, Portugal, for one month: (a) Nylon (100 µm), (b) Nylon (53 µm) and (c) Fiberglass.

According to the results presented we can conclude that for the application in question nylon fiber with 53 µm pore size is the one that best meets the requirements. However, its replacement should occur in relatively short periods (at least every two weeks).

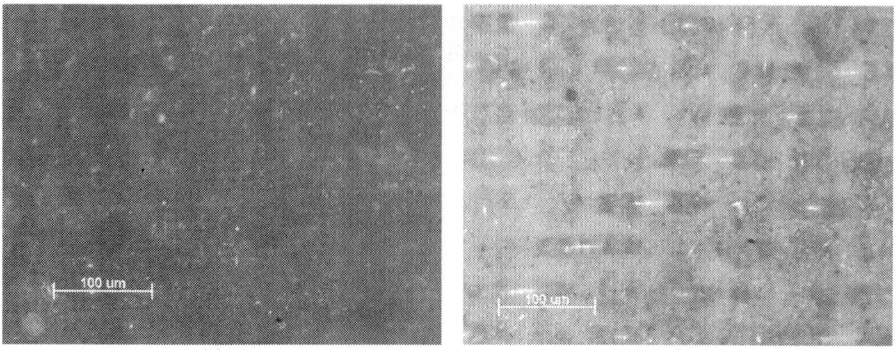

(a) (b)

Fig. 15. Images obtained from optical microscopy of the Nylon membranes, (a) 100 μm and (b) 53 μm, after immersed in the waters of the Ria de Aveiro, Portugal, for one month.

These tests also allowed analysing the performance of an aluminium structure subjected to the aqueous harsh conditions of Ria de Aveiro. In spite of its susceptibility to corrosion, the structure supported relatively well the one month "ordeal". However, the existing screw threads were clearly damaged, showing a tendency to weld the parts in contact.

4. Sensing head performance

The performance of the previously defined system is analysed in this section. In order to evaluate the influence of the structure on the measurement of environmental parameters, a comparative study is made when the LPG is fixed with and without the structure. The sensitivity of the sensing head to the refractive index change of the external medium was characterized. The thermal response of the sensing structure was also studied. The LPG used in this work were written by electric arc technique (Rego et al., 2005) in a standard telecommunication fibre (Corning SMF-28), with a period of ~ 540 μm and a length of ≈ 2 cm. To further enhance its sensitivity for refractive index changes, the fibre diameter was reduced. This was made by chemical attack with aqueous solution of fluoridic acid (HF 40%). After the etching process, the fibre was thoroughly washed with deionised water, dried, and observed under the microscope. The diameter of the fibre after suffering the chemical etch was 118.8 μm ± 1 μm, considering an initial diameter of 125 μm.

4.1 Refractive index measurement

To check the suitability of the system developed for salinity refractive index measurements, the sensitivity of the LPG based sensing head was characterized towards changes in the refractive index of the external medium. For calibration purposes, in both situations the LPGs were immersed in liquids with different refractive indexes at room temperature. The liquid was prepared by using a mixture of water and different ethylene glycol concentrations were used, resulting in samples with refractive indexes in the range [1.33, 1.43]. The refractive index of the different samples was calibrated using an Abbe refractometer. A broadband light source (Photonetics - Fiber White), with emission centred on 1550 nm illuminated the sensing element. From the transmitted broadband spectrum, the refractive index can be measured from the spectral shift of the LPG attenuation band. This shift was monitored using an optical-spectrum analyzer (ANDO AQ 6330). Fig. 16 shows the grating responses to the surrounding solutions with increasing concentration of ethylene glycol, with the device either free (without curvature) or integrated in the structure.

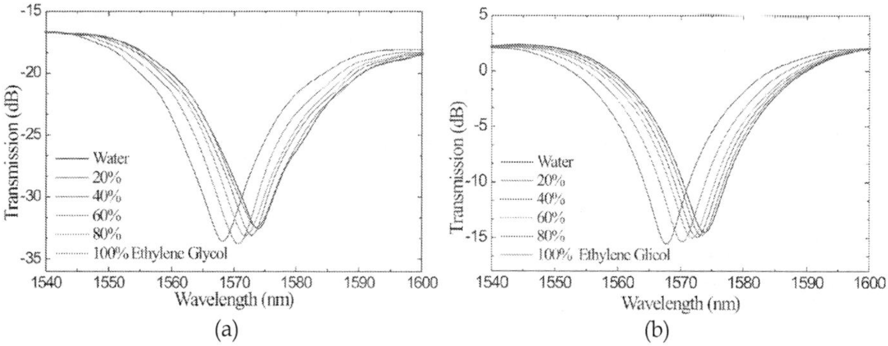

Fig. 16. Evolution of the transmission spectra of the LPG, (a) free and (b) mounted in the sensing head, as the refractive index is changed by increasing the concentration of ethylene glycol.

It can be seen that the resonance wavelength moves towards the shorter wavelength region as the concentration of ethylene glycol is increased. This can be understood through an increase of the effective refractive index of the cladding, which, according to the resonance condition, leads to lower resonant wavelengths (Lee et al., 1997). The graph presented in Fig. 17 derives directly from the analysis of the central peak of each attenuation band presented in Fig. 15. The shift is determined with reference to the spectral position considering air as the surrounding medium.

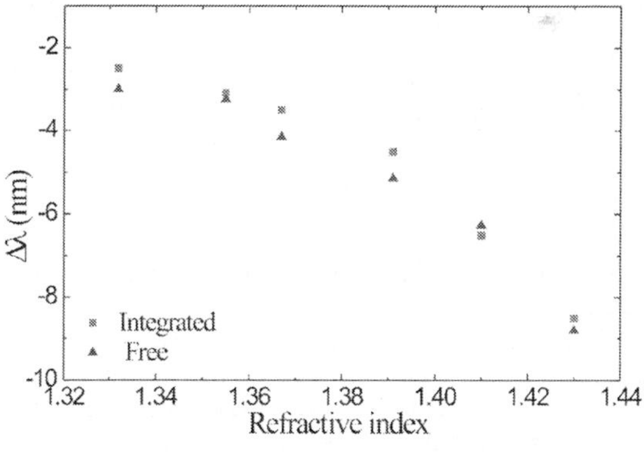

Fig. 17. Resonance wavelength shift of the LPG measured for different refractive indexes of the external medium in the situations where the device is free and integrated in the structure.

It can be seen that for both situations there is a similar response of the grating to refractive index changes. The slope of the obtained calibration curves gives an idea about the sensitivity of the refractometer. For a refractive index around 1.40, the sensitivity of the LPG is ≈ -67nm/RIU and -57nm/RIU for the cases where the device is mounted and free, respectively (RIU: refractive index unit). This difference is an indication that the pre-strain of the LPG in the structure induces a small enhancement of its refractive index sensitivity. As expected, it was also observed that the sensitivity increases substantially when the refractive index of the external medium is close to the cladding refractive index.

4.2 Salinity measurement

In order to analyse the influence of the salt concentration, the spectral responses of the LPG mounted in the structure was measured when it was immersed in water with different salt concentrations, resulting into salinity levels between 1% and 19%. The spectral response evolution of the LPG when mounted in the structure for different salinity levels is shown in Fig. 18, for salt water solutions with different weight concentrations ranging from 0% to 19%. It can be observed that the resonance wavelength moves towards the shorter wavelength region (shift of ≈ 2 nm for the salt concentration range tested). This is due to the fact that, with increasing concentration of salt, there is an increase in the refractive index of the solution. Also, it turns out that the concentration of the salt solution does not substantially affect the amplitude of the LPG loss band (variation less than 0.07 dB). The resonance wavelength shift with the salt concentration, from where it can be derived a sensitivity of ≈ -0.10 nm/ % Salt.

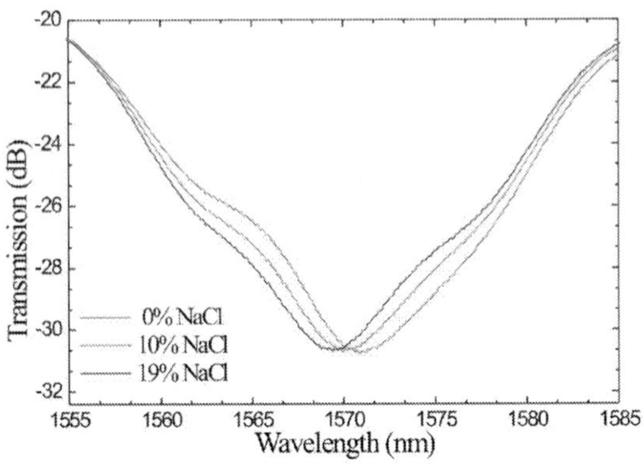

Fig. 18. Transmission spectra of the integrated LPG sensing head in an aqueous solution containing increasing concentration of salt.

4.3 Temperature measurement

To study the influence of the temperature on the developed structure, the system was placed into a thermostatic oven. The temperature was changed in increments of 5°C from 25° C to 100° C, in the air. The results are shown in Fig. 19. It is evident that the wavelength resonance shift is approximately linear up to ≈40°C, with a slope of ≈46 pm/°C. At this temperature there is a change, certainly due to some thermal induced mechanical adjustment in the sensing structure. From 45°C to 100°C, the behaviour keeps approximately linear but now with a slope of ≈ 60 pm/°C. Anyway, the sensing head was designed for application in aqueous environments where the expected temperature range is between 0 °C and 30°C. For this range, the thermal sensing behaviour is fairly linear and reversible, with a coefficient (≈ 40 pm/°C), which is different from the one found in a free LPG written in standard SMF28 fibre (≈ 70 pm/°C). This difference is certainly due to the fact that the LPG in the sensing head is under tension, and therefore the thermal induced shift of the LPG resonance is also affected by the thermal behaviour of the sensing head mechanical structure.

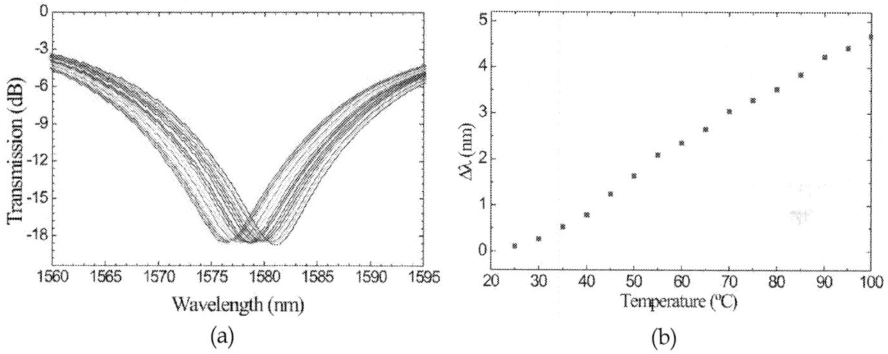

Fig. 19. Variation of the LPG resonance with temperature for the integrated sensing head.

From the results obtained it is clear that the problem of discriminating temperature induced shifts from refractive index induced wavelength shifts must be addressed in a real operational system. Nevertheless, many solutions are described in the literature that can be directly applied in our sensing head. The inscription of a single FBG in the vicinity of the LPG, for instance, would allow to univocally retrieving the temperature. Other option, was recently described [Jesus et al., 2009], using a pair of FBG that were inscribed after the LPG. However, their spectral position was chosen in such a way that each FBG was in opposite sides of the LPG resonance peak. Therefore, in this situation, the ratio of the reflected intensities of each FBG was modulated by the LPG shift being responsive to temperature and refractive index. On the other hand, the FBG wavelength was responsive only to temperature. The authors demonstrated that such a system can be operated in reflection, doubling the sensitivity to refractive index and enabling simultaneous determination of temperature and refractive index. From the practical point of view, such a system could be easily adapted into our robust sensing head protection system.

5. Conclusion

The preservation of species found in estuaries and coastal environments is achieved through constant monitoring of water quality, particularly parameters such as temperature and salinity. In the field of environmental monitoring, optical fibre refractometers (OFR) offer many advantages over conventional methods, such as immunity to electromagnetic interference, low weight, small dimensions and low insertion losses. Because of these favourable characteristics, the interest in developing fibre optic as salinity sensors has increased in the past few years, some of them based on refractive index measurements. In particular, long-period fibre gratings (LPGs) have attracted a lot of attention because of their high sensitivity to the refractive index of the surrounding environment.

This chapter focused the possible application of long-period gratings technology in environmental monitoring particularly in the measurement of surrounding refractive index or salinity. The features of this grating type were presented and it was proposed and experimentally demonstrated a prototype of a salinity sensor based on a LPG. The sensor element was written on a standard single-mode telecommunication optical fibre (Corning SMF-28) by electric arc discharge. By reducing the cladding diameter of the sensor, an improved measured resolution of 5.75×10^{-5} RIU over a measureable refractive index range from 1.33 to 1.43 (at the wavelength of 1550 nm) can be achieved, for a cladding diameter of 81 μm. It was analyzed the response of different gratings to refractive index changes with and without the structure developed, as well as its thermal response. Some important issues related with the LPG implementation in real conditions were presented, regarding the mechanical requirements and a filtration system of a housing support. The proposed robust sensing head has a filtration system for organic materials using the process of membrane separation. The developed system is easy to implement and allows the replacement of the filter without interfering with the rest of the process. Its performance analysis allowed to further underline the engineering aspects that must be taken in consideration when projecting such a system.

6. Acknowledgment

The authors gratefully acknowledge José Luis Santos and Orlando Frazão for their advice and crucial contribution. A special thanks to João Martins that designed the mechanical project for the sensing head. The acknowledgments also go to Pedro Corte-Real for his contribution with the schematics presented in this chapter. Finally, we would like to thank to José Manuel Rebordão for his encouragement.

7. References

Akiyama, M.; Nishide, K.; Shima, K., Wada, A. & Yamauchi, R. (1998). A novel long-period fibre grating using periodically releases residual stress of pure-silica core fiber. Optical Fiber Communication Conference and Exhibit, Technical Digest, pp. 276-277, ISBN 1-55752-521-8, San Jose, CA , USA, pp. 22-27.

Allsop, T.; Floreani, F.; Jedrzejewski, K.; Marques, P. & Romero, R. (2006). Spectral characteristics of tapered LPG device as a sensing element for Refractive Index and Temperature. *Journal of Lightwave Technology.*, Vol. 24, No.2, pp. 870-878.

Bhatia, V. & Vengsarkar, A. (1996). Optical fiber long-period grating sensors. *Optics Letters,* Vol.21, No.9, pp. 692-694.

Bhatia, V. (1999). Applications of long-period gratings to single and multi-parameter sensing. *Optics Express,* Vol.4, No.11, pp. 457-466.

Cardenas-Sevilla, G.; Monzon-Hernandez, D.; Torres-Gomez, I. & Martinez-Rios, A., (2009). Mechanically induced long-period fiber gratings on tapered fibers. *Optics Communications,* Vol.282, No.15, pp. 2823-2826.

Cole-Parmer catalogue (2005/06) Cole-Parmer Instrument Company, US.

Davis, D.; Gaylord,T.; Glytis, E.; & Mettler, S. (1998).Long period fibre grating fabrication with focused CO_2 laser pulses. *Electronics Letter,* Vol.34, No.3, pp. 302-303, ISSN 0013-5194.

Ding, J.; Zhang, A.; Shao, L.; Yan J. & Sailing H. (2005). Fiber-taper seeded long-period grating air as a highly sensitive refractive-index sensor. IEEE *Photonics Technology Letters,* Vol.17, No.6, pp. 1247-1249.

Dong, X., L. Su, P. Shum, Y. Chung, & C. C. Chan (2005). Photonic-Crystal-Fiber-Based Mach–Zehnder Interferometer using Long-Period Gratings. *Optics Communications,* Vol.258, No.159, pp.1379-1383.

Dong, X.; Feng, S. ; Xu, O.; Lu, S.; & Pei, L. (2009). Add/drop channel filter based on two parallel long-period fiber gratings coupler. *Optik,* Vol. 120, pp. 855-859.

Erdogan, T. (1997a). Fiber grating spectra. *Journal of Lightwave Technology.* Vol.15, No.8, pp. 1277-1294.

Erdogan, T. (1997b). Cladding-mode resonances in short- and long-period fiber grating filters. *Journal of the Optical Society of America A,* Vol.14, pp.1760-1773.

Falate, R.; Frazão, O.; Rego, G.; Fabris, J. & Santos, J. (2006). Refractometric sensor based on a phase-shifted long-period fiber grating. *Applied Optics,* Vol.45, No.21, pp. 5066-5072.

Falciai, R., Mignani, A., & Vannini, A. (2001).Long period gratings as solution concentration sensors Sensors and Actuators B: Chemical, Vol.74, No.1-3, 15 pp. 74-77. Proceedings of the 5th European Conference on Optical Chemical Sensors and Biosensors.

Fujimaki, M.; Nishihara, Y., Ohki, Y. ; Brebner, J. & Roorda S. (2000). Ion-implantation-induced densification in silica-based glass for fabrication of optical fiber gratings. *Journal of Applied Physics,* Vol.88, No.10, pp. 5534-5537, ISSN 0021-8979.

Frazão, O.; Falate, R.; Fabris, J.; Santos, J. & Ferreira, L. (2006). Optical inclinometer based on a single long-period fiber grating combined with a fused taper. *Optics Letters,* Vol. 31, No. 20, pp 2960-2962.

Gwandu B., Shu X., Allsop T., Zhang W. & Zhang L. (1999). Simultaneous refractive index and temperature measurement using cascaded long period grating in double cladding fibre. *Electronics Letters,* Vol.35, No.14, pp. 695 – 696, ISSN 0013-5194.

Han, Y.; Lee, B.; Han, W.; Paek, U. & Chung, Y (2001). Fiber optic sensing applicatios of a pair of long period fibre gratings. *Measurement Science and Technology,* Vol.12, 778-781.

Hill, O.; Malo, B.; Vineberg, K., Bilodeau, F.; Johnson, D. (1990). Efficient mode conversion in telecommunication fiber using externally written gratings. *Electronics Letters,* Vol. 26, pp. 1270-1272.

Humbert G. & A. Malki (2002). Characterizations at very high temperature of electric arc-induced long-period fiber gratings. *Optics Communications,* Vol.208, No.4-6, pp. 329-335.

Humbert, G. & Malki A. (2003). High performance bandpass filters based on electric arc-induced π-shifted long-period fibre gratings. *Optics Express*, Vol.39, No.21, pp. 1506-1505, ISSN 0013-5194.

James, S.; Rees, N.; Ashwell, G. & Tatam, R. (2002). Optical Fiber long-period Grating with a Langmuir –Blodgett thin film overlays. *Optics Letters*, Vol.9, pp. 686-688.

James, S. & Tatam, R. (2003). Optical fibre long-period grating sensors: Characteristics and application. *Measurement Science and Technology*, Vol. 14, pp. R49-R61.

Jesus, C.; Caldas, P.; Frazão, O.; Santos, J.; Jorge, P. & Baptista, J. (2009).Simultaneous Measurement of Refractive Index and Temperature Using a Hybrid FBG/LPG Configuration. *Fiber Integrated Optics*, Vol.28, pp 440–449.

Kashyp, R. (2010). Fiber Bragg Gratings. 2nd ed, Optics and Photonics Series, Academic Press, Boston.

Khaliq, S.; James, S.; & Tatam R. (2002). Enhanced sensitivity fibre optic long period grating temperature sensor. *Measurement Science and Technology*, Vol.13, No.5, pp 792–795.

Kim, B. ; Ahn, T. ; Kim, D. ; Lee, B. & Chung, Y. (2002). Effect of CO2 laser irradiation on the refractive-index change in optical fibers. *Applied Optics*, Vol.41, No.19, pp. 3809-3815.

Kondo, Y.; Nouchi, K.; Mitsuyu, T.; Watanabe, M. & Kazansky, P. (1999). Fabrication of long-period fiber gratings by focused irradiation of infrared femtosecond laser pulses, *Optics Letters*, Vol. 24, pp. 646-648.

Lee, B.; Liu, Y.; Lee, S.; Choi, S. & Jang, J. (1997). Displacement of the resonant peaks of a long-period fiber grating induced by a change of ambient refractive index. *Optics Letters*, Vol.22, pp. 1769-1771.

Lee, B.; Nishii, J. (1998). Self-interference of long period fibre gratings and its application as temperature sensor. *Electronics Letters*, Vol.34, pp. 2059-2060.

Lemaire, P.; Atkins, R.; Mizrahi, V. & Reed W. (1993). High pressure H2 loading as a technique for achieving ultrahigh UV photosensitivity and thermal sensitivity in GeO_2 doped optical fibres. *Optics Express*, Vol.29, pp. 1191.

Liu, Y.; Tu, W.; Yang, D. & Wang T. (2011), Fabrication of long-period fiber gratings by CO_2 laser in fiber under tension. *Journal of Shanghai University*, (English Edition) , Vol.15, No.1, pp. 1–6.

Murphy, R.; James, S. & Tatam, R. (2007). Multiplexing of Fiber-Optic Long-Period Grating-Based Interferometric Sensors. Journal of Lightwave Technology, Vol.25, No.3, pp. 825–829.

Mishra, V.; Singh, N.; Jain, S.; Kaur, P. & Luthra, R. (2005). Refractive index and concentration sensing of solutions using mechanically induced long period grating pair. *Optical Engineering*, Vol. 44, pp. 094402-1–094402-4.

Nikogosyan D. (2006). Long-period Gratings in a standard telecom fibre fabricated by high-intensity fentosecond UV and near-UV laser pulses. *Measurement Science and Technology*. Vol.17, pp. 960-967.

Patrick, H.; Chang, C. & Vohra, S. (1998a). Long period fiber gratings for structural bending sensing. *Electronics Letters*, Vol.34, pp. 1773 – 1775.

Patrick, H., Kersey, A. & Bucholtz, F. (1998b). Analysis of the response of long-period fiber gratings to external index of refraction. *Journal of Lightwave Technology*. Vol.16, pp. 1606–1612.

Queiroz S.; Collins, C. &; Jardim, I. (2001) .Métodos de extracção e/ou concentração de compostos encontrados em fluidos biológicos para posterior determinação cromatográfica. Quim. Nova, 24, N° 1, 68-76.

Rego G., O. Okhotnikov, E. Dianov, & V. Sulimov, (2001). High-temperature stability of long period fiber gratings produced by using an electric-arc. *Journal of Lightwave Technology*, Vol.19, pp. 1574-1579.

Rego G., Fernandes,J.; Santos, J.; Salgado, H. & Marques, P. (2003). New technique to mechanically induce long-period fibre gratings. *Optics Communications*, Vol. 220, No. 1-3, pp. 111-118.

Rego, G. ; Marques, P. ; Santos, J. & Salgado, H. (2005). Arc-induced Long-period Gratings. *Fiber and Integrated Optics*, Vol.24 (3-4), pp. 245-259.

Rego, G.; Santos, J.; & Salgado, H. (2006). Refractive index measurement with long-period gratings arc-induced in pure-silica-core fibres. *Optics Communications*, Vol.259, pp. 598–602.

Shifler D. (2005). Understanding material interactions in marine environments to promote extended structural life. *Corrosion Science*, Vol.47, pp. 2335–2352.

Shigley J. (1986). Mechanical Engineering Design - First Metric Edition, McGraw-Hill S Book Company, ISBN 0-07-056898-7.

Silva, C.; Coelho, J.; Caldas, P. ; Frazão, O.; Jorge, P.; Santos, J. (2010). Optical fibre sensing system based on long-period gratings for remote refractive index measurement in aqueous environments. *Fiber and Integrated Optics*, Vol.29, No.3, 160—169.

Swart P. L. (2004). A single fiber Michelson interferometric sensor. Proceedings of the 16 th International Conference on Optical Fiber Sensors- Japan, pp. 602-605.

Vengsarkar A.; Lemaire, P.; Judkins, J.; Bhatia, V. & Sipe, J.(1996a). Long-Period Fiber-Grating-Based Gain Equalizers. *Optics Letters*, Vol. 21, pp. 335-338.

Vengsarkar A., Pedrazzani, J.; Judkins J. B.; Lemaire, P. & Bhatia, V. (1996b). Long-period fiber gratings as band-rejection filters. *Journal of Lightwave Technology*, Vol.14, No.1, pp. 58-65.

Yokouchi T., Suzaki,Y.; Nakagawa, K.; Yamauchi, M. & Kimura (2005). Thermal tuning of mechanically induced long-period fiber grating. *Applied Optics*, Vol.44, No.24, pp. 5024-5028.

Yu, X.; Shum, P. & Dong, X. (2006). Photonic-crystal-fiber-based-Mach-Zehnder interferometer using long-period gratings. *Microwave and optical technology letters*, Vol.48, No.7, pp. 1379-1383.

Zhang, A.; Shao, L.; Ding, J. & He, S. (2005). Sandwiched long-period gratings for simultaneous measurement of refractive index and temperature. IEEE *Photonics Technology Letters*, Vol.7, No.11, pp. 2397–2399, ISSN 1041-1135.

Zanlorensi R. V., Costa, R. C. Kamikawachi, M. Muller and J. L. Fabris, (2009). Thermal characteristics of long-period gratings 266 nm UV-point-by-point induced. *Optics Communications*, Vol.282, No.5, pp 816-823.

Zhao Y., Y. Liao, B. Zhang, (2003). Monitoring Technology of Salinity in Water With Optical Fiber Sensor. *Journal of Lightwave Technology*, Vol.21, No.5, pp. 1334-1338.

Zhu Y., J. H. Chong, M. K. Rao, H. Haryono, A. Yohana, Ping Shum, Lu, C. (2003).A long-period grating refractometer: measurements of refractive index sensitivity. Proceedings SBMOIIEEE MTT-S IMOC, pp. 901-904.

14

High-Birefringent Fiber Loop Mirror Sensors: New Developments

Marta S. Ferreira, Ricardo M. Silva and Orlando Frazão
INESC Porto, Porto
Portugal

1. Introduction

The fiber loop mirror based on a ring configuration is one of the devices in optical fibers most used in communications and sensors (Mortimore, 1988). The device is formed when the two output ports of a directional coupler are spliced. In this configuration, the two waves at the coupler outputs travel in opposite directions but following the same optical paths in an optical fiber ring, which assures constructive interference as the waves re-enter the coupler. Thus, all light is reflected to the input port containing losses essentially in the fiber, the splice region and the coupler. Due to this "mirror" characteristic, the device is frequently used in the formation of resonant cavities in optical fiber lasers (Urquhart, 1989). A structure of this kind, with interesting properties, occurs when the ring contains a section of fiber with high-birefringence. When this happens, an interference pattern is generated. It depends only on the fiber birefringence and length, being independent of the remaining ring extension (Fang & Claus, 1995).

This chapter provides an overview of the state-of-the-art, birefringence concepts, and new developments of high-birefringence fiber loop mirror configurations that can be used as sensing elements.

2. State-of-the-art

In this section, a brief review about the state-of-the-art using high-birefringent fiber loop mirror (Hi-Bi FLM) configurations is described. In order to give a clear view about the evolution of Hi-Bi FLM setups, a sum up with the years of publications and the most important breakthroughs are shown in table 1.

The first Hi-Bi fiber Sagnac loop sensor for temperature measurement was proposed in 1997. The configuration, schematically shown in Figure 1 a), consisted of a 3 dB simple polarization maintaining coupler, constituted by a bow-tie fiber. By cross splicing the output ports at $\pi/2$, they guaranteed that both port lengths were different, forming an unbalanced Sagnac loop interferometer.

The temperature sensor could operate both in transmission and reflection, and its functioning did not depend on the light polarization at the input port. When compared with other temperature optical sensors (fiber Bragg gratings), this configuration presented higher sensitivity (Delarosa et al., 1997; Starodumov et al., 1997). In the same year, a strain sensor

Year	Breakthrough of Hi-Bi FLM configurations
1997	Temperature sensor (Starodumov et al., 1997); temperature sensing in NIR (Delarosa et al., 1997); strain sensor (Campbell et al., 1997)
1999	Strain sensor (Campbell et al., 1999)
2004	Temperature PCF sensor (Kim, D. H. & Kang, 2004); temperature insensitivity using PCF (Zhao et al., 2004)
2005	Interrogation system using PCF (Yang et al., 2005); displacement sensor (Liu et al., 2005)
2006	Liquid level sensor (Dong, B. et al., 2006); FBG interrogation system (Silva et al., 2011); LPG/Hi-Bi FLM (Frazao et al., 2006b)
2007	Review of Hi-Bi FLM sensors (Frazao et al., 2007a); strain PCF sensor (Dong, X. Y. et al., 2007; Frazao et al., 2007b) ; chemical etching (Frazao et al., 2007c); concatenated FLM (Frazao et al., 2007d)
2008	Temperature Erbium Hi-Bi fiber (Frazao et al., 2008a); refractive index sensor (Frazao et al., 2008b); multiparameter sensor using side-hole fiber (Frazao et al., 2008c); pressure PCF sensor (Fu et al., 2008); FBG/Hi-Bi FLM (Zhou et al., 2008); current sensor (Marques et al., 2008)
2009	Multiplexing Hi-Bi FLM (Fu et al., 2009); strain and temperature discrimination using two Hi-Bi fibers (Han et al., 2009); hollow-core PCF sensor (Kim, G. et al., 2009); holey fiber filled with metal indium (Kim, B. H. et al., 2009)
2010	Torsion PCF sensor (Kim, H. M. et al., 2010); curvature PCF sensor (Gong et al., 2010); pressure-induced SMF (Jin et al., 2010); long distance remote interrogation system (Lee et al., 2010); small core PCF (Andre et al., 2010); intensity strain sensor (Qian et al., 2010); suspended twin-core (Frazao et al., 2010); displacement PCF sensor (Zhang et al., 2010)
2011	Hi-Bi FLM with an output port probe (Frazao et al., 2011); LPG/Sagnac Loop (Kang et al., 2011); curvature PCF sensor (Hwang et al., 2011); Hi-Bi photonic bandgap Bragg fiber (Ferreira et al., 2011); transverse mechanical load (Zu et al., 2011a); PCF FLM filled with alcohol (Qian et al., 2011); PCF FLM with core offset (Dong, B. et al., 2011); magneto-optic modulator (Zu et al., 2011b); tapered FLM (Zibaii et al., 2011)

Table 1. Hi-Bi FLM configurations chronology.

based on a Hi-Bi fiber Sagnac ring configuration was investigated (see Figure 1 b) (Campbell et al., 1997). In 1999, the same authors reported similar work using a frequency-modulated continuous wave (Campbell et al., 1999).

In 2004, temperature dependence of a polarization maintaining (PM) photonic crystal fiber (PCF) in a Sagnac loop interferometer was analyzed. Comparing with a standard PM fiber, they obtained reduced temperature sensitivity (Kim, D. H. & Kang, 2004). A temperature insensitive Hi-Bi FLM using a photonic crystal fiber was demonstrated. For low-temperatures (25°C to 85°C), a wavelength spacing variation of 0.05 pm/°C was achieved (Zhao et al., 2004).

In 2005, the characteristics of a Hi-Bi FLM composed of a standard 3 dB fiber coupler with one or more sections of Hi-Bi fibers were discussed, as can be seen in Figure 1 c). Displacement and temperature sensors were theoretically and experimentally analyzed.

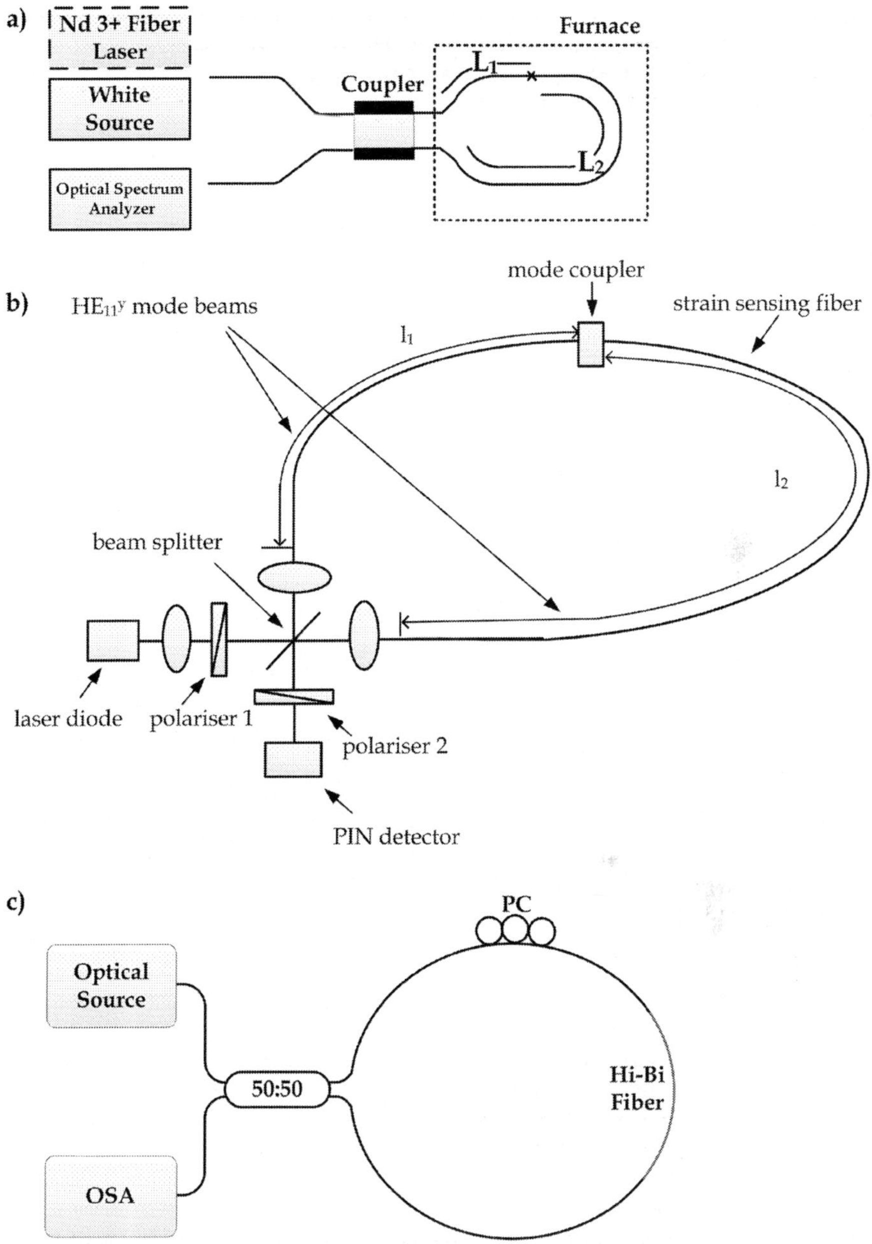

Fig. 1. a) First temperature sensor using a Sagnac interferometer, b) First strain sensor and c) conventional Hi-Bi FLM.

Using the characteristics of the Hi-Bi FLM, a strain-temperature sensor was proposed and demonstrated as an interrogation system for new applications (Liu et al., 2005). A temperature insensitive fiber Bragg grating (FBG) sensor interrogation system, which used a PCF in a Hi-Bi FLM was proposed and experimentally demonstrated (Yang et al., 2005).

In 2006, a liquid-level Hi-Bi FLM sensor based on a uniform-strength cantilever beam was presented (Dong, B. et al., 2006). The advantages of this sensor included its simple structure, high sensitivity, low-cost and repeatability. A FBG interrogation scheme based on a Hi-Bi FLM for strain-temperature discrimination was reported (Frazao et al., 2006a). The configuration improvement was the use of a Hi-Bi FLM to interrogate an array of FBGs. The same author proposed an alternative solution for simultaneous measurement of strain and temperature, using a long-period fiber grating (LPG) combined with a Hi-Bi FLM (Frazao et al., 2006b).

In 2007, a fiber-optic strain sensor was demonstrated by using a short length of PCF inserted in a Hi-Bi FLM (Dong, X. Y. et. al., 2007). This strain sensor presented temperature insensitivity. The investigation was done using an acrylate coated PCF and other uncoated. Strain and temperature measurements were made to verify their sensibilities (Frazao et al., 2007b). The uncoated PCF revealed insensitivity to temperature. Another work made by the same author (Frazao et al., 2007c) was the birefringence control of a Hi-Bi fiber (bow-tie) under chemical etching. This fiber was inserted in a FLM and the optical response was measured while chemical etching of the bow-tie fiber was taking place. A different study made, was a configuration for simultaneous strain and temperature discrimination using two concatenated Hi-Bi FLMs (Frazao et al., 2007d). By monitoring the wavelength and optical power, it was possible to obtain both measurements simultaneously.

In 2008, an optical current sensor based on Hi-Bi FLM was reported (Marques et al., 2008). An elliptical cladding fiber was coated with a metal and inserted as a sensing element in the FLM. A novel intrinsic fiber optic pressure sensor, based on a PM-PCF section inserted between the output ports of the Hi-Bi FLM was presented (Fu et al., 2008). This sensing head did not require polarimetric detection as the ones reported so far, and through wavelength shifts they obtained a sensitivity of 3.42 nm/MPa for a 58.4 cm long PM-PCF. A setup where the Hi-Bi section was combined with a FBG, enabling the simultaneous measurement of strain and temperature was presented (Zhou et al., 2008). The obtained sensing resolution was of ±1 °C for temperature and ±21 $\mu\varepsilon$ for strain. A Hi-Bi PANDA Erbium-doped FLM was presented for the first time. In this case, a temperature coefficient sensitivity of -2.22 nm/°C was achieved (Frazao et al., 2008a). Another report by the same author was a Hi-Bi D-type FLM used as refractometer (Frazao et al., 2008b). A chemical etching was applied to the D-type fiber and presented high sensitivity to refractive index. In the same year, a configuration for simultaneous measurement of multiparameters by using a PM side-hole fiber section in the FLM was proposed (Frazao et al., 2008c). Different parameters were analyzed, like torsion, temperature and longitudinal strain.

In the year of 2009, high temperature sensitivity was achieved when a section of birefringent holey fiber filled with metal indium was introduced in a FLM (Kim, B. H. et al., 2009). The analysis was done for wavelength and birefringence responses to temperature, and sensitivities of -6.3 nm/K and -3.3×10^{-6} K^{-1} were respectively obtained. On the other hand, an elliptical hollow-core photonic bandgap fiber was fabricated and tested in a FLM configuration (Kim, G. et al., 2009), measuring strain and temperature responses. Later on, simultaneous measurement of temperature and strain was performed by using a PM-PCF with Erbium-doped fiber incorporated (Han et al., 2009). Finally, three different multiplexing schemes for PCF FLMs were presented (Fu et al., 2009).

In 2010, a temperature-insensitive displacement sensor, based on Hi-Bi PCFLM was proposed (Zhang et al., 2010). This sensor presented a displacement sensitivity of 0.28286 nm/mm and good stability through a temperature range from 40 °C to 109.5 °C. A low-birefringence PCF FLM as a curvature sensor was reported for the first time (Gong et al., 2010). A sensitivity of -0.337 nm for a range between 0-9.92 m^{-1} was achieved. This sensor was also sensitive to temperature, transverse load and twist, thus being a potential multifunction sensor. A temperature sensor using a section of pressure-induced birefringent SMF FLM was firstly presented (Jin et al., 2010). A sensitivity of 0.65 nm/°C was reported. A temperature-independent strain sensor based on intensity measurements was developed (Qian et al., 2010). A small core microstructured fiber segment in the FLM for the simultaneous measurement of strain and temperature was used (Andre et al., 2010). Besides Hi-Bi, this configuration also showed intermodal interference, resulting in a complex channeled spectrum. With a sensing head composed only by a fiber section, resolutions of ±1.5 °C and ±4.7 με for temperature and strain were respectively obtained. A torsion sensor with temperature and strain independence, using a suspended twin-core fiber to form the Hi-Bi FLM was investigated (Frazao et al., 2010). The variations in amplitude caused by applying torsion on the sensing head were analyzed by the Fast Fourier Transform technique. Another torsion sensor was done by placing a PCF in a FLM and demonstrated its temperature insensitivity (Kim, H. M. et al., 2010). The variation of reflectivity for long-distance remote FGB cavity sensors using a Hi-Bi FLM was analyzed and demonstrated (Lee et al., 2010).

In 2011, a curvature sensor with a new Hi-Bi PCF design spliced inside the loop mirror was demonstrated (Hwang et al., 2011). The proposed PCF had two large air holes in the outer cladding region that gave rise to core ellipticity during the fabrication process. The curvature sensor is based on the PCF Sagnac interferometer and its sensitivity depended on the bending directions with respect to the large-air-hole alignment. A simultaneous measurement of strain and temperature based on cladding-mode-resonance in PM PCF loop mirror was reported (Dong, B. et al., 2011). By introducing a mode field diameter mismatch at a splicing joint of an inbuilt SMF-PMPCF-SMF structure connected in a Sagnac configuration, an intermodal interference was constructed by the excited cladding modes and the polarization modes of the PMPCF. An optical fiber transverse mechanical load sensor using a short section of solid core PCF which was inserted inside a FLM was reported (Zu et al., 2011a). Two new configurations of Hi-Bi FLMs with an output port probe were proposed (Frazao et al., 2011). Both configurations used two couplers spliced in between, with unbalanced arms and one output port was used as the probe sensor. The difference between them was the location of the Hi-Bi fiber section length: either between the two couplers (first new configuration) or spliced at the output port probe (second new configuration). These configurations were compared with the conventional Hi-Bi FLM when strain was applied and showed similar sensitivities. The theoretical model of these two configurations was studied by (Silva et al., 2011). A compact temperature sensor based on a

FLM combined with an alcohol-filled Hi-Bi PCF was investigated (Qian et al., 2011). A magneto-optic modulator with a magnetic fluid film inserted into an optical fiber Sagnac interferometer was proposed (Zu et al., 2011b) The magnetic fluid exhibited variable birefringence and Faraday effect under external magnetic field that led to a phase difference and polarization state rotation in the Sagnac interferometer. A Hi-Bi FLM with a LPG inscribed in PMF was proposed and experimentally demonstrated for simultaneous measurement of strain and temperature (Kang et al., 2011). This was due to the different responses of the LPG and the Sagnac interferometer observed. A Hi-Bi photonic bandgap Bragg FLM configuration for simultaneous measurement of strain and temperature was reported (Ferreira et al., 2011). The group birefringence and the sharp loss peaks were observable in the spectral response. Since the sensing head presented different sensitivities for strain and temperature measurands, these physical parameters were discriminated by using the matrix method. A single-mode non-adiabatic tapered optical fiber sensor inserted in a FLM enabling to tune its sensitivity towards refractive index was reported (Zibaii et al., 2011).

3. Birefringence concept

The Hi-Bi FLM consists in a coupler and a Hi-Bi fiber section as a sensing head. The pattern fringe is obtained through the fiber characteristics. The birefringence is guaranteed in the fabrication process through the geometrical or stress effects. The geometrical effect is easily obtained when the core is non-circular, for example, the elliptical core fiber. The stress effect can be obtained by applying mechanical tension around the core, as occurs in the case of PANDA or bow-tie fibers.

The birefringence effect in optical fibers was studied considering the phase and group modal birefringence. The phase modal birefringence is defined as:

$$B(\lambda) \equiv n_x - n_y = \frac{\lambda}{2\pi}\left(\beta_x(\lambda) - \beta_y(\lambda)\right), \tag{1}$$

where β_i (i=x, y) is the propagation constant associated to each fundamental mode of polarization. This birefringence is associated to the beat length, L_B, through the relation:

$$L_B = \frac{\lambda}{B(\lambda)} \tag{2}$$

The group modal birefringence, $G(\lambda)$, is given by:

$$G(\lambda) = \frac{d\beta_x(\lambda)}{dk} - \frac{d\beta_y(\lambda)}{dk} = B(\lambda) - \lambda\frac{dB(\lambda)}{d\lambda} \tag{3}$$

where $k = 2\pi / \lambda$. The differential phase between two polarization modes after propagating through an extension L of fiber is given by:

$$\phi = \frac{2\pi L B(\lambda)}{\lambda} \tag{4}$$

resulting in a channeled type interferometric spectral response, as illustrated in Figure 2.

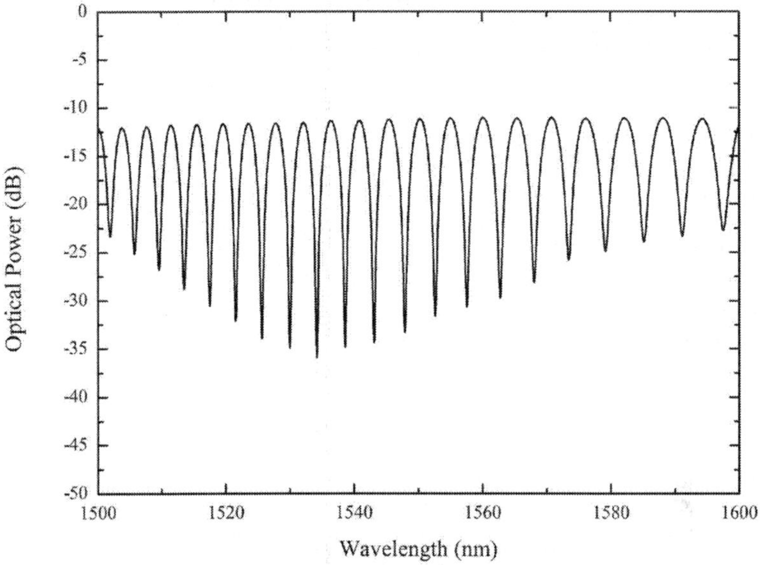

Fig. 2. Typical spectral response of an interferometric structure.

Deriving the phase ϕ with respect to λ, the following equation is obtained:

$$\frac{d\phi}{d\lambda} = \frac{2\pi L}{\lambda^2}\left(\lambda\frac{dB(\lambda)}{d\lambda} - B(\lambda)\right) = -\frac{2\pi L}{\lambda^2}G(\lambda) \tag{5}$$

If $\Delta\lambda$ is the channeled spectrum period, corresponding to a phase variation $d\phi = 2\pi$, the expression obtained for the group modal birefringence is:

$$G(\lambda) = -\frac{\lambda^2}{\Delta\lambda L}. \tag{6}$$

4. New developments

New developments have been proposed using the FLM in novel designs. Recently, three new Hi-Bi fiber loop mirror with output port probe configurations were demonstrated. These configurations present new solutions as sensing elements. One of them is to use the output port as optical refractometer without the Hi-Bi fiber section or conventional strain/temperature sensors when the probe is a Hi-Bi fiber section.

A torsion sensor with insensitivity to temperature and strain measurement is also demonstrated. In this case the conventional optical coupler is changed by a Hi-Bi coupler where the optical path difference is obtained when the two output arms are unbanlanced. These two new configurations are presented in the following sections.

4.1 Hi-Bi FLM with output probe

Three novel configurations of Hi-Bi FLMs are schematically shown in Figure 3. They consist of a loop of optical fiber formed between the output ports of two directional couplers along with an output port probe.

The first configuration proposed has a polarization controller and a Hi-Bi fiber section located between the optical couplers, (Figure 3 a)). The mode of operation is similar to the conventional FLM. In the first coupler, the light is divided in two waves that travel between the two couplers with asymmetric arms and re-couple in the output port of the second directional coupler. No interference is obtained in the output port of the second coupler because the arms between the two couplers are unbalanced. A silver mirror was fabricated in the end of the fiber tip to re-inject the light into the second coupler. Here the two waves are divided in four waves and re-injected into the first coupler.

Only the two waves that travel in different paths in the two couplers section in the downlead and uplead propagation have a path imbalance associated with the Hi-Bi fiber and, therefore, it is obtained a channeled spectrum with a periodicity similar to what is expectable when considering the standard configuration. The other two waves accumulate a substantial path imbalance and therefore the corresponding channeled spectrum has a very small periodicity, not resolvable by the OSA used. Therefore, the fringe pattern observed in the OSA comes out from the phase difference between the polarization states of the Hi-Bi fiber length. In this case, this configuration was characterized as optical refractometer (Frazao et al., 2011).

The main advantage of this configuration is that other types of interferometers (such as Fabry-Perot) can be inserted in the probe end with a reference provided by the Hi-Bi fiber section.

The second configuration proposed (Figure 3 b)) is more attractive in general. Here, the Hi-Bi fiber section is spliced to one output port of the second directional coupler. This means that the configuration can be used for remote sensing using only one fiber in the sensing zone. There is a section of an end mirrored Hi-Bi fiber at the second coupler output port, that will increase the returned optical power. It is clear that with this structure it is possible to have the sensing advantages of a FLM without the need of locating geometrically the sensing head inside the loop. A feature that increases the flexibility of the configuration shown in Figure 3 b) is the fact that the polarization controller is only needed to adjust the interferometric visibility. This configuration was compared with the conventional Hi-Bi FLM and presents similar sensitivity when the strain is applied (Frazao et al., 2011).

The last configuration (Figure 3 c)) was characterized as an interrogation setup for strain measurement. In this case, two Hi-Bi fiber sections are used as sensor and reference. The setup is formed by two simple couplers with low insertion loss, where the two output ports of the first coupler are spliced to the two input ports of the second one. In one of the splices, a Hi-Bi fiber section (reference) is inserted, while in the other arm a polarization controller is placed, in order to optimize the spectral response of the fringe pattern. The sensor is located at the output port of the second coupler and consists of the same type of Hi-Bi fiber. With the insertion of two Hi-Bi fiber sections, the configuration will have two combined unbalanced Sagnac interferometers. The Hi-Bi fiber used has a length of 1.34 m and 0.34 m for the reference and sensor signal, respectively. A signal processing based on the phase-shift quadrature of the two peaks power response of the reference signal was done in real time.

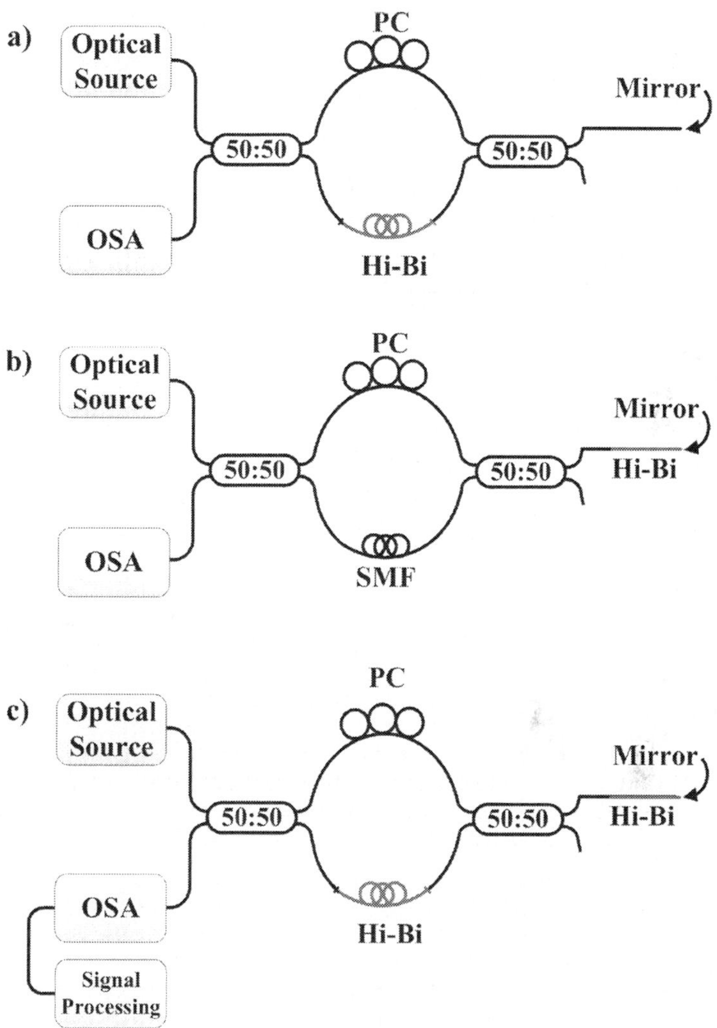

Fig. 3. Schematic setup for the: a) first; b) second; and c) third new Hi-Bi FLM configurations.

The sensitivity achieved in strain measurements is 16 mrad/με (Figure 4). The resolution of the system is 1.9 με, for a strain step variation of 29 με.

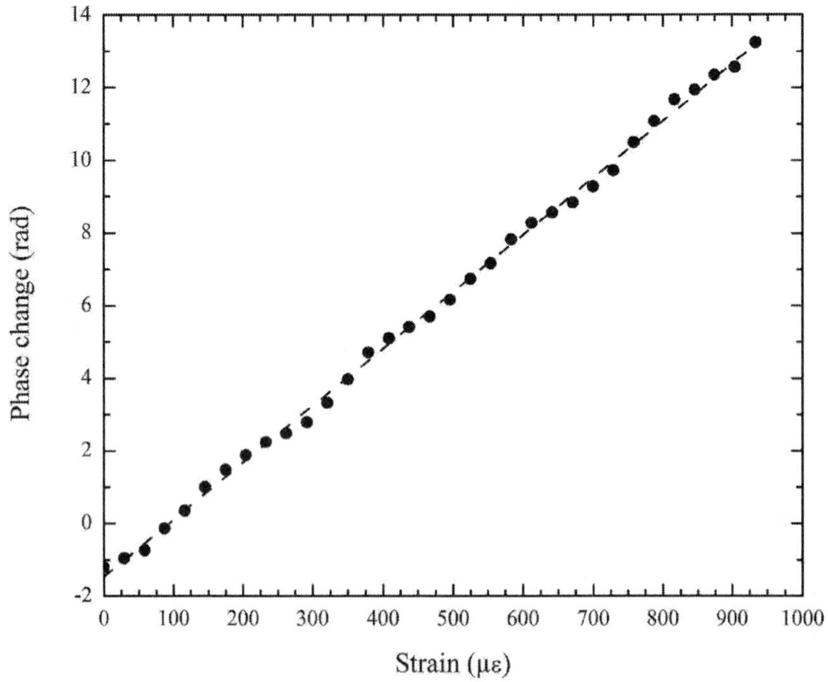

Fig. 4. Phase change versus strain.

4.2 Torsion sensor using a Sagnac interferometer

A temperature and strain independent torsion sensor is presented. In this case, the conventional optical coupler is replaced by Hi-Bi fiber coupler and the sensing head consists in a section of standard single mode fiber.

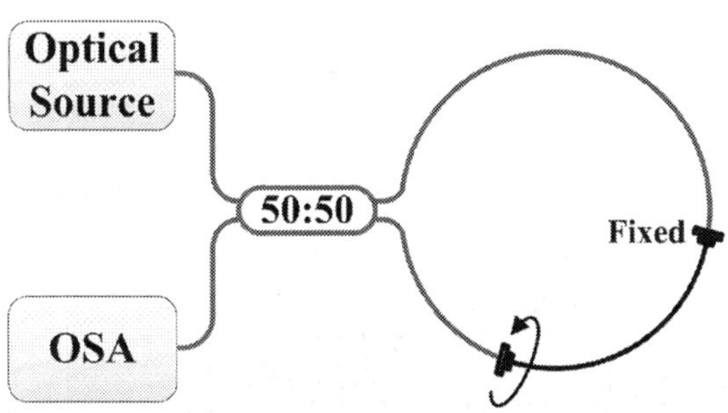

Fig. 5. Experimental setup of the torsion sensor based on Hi-Bi Sagnac loop interferometer.

Figure 5 presents the experimental setup of a Hi-Bi Sagnac interferometer sensor. Instead of using a regular fiber coupler, a 3 dB commercial Hi-Bi PANDA fiber coupler is used. This coupler will guarantee that the light will travel through the output arms with a constant polarization. A section of SMF is inserted in order to close the loop.

The interferometer is illuminated with a broadband optical source, and the transmission spectral response of the sensing head is measured with an optical spectrum analyzer (OSA).

The ports length difference of the Hi-Bi coupler will set the fringes pattern, since it introduces a phase accumulation to the travelling wave. The sensing head is located in the SMF region, and torsion measurements can be done by fixing one side of SMF section and twisting the other side, according to Figure 5. As the torsion is applied, an amplitude variation of the spectrum between maximum and minimum values is verified, which induces a change on the visibility of the fringes. Even though the amplitude changes with the torsion angle, total destructive interference is never observed, due to the existence of a beat between the two interferometers. The sensor response is periodically modulated, as the twist is applied. The sensitivity depends on the sensing head length, bigger sensing heads are more sensitive. For a 0.26 m long sensing head, sensitivities of 0.045 $degree^{-1}$ and -0.051 $degree^{-1}$ were obtained, when left and right twisting angles were respectively applied (Figure 6). The slight difference between the negative sensitivities can be due to the coating effect.

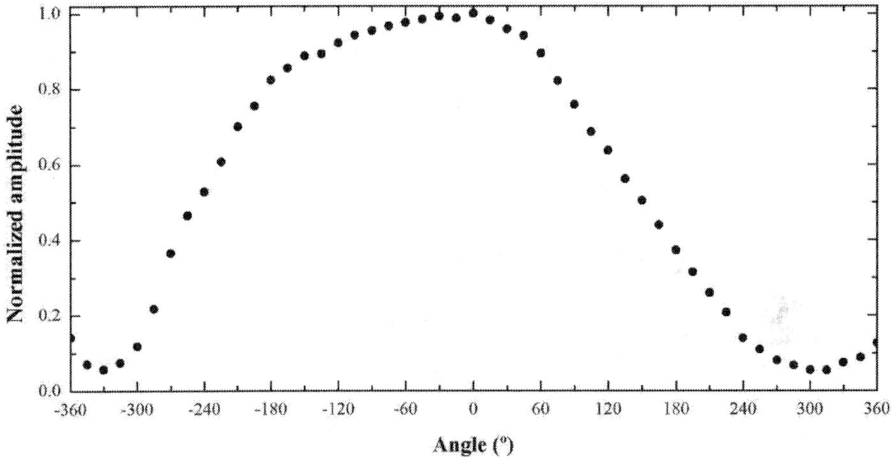

Fig. 6. Torsion measurements of the sensing head.

For temperature measurements, a sensing head with a length of 0.26 m is placed in a tubular oven, which enables the temperature to be set with an error smaller than 0.1 °C. Regarding strain characterization, the sensing head is attached to a translation stage with a resolution of 1 μm. The sensitivities obtained are 2×10^{-3} dB/°C and 7.9×10^{-9} dB/με, respectively. Due to the low sensitivities obtained, it can be stated that this sensing head is insensitive to these measurands. These results are expected because temperature and strain do not change the polarization states of the two counter propagating waves in the SMF region.

The main distinctive feature of this setup, when compared with a conventional Hi-Bi fiber loop mirror, is the polarization controller suppression.

5. Conclusion

The first optical fiber highly-birefringent Sagnac interferometer was presented in 1997 (Starodumov et al., 1997). This structure has been consistently used in the ambit of conception of new devices and systems in several areas related to the optical fiber technologies, namely in the optical fiber sensors context. By combining optical devices, it was possible to discriminate physical parameters, such as deformation and temperature. With the emerging of new fibers, like the Photonic Crystal Fibers, it is possible to measure one physical parameter with insensitivity to temperature. At last, with the development of new fiber loop mirror designs, it is possible to interrogate and measure the physical parameters with low cost and easily multiplexing with several sensing heads in series. All works recently published demonstrated the interferometric topology versatility of the Sagnac interferometer in the sensing domain. It must be pointed out that it shall be possible to obtain much better resolutions than the ones indicated if adequate signal processing techniques are applied, that allow reading the interferometer phase.

6. References

Andre, R. M., Marques, M. B., Roy, P. & Frazao, O. (2010). Fiber Loop Mirror Using a Small Core Microstructured Fiber for Strain and Temperature Discrimination. *IEEE Photonics Technology Letters*, vol. 22, No. 15, (Aug 1, 2010), pp. 1120-1122, ISSN 1041-1135

Campbell, M., Zheng, G., Holmes-Smith, A. S. & Wallace, P. A. (1999). A frequency-modulated continuous wave birefringent fibre-optic strain sensor based on a Sagnac ring configuration. *Measurement Science & Technology*, vol. 10, No. 3, (Mar, 1999), pp. 218-224, ISSN 0957-0233

Campbell, M., Zheng, G., Wallace, P. A. & Holmessmith, A. S. (1997). A distributed frequency modulation continuous wave fiber stress sensor based on a birefringent Sagnac ring configuration. *Optical Review*, vol. 4, No. 1A, (Jan-Feb, 1997), pp. 114-116, ISSN 1340-6000

Delarosa, E., Zenteno, L. A., Starodumov, A. N. & Monzon, D. (1997). All-fiber absolute temperature sensor using an unbalanced high-birefringence Sagnac loop. *Optics Letters*, vol. 22, No. 7, (Apr 1, 1997), pp. 481-483, ISSN 0146-9592

Dong, B., Hao, J. Z., Liaw, C. Y. & Xu, Z. W. (2011). Cladding-Mode Resonance in Polarization-Maintaining Photonic-Crystal-Fiber-Based Sagnac Interferometer and Its Application for Fiber Sensor. *Journal of Lightwave Technology*, vol. 29, No. 12, (Jun 15, 2011), pp. 1759-1763, ISSN 0733-8724

Dong, B., Zhao, Q. D., Feng, L. H., Guo, T., Xue, L. F., Li, S. H. & Gu, H. (2006). Liquid-level sensor with a high-birefringence-fiber loop mirror. *Applied Optics*, vol. 45, No. 30, (Oct 20, 2006), pp. 7767-7771, ISSN 0003-6935

Dong, X. Y., Tam, H. Y. & Shum, P. (2007). Temperature-insensitive strain sensor with polarization-maintaining photonic crystal fiber based Sagnac interferometer. *Applied Physics Letters*, vol. 90, No. 15, (Apr 9, 2007), pp. -, ISSN 0003-6951

Fang, X. J. & Claus, R. O. (1995). Polarization-Independent All-Fiber Wavelength-Division Multiplexer Based on a Sagnac Interferometer. *Optics Letters*, vol. 20, No. 20, (Oct 15, 1995), pp. 2146-2148, ISSN 0146-9592

Ferreira, M. S., Baptista, J. M., Roy, P., Jamier, R., Fevrier, S. & Frazao, O. (2011). Highly birefringent photonic bandgap Bragg fiber loop mirror for simultaneous measurement of strain and temperature. *Optics Letters*, vol. 36, No. 6, (Mar 15, 2011), pp. 993-995, ISSN 0146-9592

Frazao, O., Baptista, J. M. & Santos, J. L. (2007a). Recent advances in high-birefringence fiber loop mirror sensors. *Sensors*, vol. 7, No. 11, (Nov, 2007a), pp. 2970-2983, ISSN 1424-8220

Frazao, O., Baptista, J. M. & Santos, J. L. (2007b). Temperature-independent strain sensor based on a Hi-Bi photonic crystal fiber loop mirror. *IEEE Sensors Journal*, vol. 7, No. 9-10, (Sep-Oct, 2007b), pp. 1453-1455, ISSN 1530-437X

Frazao, O., Egypto, D., Bittencourt, L. A., Giraldi, M. T. M. R. & Marques, M. B. (2008a). Temperature sensor using hi-bi erbium-doped fiber loop mirror. *Microwave and Optical Technology Letters*, vol. 50, No. 12, (Dec, 2008a), pp. 3152-3154, ISSN 0895-2477

Frazao, O., Guerreiro, A., Santos, J. L. & Baptista, J. M. (2007c). Birefringence monitoring of a Hi-Bi fibre under chemical etching through a fibre loop mirror. *Measurement Science & Technology*, vol. 18, No. 12, (Dec, 2007c), pp. N81-N83, ISSN 0957-0233

Frazao, O., Marques, B. V., Jorge, P., Baptista, J. M. & Santos, J. L. (2008b). High birefringence D-type fibre loop mirror used as refractometer. *Sensors and Actuators B-Chemical*, vol. 135, No. 1, (Dec 10, 2008b), pp. 108-111, ISSN 0925-4005

Frazao, O., Marques, L. M. & Baptista, J. M. (2006a). Fibre Bragg grating interrogation based on high-birefringence fibre loop mirror for strain-temperature discrimination. *Microwave and Optical Technology Letters*, vol. 48, No. 11, (Nov, 2006a), pp. 2326-2328, ISSN 0895-2477

Frazao, O., Marques, L. M., Santos, S., Baptista, J. M. & Santos, J. L. (2006b). Simultaneous measurement for strain and temperature based on a long-period grating combined with a high-birefringence fiber loop mirror. *IEEE Photonics Technology Letters*, vol. 18, No. 21-24, (Nov-Dec, 2006b), pp. 2407-2409, ISSN 1041-1135

Frazao, O., Santos, J. L. & Baptista, J. M. (2007d). Strain and temperature discrimination using IF concatenated high-birefringence fiber loop mirrors. *IEEE Photonics Technology Letters*, vol. 19, No. 13-16, (Jul-Aug, 2007d), pp. 1260-1262, ISSN 1041-1135

Frazao, O., Silva, R. M., Kobelke, J. & Schuster, K. (2010). Temperature- and strain-independent torsion sensor using a fiber loop mirror based on suspended twin-core fiber. *Optics Letters*, vol. 35, No. 16, (Aug 15, 2010), pp. 2777-2779, ISSN 0146-9592

Frazao, O., Silva, R. M. & Santos, J. L. (2011). High-Birefringent Fiber Loop Mirror Sensors With an Output Port Probe. *IEEE Photonics Technology Letters*, vol. 23, No. 2, (Jan 15, 2011), pp. 103-105, ISSN 1041-1135

Frazao, O., Silva, S. O., Baptista, J. M., Santos, J. L., Statkiewicz-Barabach, G., Urbanczyk, W. & Wojcik, J. (2008c). Simultaneous measurement of multiparameters using a Sagnac interferometer with polarization maintaining side-hole fiber. *Applied Optics*, vol. 47, No. 27, (Sep 20, 2008c), pp. 4841-4848, ISSN 0003-6935

Fu, H. Y., Tam, H. Y., Shao, L. Y., Dong, X. Y., Wai, P. K. A., Lu, C. & Khijwania, S. K. (2008). Pressure sensor realized with polarization-maintaining photonic crystal fiber-based Sagnac interferometer. *Applied Optics*, vol. 47, No. 15, (May 20, 2008), pp. 2835-2839, ISSN 0003-6935

Fu, H. Y., Wong, A. C. L., Childs, P. A., Tam, H. Y., Liao, Y. B., Lu, C. & Wai, P. K. A. (2009). Multiplexing of polarization-maintaining photonic crystal fiber based Sagnac interferometric sensors. *Optics Express*, vol. 17, No. 21, (Oct 21, 2009), pp. 18501-18512, ISSN 1094-4087

Gong, H. P., Chan, C. C., Zu, P., Chen, L. H. & Dong, X. Y. (2010). Curvature measurement by using low-birefringence photonic crystal fiber based Sagnac loop. *Optics Communications*, vol. 283, No. 16, (Aug 15, 2010), pp. 3142-3144, ISSN 0030-4018

Han, Y. G., Chung, Y. & Lee, S. B. (2009). Discrimination of strain and temperature based on a polarization-maintaining photonic crystal fiber incorporating an erbium-doped fiber. *Optics Communications*, vol. 282, No. 11, (Jun 1, 2009), pp. 2161-2164, ISSN 0030-4018

Hwang, K. J., Kim, G. H., Lim, S. D., Lee, K., Park, J. W. & Lee, S. B. (2011). A Novel Birefringent Photonic Crystal Fiber and Its Application to Curvature Measurement. *Japanese Journal of Applied Physics*, vol. 50, No. 3, (Mar, 2011), pp. -, ISSN 0021-4922

Jin, Y. X., Chan, C. C., Zhang, Y. F., Dong, X. Y. & Zu, P. (2010). Temperature sensor based on a pressure-induced birefringent single-mode fiber loop mirror. *Measurement Science & Technology*, vol. 21, No. 6, (Jun, 2010), pp. -, ISSN 0957-0233

Kang, J. A., Dong, X. Y., Zhao, C. L., Qian, W. W. & Li, M. C. (2011). Simultaneous measurement of strain and temperature with a long-period fiber grating inscribed Sagnac interferometer. *Optics Communications*, vol. 284, No. 8, (Apr 15, 2011), pp. 2145-2148, ISSN 0030-4018

Kim, B. H., Lee, S. H., Lin, A. X., Lee, C. L., Lee, J. & Han, W. T. (2009). Large temperature sensitivity of Sagnac loop interferometer based on the birefringent holey fiber filled with metal indium. *Optics Express*, vol. 17, No. 3, (Feb 2, 2009), pp. 1789-1794, ISSN 1094-4087

Kim, D. H. & Kang, J. U. (2004). Sagnac loop interferometer based on polarization maintaining photonic crystal fiber with reduced temperature sensitivity. *Optics Express*, vol. 12, No. 19, (Sep 20, 2004), pp. 4490-4495, ISSN 1094-4087

Kim, G., Cho, T. Y., Hwang, K., Lee, K., Lee, K. S., Han, Y. G. & Lee, S. B. (2009). Strain and temperature sensitivities of an elliptical hollow-core photonic bandgap fiber based on Sagnac interferometer. *Optics Express*, vol. 17, No. 4, (Feb 16, 2009), pp. 2481-2486, ISSN 1094-4087

Kim, H. M., Kim, T. H., Kim, B. & Chung, Y. (2010). Temperature-Insensitive Torsion Sensor With Enhanced Sensitivity by Use of a Highly Birefringent Photonic Crystal Fiber. *IEEE Photonics Technology Letters*, vol. 22, No. 20, (Oct 15, 2010), pp. 1539-1541, ISSN 1041-1135

Lee, T. H., Lee, H. D., Kim, C. S. & Jeong, M. Y. (2010). Characterization of variable reflectivity of a polarization-maintaining fiber Sagnac mirror for long-distance remote fiber Bragg gratings cavity sensors. *Measurement Science & Technology*, vol. 21, No. 11, (Nov, 2010), pp. -, ISSN 0957-0233

Liu, Y., Liu, B., Feng, X. H., Zhang, W. G., Zhou, G., Yuan, S. Z., Kai, G. Y. & Dong, X. Y. (2005). High-birefringence fiber loop mirrors and their applications as sensors. *Applied Optics*, vol. 44, No. 12, (Apr 20, 2005), pp. 2382-2390, ISSN 0003-6935

Marques, B. V., Frazao, O., Mendonca, S., Perez, J., Marques, M. B., Santos, S. F. & Baptista, J. M. (2008). Optical current sensor based on metal coated Hi-Bi fiber loop mirror. *Microwave and Optical Technology Letters*, vol. 50, No. 3, (Mar, 2008), pp. 780-782, ISSN 0895-2477

Mortimore, D. B. (1988). Fiber Loop Reflectors. *Journal of Lightwave Technology*, vol. 6, No. 7, (Jul, 1988), pp. 1217-1224, ISSN 0733-8724

Qian, W. W., Zhao, C. L., Dong, X. Y. & Jin, W. (2010). Intensity measurement based temperature-independent strain sensor using a highly birefringent photonic crystal fiber loop mirror. *Optics Communications*, vol. 283, No. 24, (Dec 15, 2010), pp. 5250-5254, ISSN 0030-4018

Qian, W. W., Zhao, C. L., He, S. L., Dong, X. Y., Zhang, S. Q., Zhang, Z. X., Jin, S. Z., Guo, J. T. & Wei, H. F. (2011). High-sensitivity temperature sensor based on an alcohol-filled photonic crystal fiber loop mirror. *Optics Letters*, vol. 36, No. 9, (May 1, 2011), pp. 1548-1550, ISSN 0146-9592

Silva, R. M., Layeghi, A., Zibaii, M. I., Santos, J. L. & Frazão, O. (2011). Theoretical and Experimental Results of High Birefringent Fiber Loop Mirror with an Output Port Probe. *Journal of Lightwave Technology*, vol. -, No. -, 2011), pp. -, -(Submitted)

Starodumov, A. N., Zenteno, L. A., Monzon, D. & Delarosa, E. (1997). Fiber Sagnac interferometer temperature sensor. *Applied Physics Letters*, vol. 70, No. 1, (Jan 6, 1997), pp. 19-21, ISSN 0003-6951

Urquhart, P. (1989). Fiber Lasers with Loop Reflectors. *Applied Optics*, vol. 28, No. 17, (Sep 1, 1989), pp. 3759-3767, ISSN 0740-3224

Yang, X. F., Zhao, C. L., Peng, Q. Z., Zhou, X. Q. & Lu, C. (2005). FBG sensor interrogation with high temperature insensitivity by using a HiBi-PCF Sagnac loop filter. *Optics Communications*, vol. 250, No. 1-3, (Jun 1, 2005), pp. 63-68, ISSN 0030-4018

Zhang, H., Liu, B., Wang, Z., Luo, J. H., Wang, S. X., Jia, C. H. & Ma, X. R. (2010). Temperature-insensitive displacement sensor based on high-birefringence photonic crystal fiber loop mirror. *Optica Applicata*, vol. 40, No. 1, 2010), pp. 209-217, ISSN 0078-5466

Zhao, C. L., Yang, X. F., Lu, C., Jin, W. & Demokan, M. S. (2004). Temperature-insensitive interferometer using a highly birefringent photonic crystal fiber loop mirror. *IEEE Photonics Technology Letters*, vol. 16, No. 11, (Nov, 2004), pp. 2535-2537, ISSN 1041-1135

Zhou, D. P., Wei, L., Liu, W. K. & Lit, J. W. Y. (2008). Simultaneous measurement of strain and temperature based on a fiber Bragg grating combined with a high-birefringence fiber loop mirror. *Optics Communications*, vol. 281, No. 18, (Sep 15, 2008), pp. 4640-4643, ISSN 0030-4018

Zibaii, M. I., Latifi, H., Frazão, O. & Jorge, P. a. S. (2011). Controlling the Sensitivity of Refractive Index Measurement using a Tapered Fiber Loop Mirror. *IEEE Photonics Technology Letters*, (2011), (Accepted)

Zu, P., Chan, C. C., Jin, Y. X., Zhang, Y. F. & Dong, X. Y. (2011a). Fabrication of a temperature-insensitive transverse mechanical load sensor by using a photonic crystal fiber-based Sagnac loop. *Measurement Science & Technology*, vol. 22, No. 2, (Feb, 2011a), ISSN 0957-0233

Zu, P., Chan, C. C., Siang, L. W., Jin, Y. X., Zhang, Y. F., Fen, L. H., Chen, L. H. & Dong, X. Y. (2011b). Magneto-optic fiber Sagnac modulator based on magnetic fluids. *Optics Letters*, vol. 36, No. 8, (Apr 15, 2011b), pp. 1425-1427, ISSN 0146-9592

15

Fiber Optic Displacement Sensors and Their Applications

S. W. Harun[1,2], M. Yasin[1,3], H. Z. Yang[1] and H. Ahmad[1]
[1]*Photonic Research Center, University of Malaya, Kuala Lumpur*
[2]*Department of Electrical Engineering, Faculty of Engineering, University of Malaya, Kuala Lumpur*
[3]*Department of Physics, Faculty of Science and Technology, Airlangga University, Surabaya*
[1,2]*Malaysia*
[3]*Indonesia*

1. Introduction

Optical fiber-based sensor technology offers the possibility of developing a variety of physical sensors for a wide range of physical parameters (Nalwa, 2004). Compared to conventional transducers, optical fiber sensors show very high performances in their response to many physical parameters such as displacement, pressure, temperature and electric field. Recently, high precision fiber displacement sensors have received significant attention for applications ranging from industrial to medical fields that include reverse engineering and micro-assembly (Laurence et al., 1998; Shimamoto & Tanaka, 2001); Spooncer et al., 1992; Murphy et al., 1991). This is attributed to their inherent advantages such as simplicity, small size, mobility, wide frequency capability, extremely low detection limit and non-contact properties. One of the interesting and important methods of displacement measurement is based on interferometer technique (Bergamin et al., 1993). However, this technique is quite complicated although it can provide very good sensitivity. Alternatively, an intensity modulation technique can be used in conjunction with a multimode fiber as the probe. The multimode fiber probes are preferred because of their coupling ability, large core radius and high numerical aperture, which allow the probe to receive a significant amount of the reflected or transmitted light from a target (Yasin et al., 2009; Yasin et al., 2010; Murphy et al., 1994). For future applications, there is a need for better resolution, longer range, better linearity, simple construction and low cost unit.

In this chapter, fiber-optic displacement sensors (FODS) are demonstrated using an intensity modulation technique. This technique is one of the simplest techniques for the displacement measurement, which is based on comparing the transmitted light intensity against that of the launch light to provide information on the displacement between the probe and the target. A silicon photo-diode is used to measure the transmitted and reflected light intensity. The sensor performances are investigated for various laser sources, different probes types and arrangements and different target. The theoretical analysis and the corresponding results on various bundled fiber based sensors are also presented in this chapter. The application of the FODSs in liquid refractive index measurement is investigated theoretically and experimentally. In the last part of this chapter, a continue monitoring the liquid level is also demonstrated by using the FODS.

2. FODS with beam-through technique

The intensity based sensors can be achieved by either beam-though or reflective techniques. A change in displacement of the through-beam and reflective sensors are manifested as a variation in the transmitted light and reflected light intensity, respectively. This section demonstrates a simple design for an intensity-based displacement sensor using a multimode plastic fiber in conjunction of beam-through technique. The performances of this sensor are investigated for both lateral and axial displacements. In the sensor, light is transmitted through a transmitting fiber to a receiving fiber and the received light is then measured by a silicon detector.

Fig. 1 shows a schematic diagram of the proposed sensor, which consists of two set of fiber, one set is connected to a light source and is termed as the transmitting fiber, and the other set is connected to a silicon detector and is known as the receiving fiber. In the experiment, the transmitting fiber located opposite to the receiving fiber is moved laterally and axially. The light is scattered after travelling out from the transmitting fiber and the receiving fiber collect a portion of the scattered light to transmit into the silicon detector where its intensity is measured. The intensity of the collected light is a function of axial and lateral displacement of the fiber. The light source is a He-Ne laser with a peak wavelength of 633 nm. The light is modulated externally by chopper with a frequency of 200 Hz, which is connected to lock-in amplifier to reduce the dc drift and interference of ambient stray light. For an axial and lateral displacement, a flexible adjusting mechanism

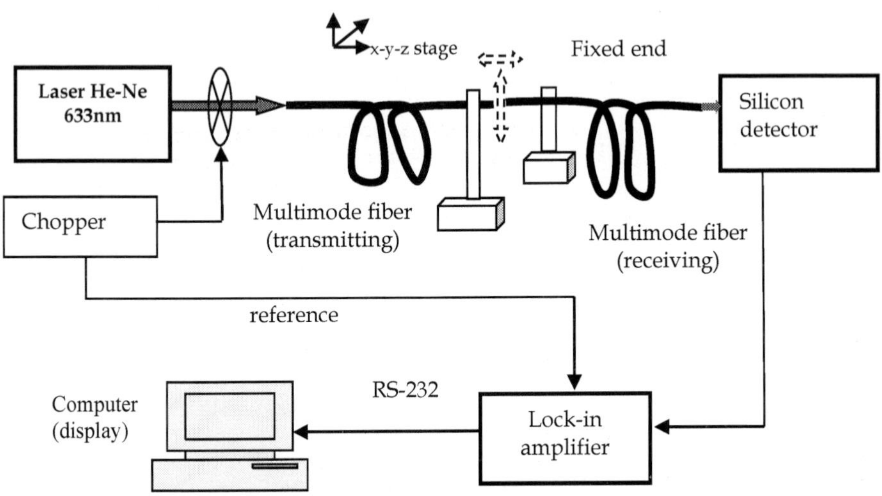

Fig. 1. Schematic diagram for lateral and axial displacement sensing using beam-through technique.

using piezoelectric is required, so the receiving fiber tip is mounted on a translational stage, which provides fine movement of the transmitting fiber surface in the axial and lateral direction. In this experiment, the axial and lateral micro distance is varied and the lock-in amplifier output voltage of the transmitted light is directly recorded by a computer automatically using Delphi software through serial port RS232. The piezoelectric micrometer can provide precise changes of about 25 and 30 nm for every positive and negative pulse, respectively, and in this experiment the displacement measurement is taken in successive steps of 45nm.

To analyze the performance of the proposed FODS, the output voltage from a receiving fiber is related to the axial and lateral displacements of the transmitting fiber. Both fibers should be mounted perpendicular to each other and positioned flush against the surface. The output voltage of the sensor should be highest in this position. The transmitting fiber is then moved away laterally and axially from the receiving fiber tip by still maintaining perpendicularity between them. Fig. 2 shows the output voltage of the lock-in amplifier against the lateral displacement between the two ends of the fibers. In the experiment, the core diameter for both ends is varied. As expected, the voltage is highest at zero displacement from the center and the lateral movements of the transmitting fiber away from the receiving fiber resulted in a reduced output voltage as shown in the figure. The power drop pattern follows the theoretical analysis by Van Etten & Van der Plaats (1991) in which the output transmission function is given by:

$$\eta = \frac{2}{\pi}\left(\left(acr\cos\left(\frac{d}{a}\right) - \frac{d}{a}\sqrt{1-\left(\frac{d}{a}\right)^2}\right)\right) \quad (1)$$

where η, d, and a are coupling efficiency, lateral displacement, and fiber core radius, respectively. η is defined as the ratio of output voltage over the maximum voltage. The sensitivity of the sensor is determined by a slope of a straight-line portion in the curves.

As shown in Fig. 2, the beam-through type of sensor has two symmetrical slopes and the sensitivity is higher at the smaller core diameter. At core diameter of 0.5 mm for both transmitting and receiving fibers, the sensitivity is obtained at around 0.0008mV/µm and the slope shows a good linearity of more than 99% within a range of 420µm. The linear range increases to around 800µm for the both slopes as the core diameter increased to 1.0 mm. The linear range can be further increased to more than 1000µm by using a larger core for the receiving fiber as shown in Fig. 2. However, the voltage is unchanged at a small lateral displacement for this sensor due to the larger receiving core, which covers the whole diverged beam from the transmitting fiber. The highest resolution of 13µm is obtained with core diameter of 0.5 mm for both fibers. In this work, the resolution is defined as the minimum displacement which can be detected by this sensor.

The performance of the sensor with lateral displacement is summarized as shown in Table 1. The output voltage from a receiving fiber is related to axial displacement of end surface of the transmitting fiber. Fig. 3 shows the output voltage of the lock-in amplifier against the axial displacement for the different core diameters. In this experiment, the end surface of transmitting fiber is moved away from the receiving fiber tip by still maintaining perpendicularity between them. As expected, the voltage is highest at zero displacement from the center and the output voltage reduces as the axial displacement increases for all

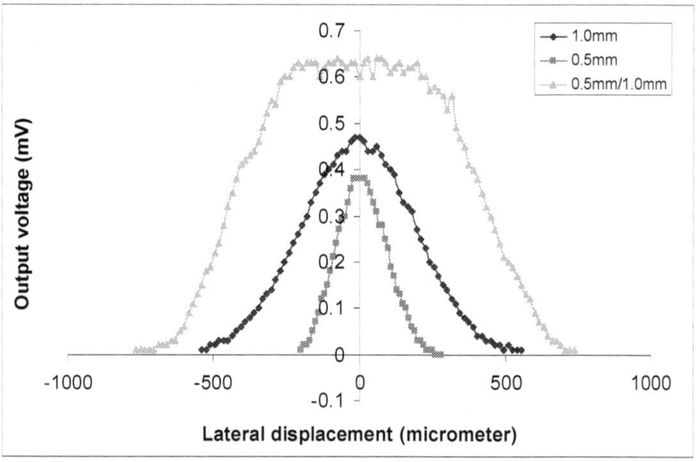

Fig. 2. The output voltage of the lock-in amplifier against the lateral displacement of the transmitting fiber.

Fiber's core diameter	The left slope		The right slope		Resolution (μm)
	Sensitivity (mV/μm)	Linear range (μm)	Sensitivity (mV/μm)	Linear range (μm)	
1.0mm /1.0mm	0.0003	855(585-1440)	0.0004	765 (1845-2610)	33
0.5mm /0.5mm	0.0008	420(120-540)	0.0007	420 (750-1170)	13
0.5mm /1.0mm	0.0005	1035(450-1485)	0.0005	1125 (3060-4185)	20

Table 1. Performance of the Lateral Displacement Sensor

core diameters used. The power drop pattern follows the theoretical analysis from (Van Etten & Van der Plaats, 1991) in which the output transmission function is given by:

$$\eta \approx 1 - \frac{z}{a}\frac{2}{\pi(NA)^2}\left(\arcsin(NA) - NA\sqrt{1-(NA)^2}\right) \tag{2}$$

where η, z, a, and NA are coupling efficiency, axial displacement, core radius, and numerical aperture, respectively.

As shown in Fig. 3, the sensors only have one slope and the sensitivity is higher at the smaller core diameter. At core diameter of 0.5 mm (for both transmitting and receiving fibers), the sensitivity is obtained at around 0.0002 mV/μm, which is the highest and the slope shows a good linearity of more than 99% within a range of 900μm. The linear range increases to 3195 m with the larger core diameter of 1.0 mm. In case of the receiving core is bigger than the transmitting core, the voltage is almost constant at small axial displacements due to the coherent light source, which has a small divergence angle. The highest resolution

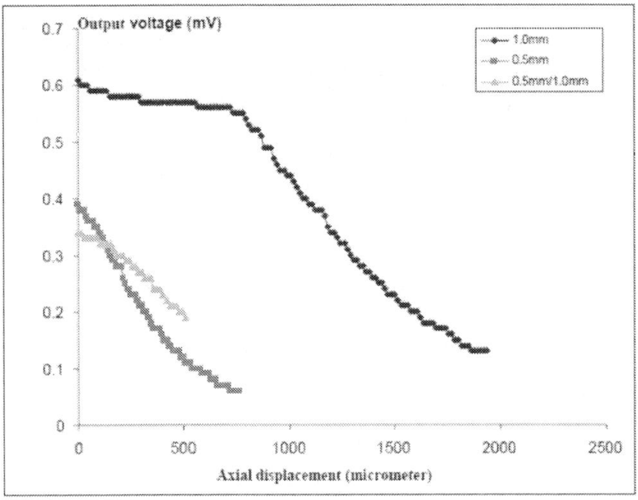

Fig. 3. The output voltage of the lock-in amplifier against the axial displacement of the transmitting fiber.

is obtained at 50 m with 0.5 mm core diameter for both fibers. The performance of the sensors with axial displacement is summarized as shown in Table 2. The stability of these sensors is observed to be less than 0.01 mV (3%). The experimental results are capable of offering quantitative guidance for the design and implementation of the displacement sensor. This sensor requires two probes, which is precisely aligned and therefore the applications are limited. In the next section, various reflective types of FODSs are proposed to improve the sensitivity, linearity and dynamic range of the sensor.

Fiber's core diameter	Sensitivity (mV/µm)	Linear range (µm)	Resolution (µm)
1.0mm/1.0mm	0.0001	3195 (2340-5535)	100
0.5mm/0.5mm	0.0002	900(0-900)	50
0.5mm/1.0mm	0.0001	1530(0-1530)	100

Table 2. Performance of the Axial Displacement Sensor

3. FODS with reflective configuration

The FODS based on intensity modulation and reflective arrangement provides a promising solution for displacement measurement in terms of wide dynamic range, with high potential for ultra-precise non-contact sensing. It also provides flexibility in incorporating the optical sensors permanently into composite structures for monitoring purpose [Wang et al., 1997). In the simplest design of reflective FODS, a probe with a pair of fibers is normally used as the media to transfer/collect the light to/from the target and its theoretical analysis is fully contributed (Faria, 1998). In the design of FODS system the sensor probe is playing a majority role comparison with the selection of laser source and reflector. Hence, the researchers are paid more attentions in the development of sensor probe to improve the performance of FODS. The FODSs described in this section are focus on the various configurations of sensor probe since they are mainly influence the performance of FODS.

3.1 FODS using a probe with two receiving fibers

In this section, a new configuration of the FODS is reported by using two receiving fibers which are bundled together. The mathematical analysis of FODS is developed to simulate the theoretical results, which is then compared to the experimental result. The performance of this FODS is also compared experimentally with the conventional FODS. The probe structure of the proposed FODS is shown in Fig. 4. It consists of one transmitting and two receiving plastic multi-mode fibers bundled together in parallel. To analyze the theory of this sensor setup, a more realistic approach - Gaussian beam is used to describe the light leaving the transmitting fiber. The irradiance of emitted light is obeying an exponential law according to

$$I(r,z) = \frac{2P_E}{\pi\omega^2(z)}\exp\left(-\frac{2r^2}{\omega^2(z)}\right) \tag{3}$$

where P_E is the emitted power from the light source, r is the radial coordinate and z is the longitudinal coordinate. $\omega(z)$ is the beam radius which is also a function of z, $\omega(z) = \omega_0\sqrt{1+(\frac{z}{z_R})^2}$. The waist radius ω_0 and Rayleigh range z_R are the important parameters in the Gaussian Beam function.

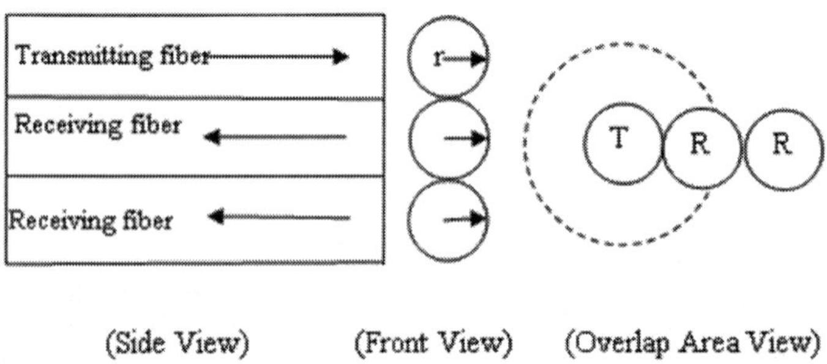

Fig. 4. Side, Front and Overlap Views of FODS probe structure with two receiving fibers

The optical power received by the receiving fiber can be evaluated by integrating the irradiance, I over the surface area of the receiving fiber end, S_r.

$$P(z) = \iint_{S_r} I(r,z)dS_r \tag{4}$$

The overlapping area of the reflected light area and the core of the receiving fibers is also illustrated in Fig 4. The power value of reflected light collected by two receiving fibers increased with the increased of displacement of probe and target mirror. Two receiving fibers collected the reflecting light results the receiving light power increased. Based on this geometrical analysis two receiving fibers collected the reflected light significantly affects the transfer function of the FODS.

The Eq. (3) can be described in other expressions in order to simulate conveniently. The power collected by the first and second receiving fibers are denoted as P_1 and P_2, respectively where P_1 is closer to the transmitting fiber. From Eq. (3), P_1 and P_2 can also be written as;

$$P_1(z) = \frac{2P_E}{\pi\omega^2(z)} \int_{y=-R_r}^{R_r} \int_{x=R_t+R_r-\sqrt{R_r^2-y^2}}^{R_t+R_r+\sqrt{R_r^2-y^2}} \exp\left(-\frac{2(x^2+y^2)}{\omega^2(z)}\right) dxdy \quad (5)$$

$$P_1(z) = \frac{2P_E}{\pi\omega^2(z)} \int_{y=-R_r}^{R_r} \int_{x=R_d+R_t+R_r-\sqrt{R_r^2-y^2}}^{R_d+R_t+R_r+\sqrt{R_r^2-y^2}} \exp\left(-\frac{2(x^2+y^2)}{\omega^2(z)}\right) dxdy \quad (6)$$

where the R_t, R_r is the radius of transmitting fiber and receiving fiber, respectively. The R_d is the diameter of the receiving fiber. The radial coordinate r is expressed by $\sqrt{x^2+y^2}$ in Cartesian coordinate system. Then the total amount of power collected by both receiving fibers is;

$$P = P_1 + P_2 \quad (7)$$

The conventional FODS only collects power P_1 using a pair of fibers bundled together. Compared to this sensor, the proposed FODS collects more reflected light due the additional P_2 power, which increases the dynamic range of the sensor.

The software simulation is programmed and implemented in MATLAB. Some important parameters are specified in the programming, wavelength of the laser source λ = 594nm, transmitting fiber core radius R_t = 0.5mm and numerical aperture value NA = 0.25. Fiber diameter R_d = 2mm. The theoretical analysis transfer function of proposed displacement sensor in Eq. (6) can be normalized by its maximum collected power P_{max}, $P_n = P/P_{max}$. The normalized distance $h_n = h/z$, while power increase to maximum for small values h from 0 to z, power decrease to zero according to the h values higher than z. The simulation results are then compared with the experimental results.

The experiment setup for the FODS with two receiving fibers is shown in Fig. 5. It consists of a light source, a chopper, a sensor probe, a flat mirror, a silicon detector, a lock-in amplifier and a PC. The sensor probe consists of one transmitting and two receiving plastic multi-mode fibers which are bundled together in parallel. The transmitting and receiving fiber length is 2m with a core diameter of 1mm. A 594nm yellow He-Ne laser is used as a light source. It has the maximum output power of 4mW and beam divergence of 0.92mRads, which is modulated by external chopper with a frequency 200Hz. The transmitting end of fiber probe radiates the modulated light from the laser to the mirror. The flat mirror is controlled by a piezoelectric motor and driver. The distance between the fiber probe and the mirror is varied in successive steps of 4μm and the light voltage which is represented the optical power is measured against the change in the mirror displacement stage. Then the mirror reflects the transmitting light into the receiving end of fiber probe. The reflected light through receiving end can be detected by the silicon detector. The photon energy collected by detector is converted into a voltage. The output of the detector transfers into the lock-in amplifier to deduce the dc drift and filter out the undesired noise. The lock-in amplifier is connected to a PC using RS232 data series line. From the PC, the output light voltage is monitored.

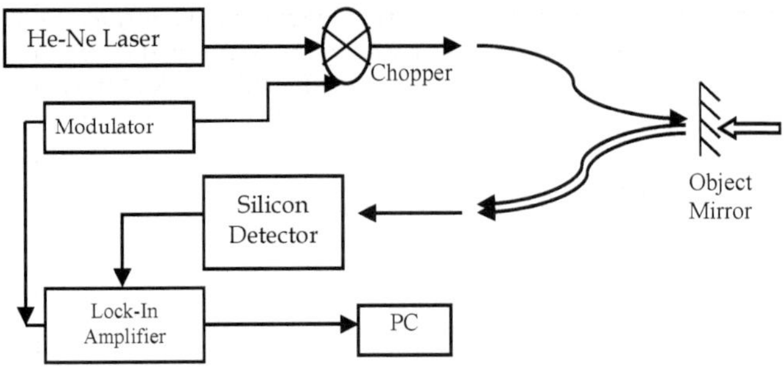

Fig. 5. Experiment setup of proposed FODS with two receiving fibers.

In both theoretical and experimental analysis, the results are processed and displayed in the normalized forms which the output power is normalized by the maximum collected optical power and the displacement is normalized by the distance parameter z. This is to make the output function a dimensionless function and eliminate the dependency of the FODS output function to the fiber core radius and divergence angle. Fig. 6 shows the comparison of the simulated and the observed response of the proposed FODS. As shown in Fig. 6, both curves exhibit a maximum with a steep linear front slope and back slope which follows an almost inverse square law relationship for the reflected light intensity versus distance of the mirror from transmitting fiber end. The signal is minimal at zero distance because the light cone does not reach the core of both receiving fibers. As the displacement increases, the size of cone of the reflected light at the plane of fiber also increases, which then starts to overlap with the receiving fiber cores leading to a small output voltage. Further increases in the displacement lead to larger overlapping which in turn results in an increase in the output voltage. However, after reaching the maximum value, the output voltage starts decreasing even though the displacement increases. This is due to the large increase in the size of the light cone and the power density decreases with the increase in the size of the cone of light.

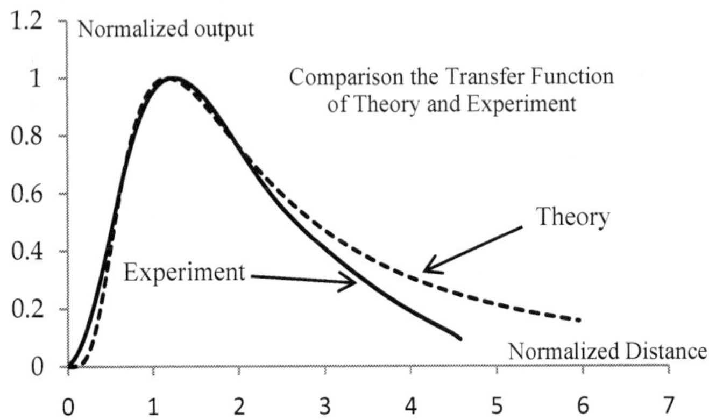

Fig. 6. Normalized collected optical power versus the normalized displacement curves for both theoretical and experimental results.

The maximum normalized output powers of 1 are obtained at the normalized displacement distances of 1.2 for the theoretical curve and 1.3 for the experimental curve as shown in Fig. 6. The close agreement of theoretical and experimental results in the Fig. 6 is quite evident. The small difference can be accounted for experimental limitations mainly due to the geometry error of the fiber used and positioning error. Table 3 summarizes the performance comparison of the simulated and the experimental result of the sensor. The slope of the response curve is the sensitivity, which is expressed in the unit of mW/µm. As shown in the table, the linearity range at the back slope is nearly 3 times that of the linearity range at the front slope. However, the sensitivity at the front slope is nearly 3~4 times higher than that of the sensitivity at the back slope.

Methods	The front slopes		The back slopes	
	Linearity range	Sensitivity	Linearity range	Sensitivity
Theoretical	0.8 µm (0.3 ~ 1.1)	0.06 mV/µm	2.5 µm (1.4 ~ 3.9)	0.015mV/µm
Experimental	0.9 µm (0.2 ~ 1.1)	0.06 mV/µm	2.5 µm (1.5 ~ 4.0)	0.017mV/µm

Table 3. The performance comparison between theoretical and experiment results.

The experiment is also repeated for the conventional sensor with one transmitting and one receiving fibers. The response of the conventional FODS is then compared to that of the proposed FODS as shown in Fig. 7. As shown in the figure, the maximum power collected by the receiver is obtained at a shorter distance for the conventional sensor. The sensor performance comparison between the proposed and the conventional sensors is summarized in Table 4. The sensitivity of both sensors is almost similar. However, the proposed sensor obviously has a better linearity range as shown in Table 4. This is attributed to the amount of the collected light intensity, which is higher in the proposed sensor compared to that of the conventional one.

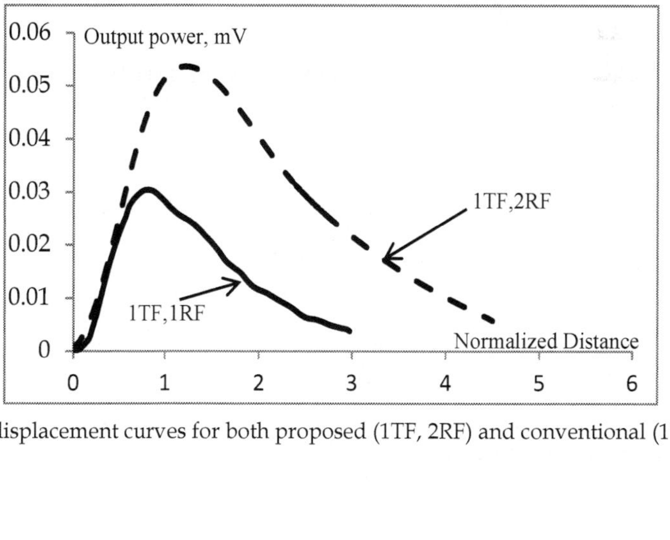

Fig. 7. The displacement curves for both proposed (1TF, 2RF) and conventional (1TF, 1RF) sensors

Methods	The front slopes		The back slopes	
	Linearity range	Sensitivity	Linearity range	Sensitivity
Conventional	0.5 μm (0.2 ~ 0.7)	0.06 mV/μm	1.4 μm (0.9 ~ 2.3)	0.016 mV/μm
Proposed	0.9 μm (0.2 ~ 1.1)	0.06 mV/μm	2.5 μm (1.5 ~ 4.0)	0.017 mV/μm

Table 4. The performance comparison between the proposed and conventional sensors

The maximum linearity ranges of 2.5 μm and 0.9 μm are obtained at the back and front slopes respectively for the proposed sensor as shown in Table 4. The linearity range of the proposed sensor is improved by about 44% for both slopes compared to the conventional sensor. As indicated by the above results, we can conclude that the employment of two receiving fibers increases the linearity range of the sensor, which is very useful for the large displacement measurements. Both theoretical and experimental results are capable of offering quantitative guidance for the design and implementation of the displacement sensor.

3.2 The FODS with two asymmetrical fibers bundled

As discussed in most of the FODSs [Ko et al., 1995; Elasar et al., 2002; Oiwa & Nishitani, 2004; Cao et al., 2007), the radius of the transmitting and receiving fibers are often made the same for the convenience of analytical study and experiments. However, there is a lack of research work on the displacement sensor using bundled fiber with different core radius. In this work, a mathematical model of displacement sensor using asymmetrical bundled fiber is developed. Some simulations were carried out based on the mathematical model and experimental results were also obtained to validate the MATLAB simulated results. The effect of different core radial ratios (CRRs) to the dynamic range, sensitivity and illumination area of bundled fiber are analyzed and discussed in this section.

Theoretical analysis

The proposed FODS consists of a transmitting and receiving fibers as well as a reflecting mirror. Both fibers are of different core radius and are bundled together in parallels as shown in Fig. 8. Let r_T and r_R denote the core radius of the transmitting fiber and the core radius of the receiving fiber. The core radial ratio, CRR is the ratio of transmitting fiber and receiving fiber core radius, as given below:

$$\text{CRR, } k = \frac{r_R}{r_T} \tag{8}$$

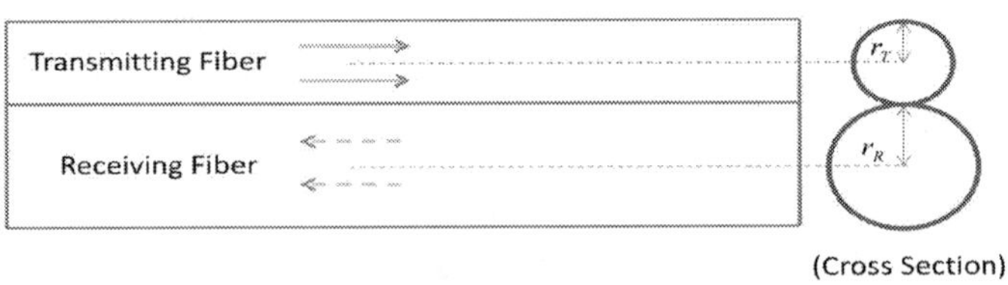

Fig. 8. Side and front views of the transmitting and receiving fiber ends.

Fig. 9 shows the geometrically illustration of the overlapping area between the reflected light circle and the core of the receiving fiber at different CRR. Based on this figure, in the same core radius of transmitting fiber; the reflected light power collected by the receiving fiber increases with the increasing core radius of the receiving fiber. The larger receiving fiber core radius and core area, the bigger fraction of reflected light can be collected by the receiving fiber. In the previous report (Faria, 1998), two major approaches have been introduced for theoretical analysis, namely geometrical and Gaussian Beam approaches. For the former approach, the simple assumption is made that the light intensity is constant within the reflected light circle. On the other hand, the light intensity outside the reflected light circle is null. This approach is apparently less accurate compared to the second approach. Gaussian Beam approach is a more realistic and more accurate method. The intensity of the light emitted from the transmitting fiber is described with Gaussian distribution as shown in Eq. (3).

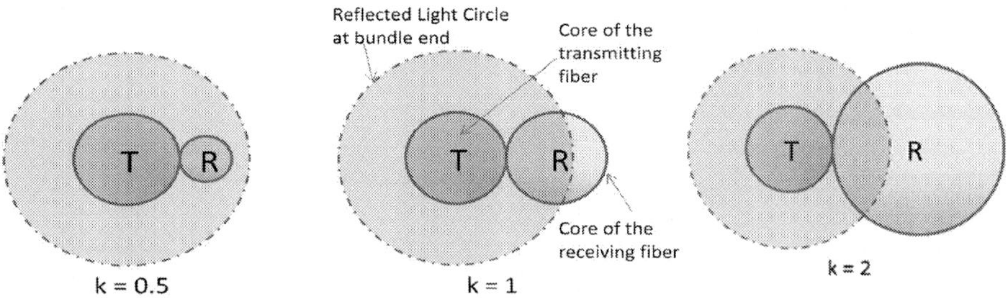

Fig. 9. Geometrical illustration for the overlapping area between the reflected light circle and the core of the receiving fiber at different CRRs.

The light power collected by the receiving fiber can be evaluated by using integral as shown in Eq. 4. However, the exact integration is tedious and impossible. Therefore, assumptions and approximation were used to dissolve the integration. For points situated in the far-field, $z >> z_R$ the following relations with the divergence angle can be obtained

$$\theta_a = \tan\theta_a = \frac{\omega(z)}{z} = \frac{\omega_0}{z_R} = \frac{\lambda}{\pi\omega_0} \tag{9}$$

Core radius of the transmitting fiber and receiving fibers are given by the approximation

$$\omega_T = z_a \tan\theta_a \approx z_a\theta_a \tag{10}$$

and

$$\omega_R = k\omega_T = kz_a\theta_a \tag{11}$$

where z_a is the distance between the beam source to the fiber end (Wang et al., 1997). The core area of the receiving fiber is computed from

$$S_a = \pi\omega_R^2 = \pi k^2 z_a^2 \theta_a^2 \tag{12}$$

The radial distance between the two core centers of the transmitting fiber and receiving fiber is determined from

$$r = \omega_T + \omega_R = \omega_T + k\omega_T = (1+k)z_a\theta_a \tag{13}$$

The path of the beam from the beam source to the bundle end after the reflection is given by

$$z_a + 2h \tag{14}$$

The displacement parameter in the normalized form is presented as

$$\zeta = \frac{z_a + 2h}{z_a} \tag{15}$$

or

$$\zeta = 1 + 2h_N \tag{16}$$

where $h_N = h/z_a$. To relate the displacement between the reflective mirror to the fiber end, h to the transfer function, with the help of the results determined above, the collected power of the receiving fiber can be expressed as

$$\begin{aligned} P(h) &= \frac{2P_E}{\pi\omega^2(z_a+2h)}\exp\left(-\frac{2r^2}{\omega^2(z_a+2h)}\right)\times S_a \\ &= \frac{2P_E k^2 z_a^2}{(z_a+2h)^2}\exp\left(-\frac{2((1+k)z_a)^2}{(z_a+2h)^2}\right) \end{aligned} \tag{17}$$

By substituting Eq. (15) into this equation, we obtain

$$P(\xi) = \frac{2P_E k^2}{\xi^2}\exp\left(-\frac{2(1+k)^2}{\xi^2}\right) \tag{18}$$

The maximum received power is achieved when $P'(\xi) = 0$, and this leads to

$$\xi_{\max} = \sqrt{2}(1+k) \tag{19}$$

Based on the relation in Eq. (3-16), the maximum h is given by

$$h_{\max} = \frac{\sqrt{2}k + \sqrt{2} - 1}{2} \tag{20}$$

The maximum power is given by

$$P_{\max} = P(\sqrt{2}(1+k)) = \frac{k^2 P_E}{(k+1)^2}\exp(-1) \tag{21}$$

In the normalized form, Eq. (3-15) is rewritten as

$$P_N(\xi) = \frac{P(\xi)}{P_{\max}} = \frac{2(k+1)^2}{\xi^2}\exp\left(1 - \frac{2(1+k)^2}{\xi^2}\right) \tag{22}$$

In the analysis, the theoretical model of the FODS is modeled based on the similar parameters used in the experiment: Wavelength of the laser source λ = 594nm, transmitting fiber core radius r_R =0.5mm and numerical aperture value NA = 0.4. Based on the same parameters, four analytical model were simulated for k = 0.5, 1, 2 and 3 which were based on the available fiber core radius combinations in the experiments.

Experiment

The experiment for the FODS with two asymmetrical bundled fibers is carried using the similar set-up as shown in Fig. 5, but using bundled fiber with different core radius. The asymmetrical bundled fiber is constructed by pairing two different plastic fibers with the core radiuses of either 0.25mm or 0.50mm or 0.75mm. Due to the limited selections of core diameters, six combinations were selected for the experiments: $[k, r_T, r_R]$ = [0.5, 0.5mm, 0.25mm], [1.0, 0.5mm, 0.5mm], [2.0, 0.25mm, 0.5mm] and [3.0, 0.25mm, 0.75mm]. *k* is the core radial ratio. A precise displacement reference between the bundle end and the reflecting mirror is imperative for the experiment. Therefore, a New Focus 9061 motorized stages which is driven by a picomotor is used to change the displacement of the reflecting mirror from the fiber probe. Each increment step in the displacement is made identical and accurate. The collected light power in the receiving fiber is converted by a silicon detector into electrical power. Lastly, the electrical signal is filtered by a lock-in amplifier and recorded in the computer.

Characteristic of the sensor

In both theoretical and experimental analysis, the results are processed and displayed in the normalized forms which the output power is normalized by the maximum output power and the displacement is normalized by the parameter z_a. This is to make the output function a dimensionless function and eliminate the dependency of the FODS output function to the fiber core radius and divergence angle. Figs. 10 and 11 show the analytical and experimental results respectively for the proposed FODS. As shown in both figures, the location of the maximum output is shifted toward the right along the axis of displacement as the value of k increases. Besides, the linear range on the front slope and back slope gets larger for every larger value of *k*. Both graphs exhibit the almost the same trend of characteristics in the curves as the value of k becomes larger. This phenomenon can be explained by the use of distinctive core radius of the two fibers. As shown in Fig. 9, at the same displacement the fraction of overlap area in receiving fiber core by the reflected light circle (percentage of shaded area in the receiving fiber core) is differ for different CRR. For a larger value of k, the fiber displacement sensor requires further displacement to achieve maximum overlap area. Adversely, the sensitivity of the fiber displacement sensor decreases as the CRR increases. On the other hand, some error in the initial displacement ($0 < h < 0.3$) is observed if the two overlaid graphs are compared. This error is accounted to the approximation used in the theoretical analysis.

The performance of the proposed FODS from the experimental results is summarized in Table 5. The results show that the magnitude of the sensitivity decreases as the CRR or *k* value becomes larger while the linear range is larger for a larger value of *k*. The sensitivity

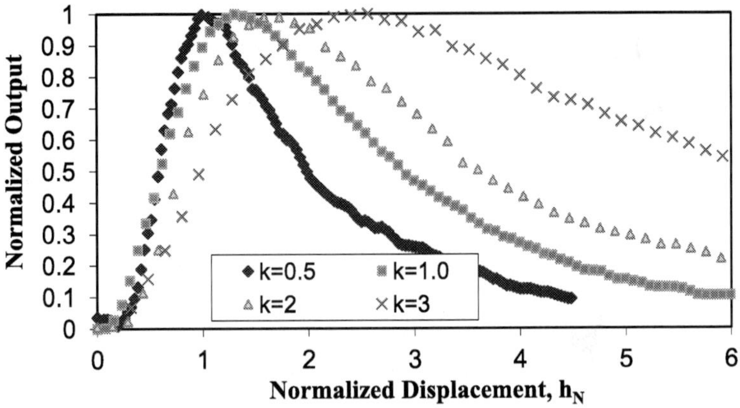

Fig. 10. The experimental result of proposed FODS model at different CRRs or k values.

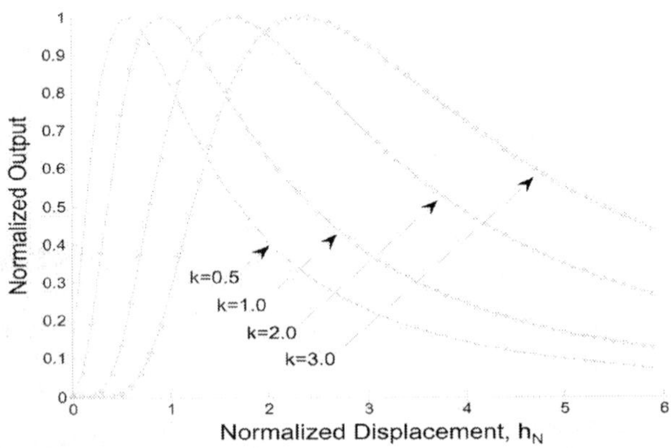

Fig. 11. The theoretical result of the fiber displacement sensor at different CRRs or k values.

CRR , *k*	Front Slope Sensitivity	Linear Range (mm)	Back Slope Sensitivity	Linear Range (mm)
0.5	1.7615	0.320 - 0.800	-0.5270	1.312 - 1.176
1	1.0984	0.230 - 0.998	-0.3378	1.536 - 3.302
2	0.9688	0.288 - 1.296	-0.2581	1.728 - 4.320
3	0.6955	0.320 - 1.440	-0.1366	2.240 - 5.400

Table 5. The sensitivity and linear range for different k values

characteristic trend is consistent with the theoretical plot as shown in Fig. 12. Fig. 12 shows the normalized sensitivity against normalized displacement at various k values. The curve width of the graph represents the linear range of the sensor. As shown in the figure, the linear range of the sensor increases with the value of k which is in agreement with the

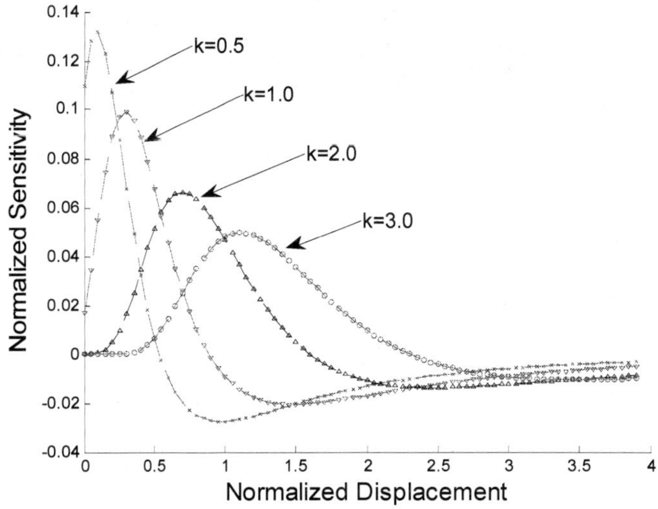

Fig. 12. Theoretical normalized sensitivity curves of the FODS at various *k* values.

experimental result in Table 5. This property provides a greater enhancement in FODS applications in terms of flexibility, wider dynamic range and high precision displacement measurement. The maximum sensitivity of 1.76 is obtained at *k*=0.5. The largest dynamic range of Fig. 10 is obtained at *k*=3. The conventional FODS which uses two fibers with identical core radius often encounters several restrictions due to limited linear range for the measurement. Besides, the limited option of sensitivity often becomes a challenge in the high-precision measurement. This restriction can be avoided using a suitable CRR or *k* value. The *k* value can be chosen in a way to provide the optimum performance.

3.3 The FODSs with two asymmetrical fibers inclined

To date, many works have been reported on the intensity modulation based FODSs (Bock et al., 2001; Cui et al., 2008; Saunders & Scully, 2007; Miclos & Zisu, 2001; Kulchin et al., 2007), which the probe consists of a pair of fibers used for transmitting and receiving the light. For instance, Buchade and Shaligram, et al. (2006) presented a FODS using two fibers inclined with a same angular angle and reported the sensitivity was enhanced compared with the conventional sensor with parallel bundled fibers. It also reported that the performances of the FODS with two fibers are depended mainly on four parameters: the offset, the lateral separation and the angle between the transmitting and receiving fiber tips, and the angle of the reflector (Buchade & Shaligram, 2007). However, there is still a lack of research work on the FODSs with different geometry of the receiving fiber. Hence, a study of this type of FODS is depicted in this section theoretically and experimentally.

Theoretical

Fig. 13 shows the geometry of the inclined displacement sensor, which consists of a transmitting fiber, receiving fiber and a reflector. The sensor performance is studied at various core radiuses of transmitting and receiving fiber. Fig. 13 (a) (Fig. 13(b)) shows the geometry of the sensor in case of the receiving core is bigger (smaller) than the transmitting core. Two asymmetrical transmitting and receiving fibers are mounted at an angle 'θ' with reference to the normal to the reflector. This ensures the receiving fiber core to collect the maximum power from cone of emitting light of the transmitting fiber. The shortest distances between the sensor probe tips and reflector are x_1 and x_2 for transmitting fiber and receiving fiber, respectively. The dash lines in receiving fiber is represented the size of diameter value same as the transmitting fiber. The image of transmitting (receiving) fiber is formed at a further distance x_1 (x_2) opposite to the transmitting (receiving) fiber beyond the reflector. The image fiber is thus seen located at $2x_1$ or $2x_2$ from the original position of the probe. Effectively the reflected light appears to form a cone and reaches the receiving fiber, which is parallel aligned in the cone as shown in Fig. 13.

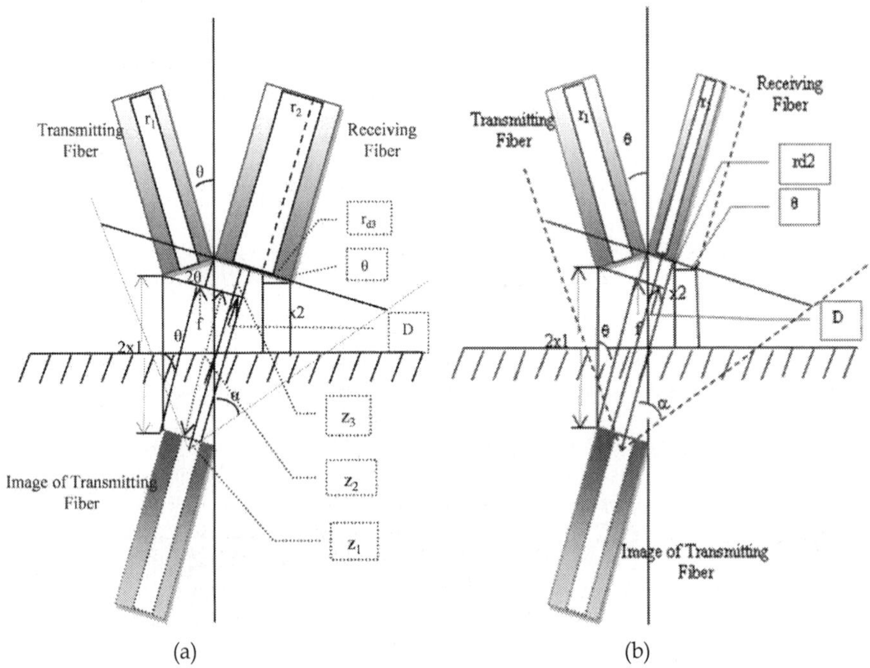

(a) $r_2 > r_1$ (b) $r_2 < r_1$

Fig. 13. The structure of sensor probes with two asymmetrical inclined fibers

As shown in Fig. 13, the core radius of the transmitting and receiving fibers are denoted as r_1 and r_2, respectively. Meanwhile, the diameters of transmitting fiber and receiving fiber are r_{d1} and r_{d2}, respectively. We assume that the ratio between the radius of two fibers is k_1, $k_1 = r_1/r_2$ and the ratio of diameter of two fibers is $k_2 = r_{d1}/r_{d2}$. From the geometry analysis of Fig. 13, the distance between the two sides of image of transmitting and receiving fibers is given by:

$$f = \left| r_{d1} \times \cos(2\theta) - 2x_1 \times \sin(\theta) \right| \tag{23}$$

Then, the distance between two fiber core, D is obtained as;

$$D = f + \frac{r_{d2} - r_{d1}}{2} = f + \frac{r_{d2}(1-k_2)}{2} \tag{24}$$

The acceptance angle of transmitting fiber α is given by $\alpha = \sin^{-1}(\frac{NA}{n})$, where NA is a numerical aperture for the transmitting fiber. The distance between emitting point of transmitted light to the receiving flat core is denoted as z, which is given by;

$$z = z_1 + z_2 + z_3 \tag{25}$$

where $z_1 = r_1 \times cot\alpha = k_1 r_2 \times cot\alpha$, $z_2 = 2x_1 \times cos(\theta)$ and $z_3 = r_{d1} \times sin(2\theta) = k_2 r_{d2} \times sin(2\theta)$ as illustrated in Fig. 13.

To analyze the power collected by the receiving fiber, we simply analysis the light inside the fiber by using a Gaussion beam approach. The irradiance of emitted light is obeying an exponential law according to Eq. (3). The optical power received by the receiving fiber can be evaluated by integrating the irradiance, I over the surface area of the receiving fiber end which is given by Eq. (4). To simulate conveniently, the Eq. (4) can be described in other expressions;

$$P_1(k_1, k_2, z) = \frac{2P_E}{\pi\omega^2(z)} \int_{y=-r_2}^{r_2} \int_{x=D-\sqrt{r_2^2-y^2}}^{D+\sqrt{r_2^2-y^2}} \exp\left(-\frac{2(x^2+y^2)}{\omega^2(z)}\right) dxdy \tag{26}$$

The $P\ (k_1, k_2, z)$ is the power collected by the receiving fiber corresponding the parameters k_1 and k_2. The radial coordinate r is expressed by $\sqrt{x^2+y^2}$ in Cartesian coordinate system.

Fig. 14 illustrates the overlap area of the reflected light area and the core of the receiving fiber. The overlap area is zero at $x_2 = 0$ (Fig. 13(a)) or $x_1 = 0$ (Fig. 13(b)) and at a very small displacement (blind area) where the jacket of the two fibers blocks the reflected. As the displacement is increased further, the overlap area increases and thus increases the total power collected by the receiving core. The total power is maxima when the receiving cone covers the entire receiving core area. After that, the received optical power starts to decay exponentially as the displacement continues to increase. The received optical power is strongly dependent on the core size of the receiving fiber. At inclination angle of 2θ between the transmitting and receiving fibers, the distance x_1 between the sensor probe tip and reflector is given by (Buchade & Shaligram, 2006)

$$x_1 = rd_1(cosec\theta - 2sin\theta)/2 \tag{27}$$

From the geometrical analysis of Fig. 13, the distance x_2 is obtained as; $x_2 = x_1 - r_{d3}sin\theta$ for $r_{d1} < r_{d2}$ (Fig. 13(a)) *or* $x_2 = x_1 + r_{d3}sin\theta$ for $r_{d1} \geq r_{d2}$ (Fig. 13(b)) where $r_{d3} = r_{d2} - r_{d1}$. Therefore, the distance between sensor probe tip and reflector mirror can be summarized as;

$$x_2 = \frac{r_{d2}}{2}\left(k_2 \cos ec\theta + 2\sin\theta(1-2k_2)\right) \qquad (rd1 > rd2)$$

$$= \frac{k_2 r_{d2}}{2}\left(\cos ec\theta - 2\sin\theta\right) \qquad \text{(rd1= rd2)}$$

$$= \frac{r_{d2}}{2}\left(k_2 \cos ec\theta - 2\sin\theta\right) \qquad \text{(rd1< rd2)} \qquad (28)$$

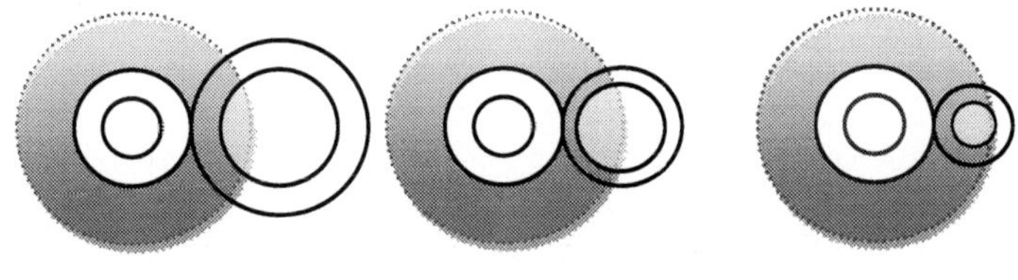

(a) (b) (c)

(a) $r_{d1} < r_{d2}$; (b) $r_{d1} = r_{d2}$; (c) $r_{d1} > r_{d2}$

Fig. 14. Overlap area view

The proposed sensor is simulated by using a MATLAB programming. To simplify the analysis, the k_1 values of 0.5, 0.667, 1, 1.5 and 2 are used. The k_2 value is set based on the availability of the fiber in our laboratory. In this simulation, the k_2 values of 0.5, 1 and 2 are used. The wavelength of the laser source λ is set at 594nm. The numerical aperture values NA_1 = 0.27, NA_2 = 0.32 and NA_3 = 0.4 are used for the core radius of 0.25mm, 0.5mm and 0.75mm, respectively.

Experiment

To verify the simulated results the FODS is constructed by mounting the transmitting and receiving fibers on the plastic board at angle with reference to the normal of the reflector as shown in Fig. 15. Separate samples with various fiber diameters and core radius are prepared for angle = 10°, 20° and 30°. Light from 594 nm He-Ne laser is modulated by an external chopper at frequency of 200 Hz and launched into the transmitting fiber. The light has an average output power of 3.0 mW, beam diameter of 0.75mm and beam divergence of 0.92 mRads. The length of transmitting and receiving fiber length is approximately 2m. The transmitting fiber radiates the modulated light from the light source to the target mirror, the displacement of sensor probe tip between mirror is controlled by a piezoelectric & driver. The reflective light from target mirror, which is mounted in the bottom of tank, is collected by the receiving fiber whose carriers the light into the silicon detector. A lock-in amplifier is connected with the detector to deduce the dc drift. The initial experiment is carried out by varying the inclination angle between the fibers.

Results and discussion

Fig. 16 compares the experimental and theoretical plots of the normalized output collected against normalized displacement between probe and reflector with air medium in between. In this study, the ratios k_1, k_2, and angle θ are set at 0.667, 0.5, 10° respectively. As shown in the figure, the theoretical curve is in good agreement with the experimental curve, verifying

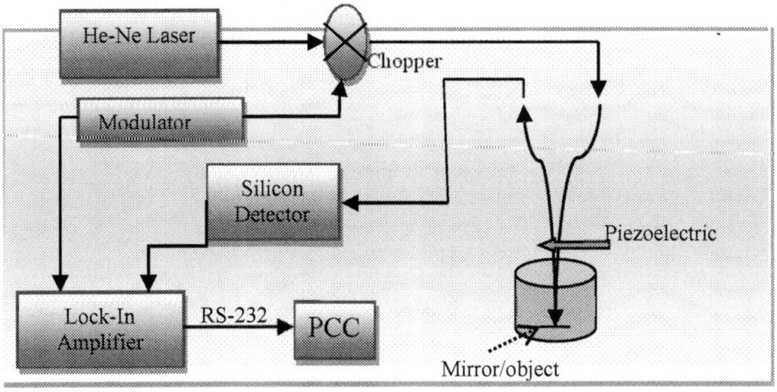

Fig. 15. Experiment setup of proposed FODS with two inclined asymmetrical fibers (Yang, et al., 2009).

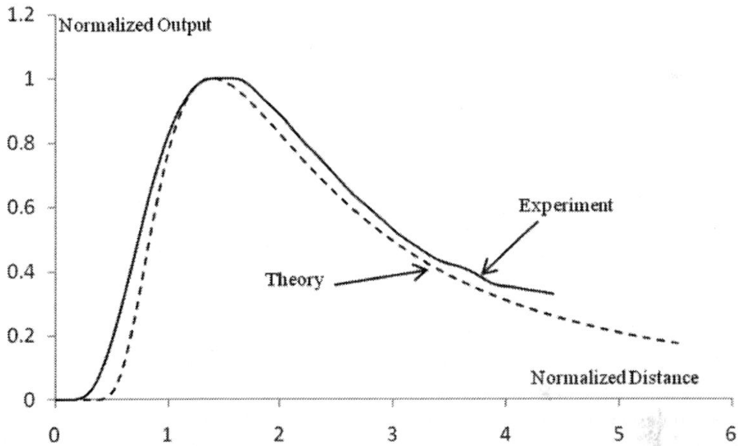

Fig. 16. Comparison between theoretical and experimental curves of the FODS with air medium in between the gap

the feasibility of our theoretical model. It is also observed that up to 0.3 (0.5) of separation for experimental (theoretical), light in transmitting fiber would be reflected back into itself and little or no light would be transferred to receiving fiber. This is then referred to as the blind region. As the distance increases, the reflected cone overlaps the receiving fiber core and hence the output intensity increases. This relation is continued until the entire face of receiving fiber is illuminated with the reflected light. This point is called optical peak and corresponds to maximum voltage. As the gap increases beyond this transition region, the intensity drops off following roughly an inverse-square law. The small discrepancy between the theoretical and experimental results is due to the noise sources such as shot noise and thermal noise, which are added to the value of the experimental results and are not calculated in theoretical analysis.

The experiments are also carried out to study the effect of k_1 and k_2 values as well as angle θ on the performance of FODS. Fig. 17 shows the normalized output power against displacement for the FODS at various k_1, k_2 and angle. Figs. 17 (a), (b) and (c) show the curves at 10°, 20°and 30° respectively with an air gap in between the displacement. By comparing the curves in Fig. 17, we understand that the performance of FODS is strongly depended on the fiber core size. The output power collected by receiving fiber is highest when the k_1 and k_2 values are set at 0.667 and 1, respectively. The inclination angle θ of two asymmetrical fiber core is also effected the sensor performance with the bigger inclined angle has a higher output sensitivity with a lower linearity range. Compared to the FODS with zero inclination angle, the sensitivity of the proposed sensor increased by 3.6, 8.5 and 16 times with the inclination angles of 10°, 20° and 30°, respectively. However, the corresponding linear range reduced by 67%, 55% and 33%, respectively. The performances of the proposed FODS are summarized as shown in Table 3.8. By using the k_1 and k_2 values of (0.667, 1), the maximum sensitivities of 0.2752 mV/mm, 0.3759 mV/mm and 0.7286 mV/mm are obtained at inclination angles of 10°, 20° and 30°, respectively. This sensitivity is higher compared to the previous work by Buchade and Shaligram (2006). The maximum linear ranges of 10.4mm, 7mm and 3mm are obtained at inclination angles of 10°, 20° and 30°, respectively for the FODS with k_1 and k_2 values of (0.667, 1).

Methods	Front slopes					
	Sensitivity (mV/mm)			Linearity Range (mm)		
(k_1, k_2)	10°	20°	30°	10°	20°	30°
(0.5, 0.5)	0.1345	0.1838	0.4761	1.5-3.5	0.4-2.1	0-0.7
(0.667,1)	0.2752	0.3759	0.7286	1 - 5.2	0.1-2.8	0 - 1
(1, 1)	0.1671	0.2224	0.5528	1.2-4.8	0.2-2.8	0 - 1
(1.5, 1)	0.1885	0.2745	0.6371	1.2-4.8	0.1-2.9	0 - 1
(2, 2)	0.0645	0.1201	0.1904	1.4-3.5	0.2- 2	0-0.8
	Back slopes					
	Sensitivity (mV/mm)			Linearity Range (mm)		
(k_1, k_2)	10°	20°	30°	10°	20°	30°
(0.5, 0.5)	0.0223	0.0479	0.1447	4.3-11.5	2.9-8.9	1.5-3.7
(0.667,1)	0.0675	0.1336	0.3224	6.6-17	3.2-10.2	1.3-4.3
(1, 1)	0.0472	0.0823	0.2296	5.6-14.8	3 - 9	1.2-4.3
(1.5, 1)	0.056	0.1155	0.2929	6.8-16	3.2-9.2	1.2-4.3
(2, 2)	0.0128	0.0389	0.0885	4.5-11.5	2.4-7.5	1 - 3

Table 6. The performances comparison of FODS with two asymmetrical inclined fibers

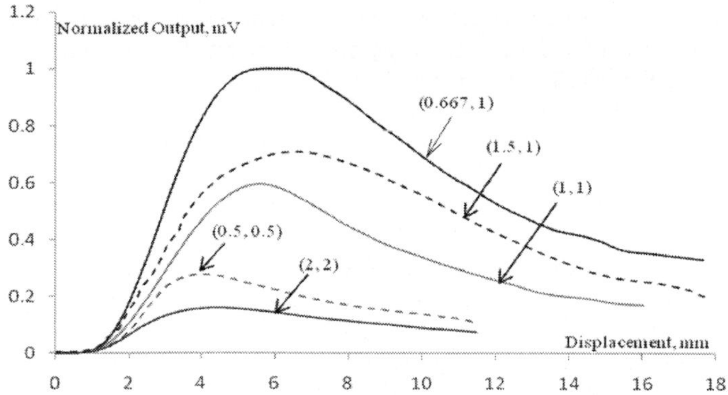

(a) Inclination angle 10°

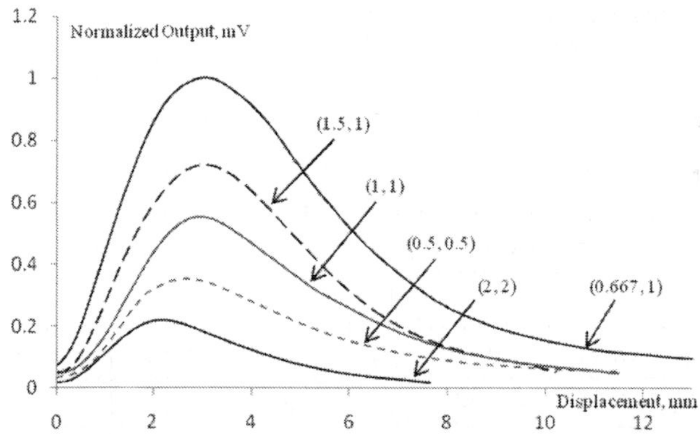

(b) Inclination angle 20°

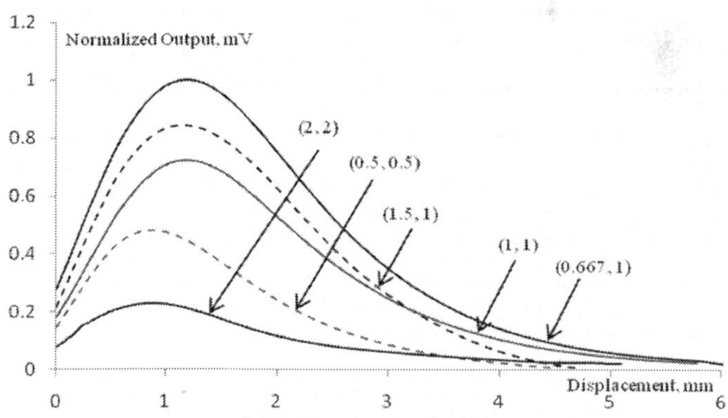

(c) Inclination angle 30°

Fig. 17. The performance output of k1, k2: (0.667, 1), (1, 1), (1.5, 1), (0.5, 0.5) and (2,2) in inclined angle (a) 10° (b) 20° (c) 30°

4. Applications of the FODS

Beside displacement measurement, the FODS can also be used to sense many other parameters such as temperature, pressure, refractive index, strain, mass and etc (Bechwith, 2000; Chang et al., 2008; Arellano-Sotelo, 2008). In this section, two different applications of FODS will be described.

4.1 Liquid refractive index detection using a FODS

In this study, the sensor probe is containing two pieces of fiber optics, one set connected to a light source and termed the transmitting fiber, and the other set connected to a photo detector (photodiode) and known as the receiving fiber. These two groups of fibers are bundled into a common probe to be used in a FODS. The FODS has a capability to measure physical quantities such as the displacement, vibration, strain, pressure, etc (Yasin et al., 2008; Yasin et al., 2009; Rastogi, 1997). However, the use of FODS sensors for detection of environmental refractive index change has not been fully explored. Refractive index sensing is important for biological and chemical applications since a number of substances can be detected through measurements of the refractive index. In the development of a liquid refractive index sensor (LRIS) (Suhadolnik et al., 1995; Chaudari and Shaligram, 2002; Yang et al., 2009; Nath et al., 2008; Kleizal & Verkelis, 2007), an intensity modulation in conjunction with multimode plastic fiber is the most suitable technique because of its non-contact sensing and many advantage properties are inherited by the multimode plastic fiber such as efficient signal coupling and being able to receive the maximum reflected light from target.

A FODS based refractive index measurement using a bundle fiber is first introduced by Suhadolnik et al. in 1995. Later on Chaudhari & Shaligram reported on study of LRIS at various types of optical sources. In our earlier work, a FODS was proposed based on two asymmetrical fibers for liquid refractive index measurement (Yang et al., 2009). In this section, a new LRIS is studied and demonstrated by using pair type of fibers bundled at various inclination angles.

The structure of proposed LRIS is shown in Fig. 18, which consists of a pair of transmitting and receiving fiber. We assume that the transmitting and receiving fibers have inclination angles of θ_1 and θ_2, respectively against the y-axis. The image of transmitting fiber is located opposite of the mirror with same distance. The central of the receiving fiber and the image of transmitting fiber are pointed as O' and O, respectively in Fig. 18. From the geometrical analysis of Fig. 18, the angle $\alpha = \sin^{-1}(NA/n)$ and $z_a = \frac{r_1}{\tan\alpha}$. Therefore, the following distances are given by;

$$AB = \sqrt{z_a^2 + r_{d1}^2}\,\sin\left(\tan^{-1}\left(\frac{r_{d1}}{z_a}\right) - \frac{\pi}{2} + \theta_1\right),$$

$$O'C = 4r_{d1}\sin\theta_1 + 2x - AB - r_{d2}\sin\theta_2,$$

$$OA = \sqrt{z_a^2 + r_{d1}^2}\,\cos\left(\tan^{-1}\left(\frac{r_{d1}}{z_a}\right) - \frac{\pi}{2} + \theta_1\right)$$

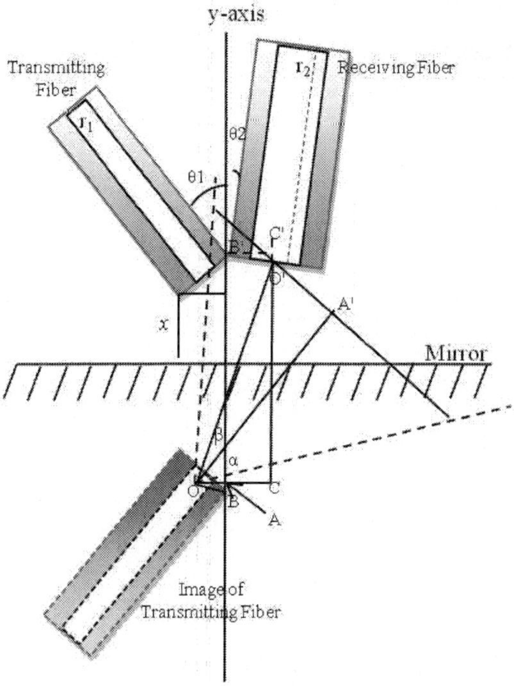

Fig. 18. Structure of sensor probe for the sensing of liquid refractive index.

$$OC = OA + r_{d2}\cos\theta_2$$

$$OO' = \sqrt{O'C^2 + OC^2}$$

where the NA is numerical aperture of transmitting fiber, n is refractive index of liquid, r_1 and r_2 are the core radius of transmitting fiber and receiving fiber while the $r_{d1}$and r_{d2} are the radius of transmitting fiber and receiving fiber, respectively and the x is the displacement between the sensor probe tip and reflector mirror.

Also from the geometrical analysis, the acceptance angle β of the light cone from the virtual point source *O*, is given by

$$\beta(z) = \tan^{-1}\left(\frac{O'C}{OC}\right) - \frac{\pi}{2} + \theta_1 \tag{29}$$

The intensity of the light emitted from the transmitting fiber can be well described with Gaussian distribution (Chang, et al., 2008) and is given by;

$$I(r,z) = \frac{2P_E}{\pi\omega^2(z)}\exp\left(-\frac{2r^2}{\omega^2(z)}\right) \tag{30}$$

where r is the radial coordinate, z is the longitudinal coordinate. $\omega(z)$ is the beam radius and expressed as a function of z, $\omega(z) = \omega_0\sqrt{1+\left(\frac{z}{z_R}\right)^2}$. The waist radius ω_0 and Rayleigh range z_R are the important parameters in the Gaussian Beam function and the detailed description can be found (Chang et al., 2008). Eq. (30) shows that the light intensity decays exponentially as it goes radially away from the center of the light circle. The radial coordinate r of Eq. (30) can be determined by;

$$r = OO'\sin\beta \tag{31}$$

The longitudinal coordinate is the distance between the sensor probe tip and the virtual laser source point O and it can be determined

$$z = OO'\cos\beta \tag{32}$$

For points situated in the far-field $z >> z_R$, the beam radius of the virtual point source can be derived as

$$\omega(z) \approx z\alpha \tag{33}$$

By the approximation,

$$r_1 = z_a \tan\alpha \approx z_a\alpha \tag{34}$$

Based on the properties above, the power harnessed by the receiving fiber, P can be evaluated by integrating the Gaussian distribution function of Eq. (30) over the area of the of receiving fiber end surface, S_r

$$P(r,z) = \int_{S_r} I(r,z)dS_r \tag{35}$$

where the core area of the receiving fiber is

$$S_r = \pi r_1^2 = \pi z_a^2\alpha^2 \tag{36}$$

By combining and substituting Eqs. (31), (32), (33) and (36) into the Eq. (35), finally the proposed LRIS response can be summarized as;

$$P_{(z,r)} = \frac{2P_E}{\pi\omega^2(z)}\exp\left(-\frac{2r^2}{\omega^2(z)}\right)\times S_r = \frac{2{z_a}^2P_E}{z^2}\exp\left(\frac{2r^2}{z^2\alpha^2}\right) \tag{37}$$

This equation shows that the liquid refractive index response of sensor is a function of displacement x and refractive index n of surrounding medium while sensor probe is design of inclination angles of θ_1 and θ_2. Therefore, based on Eq. (37), the proposed LRIS can be used to detect the liquid refractive index where the sensor probe is immersed by the liquid to be measured.

The mathematical model of proposed LRIS is simulated by MATLAB programming. In the simulation, the wavelength of the laser source λ and numerical aperture NA is set at 594nm and 0.32, respectively. The fiber core radius r_1 and r_2 are set at 0.25mm and 0.5mm while the fiber diameters r_{d1} and r_{d2} are set at 0.5mm and 0.75mm, respectively. Fig. 19 shows the experimental set-up, which consists of a 594nm yellow He-Ne laser as a light source and a bundled fiber as a probe. The emitted laser light has an average output power of 3.0mW, beam diameter of 0.75mm and beam divergence of 0.92mRads. The external chopper modulates the light at a frequency of 200 Hz before it is launched into the transmitting fiber. The transmitting fiber transfers the modulated light from laser source to reflector mirror while the receiving fiber collects the reflection light to the detector. The sensor probe is mounted on the stage controlled by NewFocus Picomotor for the displacement measurement. Silicon detector measures the received light and converts it into electrical signal which is then denoised using the lock-in amplifier. During the measurement, the room temperature is keeping in 28 °C to avoid the change of liquid refractive index.

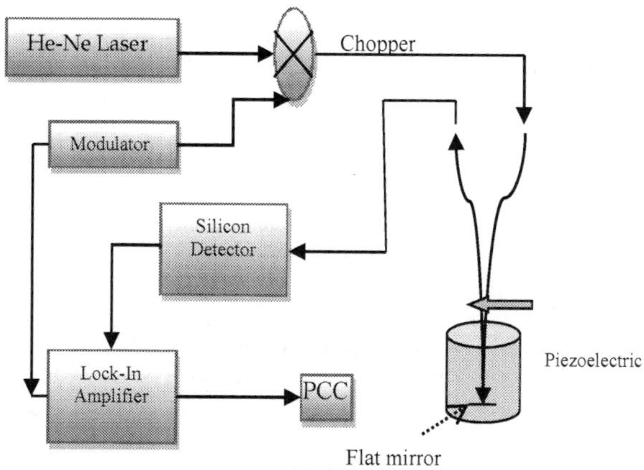

Fig. 19. Experimental set-up of the proposed liquid refractive index sensor

The simulation results are illustrated in Fig. 20. This figure shows the relationship between the sensor responses and probe inclination angles θ_1 and θ_2 in the measurement of three different refractive indices 1, 1.3 and 1.6 based on the probe inclined with the same angles of 0, 10° and 20°. In Fig. 5.8, the outputs powers are normalized in 1 and the displacements are simulated in *mm*. Three different group curves are showing three various sensors response based on the three various inclination angles. As seen in Fig. 20, it was found that the inclination angles θ_1 and θ_2 are reasonably affects the displacement curves profile and output power. The highest output power is almost 10 times of the lowest output power. The vertical dash lines are located in the displacements of 1.1 mm, 2.0 mm and 3.4 mm corresponding to the sensor probe inclination angles 20°, 10° and 0, respectively. In those positions, the sensor responses have the biggest output differentiation in the increase of refractive indices from 1.0 to 1.5. At those positions, the sensor output intensity increases almost linearly with the increase of the refractive index of the medium. From Fig. 20, we can observe that the increase the inclination angles improves the performance of the LRIS. The larger the inclination angles of θ_1 and θ_2, the better the performance of liquid refractive index response.

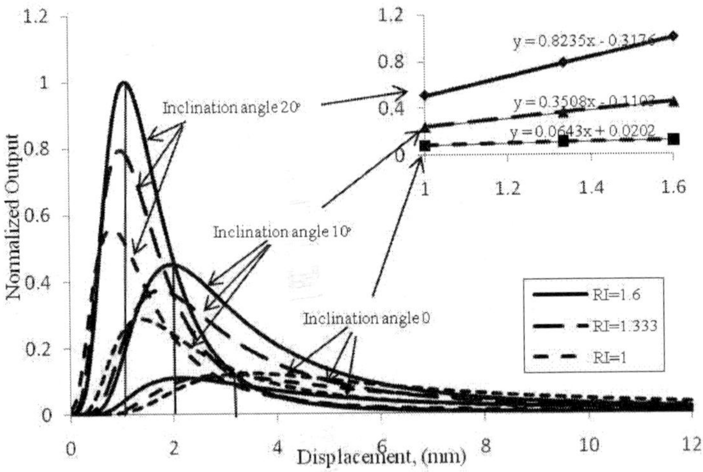

Fig. 20. Simulation results for the displacement at various inclination angles and refractive indices. Inset shows the maximum normalized output against the refractive index for inclination angles of 0°, 10° and 20°.

Inset of Fig. 20 shows the maximum normalized output of the sensor as a function of refractive index for various inclination angles. The normalized outputs were taken at the sensor probe positions of 1.1 mm, 2.0 mm and 3.4 mm for inclination angles of 20°, 10° and 0°, respectively, which is indicated by vertical dash lines in Fig. 20. It was found that the sensitivities of the sensor increases with the increment of probe inclination angle. As shown in the inset of Fig. 20, the highest sensitivity of 0.8235 is achieved by the use of probe with inclination angle of 20° which is almost 13 times higher than that in zero inclination. Fig. 21 shows the simulation curves of the LRIS at various inclination angles for the receiving and transmitting fibers when the refractive index of liquid is set at 1.3. It is clearly seen that the inclination angle of receiving fiber θ_1 has the stronger affect in the sensor output compared with the angle θ_2. As shown in Fig. 21, the highest output power is achieved by the inclination angles; θ_1 = 20° and θ_2 = 10°. The lowest output power is observed when the inclination angles of θ_1 and θ_2 are set at 0° and 10°, respectively. These results show that the sensor sensitivity can be increased by increasing the inclination angle especially for θ_1. However, increasing the inclination is very difficult to be implemented in the experiment unless we can control the position of both fibers very precisely.

In our experiment, three different liquids: isopropyl alcohol, water and methanol are used as the surrounding medium at two conditions; zero inclination for both fibers and the same inclination angles of 10° for both fibers. The refractive index values for isopropyl alcohol, water and methanol are 1.377, 1.333 and 1.329, respectively. The sensor performance with air surrounding medium is also carried out for comparison purpose. During the experiment operation, the sensor probe is mounted onto the stage and the tank is fixed in the experiment table. The liquid in the tank is changed without moving the tank to ensure the accuracy of the measurement. The room temperature was kept at 28° to ensure that the refractive index of the liquid is maintained and only displacement parameter is changed in the experiment.

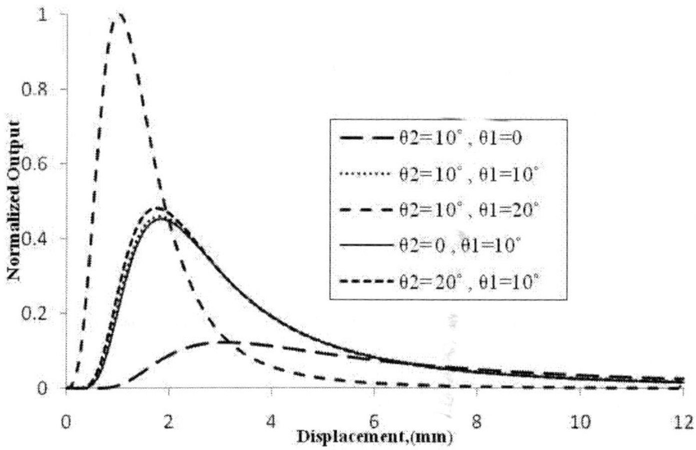

Fig. 21. The output power against displacement for various probe inclination angle combinations when the refractive index is set at 1.3.

Fig. 22 shows the displacement curve at various surrounding media when the inclination angles are set at 0° for both transmitting and receiving fibers. As shown in this figure, the normalized peak output power increases from 0.83 to 1.00 as the refractive index increases from 1.329 (methanol) to 1.377 (alcohol). It is also found that the displacement position for the peak output increases from 4.0 to 5.1 mm as the refractive index increases from 1.329 to 1.377. This is attributed the acceptance cone angle that increases with the refractive index increase. The larger acceptance angle provides a mean to collect more signal power. Fig. 23 shows the displacement curves when the inclination angles for both fibers are increased to 10°. As seen in Fig. 23, the peak power and its position increase with refractive index. The peak location of the curve also increases from 3.0 mm to 3.4 mm as the refractive index changes from 1.329 to 1.377. From these experimental results, it was found that the sensitivity of the sensor with 10° inclination of probe arrangement shows a higher sensitivity compared to that of the use of straight probe. The sensitivities are obtained at 0.11 mm^{-1} and 0.04 mm^{-1} for the sensors with 10° and 0° inclination angles respectively. This finding may be quite useful for chemical, pharmaceutical, biomedical and process control sensing applications.

4.2 The monitoring of liquid level using a FODS

Liquid level measurement is vital in many industry areas, finding the applications in such as fuel storage system, chemical processing, etc. Fiber based liquid level sensors are received in prefer of studies since they inherit many merits such as they are non-conductive, anti-erosion, and immune to electromagnetic interference. In past few years, several technical are employed to develop the fiber optic Liquid Level Sensors (LLSs). Fiber Bragg grating technology was carried out to sense the changing of liquid level by (Yun et al., 2007; Sohn & Shim, 2009; Dong & Zhao, 2010). Bending the multimode fiber for multi-points liquid level monitoring can be found in (Lomer et al., 2007). In these cases, the sensing elements are submerged inside the liquid to indicate the presence of liquid. These immersed sensing

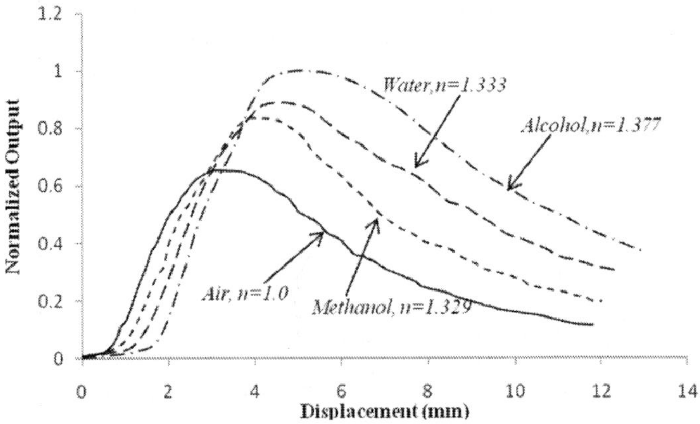

Fig. 22. Experiment results of the displacement curve at various liquid materials when the probe inclination angles are set at 0°.

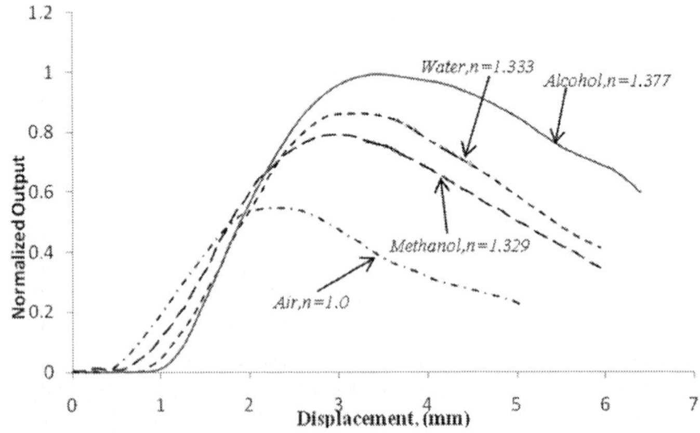

Fig. 23. Experiment results of the displacement curve at various liquid materials when the probe inclination angles for both transmitting and receiving fibers are set at 10°

elements have some limitations in their applications, such as in harsh environments, acid solutions, etc. Thereby, the fiber as sensing element for non-contact the measurement liquid seems good way to explore. There is one proposed sensor (Sohn and Shim, 2009) which employed one fiber bragg grating embedded inside the cantilever beam to connect with a float for indicating the level changing. This case is limited in its cost, surrounded temperature, and sensitivities.

In this section, a simply and cost efficiency intensity based LLS is proposed. A straightforward displacement sensor is employed; its sensing probe is engaged with a float which is contact with the measurement liquid to conduct the information of liquid. This simply setup transfers the sensing of liquid level into the measurement of displacement. It breaks down the limitations of sensitivities and measurement range which can be conquest by the selection of displacement sensors. Thus, in this study two type of displacement sensor are used to supply a flexible and compatible sensing of liquid level.

The basic sensing principle is the sensor probe displacement moving that causes from the moving of float during the liquid level change. A schematic setup of proposed fiber optic LLS is illustrated in Fig.24. It composed a He-Ne laser source light wavelength in 594nm (average output power 3.0mW, beam diameter 0.75mm and beam divergence 0.92mRads) which coupling the light into the transmitting fiber (2m length); a external chopper modulates the light at a frequency of 200Hz before launched into the transmitting fiber; a concave mirror (or flat mirror) is located at end of transmitting fiber and reflects the guided light of transmitting fiber into a receiving fiber which is bundled with the transmitting fiber. The receiving fiber guides the light into the photo detector which converts the light power into voltage. Meantime, the bundled fiber as sensor probe is fastened by an L cantilever beam which connects a float on the other side. The float is contact with the measurement liquid to indicate the information of liquid levels. A model SR-510 lock-in amplifier is connected with the chopper and photo detector. It plays the function of experiment data-acquisition system, matches the phase between the modulation light and modulator chopper and removes the noise generated by laser source, photo detector and amplifier.

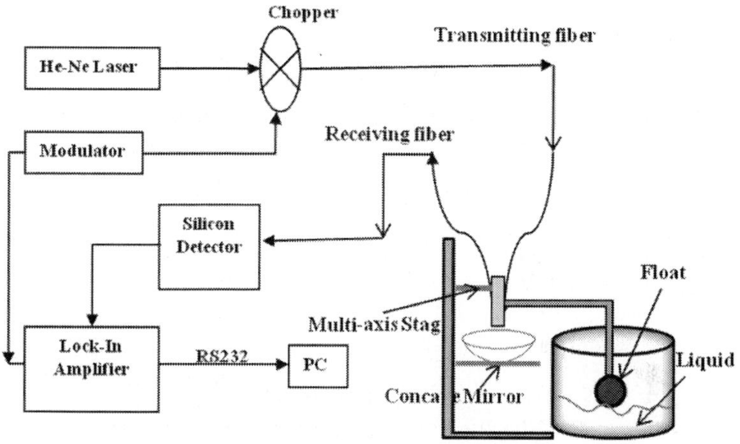

Fig. 24. Experiment setup of monitoring the liquid level

In the using of concave mirror as the reflector, the transfer function of proposed LLS is governed by parameters, namely focal length f and the diameter of the circular concave mirror D. The characteristic of the proposed sensor can be found the detailed theoretical analysis in paper (Yang et al., 2010). In consideration of the limited reflecting surface area and FL of the concave mirror, the transfer function of the proposed sensor can be given as

$$P(u)=\frac{2z_a^2P_E\left\{1-\exp\left(-\frac{D^2}{2(u+z_a)^2\theta^2}\right)\right\}}{\left[u-\frac{(u+z_a)f}{u-f+z_a}\right]^2\left(\frac{u+z_a}{f}-1\right)^2}\exp\left(-\frac{8z_a^2}{\left[u-\frac{(u+z_a)f}{u-f+z_a}\right]^2\left(\frac{u+z_a}{f}-1\right)^2}\right) \tag{38}$$

where P_E is the emitting power of laser source, u is the distance between the sensor probe tip and concave mirror, $\theta = \sin^{-1}(NA)$ and NA is the numerical aperture of the transmitting fiber. By using the flat mirror as the reflector, the transfer function of proposed sensor was analyzed in (Lim et al., 2009) detailed. It is given by

$$P(z) = \frac{2(k+1)^2}{\xi^2}\exp\left(1 - \frac{2(1+k)^2}{\xi^2}\right) \tag{39}$$

The core radial ratio k is the ratio of the transmitting fiber core radius to the receiving fiber core radius, ξ is the normalized displacement.

From the transfer functions of Eq. (38) and (39), it can be seen that the sensor response is only influenced by the focal length f and the diameter of the circular concave mirror D in the concave mirror as reflector and by the core radial ratio k for that of flat mirror. As such, the proposed LLS has the widely compatible for the variants type of FODSs, which can escapes the trade-off design in the selection of sensitivity and measurement range. Furthermore, according to the Eqs. (38) and (39), there are no more parameters can affect the sensor response, hence, the proposed LLS is independency for surrounded environments.

Fig. 25 depicts the experiment result when the liquid level is moving upward. In this experiment, the proposed LLS is composed two plastic fiber bundled parallel as sensor probe, a concave mirror with focal length 10mm and diameter 24mm as reflector. From Fig. 25, it can be seen that the sensor output intensity is modulated by the moving of liquid level in six variants slops. According to this, hence, the proposed sensor can achieve the continue monitoring of liquid level in the multi-points. Totally six monitoring points are shown in Fig. 25, each point is represented each slop which can be used as the each level of liquid measurement. These six monitoring points are located in the total measurement range of 25mm, which can provide variants sensitivities and measurement range. The performances of these monitoring points are summarized in table 7.

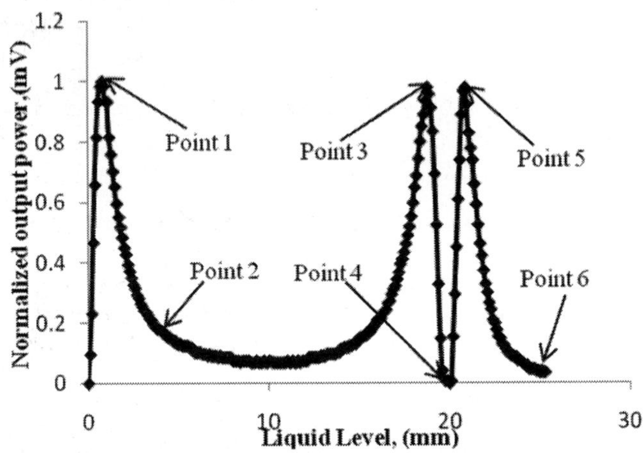

Fig. 25. The experiment result of proposed sensor for concave mirror as reflector and two plastic fibers bundled parallel

The second experimental is carried out by the sensor probe consisted with two bundled parallel and a flat mirror as reflector. On the other hand, another experimental is repeated by two asymmetrical plastic fiber bundled parallel as sensor probe and a flat mirror as reflector. Both of responses are illustrated in Fig. 26, where each of the curves are displayed two various slops in the measurement range of 3mm and 3.5mm, respectively. From the observation of Fig. 26, two various monitoring points are given from both curves as shown. They are occupied the measurement ranges of 3mm and 3.5mm, respectively. In Fig. 26, the higher sensitivity is obtained from the curve of *k*=1 and the bigger measurement range is supplied by the curve of *k*=0.5. The output curves in Fig. 26 are shifted to the right along the further of liquid level as the value of *k* decreases. Therefore, the sensitivities of front slope and back slope of the curves are decreasing with decreasing the values of *k*. Mean times, the linearity ranges of curves are enlarged with decreasing the values of *k*. The sensitivities and linearity ranges of Fig. 26 are summarized in Table 7.

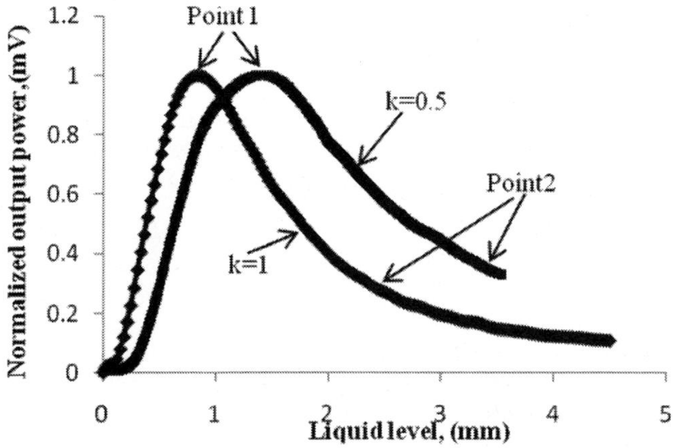

Fig. 26. The experiment results of various ratios of fibers bundled as sensor probe.

Table 7 tabulates the performances of LLS for three different experiments. From the results summarized in table 7, it can be seen that the highest sensitivity is 1.4533mV/mm when its measurement range is 0.84mm. The largest measurement range 4.8mm is obtained from the monitoring point 3 for the concave mirror as reflector. In the experiment of concave mirror as reflector, the monitoring points 1, 4, and 5 have the higher sensitivities and shorter measurement ranges than that of 2, 3, and 6. In the experiment of flat mirror as reflector, the sensitivities for k=0.5 are decreased by 52.5% and 14.8% for of front slope and back slope, respectively, if compared to k=1. The measurement ranges of front slope and back slope are increased by 46.2% and 18.8% if compared with the k=1. From the analysis, by using a particular combination of fiber core diameters and reflector, two different *k* values, ten different sensitivity and linear range selections can be achieved by using the proposed method. Thus, the proposed fiber optic LLS has wide compatibility for other type of displacement sensor to avoid the trade-off selection of sensitivity and measurement range which is always encountered in the design of conventional fiber optic LLS.

Methods	Monitoring point	Sensitivity (mV/mm)	Measurement range (mm)
Concave mirror k=1	Point 1	1.4026	0-0.8
	Point 2	0.2649	0.8-3.8
	Point 3	0.1625	14-18.8
	Point 4	1.030	18.9-20
	Point 5	1.3015	20.1-20.9
	Point 6	0.1821	21-25
Flat mirror k=0.5	Point 1	0.9529	0.1-1.4
	Point 2	0.3443	1.4-3.5
Flat mirror k=1	Point 1	1.4533	0-0.84
	Point 2	0.3951	0.84-3

Table 7. The Sensitivities and measurement ranges of proposed sensor

In the literature of (Sohn and Shim, 2009), the wavelength-shift sensitivity is around 0.15nm/cm and the resolution of intensity modulated the liquid level is approximated 0.1dB/cm are achieved. However, these sensitivities cannot satisfy the high precision measurements and the LLS proposed in (Sohn and Shim, 2009) is high costly. In this paper, the highest sensitivity is achieved by using a probe with two fibers bundled and flat mirror as reflector, it is around 100 times higher than that of (Sohn and Shim, 2009) and is 20 times lower in the cost.

5. Conclusion

In this chapter, the performance of various FODSs is investigated theoretically and experimentally. A beam-through FODS has been demonstrated for sensing of lateral and axial displacements, which shows that the usage of smaller core normally shows a better sensitivity and resolution with the expense of the smaller linear range. Various reflective FODSs are then proposed to provide a possibility for the development of the cheap, simple, sensitive and wide dynamic range sensor systems. Compared with the conventional sensor with only one receiving fiber, FODS with two receiving fibers is observed to have a better linearity range. The performance of FODS with asymmetrical bundled fiber is also theoretically and experimentally demonstrated in this chapter. The effect of different core radial ratios (CRRs) on the performance of the sensor is investigated in terms of dynamic range and sensitivity. The experimental results are almost in agreement with the theoretical results. The location of the maximum output is shifted toward the right along the axis of displacement as the value of k increases. Besides, the linear range for both front and back slopes increases with the value of k. To improve the sensitivity, a FODS with two inclined asymmetrical fiber has been proposed. This FODS has been applied in liquid refractive index sensor. An extra-low cost, ultra-high sensitivity, and wide compatibility liquid level fiber optic sensor has been demonstrated as another application for the FODS. A float is touched with the measurement liquid to translate the information of liquid change to the sensor probe displacement moving where an L cantilever beam is fixed in between, the moving of sensor probe finally results the output intensity is modulated. All the theoretical and experimental results are capable of offering quantitative guidance for the design and implementation of a practical FODS.

6. References

Arellano-Sotelo, H.; Barmenkov, Yu. O.; Kir'yanov, A.V. (2008). The use of erbium fiber laser relaxation frequency for sensing refractive index and solute concentration of aqueous solutions. *Laser Physics Letters*, Vol. 5, No. 11, pp. 825-829.

Bechwith, M.L. (2000). Mechanical measurement. Addison Wesley Longman.

Bergamin; Cavagnero, G.; Mana, G. (1993). A displacement and angle interferometer with subatomic resolution. *Rev Sci Instrum*, 64, 3076-3081.

Bock, W.J.; Nawrocka, M.S.; Urbanczyk, W. (2001). Highly sensitive fiber optic sensors for dynamic pressure measurement. *IEEE Trans. Instrum. Meas.*, Vol. 50, No. 5, pp. 1085-1088.

Buchade, P.B. & Shaligram, A.D. (2006). Simulation and experimental studies of inclined two fiber displacement sensor. *Sensor and Actuators A*, Vol. 128, pp. 312- 316.

Buchade, P.B. & Shaligram, A.D. (2007). Influence of fiber geometry on the performance of two-fiber displacement sensor. *Sensors and Actuators A*, 136, 199- 204.

Cao, M.H.; Chen, Y.; Zhou, Z.; Zhang, G. (2007). Theoretical and experimental study on the optical fiber bundle displacement sensors. *Sensor and Actuators A*, Vol. 136, pp. 580-587.

Chang, J.; Wang, Q.; Zhang, X.; Ma, L.; Liu, T.; Wang, Q.; Liu, Z.; Zhang S.; Ding, S. (2008). Single-end vibration sensor based on an over-coupled fiber-loop reflector. *Laser Physics*, Vol. 18, pp. 452-454.

Chaudhari, A.L. & Shaligram, A.D. (2002). Multi-wavelength optical fiber liquid refractometry based on intensity modulation", *Sensor and Actuators A*, Vol.100, pp.160- 164.

Cui, C.; Wang, Q.; Hu, X.; Yu, Y.; Zhao, S. (2008). Simulation of spin-axis position measurement of superconducting sphere rotor by fiber optic sensor. *IEEE Trans Appl Supercond.*, Vol 18, pp. 836-839.

Dong, X.W. & Zhao, R.F. (2010). Detection of liquid level variation using a side polished fiber bragg grating. *Optics & laser technology*, 42, 214-218.

Elasar, J.; Selmic, S.; Tomic, M.; Prokin, M. (2002). A fiber optic displacement sensor for a cyclotron environment based on a modified triangulation method. *J. Opt. A: Pure Appl. Opt.*, Vol. 4, pp. 347–355.

Faria, J.B. (1998). A theoretical analysis of the bifurcated fiber bundle displacement sensor. *IEEE Transactions on instrumentation and measurement*, Vol. 47, pp. 742-747.

Kleizal, V. & Verkelis, J. (2007). Some Advanced Fiber-Optic Amplitude Modulated Reflection Displacement and Refractive Index Sensors. *Nonlinear Analysis: Modelling and Control*, Vol. 12, No. 2, pp. 213–225.

Ko, W.H.; Chang, K.M.; Hwang, G.J. (1995). A fiber optic reflective displacement micrometer. *Sensor Actuators A*, Vol. 49, pp. 51–55.

Kulchin, Y.N.; Vitrik O.B.; Dyshlyuk A.V.; Shalagin, A.M.; Babin, S.A.; Vlasov, A.A. (2007). Applications of optical time domain reflectometry for the interrogation of fiber bragg sensors. Laser Physics, Vol. 17, pp. 1335-1339.

Laurence Bergougnous, Jacqueline Misguich-Ripault, and Jean-Luc Firpo, (1998). Characterization of an optical fiber bundle sensor. *Rev. Sci. Instrum.* 69, 1985–1990.

Lim, K.S.; Harun, S.W.; Yang, H.Z.; Dimyati, K.; Ahmad, H. (2009). Analytical and experimental studies on asymmetric bundle fiber displacement sensors. *J. of Modern Optics*, Vol. 56, pp.1838-1842.

Lomer, M.; Arrue, J.; Jauregui, C.; Aiestaran, P.; Zubia, J.; Lopez-Higuera, J.M. (2007). Lateral polishing of bends in plastic optical fibers applied to a multipoint liquid level measurement sensor. *Sensors and Actuators A*, 137, 68-73.

Miclos, S. & Zisu, T. (2001). Chalcogenide fibre displacement sensor. J. Optoelectron Adv. Matter., Vol. 3, pp. 373-376.

Murphy, A.M.; Gunther, M.F.; Vengsarkar, A.M.; Claus, O.R. (1991). Quadrature phase-shifted extrinsic Fabry-Perot optical fiber sensor. *Opt. Lett .*, 16, 273.

Murphy, M.M. & Jones, G.R. (1994). A variable range extrinsic optical fiber displacement sensor. *Pure Appl. Opt.*, Vol. 3, pp. 361–369.

Nalwa, S. (2004). Polymer optical fibers. California: American Scientific Publishers.

Nath, P.; Singh, H.K.; Datta, P.; Sarma, K.C. (2008). All-fiber optic sensor for measurement of liquid refractive index. *Sensor and Actuators A*, Vol.148, pp.16- 18.

Oiwa T. & Nishitani, H. (2004). Three dimensional touch probe using three fiber optic displacement sensor. *Meas. Sci. Technol.*, Vol. 15, pp. 84–90.

Rastogi, P.K. (1997). *Optical Measurement Techniques and Applications*. Artech House, Inc. Boston, London.

Saunders, C. & Scully. (2007). Sensing applications for POF and hybrid fibres using a photon counting OTDR. Meas. Sci. Technol., Vol. 18, pp. 615-622.

Shimamoto, A. & Tanaka, K. (2001). Geometrical analysis of an optical fiber bundle displacement sensor. Applied Optics, Vol. 35, Issue 34, pp. 6767-6774.

Sohn, K.R. & Shim, J.H. (2009). Liquid level monitoring sensor system using fiber bragg grating embedded in cantilever. *Sensors and Actuators, A*, 152, 248-251.

Spooncer, R.C.; Butler, C.; Jones, B.E. (1992). Optical fibre displacement sensors for process and manufacturing applications. Opt. Eng. 31, 1632-1637.

Suhadolnik; Babnik, A.; Mo2ina,J. (1995). Optical fiber reflection refractometer. *Sensor and Actuators B*, Vol. 29, pp.428- 432.

Van Etten W. & Van der Plaats J. (1991). Fundamentals of optical fiber communi-cations. Prentice-Hall, London.

Wang, D.X.; Karim, M.A.; Li, Y. (1997). Self referenced fiber optic sensor performance for micro displacement measurement. Opt. Eng., Vol. 36, pp. 838–842.

Yang, H.Z.; Lim, K.S.; Harun, S.W.; Dimyati, K.; Ahmad, H. (2010). Enhanced Bundle Fiber Displacement Sensor Based on Concave mirror. *Sensors and Actuators A*, 162, 8-12.

Yang, H.Z.; Harun, S.W.; Ahmad, H. (2009). Fiber Optic Displacement and Liquid Refractive Index Sensors with Two Asymmetrical Inclined Fibers. *Sensor and Transducer*, Vol.108, pp. 80-88.

Yasin,M.; Harun, S.W.; Pujiyanto, Ghani, Z.A.; Ahmad H. (2010). PerformanceComparison between Plastic-Based Fiber Bundle and Multimode FusedCoupler as Probes in Displacement Sensors. *Laser Physics*, Vol. 20, No. 10, pp. 1890–1893.

Yasin, M.; Harun, S.W.; Abdul-Rashid, H.A.; Kusminarto; Karyono; Ahmad, H. (2008). The performance of a fiber optic displacement sensor for different types of probes and targets. *Laser Physics Letters*, Vol. 5, No. 1, pp. 55-58.

Yasin, M.; Harun, S.W.; Samian; Kusminarto; Ahmad, H. (2009). Simple design of optical fiber displacement sensor using a multimode fiber coupler. *Laser Physics*, Vol. 19, No.7, pp. 1446-1449.

Yun, B.F.; Chen, N.; Cui, Y.P. (2007). Highly sensitive liquid level sensor based on etched fiber bragg grating. *IEEE Photonic Technology Letters*, 19, 1747-1749.

16

Sensing Applications for Plastic Optical Fibres in Civil Engineering

Kevin S. C. Kuang
National University of Singapore,
Department of Civil and Environmental Engineering
Singapore

1. Introduction

Monitoring of the performance and integrity of structures using sensors attached or embedded to them has been of great interest in view of the possibility of predicting potential structural failures, optimising the operational efficiency of the structure, extending the useful lifespan of the host structure as well as tracking the propagation of cracks, amongst other benefits. Indeed, the field of structural health monitoring has received considerable attention and investments from private and public institutions including academia, professional institutes, industrial research laboratories, government agencies as well as the industry. A wide range of companies involved in the designing, fabrication, integration, usage, servicing and maintenance of structural components, assemblies and systems have a common need to monitor, understand and manage the performance and integrity of the structures in their possession. These companies hail from industries such as the offshore, oil-and-gas, power generation, building construction, marine, aerospace, civil and environmental.

Monitoring the performance and integrity of a host structure is often translated into the measurement of key parameters or health signatures of the structure as indicator of the integrity and level of performance of the host. Common parameters of interest include strain, load, deflection, displacement, pH-level, temperature, pressure, vibration frequency, liquid level, detection of cracks and monitoring of their rates of propagation, moisture, detection of events such as impact, overload, fracture and others. Several review papers are available in the open literature outlining various examples of optical fibre sensor applications [1-3].

In recent years, optical fibre sensor technology has gained significant attraction as the sensor of choice in many structural health-monitoring applications. Several optical fibre sensing techniques have been commercialised whilst others are being developed. These sensors in general, offer several distinctive benefits and advantages over conventional sensors - these include their insensitivity to electromagnetic and microwave radiation (occurring near power-generating equipment), being spark-free (hence suitable for use in areas where risk of explosion is a concern), the fact that they are intrinsically safe (inert and passive), non-conductive, lightweight and potentially suitable for embedding into structures such as fibre composite laminates during the fabrication stage.

There are several unique features associated with optical fibre sensors. Certain types of optical fibre sensors (e.g. fibre Bragg gratings) possess the possibility of having multiple sensing regions in a single strand of fibre - this capability where multiple sensors could be interrogated simultaneously in quasi-real time using a single channel is a significant advantage over conventional sensors such as strain gauges where a single channel monitors only a single sensor. In addition, recent advancement in the development of interrogators based on optical time-domain reflectometry (OTDR) principle has opened up the possibility of performing distributed monitoring of strain and temperature along a single fibre sensor. Also, although optical fibre sensing techniques, in general, are known to be capable of obtaining high measurement resolution (e.g. sub-microstrain measurement is possible with fibre Bragg grating sensor), their ability to monitor large strains (e.g. 40% strain or more) has also been reported [4]. In applications requiring the measurement of parameters that modulate over time (e.g. as in fluctuating liquid level [5]), optical fibre sensors offer a range of dynamic characteristics or responses which could be tailored to suit specific needs.

There are two main types of optical fibres based on the material from which they are made and these can be classified as either glass-type optical fibres (GOFs) or polymer-type optical fibres (POFs). Within each class of fibres, a variety of materials have been used, developed, doped and purified to achieve low transmission attenuation. Historically, optical fibre sensors were developed using GOFs due to their availability, excellent transmission properties, and other optical characteristics which make them amendable to sensor fabrication – e.g. the ease of changing the refractive-index of a section of a specially-doped glass fibres via light radiation has led to the development of fibre Bragg grating sensors. In recent years, however, interest in the use of POFs for short distance networking (e.g. local area networking), significant improvements in transmission properties, lower cost and various practical advantages of POFs over GOFs have spurred intense research and development in sensor technologies based on polymer fibres. Whilst many of the sensing methodologies developed based on GOFs could also be applied to POFs, the unique features and advantages of POFs opened up new possibilities and opportunities for its applications in the field of civil engineering.

Plastic optical fibres usually come with core sizes ranging from diameters of 0.25mm to 1mm although fibres of larger diameters could also be purchased - a search on the Internet will reveal many suppliers of POFs. The core of the fibre could be made from poly-methymethacyrlate (PMMA), polycarbonate (PC), polystyrene (PS) and cyclic-transparent optical polymer (CYTOP), this being a recent addition to the list. Although the transmission attenuation of POF based on PMMA (160 dB/km) is inferior to the CYTOP-type fibre, it is the most readily available POF. Being inexpensive, many cost-effective sensors have been developed based on PMMA step-index fibre for a variety of applications in civil engineering.

POF sensors have attracted increasing interest for use in health-monitoring of civil structures due to a number of advantages over their silica counterpart. These include their lower cost (the fibre itself and the associated accessories and hardware), less susceptible to fracture and flexibility compared to bare GOFs. They also offer ease of termination (i.e. cleaving can be done with a razor blade and connected easily to the ferrule), safe disposability and ease of handling during sensor fabrication and installation to the host structure. Compared to GOFs of the same diameter, POFs are lighter, cheaper and less prone to flexural damage. In addition, POF has an elastic limit of an order higher than GOF [6,7] and it has been shown that standard step-index PMMA POF can be integrated into geotextile materials to measure strains up to 40%. When it is desirable to embed sensors within concrete structures, POF sensors offer a possible solution since the extremely alkaline (pH 12) environment of the concrete mixture is known to be corrosive to standard glass fibres [8]. Furthermore, the presence of moisture can weaken the glass core and accelerates crack growth in the fibre. Although a polymer coating may be applied in order to protect the glass fibre from the corrosive environment, this will incur additional cost and change the sensing properties of the sensor.

The brief introduction above serves as the background information to the following examples of applications encountered in civil engineering. Initially, examples of POF sensors based on intensiometric principles will be highlighted, followed by techniques based on OTDR. This chapter will focus on these two particular techniques in view of the author's experience in working with these in many projects over the last ten years. The principle of operation of each technique will be outlined in each category prior to a showcase of some examples to assist the reader in appreciating the subject matter.

The aim of this chapter is to present to the reader the potential of POF sensing technique as an attractive option for various applications in civil engineering and is not intended to be a comprehensive review of the various studies published in the literature in the area of POF sensor. Through illustrative examples of applications, it will be shown that POF sensors are ideal candidates in numerous civil engineering projects, either complementing or replacing conventional sensors. In some cases, POF sensors will be shown to offer a solution not possible with existing commercial sensors in the same class. With further development aimed at exploiting the unique features of POF-based sensors they may well prevail as the sensor of choice in many civil engineering applications where their cost-effectiveness, reliability, simplicity in design and ability to perform under demanding conditions are the key considerations in the sensor selection process.

2. POF sensors for civil engineering applications

To date, there are several sensing methodologies developed for the POF. These techniques differ in the way which the optical signal (which contains information on the measured quantity) is interrogated. Depending on which optical characteristic was monitored, the technique could be classified as intensity-based, interference-based, polarisation-based, wavelength-based or OTDR-based.

Of the various types listed, intensity-based sensing technique represents one of the earliest and perhaps the most direct and inexpensive way of interrogating optical fibre sensors for structural health monitoring applications. The sensing principle is straightforward and relies on the monitoring of the intensity level of the optical signal as it modulates in response to the measured quantity. Although intensity fluctuation in the optical signal due to possible power variation at the source as well as influence of external parameters unrelated to the measured quantity (e.g. micro- and macro-bending along the fibre length) can occur, these could be overcome. For example, with the availability of stable and inexpensive light sources and low bend-sensitivity fibres, the intensity-based approach offers excellent technical and commercial prospects for high volume, large-scale applications from a cost-effectiveness point of view.

Intensity-based technique is also well-suited for vibration measurement since absolute strain values are not required for this quantity - provided that the oscillation is within the sensitivity of the sensor. In recent years, POF-based accelerometers have been developed with vibration monitoring capability and it has been demonstrated that these optical fibre-based vibration sensors compare well with conventional capacitive accelerometers for vibration up to 1 kHz.

In addition, intensity-based technique can be integrated to wireless modules with ease for monitoring of static and dynamic loads. This can be carried out at a low-cost and with ease in view of the availability of miniaturised solid-state devices which could be used as light sources and detectors for the POF sensor. An example of this concept has been demonstrated for a flood monitoring application [5]. These features highlight the potential of intensity-based technique for field deployment where their sensing capabilities could be exploited whilst achieving cost-effectiveness. Instead of the common approach which is to use an optical fibre technology based on best-in-class for their measurement needs (e.g. high strain resolution of fibre Bragg grating sensors, which often entails a high up-front cost) potential users need to be clear in their requirements and be open to evaluate intensity-based optical techniques which to provide a more cost-effective solution.

In this section, the principle of operation of each type of interrogation method covered in this chapter will be outlined accordingly. The two main types considered in this chapter are (1) Intensity-based sensing and (2) OTDR-based sensing.

2.1 Intensity-based POF sensors

2.1.1 Operating principles and background of intensity-based POF sensors

The ease of monitoring the intensity of the optical signal in POF naturally leads to the investigation of their potential as intensity-based sensors. Light sources are easily available and inexpensive and could come in the form of high-intensity light-emitting diodes (LEDs), typically those operating at the red region (centred at 625 nm) region of the visible light spectrum. Light detectors are also widely availably and those that are commonly used in association with POF are light-dependant resistors (LDRs) or solid-state photodetectors. The interrogation technique is straightforward and involves monitoring the modulation of the light intensity as a function of an external perturbation of interest which include strain, displacement, load, pressure and others. Indeed the simplicity in design of the sensor as well as the low cost of the light source and detector associated with intensity-based technique have led to numerous applications not just for civil engineering but also for many other fields where detection and monitoring are required.

In general there are two categories of intensity-based POF sensors and these are classified as either (a) intrinsic or (b) extrinsic sensors. In the former category, the modulation of the intensity of the optical signal in the fibre is a result of the physical change in the optical fibre due to an external perturbation, such as load, strain or pressure acting on the fibre itself. This phenomenon can be observed when a sensitised region of the POF is bent resulting in the modulation of the light intensity. In the latter category of sensor, the modulation of the optical signal in the fibre is a result of a change in the physical-optic state of a transducer attached to the fibre resulting in the modulation of light intensity.

The optical fibre in this case serves as a light conduit and is not physically affected by the external perturbation itself. An example of this effect is when the light intensity modulates as a result of a change in the distance between two cleaved fibres resulting from an external perturbation.

2.1.2 Applications of intensity-based POF sensors

Intensity-based POF sensors have been demonstrated for a number of structural applications ranging from monitoring of flexural loading to detection of cracks in concrete specimens. In one of their earlier works, Kuang et al. [9] investigated the use of such type of POF sensor to monitor the mechanical response of a number of plastic specimens. The POF utilised in their studies was a step-index 1mm diameter PMMA POF supplied by Mitsubishi Rayon Co. Ltd. The flexural sensitivity of the fibre used (ESKA CK40) was enhanced by removing the cladding layer of a segment of the POF over a pre-determined length, the aim being to promote light loss in this region due to the reduction in the number of modes undergoing total internal reflection when the fibre was bent. In their studies, the sensitised regions reported range from 70mm for smaller specimens to 300mm for larger specimens were shown to be appropriate although it may be postulated that the length of choice will depend largely on the sensitivity required and the size of the host structure of the actual application. The study also demonstrated that the POF sensor based on the removal of cladding layer possess directional sensitivity in that highest flexural sensitivity could be achieved when the plane of cut resulting from the removal procedure was perpendicular to the direction of flex. The flexural strain-optic coefficient was reported to be approximately $1.8 \times 10^{-5}/\mu\varepsilon$, implying that at 1% flexural strain, a normalised loss of 18% in light intensity could be expected.

Since the removal of the cladding layer was done manually by scrapping the POF with a razor blade, the light intensity transmitted through to the fibre end would have decreased considerable with each subsequent scrapping process since with each cut, more of the POF core was removed. To achieve consistency in the amount of material removed, the procedure should be done with the POF placed in grooves machined to specific depths. The scrapping of the POF material would then be carried out in steps of decreasing groove depth as desired. A photograph showing the cross-section of the sensitised POF is given in Figure 1.

To test the response of the POF sensor under flexural load and tensile load, the authors mounted the POF onto its respective specimen using an acrylate-based adhesive. For the flexural load condition, a section of a straight fibre was aligned along the longitudinal axis of the specimen and attached to the bottom surface of the plastic specimen. The specimen was tested under a three-point bend set-up to evaluate the response of the sensor. To sensitise the POF to the flexural load, a segment of the fibre attached to the specimen was removed as described above. To test the POF response to tensile loading, the sensing region of the POF was curved slightly such that the fibre would tend to straighten out as the tensile load was applied.

Figure 2(a) shows the repeatability of the POF sensor response under a cyclic flexural loading, demonstrating the potential of the sensor to monitor the flexural displacement of the specimen reliably. The response of the POF sensor when tested in the tensile direction is

Fig. 1. Photograph showing the cross-sectional view of the sensitised POF where a segment of the POF (cladding and core) was removed (after [9]).

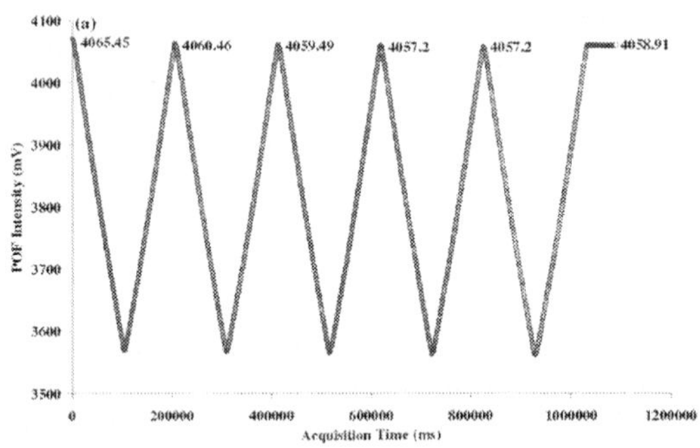

(a)

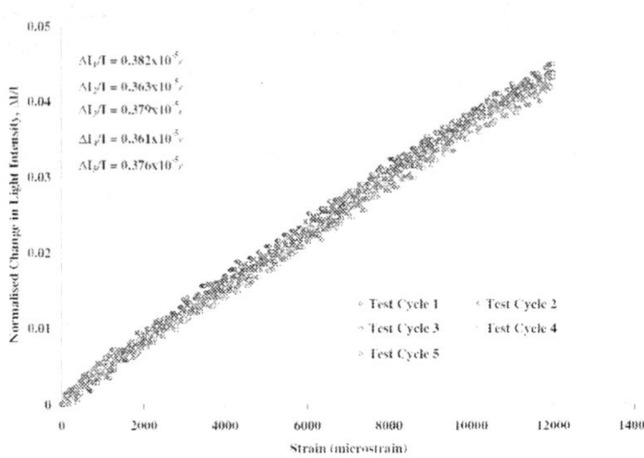

(b)

Fig. 2. Plot showing the response of the POF sensor during (a) cyclic flexural loading, (b) tensile loading (after [9]).

shown in Figure 2(b). Here the result also illustrates the high degree of repeatability of the response of the POF. It was highlighted that although the tests stopped at flexural and tensile strains of 0.7% and 1.2% respectively, the ability of the POF sensor design used in this study to measure strain values higher than that observed in the tests can be achieved. This may be supported by the fact that at those levels of strain observed in the study, the loss of intensity in each case did not show evidence of a plateau. In addition, studies conducted by other workers up to 45% [6,7] can be expected while other studies have shown that POF could endure stretching of more than 80% strain under certain conditions [10].

The performance of the POF was also investigated under dynamic loading conditions to assess their potential for vibration measurement, a technique frequently employed in health monitoring of civil engineering structures. Here, Kuang et al. [11] used the same POF sensor design described above in a set-up involving monitoring the natural frequencies of a series of increasingly damaged composite beams subjected to various degrees of damaged induced by low-velocity impact. The ability to detect the change in the frequency response of the host beam as a result of impact-induced damage is shown in Figure 3(a). In addition, the damping ratio of the cantilever set-up was also monitored. The ability of the sensor to monitor the damping ratio as a function of impact energy (inferring degree of damage) is shown in Figure 3(b). The changes in natural frequency and the decrease in the damping ratio were used to characterise the reduction in the post-impact flexural stiffnesses and the residual strengths of the composite beams with increasing level of impact damage. The study showed that the POF sensor was evidently able to detect a change in the damping ratio as small as 2.5%. The natural frequencies of the beams at various stages of damage based on the results of the POF sensor were found to compare well with the predicted theoretical values as well as the readings obtained from an electrical strain gauge. The authors also discovered in their study that impact tests on simply supported carbon-fibre reinforced beams also showed that the sensor was able to monitor out-of-plane deflections during the impact event. A laser Doppler velocimeter was also used in conjunction with a piezoelectric load cell as a means to validate the results of the POF sensor.

The ability to monitor the dynamic response of the host beam encouraged the authors to further apply the technique to monitor the morphing response of a nickel-titanium fibre- metal laminate [12]. A collocated electrical strain gauge provided a reference measurement to compare the response of the POF sensor. It was reported that following activation of the smart fibre metal laminate using a heat gun, the nickel-titanium deformed as predicted and the POF sensor was found to monitor the flexural response of the smart composite accurately. The response of the POF was compared to the electrical strain gauge and this is summarised in Figure 4 showing the high degree of similarity in the responses of both sensors.

This type of intensity-based POF sensor also found applications in crack detection and monitoring of deflection of civil structures. Kuang et al. [13] conducted a series of flexural tests on scaled-specimens where the POF sensor was attached to the bottom surface of the concrete beams and demonstrated in their study that the POF sensor design used were of sufficient sensitivity to detect the presence of hair-line cracks. Figure 5 shows photomicrographs of the size of the cracks in the concrete relative to the POF. The smallest crack width measured approximately 0.04 mm was successfully detected using the POF sensor. In order to improve the sensitivity of the POF to the beam deflection and crack initiation, a segment of the POF was

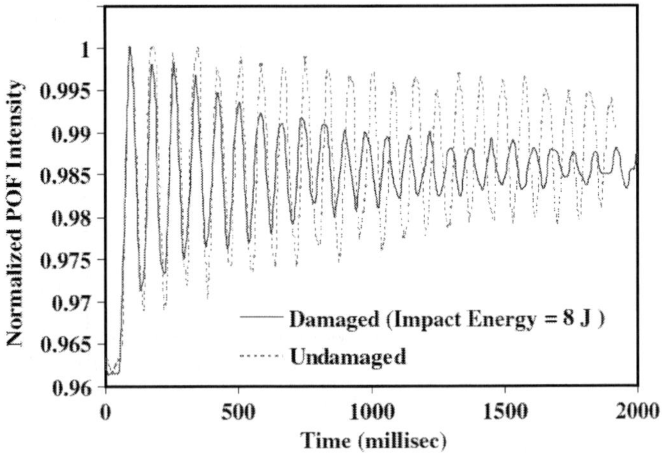

(a)

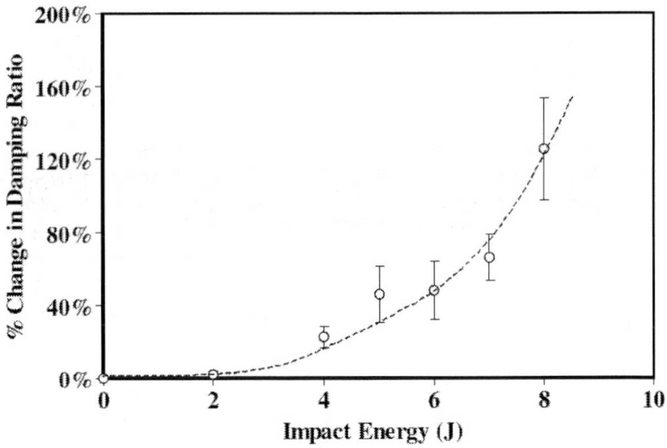

(b)

Fig. 3. Plots showing (a) a comparison of the dynamic response of undamaged and damaged specimen as measured by the POF sensor, (b) the ability of the POF sensor to monitor the change in the damping ratio response for an increasingly damaged specimen (after [11]).

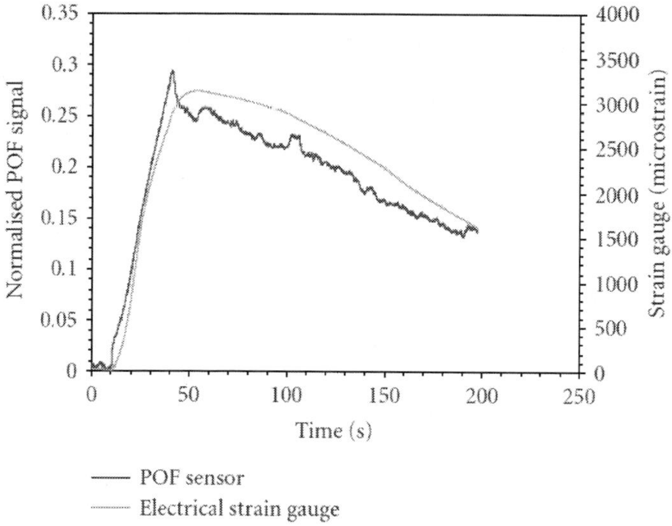

Fig. 4. Plots showing response of the POF in monitoring the fibre metal laminate upon activation of the nickel-titanium (after [12]).

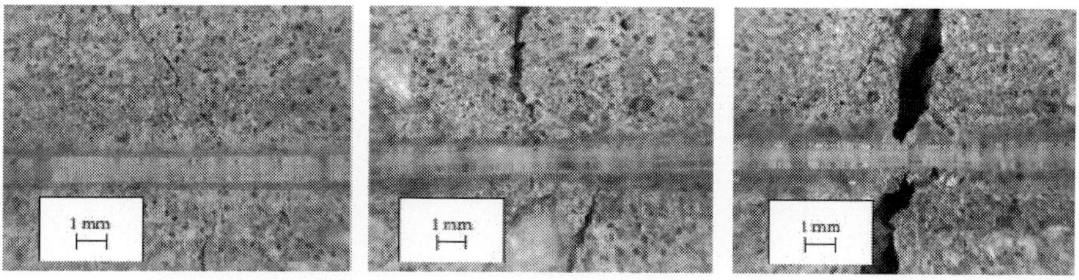

Fig. 5. Photographs showing intersection of the crack and the POF sensor for the concrete specimens at difference degrees of crack width (after [13]).

removed as described earlier over a predetermined length (70 mm for a series of scaled un-reinforced concrete specimens and 300 mm for full-scale reinforced concrete specimens). For the scaled specimens, the authors showed that under three-point bend loading, the POF sensor was able to detect the presence of crack once it intersected the sensor and this observation was repeatable for the entire series of specimens as shown in Figure 6. Similarly, for the full-size specimens the test results affirmed the potential of the POF sensor to monitor the response of the concrete beam under load. Initial cracks in the concrete beam were successfully detected and the POF continued to monitor the post-crack vertical deflection through to finally detecting the failure cracks in the specimen.

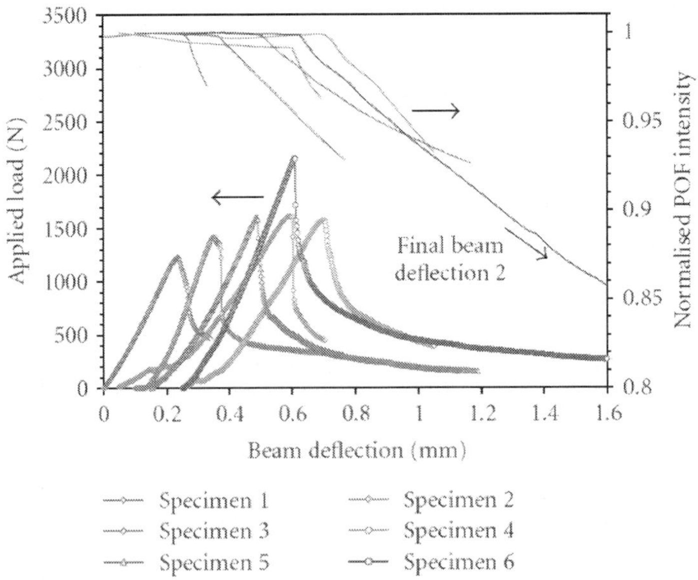

Fig. 6. Plots showing the ability of the POF sensors in detecting the crack for a series of six scaled concrete specimens (after [13]).

In a different area of application within the field of civil engineering, POF sensor has been demonstrated to have potential to be used as flood sensors. Monitoring of the water level in flood-prone areas could provide early warning of flash floods which devastated various regions in ASEAN (Association of the Southeast Asian Nations) in recent times, both in the urban and rural areas. The monitoring of liquid level is also an important activity in many industries and the special requirements to monitor a large number of multiple containers of volatile fluids demands the implementation of cost-effective and safe sensor systems. Other applications where liquid level monitoring is desirable include monitoring of the water level in catchment areas, monitoring of illegal dumping of trash and pollution in canals (which can be inferred by an unexpected rise of water level) and early warning of impending drought (when water levels drops to a minimum level). Information collected by a network of sensors could also be used to complement and update flood monitoring and prediction software models.

The intensity-based POF sensor proposed for flood monitoring by Kuang et al. [5] was simplified from earlier designs applied to glass-type fibres. In the simplified design, neither fusion-splicing nor heating were used in the sensor construction as the POF could be readily bent to shape without any risk of fracture. No prismatic element was used in the proposed design, thus rendering its construction significantly easier whilst exhibiting excellent optical response. It was argued that the simplicity in design of the proposed sensor could lead to an inexpensive, rugged, easy-to-use and effective optical fibre-based sensor for mass deployment in a flooding monitoring application. In addition, the authors demonstrated the ease of integrating the POF sensor to a commercial wireless mote, utilizing the on-board light dependent resistor to monitor the optical intensity transmitted through the fibre. Figure 7 shows the assembled unit consisting of the POF sensor and the wireless mote. The convenience and ease of integration the POF to off-the-shelve wireless motes would not be possible with other interrogation techniques, highlighting the unique advantage of intensity-based approach.

Fig. 7. Photograph of the wireless POF prototype and the MICA2DOT unit used in the flood monitoring experiment (after [5]).

The principle of operation of the POF sensor used in the flood monitoring study was based on the loss of total internal reflection or absorption of evanescence wave when the probe was immersed into the liquid medium. Before the sensitized probe was immersed into the liquid (i.e. in air which has a refractive index of ~1), the majority of the light ray will undergo total internal reflection and will be guided along the optical fibre to the output end of the fibre. However, when the probe was immersed into a liquid with a higher refractive index (e.g. water, which has a refractive index of 1.3), light traveling as cladding-mode will be absorbed by the liquid as the condition for total internal reflection could not be satisfied at the fibre–water interface. The intensity of the optical signal arriving at the output end of the fibre will correspondingly be reduced. It follows that when the probe was immersed in a liquid with higher refractive index, more light will be lost from the optical fibre into the liquid. This being a point sensing system, several such sensors may be deployed concurrently to achieve a multi-level discrete measurement of liquid level.

The response of the sensor was also tested and Figure 8 shows the repeatability of the liquid detection capability of the proposed sensor. From the sensor response, it is clear that it exhibits excellent repeatability showing no sign of hysteresis. At the circle marked A, corresponding to the moment the tip of the probe touched the surface of the liquid, the displacement transducer reading was initialized and set to zero. At that moment, the sensor signal decreased sharply as light escaped into the test liquid. When the probe was being drawn out, the sensor signal began increasing gradually starting from time B when the probe was beginning to leave the surface of the liquid. This occurred at approximately the same probe position corresponding to that of circle A. As a result of the surface tension of the liquid, a cusp formed at the tip of the probe as it past the zero position. The wetted area at the tip decreased gradually as the probe was being drawn out and this corresponds to the increasing signal reading as more light rays undergo total internal reflection. As the cusp broke at time corresponding to circle C, the sensor signal reaches maximum value at approximate probe position of 0.37mm. From the chart, the response of the sensor when withdrawn from the liquid lags behind the physical position of the probe as defined by the difference in the distance between positions A and C, this interval being approximately 0.37mm for the liquid used. Clearly, the value of this interval, which is a function of the surface tension, will vary accordingly with liquid used. However, level of accuracy better than 1mm may not be at all necessary for many applications in civil engineering such as flood monitoring or water level in reservoirs. This example provides further evidence of the cost-effectiveness of the intensity-based approach in meeting the sensing needs of an important area in civil engineering.

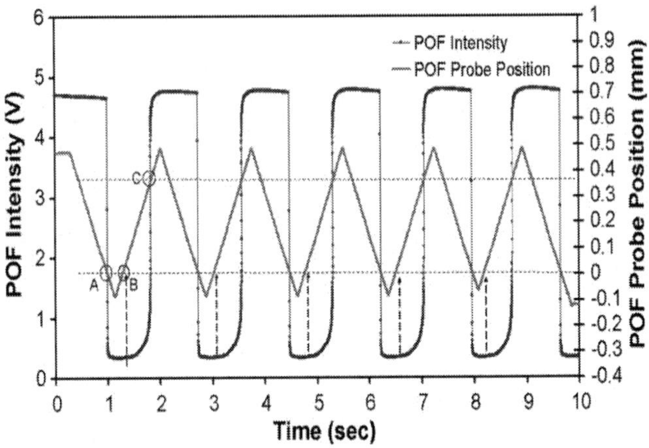

Fig. 8. Chart showing the repeatability of the POF sensor as the POF was immersed in and removed from water (after [5]).

Another significant design based on intensity sensing technique relies on the principle that a change in the gap between two cleaved POF surfaces resulting from an applied strain will leads to a corresponding chance in the light intensity of the sensor [14]. The operating principle of the POF sensor is straightforward- the sensor relies on the monitoring of the optical power transmitted through an air-gap between two cleaved optical fibre surfaces. The two fibres are aligned within a housing in which the fibre could slide smoothly. In this case, the housing itself is strain-free while the POFs are displaced relative to each other during the bending of the beam. It is important to note that the two fibres on either ends of the housing do not undergo any strain since they are free to slide relative to the housing. The gap changes in proportion to the applied strain leading to the modulation of the transmitted optical power.

Two versions of the sensor design are shown schematically in Figure 9. The authors showed that by adding a suitable opaque liquid, the strain sensitivity of the sensor could be improved significantly compared to one without the inclusion of the liquid. A comparison of the strain (or extension) sensitivity of the sensor is shown in Figure 10, highlighting the effect of different liquid opacity on its sensitivity. In the study, four POF sensors and an electrical strain gauge were attached to the bottom surface of a reinforced scaled concrete

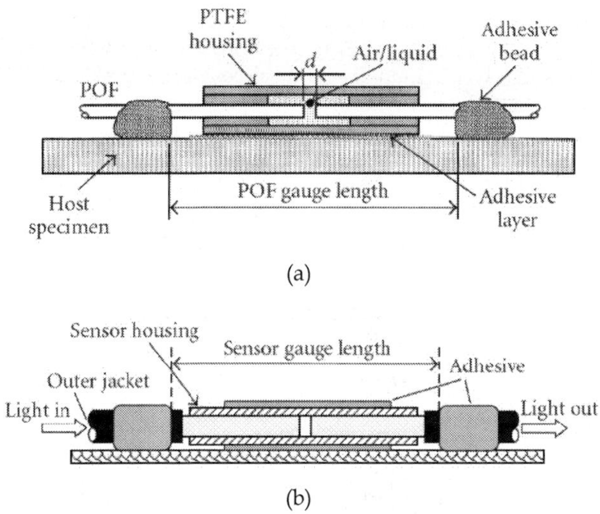

Fig. 9. Schematic showing the two versions of the POF sensor design based on changes in the gap between two cleaved fibres (a) (after [14]), (b)(after [4]).

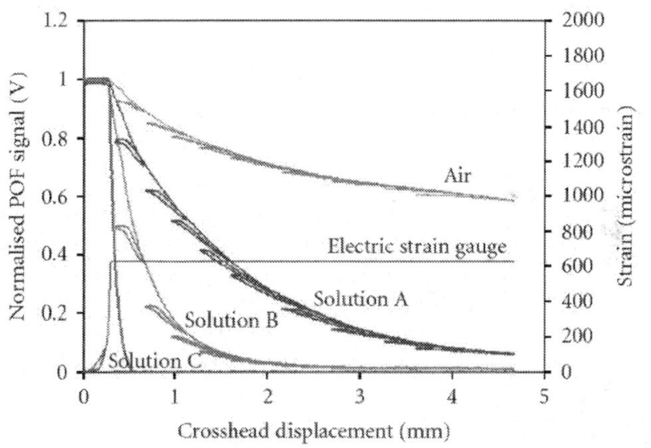

Fig. 10. Plot showing the different responses of the four POF sensors and the electrical strain gauge under a cyclic loading test (after [14]).

beam. The specimen was subjected to a flexural cyclic loading in a three-point bent set-up. When the failure load was reached, a crack developed across the beam which also damaged the electrical strain gauge. The four POF sensors were found to have survived and were able to continue monitoring the loading cycle as shown in Figure 11.

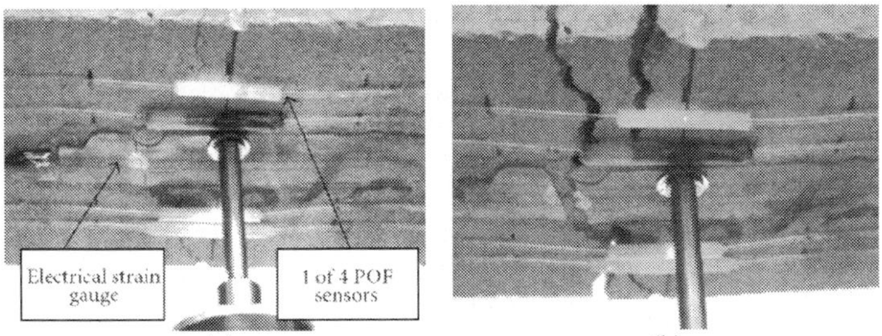

(a) (b)

Fig. 11. Photos showing (a) the damage of the electrical strain gauge due to an intersecting crack (b) the relatively better survivability of the four POF sensors (no visible damage) even under severe crack (after [14]).

Building on the work conducted earlier, here the authors made further progress in demonstrating the ease of integrating the signal of the intensity-based POF sensor as a feedback to a control system to direct the amount of deflection in a smart fibre laminate beam [15]. Figure 12 shows a schematic of the set-up used in the experiment. Here, in order to activate the morphing of the smart beam, a thin-film heating source was attached to the beam. The POF sensor, attached to the opposite surface provided the feedback signal to the controller such that the appropriate amount of power was supplied to the heater to allow the smart beam to deflect to the desired position. The deflection of the beam was monitored continuously by the POF sensor and used as feedback to the controller to achieve the pre-set deflection position accurately. Figure 13 shows the POF sensor readings at three different beam deflections (A) to (C). The result illustrates that the POF-instrumented beam could be controlled accurately to within 3% of the desired deflection using the feedback data from the sensor with very little overshoot.

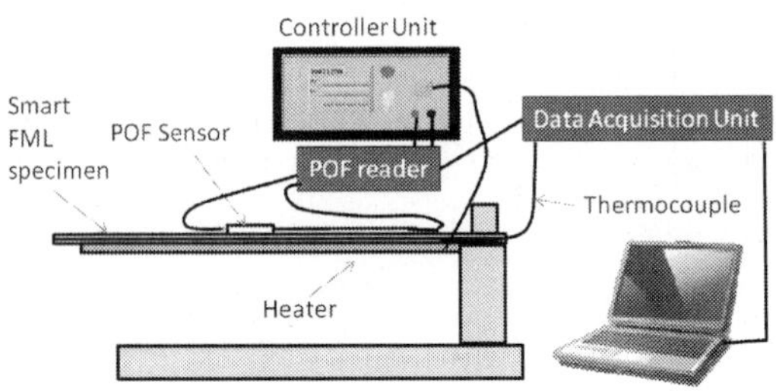

Fig. 12. Experimental set-up used to evaluate the potential use of the POF as feedback sensor to control amount of deflection of the smart composite beam (after [15]).

It was also of interest to gain an insight of the ability of the POF sensor to detect external disturbances acting on the beam and the ability of the control system to use the feedback data from the POF sensor to self-correct the amount of deflection of the smart beam. When a weight was added to the free end of the specimen after the smart beam had achieved steady-state, the POF immediately detected the change in the deflection of the beam which in turn resulted in a decrease in the POF sensor output triggering the controller to activate the

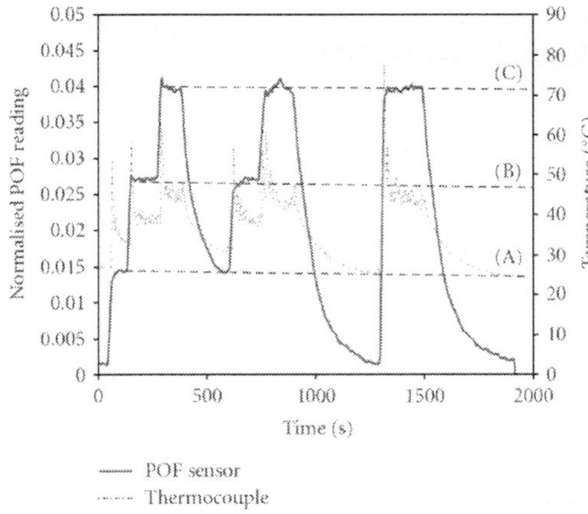

Fig. 13. Plot showing the response of the POF sensor demonstrating the potential of using the POF sensor to provide feedback signal achieve the amount of deflection desired in the smart composite beam (after [15]).

shape memory alloy via the heater. On heating to its activation temperature, the smart beam position was restored as evidenced by the steady state values of the POF sensor at the original controlled position. When the weight was removed, the position of the beam was again perturbed. Here the POF sensor triggered a corresponding series of action similar as before resulting in the beam settling to the controlled position as illustrated in the POF trace in Figure 14.

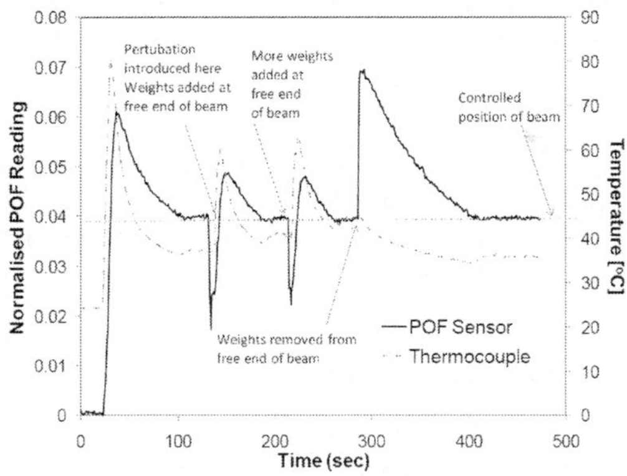

Fig. 14. Plot showing the self-correction capability of the control system based on the POF sensor output (after [15]).

The ability of the above POF sensor to monitor relatively large extension was further studied in a separate application [4, 16]. Here, the POF sensor based on Figure 9(b) was attached to two types of polypropylene geotextile specimens as shown in Figure 15, one being woven whilst the other being non-woven. A series of specimens, instrumented with

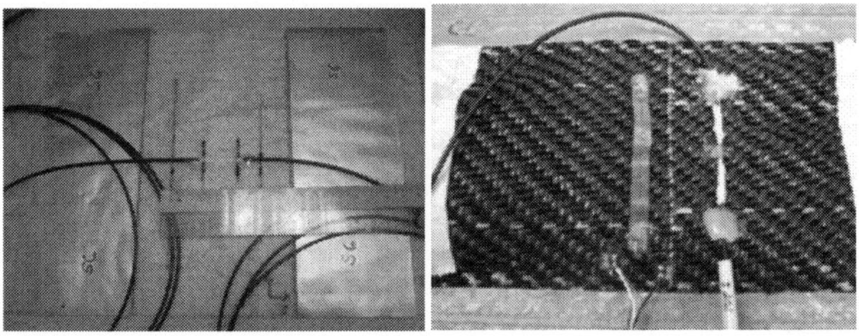

Fig. 15. Photographs showing the two types of POF-instrumented geotextile specimens (after [16]).

the POF sensor, were tested under quasi-static tensile load as well as being subjected to an acceleration field of 100g in a geotechnical centrifuge machine in a submerged condition.

In the quasi-static tensile test, the geotextile specimens were loaded to failure and the typical results are shown in Figure 16. The results show the ability of the POF sensor to monitor strains exceeding 35% for the non-woven specimens and approximately 20% for the woven specimens. The responses of the POF sensors in both cases compared very well with other reference measurement techniques. The POF sensor is expected to be able to monitor even higher strains since the fibre itself does not undergo any strain. It is noteworthy that the sensing principle relies on the changes in the distance of the gap between two fibres and the adhesive to attach the sensor to the host substrate was applied on the fibre, not on the housing itself, hence, neither the fibre nor the housing experience any significant strain, since the strain limit is a function of the gauge length (i.e. distance between the adhesives applied on both sides of the fibre) and the provision of sufficient fibre length within the housing is important such that at maximum extension, the fibre does not protrude from the housing.

In the centrifuge test, it was reported that the POF sensors survived the acceleration of 100g whilst submerged under water. Here, the instrumented geotextile tube was released from a hopper and dumped to the bottom of the centrifuge box to study the strain development of the geotextile tube throughout the whole of this process. A study of the sensors output showed that the POF sensor was able to monitor the strain experienced by the model geotextile tube reasonably well, following a similar trend and magnitude of measurement as obtained using the electrical strain gauge as illustrated in Figure 17. Towards the end of the test, it was noted that the two sensors gave different responses; the POF sensor indicated a sharp increase in strain while the strain gauge recorded a decrease. The resistance type strain gauge was deemed to have correctly recorded the decreasing strain trend. It is postulated that since the geotextile tube came to its resting position on the base of the strong box, without any external loading, the strain developed on the base centre portion of the geotextile would be insignificant, as registered by the strain gauge. On inspection of the POF sensor, it was noted that the fibre ends of POF sensor had protruded out of its housing which led to the sudden loss of intensity at model time 6 s as indicated in Figure 17. It was likely that the POF was tugged as the geotextile tube was being released resulting in the separation of the fibre from its housing. Although the POF sensor was damaged during the last phase of the test, the centrifuge experiment was encouraging, demonstrating the potential of the POF sensor for measuring large strains in geotextile tubes.

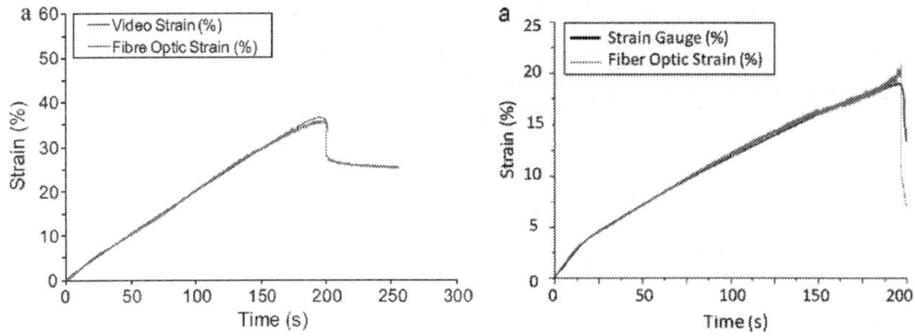

(a) (b)

Fig. 16. Plot showing the response of the POF sensor during a quasi-static tensile loading for (a) a non-woven specimen, (b) a woven specimen (after [16]).

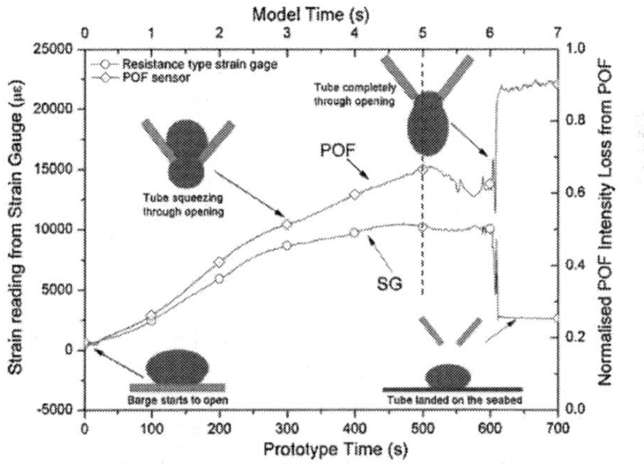

Fig. 17. Plot showing the excellent response of the POF sensor in the centrifuge test compared to the electrical strain gauge (after [16]).

2.2 OTDR-based POF sensors

OTDR-based POF sensors have received some attention recently partly in view of the recent advancement in the field of optical time-domain reflectrometry for POF. Developmental work in the area of the interrogator has resulted in the availability of high-resolution photon-counting systems allowing distributed monitoring along a single POF fibre. OTDR is a well-known technique in telecommunication for fault analysis and crack detection in silica fibres. The ability to detect the very low intensity light levels and high detection speed whilst achieving a high level of spatial resolution is not possible with standard OTDRs. The use of graded-index fibre in conjunction with v-OTDR also reduces the modal dispersion encountered in step-index PMMA fibres and hence graded index fibre is also used for monitoring of fibre over longer distances compared to step-index fibres. One of the most significant advantages of the OTDR technique is the possibility of distributed sensing, i.e. monitoring along the entire length of the fibre is in principle achievable.

2.2.1 Operating principles and background of OTDR-based POF sensors

OTDR sensing exploits the monitoring of the backscatter light in an optical fibre following the launched of a short optical pulse at one end of the fibre. The ability of the OTDR to detect the position of a defect along its length is based on the principle of time-of-flight. The time it takes for the pulse to be reflected back to the detector as backscatter signal is measured very precisely i.e. the backscatter signal is recorded as a function of time which is then converted to a distance measurement. Perturbations, such as strain or defects along the length of the fibre will result in either a peak Fresnel reflection or loss in the backscatter signal at the location of the perturbation. Specifically for POF use, a few companies have developed and commercialized v-OTDR to detect the very weak backscatter light signal due to the high attenuation and large dispersion in POFs. The photon-counting technology, based on an avalanche photodiode allows detection of very short optical pulses (~1 ns) necessary to resolve the location of defects accurately (to within 10 cm resolution).

An OTDR trace consists of a number of reflection peaks corresponding to specific physical junctions along the fibre such as splices, connectors, fibre terminations or defects. The trace is typically plotted on an intensity-distance graph where the location of the peaks along the x-axis refers to their positions of the event or physical entity (splices, connectors, defects etc.) along the fibre. Hence the position of a crack along the fibre will be shown as a reflection peak on the OTDR trace. In order to match the peaks in the OTDR trace to their respective event, the position of a physical reference along the fibre such as a connector, should be defined. Unknown positions of defects or cracks anywhere along the fibre can then be determined accurately with reference to the pre-defined position.

When used to detect damage in structures, the POF should clearly be sufficiently sensitive to detect the parameter of interest. Since detection of cracks along the fibre is an inherent capability of the OTDR-technique, a possible application of the technique is to apply it for crack detection in structures such as steel tubes. Other applications, which will not be included in this chapter, include large strain measurement in geotextile, exploiting the high yield strain of the POF compared to its glass counterpart which has been reported elsewhere [10]. Here, the segment of the fibre where stress is concentrated will result in the formation of a growing peak in the OTDR trace as the fibre experienced increasing strain.

2.2.2 Applications of OTDR-based POF sensors

Graded-index POF sensors in conjunction with the v-OTDR could be used for detection of cracks in tubular steel specimens encountered in the offshore oil platform. Small hairline cracks along weld lines are frequently the precursor to structural failure under the cyclic loading condition in which the structure is subjected to. Typically, offshore platforms are constructed using steel jacket structures. The steel jacket or 'substructure' is fabricated by joined together tubular steel members using welding and pin-jointed to the seabed. The steel jacket is a massive structure which can weigh up to 20,000 tonnes. In a complex structure such as that of an offshore platform, potential failure hot spots are many which may result from fatigue due to cyclic stresses. Cracking in the steel tubular members may occur due to the high stress intensity experienced at the intersections of brace and chord members. At certain critical sections, stress concentration would be as high as 20 times that of continuous non-welded sections. The welded joints and the abrupt change in geometry are likely locations where cracks would initiate and grow. An array of conventional non-destructive techniques such as ultrasonic scanning, X-radiography and acoustic emission exists for damage detection purposes but these are not suitable for real-time monitoring of large structures since these techniques, in general, require an experienced personnel to conduct the assessment in a localized manner. Hence a distributed sensing based on optical fibre sensor technology for crack detection offers a potential solution for health monitoring of large structures.

In a recent study conducted by Kuang et al., [17] scaled-size tubular steel specimens loaded under flexure was tested with the intention of using POF to monitor the crack growth in the specimen. A pre-crack was initially introduced and subsequently propagated under a three-point bend set-up. Graded-index POF was attached to the steel tube with the aim to detect and monitor the crack propagation as the applied load was increased in a quasi-static manner. In order to achieve this aim, a single POF was attached to the specimen in a serpentine manner such that as the crack propagated, different segments along the length of the fibre will be affected and will be reflected in the OTDR trace. The schematic of the specimen and photo of the set-up are shown in Figure 18. The POF (GigaPOF-50SR) tested had core and cladding diameters of 50 μm and 490 μm respectively. A 5 m long fibre was prepared in addition to a spacer fibre of 2.3 m. A pre-determined segment of the 5 m fibre was attached onto the polished section of the tube using a cynoacrylate-based adhesive. The fibre length from one position of crack to the adjacent one was pre-determined to be approximately 0.5 m (i.e. distance for each bent segment mid-point to mid-point), hence if the POF was able to detect the crack, a peak should appear at intervals of 0.5 m sequentially as the crack propagated.

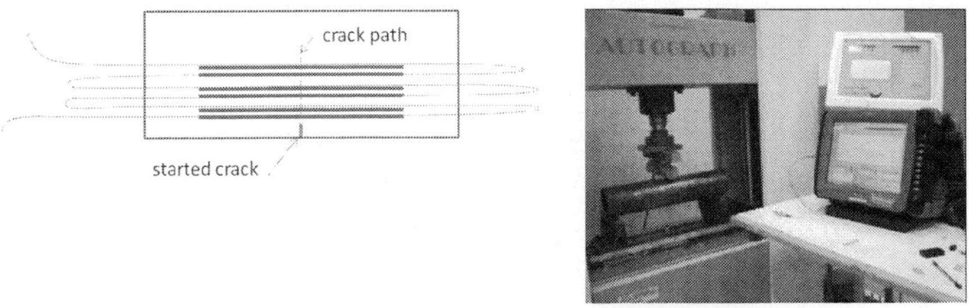

(a) (b)

Fig. 18. (a) Schematic of the specimen, the bold lines indicating the segment of the fibre attached to the specimen (b) Photo showing the three-point bend set-up and the OTDR used in the study (after [17]).

The specimen was loaded transversely to propagate the pre-crack until the crack front reaches the vicinity of the first segment of the POF fibre. Any degradation or fracture of the plastic fibre was noted and compared to the OTDR plot. The OTDR was set to monitor the POF sensor continuously and updates the OTDR. Initially the measurement time of the OTDR was set to a minimum of 1 sec without signal averaging to assess whether this setting is sufficient to detect and locate the crack accurately to within 10 cm. Following the fracture of the first segment of the POF (i.e. closest to the crack front), the loading was momentarily stopped to allow the OTDR data to be saved for further inspection. The loading was continued subsequently until the crack propagates through the entire segment of the POF sensor. The locations of the fracture where they were expected to occur along the fibre were recorded and then compared to the OTDR results. The objective of the mechanical test was to observe the response of the POF and its potential as crack sensors on pre-cracked tubular steel tubes under transverse loading.

A close-up shot of the specimen following fracture of the POF sensors in all segments is shown in Figure 19. The photo reveals that the POF sensor underwent brittle fracture with little signs of necking and stress whitening. The sensing fibre was also noted to experience damage at approximately the same instance as when the crack front intersected the fibre. The damage of the sensing fibre was corresponded to a new peak in the OTDR trace. All the traces were consolidated and presented in Figure 20.

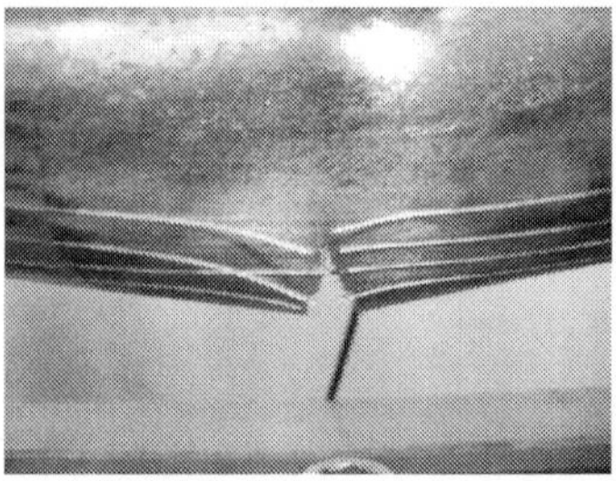

Fig. 19. Photo showing the fracture of the POF sensor along the crack path (after [17])

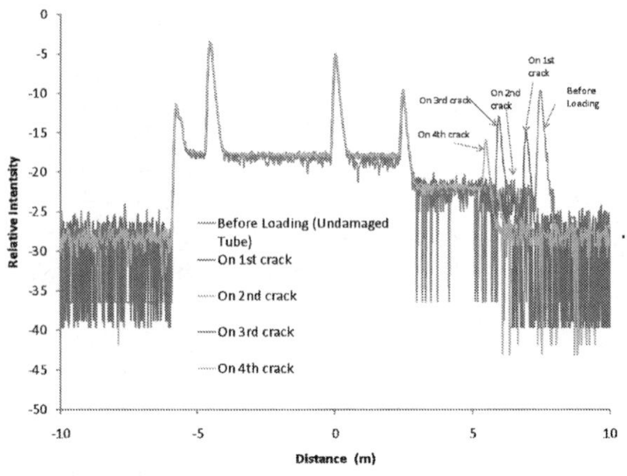

Fig. 20. Plot showing the consolidated OTDR traces highlighting the potential of the POF sensor in detecting and locating the position of the crack along the fibre (after [17]).

All the damage locations along the sensing fibre corresponding to the intersection of the crack front and the fibre were successfully located to within 20 mm by the v-OTDR except the second crack position. The authors argued that the POF segment at location 2 was not sufficiently sensitive i.e. the fibre at that particular location did not fracture when the crack intersected it. The sensitization of the fibre was clearly an important process to ensure that all cracks are detected. The adhesive used was also an important consideration as it is known that cynoacrylate-based adhesive embrittles the POF sensor and thereby improving its sensitivity to fracture.

Although the crack and location 2 was not detected in the particular test, it is evident that the technique shows excellent potential for detecting the location of cracks along the fibre and when configured in a serpentine shape, the sensor is capable of monitoring crack propagation accurately. From Table 1, it is clear that the v-OTDR possess the accuracy in determining the location of any crack along the POF given its high spatial resolution. The accuracy of locating the exact position of the crack along the fibre sensor relies on a proper calibration of the OTDR and determination of an accurate average refractive index of the graded-index POF. The results clearly showed that the calibration of the system was appropriate for the length of fibre used in the study.

Position	Expected (m)	Actual (m)
Location of 1st crack	6.964	6.95
Location of 2nd crack	6.464	-
Location of 3rd crack	5.964	5.95
Location of 4th crack	5.464	5.464

Table. 1. Accuracy of the crack locations detected by the OTDR versus actual positions (after [17]).

3. Conclusion

Plastic optical fibre sensors represent an emerging alternative for various applications in engineering. In this chapter, various examples of applications of POF sensors in civil engineering have been demonstrated e.g. for monitoring strain, deflection, liquid level, vibration, detection of cracks, and these have been applied to different materials including concrete, steel, geotextile and laminate composites. POF sensors offer many unique features that could be exploited to achieve a cost-effective sensing system. Potential users need to have an appreciation of these benefits and the capability of these sensors. Following an objective assessment of the capability of POF sensors together with a clear understanding of the sensing specification required in any particular project, POF sensors may proved to be the ideal candidate for the job.

4. Acknowledgment

Funding by A*STAR-MPA through the National University of Singapore CORE centre under joint Grants no. R-264-000-226-490 and no. R-264-000-226-305 is acknowledged.

5. References

[1] K. S. C. Kuang and W. J. Cantwell, "The use of conventional optical fibres and fibre Bragg gratings for damage detection in advanced composite structures—a review," Applied Mechanics Reviews, vol. 56, pp. 493–513, 2003.

[2] G. Zhou and L. M. Sim, "Damage detection and assessment in fibre-reinforced composite structures with embedded fibre optic sensors-review," Smart Materials and Structures, vol. 11, pp. 925–939, 2002.

[3] K. Peters "Polymer optical fiber sensors—a review" Smart Materials and Structures, vol. 20, Article ID 013002, 17 pages, 2011.
[4] K. S. C. Kuang, S. T. Quek, C. Y. Tan, and S. H. Chew," Plastic optical fiber sensors for measurement of large strain in geotextile materials," Advanced Materials Research, vol. 47-50, pp. 1233-1236, 2008.
[5] K. S. C. Kuang, S. T. Quek, M. Maalej, "Remote flood monitoring system based on plastic optical fibres and wireless motes" Sensors and Actuators A: Physical A, vol. 147 pp. 449-455, 2008.
[6] S. Kiesel, K. Peters, T. Hassan, and M. Kowalsky, "Calibration of a single-mode polymer optical fiber large-strain sensor," Measurement Science and Technology, vol. 20, Article ID 034016, 7 pages, 2009.
[7] S. Kiesel, K. Peters, T. Hassan, and M. Kowalsky, "Behaviour of intrinsic polymer optical fibre sensor for large-strain applications," Measurement Science and Technology, vol. 18, pp. 3144-3154, 2007
[8] K. Murphy, S. Zhang, and V. M. Karbhari, "Effect of concrete based alkaline solutions on short term response of composites," in Proceedings of the 44th International Society for the Advancement of Material and Process Engineering Symposium and Exhibition, vol. 44, pp. 2222-2230, 1999.
[9] K. S. C. Kuang, W. J. Cantwell, and P. J. Scully, "An evaluation of a novel plastic optical fibre sensor for axial strain and bend measurements," Measurement Science and Technology, vol. 13, pp. 1523-1534, 2002.
[10] P. Lenke, S. Liehr, K. Krebber, F. Weigand, and E. Thiele, "Distributed strain measurement with polymer optical fiber integrated in technical textiles using the optical time domain reflectometry technique," in Proceedings of the 16th International Conference on Plastic Optical Fibre, pp. 21-24, 2007
[11] K. S. C. Kuang and W. J. Cantwell, "The use of plastic optical fibre sensors for monitoring the dynamic response of fibre composite beams," Measurement Science and Technology, vol. 14, pp. 736-745, 2003.
[12] P. Cortes,W. J. Cantwell, K. S. C. Kuang, and S. T. Quek, "The morphing properties of a smart fibre metal laminate," Polymer Composites, vol. 29, pp. 1263-1268, 2008.
[13] K. S. C. Kuang, Akmaluddin, W. J. Cantwell, and C. Thomas, "Crack detection and vertical deflection monitoring in concrete beams using plastic optical fibre sensors," Measurement Science and Technology, vol. 14, pp. 205-216, 2003.
[14] K. S. C. Kuang, S. T. Quek, and M. Maalej, "Assessment of an extrinsic polymer-based optical fibre sensor for structural health monitoring," Measurement Science and Technology, vol. 15, pp. 2133-2141, 2004.
[15] K. S. C. Kuang, S. T. Quek, and W. J. Cantwell, "Morphing and control of a smart fibre metal laminate utilizing plastic optical fibre sensor and Ni-Ti sheet," in Proceedings of the 17th International Conference on Composite Materials, 2009.
[16] K. S. C. Kuang, C. Y. Tan, S. H. Chew and S. T. Quek, "Monitoring of large strains in submerged geotextile tubes using plastic optical fibre sensors" Sensors and Actuators A: Physical A, vol. 167, pp. 338-346, 2011.
[17] K. S. C. Kuang and C. G. Koh, "Structural damage detection of offshore platforms using plastic optical fibre sensors" Proceedings of the 7th Asian-Australasian Conference on Composite Materials (ACCM-7), 2010.

17

Plastic Optical Fiber pH Sensor Using a Sol-Gel Sensing Matrix

Luigi Rovati[1], Paola Fabbri[2], Luca Ferrari[1] and Francesco Pilati[2]
[1]Department of Information Engineering,
[2]Department of Materials and Environmental Engineering,
University of Modena and Reggio Emilia
Italy

1. Introduction

Because it is the most ubiquitous species encountered in chemical reactions, hydrogen ion occupies a very special place in chemistry and biology, most of the chemical and biological processes being dependent on its activity. From the analytical point of view, the abundance of hydrogen ions is quantified in terms of pH, the negative logarithm of its activity. Its importance is evident by considering that, if the pH of the human blood changes as little as 0.03 pH units or less, the functioning of the body will be greatly impaired; also, brain pH decreases from normal pH of 7.4 to a pH of 6.75 during the brain insult and a continuous monitoring system would be beneficial in the treatment of comatose neurosurgical patients and those who have suffered traumatic brain injury, ischemic brain insult and so forth (Zauner, 1995). Furthermore, the kind of animals and plants living in lakes, rivers and oceans depends on pH values, as well as pH of soil affects the livability of plants. For this reason, the use of pH sensors is widely diffused in various fields to monitor chemical and biological processes and it is finding an increasing number of applications in medicine, biomedicine, industry and environmental monitoring.

The earliest methods of pH measurement fall roughly into four categories: indicator reagents, pH test strips, amperometric or potentiometric devices. Since a long time, the glass electrode pH sensor has been the most popular due to its ideal Nernstian response, independent of redox interferences, and because of its high selectivity for hydrogen ions in a solution, short balancing time of electrical potential, reliability and high reproducibility, and long lifetime. For these reasons, it is still considered to be the standard measuring method for the pH measurements.

However, glass electrodes have several drawbacks for many applications. Firstly, they exhibit a sluggish response and are difficult to miniaturize. Moreover they can generate electromagnetic interferences with other devices and cannot be used in food or in vivo applications due to their brittle nature. For these reasons, there is increasing need for alternative pH sensors and, in this respect, optical methods offer many advantages over the common glass electrode or other electrochemical devices. Among these, optical fiber colorimetric pH sensors monitoring the change in the optical response of proton-sensitive indicators are the most popular systems. Other pH-sensing devices, based on different pH-related properties, have also been proposed; they exploit mass-changing in pH-responsive hydrogel (Ruan, 2003), pH-sensitive changes of refractive-index in polymers (Zamarreno, 2010), pH sensing based on the interferometric response of nanofilms built from layer-by-layer electrostatic self-assembly of polyelectrolytes (Goicoechea, 2009), pH sensors based on conducting polymers (including polypirrole, polyaniline, etc.) (Talaie, 2000), ultrasensitive microcantilever structures (Bashir, 2002; Fritz, 2000), to mention only someone. However, these devices suffer from various drawbacks such as instability or drift or their use is not straightforward or are too expensive to be mass-produced at industrial level.

For a more detailed discussion about pH sensors, one may refer to several reviews appeared in the last decade (Lin, 2000; Wolfbeis, 2004; Yuqing, 2005; Korostynska, 2007; Baldini, 2008; Wolfbeis, 2008; McDonagh, 2008).

As said above, optical fiber methods are the most interesting alternative to the glass electrode for pH as they enable optical spectroscopy to be performed on sites inaccessible to conventional spectroscopy and over long distances. This is particularly interesting in medicine where the high degree of miniaturization of optical fibers, their considerable geometrical versatility, and their extreme handiness enable unique performances in invasive applications. With respect to glass electrode pH sensors, optical fiber methods take also advantage of the immunity to electromagnetic interferences, and from the absence of electrical contacts that make them useful also in the case of non-invasive applications. For these reasons, optical fiber sensors are in continuous development and offer to the physicians efficient tools for prompt and reliable diagnosis, in particular when an in-vivo continuous monitoring has to be carried out (Baldini, 2008).

The basic concept of the optical fiber methods of pH measurements relies on the fact that the incident beam of light is passed through a light guide to the active end of the optical fiber where it interacts with a chemical indicator, which alters the beam's intensity, usually by absorption or fluorescence, of the modified optical signal which is guided back to the detector. In this way, remote sensing can be achieved since the optical signal can be carried over long distances.

The indicator is usually an organic dye able to change its color over a given pH range (typically 3-4 pH units centered at the pKa of a given indicator) and is generally confined to the surface of the optical fiber or immobilized in an adjacent space according to various possible configurations of the sensing element (Lin, 1997; McDonagh, 2008).

In order to record reliable pH values, the hydrogen ion concentration must equilibrate between the bulk of the sample and the space where the indicator is confined, and the sensor response in terms of precision, sensitivity, stability and response-time will therefore depend on the conditions existing within this space.

The most used method exploited to prepare an optical fiber responsive to pH changes is the immobilization of pH indicators in/on a suitable material. This is a key step that will largely influence the characteristics of the pH sensor.

Photopolymerization (Barnard, 1991; Bronk, 1994; Ferguson, 1997; Healey, 1997; Song, 1997) and sol-gel chemistry (Lin, 1997 and 2000; McDonagh, 2008) have been the mainly exploited reactions, and adsorption, covalent binding and entrapment are the most widely used approaches to immobilize indicator dyes.

In the adsorption method, a pH indicator is adsorbed, physically or chemically, on a solid support attached to the optical fiber; it is relatively simple, but the amount of the adsorbed indicator may be small, resulting in a weak signal, and the indicator may leach out, resulting in a pH response that changes with time. Commercial ion-exchange resins, sulphonated polystyrene (Igarashi, 1994) and polyelectrolyte containing silica (Shi, 1997) have been proposed, among the other, for electrostatic immobilization, while hydrophobic organic products (polymers or organically modified alkoxysilanes) have been used to improve hydrophobic chemical adsorption of indicators (Korostynska, 2007; Kowada, 2005; Wu, 2006).

Adsorption of the indicator may be slow in non-porous monolithic solids and in solids with random microporosity, and the development of mesoporosity, having large and controlled open pores (Cagnol, 2004; Grosso, 2004), may result in a shorter response time of the pH sensor (enhanced diffusion and accessibility for protons and therefore faster equilibration) and a larger amount of pH indicator to be immobilized. However indicator leaching may also become a more relevant problem.

The covalent method can overcome leaching problems as the pH indicator is strongly bound to the solid substrate (Baldini, 2008; Barnard, 1991; Bronk, 1994; Ferguson, 1997; Healey, 1997; Song, 1997). However, the immobilization by covalent attachment requires a previous surface modification of the optical fibers and the control of the surface reaction of the modified substrate with a suitable indicator. The resulting procedure is often complicated and time consuming and may lead to loss of indicator sensitivity or result in poor fluorescent properties (Bacci, 1991; Lobnik, 1998). In general, this leads to a costly technology unsuitable for the preparation of mass-produced disposable devices.

In the entrapment method, the pH indicator is entrapped in a substrate, often starting from a polymer solution or a polymerizable liquid. The method is relatively easy, and a suitable choice of the indicator/polymer pair can strongly reduce or completely overcome the problem of indicator leaching. In addition, polymers present the advantage to be less brittle than inorganic glasses, and a suitable design of the molecular architecture can lead to a good balance of hydrophobicity and water swelling. This could allow combining a faster penetration of water and ions with a minimized indicator leaching. Various polymers have been used for entrapment of pH indicators by exploiting either their thermodynamic affinity or by covalent binding of pH indicators (Korostynska, 2007; McDonagh, 2008).

Combination of adsorption and entrapment methods is expected to occur in polymer-containing organic-inorganic hybrid materials, where highly interpenetrated organic and inorganic nanosized domains are formed during the sol-gel process. This could result in improved sensors, when, in addition to the control of water swelling by a suitable polymer choice, the presence of an extended interfacial surface between organic and inorganic domains, can favor the physical adsorption of the indicator. This can allow to optimize the mobility of the hydrogen ions, from the sample into the sensing element, resulting in a reduced response time, and to minimize indicator leaching.

Optical fiber pH sensors can also be prepared in the absence of indicators by exploiting the change of color as a function of pH of some conductive polymers such as polyanilines (Pringsheim, 1997; Sotomayor, 1997; Grummt, 1997) and polypirrole (de Marcos, 1996). Even though they present some intrinsic advantages, such as simple fabrication and measurements in the near-IR region, they suffer from shortcomings like long response time (slow diffusion of water and hydrogen ions into the glassy polymers), interference from other ions, oxidation by oxidizing agents and the need for pre/reconditioning before each measurement.

In the last two decades sol-gel chemistry has become one of the most widely used method for the development of optical fiber pH sensors due to its versatility and to the attractive optical properties of inorganic sol-gel materials (Lin, 1997, 2000; Wolfbeis, 2004, 2008; McDonagh, 2008). Indeed, either inorganic glasses or organic-inorganic hybrid materials can be easily obtained under mild conditions ($T < 100$ °C) starting from a low-viscosity solution containing metal alkoxides alone or mixed with organically modified alkoxysilanes or with suitable polymers. Sol-gel process key parameters like reactant formulation, catalyst, water content and reaction temperature, can be exploited to produce a large variety of materials whose properties can be finely adjusted to tailor the physicochemical properties of the final solid in order to optimize sensor performances (Wen, 1996; Cagnol, 2004; Grosso, 2004; Estella, 2007).

Various sol-gel material configurations, like monoliths, thin films, capillaries and droplets including miniaturized configurations, have been proposed to generate the sensing element at the optical fiber tip (Lin, 1997; McDonagh, 2008).

Alkoxysilanes are most frequently used as precursors of inorganic phases, and the resulting inorganic silica glasses are porous matrices, that can be characterized by tailored porosity (Wen, 1996; Cagnol, 2004; Grosso, 2004; Estella, 2007). Compared with pure organic polymers, sol-gel inorganic glasses can offer some advantages: higher stability, optical transparency down to 250 nm, feasibility of direct coating on glass and silica fibers. However, they also may suffer from some shortcomings, and in particular, slow response to pH changes (typically several tens of seconds) (Lin, 1997; Ben-David, 1997; Seki, 2007), indicator leaching, difficult adhesion to plastic optical fibers.

Sol-gel materials derived from organically modified alkoxysilane (Ormosil) can be prepared under the same reaction conditions used for the preparation of all inorganic sol-gel glasses. The resulting material can have a better surface interaction with organic indicators reducing the shortcomings deriving from leaching (Kowada, 2005; Wu, 2006).

The inclusion of suitable organic polymers in the initial solution with alkoxysilanes leads to the formation of silica-based hybrid materials consisting of highly interpenetrated organic and inorganic domains. A suitable choice of the organic polymers will result in the formation of organic domains that can allow to immobilize higher amounts of indicator and to optimize indicator performances (Cjlakovic, 2002; Aubonnet, 2003) versus water adsorption, while preserving good transparency. As it will be discussed in the following, this can result in fast and reliable cheap sensors.

A lot of progress, improvement and innovation have been made in the field of optical fiber sensors in the last two decades, however notwithstanding the high number of optical fiber chemical and biochemical sensors described in the literature, only a few have reached the market, and the goal of mass production and commercialization of optical fiber pH sensors has not really been accomplished, and research development and application of optical fiber pH sensors is still very challenging and demanding interdisciplinary field. To reach this goal issues such as low cost, user-friendliness, sensitivity, robustness, long-term stability and shelf lifetime of the sensors have to be addressed adequately.

Within this context, this paper aims at illustrating a facile method to develop plastic optical fiber pH sensors with a tip-based sensing element prepared by a sol-gel process and consisting of phenol red indicator entrapped in a PEO-silica organic-inorganic hybrid material.

2. Principle, theory, design of the POE pH sensor

Current trends in optical sensors, such as miniaturization, flexibility and enhanced sensitivity, are indicating a new chemical route for the development of advanced multifunctional materials for optical applications. Those chemical technologies, which can be more easily customized and allows the inclusion of multiple functionalities within a unique preparation step, are bound to be progressively more and more applied to the preparation of optical materials.

In this perspective, the sol-gel technology certainly represents one of the most promising chemical strategies, thanks to numerous advantages mainly related to simplicity and mild operative conditions. It enables creating a glass-like porous structure at room temperature by a two-step acid or base catalyzed reaction involving hydrolysis and condensation, starting with metal alkoxides $M(OR)_4$, which transforms into a rigid three-dimensional metal-oxide network (Brinker, 1990). The sol–gel process has been proved to be flexible enough for an efficient incorporation of organic polymer chains that can behave as flexible links between the metal-oxide domains in the inorganic network, in particular when they are bearing reactive groups that can be involved in the hydrolysis–condensation reactions. The resulting materials are known as organic–inorganic hybrids (Schmidt, 2000), also commonly designated as ceramers due to the combination of the properties of ceramics (high modulus, thermal stability and low coefficient of thermal expansion) with those of organic polymers (high ductility, molecular flexibility and low temperature processing). These materials are often also known as phase-interconnected nanocomposites because of the high level of interconnection between the two phases with domain phase sizes approaching the nanometer scale. Ceramers have a huge potential for application in a variety of advanced technologies (Eckert, 2001; Sanchez, 2011; Kickelkick, 2006), both as structural materials and functional materials, such as catalyst supports, protective coatings (Messori, 2003, 2004a); Toselli, 2007; Fabbri, 2008), sensors (Rovati, 2011; Fabbri, 2011), and active glasses.

Optical fiber sensors are traditionally obtain by fully-inorganic sol-gel process that allows the creation of Si-O-Si linkages between the silica core of the optical fiber and the silica porous matrix deriving from the jellification of the sensitive dye-doped colloidal suspension (Cao, 2005). However, this approach cannot be easily applied in the case of plastic optical fibers, due to the ineffective interaction between the organic PMMA optical fiber core (Lin, 2000).

The approach proposed in this work consists in the fabrication of a pH sensor based on an organic-inorganic hybrid matrix obtained by a sol-gel process, doped with a pH sensitive indicator, to be applied at the tip of plastic optical fibers. Inside the sensitive element, the organic part of the hybrid glass, polyethylene oxide (PEO), plays a multiple role: (i) it allows good adhesion between the plastic optical fiber and the whole sensitive element; (ii) its weak hydrophilicity permits to tune the kinetic of response of the sensor by influencing the diffusion rate of the analyte inside the porous matrix and its interaction with the indicator; (iii) its nature of organic compound allows better physical and chemical interactions with the organic pH indicator dispersed in the hybrid matrix, thus reducing problems of leaching and enhancing the response rate of the sensor.

2.1 Polymer-silica sol-gel hybrids for pH indicators entrapment

Several studies have been performed on the preparation of pH sensors obtained by immobilization of a pH sensitive indicator onto the tip or onto the sides of traditional silica core optical fibers by means of the sol-gel technology (MacGraith, 1991; Gupta, 1998; Alvarado-Mendez, 2005; Rayss, 2002; Miled, 2002; Dong, 2008; Lee, 2001), thanks to the outstanding flexibility of this chemical technique. For this application, silica glass is usually prepared by sol-gel via hydrolysis and condensation reactions of the precursor tetraethylorthosilicate (TEOS), in the presence of water (to promote hydrolysis), ethanol (to enhance the miscibility of water and TEOS), and an acid or base catalyst. The resulting silica-based glass can be attached to the silica core of traditional optical fibers by uncladding of a portion of the optical fiber and activation of the silica core by immersion in nitric acid in order to form silanol groups onto the optical fiber surface, followed by deposition of the silica sol containing the pH indicator on the naked part of the optical fiber (Sharma, 2004). This procedure allows the creation of Si–O–Si linkages between the silica core of the optical fiber and the silica porous matrix deriving from gelification of the sensitive dye-doped colloidal suspension, thus achieving a permanent bonding between the sensitive element (sol-gel glass doped with indicator) and the optical fiber itself which will transmit the optical signal.

The nature of organic compound of the pH sensitive indicators can cause a scarce affinity with the inorganic silica matrix (Wu, 2006). This induces leaching out of the sensitive species from the support, and the only effective way to overcome this problem is to create covalent bonding or stabilizing interactions between the pH indicator and the embedding matrix; nevertheless, this approach demonstrated to strongly reduce the indicator's sensitivity because of the reduction of its mobility within the solid matrix, and to cause slow response.

If the same approach described above for the preparation of pH optical sensors applied to silica-core optical fibers would be applied to the new generation of plastic optical fibers, an ineffective interaction between the organic PMMA core and the inorganic silica sensitive element would be found, resulting in an unstable device with physical discontinuity between the optical fiber and the sensitive head applied onto its tip.

2.2 Plastic optical fibers and sensing probe

The developed sensor is based on a low-cost plastic optical fiber terminated with a sensing element. The sensing matrix takes advantage of the use of PEO/silica organic-inorganic hybrids obtained from sol-gel technique, and in particular of the weak hydrophilicity of PEO chains crosslinked by sol-gel reactions with TEOS. In the presence of aqueous solutions, the PEO domains can swell allowing a faster ion exchange with the surrounding solution. The extent of swelling depends on the PEO molecular weight and, in addition to the beneficial effect on ion diffusion rate, it leads to changes of the volume of the sensing element that can causes problems to the extraction of the pH information. A good balance of ion diffusion rate, adhesion to PMMA fibers and volume change was obtained using a PEO having a molecular weight 8000. A schematic diagram of the probe configuration is reported in Fig. 1. Basic working principle is based on a classic colorimetric approach. The white interrogation light $I_i(\lambda)$ is scattered and absorbed by the sensing matrix that include a pH sensitive indicator (indicator). The change in color of this indicator induces a change in the optical spectrum of the outgoing light $I_o(\lambda, pH)$ collected by the fiber. Analyzing the spectrum of this optical signal the pH of the solution can be recovered.

Absorption operated by the indicator, which depends on the pH value of the solution, induces a change in the optical spectrum of the outgoing light collected by the fiber.

As the pH varies, the relative fractions concentrations of the dissociated $[A^-]$ and protonated $[HA^+]$ forms of the indicator are changed according to the Henderson-Hasselbalch equation:

$$[HA^+] = \frac{C}{1 + 10^{pH - pK_a}}$$
$$[A^-] = \frac{C}{1 + 10^{pK_a - pH}} \tag{1}$$

where Ka is the acid dissociation constant, pKa is −logKa, and C is the concentration of the indicator. According to Fig. 1, interrogation light $I_i(\lambda)$ is absorbed and scattered by the sensing matrix generating the outgoing light $I_o(\lambda, pH)$. According to the modified Beer-Lambert law, the spectrum of the outgoing light $I_o(\lambda, pH)$ can be estimated from the spectrum of the entering light $I_i(\lambda)$ (Splinter, 2006):

$$OD(\lambda, pH) = Log\frac{I_i(\lambda)}{I_o(\lambda, pH)} = \alpha_A(\lambda)\cdot d(\lambda, pH)\cdot[A^-] + \alpha_{HA}(\lambda)\cdot d(\lambda, pH)\cdot[HA^+] + G$$
$$OD(\lambda, pH) = C\cdot d(\lambda, pH)\cdot\left\{\frac{\alpha_A(\lambda)}{1 + 10^{pK_a - pH}} + \frac{\alpha_{HA}(\lambda)}{1 + 10^{pH - pK_a}}\right\} + G \tag{2}$$

where $OD(\lambda, pH)$ is the optical density, $\alpha_A(\lambda)$ and $\alpha_{HA}(\lambda)$ the specific extinction coefficients of the two forms, $d(\lambda, pH)$ the mean optical path of the photons through the probe, and G a parameter related to the "geometry" of the probe. Hence, we consider the relative variations of the outgoing light as the sensor response:

$$\Re(\lambda, pH) = \frac{I_0(\lambda, pH_{ref}) - I_0(\lambda, pH)}{I_0(\lambda, pH)} = -\frac{\Delta I_0(\lambda, pH)}{I_0(\lambda, pH)} = \frac{\Delta OD(\lambda, pH)}{Log(e)}$$
$$\Re(\lambda, pH) = \frac{C\cdot d(\lambda, pH)}{Log(e)}\cdot\left\{\frac{\alpha_A(\lambda)}{1 + 10^{pK_a - pH}} + \frac{\alpha_{HA}(\lambda)}{1 + 10^{pH - pK_a}}\right\} - Bl(\lambda, pH_{ref}) \tag{3}$$

where baseline Bl is

$$Bl(\lambda, pH) = \frac{C\cdot d(\lambda, pH_{ref})}{Log(e)}\cdot\left\{\frac{\alpha_A(\lambda)}{1 + 10^{pK_a - pH_{ref}}} + \frac{\alpha_{HA}(\lambda)}{1 + 10^{pH_{ref} - pK_a}}\right\} \tag{4}$$

and pH_{ref} is the reference value of pH that annuls response $\Re(\lambda, pH)$.

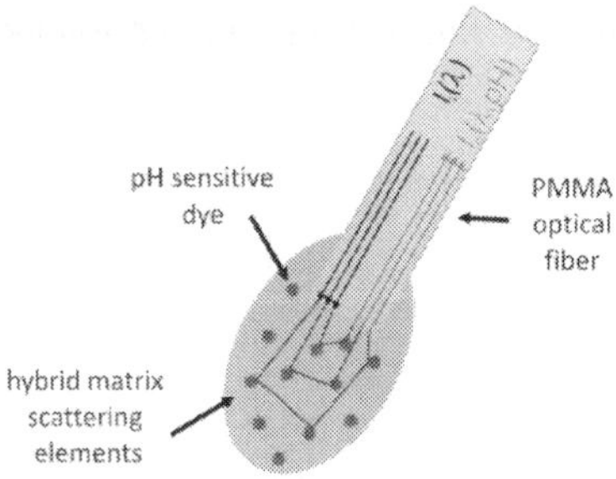

Fig. 1. Drawing of the probe configuration: $I_i(\lambda)$ is the interrogation light absorbed and scattered by the sensing hybrid matrix; $I_0(\lambda, pH)$ is the outgoing light containing the information of interest.

3. Prototype sensors

The developed sensor is based on a low-cost plastic (PMMA) optical fiber (Fort Fibre Ottiche, Italy) with core diameter 1 mm and numerical aperture 0.22. The total length of the taken optical fiber is 35 cm. The pH sensitive indicator (phenol red) was immobilized on the tip of the optical fiber taking advantage of the use of PEO to make hybrid gels. In the next sections the details about the prototype preparation are provided.

3.1 Chemicals

α,ω-Hydroxy-terminated poly(ethylene oxide) (PEO; purchased by Fluka, Milan, Italy) having a number-average molecular weight of 8000 g/mol, 3-isocyanatopropyltriethoxysilane (ICPTES; Fluka, Milan, Italy), tetraethylorthosilicate (TEOS; Aldrich, Milan, Italy), hydrochloric acid at a 37% concentration (HCl, Carlo Erba, Milan, Italy), and ethanol (EtOH; Carlo Erba, Milan, Italy) were high-purity reagents and were used as received without further purification. Phenol red (PR; Aldrich, Milan, Italy) was used as received as pH-sensitive indicator. α,ω-Triethoxysilane-terminated poly(ethylene oxide) (called PEOSi in the following) was prepared by the bulk reaction of the corresponding α, ω-hydroxy-terminated poly(ethylene oxide) with ICPTES (molar ratio of 1:2). The reaction was carried out in a 50 mL glass flask equipped with a calcium chloride trap and under magnetic stirring at 120 °C for 3 h, following a reported procedure (Messori, 2004b).

3.2 Preparation of the sensitive matrix

PEOSi/SiO_2 hybrids were prepared by dissolving TEOS and PEOSi in EtOH at a concentration of about 60% (w/v), and then adding water (for the hydrolysis reaction) and HCl in the following molar ratios with respect to the ethoxide groups of both the PEOSi and TEOS: $EtO^-/H_2O/HCl$ 1/1/0.05. Then pH indicator PR was dissolved into the mixture (approx. 4 mg/50 mL) and left under stirring at room temperature for 15 min. The homogeneous indicator-doped mixture was finally heated at 60 °C for 15–30 min (in order to approach the gel point) before application of a drop onto the tip of the optical fiber.

The hybrid material was characterized by final organic:inorganic weight ratio of 8:2 (assuming the completion of the sol-gel reactions). Samples were coded as follows: PEO-PR where PEO refers to the polymer phase within the hybrid and PR indicates the pH indicator entrapped in the hybrid.

3.3 Preparation of the sensing probe

The polymer-silica hybrid matrix, doped with pH indicator, was prepared as above reported; after the thermal treatment at 60 °C for 15–30 min, condensation reactions involving ethoxysilane groups induced an increase in the viscosity of the solution, leading to a gel-like material. A drop of hybrid gel was then manually deposited onto the tip of the plastic optical fiber, and the optical fiber allowed to dry at atmospheric pressure and room temperature for 2 h and then cured at 105 °C for 24 h to increase the extent of the sol-gel reaction and to induce a strong interfacial connectivity between the sensing matrix and the POF. The dimension of the hybrid sensitive element was optimized looking for the best compromise between the rate of the response of the optical sensor and its physical and mechanical robustness. The final appearance of the sensing probe is reported in Fig. 2.

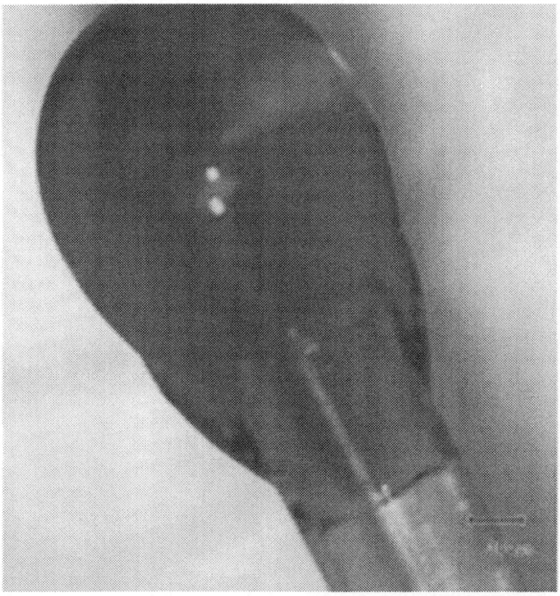

Fig. 2. Picture of the sensing probe: polymer-silica sol-gel hybrid deposited onto the tip of the POF.

4. Instruments and methods

In this section the instruments and methods used in the experimental activity are presented. To evaluate the performance of the developed probes a custom optical setup has been realized and characterized. Afterwards, ahead of the sensor characterization, the spectral properties of the sensing indicator have been determined.

4.1 Design and characterization of the optical setup

The instrumental setup realized to interrogate the sensing probe is shown in Fig. 3. It performs the illumination of the tip of the optical fiber and the collection of the reflected beam.

It consists of a white light source WLS (tungsten halogen lamp, HL-2000-HF, Avantes, Netherlands), two fiber-optic collimators (PAF-SMA-5-550, OFR, USA), a cube beam splitter (039-1130, Optosigma, USA), a glass electrode pH-meter (Eutech Instrument, pH 700 Netherlands) and a grating spectrometer OSA (PMA-11 C5966, Hamamatsu, Japan) equipped with a charge-coupled device detector. The beam, generated by the WLS, is delivered to the first collimator through a 0.5 m long multi-mode optical fiber (1 mm core diameter, NA 0,51, Fort Fibre Ottiche, Italy). The beam splitter, fed by the collimated light, transmits the beam to the second collimator that re-focuses it into the optical fiber sensor, whose sensitive tip is submerged into the testing solution. The glass electrode pH-meter monitors the testing solution.

The colored light reflected by the optical fiber sensor goes back over the optical fiber and reaches the entrance optical fiber bundle of the grating spectrometer OSA, after being deflected by the beam splitter. The measuring spectral range of our spectrometer covers the visible spectrum from 400 to 800 nm. The maximum resolution achievable is 0.487 nm; it has been computed dividing the spectral range by the number of acquisition spectral channels (400nm/820nm). In fact, according to the Rayleigh criterion, which points out how wide the spectral distance between two lines must be to allow their recognition as separate lines, the spectral width of the individual lines must be markedly smaller than their spacing. The theoretical limit computed for our spectrometer does not take into account channels crosstalk and other non-idealities.

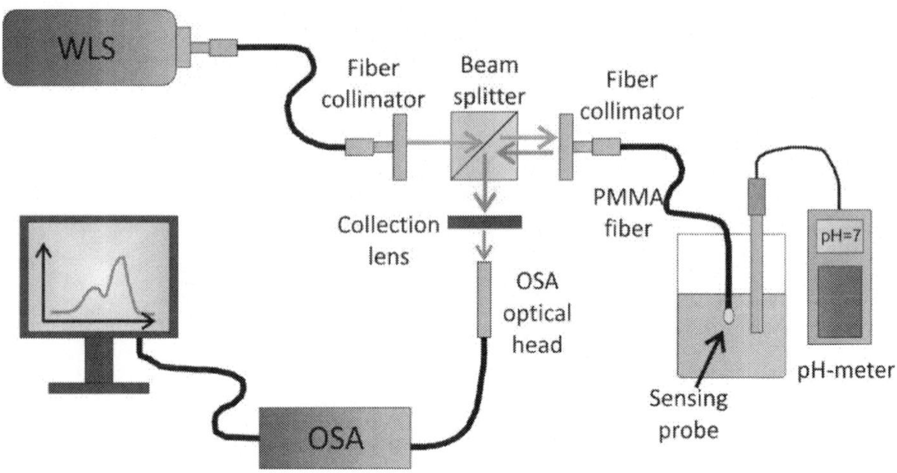

Fig. 3. Schematic design of the optical setup realized. WLS: tungsten halogen lamp, OSA: grating spectrometer.

In order to avoid the corruption of the spectral properties of the outgoing light because of the response of the measuring system and the transmission of the sensing POF, the spectral response of the interrogation system was measured using the experimental setup depicted in Fig. 4.

For this purpose, two measurements were performed; in particular, firstly, as shown in Fig. 4a, the light spectrum at the output of a 35 cm-long POF without sensing matrix was measured using spectrometer OSA. Afterwards, the spectrometer was reassembled as in Fig. 4b using the same 35 cm-long POF.

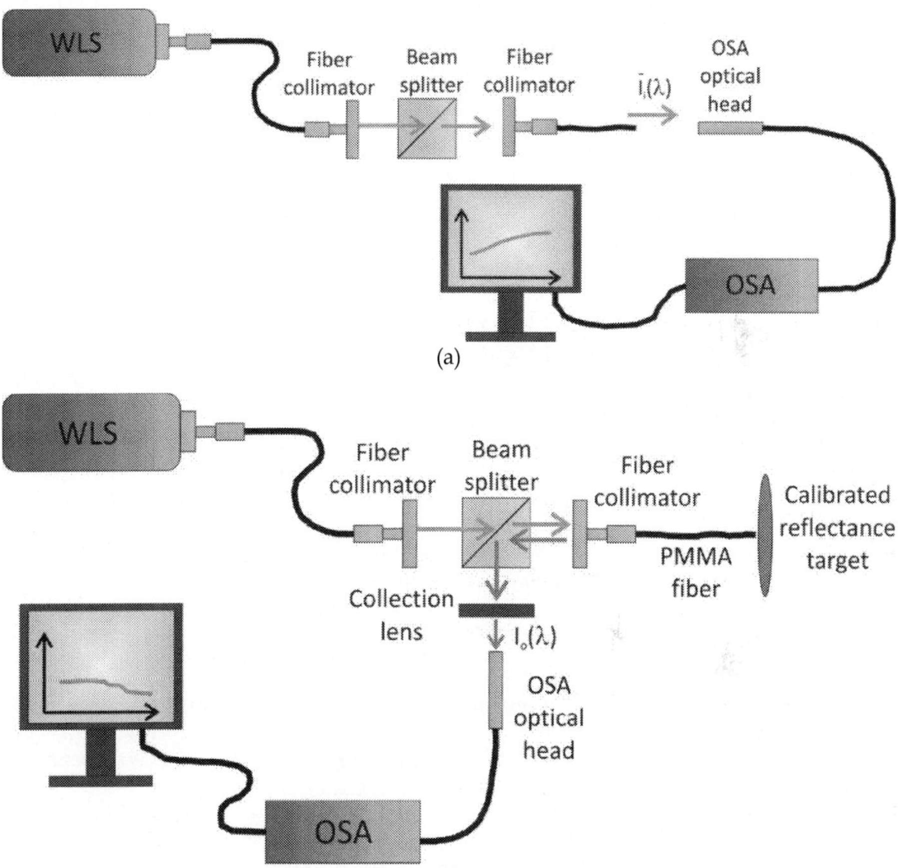

(b)

Fig. 4. Experimental setup used to measure the spectral response of the interrogation system. (a) The light spectrum at the output of a 35 cm-long POF without sensing matrix was measured. (b) Afterwards, the spectrometer was reassembled and in front of the POF a calibrated reflectance target acted as a diffuser.

A NIST (National Institute of Standards and Technology) calibrated reflectance target ($\Im$=99%, Labsphere, USA) acted as a lambertian diffuser in front of the POF. This target has a spectrally flat (± 4%) reflectance over the UV-NIR spectrum. Spectrum $I_o(\lambda)$ acquired by OSA can be calculated as:

$$I_0(\lambda) = k \cdot \Im \cdot S_r(\lambda) \cdot \tilde{I}_i(\lambda) \tag{5}$$

where k is a coupling constant, $\Im$ the calibrated target reflectance assumed not dependent on the wavelength and $S_r(\lambda)$ the interrogation system response. Therefore, the ratio $\frac{I_0(\lambda)}{\tilde{I}_i(\lambda)}$ has been used to study the spectral trend of the interrogation system response.

In particular, the transmission features of the POF have been determined from the comparison between the spectrum of the halogen lamp WLS and $\tilde{I}_i(\lambda)$; as reported in literature (Feng, 2011), the POF exhibits an attenuation peak at about 750 nm.

4.2 Indicator optical characterization

Subsequent to the characterization of the optical properties of the interrogation system, in this section we are going to report the optical measurements performed to determine the spectral properties of the sensing indicator.

The absorbance spectra of phenol red in distilled water at different pH values were measured using a commercial spectrometer (Lambda 19 Spectrometer, Perkin Elmer USA); thus the absorption values at 560 nm were normalized and fitted by the Boltzmann (sigmoidal) function according to Eq. (6).

$$S = S_0 - \frac{a}{1+e^{-\frac{pH-pK}{b}}} \tag{6}$$

where a, b and S_0 are constants.

4.3 Sensors characterization

In our disposable sensor, the spectrum of the outgoing light $I_0(\lambda, pH)$ and thus the response of the sensor $\Re(\lambda, pH)$ is determined by two phenomena: the photon diffusion process in the hybrid-material drop applied to the optical fiber tip and the absorption of the indicator. Both these processes are pH-dependent; indeed, the size of the hybrid drop depends on the pH level of the solution and so, varying the pH, different photon optical paths are possible.

In order to characterize the sensing probe, measurements were performed at room temperature in standard pH buffers (Fisher Scientific) evaluating the response of the disposable sensor $\Re(\lambda, pH)$. According to the schematic representation in Fig. 5, the sensitive tip of the POF was immersed in 20 ml of sample solution; measurements were taken at 5 min after changing the pH level, monitored with the reference pH meter, which has a measuring uncertainty of 0.05 units. The system response $\Re(\lambda, pH)$ was calculated acquiring a reference spectrum $I_0(\lambda, pH_{ref})$ at pH_{ref}=10 and varying the pH of the solution from 9 to 4.

To evaluate the temporal response of the probe to a rapid change in pH, the pH value of the solution has been staircase decreased from 8 to 5 with unitary steps of duration 10 s each. The same protocol has been applied for the reverse direction of the staircase, i.e. from 5 to 8. Also during these measurements the testing solution was continuously monitored with the reference pH meter. The temporal response has been investigated firstly visually inspecting the color changes and then acquiring continuously the spectrum $I_0(\lambda, pH)$ and calculating $\Re(\lambda, pH)$ as a function of time.

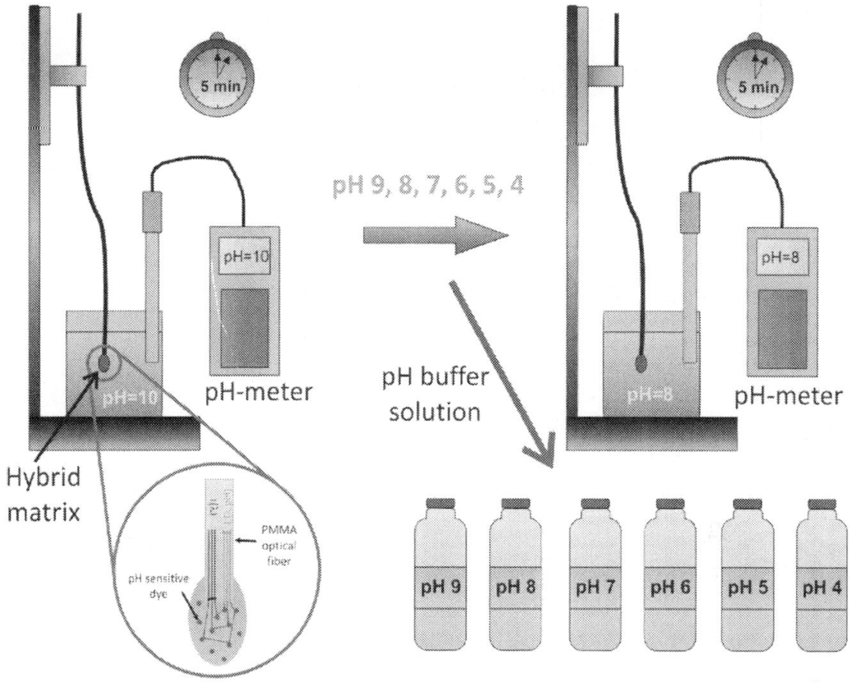

Fig. 5. Schematic representation of the measurement procedure. Firstly, a reference spectrum $I_0(\lambda, pH_{ref})$ at pH_{ref}=10 was acquired. Then, the pH has been changed from 9 to 4, using standard pH buffers and the response of the disposable sensor has been taken at 5 min after changing the pH level, monitored with the reference pH meter.

5. Results and discussion

The optical properties of the interrogation system have been determined according to the measurement procedures reported in section 4.1. Spectra $I_0(\lambda)$, $\tilde{I}_i(\lambda)$ and their ratio are shown in Fig. 6 together with the emission spectrum of the halogen lamp WLS. Other distortions of the halogen lamp spectrum are due to the transmission properties of the optical components used in the interrogation setup.

Figure 6b shows the ratio $\frac{I_0(\lambda)}{\tilde{I}_i(\lambda)}$ that, according to Eq. (5), is proportional to the spectral response of the interrogation system. The spectral region of interest in our application is 500-600 nm, i.e. green-yellow-red; since the response of the interrogation system is satisfactorily flat in this interval, as shown in Fig. 6b, the information of interest is not corrupted.

The spectral properties of the sensing indicator have been measured according to the procedure reported in section 4.2. The absorbance spectra of phenol red at different pH values are shown in Fig. 7; the maximum of all the spectra is well visible at the wavelength of about 560 nm. According to literature (Wang, 2003), at this wavelength the phenol red

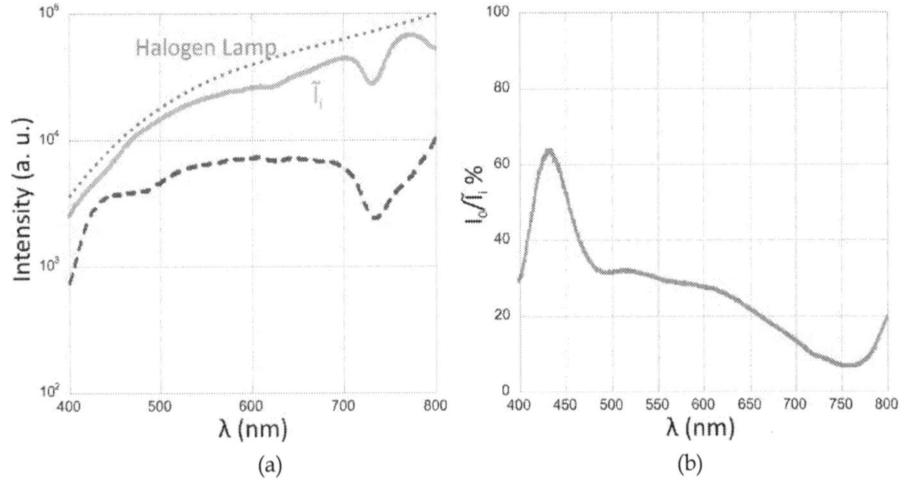

Fig. 6. Spectra of $\tilde{I}_i$ (λ) and I_o(λ) (a), and spectral response of the interrogation system (b). In (a), the upper dashed blue line represents the emission spectrum of the halogen lamp WLS. The spectral region of interest for our application is 500-600 nm.

molecules exhibit a very low absorption that can be significantly enhanced by increasing the pH of the solution. Note that the absorbance at 560 nm showed a sharp transaction when the pH of the solution changed from 5 to 8.

As shown in Fig. 8, the point of inflection (sigmoidal fitting function Eq. (6) equal to 0.5) corresponds to the apparent pKa value of 7.9. This value is in good agreement with the pKa of phenol red in aqueous solution reported in literature (Budavari, 1989).

The sensor characteristics have been measured according to the procedure reported in section 4.3. As shown in Fig. 9, a visual inspection of the probe at two different pH values of the solution underlines the swelling described in the previous sections. The spectra $\Delta I_0(\lambda, pH) = I_0(\lambda, pH) - I_0(\lambda, pH_{ref})$ acquired at different values of pH are shown in Fig. 10.

Observing Fig. 10, it is clear that the swelling/shrinking processes induce a change in the amplitude of the whole spectrum. As far as these phenomena are concerned, parameters A and S were considered to quantify the response of the sensor. In particular, as shown in Fig. 11, given a typical spectrum $\Delta I_0(\lambda)$, parameter A and S were calculated as follow: (i) the two maxima $M_1 = (\lambda_1, I_1)$, $M_2 = (\lambda_2, I_2)$ and the minimum $m = (\lambda_3, I_3)$ of $\Delta I_0(\lambda)$ have been calculated using a standard Matlab© routine; (ii) the equation of the line joining M_1 and M_2 has been calculated as $L(\lambda) = \frac{I_2 - I_1}{\lambda_2 - \lambda_1}(\lambda - \lambda_1)$; (iii) A has been defined as $L(\lambda_3) - I_3$ and S as $L(\lambda_3)$.

As shown in Fig. 11, parameter S is correlated to the spectrum amplitude and thus to swelling/shrinking processes while A is strictly related to the absorption (at about λ= 560 nm) performed by the indicator. In Fig. 12, parameters A and S were calculated for each spectrum acquired and the normalized results are plotted against pH.

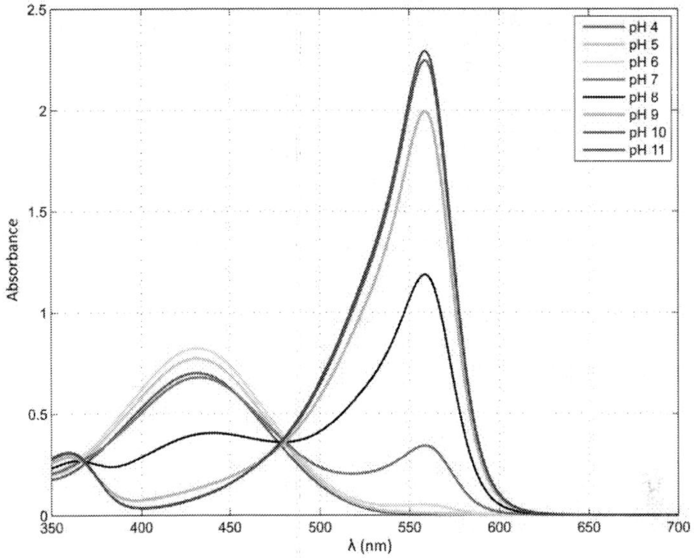

Fig. 7. Absorbance spectra of phenol red at different pH values.

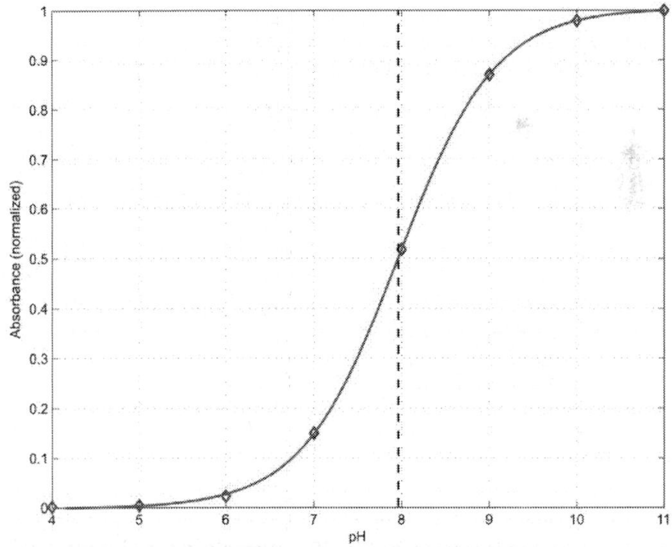

Fig. 8. Absorbance versus pH and Boltzmann interpolation function.

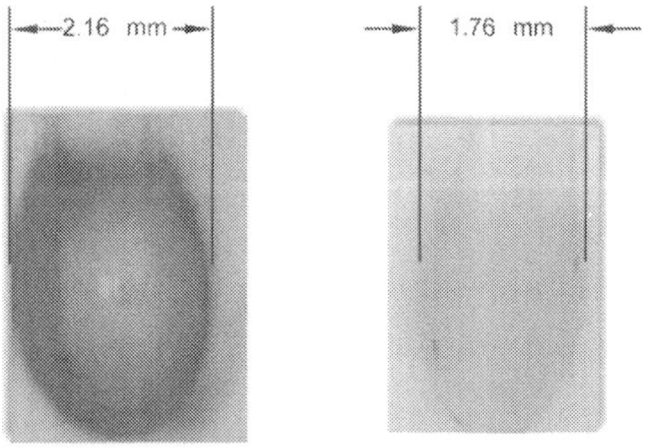

Fig. 9. Two shapes appearance of the probe PEO8000-8:2-PR after 5 min of immersion in basic (left) and acid (right) solution.

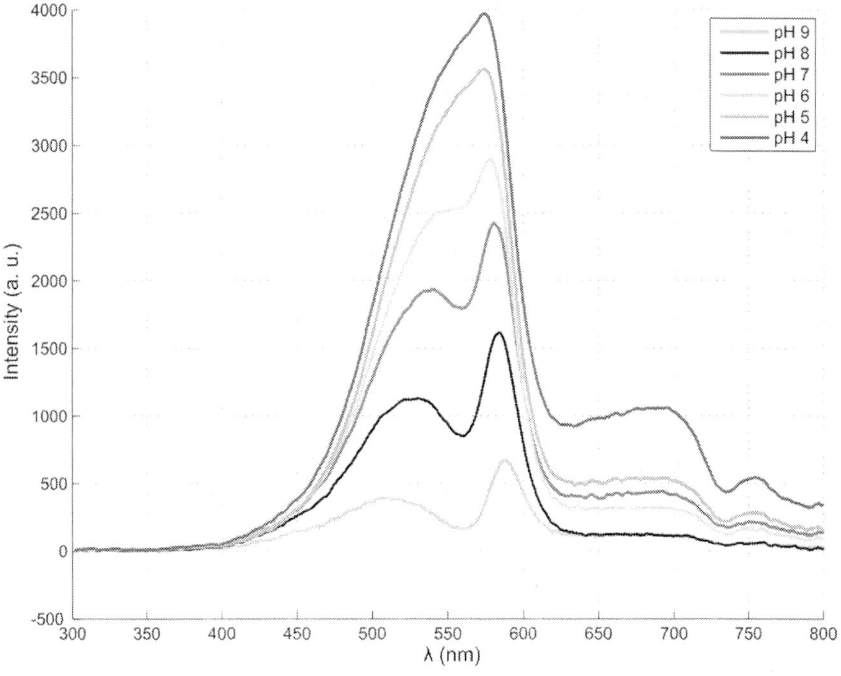

Fig. 10. Spectra of $\Delta I_0(\lambda)$ acquired at different values of pH. The light absorption of phenol red is well visible at the wavelength of 560 nm. Swelling/shrinking processes induce a change in the amplitude of the whole spectrum.

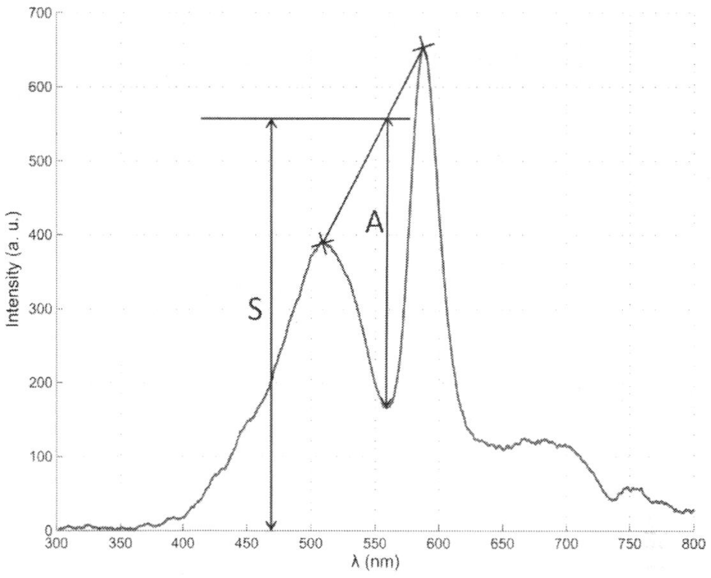

Fig. 11. Graphical definition of the two characteristic parameters. Parameter A is strictly related to the absorption performed by the indicator, whereas S is correlated to the swelling/shrinking processes.

Parameter A exhibits a sharp transaction changing the pH of the solution from 5 to 8. The experimental data was fitted by the Boltzmann (sigmoidal) function according to Eq. (6). The point of inflection (sigmoidal fitting function equal to 0.5) corresponds to the apparent pKa value $pK_a' = 6.8$. This value is lower than the measured value of pKa=7.9 of phenol red in aqueous solution (Fig. 8). However, the pKa value can be shifted to lower pH with increasing ionic strength as observed by Holobar et al. (Holobar, 1993), and this phenomenon could occur also in our hybrid matrix.

An increase in the value of pH determines a higher swelling of the hybrid drop, as shown in Fig. 9, which leads to an increase in the distance between the optical fiber tip and the probe-liquid interface and a consequent decrease in the mean optical path of the photons through the probe. As a consequence, according to Eq. (3), also the sensor response diminished. Coherently, as shown in Fig. 12b, parameter S decreased linearly with the pH of the solution. The linear fit of the experimental data exhibits a slope of 0.17.

The response time of the probe PEO8000-8:2-PR was measured. A trade-off between the response time and the indicator leaching together with the adhesion between the sensitive matrix and the POF is required. In fact, high molecular weight PEO improves the adhesion and diminishes the indicator leaching, but causes long response time of the sensor. In general there are two different time constant; in fact, when pH is changed from 10 to 3, the sensor was found to reach 90% of the regime signal intensity (τ_{90}) between 1 and 2 seconds, while changing the pH from a low value to a high one resulted in longer response times that can be quantify in the order of some minutes, i.e. the kinetics depends on the direction of the

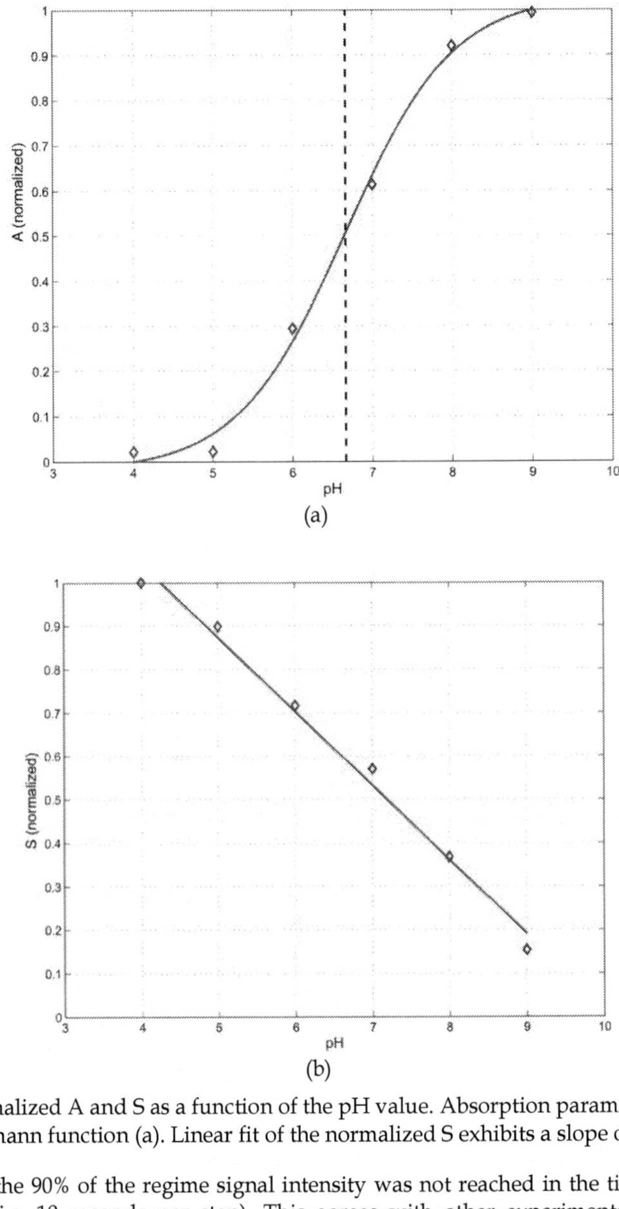

Fig. 12. Normalized A and S as a function of the pH value. Absorption parameter A is fitted by the Boltzmann function (a). Linear fit of the normalized S exhibits a slope of 0.17 (b).

pH change (the 90% of the regime signal intensity was not reached in the time scale of the experiment, i.e. 10 seconds per step). This agrees with other experimental observations reported in literature (Ismail, 2002; Badini, 1995). Empirical quantification of this sensor behavior is shown in Fig. 13.

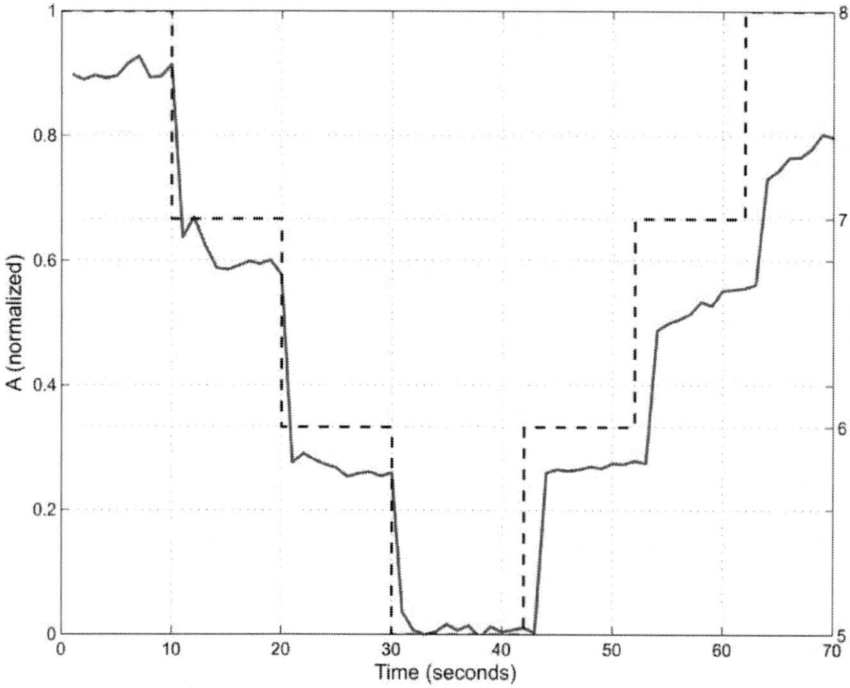

Fig. 13. Normalized A as a function of time for different values of pH. The pH value of the solution has been staircase changed from 8 to 5 and then backs from 5 to 8 with unitary steps of duration 10 s each.

In Fig. 13, the parameter A (blue line) is represented as a function of time together with the value of the reference pH (dashed line).

We think that the slow response time observed changing the pH from 5 to 8 is strictly associated with the swelling/shrinking kinetics of the sensitive drop. Finally, the kinetics of swelling/shrinking has shown to be dependent on the direction of the pH change. This complex mechanism meddles in the photons propagation in the sensitive drop and thus in the response of the probe.

6. Characteristics of the developed pH sensor

The mean lifetime of disposable sensors, checked in the continuous mode operation at a pH level in the middle of the range, is two days. The reusability, tested with a series of probes checked once a day at a pH level in the middle of the range and preserved dry, has a mean life of one month. Table 1 shows these and other analytical parameters of the sensor.

Parameter	Value
Optical interrogation wavelength	560 nm
Probe diameter	2 mm (typ.)
Measuring pH Range	5-8
Sensitivity	0.5 a.u./pH unit (max.)
Response time pH 8 to pH 5	2 s (max.)
Response time pH 5 to pH 8	4 min (typ.)
Reusability	1 month
Mean lifetime	2 days
Cost (few probes)	less than 0.7 Euro

Table 1. Characteristics of the developed disposable pH probe.

7. Conclusions

We present a disposable sensor based on a low-cost plastic optical fiber that covers an important pH range for physiological applications, e.g. in medicine and biotechnology, with excellent sensitivity. Some other advantages of the proposed sensor are: simple realization process, small size, high resistance to aggressive environments and electrical isolation.

The sensor characterization has been performed using a simple optical setup based on a single wavelength measurement. Nevertheless according to Lei (Lei, 2010), a ratiometric approach could improve the sensing overall performance. To this purpose, the isosbestic wavelength, i.e. 480 nm, highlighted in the spectra shown in Fig. 7, could be conveniently exploited.

For practical applications, swelling/shrinking processes could make not trivial the extraction of the pH information and deteriorate the sensor response time. Nevertheless, this phenomenon can be practically cancelled minimizing the size of the sensing element. Large drop volume allows to make the probe fabrication simple and reproducible and to obtain high signal-to-noise ratio but emphasizes the swelling/shrinking effects. Depending on the application, the trade-off among these aspects can establish the optimal drop size.

8. References

Alvarado-Mendez E., Rojas-Laguna R., Andrade-Lucio J.A., Hernandez-Cruz D., Lessard R.A. & Avina-Cervantes J.G., (2005). Design and characterization of pH sensor based on sol–gel silica layer on plastic optical fiber. Sens. Actuator B-Chem.: Vol. 106, pp. 518–522, ISSN 0925-4005

Aubonnet S., Barry H.F., von Bueltzingsloewen C., Sabbatie J.M. & MacCraith B.D., (2003). Photo-patternable optical chemical sensors based on hybrid sol-gel materials. Electron. Lett., Vol.39, No.12, pp.913-914, ISSN 0013-5194

Bacci M., Baldini F.& Bacci S. (1991). Spectroscopic behavior of acid-base indicators after immobilization on glass supports. Appl. Spectrosc., Vol.45, No.9, pp.1508-1515, ISSN 0003-7028

Badini G.E., Grattan K.T.V. & Tseung A.C.C. (1995). Impregnation of a pH-sensitive dye into sol-gels for fibre optic chemical sensors. Analyst, Vol.120, No.4, pp. 1025-1028, ISSN 0003-2654

Baldini F., Giannetti A., Mencaglia A.A. & C. Trono. (2008). Fiber-optic sensors for biomedical applications. Curr. Anal. Chem., Vol.4, pp. 378-390; ISSN 1573-4110

Barnard S.M.& Walt D.R. (1991). A fibre-optic chemical sensor with discrete sensing sites. Nature, Vol.353, pp. 338-340, ISSN 0028-0836

Bashir R., (2002), Micromechanical cantilever as an ultrasensitive pH microsensor. Appl. Phys. Lett. , 81, 3091-3093; ISSN 0003-6951

Ben-David O., Shafir E., Gilath I., Prior Y. & Avnir D. (1997). Simple absorption optical fiber pH sensor based on doped sol-gel cladding material. Chem. Mat.,Vol.9, pp. 2255-2257, ISSN 0897-4756

Brinker C.J. & Scherer G.W. (1990). Sol-gel science. The physics and chemistry of sol-gel processing. Academic Press, ISBN: 978-0121349707, New York

Bronk K. & Walt D.R. (1994). Fabrication of patterned sensors arrays with aryl azides on a polymer-coated imaging optical fiber bundle. Anal. Chem, Vol.66, No.20, pp. 3519-3520, ISSN 1520-6882

Budavari S., O'Neil M.J., Smith A. & Heckelman P.E. (1989). The Merck Index, 11th ed. Merck & Co., ISBN: 0-911910-28-X, Inc. Rahway, NJ

Cagnol F., Grosso D. & Sanchez C. (2004). A general one-pot process leading to highly functionalised ordered mesoporous silica films. Chem. Commun., pp. 1742-1743, ISSN 1359-7345

Cao W. & Duan Y. (2005). Optical fiber-based evanescent ammonia sensor. Sens. Actuator B-Chem., Vol.110, No.2, pp. 252-259, ISSN 0925-4005

Cjlakovic M., Lobnik A. & Werner T. (2002). Stability of new optical pH sensing material based on cross-linked poly(vinyl alcohol) copolymer. Anal. Chim. Acta, Vol.455, pp. 207-213, ISSN 0003-2670

de Marcos S. & Wolfbeis O.S. (1996). Optical sensing of pH based on polypyrrole films. Anal. Chim. Acta, Vol.334, pp. 149-153, ISSN 0003-2670.

Dong S., Luo M., Peng G. & Cheng W. (2008). Broad range pH sensor based on sol-gel entrapped indicators on fiber optic. Sens. Actuator B-Chem., Vol. 129, pp. 94-98, ISSN 0925-4005

Eckert H. & Ward M. (Eds.). (2001) Nanostructured and functional hybrid organic-inorganic materials. Chem. Mat., Vol. 13, No. 10. ISSN 0897-4756

Estella J., Echeverria J.C., Laguna M. & Julian J. (2007). Silica xerogels of tailored porosity as support matrix for optical chemical sensors. Simultaneous effect of pH, ethanol:TEOS and water:TEOS molar ratios, and synthesis temperature on gelation time, and textural and structural properties. J. Non-Cryst. Solids, Vol.353, No.3, pp. 286-294, ISSN 0022-3093

Fabbri P., Leonelli C., Messori M., Pilati F., Toselli M., Veronesi, P., Morlat-Therias S., Rivaton A. & Gardette J.L. (2008). Improvement of the surface properties of polycarbonate by organic-inorganic hybrid coatings. J. Appl. Polym. Sci., Vol. 108, No. 3, pp. 1426-1436. ISSN 0021-8995

Fabbri P., Pilati F., Rovati L., McKenzie R. & Mijovic J. (2011). Poly(ethylene oxide)-silica hybrids entrapping sensitive dyes for biomedical optical sensors: molecular dynamics and optical response. Opt. Mater., 33(8), 1362-1369 ISSN 0925-3467

Feng C. & Zhongyang R. (2011). Research on Measurement of POF Attenuation Spectrum. Intelligent Computation Technology and Automation (ICICTA), 2011 International Conference on , Vol.2, pp.672-674.

Ferguson J.A., Healey B.G., Bronk K.S., Barnard S.M. & Walt D.R. (1997). Simultaneous monitoring of pH, CO_2 and O_2 using an optical imaging fiber. Anal. Chem, Vol.340, pp 123-131; ISSN 1520-6882

Fritz J., Baller M.K., Lang H.P., Strunz T., Meyer E., Guntherodt H.J., Delamarche E., Gerber Ch. & Gimzewski J.K. (2000). Stress at the solid–liquid interface of self-assembled monolayers on gold investigated with a nanomechanical sensor. Langmuir, Vol.16, (November 2000), pp 9694-9696, ISSN 0743-7463

Goicoechea J., Zamarreno C.R., Matias I.R. & Arregui F.J. (2009). Utilization of white light interferometry in pH sensing applications by mean of the fabrication of nanostructured cavities. Sens. Actuator B-Chem., Vol.B138, pp 613-618, ISSN 0925-4005

Grosso D., Cagnol F., Soler-Illia G.J., Crepaldi E.L., Amenitsch H., Brunet-Bruneau A., Bourgeois A. & Sanchez C. (2004). Fundamentals of mesostructuring through evaporation-induced self-assembly. Adv. Funct. Mater. Vol.14, No.4, pp. 309-322, ISSN 1616-301X

Grummt U.W., Pron A., Zagorska M. & Lefrant S. (1997). Polyaniline based optical pH sensor. Anal. Chim. Acta, Vol.357, pp. 253-259; ISSN 0003-2670

Gupta B.D. & Sharma S. (1998). A long-range fiber optic pH sensor prepared by dye doped sol–gel immobilization technique. Opt. Commun., Vol. 154, pp. 282-284, ISSN 0030-4018

Healey B.G. & Walt D.R. (1997). Fast temporal response fiber-optic chemical sensors based on the photodeposition of micrometer-scale polymer arrays. Anal. Chem, Vol.69, No.11, pp. 2213-2216; ISSN 1520-6882

Holobar A., Weigl B.H., Trettnak W., Benes R., Lehmann H., Rodriguez N.V., Wollschlager A., O'Leary P., Raspor P. & Wolfbeis O.S. (1993). Experimental results on an optical pH measurement system for bioreactors. Sens. Actuator B-Chem., Vol.11, No.1-3, pp. 425-430, ISSN 0925-4005

Igarashi S., Kuvae K. & Yotsuyanagi T. (1994). Optical pH sensor of electrostatically immobilized porphyrin on the surface of sulfonated-polystyrene. Anal. Sci., Vol.10, (October 1994), pp. 821-822, ISSN 0910-6340

Ismail F., Malins C. & Goddard N.J. (2002). Alkali treatment of dye-doped sol-gel glass films for rapid optical pH sensing. Analyst, Vol.127, No.2, pp. 253-257, ISSN 0003-2654.

Kickelkick G. (2006). Hybrid materials. Synthesis, characterization, and application. Wiley-VCH, , ISBN 3527312994, New York

Korostynska O., Arshak K., Gill E. & Arshak A. (2007), Review on state of the art in polymer based pH sensors. Sensors, Vol.7, pp 3027-3042, ISSN 1424-8220

Kowada Y., Ozeki T & Minami T. (2005). Preparation of silica-gel film with pH indicators by the sol-gel method. J. Sol-Gel Sci. and Techn., Vol.33, pp. 175-185, ISSN 0928-0707

Lee S.T., Gin J., Nampoori V.P.N., Vallabhan C.P.G., Unnikrishnan N.V. & Radhakrishnan P. (2001). A sensitive fiber optic pH sensor using multiple sol–gel coatings. J. Opt. A-Pure Appl. Opt., Vol. 3, pp. 355–359, ISSN 1464-4258

Lei J., Wang L. & Zhang J. (2010). Ratiometric pH sensor based on mesoporous silica nanoparticles and Förster resonance energy transfer. Chem. Commun., Vol. 46, pp 8445-8447, ISSN 1359-7345

Lin J. (2000). Recent development and applications of optical and fiber-optic pH sensors. Trends Anal. Chem., Vol.19, No. 9, pp 541-552, ISSN 0167-2940

Lin J. & Brown C.W. (1997). Sol-gel glass as a matrix for chemical and biochemical sensing. Trac-Trends Anal. Chem.Vol.16, No.4, pp. 200-211, ISSN 0167-2940

Lobnik A., Oehme I., Murkovic I. & Wolfbeis O.S. (1998). pH optical sensors based on sol-gels: chemical doping versus covalent immobilization. Anal. Chim. Acta, Vol.367, pp. 159-165, ISSN 0003-2670

MacGraith B.D., Ruddy V., Potter C., O'Kelly B. & McGilp J.F. (1991). Optical waveguide sensor using evanescent wave excitation of fluorescent dye in sol-gel glass. Electron. Lett., Vol. 27, pp. 1247–1248, ISSN 0013-5194

McDonagh C., Burke C.S. & MacCraith B. D. (2008). Optical chemical sensors. Chem. Rev., Vol.108, pp. 400-422, ISSN 0009-2665

Messori M., Toselli M., Pilati F., Fabbri E., Fabbri P., Busoli S., Pasquali L. & Nannarone S. (2003). Flame retarding poly(methyl methacrylate) with nanostructured organic-inorganic hybrids coatings. Polymer, Vol. 44, No. 16, pp. 4463-4470, ISSN: 0032-3861

Messori M., Toselli M., Pilati F., Fabbri E., Fabbri P. & Busoli S. (2004a). Poly(caprolactone)/silica organic-inorganic hybrids as protective coatings for poly(methyl methacrylate) substrates. Surf. Coat. Int., Vol. 86, No. 3, pp. 183-186, ISSN 1754-0925

Messori M., Toselli M., Pilati F., Fabbri E., Fabbri P., Pasquali L. & Nannarone S. (2004b). Prevention of plasticizer leaching from PVC medical devices by using organic-inorganic hybrid coatings. Polymer Vol. 45, pp. 805–813, ISSN 0032-3861

Miled O.B., H.B. Ouada & J. Livage (2002). pH sensor based on a detection sol–gel layer onto optical fiber. Mater. Sci. Eng. C-Mater. Biol. Appl., Vol. 21, pp. 183–188, ISSN 0928-4931

Pringsheim E., Terpetschnig E. & Wolfbeis O.S. (1997). Optical sensing of pH using thin films of substituted polyaniline. Anal. Chim. Acta, Vol.357, pp. 247-252, ISSN 0003-2670

Rayss J. & Sudolski S. (2002). Ion adsorption in the porous sol–gel silica layer in the fiber-optic pH sensor. Sens. Actuator B-Chem., Vol. 87, pp. 397–405, ISSN 0925-4005

Rovati L., Fabbri P., Ferrari L. & Pilati F. (2011). Construction and evaluation of a disposable pH sensor based on a large core plastic optical fiber. Rev. Sci. Instrum., Vol. 82, p. 023106, ISSN 0034-6748

Ruan C., Zeng K. & Grimes C.A. (2003). A mass-sensitive pH sensor based on a stimuli-responsive polymer. Anal. Chim. Acta, Vol.497, pp 123-131, ISSN 0003-2670

Sanchez C., Shea K.J. & Kitagawa S. (2011). Recent progress in hybrid materials science. Chem. Soc. Rev., Vol. 40, pp. 471-472, ISSN: 0306-0012

Schmidt H. (2000). Sol-gel derived nanoparticles as inorganic phases in polymer-type matrices. Macromol. Symp., Vol. 159, pp. 43–55, ISSN 1521-3900

Seki A., Katakura H., Kai T., Iga M. & Watanabe K. (2007). A hetero-core structured fiber optic pH sensor. Anal. Chim. Acta, Vol 582, No.1, pp. 154-157, ISSN 0003-2670

Sharma N.K. & Gupta B.D. (2004). Fabrication and characterization of a fiber-optic pH sensor for the pH range 2 to 13. Fiber Integrated Opt., Vol. 23, pp. 327–335 ISSN 0146-8030

Shi Y. & Seliskar C.J. (1997). Optically transparent polyelectrolyte-silica composite materials: preparation, characterization, and application in optical chemical sensing. Chem. Mater., Vol.9, pp. 821-829, ISSN 0897-4756

Song A., Parus S. & Kopelman R. (1997). High-performance fiber-optic pH microsensors for practical physiological measurements using a dual-emission sensitive dye. Anal. Chem, Vol.69, No.5, (March 1997), p. 863867, ISSN 1520-6882

Sotomayor M.D.P.T., De Paoli M.A. & Oliveira W.A.D. (1997). Fiber-optic pH sensor based on poly(o-methoxyaniline). Anal. Chim. Acta, Vol.353, pp 275-280, ISSN 0003-2670

Splinter R., Brett A. & Hooper (2006). An introduction to biomedical optics. Taylor & Francis Ed, ISBN: 0750309385 9780750309387, New York

Talaie A., Lee J.Y., Jang J., Romagnoli J.A. & Taguchi T. (2000). Dynamic sensing using intelligent composite: an investigation to development of new pH sensors and electrocromic devices. Thin Solid Films, Vol. 363, pp. 163-166, ISSN 0040-6090

Toselli M., Marini M., Fabbri P., Messori M., & Pilati F. (2007). Sol-gel derived hybrid coatings for the improvement of scratch resistance of polyethylene. : J. Sol-Gel Sci. Technol., Vol. 43, No. 1, pp. 73-83, ISSN 0928-0707

Wang E., Chow K.F., Kwan V., Chin T., Wong C. & Bocarsly A. (2003). Fast and long term optical sensors for pH based on sol–gels. Anal. Chim. Acta, Vol. 495, No.1-2, pp. 45-50, ISSN 0003-2670

Wen J. & Wilkes G.L. (1996). Organic/inorganic hybrid network materials by the sol-gel approach. Chem. Mater., Vol.8, No.8, pp. 1667-1681, ISSN 0897-4756

Wolfbeis O.S. (2004). Fiber-optic chemical sensors and biosensors. Anal. Chem., Vol.76, No.12, pp. 3269-3284, ISSN 1520-6882

Wolfbeis O.S. (2008). Fiber-optic chemical sensors and biosensors. Anal. Chem., Vol.80, No.12, pp. 4269-4283, ISSN 1520-6882

Wu Z., Jiang Y., Xiang H., You L. (2006). Understanding the mechanisms of reaction and release of acid–base indicators entrapped in hybrid gels. J. Non-Cryst. Solids, Vol. 352, pp. 5498–5507, ISSN 0022-3093

Yuqing M., Jianrong C., & Keming F. (2005). New technology for the detection of pH. J. Biochem. and Biophys. Methods, Vol.63, pp. 1-9, ISSN 0165-022X

Zamarreno C.R., Hernaez M., Del Villar I., Fernandez-Valdivielso C., Arregui F.J. & Matias I.R. (2010). Optical fiber pH sensor fabrication by means of Indium tin oxide coated optical fiber refractometers. Phys. Status Solidi C: Current Topics in Solid State Phys., Vol.7, No. 11-12, pp. 2705-2707, ISSN 1862-6351

Zauner A., Bullock R., Di X. & Young H.F. (1995). Brain oxygen, CO_2, pH and temperature monitoring: evaluation in the feline brain. Neurosurgery, Vol.37, No.6, pp. 1176-1177, ISSN 0148-396X

18

Mechanical Property and Strain Transferring Mechanism in Optical Fiber Sensors

Dongsheng Li, Liang Ren and Hongnan Li
Dalian University of Technology
China

1. Introduction

Optical fiber sensors (OFSs) have attracted considerable interests for their superior sensing abilities, especially due to their electromagnetic inference immunity and high sensitivity. However, OFS sensors based on bare fibers are fragile and easily damaged. For their safe use in engineering sensing, the glass core of optical fibers has to be coated with protective coatings, or to be bonded with adhesive materials, for instance, epoxy. Low elastic modulus of the soft coatings or bonding layer results in various shear stresses along the middle layer between the fiber core and the host structure to be measured. A portion of the host material strain is absorbed by the protective coatings when the strain transfers from the host material to the fiber core, and hence only a small fraction of structural strain is sensed. Consequently, the strains directly sensed by OFSs are different from actual structural strains. This is the main issue to be discussed in the chapter.

For the purpose to discuss the strain transferring problem, fundamental mechanical properties of OFSs are first presented. The chapter then develops an analytical model to derive the relationship between OFS sensed strains and actual structural strains. It is discovered that the strains, directly sensed by an OFS, have to be magnified by a factor (strain transfer rate) to reflect actual structural strains. This is, of course, of paramount interests for the application of optical fiber sensors. The factors that affect the efficiency of strain transfer on the optical fiber sensor are deduced and discussed in details based on the theoretical analysis. Critical adherence length is proposed as a measure to guarantee sufficient strain transfer rate based on the analytical model. Detailed analysis shows that the critical adherence length is the minimum length with which the fiber has to be tightly embedded to a structure for adequate sensing. Furthermore, theoretical results are verified through laboratory experimentation with the fiber Bragg grating sensors. Measured average strain transfer rates agree well with those calculated from the analytical model. Furthermore, the strain transfer rate of an OFS embedded in a multi-layered structure is consequently developed and the shear modulus of the host material on the influence of strain transmission is discussed.

For surface-bonded OFSs, optical fibers are adhered on the surfaces of structural members. Adhesive material still exists in-between and optical fibers are only partial buried. The strain readings of OFSs can be heavily affected by the thickness and mechanical properties of the adhesive and therefore it is important to determine how these factors affect the measurements. Numerical finite element simulations are reviewed for strain transferring mechanism of surface-bonded OFSs. In the end, the chapter summarizes current investigations on strain transferring mechanism of optical fiber sensors and practical packaging issues.

2. Fundamental mechanical properties of optical fibers

An optical fiber is a flexible, transparent fiber made of very pure glass (silica) not much wider than a human hair that acts as a waveguide, or "light pipe", to transmit light between the two ends of the fiber (Thyagarajan & Ghatak, 2007). Optical fiber typically consists of a transparent core surrounded by a transparent cladding material with a lower index of refraction.

Materials Properties (1)	Symbols (2)	Values (3)	Unit (4)
Young's modulus of silica fiber	E	7.3×10^{10}	Pa
Young's modulus of silicon coating	E	2.55×10^{6}	Pa
Poisson's ratio of silicon coating	ν	0.17~0.48	-
Shear modulus of silicon coating	G	8.5×10^{5}	Pa
Strength of silica fiber	σ	$0.35\text{~}4.8 \times 10^{10}$	Pa
Breaking strain of silica fiber	ε	~10%	-
Density	ρ	2200	Kgm^{-3}
Coefficient of thermal expansion	*f*	0.54×10^{-6}	K^{-1}

Table 1. Typical mechanical properties of the optical fiber (Fernando et al., 2002; Measures, 2001)

Optical fibers are drawn from a molten glass perform. Although fluorozirconate, fluoroaluminate, and chalcogenide glasses as well as crystalline materials like sapphire, are used for optical fibers, silica is the most common material, especially for optical fiber sensing applications. Optical fibers made of fused silica are perfectly elastic until their breaking point. They are brittle and do not yield as do metals when overstressed. For the interests of sensing applications, key mechanical properties of silica fiber are listed in Table 1. Due to variation of manufacturing processes, mechanical properties of optical fibers may fluctuate around the nominate values in Table 1.

Although the strength of silica fiber is in excess of 4.8 Gpa (~7% strain), the practical stress limit is much lower due to the presence of many microscopic flaws. Proof tests are then carried out by manufactures to guarantee a safe operating stress and a predictable life. Typically, optical fibers are proof tested to 0.35 Gpa, corresponding to about 0.5% strain (Measures, 2001). For most sensor applications, the strength of silica fiber is much higher than that of structural members to be measured. Fatigue tests showed that optical fibers have a long working life when its operating stress is a small fraction of the proof stress. For most long term sensing application, OFSs can survive and are adequate.

Furthermore, fiber Bragg gratings (FBG) are the most common sensing elements among OFSs. Their strength is of considerable interests. Experiments showed that the formation of FBGs with UV light leads to a substantial weakening of the optical fiber (Feced et al., 1997). Controversy exists. Varelas et al. (1997) have demonstrated that producing FBGs with a continuous wave caused only minimal reduction in the strength of the optical fiber. Further investigations are needed to study the effects of fabrication of FBGs on the fiber strength.

3. Strain transferring mechanism for embedded OFSs

Conventionally, the values measured by optical fiber sensors were assumed to be actual structural strains (Baldwin et al., 2002). In fact, the strain measured by an OFS is different from actual host structure strain because of the difference between the optical fiber core modulus and the modulus of the fiber coating or the adhesive. Such strain difference depends on detailed packing measures of OFSs. Generally, there are three methods to integrate OFSs with host structures in terms of packing strategies: (1) direct integration, in which OFSs are directly embedded in or surfaced bonded to a host structure; (2) sensor-packaging integration, by which OFSs are first fixed in a small tube or bonded on the surface of a plate, and then the tube or the plate is anchored in the host structure; (3) clamping integration, in which an OFS is gripped at two ends by a bracket fixed on the host structure. No matter how the OFSs are packaged and integrated, the fiber core is brittle and has to be protected by adhesives or a coating layer in most sensing applications to avoid fiber breakage and to ascertain its long term stability. However, such a protection results in inconsistency between the fiber strain and the structural strain. Such discrepancies are neglected in many applications of OFSs by simply assuming that the fiber strain is consistent with the host structural strain (Friebele et al., 1999; Udd, 1995). Such assumption gives acceptable measurement results for the OFSs with long gauge length, in which the peak host strain can be fully transferred into the fiber strain, but cannot provide good measurement strains for short gauge OFSs, for instance, FBG sensors, in which the effect of the bonded fiber length on strain transfer between the fiber and host structure is significant (Galiotis et al., 1984). It is, thus, of primary importance to have a detailed knowledge of the relationship between the host structural strain and the fiber strain in order to correctly interpret structural strain from the strains sensed in the fiber.

Since the Young's modulus of the fiber is typically much larger than that of the adhesive or the coating as shown in Table 1, the axial elastic displacements of the fiber and the host material are different. In order to measure the strain of host materials accurately, there is a need to develop the relationship between the strain sensed by an OFS and the actual strain of the host material. Many researchers all over the world have made great efforts in this field and obtained notable achievements. Pak (1992) analyzed the strain transfer of a coated optical fiber embedded in a host composite, which is strained by a far-field longitudinal shear stress parallel to the optical fiber. In this case, the maximum shear transfer occurs when the shear modulus of the coating is the geometric mean of the shear moduli of the fiber and the host material. Ansari & Libo (1998) developed a simply model for evaluating the strain transfer percentage from the surrounding matrix to a length of optical fiber embedded in it under the assumption that the strain at the middle of the bonded fiber is equal to that of the host structure at the same position. Experiments were performed with a white-light fiber-optic interferometer to confirm the theory results. Li et al (2002) considered the coating as an ideal elasto-plastic material and deduced the strain transfer coefficients when the host material experiences tensile stress and compression stress respectively. Galiotis (1984) designed a polydiacetylene single fiber in an epoxy resin host material subjected to tensile strain along the fiber direction, and measured the strains at all points along the length of the fiber by the method of resonance Raman. They found that the axial strain in the fiber rises from a finite value at the end of the fiber to a fairly constant value at the central portion of the fiber, and that the axial strain at the midpoint of the fiber is lower than that applied to the host material, which can be approximately explained by the shear-lag model of Cox (1952).

This chapter mainly concerns the study of a single optical fiber embedded in a finite host structure, which is subjected to uniform axial stress. To derive the relationship between the strain measured by an optical fiber and the actual one experienced by the structure, a more realistic hypothesis is proposed in this chapter.

3.1 Theoretical approach

An optical fiber usually consists of three layers: fiber core, cladding and coating. The core diameter of a single mode fiber is usually 9μm, and the value for a multiple-mode fiber is 50μm (Fig.1). The coating, i.e., the outer layer of an optical fiber, often has an outer diameter of 250μm and an inner diameter of 125μm. The coating is usually made of plastics.

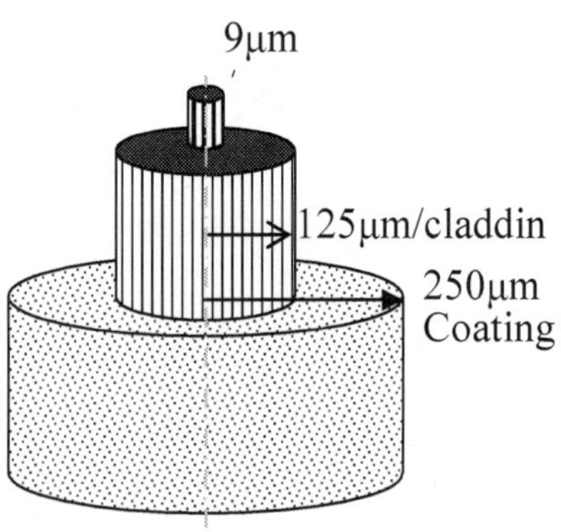

Fig. 1. Structure of a typical single mode fiber

We will deal with a concentric model of OFSs, which is similar to the fiber structure in Fig. 1. The concentric model is just a simplification of direct integration strategy. This model is suitable for two different cases. In the first case, a fiber with coating is directly embedded in the host structure. In the second case, the coating of a fiber is first stripped off, and then the bare fiber (including the core and cladding only) is attached to the inner surface of a steel tube by adhesives. In these two cases, there is an adhesive layer, often called middle layer, between the bare fiber and the host material. The three layers are concentric and the strain of the outer host material is transmitted to the inner fiber through the deformations of the middle layer.

For the concentric model, the strain transfer process from host material to the fiber core can be described as follows: as the host material generates strain under external loads, the deformation makes the shear strain appear in the interface between the host material and interlayer, and the shear strain is transferred to the interior of the interlayer to an extent which depends on the effective bonding of the interface. The interlayer generates axial strain at the same time. The axial strain in the interlayer is transferred to the fiber core through the shear strain in the interface between the interlayer and fiber core. The fiber core generates axial strain and senses the strain of the host material.

In the analysis of strain transfer to the optical fiber from the host material quantitatively, following assumptions are adopted for the concentric model.

Assumption 1) All the materials pertinent to the model remain elastic, and only the outer host material is subjected to axial stress and is uniformly strained, whereas the bare fiber and the middle layer do not directly bear any external loadings.

Assumption 2) Mechanical properties of the core and cladding of the fiber are the same. In reality, their properties are slightly different owing to their difference in some chemical components and the writing process of Bragg gratings. The core and cladding, hereinafter, are referred collectively as fiber for simplicity.

Assumption 3) There are no strain discontinuities across the interfaces, including the one between the host material and the middle layer and the one between middle layer and fiber interfaces, i.e. the bond between all the interfaces is perfect and no de-bonding exists.

In the concentric model, the Young's modulus of the single optical fiber is E_g with a radius of r_g embedded in a host material, separated by a middle layer of Young's modulus E_c and radius r_m and Poisson's ratio v, as illustrated in Fig.2. The host material, assumed to be infinite in all directions, is subjected to a uniform axial stress. L is the half gauge length of an optical fiber sensor, and $2L$ is the total length that the fiber is bonded to the host material through the middle layer. $\tau(x,r)$ is the shear stress in the middle layer a distance r above a given x coordinate along the center of the fiber. $\tau(x,r_g)$ is the shear stress at a given x coordinate along the fiber-middle layer interface. σ_g, σ_c and σ_m are the axial stress in the fiber, middle layer and the host material, respectively.

Based on the stress equilibrium for a small segment of the fiber and assumptions 1-3, the longitudinal stress along the fiber σ_g, is related to the interfacial shear stress at the fiber/middle layer interface through,

$$\frac{d\sigma_g}{dx} = -\frac{2\tau(x,r_g)}{r_g} \tag{1}$$

In reference to the free body diagram pertaining to the middle layer (Fig.2c), the equilibrium in the x direction results in the following relationship by balancing the shear stresses $\tau(x,r_g)$ and $\tau(x,r)$ with the axial stress σ_c,

$$\tau(x,r) = \frac{r_g}{r}\tau(x,r_g) - \frac{r^2 - r_g^{\,2}}{2r}\frac{d\sigma_c}{dx} \tag{2}$$

Substitution of (1) into (2) leads to,

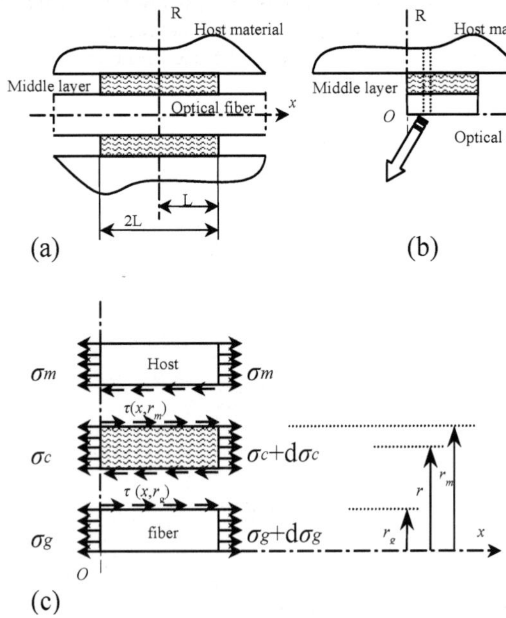

Fig. 2. Coordinate system and free body diagram of symmetrical section of optical fiber: (a) Optical fiber of gauge length 2*L*; (b) One quarter of the fiber; (c) Stress distribution of the fiber and the coating.

$$\tau(x,r) = -\frac{r_g^2}{2r}\frac{d\sigma_g}{dx} - \frac{r^2 - r_g^2}{2r}\frac{d\sigma_c}{dx} \tag{3}$$

Since the shear modulus of deformation predominates when it comes to the load transferring between the host material and the fiber, lateral motions perpendicular to the x coordinate are of secondary importance for the present study, and the Poisson effect can be neglected. Using the constitutive equation relating stress to strain, $\sigma_g = E_g \varepsilon_g$, where $\varepsilon_g = \frac{du}{dx}$ is the axial strain along the fiber and $u \equiv u(x)$ is the displacement, one obtains,

$$\begin{aligned} \tau(x,r) &= -\frac{r_g^2}{2r}E_g\frac{d\varepsilon_g}{dx} - \frac{r^2 - r_g^2}{2r}E_c\frac{d\varepsilon_c}{dx} \\ &= -\frac{E_g r_g^2}{2r}\left(\frac{d\varepsilon_g}{dx} - \frac{r^2 - r_g^2}{r_g^2}\frac{E_c}{E_g}\frac{d\varepsilon_c}{dx}\right) \end{aligned} \tag{4}$$

Since the fiber is strained together with the middle layer, the strain gradients are expected to be of the same order,

$$\frac{d\varepsilon_g}{dx} \cong \frac{d\varepsilon_c}{dx} \tag{5}$$

Thus, the important factor that determines $\tau(x,r)$ in (4) is the ratio of the stiffness between the fiber and the middle layer. The Young's modulus of the optical fiber coatings, in the case of a coated fiber, or the Young's modulus of typical structural epoxies, in the instance of a bare embedded fiber is, as usual, ten percent or less than that of the glass fiber, and r is not much larger than r_g (the middle layer is typically very thin to admit efficient strain transfer between the fiber and the host material), the second part of the right hand side of (4) is , therefore, insignificant compared to the first part,

$$\frac{r^2 - r_g^{\ 2}}{r_g^{\ 2}} \frac{E_c}{E_g} \frac{d\varepsilon_c}{dx} \cong o(\frac{d\varepsilon_g}{dx}) \tag{6}$$

Substituting (5) and (6) into (4) results in,

$$\tau(x,r) = -\frac{r_g^{\ 2}}{2r} \frac{d\sigma_g}{dx} = -\frac{r_g^{\ 2}}{2r} E_g \frac{d\varepsilon_g}{dx} \tag{7}$$

The shear stress term in (7) is determined by using Hooke's law as follows,

$$\tau(x,r) = G_c \gamma(x,r) = G_c(\frac{\partial u}{\partial r} + \frac{\partial w}{\partial x}) \cong G_c \frac{\partial u}{\partial r} \tag{8}$$

where $u \equiv u(x)$ and $w \equiv w(x)$ are the axial and radial displacements in the middle layer, respectively. Substituting (8) into (7) and integrating the resulting expression over r_g from the fiber and middle layer interface to the middle layer and host material interface radius r_m gives,

$$\int_{r_g}^{r_m} G_c \frac{\partial u}{\partial r} dr = \int_{r_g}^{r_m} \left[-\frac{r_g^2}{2r} E_g \frac{d\varepsilon_g}{dr} \right] dr \tag{9}$$

The integration result of (9) is,

$$u_m - u_g = -\frac{E_g}{2G_c} \frac{d\varepsilon_g}{dx} r_g^{\ 2} \ln(\frac{r_m}{r_g}) = -\frac{1}{k^2} \frac{d\varepsilon_g}{dx} \tag{10}$$

where u_m and u_g respectively denote the axial displacement from the x coordinate origin in the host material and fiber. The strain lag parameter, k, containing both effects of the geometry and the relative stiffness on the system components, can be written as,

$$k^2 = \frac{2G_c}{r_g^{\ 2} E_g \ln(\frac{r_m}{r_g})} = \frac{1}{(1+\mu)\frac{E_g}{E_c} r_g^{\ 2} \ln(\frac{r_m}{r_g})} \tag{11}$$

where, G_c, is the shear modulus of the middle layer. Differentiation of (10) with respect to x yields,

$$\frac{d^2 \varepsilon_g(x)}{dx^2} - k^2 \varepsilon_g(x) = -k^2 \varepsilon_m \tag{12}$$

The general solution to (12) takes of the following form,

$$\varepsilon_g(x) = c_1 e^{kx} + c_2 e^{-kx} + \varepsilon_m \tag{13}$$

where c_1 and c_2 represent the integration constants determined by boundary conditions. Since the host material does not contact the fiber beyond the ends of the interface between the fiber and the middle layer, the fiber is assumed to be free from axial stress at both ends. This assumption means that the strain transferred to the fiber is equal to zero at both ends of the OFS, and is given by,

$$\varepsilon_g(L) = \varepsilon_g(-L) = 0 \tag{14}$$

The boundary conditions established by (14) are identical to the following equations,

$$\varepsilon_g(L) = 0,\ \dot{\sigma}_g(0) = 0\, , \dot{\varepsilon}_g(0) = 0 \tag{15}$$

The relationship (15) states that there is no shear stress in the fiber at the midpoint of the fiber (refer to (1)) due to the symmetric nature of the structure. The boundary conditions established in (14) and (15) will lead to the same solution as (13). Hence, by imposing these boundary conditions on (13), c_1 and c_2, are evaluated and obtained as

$$c_1 = c_2 = -\frac{\varepsilon_m}{2\cosh(kL)} \tag{16}$$

Thus, the final solution to (13), i.e. the strain distribution along fiber and its relationship to the strain of the host material at a given x coordinate is,

$$\varepsilon_g(x) = \varepsilon_m\left(1 - \frac{\cosh(kx)}{\cosh(kL)}\right) \tag{17}$$

Eq.(17) is the governing equation that describes the strain distribution along the fiber. The effects of the Young's moduli of the fiber and the middle layer, as well as the effects of the radii of the fiber and the middle layer, on the strain transfer, are all included in the strain lag parameter k defined in (11).

Fig.3 shows the strain transfer rate along an optical fiber. The mechanical properties of the optical fiber employed in the investigation, are given in Table 2. These properties are the same as those used by Ansari & Libo(1998) for comparison purposes. The strain difference between the fiber and the host material at a given x coordinate, is determined by the strain lag parameter k. Fig.3 also demonstrates that the strain sensed by the fiber at the midpoint is not equal to the strain in the host material in this instance.

3.2 Influence of the host material

Previous research on embedded OFSs have been carried out under the assumption that the strain distribution of the host material is not influenced by the embedding of the OFS and equals to the actual strain of the host material on the outer surface of the interlayer. According to Newton's third law, when strain in the host material causes deformations in

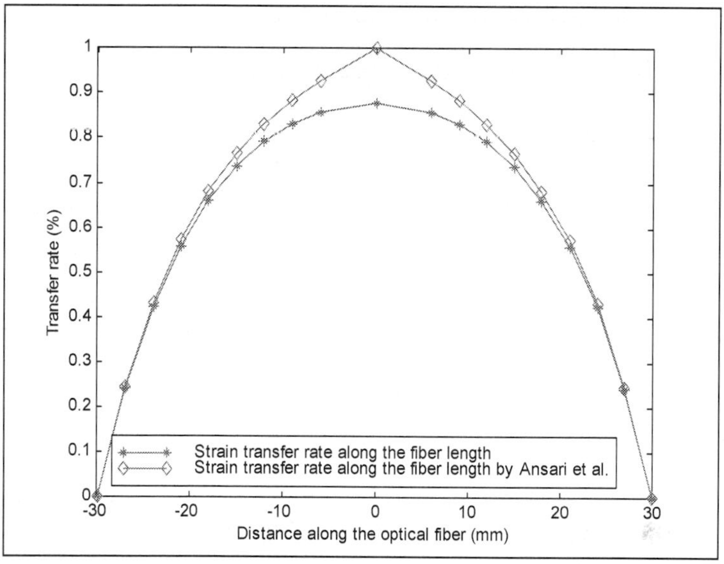

Fig. 3. Distribution of normal strain in fiber along the length

Materials Properties (1)	Symbols (2)	Values (3)	Unit (4)
Young's modulus of the fiber	E_g	7.2×10^{10}	Pa
Young's modulus of silicon coating	E_c	2.55×10^{6}	Pa
Poisson's ratio of silicon coating	μ	0.48	-
Shear modulus of silicon coating	Gc	8.5×10^{5}	Pa
Radius of outer boundary of silicon coating	r_m	102.5	μm
Radius of the fiber	r_g	62.5	μm

Table 2. Mechanical properties of the optical fiber

the interlayer, the interlayer reacts causing opposite deformations in the host material. Thereby, the strain level of some of the host material surrounding the embedded OFSs is lower than the actual strain of host material. The effects of host material properties in strain transfer are often ignored. However, many experiments and interrelated investigations reveal that this will result in large errors of strain transfer in the calculation especially when the stiffness of host materials is much lower than that of the fiber core.

In the field of short-fiber composites, Cox(1952) first developed the theoretical analysis of load transfer from the matrix to the fiber and this theory is referred to as the shear-lag analysis. The Young's modulus of matrix is considered as an important parameter. Subsequently, Monette et al. (1992), Rosen et al. (1965) and Tsai et al. (1990) deduced different improved strain transfer formulae using different assumptions based on the principle of shear-lag theory. In their formulae, the Young's modulus of matrix was introduced as a variable. All the results indicated that the strain transfer depends on the characteristics of the matrix. Monette et al (1992) investigated the strain transfer employing both analytical theory and computer simulation. The conclusions imply that the strain distribution in the fiber is expressed as a function of E_f/E_m, here E_f and E_m are the Young's modulus of fiber and matrix respectively; the stress transfer efficiency increases as the value of E_f/E_m increases. In the same paper (Monette et al. 1992), a general theory was outlined based on a shear-lag type approach for a three-phase composite material made of a single fiber plus an interphase region embedded in a soft matrix. The shear strain in the matrix exists from the interphase to a distance chosen far away in the matrix. The computer simulation results infer that the strain transfer is deeply influenced by the ratio, E_f/E_m. Grubb and Li (1994) used Raman spectroscopy to measure the axial stress distribution along a fiber during a quasi-static single fiber pull-out test. They obtained that the shear-lag constant can be expressed as a function of the matrix shear modulus and geometric terms. One of these terms is the effective interfacial radius, r_g, which is the radius for which the strain in the matrix equals the average matrix strain. Raman measurements indicate that r_g is small, only four times the fiber radius. This result is supported by polarizing microscopy. The model of Greszczuk (1969), which assumes a uniform shear within an effective interaction thickness, gives a similar result. Jiang et al. (1998) concluded that the conventional shear-lag theory has many drawbacks. Despite considering the strain transfer from matrix to fiber, the influence of the embedded fiber on the matrix is not taken into account by the shear-lag theory. They developed the theoretical formulae of fiber and matrix strain field distributions employed both shear-lag theory and elastic theory. The significance of the matrix is emphasized sufficiently in their deformation model. They not only considered the strain transfer from the matrix to the fiber, but also considered the influence of the fiber on the behavior of the matrix.

All the research findings indicate that the strain distribution of the host material surrounding the fiber is influenced by the fiber embedding; the character of the host material have important effects on the strain transferring from host material to fiber. Therefore, the mechanical properties of the host material must be taken into account when exploring the strain transferring from host material to fiber.

In our previous investigations(Li et al., 2006; LI et al., 2009), an improved strain transfer model is developed considering the influence of the host material. An additional assumption is that the normal stress and the shear stress are both existent in some range of the host materials and the influencing radius is equal to four times the external radius of the interlayer. The derivation is similar to the simple model and the final solution is in the same form as in Eq.(17) with the only exception that the shear lag parameter, k, takes a different expression as follows,

$$k^2 = \frac{2}{r_f^{\ 2} E_f \{\frac{1}{G_c}\ln(\frac{r_c}{r_f}) + \frac{1}{G_m}\left[\frac{r_m^2}{r_m^2 - r_c^2}\ln(\frac{r_m}{r_c}) - \frac{1}{2}\right]\}} \tag{18}$$

3.3 Average strain transfer rate

The strain transferred from the host material to an optical fiber varies with the different points along the gauge length of the fiber. Strain transfer rate (STR) $\alpha(x)$ can be defined as the ratio of the strain measured by an OFS sensor and the strain actually experienced by the host material at a given point of the fiber (a given x coordinate) as following,

$$\alpha(x) = \frac{\varepsilon_g(x)}{\varepsilon_m} = (1 - \frac{\cosh(kx)}{\cosh(kL)}) \tag{19}$$

The maximum STR $\alpha_m(x)$ occurs at the midpoint of the fiber, i.e. at the point where x is equal to zero.

$$\alpha_m(0) = \frac{\varepsilon_g(0)}{\varepsilon_m} = (1 - \frac{1}{\cosh(kL)}) \tag{20}$$

Generally, the strain measured by an OFS sensor, is the average strain over the gauge length of the fiber. Average STR (ASTR) can be expressed in the following form,

$$\bar{\alpha} = \frac{\bar{\varepsilon}_g(x)}{\varepsilon_m} = \frac{2\int_0^L \varepsilon_g(x)}{2L\varepsilon_m} = 1 - \frac{\sinh(kL)}{kL\cosh(kL)} \tag{21}$$

The ASTR determined by (21) depends on the gauge length and the mechanical properties of the optical fiber, the middle layer and the host material. Therefore, it can be readily employed for correct interpretation of structural strains from the optical fiber measurement values. The results from the developed analytical model can be used as a complement to experiments especially where calibration tests are difficult to perform or where qualitative interpretation of a measurement system is required, for instance, in embedment applications of OFSs.

Fig.4 shows the variation of ASTRs $\bar{\alpha}$ over the fiber length and maximum STRs at the midpoint of the fiber in terms of the gauge length of the optical fiber, and also the experimental values by Ansari and Yuan (1998). It can be seen that the ASTRs and maximum STRs for the optical fiber in our study differs from those derived by Ansari and Yuan (1998), especially for the optical fiber with half-gauge length shorter than 60mm. When the fiber gauge length increases, the ASTRs and the maximum STRs determined by Ansari and Yuan (1998) are close to the values presented in this paper since sinh(kL) is nearly equal to cosh(kL) for a larger fiber half gauge length L. It should be emphasized that (21) is a more accurate solution for the strain transfer problem in the concentric cylinder model.

As indicated in Fig.4, the ASTR experimental values by Ansari and Yuan (1998) agree quite well with those evaluated by (21). Although the model has been verified for the simple case, it needs further experiments to validate its generalization. In any case, it facilitates a simple and direct qualitative interpretation of structural strains from measurement values made by OFSs.

Since the STRs at all the points within the fiber gauge length varies and a Bragg grating demands uniform axial deformations to avoid multiple-peak reflection and light spectrum expansion, it is required that an optical fiber be evenly stressed. However, only a short portion of FBG sensor is usually bonded, and this will lead to inadequate strain transfer, i.e. the strain sensed by a FBG sensor is only a fraction of the host material strain. Fig.5 shows

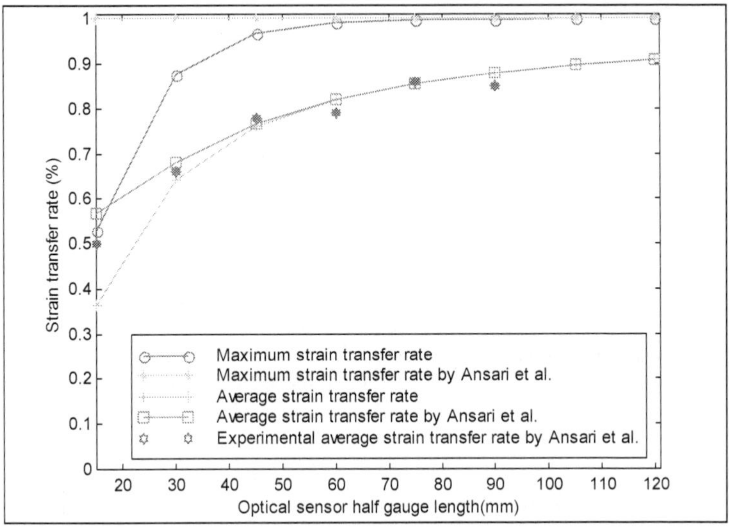

Fig. 4. Comparison of maximum STR and ASTR with experimental data for various gauge length sensors

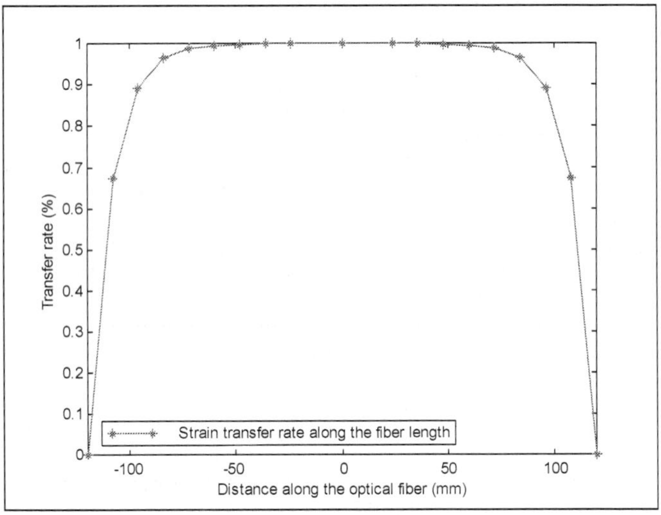

Fig. 5. Strain transfer rate of an 120mm gauge length fiber

the variation of ASTRs $\bar{\alpha}$ along the fiber with a gauge length 120mm. The figure indicates that the STRs near the center of the fiber approach unity. Therefore, an adequate fiber length has to be bonded to ensure the correct measurement of the host structural strain. Critical adherence length (CAL) can be defined as the minimum length to be bonded so that the STRs, at least over the middle half-length of the fiber, are larger than 0.9, i.e.

$$\alpha(l_c / 2) \geq 0.9 \tag{22}$$

Substituting (19) into (22) gives,

$$\frac{\varepsilon_g(l_c / 2)}{\varepsilon_m} = 1 - \frac{\cosh(kl_c / 2)}{\cosh(kl_c)} \geq 0.9 \tag{23}$$

The solution of (23) is,

$$\mathbf{CAL} = \boldsymbol{l_c} = \mathbf{9.24}/\boldsymbol{k} \tag{24}$$

where l_c is only computed over the half gauge length. Critical adherence length means that an OFS can be bonded over a longer portion, and the effective gauge length is located in the middle. For the case of the OFS shown in Table 2, its CAL is 99.78mm. This implies that the STRs along the middle half-length of the fiber are larger than 0.9 as long as the minimum bonded fiber length is beyond 99.78mm. **CAL** indicates that we can bond an FBG over a longer length, for instance 80mm, to locate the FBG just in the middle of the adhered length for efficient strain transferring.

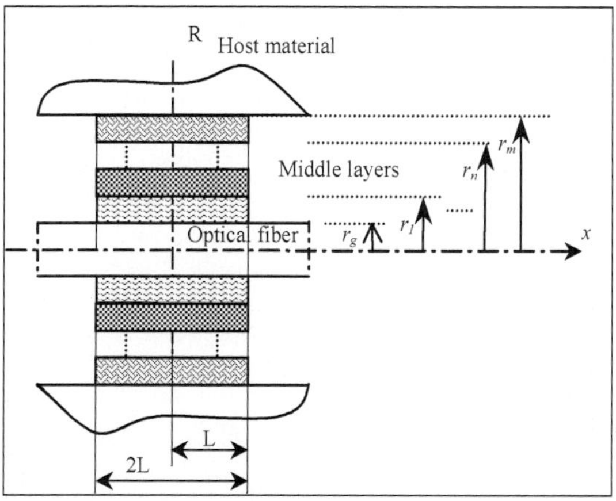

Fig. 6. Cross section of a multi-layer adhered fiber

3.4 Strain transferring for a multi-layered concentric model

In many cases, there are several middle layers between the fiber and host material, for instance, the fiber is first coated with an adhesive that solidifies quickly, and then bonded to the structure by an epoxy that solidifies slowly to ensure a uniform stress distribution. One typical case is to bond the fiber to a structure by directly applying adhesives on the fiber coating, and the coating and the adhesive form two separate layers between the fiber and host material. In some sensing applications, specialized coatings are required to enhance an optical fiber's measurement sensitivity and to accommodate the host structure. The OFSs maybe coated in this way with two different layers of coatings to employ their advantages of mechanical properties.

Fig.6 shows a multiple layer model available for the strain transfer analysis. Additionally, r_i is the outer radius of the ith layer (i =1~n), r_g is the outer radius of the fiber layer and r_m is the inner radius of the host materials (i.e. the outer radius of the (n+1)th layer). G_i is the shear modulus of the ith layer (i =2~n), and G_c is the shear modulus of the 1st layer.

Equation (9) can be rewritten as,

$$\int_{r_g}^{r_m}[\frac{dx}{dr}]dr=\int_{r_g}^{r_m}[-\frac{1}{G_c}\frac{r_g^{\ 2}}{2r}E_g\frac{d\varepsilon_g}{dx}]dr \tag{25}$$

The resulting expression is integrated, over r_g, from the fiber and first layer interface, to the outmost layer and host material interface radius r_m through all the middle layers as follows,

$$(\int_{r_g}^{r_1}+\int_{r_1}^{r_2}+...+\int_{r_n}^{r_m}[\frac{dx}{dr}])dr=(\int_{r_g}^{r_1}+\int_{r_1}^{r_2}+...+\int_{r_n}^{r_m}[-\frac{1}{G_c}\frac{r_g^{\ 2}}{2r}E_g\frac{d\varepsilon_g}{dx}]dr \tag{26}$$

The integration result of (25) is given by,

$$u_m-u_g=-\frac{1}{k_m^{\ 2}}\frac{d\varepsilon_g}{dx}=-r_g^{\ 2}\frac{E_g}{2}\{\sum_{i=2}^{n}\frac{1}{G_i}\ln(\frac{r_i}{r_{i-1}})+\frac{1}{G_c}\ln(\frac{r_1}{r_g})\}\frac{d\varepsilon_g}{dx} \tag{27}$$

where u_m and u_g denotes the axial displacement from the x coordinate origin in the host material and fiber, respectively.

$$k_m^{\ 2}=\frac{2}{r_g^{\ 2}E_g\{\sum_{i=2}^{n}\frac{1}{G_i}\ln(\frac{r_i}{r_{i-1}})+\frac{1}{G_c}\ln(\frac{r_1}{r_g})\}} \tag{28}$$

The strain lag parameter k_m, similar to the formerly discussed case, is determined by Young's moduli of the fiber and the middle layers, and the diameters of the fiber and the middle layers.

Thus, (19) and (28) can be directly used to compute the strain transfer rate for a fiber embedded in host material with multiple layers. Consequently, the critical adherence length and the average strain transfer rate within a gauge length for an optical fiber sensor can be computed by (19) and (28).

3.5 Test validation

The experiment is conducted on a universal material testing machine to study the strain transmission characteristics of an FBG sensor on different material sheets through a comparison of the strain values measured with the bare FBG and the electric strain gauge. A steel plate and a plexiglass plate are used in the experiments. Both plates are in rectangular shapes with uniform thickness. Two bare FBG sensors were bonded directly on one side of the plate with the two-component epoxy resin. One is located in the middle of the specimen and the other is located near the end of the specimen. Two electric strain gauges are fixed at the same positions on the other side of the specimen. The experimental specimen is shown in Figs. 7 and 8.

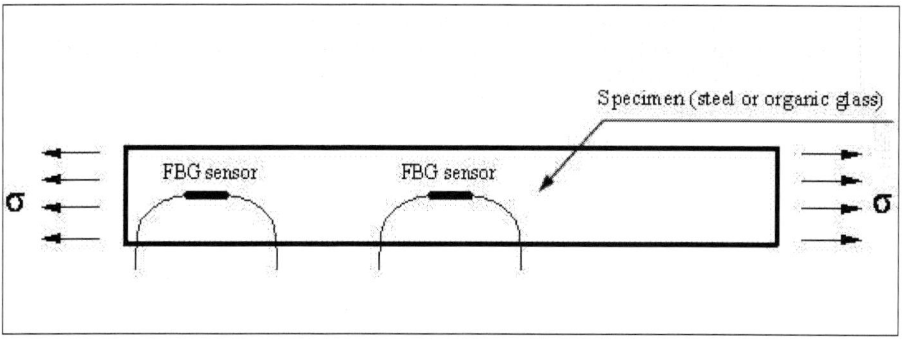

Fig. 7. The plan figure of the FBG sensors distribution.

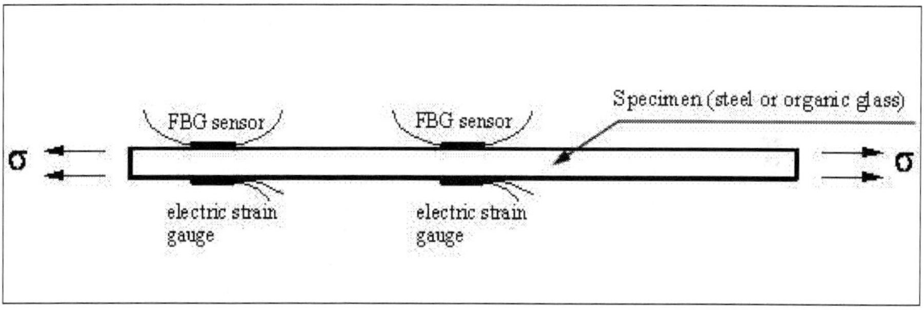

Fig. 8. The profile chart of sensor distribution.

Tensile experiments were carried out on the specimens. The load is controlled in the linear elastic range of the specimens. Therefore, each point of the specimens is under the same strain presumably from the point of view of elastic behaviour. The strain measured by the strain gauge is the actual strain of the plate. The ratio of the strain sensed by the FBG sensor to the strain measured by the strain gauge is considered to represent the strain transfer rate from the host material to the fiber core. As shown in Fig. 9, the steel plates is loaded step by step and continuously from 0 με to 250 με, then unloaded to 0 με. It is seen that the linearity of strain with the wavelength is very good, and that the coefficient of linear correlation is larger than 0.999. As shown in Fig. 10, the plexiglass sheet is loaded step by step and continuously from 0 μεto 500 με, and the coefficient of linear correlation is larger than 0.999 as well.

Based on the theory developed in this chapter, the average strain transfer ratios in steel and plexiglass as shown in Eq.(21) are 0.82330 and 0.7892 respectively. The ratio of average strain transfer with two different host materials is 0.9586. The experiment values can be calculated using test data. The experiments were repeated three times with three specimens and the data are listed in Table 3. All differences between the test and the theoretical values are less than 5%. Thus, the theoretical model is verified and the model is suitable for the analysis of fiber optic sensor strain transfer with different shear modulus of host materials.

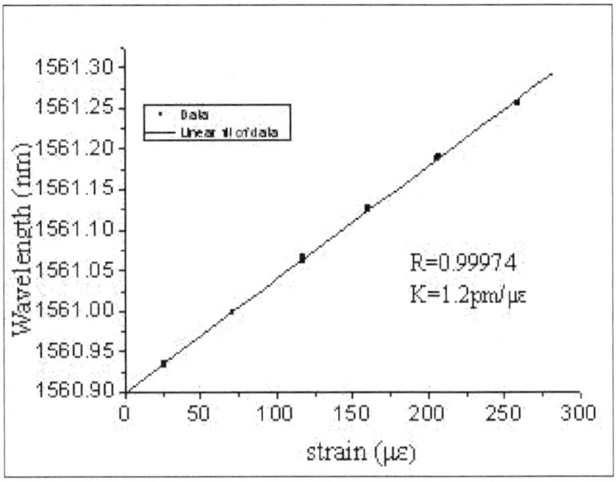

Fig. 9. the relationship between FBG wavelength and strain on the steel plate.

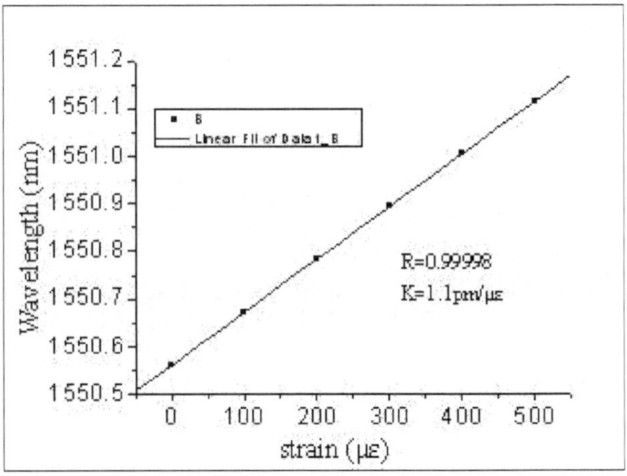

Fig. 10. the relationship between FBG wavelength and strain on the plexiglass plate.

Average strain transfer rate	First specimen	Second specimen	Third specimen
Experiment values (c_1)	0.9602	0.9596	0.9543
Theoretical values (c_2)	0.9586	0.9586	0.9586
Error $((c_1-c_2) / c_2)$	0.2%	0.1%	0.5%

Table 3. Average strain transfer rates of bonded fiber Bragg grating sensors.

When the axis of the optical fiber has an angle with the direction of the applied principal strain/ stress, there would be a different strain/stress transfer rate existence because of the interlayer between the fiber optic sensor and the surrounding matrix. Li et al.(2007) discussed this aspect. Jiang & Peters (2008) developed a shear-lag model for unidirectional multi-layered structures whose constituents vary throughout the cross section through the extension of an existing optimal shear-lag model suitable for two-dimensional planar structures and discussed solution algorithms for a variety of boundary conditions.

4. Strain transferring mechanism for surface-bonded OFSs

Apart from embedded FBG sensors, bonded FBG sensors are used more frequently. Because of the asymmetry of the bonded sensors system, the theoretical analysis is difficult. The research methods concentrated on computer simulation with finite element model (FEM) mostly. Betz et al. (2003) studied strain transferring of patch bonded FBG sensors with finite element analysis and experiments. The fiber is first placed on a backing patch and this patch is simply glued to the surface of a structure. The thickness and the modulus of the backing material are varied in the model. When the structure is strained 0.3% parallel to the direction of the fiber axis, the strain level in the fiber core is found to vary between 0.26%~0.28%. It is concluded that the structural strain has not been completely transferred due to the fiber due to the presence of the backing patch, and that the degree to which the strain is transferred depends on the thickness and Young's modulus of the backing. Lin et al., (2005) investigated three packaging methods of FBG sensors and found that strain transmission rates decrease with the increase in thickness of the steel tube. Their experiments show, however, that the thickness and Young's modulus of the glues have little influence on the strain transmission.

Li et al. (2008) and Zhou et al. (2010) found also that the strain transmission loss is small when the substrate is thick and stiff as compared to the bonding layer and the FBG. However, it becomes large when the substrate is thin and made by low modulus materials. Wan et al. (2008) studied strain transfer with stiff adhesive case with a 3D-FEM for a surface-mounted strain sensor and verified it by experiments. They found that the bond length and the bottom thickness are dominant factors besides side width and top thickness. Jahani & Nobari (2008) demonstrated that both Young's and shear moduli of adhesive are frequency dependent in dynamic testing because adhesive materials show viscoelastic behaviour.

5. Conclusion

The strain transfer mechanism from host material to the fiber core can be described as follows: as the host material generates strain under the external load, the non-uniform deformation makes the shear strain appear in the interface between the host material and interlayer, and the shear strain is transferred to the interior of the interlayer to an extent which depends on the effective bonding of the interface. The interlayer generates axial strain at the same time. The axial strain in the interlayer is transferred to the fiber core through the shear strain in the interface between the interlayer and fiber core. The fiber core generates axial strain and senses the strain of the host material. In the course of strain transferring from host material to fiber core, a portion of strain is absorbed by the interlayer; the fiber core senses a partial strain of the host material only. Theoretical studies and experiments found that the structural strain may not be completely transferred due to the fiber core due to the presence of the coatings or packaging (middle layer). The strain transfer rate depends not only on geometry and mechanical properties of the middle layer or the material between the fiber and host surface, but on the Young's modulus of the host material. Naturally, packaging methods have significant influences on the strain transfer rate.

In practical applications, primitive protective coatings by fiber manufactures have to be stripped off for sufficient strain transfer rate, or at least its low Young's modulus has to be considered in experiments. Moreover, the aging problem of the adhesive should be considered since the service life of most structures to be measured is generally designed for 50 years or longer and adhesive must be adequate.

6. Acknowledgment

This research work was supported by the Science Fund for Creative Research Groups of the National Natural Science Foundation of China (Grant No. 51121005).

7. References

Ansari, F., & Libo, Y. (1998). Mechanics of Bond and Interface Shear Transfer in Optical Fiber Sensors. *Journal of Engineering Mechanics*,Vol.124, No.4,pp.385-394,ISSN 0733-9399.

Baldwin, C.;Salter, T.;Niemczuk, J.;Chen, P., & Kiddy, J. (2002). Structural monitoring of composite marine piles using multiplexed fiber Bragg grating sensors: In-field applications. *Smart systems for bridges, structures, and highways*,Vol.4696,pp.82,ISSN 0819444448.

Betz, D. C.;Thursby, G.;Culshaw, B., & Staszewski, W. J. (2003). Acousto-ultrasonic sensing using fiber Bragg gratings. *Smart Materials & Structures*,Vol.12, No.1,pp.122-128,ISSN 0964-1726.

Cox, H. (1952). The Elasticity and Strength of Paper and other Fibrous Materials. *British Journal of Applied Physics*,Vol.3, No.MAR,pp.72-79,ISSN 0022-3727.

Feced, R.;Roe-Edwards, M.;Kanellopoulos, S.;Taylor, N., & Handerek, V. (1997). Mechanical strength degradation of UV exposed optical fibres. *Electronics Letters*,Vol.33, No.2,pp.157-159,ISSN 0013-5194.

Fernando, G.;Webb, D., & Ferdinand, P. (2002). OPTICAL-FIBRE SENSORS. *MRS Bulletin*,Vol.27, No.5,pp.359-361,ISSN 0883-7694.

Friebele, E.;Askins, C.;Bosse, A.;Kersey, A.;Patrick, H.;Pogue, W., et al. (1999). Optical fiber sensors for spacecraft applications. *Smart Materials and Structures*,Vol.8, No.6,pp.813-838,ISSN 0964-1726.

Galiotis, C.;Young, R.;Yeung, P., & Batchelder, D. (1984). Study of model polydiacetylene/epoxy composites. Part 1- the axial strain in the fibre. *Journal of Materials Science*,Vol.19, No.11,pp.3640-3648,ISSN 0022-2461.

Greszczuk, L. Theoretical Studies of the Mechanics of the Fiber-Matrix Interface in Composites.ISSN 0091-3286.

Grubb, T., & Li, T. (1994). Single-fiber polymer composites: Part I Interfacial shear strength and stress distribution in the pull-out test. *J. Mater. Sci*,Vol.29,pp.189¨C202,ISSN.

Jahani, K., & Nobari, A. (2008). Identification of dynamic (Youngs and shear) moduli of a structural adhesive using modal based direct model updating method. *Experimental Mechanics*,Vol.48, No.5,pp.599-611,ISSN 0014-4851.

Jiang, G., & Peters, K. (2008). A shear-lag model for three-dimensional, unidirectional multilayered structures. *International Journal of Solids and Structures*,Vol.45, No.14-15,pp.4049-4067,ISSN 0020-7683.

Jiang, Z.;Lian, J.;Yang, D., & Dong, S. (1998). An analytical study of the influence of thermal residual stresses on the elastic and yield behaviors of short fiber-reinforced metal matrix composites. *Materials Science and Engineering: A*,Vol.248, No.1-2,pp.256-275,ISSN 0921-5093.

Li, D. S.;Li Sr, H.;Ren, L., & Song, G. (2006). Strain transferring analysis of fiber Bragg grating sensors. *Optical Engineering*,Vol.45, No.2,pp.4402,ISSN 0091-3286.

LI, H. N.;Zhou, G. D.;Ren, L., & LI, D. S. (2009). Strain Transfer Coefficient Analyses for Embedded Fiber Bragg Grating Sensors in Different Host Materials. *Journal of Engineering Mechanics*,Vol.135, No.12,pp.1343-1353,ISSN 0733-9399.

Li, H. N.;Zhou, G. D.;Ren, L., & Li, D. S. (2007). Strain transfer analysis of embedded fiber Bragg grating sensor under nonaxial stress. *Optical Engineering*, Vol.46 ,pp.054402,ISSN 0091-3286.

Li, Q.;Li, G.;Wang, G.;Ansari, F., & Liu, Q. (2002). Elasto-Plastic Bonding of Embedded Optical Fiber Sensors in Concrete. *Journal of Engineering Mechanics*,Vol.128, No.4,pp.471-478,ISSN 0733-9399.

Li, W. Y.;Cheng, C. C., & Lo, Y. L. (2008). Investigation of strain transmission of surface-bonded FBGs used as strain sensors. *Sensors & Actuators: A. Physical*,ISSN 0091-3286.

Lin, Y. B.;Chang, K. C.;Chern, J. C., & Wang, L. A. (2005). Packaging methods of fiber-Bragg grating sensors in civil structure applications. *Sensors Journal, IEEE*,Vol.5, No.3,pp.419-424,ISSN 1530-437X.

Measures, R. M. (2001). *Structural monitoring with fiber optic technology* ISBN 0124874304: Academic.

Monette, L.;Anderson, M.;Ling, S., & Grest, G. (1992). Effect of modulus and cohesive energy on critical fibre length in fibre-reinforced composites. *Journal of Materials Science*,Vol.27, No.16,pp.4393-4405,ISSN 0022-2461.

Pak, Y. (1992). Longitudinal Shear Transfer in Fiber Optic Sensors. *Smart Materials and Structures(UK)*, No.1,pp.57-62,ISSN 0964-1726.

Rosen, B. W. (1965). Mechanics of composite strengthening. *Fiber composite materials*,Vol.2,pp.37-75,ISSN 0091-3286.

Thyagarajan, K., & Ghatak, A. K. (2007). *Fiber optic essentials* ISBN 0470097426: Wiley-IEEE Press.

Tsai, H. C.;Arocho, A. M., & Gause, L. W. (1990). Prediction of fiber-matrix interphase properties and their influence on interface stress, displacement and fracture toughness of composite material. *Materials Science and Engineering: A*,Vol.126, No.1-2,pp.295-304,ISSN 0921-5093.

Udd, E. (1995). *Fiber optic smart structures* ISBN 0471554480: Wiley-Interscience.

Varelas, D.;Limberger, H., & Salathe, R. (1997). Enhanced mechanical performance of singlemode optical fibres irradiated by a CW UV laser. *Electronics Letters*,Vol.33, No.8,pp.704-705,ISSN 0013-5194.

Wan, K. T.;Leung, C. K. Y., & Olson, N. G. (2008). Investigation of the strain transfer for surface-attached optical fiber strain sensors. *Smart Material Structures*,Vol.17, No.3,pp.5037,ISSN 0964-1726.

Zhou, J.;Zhou, Z., & Zhang, D. (2010). Study on Strain Transfer Characteristics of Fiber Bragg Grating Sensors. *Journal of Intelligent Material Systems and Structures*,ISSN 1045-389X.

19

High-Sensitivity Detection of Bioluminescence at an Optical Fiber End for an ATP Sensor

Masataka Iinuma, Yasuyuki Ushio, Akio Kuroda and Yutaka Kadoya
Graduate School of Advanced Sciences of Matter, Hiroshima University
Japan

1. Introduction

In biological studies, the luminescence from fluorescent proteins or luminescent enzymes is widely used for monitoring a change of environment at a cell. Biomolecules used for this probing, such as Green Fluorescence Protein(GFP) or luciferase molecules can respond to the existence of specific molecules or ions and subsequently emit a photon. The detection of a specific molecule can then be confirmed by detecting the emitted photons efficiently with a photon detector. A highly efficient detection of the luminescence is normally essential to a high sensitivity to the specific molecules or ions. An improvement of the sensitivity can upgrade the capability of detection in a low concentration of sample solution. Therefore, there are many efforts to improve the efficiency of the collection of emitted photons and of the optical coupling to the photon detector (Yotter, 2004).

A straightforward approach is to directly detect the luminescence from the whole sample solution in a test tube as shown in Fig. 1 (a). However, to realize high efficiency detection, this method needs a single photon detector with a wide photon-sensitive area, which is ideally larger than the photon-emission area in the test tube. The reason is that it is difficult to image incoherent light such as natural light to a smaller area than the emission area. Here, we are introducing an alternative method, where the luminescent biomolecules are immobilized at an optical fiber end and the luminescence is detected by a photon detector which is optically coupled to the other optical fiber end. The sketch of the optical fiber-based systems is shown in Fig. 1 (b). This method has been investigated for application to a fiberoptic biosensor, which is constructed by immobilizing either an enzyme or an antibody. A review is given in (Arnold, 1991), (Blum & Gautier, 1991).

This method has three merits. The first one is to permit a local detection within the sample solution, because the optical fiber end functions as a needle-like probe in the solution. The second one is that the detection scheme does not require that the photon detector is very close to the sample solution. This feature makes it easier to mount the sensing parts in integrated bioengineering, such as μ-TAS. The third merit is that single photon detectors with a small sensitive area can be used, because the photon-emission area, which is almost identical to the cross section of the core part in the optical fiber, is small. In general, single photon detectors have lower dark counts for smaller sensitive area. Low dark counts are very significant, because it essentially gives the upper limit of the sensitivity of photon detection. Recently, single photon detectors using avalanche photodiodes (APDs) have become widely available with good performance, but their sensitive area is small and has a typical size of 0.1 mm. The

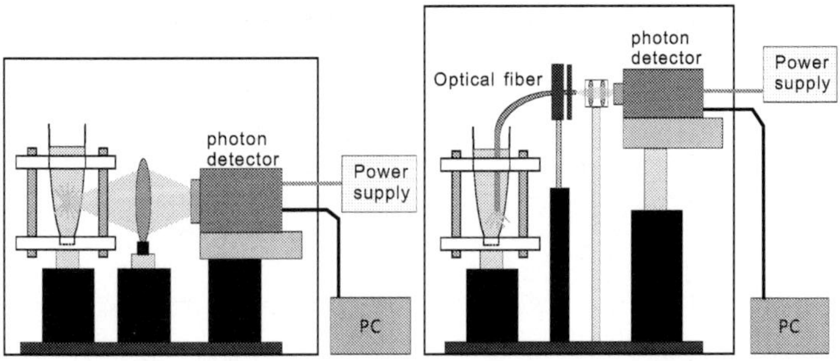

(a) Direct detection system (b) Optical fiber-based detection system

Fig. 1. Schematic figure of two detection systems

luminescence detection by the optical fiber-based system allows us to fully use the merits of compactness, high quantum efficiency, and low noise of these APD detectors.

We have built a detection system of bioluminescence at an optical fiber end and investigated the sensitivity of Adenosine triphosphate (ATP) detection by using an APD-type photon detector (Iinuma et al., 2009). ATP is a good indicator of biochemical reaction or life activity, since ATP is considered as the universal currency of biological energy for all living things. Therefore, there are a lot of efforts to develop ATP-sensing techniques for compact and efficient ATP detection (Stanley, 1992), (Andreotti & Berthold, 1999), (Gourine et al., 2005). In particular, high-sensitivity detection of ATP can indicate the existence of microorganisms even in low numbers. Thus, a compact and simple detection system with extremely high sensitivity is very desirable.

One powerful method for highly sensitive ATP detection is to use the chemical reaction involved in the bioluminescence, the luciferin-luciferase reaction (Fraga, 2008). In this reaction, after one ATP molecule and one luciferin molecule are bound to one luciferase molecule, the luciferin molecule is oxidized using the energy of ATP. As a consequence, one photon is emitted during the transition from the excited state to the ground state of the oxidized luciferin molecule bound to the luciferase molecule. The emission of one photon indicates the use of the energy of one ATP molecule. In the method using the luciferin-luciferase reaction, the efficient detection of the bioluminescence is essential for high-sensitivity detection of ATP.

Since the oxidation of luciferin is catalyzed by the enzyme luciferase, the immobilization of luciferase molcules on solid probes of various sizes allows highly sensitive and local measurements of ATP. Three types of immobilization have been used: firstly attachment to the cell surface (Nakamura et al., 2006), secondly attachment to small particles, such as nanoparticles (Konno et al., 2006) and glass beads (Lee et al., 1977), thirdly attachment to extended objects with a size in the centimeter range, such as strips (Blum et al., 1984), (Ribeiro et al., 1998) and films (Worsfold & Nabi, 1986). For the ATP-detection on an intermediate scale below 1 millimeter, a fiberoptic probe employing immobilized luciferase (Blum & Gautier, 1991) as well as microchips (Tanii et al., 2001), (Tsuboi et al., 2007), can be utilized. The detection system of bioluminescence at an optical fiber end can achieve local detection of ATP within the sample solution. Realization of high sensitivity potentially provides the local detection of extremely low number of microorganisms. Thus, it is desirable to construct a highly efficient detection system of the bioluminescence at an optical fiber end and to evaluate the detection limit with the system.

The rest of this chapter is organized as follows. In sec. 2, we describe a concept for the construction of the optical fiber-based system for efficient detection of a fluorescence at the optical fiber end. In sec. 3, we show how to optimize the parameters of the optical fiber and the coupling optics so as to realize high photon-collection efficiency. In sec. 4, we describe the application of the constructed detection system to ATP sensing. By immobilizing luciferase molecules at the optical fiber end, the bioluminescence by luciferin-luciferase reaction can be detected using the optical fiber-based system. We evaluated the sensitivity of ATP with this system. Sec. 5 summarizes present results and problems.

2. General concept for construction of the optical fiber-based system

For a luminescence detection system using an optical fiber with a core diameter ϕ_0 and a numerical aperture NA_0, a collection efficiency of the luminescence η at the optical fiber end depends only on NA_0 as shown in Fig. 2. From a simple calculation based on the solid angle

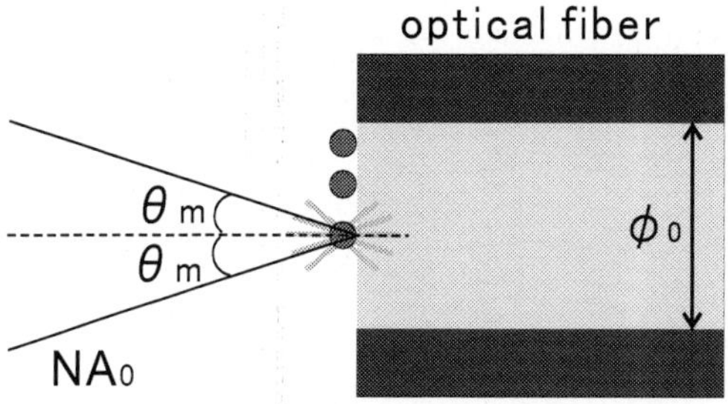

Fig. 2. Fluorescence at the optical fiber end. θ_m is a maximum opening angle for light propagation in the optical fiber.

with a maximum opening angle θ_m, $\eta(NA_0)$ can be expressed as,

$$\begin{aligned} \eta\left(NA_0\right) &= \frac{1}{2}\left(1-\cos\theta_m\right) \\ &= \frac{1}{2}\left[1-\sqrt{1-\left(\frac{NA_0}{n_w}\right)}\right], \end{aligned} \tag{1}$$

where n_w is the refraction index of the substance surrounding the optical fiber end. In immersing the optical fiber end into water, its value should be identical to the value of water, which is about 1.33. Fig. 3 shows the calculated values of the collection efficiency $\eta(NA_0)$ as a function of NA_0 using Eq. (1) at $n_w = 1.33$. One can easily see that $\eta(NA_0)$ increases with NA_0.

In the following, let us consider the situation where the other optical fiber end is optically coupled to a photon detector with a circular sensitive area having a diameter ϕ_3 and a numerical aperture NA_3. The coupling efficiency ϵ between the optical fiber end and the photon detector depends on ϕ_0, NA_0 of the optical fiber and ϕ_3, NA_3 of the photon detector

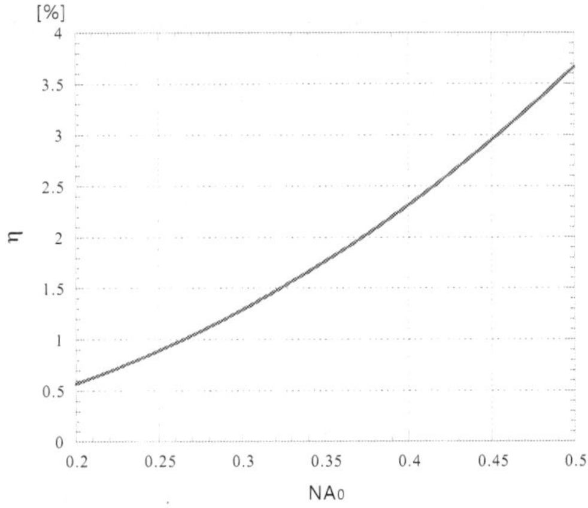

Fig. 3. Calculated values of collection efficiency as a function of NA_0 at n_w=1.33

used. Since the number of emitted photons is proportional to the square of ϕ_0 and $\eta(NA_0)$ increases with NA_0 as expressed by Eq. (1), the number of transmitted photons to the other optical fiber end is proportional to the square of ϕ_0 and $\eta(NA_0)$. On the other hand, the coupling efficiency ϵ generally decreases as ϕ_0 or NA_0 increases. Thus, we can define the following formula for a figure of merit (FOM) and optimize ϕ_0, NA_0 and parameters of the coupling optics x_i to maximize this FOM:

$$FOM = \phi_0^2 \cdot \eta(NA_0) \cdot \epsilon(\phi_0, NA_0, x_i; \phi_3, NA_3) \tag{2}$$

It should be noted that ϵ can ideally be 100 % under the conditions of $\phi_0 \leq \phi_3$ and $NA_0 \leq NA_3$. In many cases, the condition $NA_0 \leq NA_3$ is satisfied when using typical photon detectors. Thus, under the condition of $NA_0 \leq NA_3$, we can classify two cases: case (1) is $\phi_0 \leq \phi_3$ and case (2) is $\phi_0 > \phi_3$. In case (1), ϵ is constant and can ideally be 100 % and the total detection efficiency is limited only by $\eta(NA_0)$. Therefore, optimization of the coupling optics is not necessary. The conditions $\phi_0 = \phi_3$ and $NA_0 = NA_3$ both maximize the FOM and the sensitivity becomes highest. In case (2), however, the optimization of ϕ_0, NA_0, and the parameters x_i for a design of the coupling optics are necessary for given values of NA_3 and ϕ_3, because ϵ decreases as ϕ_0 or NA_0 increases.

3. Design of coupling optics

3.1 Photon detectors

Photon detectors generally have two significant factors contributing to the sensitivity of detection for weak light: the efficiency and the dark counts of the photodetector. A cooled APD which can detect for single photons is mostly used because of its very low dark counts. The sensitive area must be very small for realizing a large reduction of the dark counts, but the quantum efficiency is several times larger than that of a photomultiplier tube(PMT). Furthermore, the APD has the useful characteristics of compactness, easy operation, and durability in comparison with a PMT. To construct an optical fiber-based system with high sensitivity, we chose an APD-type photon counting module (SPCM-AQR-14) manufactured by Perkin Elmer Co. Ltd., which has a quantum efficiency η_{qe} of 55 % at 550 nm and dark counts of about 100 s^{-1}. The APD has a circular sensitive area, where ϕ_3 is 0.175 mm and NA_3 is 0.78, as calculated from the geometrical structure between the sensitive area and the photon detection window. If we use an optical fiber with $\phi_0 > \phi_3$ to increase the number of emitted photons, it is necessary to optimize ϕ_0, NA_0, and to design the coupling optics for maximal sensitivity.

3.2 Design concept and procedure

We can consider the coupling optics between the optical fiber end and the APD as an optical system imaging a light source with NA_0 and ϕ_0 onto the APD with NA_3 and ϕ_3. The basic design of the coupling optics is shown in Fig. 4.

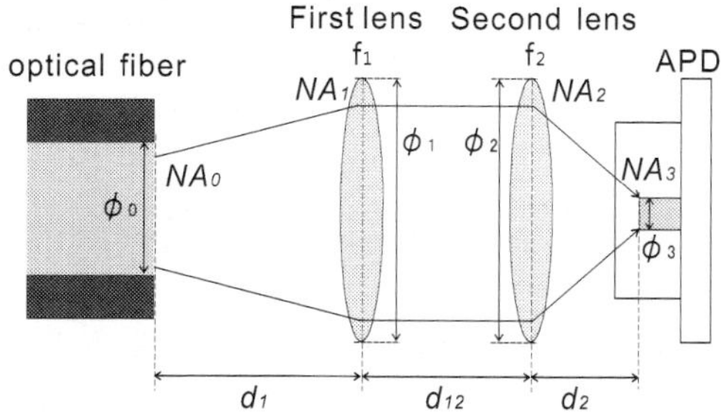

Fig. 4. Design of the coupling optics between the fiber output and the APD

Among the parameters shown in Fig. 4, we firstly determine the parameters of optical components, the focal length f_1 and the numerical aperture NA_1 of the first lens, the focal length f_2 and the numerical aperture NA_2 of the second lens, and the distance d_{12} between the first lens and the second lens. The second lens is selected among available lenses to make NA_2 as large as possible while staying below NA_3. We also select the first lens among available lenses to make NA_1 as large as possible while staying below NA_2. As the result, f_1, NA_1, f_2, NA_2 were determined as shown in Table 1.

	focal length	NA	diameter
first lens	8.0 mm	0.50	8.68 mm
second lens	3.1 mm	0.68	5.1 mm
APD sensitive area		0.78	0.175 mm

Table 1. Parameters of optical components

Taking into account of geometrical structure of lens mounts and fixing both of the first and second lens, the distance d_{12} should be longer than 3 mm. In this case, we determined $d_{12} = 5$ mm. The remaining parameters in this optical system were NA_0, ϕ_0, the distance d_1 between the fiber end and the first lens, and the distance d_2 between the second lens and the APD. These parameters can be determined in two steps as follows. In the first step, NA_0 and ϕ_0 are optimized to maximize the FOM expressed as Eq. (2) under the conditions $f_1 = d_1$ and $f_2 = d_2$. In the second step, d_1 and d_2 are optimized to maximize $\epsilon(d_1, d_2, \phi_0, NA_0)$ for given values of NA_0 and ϕ_0 obtained in the first step.

3.3 Determination of parameters in the optical system

The optimization procedure requires specific values of $\eta(NA_0)$ and $\epsilon(NA_0, \phi_0)$ for obtaining the FOM. Since the values of $\eta(NA_0)$ were given from Eq. (1) as shown in Fig. 3, it is necessary to calculate the values of $\epsilon(NA_0, \phi_0)$ at $d_1 = f_1$ and $d_2 = f_2$. The simple method for obtaining the efficiency is a statistical simulation by ray tracing.

The optical fiber end can be considered as a light source with ϕ_0 and NA_0. To calculate the efficiency, an event of light emission is randomly generated at an arbitrary position within ϕ_0 and at an arbitrary direction within NA_0. Subsequently, the final state of light at the APD sensitive area is calculated by the transformation of the initial state based on ray tracing. We repeat a procedure containing the generation of one event in the light emission and the subsequent transformation to the final state at the APD via intermediate states at the first lens and the second lens, and count the number of the events, where the conditions at the first lens NA_1, ϕ_1, at the second lens NA_2, ϕ_2, and at the APD NA_3 and ϕ_3 are all fulfilled. The efficiency can be obtained as the ratio of number of counts to the total number of event generation.

The propagation of light can be described with matrix formalism in paraxial optics (Yariv, 1997). The light at the initial state (r_i, r'_i), where r_i is a distance from an optical axis of optics and r'_i is a slope of the light direction, can be transferred by the following matrices, for example,

$$M_{free} = \begin{bmatrix} 1 & d \\ 0 & 1 \end{bmatrix}$$

$$M_{lens} = \begin{bmatrix} 1 & 0 \\ -\frac{1}{f} & 1 \end{bmatrix},$$

where M_{free} is the transfer matrix of free-space propagation far from the distance d and M_{lens} is the transfer matrix of thin lens with the focal length f. Any state of light expressed as the vector form (r, r') can be transferred by any combination of the transfer matrices. Therefore, the transfer matrix $M_{coupling}$ describing the optical system shown in Fig. 4 can be expressed as follows,

$$M_{coupling} = \begin{bmatrix} 1 & d_2 \\ 0 & 1 \end{bmatrix} \cdot \begin{bmatrix} 1 & 0 \\ -\frac{1}{f_2} & 1 \end{bmatrix} \cdot \begin{bmatrix} 1 & d_{12} \\ 0 & 1 \end{bmatrix} \cdot \begin{bmatrix} 1 & 0 \\ -\frac{1}{f_1} & 1 \end{bmatrix} \cdot \begin{bmatrix} 1 & d_1 \\ 0 & 1 \end{bmatrix}$$

$M_{coupling}$ can transfer the initial state (r_i, r'_i) at the optical fiber end to the final state (r_f, r'_f) at the APD sensitive area.

Thus, the values of ϵ were obtained by the statistical method with the transfer matrix $M_{coupling}$, where random numbers for setting initial states were generated with the software package based on algorithm of Mersenne Twister (Matsumoto & Nishimura, 1998). The calculated results are shown as a function of NA_0 in Fig. 5.

Fig. 5 shows that ϵ is decreasing with increasing NA_0 and ϕ_0. As a consequence, ϵ becomes 100 % at $\phi_0 = 0.4$ mm and $NA_0 \leq 0.25$. However, the reduction of ϕ_0 and NA_0 makes both η

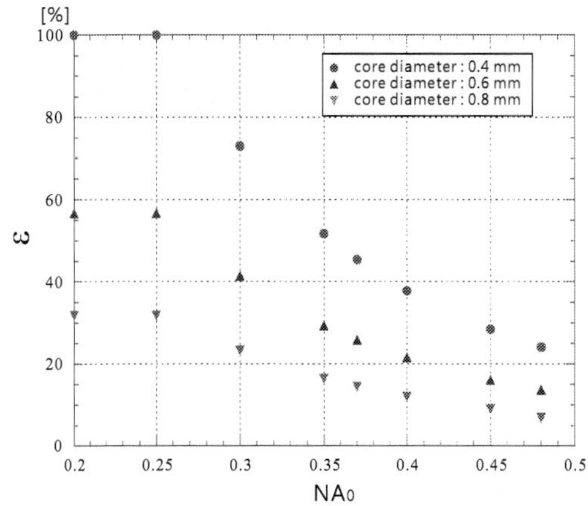

Fig. 5. Calculated values of coupling efficiency ϵ as a function of NA_0. The solid circles indicate the values at $\phi_0 = 0.4$ mm, the solid triangles are the values at $\phi_0 = 0.6$ mm, and the solid inverse triangles are the values at $\phi_0 = 0.8$ mm.

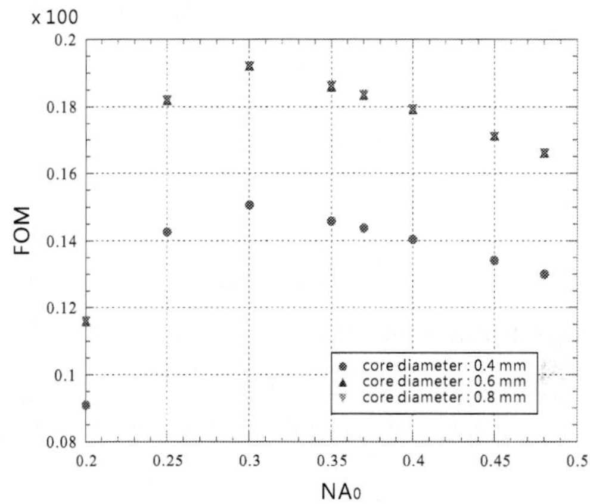

Fig. 6. Calculated values of FOM as a fuction of NA_0. The solid circles indicate the values at $\phi_0 = 0.4$ mm, the solid triangles are the values at $\phi_0 = 0.6$ mm, and the solid inverse triangles are the values at $\phi_0 = 0.8$ mm.

and the number of emitted photons smaller. Therefore, the calculation of the FOM is necessary for the optimization of NA_0 and ϕ_0.

The plot of the FOM as a function of NA_0 is shown in Fig. 6. One can easliy see that the FOM is maximal at $NA_0 = 3.0$. In addtion, the value of the FOM for $\phi_0 = 0.6$ mm is almost same as the one for $\phi_0 = 0.8$ mm. This means that the FOM is saturated for larger diameters than $\phi_0 = 0.6$ mm because the size of the transferred image at the APD is larger than the APD sensitive area.

Optical fibers with $NA_0 = 3.0$ are not easily obtainable, whereas optical fibers with $NA_0 = 2.5$ and $NA_0 = 3.7$ are readily available. Fig. 6 shows that the slope above $NA_0 = 3.0$ is flatter than the one below. In the upper part, the misalignment or imperfection of the optical system has less influence on the coupling efficiency. Therefore, we selected $\phi_0 = 0.6$ mm and $NA_0 = 3.7$.

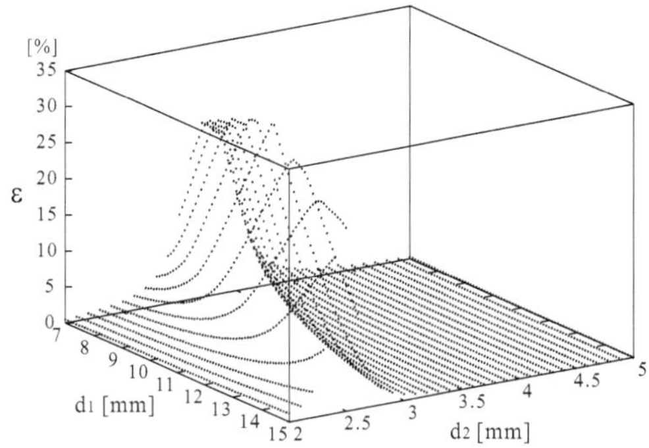

Fig. 7. Calculated values of ϵ as fuctions of d_1 and d_2 for given values of $NA_0 = 0.37$ and $\phi_0 = 0.6$ [mm].

In a second step, d_1 and d_2 were optimized to maximize $\epsilon(d_1, d_2)$ for $NA_0 = 0.37$ and $\phi_0 = 0.6$ mm. Fig. 7 shows the calculated values of ϵ as functions of d_1 and d_2. From the peak of ϵ in Fig. 7, the results of $d_1 = 11.6$ mm and $d_2 = 2.7$ mm were obtained for $NA_0 = 0.37$ and $\phi_0 = 0.6$. These parameters provide the maximum $\epsilon(d_1, d_2)$ of 33.33 %, $\eta(NA_0)$ of 1.95 % and FOM $\times 100$ is 0.234 which value is higher than the maximum in Fig. 6.

4. Application to the ATP sensing system

4.1 Luciferin-luciferase reaction

Bioluminescence in living organisms, such as fireflies and some marine bacteria, typically occurs due to the optical transition from the excited state to the ground state of oxidized luciferin molecules produced by the luciferin-luciferase reaction under the catalytic activity of luciferase molecules. This reaction can be expressed by the following sequence of reaction steps:

$$\begin{aligned} &\mathrm{E} + \mathrm{LH_2} + \mathrm{S} \rightleftharpoons \mathrm{C} + \mathrm{PP_i} \\ &\mathrm{C} + \mathrm{O_2} \rightarrow \mathrm{E} + \mathrm{P}^* + \mathrm{AMP} + \mathrm{CO_2} \\ &\mathrm{P}^* \rightarrow \mathrm{P} + \gamma \end{aligned}$$

where E indicates luciferase, LH_2 luciferin, PP_i pyrophoric acid, C is an enzyme-substrate compound $E \cdot LH_2$-AMP, AMP adenosine monophosphate, P oxidized luciferin, and γ a photon (DeLuca, 1976). The emission of one photon at the position of luciferase molecule indicates the use of the energy of one ATP molecule.

The immobilization of luciferase molecules at the optical fiber end enables us to sense the presence of ATP around the fiber end using single photon detection. For this purpose, we used a compound protein containing a silica-binding protein (SBP) molecule and a luciferase molecule (SBP-luciferase), which were recently synthesized by Taniguchi and co-workers (Taniguchi et al., 2007). This protein makes it possible to immobilize a luciferase molecule on the optical fiber end via a SBP molecule while retaining its activity. The spectrum of the emitted photons shows a central wavelength of 550 nm and a width of about 100 nm (Denburg et al., 1969), (Ugarova & Brovko, 2002). Since the APD module has the large quantum efficiency for the photons from the luciferin-luciferase reaction, the APD photon counting module is suitable for ATP sensing.

4.2 Reaction-diffusion differential equation

Fig. 8 shows the enzyme reaction which describes the sequence of reactions.

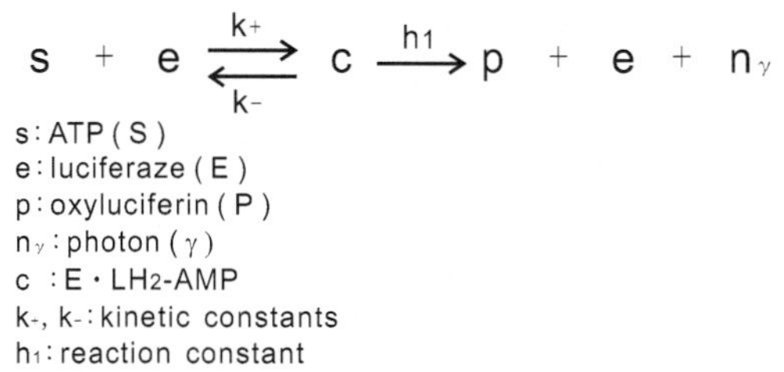

Fig. 8. Enzyme reaction describing luciferin-luciferase reaction

Here, s, e, c, p, n_γ in Fig. 8 represents a concentration of ATP, luciferase, $E \cdot LH_2$-AMP, oxidized luciferin, and emitted photon, respectively. In addition, k_+, k_- are kinetic constants for equilibrium and h_1 is a reaction constant.

In a solution containing nonlocalized homogeneously dispersed luciferase and ATP, the Michaelis-Menten theory can be simply applied to the above reaction. In the presence of enough luciferin molecules in the solution, a rate of emitted photons at steady state v_γ can be expressed as the Michaelis-Menten formula,

$$v_\gamma = \frac{V_0 S}{S + K_M}, \tag{3}$$

where V_0 is a maximum reaction rate which is equivalent to a product of the number of luciferase molecule and h_1, K_M is the Michaels constant expressed as $K_M = \frac{k_- + h_1}{k_+}$ and S is the ATP concentration.

In the fiber-based system for sensing dispersed ATP molecules, on the other hands, an ATP-flow generated by a gradient of ATP concentration around the luciferase-terminated fiber end carries ATP molecules to the vicinity of immobilized SBP-luciferase molecules. The ATP molecule is bound to the immobilized SBP-luciferase molecule near them and subsequently contributes the luciferin-luciferase reaction at this fiber end. To calculate the rate of emitted photons, therefore, it is necessary to consider not only a reaction rate but also an ATP diffusion rate.

The series of reaction-diffusion equations describing the enzyme reaction shown in Fig. 8 can be expressed as,

$$\begin{aligned}\frac{ds}{dt} &= D\left(\frac{\partial^2 s}{\partial x^2} + \frac{\partial^2 s}{\partial y^2}\right) + R(s,e,c) \\ \frac{de}{dt} &= -k_+ se + k_- c + h_1 c \\ \frac{dc}{dt} &= k_+ se - k_- c - h_1 c \\ \frac{dn_\gamma}{dt} &= h_1 c, \end{aligned} \tag{4}$$

where D is a diffusion constant of ATP in water and $R(s,e,c)$ is expressed as follows.

$$R(s,e,c) \equiv \begin{cases} -k_+ se + k_- c & (x, y \in \Gamma_{fiber}) \\ 0 & (\text{otherwise}) \end{cases}$$

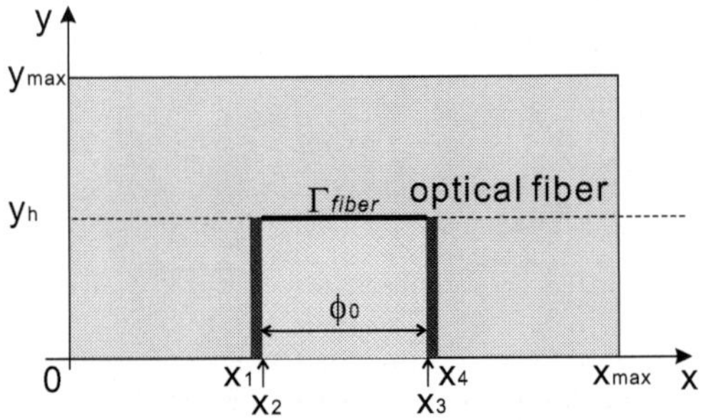

Fig. 9. Coordinate system of x and y around the fiber end.

Fig. 9 shows the definition of the coordinate system of x and y . Here, it should be noted that $s(x,y,t)$ is a function of x, y, t and $e(t)$ and $c(t)$ are functions of only t. Γ_{fiber} is defined as an area where the reaction occurs and equivalent to the core part ϕ_0 in the optical fiber. In this coordinate system, it can be represented as the interval of $x_2 \leq x \leq x_3$ with $y = y_h$. The both intervals of $x_1 \leq x \leq x_2$ and $x_3 \leq x \leq x_4$ with $y = y_h$ represent the parts of clad in the optical fiber.

By dividing the space around the fiber end into finite spatial steps and also dividing the time into finite time steps, we can numerically solve the series of reaction-diffusion equation Eq. (4) under the boundary conditions presented in Table 2. Here, s_0 in Table 2 is an initial ATP concentration that should be uniform into the whole solution before starting the reaction. The numerical solutions can be given as time evolution of spatial distribution of concentration $s(x_i, y_i, t_i)$, $e(t_i)$, and $c(t_i)$ and the rate of emitted photons can be obtained from $h_1 c(t_i)$. The peak of the photon-emission rate in time corresponds to the reaction rate given by the Michaelis-Menten formula.

In order to check the possibility that the emission rates are limited by the diffusion rate of ATP, we numerically obtained the peak values of the photon-emission rate at various ATP concentration and compared to the reaction rate calculated by Michaelis-Menten formula. To obtain numerical solutions at each time step, spatial segmented equations derived from Eq.

region	$x = 0$	$0 \le x < x_1$, $y = 0$	$x = x_1$, $0 \le y < y_h$	$x_1 \le x < x_2$, $y = y_h$	$x_2 \le x \le x_3$, $y = y_h$
condition	$s = s_0$	$s = s_0$	$\left.\frac{ds}{dx}\right\vert_{x=x_1} = 0$	$\left.\frac{ds}{dy}\right\vert_{y=y_h} = 0$	$\left.\frac{ds}{dy}\right\vert_{y=y_h} = 0$
region	$x_3 < x \le x_4$, $y = y_h$	$x = x_4$, $0 \le y < y_h$	$x_4 < x \le x_{max}$, $y = 0$	$x = x_{max}$	$y = y_{max}$
condition	$\left.\frac{ds}{dy}\right\vert_{y=y_h} = 0$	$\left.\frac{ds}{dx}\right\vert_{x=x_4} = 0$	$s = s_0$	$s = s_0$	$s = s_0$

Table 2. boundary conditions for solving Eq. (4). s_0 is the initial ATP concentration.

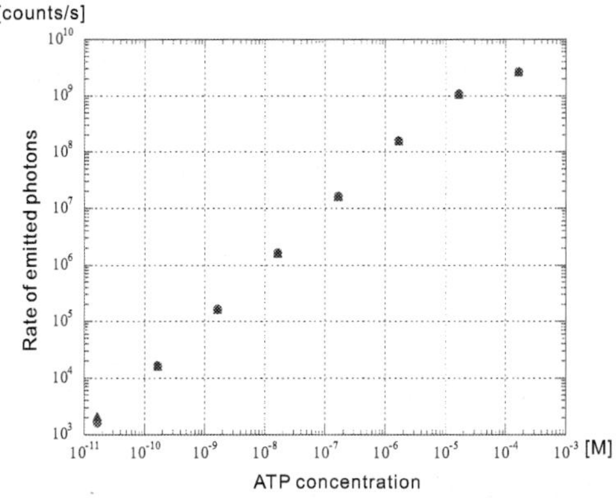

Fig. 10. Comparison of the results calculated from numerical solutions of reaction-diffusion equations to the results of Michaelis-Menten formula. Closed triangles indicate values of ATP diffusion process using typical values of kinetic constants, closed circles are values given by Michaelis-Menten formula.

(4) were solved in terms of time using the software package of ordinary differential equations DASKR (Brown et al., 1998), where the ATP diffusion constant was $D = 0.5 \times 10^{-5}$ cm^2/s (Aflao & DeLuca, 1987), the geometrical parameters were $x_{max} = 1.4$ mm, $y_{max} = 2$ mm, $y_h = 1$ mm, $x_1 = 0.38$ mm, $x_4 = 1.02$ mm, the kinetic constants were $k_+ = 20000$ M^{-1}s^{-1}, which was estimated from the typical buildup time of 0.3 s (DeLuca & McElory, 1974), and $k_- = 0.515$ s^{-1}, which was calculated with the relational form $K_M \cdot k_+ - h_1$. As other parameters, we used $h_1 = 0.125$ s^{-1} (Branchini et al., 2001), $K_M = 3.2 \times 10^{-5}$ M, and the surface density of luciferase molecule $\sigma_0 = 9.03 \times 10^{10}$ mm^{-2} (Taniguchi et al., 2007), which were also used for Michaelis-Menten formula. The Michaelis constants $K_M = 3.2 \times 10^{-5}$ M was obtained from data analysis of counts of detected photons, which describes in Sec. 4.

Fig. 10 shows the results of two kinds of calculations. The closed triangles indicate the values obtained from numerical solutions of Eq. (4), whereas the closed circles are the values deduced from Michaelis-Menten formula. The results of ATP diffusion process have good agreements with ones given by the Michaelis-Menten theory. Therefore, we can conclude that the ATP diffusion is not a rate-limiting process for the present rate of the chemical reaction.

4.3 Measurement of the sensitivity

4.3.1 Immobilization of luciferase

Before immobilizing the luciferase molecules, we cut optical fiber and cleaned the cut surface with ethanol and Tris buffer (0.25 mM Tris-HCl with 0.15 M NaCl). After cleaning, the surface was immersed in a solution of SBP-luciferase and was left at a temperature of 3°C to 6°C for a period of about two hours.

4.3.2 Sample solutions

The samples were a 1:4:4:31 mixture of 20 mM D-luciferin solution, Tris buffer solution(250 mM Tris-HCl mixed with 50 mM $MgCl_2$), ATP solution, and distilled water. Several solutions of ATP with different ATP concentrations were made by diluting the ATP standard in ATP Bioluminescence Assay Kit CLS II manufactured by Roche Co. Ltd. Thus, a series of sample solutions with different ATP concentrations were prepared in advance. An additional sample without ATP was also produced by mixing distilled water instead of the ATP solution. This sample was measured in order to obtain a background before the ATP measurements.

4.3.3 Data taking system

TTL pulses outputted from the APD photon counting module were counted by a PC card installed in a personal computer(PC). The number of pulses occurring during 10 s were recorded every 10 s by the PC

4.3.4 Results

The time dependence of photon counts per 10-s interval were measured after the luciferase-terminated end were immersed in the sample solutions with various ATP concentration. The results for 100 $\mu\ell$ solution from 1.65×10^{-4} M to 1.65×10^{-9} are shown in Fig. 11.

The photon counts increase and reach a maximum at about 150 s after immersion. Then, they decrease very slowly to the background level. Therefore, for obtaining high sensitivity, it is practical to continue the measurement for about 300 s after the luciferase-terminated end is immersed in the solution. A background level of approximately 120 s^{-1} was determined as an average of counts for background data. It was found to correspond well to the dark counts of the photon counting module.

ATP concentration	integrated counts	BG-subtracted counts	statistical errors
no ATP (background)	34490		
1.65×10^{-11} M	34664	174	262.97
1.65×10^{-10} M	35346	856	264.27
1.65×10^{-9} M	46770	12280	285.06
1.65×10^{-8} M	139300	104810	416.88

Table 3. Integrated counts for ATP sample solutions

Table 3 presents integrated counts of detected photons during 300 s from the origin of time. Statistical errors are estimated as one standard deviation assuming Poisson distribution. From Table 3, the sensitivity in this optical fiber-based system is limited to 1.65×10^{-10} M, which

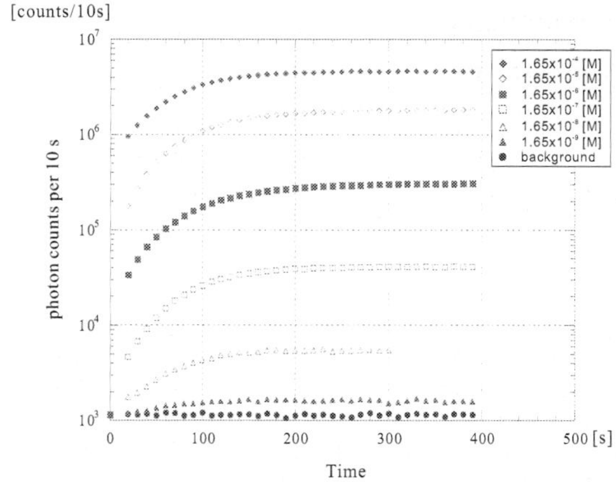

Fig. 11. Measured photon counts per 10 s as a function of time at ATP concentration 1.65×10^{-4} M, 1.65×10^{-5} M, 1.65×10^{-6} M, 1.65×10^{-7} M, 1.65×10^{-8} M, 1.65×10^{-9} M, which are represented by closed diamonds, open diamonds, closed squares, open squares, open triangles, closed triangles, respectively. Closed circles indicate background counts.

corresponds to a number of ATP molecules of about 10^{-14} mol in the 100 $\mu\ell$ solution. We also ascertained that our system is sensitive to the 10^{-10} M level even with a $10\mu\ell$ solution, in which the absolute ATP concentration is 10^{-15} mol.

In order to check the ATP concentration dependence of the photon counting rate at maximum, the average of counts in six 10-s intervals around the time at which the counting rate became maximal was calculated for each ATP concentration. The results are indicated by solid circles in Fig. 12. This figure shows that the lower limit of the sensitivity is 1.65×10^{-9} M taking into account of statistical errors. If the statistical errors reduce to half of the present ones, which means that statistics at each point becomes four times as large as 60-s interval, the detection of 1.65×10^{-10} M also becomes feasible. This is consistent with the results presented in Table 3. The analysis of fitting data points in Fig. 12 provided the Michaelis constant of 3.2×10^{-5} M.

4.3.5 Discussion

In Fig. 12, the open squares show predictions of the counting rate estimated from the results of numerical solutions shown in Fig. 10 with the total detection efficiency in our detection system, which can be expressed as $\epsilon_{total} = \eta \cdot \epsilon \cdot \eta_{qe}$. Its value was calculated to be 0.00354 for photons with the wavelength 550 nm. Comparing the experimental data with the predictions, there are disagreements amounting to roughly one order of magnitude.

The results suggest the ATP diffusion does not limit the detection limit. Therefore, the possible reasons for the discrepancies may be the following.

(1) The rate of the chemical reaction per luciferase molecule itself becomes small.
(2) The number of immobilized active molecules is lower than expected.

We have also confirmed that the buildup time in Fig. 11 is about 30 times longer than that in a solution containing only nonlocalized homogeneously dispersed SBP-luciferase

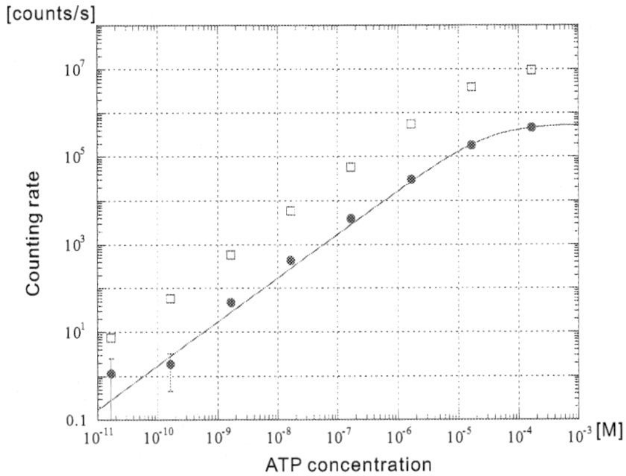

Fig. 12. Measured photon counting rate as a function of ATP concentration. Solid line is a curve obtained by fitting data with the Michaelis-Menten formula. Open squares are values estimated from the results of numerical solutions shown in Fig. 10 with total detection efficiency $\epsilon_{total} = 0.00354$.

molecules. There are also other experimental results claiming the decrease in acitivity of immobilized luciferase moleclues (Nishiyama et al., 2008), (Tanaka et al., 2011). The reason of this discrepancy is still an open question. These results suggest that it will be necessary to study the reaction at the end of the optical fiber in more detail in order to improve the sensitivity.

5. Summary

We introduced a method of high-sensitivity detection of bioluminescence at an optical fiber end for an ATP sensor as an efficient alternative to direct detection of bioluminescence from a sample solution. For efficiently detection of the bioluminescence, we have constructed an optical fiber-based system, where the luciferase molecules are immobilized on the optical fiber end and the other end is optically coupled to an APD-type photon counting module. We have evaluated the sensitivity of the ATP detection of this system and found that the optimization of the optical coupling system and the use of the SBP-luciferase molecules provide the detection limit of 10^{-10} M, which allows us to detect the absolute ATP concentration of 10^{-15} mol with a $10\mu\ell$ solution. To improve the sensitivity, it is necessary to study the details of the reaction at luciferase-terminated end of the optical fiber.

6. Acknowledgements

We are grateful to Dr. Kazutaka Nomura for providing the SBP-luciferase molecules and Prof. Holger F. Hofmann for reading the manuscript. We also express our thanks to Dr. Tetsuya Sato, Dr. Koichiro Sadakane, and Dr. Shuhei Miyoshi for their help in setting up the experiments. This work has been partially supported by the International Project Center for Integration Research on Quantum, Information, and Life Science of Hiroshima University and the Grant-in-Aid for Scientific Research (C)(19560046) of Japanese Society for the Promotion of Science, JSPS.

7. References

Andreotti P. E. & Berthold F. (1999). Application of a new high sensitivity luminometer for industrial microbilogy and molecular biology, *Luminescence*, Vol. 14: 19-22, ISSN 15227235

Arnold M. A. (1991). Fluorophore- and Chromophore-Based Fiberoptic Biosensors, In: *Biosensor principles and applications*, Blum L. J. & Coulet P. R. (Ed.), 195-211, Marcel Dekker, 1991, ISBN 0824785460, New York

Aflao C. & DeLuca M. (1987). Continuous Monitoring of Adenosine 5′-Triphosphate in the Microenvironment of Immobilized Enzymes by Firefly Luciferase, *Biochemistry*, Vol. 26: 3913-3920, ISSN 00062960

Blum L. J.; Coulet P. R.; Gautheron D. C. (1984). Collagen Strip with Immobilized Luciferase for ATP Bioluminescent Determination, *Biotechnol. Bioeng.*, Vol. 27: 232-237, ISSN 00063592

Blum L. J.; Gautier S. M. (1991). Bioluminescence- and Chemiluminescence-Based Fiberoptic Sensors, In: *Biosensor principles and applications*, Blum L. J. & Coulet P. R. (editors), 213-247, Marcel Dekker, 1991, ISBN 0824785460, New York

Branchini B. R.; Magyar R. A.; Murtiashaw M. H.; Portier N. C. (2001). The Role of Active Site Residue Arginie 218 in Firefly Luciferase Bioluminesence, *Biochemistry*, Vol. 40: 921-925, ISSN 00062960

Brown P. N.; Hindmarsh A. C.; Petzold L. R. (1998). Consistent initial condition calculation for differential-algebraic systems, *SIAM J. Sci. Comput.*, Vol. 19: 1495-1512, ISSN 10648275. URL: http://www.netlib.org/ode/

DeLuca M. & McElory W. D. (1974). Kinetics of the Firefly Luciferase Catalyzed Reactions, *Biochemistry*, Vol. 13: 921-925, ISSN 00062960

DeLuca M. (1974). Firefly luciferase, In: *Advances in Enzymology*, Meister A. (Ed.), Vol. 44: 37-68, Wiley, 1976, ISBN 9780471591795, New York

Denburg J. L.; Lee R. T.; McElory W. D. (1969). Substrate-Binding Properties of Firefly Luciferase 1. Luciferin-binding Site, *Arch. biochem. biophys.*, Vol. 134: 381-394, ISSN 00039861

Fraga H. (2008). Firefly luminescence: A histrical perspective and recent developments, *Photochem. Photobio. Sci.*, Vol. 7: 146-158, ISSN 1474905

Gourine A. V.; Laudet E.; Dale N.; Spyer K. M. (2005). ATP is a mediator of chemosensory transduction in the central nervous system, *Nature*, Vol. 436: 108-111, ISSN 00280836

Iinuma M.; Ushio Y.; Kuroda A.; Kadoya Y. (2009). High Sensitivity Detection of ATP Using Bioluminescence at An Optical Fiber End, *Electronics and Communications in Japan*, Vol. 92: 53-59, ISSN 03854221

Konno T.; Ito T.; Takai M.; Ishihara K. (2006). Enzymatic photochemical sensing using luciferase-immobilized polymer nanoparticles covered with artificial cell membrane, *J. Biometer. Sci. Polymer. Edn.*, Vol. 17: 1347-1357, ISSN 09205063

Lee Y.; Jablonski I.; DeLuca M. (1977). Immobilization of Firefly Luciferase on Glass Rods: Properties of the Immobilized Enzyme, *Anal. Biochem.*, Vol. 80: 496-501, ISSN 00032697

Matsumoto M. & Nishimura T. (1998). Mersenne Twister: A 623-dimensionally equidistributed uniform pseudorandom number generator, *ACM Trans. model. comput. simul.*, Vol. 8: 3-30, ISSN 10493301. URL: http://www.math.sci.hiroshima-u.ac.jp/ m-mat/MT/emt.html

Nakamura M.; Mie M.; Funabashi H.; Yamamoto K.; Ando J; Kobatake E. (2006). Cell-surface-localized ATP detection with immobilized firefly luciferase, *Anal. Biochem.*, Vol. 352: 61-67, ISSN 00032697

Nishiyama K.; Watanabe T.; Hoshina T.; Ohdomari I. (2008). The main factor of the decrease in activity of luciferase on the Si surface, *Chem. Phys. Lett.*, Vol. 453: 279-282, ISSN 00092614

Ribeiro A. R.; Santos R. M.; Rosário L. M.; Gil M. H. (1998). Immobilization of Luciferase from a Firefly Lantern Extract on Glass Strips as an Alternative Strategy for Luminescent Detection of ATP, *J. Biolumin. Chemilumin.*, Vol. 13: 371-378, ISSN 08843996

Stanley P. E. (1992). A survey of more than 90 Commercially Available Luminometers and Imaging Devices for Low-light Measurements of Chemiluminescence and Bioluminescence, Including Instruments for Manual, Automatic and Specialized Operation, for HPLC, LC, GLC and Microtitre Plates, *J. Biolumin. Chemilumin.*, Vol. 7: 77-108, ISSN 08843996

Tanaka. R.; Takahama E.; Iinuma M.; Ikeda T.; Kadoya Y.; Kuroda A. (2011). Bioluminescent Reaction by Immobilized Luciferase, *IEEJ Trans. EIS*, Vol. 131: 23-28, ISSN 03854221

Taniguchi K.; Nomura K.; Hata Y.; Nishimura Y.; Asami Y.; Kuroda A. (2007). The Si-tag for immobilizing proteins on a silica surface, *Biotechnol. Bioeng.*, Vol. 96: 1023-1029, ISSN 00063592

Tanii T.; Goto T.; Iida T.; Koh-Masahara M.; Ohdomari I. (2001). Fabrication of Adenosine Triphosphate-Molecule Recognition Chip by Means of Bioluminous Enzyme Luciferase, *Jpn. J. Appl. Phys.*, Vol. 40: L1135-L1137, ISSN 00214922

Tsuboi Y.; Furuhata Y.; Kitamura N. (2007). A sensor for adenosine triphosphate fabricated by laser-induced forward transfer of luciferase onto a poly(dimethylsiloxane) microchip, *Appl. surf. sci*, Vol. 253: 8422-8427, ISSN 01694332

Ugarova N. N.; Brovko L. Y. (2002). Protein structure and bioluminescent spectra for firefly bioluminescence, *Luminescence*, Vol. 17: 321-330, ISSN 15227235

Worsfold P. J.; Nabi A. (1986). Bioluminescent assays with immobilized firefly luciferase based on flow injection analysis, *Anal. Chim. acta*, Vol. 179: 307-313, ISSN 00032670

Yariv A. (1997). *Optical Electronics in Modern Commnications Fifth Edition*, Oxford University Press, ISBN 0-19-510626-1, New York.

Yotter R. A.; Lee L. A.; Wilson D. M. (2004). Sensor Technologies for Monitoring Metabolic Activity in Single Cells - Part I: Optical Methods, *IEEE Sensors Journal*, Vol. 4: 395-411, ISSN 00032670

20

Fiber Optics for Thermometry in Hyperthermia Therapy

Mario Francisco Jesús Cepeda Rubio,
Arturo Vera Hernández and Lorenzo Leija Salas
Centro de Investigación y de Estudios Avanzados del Instituto Politécnico Nacional, Department of Electrical Engineering/Bioelectronics
Mexico

1. Introduction

A review of electromagnetic hyperthermia ablative therapies, a temperature measurements during microwave ablation using optical fiber temperature sensors and a fiber optic temperature sensor developed for a hyperthermia laboratory are summarized in this chapter. Hyperthermia also called thermotherapy or thermal therapy is a type of cancer treatment in which body tissue is exposed to high temperatures. All methods of hyperthermia include transfer of heat into the body from an external energy source and are currently under study, including local, regional, and whole-body hyperthermia. The application of heat to treat patients with a malignant tumor is not a novel concept. The Edwin Smith papyrus explains the topical application of heated metallic implements or hot oil that were used approximately 5000 years ago to treat patients with tumors (Izzo et al., 2001). Local hyperthermia is used to heat a small area. It involves creating very high temperatures that destroy the cells that are heated. Research has shown that high temperatures can damage and kill cancer cells, usually with minimal injury to normal tissues (van der Zee, 2002). By killing cancer cells and damaging proteins and structures within cells, hyperthermia may shrink tumors (Hildebrandt et al., 2002). Numerous clinical trials have studied hyperthermia in combination with radiation therapy and/or chemotherapy. Many of these studies, have shown a significant reduction in tumor size when hyperthermia is combined with other treatments (Wust et al., 2002). Otherwise, ablation or high temperature hyperthermia, including lasers and the use of radiofrequency, microwaves, and high-intensity focused ultrasound, are gaining attention as an alternative to conventional surgical therapies (van Esser et al., 2007). Each of these techniques works differently, the objective is to heat tissue to a temperature 50°C above to destroy cells within a localized section of a malignant tumor. Frequently, temperature monitoring during thermal therapy has been achieved using interstitial thermocouples, thermistors, or fiber optics probes. Fiber optics technology should exploit one or more of electromagnetic immunity, noninvasiveness, chemical immunity, small size, or the capacity for distributed measurement. Fiber optic thermometers are used when electrical insulation and EMI immunity are necessary. The most relevant application involves tissue-heating control during microwave ablation (MWA) or radiofrequency ablation (RFA) hyperthermia therapy for cancer treatment (Mignani & Baldini, 1996).

1.1 Radiofrequency ablation

RFA works by converting radiofrequency waves into heat through ionic vibration. Alternating current passing from an electrode into the surrounding tissue causes ions to vibrate in an attempt to follow the change in the direction of the rapidly alternating current. It is the ionic friction that generates the heat within the tissue and not the electrode itself. The higher the current, the more vigorous the motion of the ions and the higher the temperature reached over a certain time, eventually leading to coagulation necrosis and cell death. The ability to efficiently and predictably create an ablation is based on the energy balance between the heat conduction of localized radiofrequency energy and the heat convection from the circulation of blood, lymph, or extra and intracellular fluid. The amount of radiofrequency produced heat, is directly related to the current density dropping precipitously away from the electrodes, thus resulting in lower periphery temperatures. The goal of radiofrequency ablation is to achieve local temperatures that are lethal to the targeted tissue. Generally, thermal damage to cells begins at 42°C; and once above 60°C, intracellular proteins are denatured, the lipid bilayer melts, and irreversible cell death occurs (Simon et al., 2005).

1.2 Microwave ablation

Water molecules are polar; the electric charges on the molecules are not symmetric. The orientation and the charges on the atoms are such that the oxygen side has a negative charge, and the hydrogen side of the molecule has a positive charge. When an oscillating electric charge from microwave radiation interacts with a water molecule, it causes the molecule to flip. Microwave radiation is specially tuned to the natural frequency of water molecules to maximize this interaction. As a result of the radiation hitting the molecules, the electrical charge on the water molecule flips back and forth 2-5 billion times a second depending on the frequency of the microwave energy. Temperature is a measure of how fast molecules move in a substance, and the vigorous movement of water molecules raises the temperature of water. Thus, electromagnetic microwaves heat matter by agitating water molecules in the surrounding tissue, producing friction and heat, thus inducing cellular death via coagulation necrosis (Goldberg & Gazelle, 2001).

2. Fiber optics thermometer

One of the first commercial fiber optic temperature probe was a fluorescence-based temperature sensor introduced in the early 1980s by the Luxtron Corporation of Mountain View, CA. Successors to these sensors are commercially available and are effective, approach to solving measurement problems (Culshaw, 1999). These include monitoring temperature patterns, examining temperature distribution in power transformer oils, and similar areas where the issue is the operation of a reasonably precise temperature probe within very high electromagnetic fields. In such circumstances, a metallic temperature sensor either distorts the electromagnetic field significantly or is subjected to very high levels of interference, producing spurious evaluations. Other applications sectors exploit the small size or chemical passivity of the device, including operation within corrosive solvents or examination of extremely localized phenomena such as laser heating or in determining the selectivity of radiation and hyperthermia treatments.

2.1 Fiber optics for thermometry in hyperthermia

Temperature probes that employ optical fibers were introduced for clinical hyperthermia treatment in the 1980s. Clini-therm Corporation, Dallas, TX founded in 1978 (no longer in business), at that time the company was working to develop and manufacture a system for hyperthermia therapy. The system incorporates fiber optic temperature probes which are inserted into a malignant tumor and surrounding healthy tissue through catheters. The probes detect changes in temperature during hyperthermia treatment.

(Vaguine et al., 1984) developed another probe. It uses a small crystal of the semiconductor gallium arsenide (GaAs) whose optical absorption at a specifically chosen wavelength is sensitively related to the crystal's temperature. The amount of optical signal returned after passage through the sensor may be detected and electronically translated, after calibration, into an indication of the probe's temperature. The system has up to 12 temperature sensors which are packaged in two basic probe configurations: a single-sensor probe with a length of 1.2 m and a diameter of 0.6 mm; and a four-sensor linear array probe with a length of 1.2 m, diameter of 1.1 mm, and spacing of 1.5 cm between adjacent sensors.

(Katzir et al., 1989) reported a fiber optic radiometer system, which is based on a nonmetallic, infrared fiber probe, which can operate either in contact or noncontact mode. In preliminary investigations, the radiometer worked in a strong microwave or radiofrequency field, with an accuracy of ± 0.5°C. The fiber optic thermometer was used to control the surface temperature of objects within ± 2°C.

(Wook et al., 2009) reported a measured temperature distribution using the infrared optical fibers during radiofrequency ablation. Infrared radiations generated from the water around inserted single-shaft plain electrode are transferred by the three silver halide optical fibers and are measured by a thermopile sensor array. Also, the output voltages of the thermopile sensor array are compared with those of the thermocouple recorder. It was concluded that is expected that a non-contact temperature sensor using the infrared optical fibers can be developed for the real time temperature monitoring during RFA treatments based on the results of this study.

Intelligent Optical Systems, Inc., has reported in 2009 an Optical Fiber Temperature (OFT) probe that will provide 10 points of high-resolution temperature data in a single fiber. The OFT probe is an alternative to thick bundles of arrayed single point temperature. According to the company, this new probe will provide a realistic, cost effective, and efficient technique for distributed temperature monitoring and profiling in treated tissues, and use of the OFT probe can reduce complications as a result of excessive heat exposure and damage to the healthy tissue surrounding a tumor. As microwave hyperthermia becomes increasingly popular for the treatment of benign prostatic hyperplasia and other benign or malignant tumors. The OFT provides multipoint and self-calibrating temperature sensing. Measures tumor temperatures at multiple points along a 5 to 10 cm sensor length, with the ability to pinpoint target areas within 0.5 cm. Monitors these temperatures over a large range, including 35°C to 55°C range, with a 0.1°C temperature resolution. Readily interfaces with clinical hyperthermia catheters because of its size (0.25mm thick) and thus will be adaptable to a variety of clinical conditions.

(Saxena & Hui, 2010) reported a polymer-coated fiber Bragg grating (PFBG) technology that provides a number of FBG thermometry locations along the length of a single optical fiber.

The PFBG probe was tested in an environment designed to approximate the microwave exposure that might be encountered during clinical hyperthermia treatments. According to the authors, this offers an enabling alternative to either scanning or bundled single point temperature probes for distributed thermometry in clinical applications.

3. Fiber optics thermometry system for a hyperthermia laboratory

In this section, we present a fiber optic temperature sensor developed for a hyperthermia laboratory. Many benefits coming from hyperthermia depends on the ability to induce and to measure high temperatures locally. The heating is carried out via local irradiation of energy, mainly by means of ultrasonic or electromagnetic waves. To be able to manage the irradiation parameters it is necessary to know the temperature to which the radiated tissue rises, that makes practically indispensable to have a highly reliable system of temperature measurement. The measurement of temperatures is carried out with thermometry systems that can be invasive, or non invasive. The use of the conventional sensors (thermistors, thermocouples, etc.) it is not satisfactory in some applications, just as it is the case of the therapies for hyperthermia with microwaves. This is due to that currents and voltages are induced by electromagnetic interference in the metallic elements and a self-heating by induction appears. Both factors produce erroneous readings as a result in the measurements when using these sensors. When these conditions are presented it is necessary to build sensors denominated, as non-disturbing electromagnetic field, like those based on optical fibers. Among them we find fluorescent sensors, interferometric sensors, those based on the variation of intensity caused by absorption or reflection and the evanescent sensors. At the present time, there are in the world diverse laboratories dedicated to the experimental investigation in hyperthermia, where the electromagnetic radiation is used inside a controlled and safety ambient. In the Section of Bioelectronics of the CINVESTAV-IPN there is an automated laboratory (Chong et al., 2000), in which controlled radiation experiments with microwaves are carried out in the interval from 900 MHz to 8 GHz and pulsating wave ultrasound at 1 MHz. This radiation has an impact on a biological tissue substitute material, where heating is obtained by the interaction of radiation with the radiated material. The experiments are carried out inside an anechoic chamber to avoid the external electromagnetic interferences and at the liberation of the energy radiation to the environment. The thermometry system that is described is part of this automated laboratory (Pennisi et al., 2002). This system measures the temperature inside the phantom, which is heated through the electromagnetic and ultrasonic fields radiations that are being carried out.

3.1 Methodology

The system can be divided in four parts: the stage of temperature measurement using optical fiber sensors and the analog signal conditioning; the stage of automated positioning of the temperature sensors inside the phantom; the communication and control stage using two personal computers; and a general software for the visualization of the temperature distribution inside the phantom. Figure 1 shows a block diagram of the system.

3.1.1 Temperature sensor

This model is an approach based on the theory of the weakly guiding fibers, and don't take into account the length of the sensor. However, we can see that the experimental results

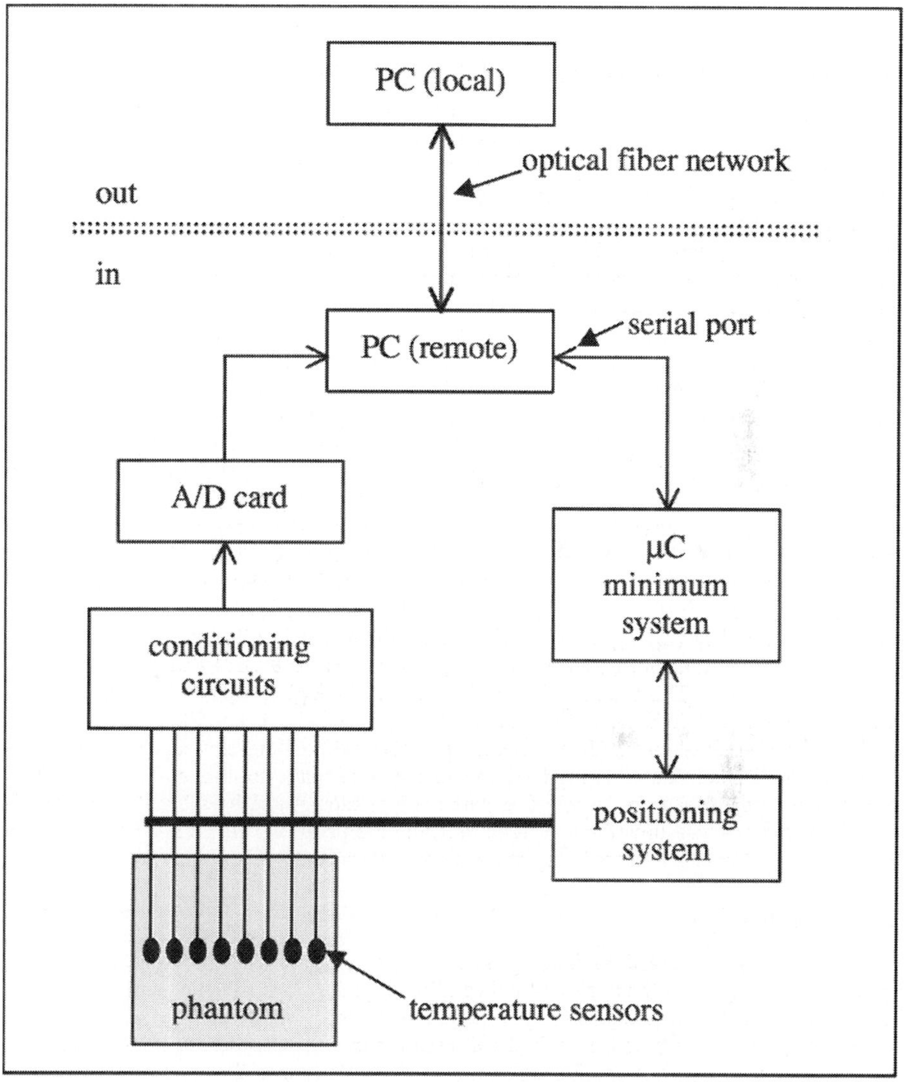

Fig. 1. Diagram of the hyperthermia automated laboratory. In the inner of the anechoic chamber, a PC controls the temperature sensors and the positioning system. A PC located outside the chamber for the visualization of the temperature distribution.

match with the response predicted by the model. Assuming that there is an allowable power flow in both core and cladding, the combined fiber power flow is:

$$P = P_1 + P_2 \text{ (Watts)} \tag{1}$$

where the subscripts 1 and 2 refer to the core and cladding respectively. The ratio of optical powers in core and cladding is:

$$\frac{P_1}{P_2} = \frac{3}{4} N_m^{1/2} - 1 \tag{2}$$

where N_m, represents the total number of free-space modes accepted and transmitted by a step-index fiber. The number of modes is related with the so-called v value by the relationship:

$$N_m = \frac{1}{2} v^2 \tag{3}$$

The v value, sometimes referred to the normalized frequency, is an important fundamental fiber property, because it contains the most important fiber parameters, and is defined by:

$$v = 2\pi \left(a_1 / \lambda \right) \left(n_1^2 - n_2^2 \right)^{1/2} \tag{4}$$

where a_1 is the fiber core radius, λ is the free space wavelength of operation and n is the refractive index. The normalized power at the end of the fiber is then:

$$P_{norm} = 1 - \frac{P_2}{P} = 1 - \frac{2\sqrt{2}}{3\pi \left(a_1 / \lambda \right) \left(n^2{}_1 - n_2 (T)^2 \right)^{1/2}} \tag{5}$$

From the latter expression, we can see that the power transmitted along the fiber is dependent on the temperature, with an inverse square root relationship. Due to the refraction index wavelength dependence, we have evaluated three different wavelengths for the sensor operation (660 nm, 850 nm and 1300 nm), as well as several different types of oils, with a slightly different refraction index. For our range of interest the best results we have obtained were with vegetable oils, operating in the infrared range. The sensor was fabricated using a multimode patch-cord with ST connectors on both ends, which is commercially available. First, the patch-cord is divided in halves. A 30 mm portion of jacket and plastic buffers are mechanically removed from both ends. After that, a 20 mm glass tube with an external diameter of 1.1 mm is introduced through one of the ends, and both ends of the fiber are joined together with a fusion splicer (RXS-X74). Then the fiber without buffer of this portion is treated with an acid, to remove partially the glass cladding. The acid attack is stopped when the transmitted power falls 30% from its initial value. The glass tube is then located in the central part of the fiber and is filled with the oil. Finally, both ends of the glass tube are sealed with epoxy glue. Figure 2 shows a detailed outline of the probe described above.

The basic setup is showed on Figure 3. There is a lightwave multimeter (HP 8153A) configured with an optical source module (Hp8155IMM) and a power sensor module (HP

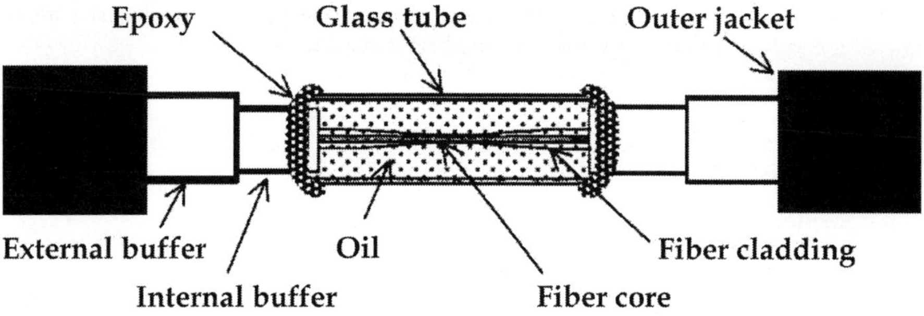

Fig. 2. Detailed outline of the constructed sensor (not in scale).

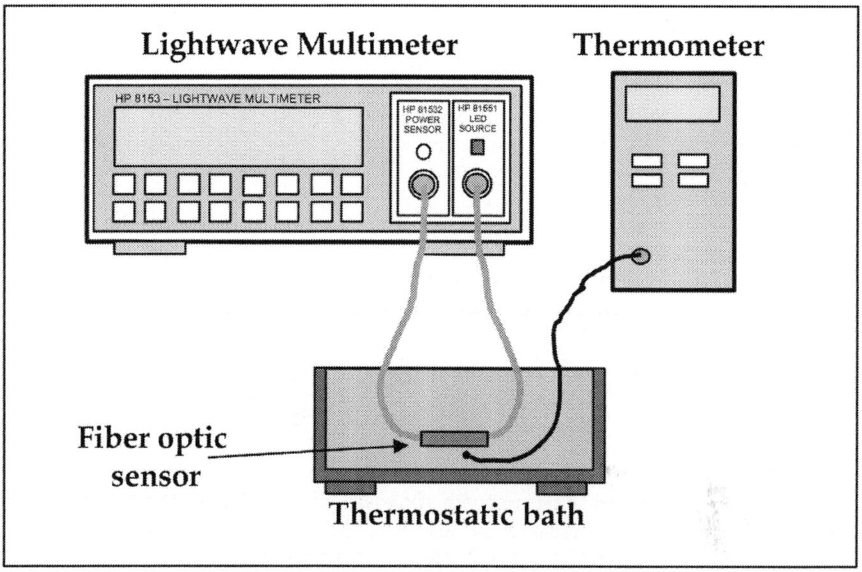

Fig. 3. Experimental setup.

81532A). The stability of the source was evaluated, and was +0.03 dB in a 6 h interval. The fiber optic sensor is connected between the source and the power sensor, and is immersed in a thermostatic bath. The wavelength of operation selected is 850 nm.

3.1.2 Positioning stage

This stage is constituted by a personal computer (PC), a microcontroller and a PC controlled automated positioning system. The PC is IBM compatible. It sends the data that correspond to a position selected by the user toward the minimum system. This PC is located inside the anechoic chamber and it is externally operated through an optical fiber communication system. The communication between the PC and the minimum system is carried out through the communications port COM1. The temperature sensors are displaced in an axis through the phantom by means of an automated positioning system. Two DC motors controlled by the minimum system by means of the feedback signal that provides an optical encoder impel this. The maximum lineal travel distance is of 35 cm, with a resolution of 2.5 mm. The physical construction of the positioning system is built on a square base of acrylic of 65 x 120 cm of external area, of 1 cm thick. Bars of Nylamid were also used for support. The used materials minimize the interaction with the electromagnetic fields that could be present.

3.1.3 Communication and control stage

The communication system is based on the use of two IBM compatible computers, one of them inside the anechoic chamber (remote) and the other one in the exterior (local). The communication between both computers was carried out through an optical fiber cable; the local PC operates the remote PC. The local PC controls all the equipment inside the chamber and captures all the acquired information. An optical fiber based net card was used to communicate the two computers (3C905B-FX(SC), 3Com 100Base-FX). It possesses a speed of transmission of 100 Mbps. The protocol used to communicate both computers was the TCP/IP. The package used for the communication was the Remote Administrator v2.0, which allows control all the devices and programs of the remote computer from the local computer. The analog signals coming from the eight conditioning circuits were digitized by means of a 12 bits analog to digital card (Lab PC-1200 A/I, National Instruments) that is installed in the remote computer. The communication system is based on the use of two IBM compatible computers, one of them inside the anechoic chamber (remote) and the other one in the exterior (local). The communication between both computers was carried out through an optical fiber cable; the local PC operates the remote PC. The local PC controls all the equipment inside the chamber and captures all the acquired information.

3.1.4 Visualization software

Software was developed in the LabWindows CVI programming language (National Instruments). By means of this software the user can control all the controllable devices that conforms the automated laboratory. Data like the radiation time and frequency, the distance between the radiator and the phantom, the quantity of temperature measurements in a plane of the phantom, the quantity of phantom mappings, sensor thermal response time, the elapsed time between each temperature mapping, among other, should be provided by the user. Once the experiment begins, the laboratory operates automatically and allows the user to visualize the results obtained in each temperature mapping. In Figure 4 the main screen of the program is shown.

3.2 Results

Calibration of the sensor was done using a thermostatic bath and a thermocouple thermometer with an accuracy of 0.1"C (TES-1310). Several calibration curves in different days were taken for the sensor, as shown in Figure 5. With these curves we evaluated the repeatability and the stability of the sensor. We also evaluated the thermal time constant of the sensor, resulting in a value of 1.9 s. As we see, the transmitted power increases with the temperature. While the temperature increases, the refraction index of the cladding decreases in a linear way. At low temperatures (below 5"C), the refraction index of the oil is higher than the refraction index of the core, and no power is transmitted through the fiber, because

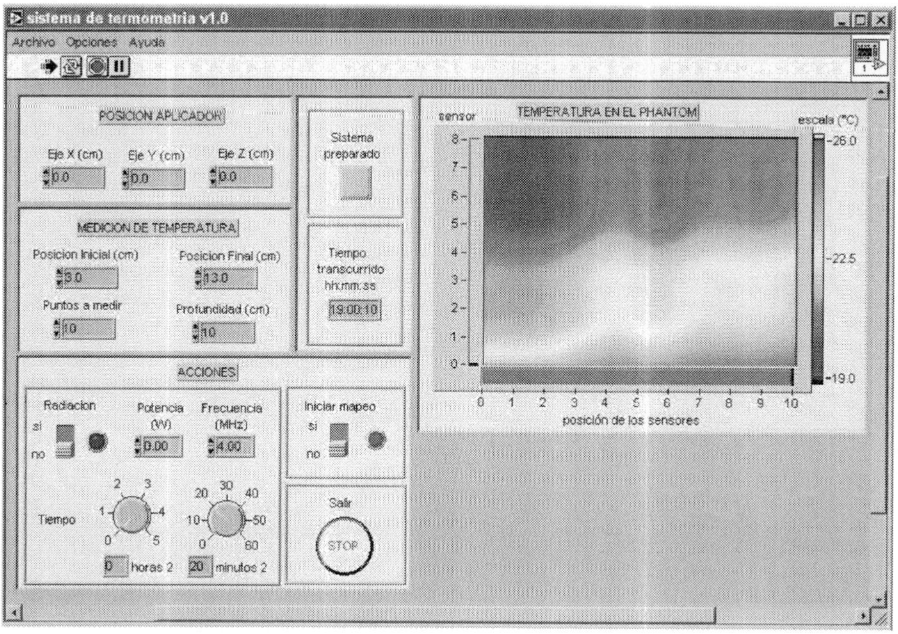

Fig. 4. Software main screen.

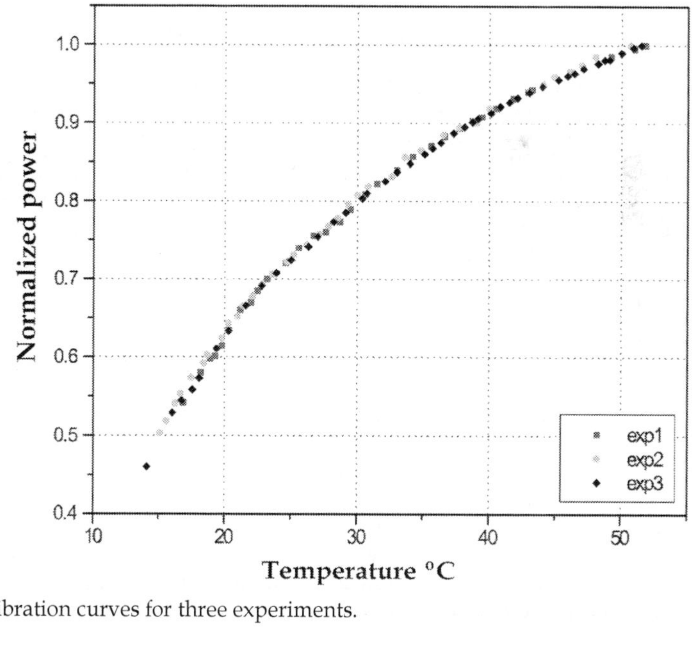

Fig. 5. Calibration curves for three experiments.

waveguide conditions are not satisfied. As the refraction index decreases the power transmitted increases, due to the increasing number of guided modes in the fiber. When the oil index equals the refraction index of the cladding, the transmitted power reaches a maximum value, and the curve shows a saturation effect. The sensor performance using microwaves was tested using a domestic microwave oven that operates at 2.45 GHz. The experimental setup was similar to the previously described. Instead the thermostatic bath, we have used the microwave oven, to heat water inside a plastic container. Figure 6 shows the achieved results, compared with a calibration curve obtained with the thermostatic bath.

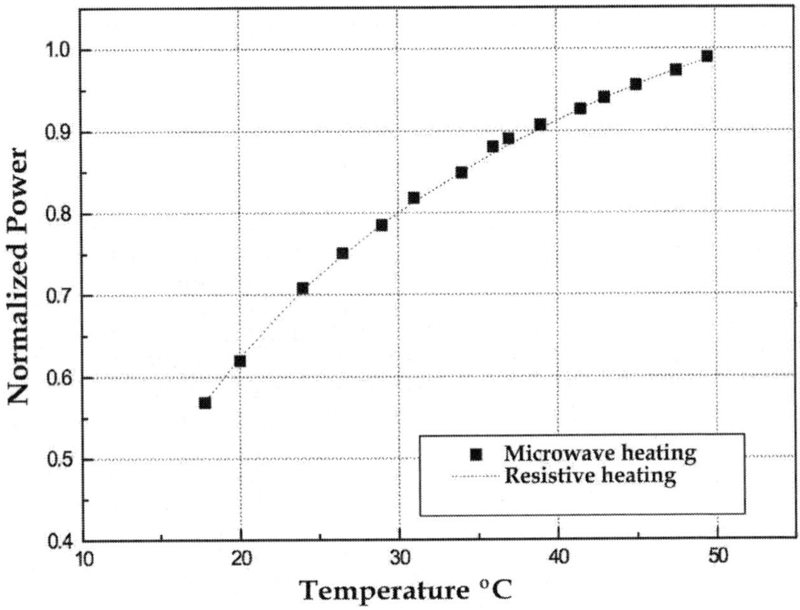

Fig. 6. Calibration curves for three experiments.

Figure 7 shows the fiber optic temperature sensor constructed for the hyperthermia laboratory.

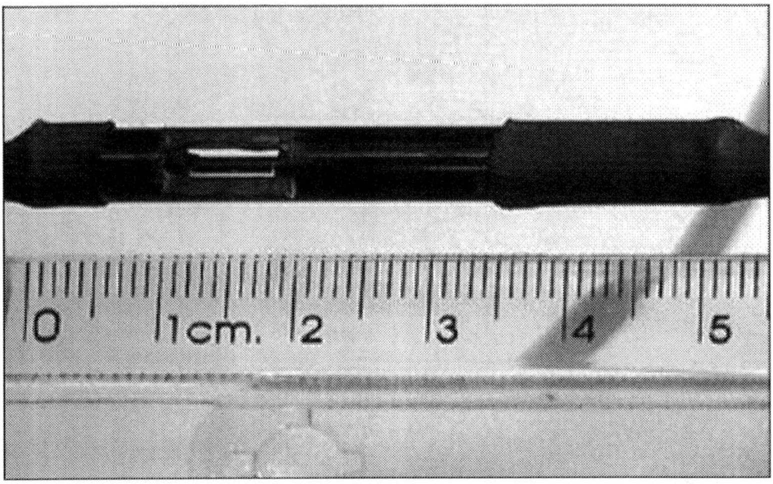

Fig. 7. Fiber optic temperature sensor constructed.

3.3 Discussion

A reasonable working range for the thermometry system is between 20°C to 35°C, as the phantom is heated about 3°C to 5°C from ambient temperature. According to the required precision, the thermometer should be accurate to within ±0.2°C at the: time of calibration. Additionally, a low thermal time constant and a low drift are required. The results obtained with the developed sensor show an appropriate performance for the selected application. The sensor was evaluated using microwaves to heat the bath in the operational interval of temperature. In this situation, we found that the sensor response still remains between the confidence band. The observed non-linearity is not an important issue, since we think to digitize the readings of multiple sensors with a personal computer. That implies the use of a calibration table to obtain the temperature values relative to the optical power detected. The reported results were obtained by means of optical instruments, but the idea is to use specifically designed circuits. For this reason, we are developing an optical source and optical detector circuits, to construct a thermometer with the presented sensor. As preliminary results, we have an optical source with an excellent stability (±0.0005 dB), thanks to the use of an optic feedback technique. That grants the required long term drift requirements of the thermometer. The optical components that we are using are the pair OPF1414-OPF2414 (Optek Inc.), which operate at a wavelength of 840 nm and possess ST ports for direct connection of the sensor. The procedure of construction of the sensor is simple and the required materials are relatively inexpensive. For example, the used fiber is a standard telecommunications grade patch-cord. The sensor could be used in other applications involving temperature measurements under strong electromagnetic fields, such as power transformers, microwave antennas, etc. The measurement range could be changed simply using oil with an appropriate refraction index. One of the drawbacks is the limited range within the sensor can make the measurements.

4. Fiber optics thermometry in microwave ablation

In this section, we present an experimental setup and results of temperature measurements during MWA using fiber optics thermometry in *ex vivo* tissue (Cepeda et al., 2011). Luxtron STB MAR'05 fiberoptic thermal probes were placed 5mm above the antenna slots longitudinally to measure real-time temperature during high hyperthermia experiment (Figure 8). The fiberoptic thermal probes are connected to a Luxtron 3300 Fluoroptic thermometer, which monitors the temperature and saves the data to a personal computer via a RS-232 serial cable.

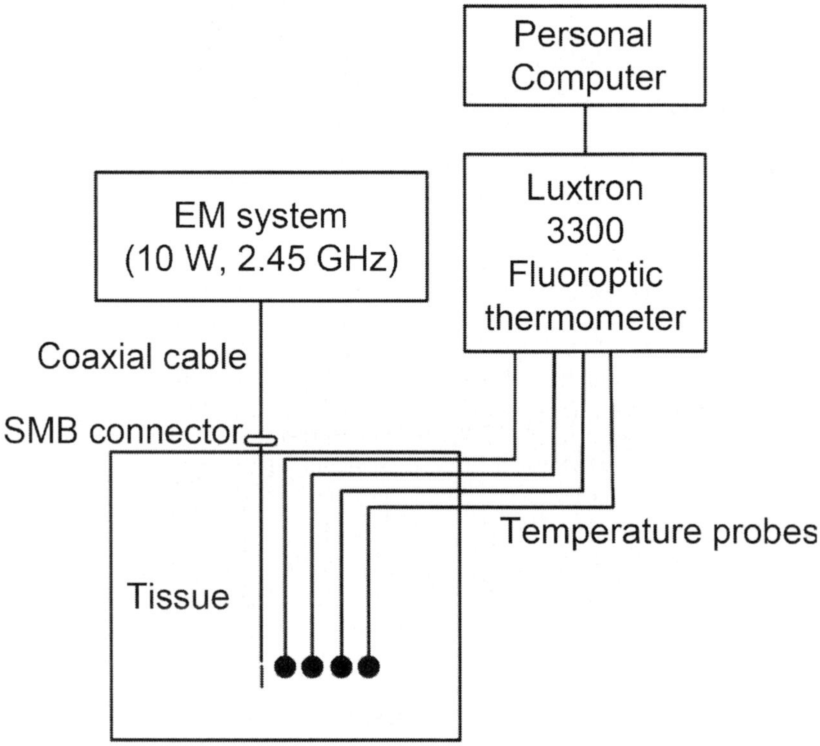

Fig. 8. Schematic of equipment and experimental setup. The antenna was completely inserted into homogeneous muscle tissue and connected to the EM system. Temperature sensors were 5 mm above the antenna slots longitudinally.

High temperature hyperthermia experiments were performed at 10 W for 3 min. The initial tissue temperature was 18 to 19 °C. Figure 9 shows the tissue with an ablated zone.

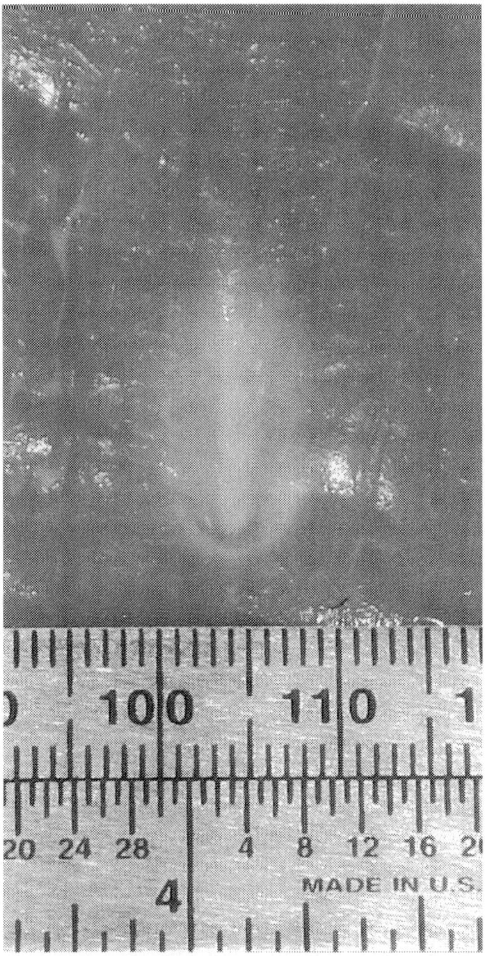

Fig. 9. A lesion created in an ex-vivo experiment at the end of a 180 s hyperthermia procedure.

The measured thermal histories at the specified locations during high hyperthermia therapy were shown in Figure 10. A fluoroptic thermometer was selected because its fiber optic temperature sensors have minimal disturbance on the antenna Specific Absorption Rate (SAR) and are unaffected by microwave radiation. Figure 11 shows a photo of *ex vivo* swine breast tissue temperature measurement, using one fiber optic during MWA.

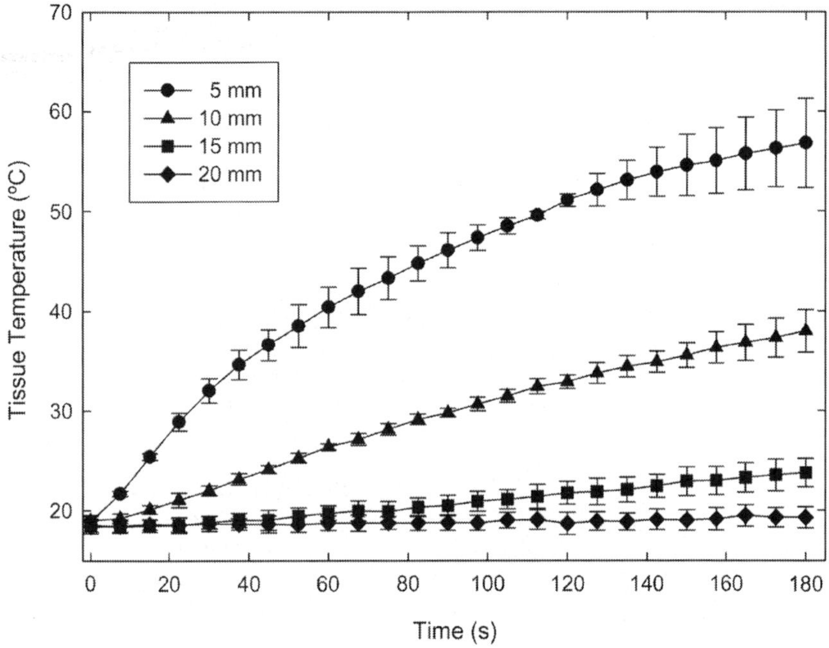

Fig. 10. A lesion created in an ex-vivo experiment at the end of a 180 s hyperthermia procedure.

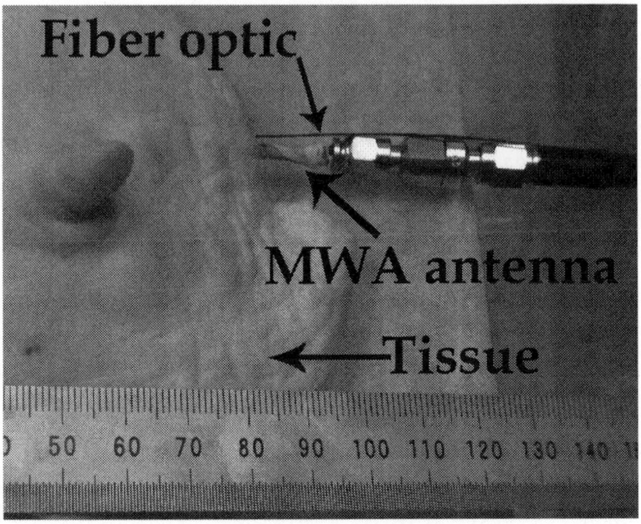

Fig. 11. Ex vivo temperature measurement in swine breast tissue.

5. Conclusion

All methods of hyperthermia involve transfer of heat into the body from an external energy source and are currently under study, including local, regional, and whole-body hyperthermia. Local hyperthermia is used to heat a small area. It involves creating very high temperatures that destroy the cells that are heated. Microwave ablation, like radiofrequency ablation, uses heating to cause tissue necrosis. RFA heats the tissue by electrical resistive heating but MWA works on a different principle. An antenna emits microwave radiation into the tissue; this results in excitation and oscillation of polar molecules and causes frictional heating. Frequently, temperature monitoring during hyperthermia cancer therapy has been achieved using interstitial thermocouples, thermistors, or fiber optic probes. The thermocouples and other conventional probes possess numerous advantages that make them attractive to be used in the characterization of temperature in hyperthermia. Among these advantages are a small size, accuracy, excellent reliability and low cost. Furthermore, this technology also possesses disadvantages, like metallic conductive components, shielding and plug wires. These metallic parts are often the reason of considerable errors in measurement of temperature when the probes are used for characterizing equipment of hyperthermia in phantoms exposed to electromagnetic radiations. These errors of measurement are caused by three phenomena, which can be presented alone or acting in combination: sensor heating due to the induced currents, disturbance of the electromagnetic field and electromagnetic interference. Thermometers based on optical fibers offer the advantage of not possessing metallic components, and therefore they do not to disturb the electromagnetic field. Fiber optic thermometers are used when electrical insulation and electromagnetic immunity are necessary. The most relevant application includes tissue-heating temperature control during radiofrequency or microwave hyperthermia cancer treatment.

6. References

Cepeda, M., Vera, A., Leija, L., Avila-Navarro, E. & Navarro, E. (2011). Coaxial Slot Antenna Design for Microwave Hyperthermia using Finite-Difference Time-Domain and Finite Element Method. *The Open Nanomedicine Journal,* Vol. 3, No. 1, (2011), pp. (2-9), ISSN 1875-9335

Chong, J., Leija, L., Posada, R., Fonseca, W. & Vera, A. (2000). Design and construction of an automated laboratory for electromagnetic and ultrasonic radiation in the study of hyperthermia in biological systems, *Proceedings of the 2000 Annual Fall Meeting of the Biomedical Engineering Society,* Seattle, WA, October 2000

Culshaw, B. (1999). Fiber-Optic Thermometers, In: *Measurement, Instrumentation, and Sensors Handbook,* John G. Webster, 32.11, CRC Press LLC, ISBN 978-0-8493-2145-0, USA

Goldberg, S. & Gazelle, D (2001). Radiofrequency tissue ablation: physical principles and techniques for increasing coagulation necrosis. *Hepatogastroenterology,* Vol. 48, No. 38, (2001), pp. (359-367), ISSN 0172-6390

Hildebrandt, B., Wust, P., Ahlers, O., Dieing, A., Sreenivasa, G., Kerner, T., Felix, R. & Riess H. (2002). The cellular and molecular basis of hyperthermia. *Critical Reviews in Oncology/Hematology,* Vol. 43, No. 1, (July 2002), pp. (33-56), ISSN 1040-8428

Izzo, F., Thomas, R., Delrio, P., Rinaldo, M., Vallone, P., DeChiara, A., Botti, G., D'Aiuto, G., Cortino, P. & Curley, S. (2001). Radiofrequency ablation in patients with primary breast carcinoma: a pilot study in 26 patients. *Cancer,* Vol. 92, No. 8, (October 2001), pp. (2036-2044), ISSN 1097-0142

Katzir, A., Bowman, H., Asfour, Y., Zur, A., & Valeri, C. (1989). Infrared fibers for radiometer thermometry in hypothermia and hyperthermia treatment. *IEEE Transactions on Biomedical Engineering,* Vol. 36, No. 6, (June 1989), pp. (634-637), ISSN 0018-9294

Mignani, A. & Baldini, F. (1996). Biomedical sensors using optical fibres. *Reports on Progress in Physics,* Vol. 59, No. 1, (January 1996), pp. (1-28), ISSN 0034-4585

Pennisi, C., Leija, L., Fonseca, W. & Vera, A. (2002). Fiber optic temperature sensor for use in experimental microwave hyperthermia, *Sensors, 2002. Proceedings of IEEE,* ISBN 0-7803-7454-1, Orlando, Florida, USA, June 2002

Saxena, I. & Hui, K. (2010). Polymer coated fiber Bragg grating thermometry for microwave hyperthermia. *Medical physics,* Vol. 37, No. 9, (September 2010), pp. (4615-4619), ISSN 0094-2405

Simon, C., Dupuy, D. & Mayo-Smith, W. (2005). Microwave Ablation: Principles and Applications. *RadioGraphics,* Vol. 25, No. suppl 1, (October 2005), pp. (S69-S83), ISSN 0271-5333

Vaguine, V., Christensen, D., Lindley, J. & Waltson, T. (1984). Multiple Sensor Optical Thermometry System for Application in Clinical Hyperthermia. *IEEE Transactions on Biomedical Engineering,* Vol. 31, No. 1, (January 1984), pp. (168-172), ISSN 0018-9294

van der Zee, J. (2002). Heating the patient: a promising approach?. *Annals of Oncology,* Vol. 13, No. 8, (August 2002), pp. (1173-1184), ISSN 1569-8041

van Esser, S., van den Bosch, M., van Diest, P., Mall, W., Borel Rinkes, I. & van Hillegersberg, R. (2007). Minimally Invasive Ablative Therapies for Invasive Breast Carcinomas: An Overview of Current Literature. *World Journal of Surgery,* Vol. 31, No. 12, (December 2007), pp. (2284-2292), ISSN 0364-2313

Wook, J., Jeong, K., Dong, H., Kyoung, W., Ji, Y., Jae, H., & Bongsoo L. (2009). Infrared Fiber-Optic Temperature Array Sensor for Radiofrequency Ablation, *Proceedings of Biomedical Engineering and Informatics, 2009. BMEI '09. 2nd International Conference on,* ISBN 978-1-4244-4132-7, Tianjin, October 2009

Wust, P., Hildebrandt, B., Sreenivasa, G., Rau, B., Gellermann, Riess H., Felix, R. & Schlag P. (2002). Hyperthermia in combined treatment of cancer. *The Lancet Oncology,* Vol. 3, No. 8, (August 2002), pp. (487-497), ISSN 1470-2045

21

White Light Sensing Systems for High Voltage Measuring Using Electro-Optical Modulators as Sensor and Recovery Interferometers

Josemir C. Santos[1], José C. J. Almeida[2] and Luiz P. C. Silva[1]
[1]*University of São Paulo*
[2]*Optsensys*
Brazil

1. Introduction

There are several known issues related to the use of electro-optical sensors as measuring devices. In applications intended for use in measuring currents and potentials in high voltage electric power systems, optical sensor systems must attend severe specifications, such as: large withstand voltage of the primary sensor, high accuracy (better than 0,5%, in case of revenue metering), long term stability of scale factor, large dynamic range (>50 dB, in some applications for protection purpose) and relatively wide frequency response range (of several kHz, in case that harmonics content evaluation is needed). However, the potential advantages arising from the use of optical fiber sensors as instrument transformers (ITs) in substitution to conventional electromagnetic transformers, may explain the intensive efforts that have been done during the last 30 years to develop optical ITs.

In a previous work (Santos, 2009), focused on the development of electro-optical sensors applied to direct measuring high voltages, in order to fulfill the requirements for ITs previously described, various sensing systems linked by optical fibers were presented. The progress in this field is commonly divided into three main approaches.

In the first approach, conventional transformers connected to primary electronic converters and communicated by means of optical fiber links to secondary converters have been used to improve the EMI toughness and to provide complete electrical isolation between the measured and the measuring systems. Typical examples of such approach (Crotti et al., 2006; Rahmatian, 2010) describe conventional ITs connected to primary electronic converters that generate modulated optical signals and provide their adequate coupling to optical fiber communication links.

In the second aproach, non-conventional transformers, also called as electronic transformers, have been developed using different types of sensors or transducers as measuring devices that are connected by optical fiber links to secondary converters. In representative examples of this approach (Bull et al., 2005; Mariscotti, 2009) Rogowski coils are used as sensor element.

The third approach consists in using optical sensors to directly measuring of voltage or current in high voltage systems. Passive optical sensors, such as electro-optical and magneto-optical modulators, are attractive for such purpose because all their components can be made with dielectric materials and have small dimensions. Since these sensors, generally, show low sensitivity to EMI and wide frequency response, they usually have much superior performance characteristics than the conventional transformers, and are capable to make better use of potentially excellent quality of the optical fiber links.

The linear electro-optic effect, also called Pockels effect, has been largely used to build electro-optical modulators that have many applications in communications, signal processing and sensing systems. These modulators can be used as sensors for measuring of electric field, current and voltage, among other quantities. The optical sensors used as ITs for measuring voltage and current are called Optical Voltage Transformers (OVTs) and Optical Current Transformers (OCTs), respectively.

The main problem in trying to apply electro-optical sensors to perform direct measuring of potentials at high voltage levels is their sensitivity to applied voltage, commonly too high in comparison with the voltage to be measured. Since in typical Pockels modulators the transmitted light intensity has a sinusoidal dependency with applied modulation voltage, as shown later in this chapter, there is a limit for holding a univoque relationship between the transmitted light intensity and the applied voltage, imposing a restriction in maximum value of applied voltage that can be actually measured.

A solution that has been adopted to allow measuring of higher voltages relies in using capacitive or resistive potential dividers for obtaining a sample of the high voltage to be measured and to apply this low voltage signal to the optical sensor. This solution limits the performance of the OVT to that of the potential divider, which is not always satisfactory. The most favorable solution, however, should be the direct application of high voltages to a properly designed Pockels modulator, avoiding the use of dividers and, therefore, making the best use of superior characteristics of optical sensors. To reach this goal, the Pockels modulator should be built in an appropriate configuration, in order to increase its dielectric withstand and reduce its sensitivity to suitable values. Several alternatives have been proposed to build optical fiber sensors appropriate for high voltage measuring systems, but a small number of them could lead to practical devices able to be used in field applications.

Another approach to solve the posed problem is the application of a different interrogation technique, instead of the polarimetric intensity modulation normally used in Pockels modulators, capable to extend the dynamic range of the sensor to a sufficiently high value.

Recently, the White Light Interferometry (WLI) technique has been used as an alternative to implement optical fiber sensor systems that can meet severe specifications, similar to the requirements of applications as in ITs. White light interferometric systems based on interferometers linked by optical fibers have been largely used for implementation of various high performance optical sensors on trial basis. The design of a specific sensor depends upon the particular requirements of its application as well as the physical quantity to be measured, which may be, for example, displacement, strain, temperature, etc. In such a kind of WLI sensor systems, a broadband light source is used to illuminate two interferometers that are connected in series by an optical fiber link. The application of this method can add a large dynamic range to the high accuracy, which is a typical advantageous characteristic of conventional interferometric sensor systems.

Following this approach, it is shown in this chapter how WLI technique can be applied to a Pockels voltage sensor, enhancing its performance while keeping its linearity and accuracy.

At first, the general concepts related to WLI are presented.

Next, it is introduced an application, originally proposed by the first author, in which the WLI is used together with two electro-optical modulators, based on the Pockels effect, for measuring high voltages directly. Since this approach has no moving parts, the frequency bandwidth achievable can be much higher than any mechanically or electronically scanned WLI system previously proposed.

Finally, practical realizations of ITs based on WLI sensor systems using electro-optical modulators as sensor and recovery interferometers are presented. Special attention is given to electronic signal processing techniques used to demodulate the optical signal available at the output of the sensor system. Two different techniques are depicted; both capable to perform measuring with compensation for fluctuations on optical average power delivered at the output of sensor system. Such fluctuations normally arise from variations in attenuations or variations in the output power of optical source, for example. Experimental investigations carried out with the sensor systems developed are presented, including experimental results obtained with both electronic signal processing techniques. At the end of chapter, a brief discussion on the future perspectives and under developing initiatives related to this field is addressed.

2. Principles of White Light Interferometry (WLI) sensing

In typical interferometric optical sensors, highly monochromatic (or long coherence length) light sources are commonly used, therefore the unambiguous measurement range is limited to one period of the interferometer transfer function and the basic unit of measurement is the source wavelength.

By the other hand, white light interferometry has, potentially, the capability of identifying the interference fringe order from the output fringe pattern by using a broadband optical source (Chen, 1992). As a result, the unambiguous range of the output signal is no longer limited to within half of a fringe, and an absolute phase measurement over a large operating range can be achieved. The output intensity pattern of a white light interferometer has a visibility profile determined by the low-coherence property of the broadband source used.

Since the output of an interferometer is, in theory, the Fourier transform of the source spectrum (Françon, 1967) and the Fourier transform a Gaussian function is also a Gaussian function, it follows that the normalized output intensity pattern of a white light interferometer is a cosine function modified by a Gaussian visibility profile, as shown in Fig. 1. This output intensity pattern is commonly also called as fringe pattern in optical engineering field.

The principle of WLI sensing may be explained by using Fig. 2, which shows a block diagram of a generic WLI sensor system.

In Fig. 2 it is shown that light emitted from a spectrally broadband source, for example a superluminescent diode (SLD), is coupled via a beam splitter into an optical fiber link connected to the sensor interferometer. The optical path difference (OPD) introduced by the

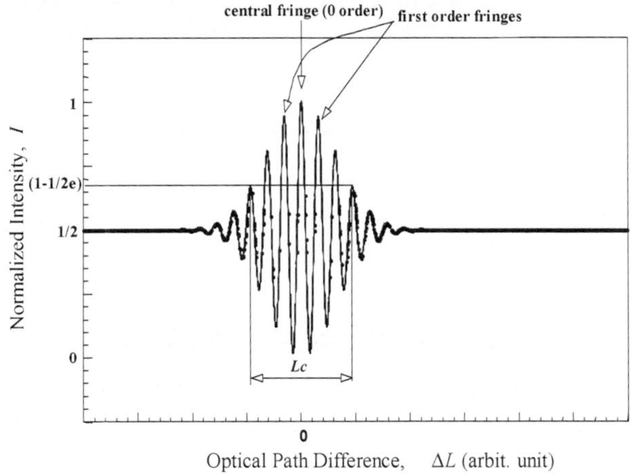

Fig. 1. Theoretical output intensity pattern of a white light interferometer.

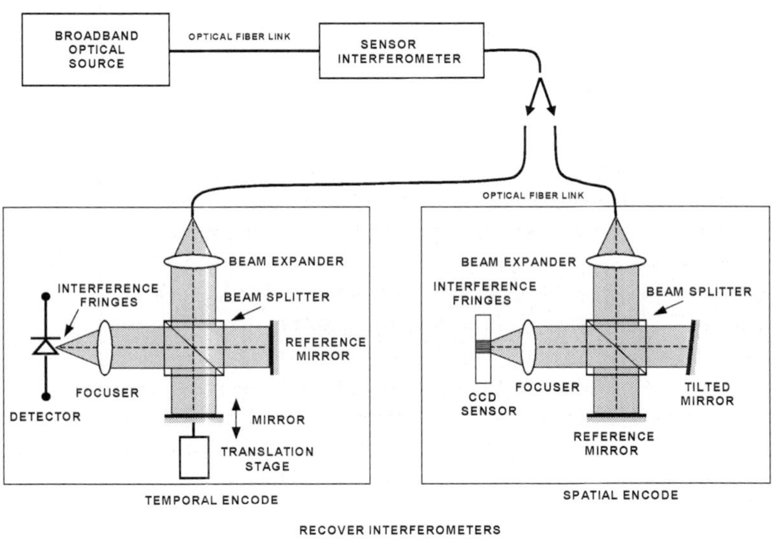

Fig. 2. Principle of operation of a white light interferometric sensor.

sensor interferometer, $\Delta L_s(X)$, is linearly sensitive to the external measurand X. If the OPD introduced by the sensing interferometer is greater than the coherence length of the source, L_c, no interference is observed at the output light of the sensor interferometer. When the output of sensor interferometer is linked to the input of a second interferometer, called as "recovery" interferometer, scanning of the path difference introduced by this interferometer, $\Delta L_r(Y)$, results in interference being observed at its output when $\Delta L_r(Y)$ approaches to $\Delta L_s(X)$.

Neglecting the spectral dependency of OPD, the behavior of light intensity at the output of sensor interferometer, I_s, as a function of light wavenumber, σ, can be described by the following relation (Lequine, 1990):

$$I_s(\sigma) = T_1\, T_s\, I_o(\sigma)\left\{1 \pm K_s \cos\left[2\pi\sigma\Delta L_s(X)\right]\right\} \tag{1}$$

where T_1 is the transmission factor of the forward fiber link, T_s and K_s are the transmission factor and the visibility of sensor interferometer, respectively, and $I_0(\sigma)$ is the spectral intensity distribution of the light source. Analogously, the behavior of the recovery interferometer can be written as:

$$I_r(\sigma) = T_2\, T_r\, I_s(\sigma)\left\{1 \pm K_r \cos\left[2\pi\sigma\Delta L_r(Y)\right]\right\} \tag{2}$$

where T_2 is the transmission factor of return fiber link and T_r and K_r are the transmission factor and the visibility of recovery interferometer, respectively. The OPD introduced by the recovery interferometer, $\Delta L_r(Y)$, is linearly sensitive to scanning quantity Y. The total intensity available for detection at the output of WLI system is calculated by integrating the previous equation over all the wavenumbers, and, if the source has a Gaussian spectral distribution, this integral becomes:

$$I = T_1 T_2\, T_s T_r\, I_0 \left\{ 1 + K_{0s}\, K_{0r}\, e^{-\left[\pi \frac{(\Delta L_s - \Delta L_r)}{L_c}\right]^2} \cos\left[2\pi \frac{(\Delta L_s - \Delta L_r)}{\lambda_0}\right] \right\} \tag{3}$$

The resulting intensity signal, I, appears in eq. (3) as a cosine function modified by a visibility function, both dependent of the OPD difference, exactly as shown in Fig. 1. Both eq. (3) and Fig. 1 show that, at the matching point, the interferometer output is at the center of the brightest fringe. The order of this fringe is zero and it is called *central fringe*. The fringe-order identification is dependent on the identification of central fringe.

To achieve a precise measurement in a large dynamic range, the sensor system must, at first, correctly identify the central fringe and then measure accurately the OPD introduced by the recovery interferometer, $\Delta L_r(Y)$, necessary to lead the system to the matching point, i.e., the center of central fringe. There are several proposed methods to reach this aim, using different topologies and signal processing techniques. The two most commonly used OPD scanning techniques, the temporal encode and the spatial encode, are illustrated in Fig. 2.

Temporal encode techniques make use of mechanically tunable OPD interferometers to scan the fringe pattern and record it for further analysis. This method allows high accuracy and low signal-to-noise ratio (SNR) level, but it is normally very slow, since it uses moving parts.

In a system using electronically scanned (spatial encode) WLI, the interference fringes pattern is produced by superposing two expanded light beams. The resulting pattern is imaged over a CCD array. This method is much faster than the temporal encoding, but the fringe visibility is reduced due to spatial coherence mismatches and intrinsic noises of beam profile and CCD array (Ning et al., 1995).

In both cases, the signal processing techniques normally include digital treatment of the signal and the computation time may limit the response frequency bandwidth of the sensor system. Besides these two techniques, other electronically scanned methods using dual wavelength light sources and no moving parts can be applied to WLI sensors.

In this chapter a new approach for temporal encoded WLI sensor systems with no moving parts is proposed by using a Pockels electro-optic modulator as recovery interferometer. In such approach, the frequency bandwidth of the system can be significantly enhanced while keeping a large measuring range of operation.

3. WLI method applied to fiber optical linked electro-optical sensor systems

Recently, several WLI sensor systems have been proposed to measure different physical quantities (measurands). In Table 1 some examples are listed and information are given about the measurand, the encoding technique, the bandwidth and the dynamic range for each sensor.

Measurand	Encoding Technique	Bandwidth	Dynamic Range	Reference
Temperature	Temporal (fiber stretch)	30 Hz	20 ~ 70 °C	(Kim, 2008)
Position	Temporal (moving mirror)	< 1 Hz	0 ~ 7.2 mm	(Chen, 2010)
Displacement	Temporal (moving mirror)	~20 Hz	0 ~ 1050 nm	(Liao, 2010)
Micro-Strain	Temporal (moving lens)	1 Hz	0 ~ 350 με	(Velosa, 2011)
Temperature	Temporal (moving lens)	> 60 Hz	35 ~ 75 °C	(Yu, 2005)
Displacement	Spatial (electronically scanned)	~500 Hz	0 ~ 20 μm	(Murtaza, 1991)

Table 1. Examples of WLI sensor systems

From table 1, it is possible to observe that typical WLI sensor systems presented show dynamic ranges that can be considered larges, when compared with the ones achievable by normal interferometric systems in same conditions. However, in all cases, relatively narrow bandwidths are observed, preventing the use of mentioned systems in applications intended for measurement of fast varying measurands.

In alternate current (ac) electrical power systems the main electrical measurands of interest are voltage and current, and the lowest nominal frequency of these quantities is 50 Hz. Therefore, only the example shown in last line of table 1 could be applied to measure such quantities with significant information about their harmonic content. Furthermore, larger bandwidth may be required from a sensor system to be applicable as IT for protection purposes, since they must to respond to transitory phenomena, such as short circuits or lightning and switching impulses, which superpose overvoltage and overcurrent variations much faster than the period associated to the nominal frequency. Consequently, practical application of WLI technique in optical ITs demands advances in the fields of encoding techniques and signal processing schemes, to allow production of sensor systems able to provide bandwidths sufficiently larges to attend requirements of defined accuracy classes, which will depend on the foreseen use of IT for metering or protection.

Recently a new encoding technique based on electro-optical interferometers has been developed to overcome both mentioned limitations, in dynamic range and bandwidth, for the specific case of OVT to be used for measuring in high voltage electric power systems.

Most part of electro-optical sensors utilized in OVTs makes use of electro-optic modulators based on Pockels effect configured as birefringence interferometers. In this case, there is a limitation in dynamic range of measured voltage, which is a typical disadvantage arising from the polarimetric method used for measuring the electrically induced phase retardation due to the Pockels effect, since normal interferometric methods are useful only to measure OPD in the range of one wavelength. However, as described earlier, WLI sensor systems are based on interferometers linked in series by optical fibers. Therefore, by linking in series two properly designed Pockels modulators configured as birefringence interferometers, one as voltage sensor and other as recovery interferometer, it is possible to set up a WLI sensor system.

The working principle of a Pockels modulator, represented as shown in Fig. 3, can be described as a two-beam birefringence interferometer by assuming that the components of incident light, which are polarized parallel to x and y axes, are two independent beams having intensities I_x and I_y, respectively. When light is crossing the Pockels crystal, both components travel along a path of same length, L, which is the length of crystal, but each one is affected by a different refractive index. The difference between these two refractive indices, $\Delta n = n_y - n_x$, is the birefringence of medium. The light emerging from the crystal crosses a second polarizer, called as analyzer, which is oriented at 45° with respect to

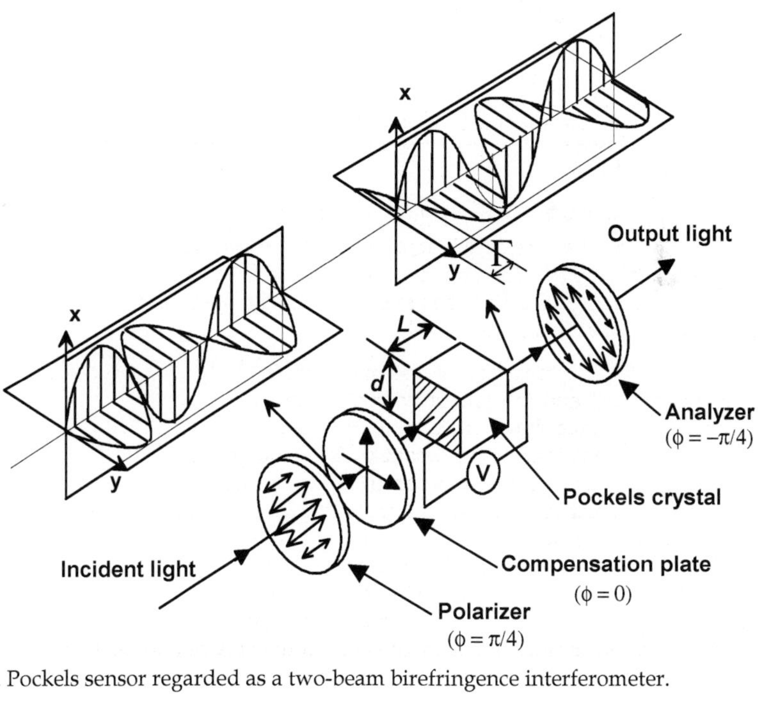

Fig. 3. Pockels sensor regarded as a two-beam birefringence interferometer.

polarization directions of both components. Since the analyzer transmits only the part of light intensity of each component that is parallel to its polarization direction, interference of the two light components occurs at its output.

The key part in designing a Pockels modulator is the calculus of its birefringence. The general procedure involved in this task depends on several parameters, such as: the specific crystal selected (including its optical and electrical characteristics, geometry and dimensions), the configuration chosen (longitudinal or transverse) and the relative orientation among the propagation direction, the electric filed direction and the crystal's optical axes directions.

For example, using an electro-optical crystal with cubic lattice (of $\bar{4}3m$ symmetry group, for instance), in a polarimetric modulator in longitudinal configuration as shown in Fig. 3, where polarizer and analyzer polarization directions are crossed to each other, the propagation direction is parallel to an optical axis of crystal, and considering that applied voltage V generates an uniformly distributed electric field, E, inside crystal, which implies that $E = V/L$, the electricaly induced birefringence, Δn_e, becomes (Yariv & Yeh, 1984):

$$\Delta n_e(V) = n_o^3 r_{41} E = n_o^3 r_{41} \frac{V}{L} \quad (4)$$

where n_0 is the ordinary refractive index and r_{41} is, in this case, the only relevant electro-optic coefficient of crystal.

The electrically induced phase retardation, Γ_e, between the two orthogonal light components is:

$$\Gamma_e = \frac{2\pi}{\lambda} \Delta n_e \times L \quad (5)$$

where λ is the wavelength of light inside crystal when there is no applied voltage.

In this case, the half-wave voltage, which is defined as the value of V that makes $\Gamma_e = \pi$, is given by (Almeida, 2001):

$$V_\pi = \frac{\lambda}{2 n_0^3 r_{41}} \quad (6)$$

The total phase retardation, Γ_t, is the summation of the electrically induced phase retardation, Γ_e, with a fixed contribution of compensation plate, $\Delta\phi_c$, and can be written as:

$$\Gamma_t = \Delta\phi_c + \frac{2\pi}{\lambda} \Delta n_e \times L = \Delta\phi_c + \frac{2\pi}{\lambda} n_0^3 r_{41} V \quad (7)$$

The total optical phase retardation can be also given in terms of V_π by:

$$\Gamma_t = \Delta\phi_c + \pi \frac{V}{V_\pi} \quad (8)$$

By using eq. (8) the optical path difference, ΔL, correspondent to this optical phase retardation is given by:

$$\Delta L = \frac{\lambda}{2\pi}\Gamma_t \tag{9}$$

Applying eq. (7) to this relation, the total path difference becomes

$$\Delta L = \Delta L_c + n_o^3 r_{41} V \tag{10}$$

where $\Delta L_c = \dfrac{\lambda}{2\pi}\Delta\phi_c$ is the OPD introduced by the compensation plate.

From eq. (10), it is found that the total OPD in this interferometer is independent of wavelength and is linearly dependent on applied voltage. Since the two beams have crossed polarization directions they cannot interfere with other. The recombination of beams is made by the analyzer, which actually is a polarizer with its polarization direction oriented at -45° with respect to x axis. Since these two beams have the same intensity ($I_x = I_y$) and only half of intensity of each beam is transmitted by the analyzer, the visibility of its central fringe is $K_{0s} = 1$, and the transmission factor of this sensing interferometer is:

$$T_s = \frac{1}{2}\alpha_s \tag{11}$$

where α_s is the attenuation introduced in the transmitted light by the sensing interferometer due to factors such as reflections, scattering, misalignments, absorption, etc.

The behavior of this interferometer is given, from eq. (2), by:

$$I_s(\sigma) = T_1\,\alpha_s \frac{1}{2} I_o(\sigma)\left\{1 \pm \cos\left[2\pi\sigma\left(\Delta L_c + n_o^3 r_{41} V\right)\right]\right\} \tag{12}$$

If a second Pockels birefringence interferometer is linked in series with the sensing interferometer described above, the overall behavior of the white light interferometric sensor system obtained is given, analogously to eq. (3), by:

$$I = \frac{1}{4}\alpha_s\alpha_r T_1 T_2 I_0 \left\{1 + \frac{1}{2} e^{-\left\{\left[\left(\Delta L_{cs} - \Delta L_{cr}\right) + n_o^3 r_{41}\left(V_s - V_r\right)\right]\frac{\pi}{L_c}\right\}^2} \cos\left[2\pi \frac{\left(\Delta L_{cs} - \Delta L_{cr}\right) + n_o^3 r_{41}\left(V_s - V_r\right)}{\lambda_0}\right]\right\} \tag{13}$$

where indices s and r correspond to values related to sensor and recovery interferometer, respectively, and L_c is the coherence length of source, given by (Yu, 2005):

$$L_c \cong \frac{\lambda_0^2}{d\lambda} \tag{14}$$

where λ_0 is the central wavelength of low-coherence source in vacuum and $d\lambda$ is the spectral width of source.

In order to operate the sensor system near to central fringe, both OPDs introduced by retardation plates must to be equal. In this situation, $\Delta L_{cs} - \Delta L_{cr} = 0$ and eq. (13) becomes:

$$I = \frac{1}{4}\alpha_s \alpha_r T_1 T_2 I_0 \left\{ 1 + \frac{1}{2} e^{-\left\{\left[n_o^3 r_{41}(V_s - V_r)\right]\frac{\pi}{L_c}\right\}^2} \cos\left[2\pi \frac{n_o^3 r_{41}(V_s - V_r)}{\lambda_0}\right] \right\} \quad (15)$$

It is found from last equation that, for a system built as described, the voltage applied to recovery interferometer, V_r, must to be equal to the voltage to be measured, V_s, in order to reach the maximum light intensity at system output. Therefore, by observing the output light intensity, I, meanwhile V_r is changed and measuring V_r when I is at maximum, it is possible to know directly the value of V_s at that moment.

The measuring method just described corresponds to a hybrid encoding technique, since it can be considered as temporal, once it could be necessary to scan periodically V_r to get one measured value of V_s per period and, also, it can be seem as electronically scanned, once there are no moving parts in the system and the OPD scanning is performed by means of an electro-optical device.

The exact classification of this technique, in fact, depends on how the electronic signal processing of the sensor system is implemented. The behavior and performance characteristics of the sensor system are ultimately defined by the association of encoding technique with the signal processing technique.

4. Application of WLI technique to a high voltage Pockels sensor system

Considering the application of a WLI sensor system as high voltage OVT, the method just described is not useful because it imposes the need to apply to recovery interferometer the same high voltage that is applied to the sensor interferometer, which is meaningless.

There are, at least, two possibilities for introducing a measurement scale factor in the sensor system, in order to reduce the maximum value of scanning voltage necessary to find the central fringe position. The first alternative consists of using different crystals in sensor and recovery interferometers. Choosing a crystal to be used in recovery interferometer with a product $n_{or}^3 r_{41r}$ greater than the product $n_{os}^3 r_{41s}$ of crystal used in sensor interferometer, the voltage V_r necessary to matches the OPD introduced by sensor interferometer will be smaller. In this case, eq. (15) becomes:

$$\begin{aligned} I &= \frac{1}{4}\alpha_s \alpha_r T_1 T_2 I_0 \left\{ 1 + \frac{1}{2} e^{-\left[\left(n_{os}^3 r_{41s} V_s - n_{or}^3 r_{41r} V_r\right)\frac{\pi}{L_c}\right]^2} \cos\left[2\pi \frac{n_{os}^3 r_{41s} V_s - n_{or}^3 r_{41r} V_r}{\lambda_0}\right] \right\} = \\ &= \frac{1}{4}\alpha_s \alpha_r T_1 T_2 I_0 \left\{ 1 + \frac{1}{2} e^{-\left\{\left[n_{os}^3 r_{41s}(V_s - f V_r)\right]\frac{\pi}{L_c}\right\}^2} \cos\left[2\pi \frac{n_{os}^3 r_{41r}(V_s - f V_r)}{\lambda_0}\right] \right\} \end{aligned} \quad (16)$$

where $f = \frac{n_{or}^3 r_{41r}}{n_{os}^3 r_{41s}}$ is the scale factor.

The second alternative to introduce a scale factor is making use of a configuration in the Pockels modulator applied as sensor interferometer different than the configuration used in the modulator used as recovery interferometer. For example, a transverse modulation configuration can be applied in recovery interferometer while longitudinal modulation configuration is applied in sensor interferometer. In such a case, eq. (15) becomes:

$$I=\frac{1}{4}\alpha_s\alpha_r T_1T_2 I_0\left\{1+\frac{1}{2}e^{-\left\{\left[n_o^3 r_{41}\left(V_s-\frac{L}{d}V_r\right)\right]\frac{\pi}{L_c}\right\}^2}\cos\left[2\pi\frac{n_o^3 r_{41}\left(V_s-\frac{L}{d}V_r\right)}{\lambda_0}\right]\right\} \quad (17)$$

where d is the thickness of Pockels crystal used in the recovery interferometer and the scale factor is L/d. In order to describe in a general form the scale factor arising from the use of different electro-optic modulators in WLI sensor systems, eq. (15) may be represented in terms of V_π as follows:

$$I=\frac{1}{4}\alpha_s\alpha_r T_1T_2 I_0\left\{1+\frac{1}{2}e^{-\left[\left(\frac{\lambda_0}{2V_{\pi s}}V_s-\frac{\lambda_0}{2V_{\pi r}}V_r\right)\frac{\pi}{L_c}\right]^2}\cos\left[\pi\left(\frac{1}{V_{\pi s}}V_s-\frac{1}{V_{\pi r}}V_r\right)\right]\right\}=$$
$$=\frac{1}{4}\alpha_s\alpha_r T_1T_2 I_0\left\{1+\frac{1}{2}e^{-\left\{\left[\frac{\lambda_0}{2L_c}\frac{\pi}{V_{\pi s}}\left(V_s-f_\pi V_r\right)\right]\right\}^2}\cos\left[\frac{\pi}{V_{\pi s}}\left(V_s-f_\pi V_r\right)\right]\right\} \quad (18)$$

where $V_{\pi r}$ and $V_{\pi s}$ are half wave voltages of Pockles modulators used as sensor and recovery interferometers, respectively, and the scale factor, expressed in terms of V_π, is: $f_\pi=\frac{V_{\pi s}}{V_{\pi r}}$.

The advantage in using f_π is that, regardless of all characteristics of any specific electro-optical modulator selected to be used as sensor or recovery interferometer, it is enough to calculate $V_{\pi r}$ and $V_{\pi s}$ to determine the optical behavior of WLI sensor system. Therefore, the procedure to design a high voltage OVT using a WLI sensor system as described must include the selection of a scale factor in such a way that the scanning voltage becomes suitable to be driven and measured by an electronic signal processing system.

4.1 Pockels modulators as high voltage sensor interferometers

Recently, different possibilities for development of Pockels modulators specially designed for direct measurement of high voltages have been investigated. Among these possibilities, a particular approach comprises an unusual configuration of Pockels modulator in which the electro-optic crystal is divided in several slices (Santos et al., 2000). This kind of modulator, which is called multi-segmented Pockels sensor, allows V_π to be defined to arbitrarily high value. Several prototype versions of devices based on such approach were already built and tested, demonstrating their suitability to be applied as primary sensor in high voltage OVTs.

A basic version of high voltage Pockels sensor in longitudinal modulation configuration consists in an arrangement of one piece of electro-optic Pockels crystal inserted between two aluminum electrodes, which are separated and supported by acrylic tubes. In a prototype formerly built, the electro-optic crystal used was a piece of $Bi_3Ge_4O_{12}$ (BGO) in parallelepiped shape (1x5x100 mm) encapsulated in an acrylic cylinder. Since BGO exhibits $n_o = 2.098$ and $r_{41} = 1.03\times10^{-12}$ (m/V), and using a SLD operating in $\lambda = 1.321$ μm as light source in the high voltage sensor described, the half-wave voltage calculated using eq. (6) is: $V_\pi \cong 69.4$ kV.

In order to increase V_π value, a high voltage Pockels sensor using longitudinal modulation in a multi-segmented configuration was developed to be used as sensor interferometer in an OVT prototype (Santos, 1996). Such sensor is composed by electro-optic crystals and pieces of dielectric material acting as mechanical supports that maintain the crystals spaced by gaps, as shown in Fig.4.

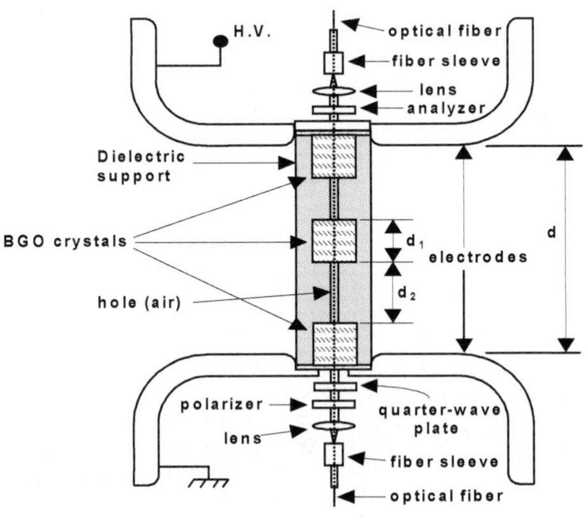

Fig. 4. Multi-segmented high voltage electro-optical sensor.

A simple method to make an estimation of V_π voltage for multi-segmented sensors is achieved by using the expressions (Santos, 1996):

$$V_\pi = \frac{(\lambda.K)}{2.n_0^3.r_{41}\,n.d_1} \quad (19)$$

$$K = \left[n.d_1 + \frac{\varepsilon_1}{\varepsilon_2}(n-1).d_2 \right] \quad (20)$$

where n is the number of crystal pieces, d_1 is their thickness, d_2 *is* the thickness of gaps, ε_1 is the permittivity of crystal and ε_2 is the permittivity of gas medium filling the gaps.

Although eq. (20) is a good approximation when crystal pieces are thin (≅ 1 mm), it does not remain valid when the thickness d_1 increases, due to non-uniformities of electric field inside the pieces. An equivalent value of K that allows a better estimation of V_π can be found from the voltage applied to electrodes and the average value of electric field in crystal pieces by:

$$K = \frac{V}{\overline{E}} \tag{21}$$

$$\overline{E} = \frac{1}{n}\sum_{i=1}^{n} \overline{E_i} \tag{22}$$

The average values of electric fields inside the crystals, $\overline{E}_i$, were computed by applying a 2D multiphysics simulation software based on finite elements method (FEM), which allows the exploration of axial symmetry presented by the sensor (Rubini et al., 2004). This kind of software has been used in analysis by FEM simulation of several electro-optic components, including integrated optics devices (Franco et al., 1999).

For example, using 3 pieces of BGO having d_1 = 6,26 mm, placed in a sensor as shown in Fig. 4, the calculated value using FEM simulation is: $V_\pi \cong 421{,}9$ kV. The experimentally measured value, using a prototype built with same configuration and materials above described, was: $V_\pi \cong 439{,}6$ kV. The good agreement between calculated and measured V_π values indicates that multi-segmented approach is suitable for obtaining Pockels modulators appropriate for direct measurement of high voltages.

4.2 Recovery interferometer

In order to allow the identification of the central fringe, a Pockels modulator suitable to be used as recovery interferometer in a WLI sensor system must to be able to reconstruct a significant part of the fringe pattern. Since each fringe is within an interval of 2π radians of phase difference and V_π is the voltage necessary to produce π radians of phase difference in a Pockels modulator, it follows that a Pockels recovery interferometer must have $V_{\pi r}$ as small as possible. Even though in a particular WLI sensor system multi fringe recovering is not required, a small $V_{\pi r}$ value is favorable for the electronic signal processing system, which is responsible for driving the scanning voltage V_r.

Reduction of V_π in electro-optic modulators can be achieved, for example, by adopting a transverse modulation configuration. In this case, the relation between electric field and applied voltage is given by $E = V/d$, and, from eq. (4) and eq. (5), V_π becomes:

$$V_\pi = \frac{\lambda}{2n_o^3 r_{41}} \frac{d}{L} \tag{23}$$

To achieve smaller values of V_π, the ratio d/L must be minimized. Using a piece of BGO, in parallelepiped shape with dimensions of d = 1 mm and L = 100 mm, as electro-optic crystal in a prototype of recovery interferometer, and the same light source described before, the half-wave voltage calculated using Eq. (23) is $V_\pi \cong 694$ V. Although such value can be considered satisfactory for some applications, it is still relatively high.

An alternative to reduce $V_{\pi r}$ even more, as mentioned before, is to make use of materials with a product $n_{or}^3 r_{41r}$ greater than the product $n_{os}^3 r_{41s}$ of crystal used in sensor interferometer. It is possible to find crystals like Barium Strontium Niobate (BSN) that presents Pockels coefficients many times greater than that of BGO, which may allow obtaining much smaller $V_{\pi r}$, as is desirable. Another possibility to obtain electro-optical modulators with low $V_{\pi r}$ is using integrated optical devices, which typically show half-wave voltages around 5 V. However, since there are no integrated optical devices suitable for WLI applications commercially available, development efforts must be done to make feasible this possibility.

5. Signal processing techniques applied to WLI high voltage sensor system

As stated earlier, behavior and performance characteristics of a WLI sensor system, as dynamic range, accuracy, resolution, bandwidth, thermal stability, etc., are all dependent on the association of an encoding technique with a signal processing technique.

The presented encoding technique based on Pockels modulators linked by optical fibers may, potentially, allow a WLI sensor system applied as OVT to have large bandwidth, high accuracy and fine resolution, among other desirable characteristics. However, since Pockels modulators can be seen as polarimetric interferometers, high voltages measured are traduced in terms of optical intensity at their outputs. Therefore, the measurement becomes dependent on link losses, which can change unpredictably in time.

Recently, applications of specially designed electronic signal processing systems have been proposed as options to overcome such problem (Almeida, 2001; Santos, 1996). The main idea behind these signal processing techniques is to take advantage of a particular property arising in WLI sensor systems when fixed OPDs introduced by compensation plates are larger than coherence length of the light source. In this case, from eq. (12) it is possible to observe that, since visibility approaches to zero, the information in the output of sensor interferometer is encoded only in the spectrum of light, suggesting that a measurement can be done independently of optical intensity present in the optical fiber link output.

In WLI systems using sensor and recovery interferometers connected in series, both introducing identical fixed OPDs, the information is translated back to the amplitude domain, as shown by eq. (13), and measuring can be done in a large bandwidth, but becomes again dependent on the optical intensity at output of system. To achieve a WLI sensor system that provides measurements independent on optical intensity it is necessary to develop electronic signal processing techniques specially designed to compensate optical intensity fluctuations not related to the measurand.

Following, two topologies of WLI sensor systems, using different encoding techniques associated with their specific signal processing techniques, both able to offer a.c. high voltage measurements compensated for optical intensity fluctuations, are presented. Prototypes were built and tested, and experimental results are presented and commented.

5.1 Peak and valleys values detection technique

5.1.1 Principle of operation

The strategy adopted in this signal processor approach is based on the precise detection of peaks and valleys values present in the output signal of a modulated recovery interferometer, which contains desired information about the ac high-voltage applied to the specially designed Pockels modulator used as sensor interferometer.

The analysis of this patent pending signal processing technique may start by the photo-current, $I_d(t)$, detected at output of optical fiber emerging from recovery interferometer, which is proportional to optical intensity, $I(t)$, as shown in Fig.5 (Almeida & Filho, 2003).

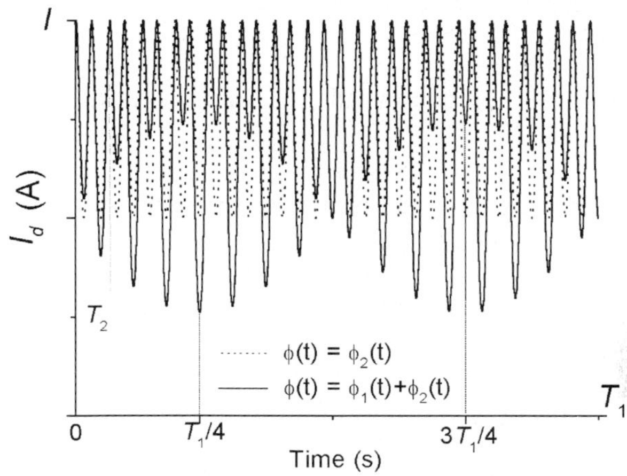

Fig. 5. Photo-current, I_d (t), for a period T_1 of the phase of modulation voltage applied to the sensor interferometer.

The behavior of $I_d(t)$ is shown in Fig. 5 for a particular condition, characterized by the phases of modulating voltages $\varphi_1(t) = \varphi_1 \sin\omega_1 t$ and $\varphi_2(t) = \varphi_2 \sin\omega_2 t$, applied, respectively, to sensor and recovery interferometers. To better explain the principle of operation, it were adopted $\omega_2 = 16\omega_1$, $\varphi_1(t) = 0{,}5$ radian, $\varphi_2(t) = \pi/2$ radians. $T_1 = 1/f_1$ is the period of fundamental frequency f_1 of $\varphi_1(t)$ and $\omega_2 = 1/T_2$ ($T_2 = T_1/16$).

In Fig. 5 it is illustrated that:

- the valley values change with the presence of phase modulation $\varphi_1(t)$ due to the ac high voltage applied to the sensor interferometer,
- these changes occur at the instants $T_2/4$ and $3T_2/4$ of each cycle of phase modulation $\varphi_2(t)$ applied to the recovery interferometer. This deviation of valley values occurs alternately bellow and above the reference value obtained without the phase modulation, $\varphi_1(t)$,
- the peak values don't change with the presence of phase modulation $\varphi_1(t)$ due to the ac high-voltage applied to the sensor interferometer. However, their positions change and occur symmetrically to the instant $T_2/2$ at each cycle of phase modulation, $\varphi_2(t)$.

In order to clarify the topics above, in Fig.6 it is shown a zoom in time, detailing one cycle of the voltage $V_d(t)$, present at the output of a transimpedance amplifier, used to convert the photocurrent $I_d(t)$ to voltage. An ideal transimpedance amplifier built in a classical configuration presents an output voltage $V(t)$ given by:

$$V(t) = R_f\, I(t) \tag{24}$$

were R_f is the feedback resistance, which is numerically equal to the transimpedance gain, G.

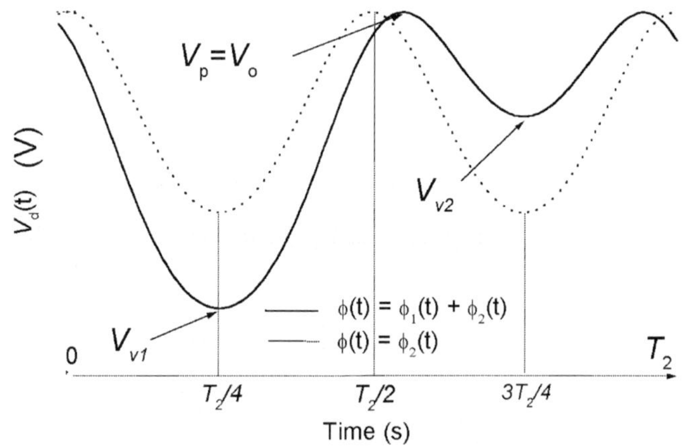

Fig. 6. Magnification in time showing output voltage at the output of transimpedance amplifier in a period of T_2.

In Fig.6 it is shown the occurrence of:

- a valley value V_{v1} and a valley value V_{v2}, bellow and above, respectively, reference value, represented by the dotted line, obtained for $\varphi_1(t) = 0$ radian,
- a peak value occurring after $T_2/2$.

The photo current, can be expressed by (Santos et al., 2002):

$$I_d(t) = \frac{I_o}{2}\left\{1 + \cos\left[\varphi_1(t) + \varphi_2 \sin \omega_2 t\right]\right\} \tag{25}$$

where I_o is its maximum value. $I_d(t)$ can also be expressed using Bessel functions as:

$$\begin{aligned} I_d(t) = I_o\left\{1 + \frac{1}{2}\left[J_0(\varphi_2) + 2\sum_n J_{2n}(\varphi_2)\cos(2n\omega_2 t)\right]\cos\left[\varphi_1(t)\right]\right\} + \\ - I_o\left\{\sum_n J_{2n-1}(\varphi_2)\sin\left[(2n-1)\omega_2 t\right]\sin\left[\varphi_1(t)\right]\right\} \end{aligned} \tag{26}$$

where $n \in N^*$, and J_n denotes the nth order Bessel function of first kind (Almeida, 2001).

From eq. (25) and eq. (26) it is obtained the expression for the low frequency components of $I_d(t)$, labeled as $I_{lf}(t)$ and characterized by $\omega \langle\langle\, \omega_2 - \omega_1$, as:

$$I_{lf}(t) = I_o\left\{1 + \frac{1}{2}J_0(\varphi_2)\cos\left[\varphi_1(t)\right]\right\} \tag{27}$$

Likewise, from eq. (24) and eq. (27) the low frequency components of $V_d(t)$, termed as $V_{lf}(t)$, can be expressed as:

$$V_{lf}(t) = V_o\left\{1 + \frac{1}{2}J_0(\varphi_2)\cos\left[\varphi_1(t)\right]\right\} \tag{28}$$

Based on eq. (24) and eq. (26) the high frequency components of $V_d(t)$ labeled as $V_{hf}(t)$ and identified by $\omega_2 - \omega_1 \rangle\rangle \omega_1$ can be written as:

$$V_{hf}(t) = V_o\left(\frac{1}{2}\left\{\cos\left[\varphi_1(t) + \varphi_2 \sin\omega t\right] - J_0(\varphi_2)\cos\left[\varphi_1(t)\right]\right\}\right) \tag{29}$$

From eq. (29) results that, the valley values of $V_{hf}(t)$, from now on termed $V_{v1}(t)$ and $V_{v2}(t)$, can be expressed in each cycle of $\varphi_2(t)$ as:

$$V_{v1} = \frac{V_o}{2}\left\{\cos\left[\varphi_2 + \varphi_1(t)\right] - J_0(\varphi_2)\cos\left[\varphi_1(t)\right]\right\} \tag{30}$$

$$V_{v2} = \frac{V_o}{2}\left\{\cos\left[\varphi_2 - \varphi_1(t)\right] - J_0(\varphi_2)\cos\left[\varphi_1(t)\right]\right\} \tag{31}$$

A straightforward substitution of trigonometric identities in eq. (30) and eq. (31), gives:

$$V_{v1}(t) = \frac{V_o}{2}\left\{\cos\varphi_2 \cos\left[\varphi_1(t)\right] - \sin\varphi_2 \sin\left[\varphi_1(t)\right]\right\} - \frac{V_o}{2}\left\{J_0(\varphi_2)\cos\left[\varphi_1(t)\right]\right\} \tag{32}$$

$$V_{v2}(t) = \frac{V_o}{2}\left\{\cos\varphi_2 \cos\left[\varphi_1(t)\right] + \sin\varphi_2 \sin\left[\varphi_1(t)\right]\right\} - \frac{V_o}{2}\left\{J_0(\varphi_2)\cos\left[\varphi_1(t)\right]\right\} \tag{33}$$

In order to maximize the influence of $\varphi_1(t)$ in the difference between $V_{v1}(t)$ and $V_{v2}(t)$, one should set $\varphi_2 = \pi/2$ radians as peak value of phase modulation $\varphi_2(t)$. Such a substitution in eq. (32) and eq. (33) results in:

$$V_{v1}(t) = \frac{V_o}{2}\left\{-\sin\left[\varphi_1(t)\right] - J_o(\pi/2)\cos\left[\varphi_1(t)\right]\right\} \tag{34}$$

$$V_{v2}(t) = \frac{V_o}{2}\left\{\sin\left[\varphi_1(t)\right] - J_o(\pi/2)\cos\left[\varphi_1(t)\right]\right\} \tag{35}$$

From eq. (34) and eq. (35), $\varphi_1(t)$ can be written as:

$$\varphi_1(t) = \arcsin\left[\frac{V_{v2}(t) - V_{v1}(t)}{(V_o/1{,}5)}\right] \tag{36}$$

Strictly speaking, the parameter V_o changes with fluctuations in the intensity of the light incident on the photodiode and should be written as $V_o(t)$. In spite of that, eq. (36) shows that, from a theoretical point of view, these fluctuations have no influence on the measured value of $\varphi_1(t)$. From (36), in the interval $0 \leq \varphi_2 \leq \pi$, $\varphi_2 = \pi/2$ radians is the peak value of the phase modulation, $\varphi_2(t)$, which:

- maximizes the differential sensitivity between the valley values $V_{v1}(t)$ and $V_{v2}(t)$ as regard to $\varphi_1(t)$;
- minimizes the dependence of scale factor with respect to time fluctuations in φ_2, since that $\partial\varphi_2(t)/\partial t = 0$ for this point of operation;
- maximizes the dynamic range for the measurement of $\varphi_1(t)$, i.e., $|\varphi_1(t)|_{max} = \pi/2$ radians.

The presented technique allows keeping control of phase modulation depth, φ_2, by means of:

$$\varphi_2 = \arcsin\left[\frac{V_{v2}(t) + V_{v1}(t)}{(V_o/1{,}5)}\right] \tag{37}$$

likewise as provided by other techniques proposed for processing interferometric signals, based on division of RMS values of some of its harmonics or zero-crossings (Burns, 1994; Tselikov et al., 1998).

A blocks diagram of electronic signal processor developed for demodulating ac high voltages applied to OVT prototype is shown in Fig.7.

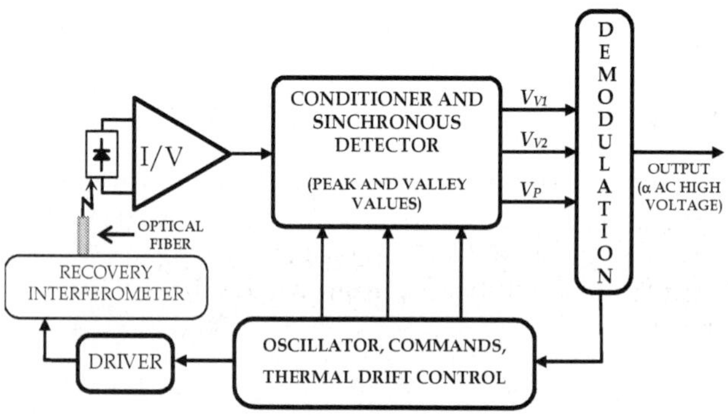

Fig. 7. Blocks diagram of the signal processor applied to the OPT.

The synchronous detection of consecutive peak and valleys values the transimpedance amplifier output yields, according to eq. (37), the demodulation of ac high voltage applied to WLI high voltage sensor system.

5.1.2 Experimental results

Aiming the development of a highly precise OVT based on WLI technique, a prototype of optical fiber sensor system using a Pockels modulator as recovery interferometer was built (Rubini et al., 2004), as shown in Fig. 8.

The sensor system was integrated to the electronic signal processor and the whole set was subjected to ac high-voltage tests conducted to determine the overall performance with

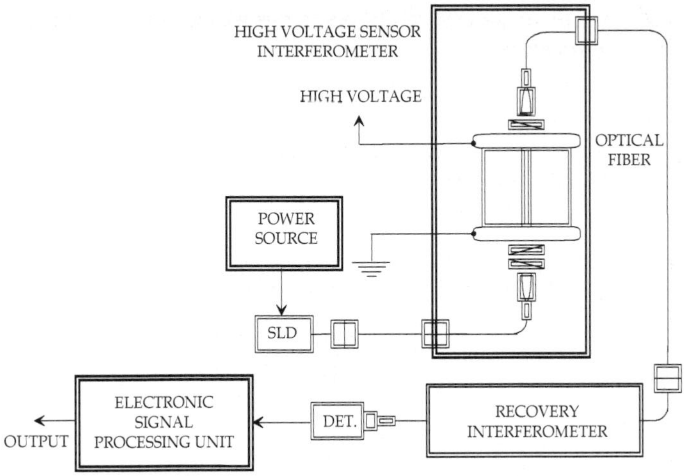

Fig. 8. Blocks diagram of OVT prototype using WLI configuration.

respect to the voltages applied. In Fig.9(a) it is illustrated the OVT used for the essays up to 10 kV_{RMS} applied to the sensor at the Optical Sensors Laboratory (LSO) of University of São Paulo (USP). The sensor interferometer is connected to a conventional instrument voltage transformer, rated 80.5 kV primary, 115/230V secondary, 7.5 kVA. Fig. 9(b) illustrates the OVT prototype at

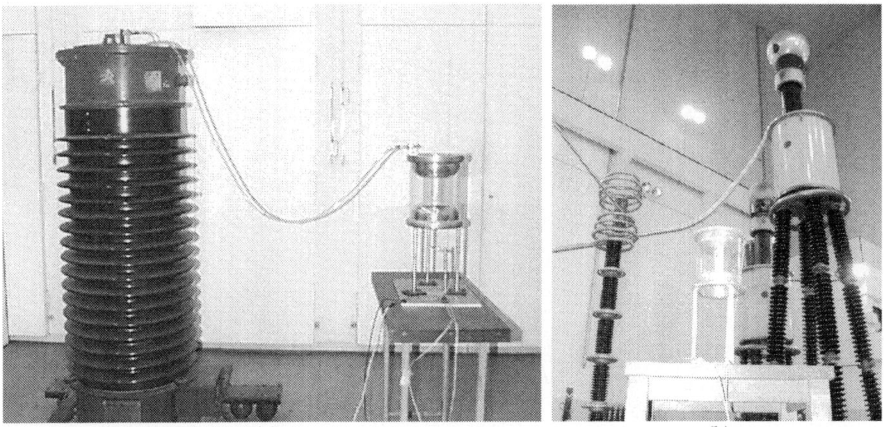

(a) (b)

Fig. 9. Sensor interferometer of the OVT prototype: (a) connected to the voltage transformer at Optical Sensors Laboratory (LSO-USP) and (b) connected to the cascade transformer at High Voltages Laboratory (LAT-IEE/USP).

the High Voltages Laboratory (LAT) facilities of Electrotechnical and Energy Institute (IEE/USP) used for the essays from 10 kV_{RMS} up to 100 kV_{RMS}. The sensor interferometer connected to a high voltage cascade transformer, rated 1000 kV secondary, 150 kVA.

High-voltages at power frequency (60 Hz) up to 80 kV_{RMS} were applied to the sensor interferometer. The experimental results from these tests and related linear regression curve are shown in Fig. 10.

From results shown in Fig. 10, it is observed that a linearity better them +/- 1,5% is obtained.

In addition, from other tests, it was possible to observe good temperature and time stabilities showed by the OVT, lower than 1000 ppm/°C in an observation period of half an hour.

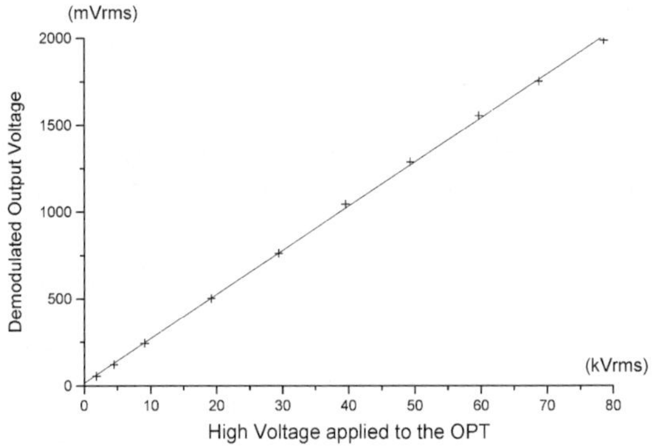

Fig. 10. Demodulated output voltage vs. ac input voltage applied to the sensor interferometer.

Such behaviors are due to the optical encoding technique as well as to the peculiar design of the electronic signal processing technique, which was specially developed for this application. The demodulation bandwidth of 1 kHz used makes possible to efficiently demodulate the firsts 16 harmonics of ac voltage applied to the sensor.

In Fig. 11 it is shown time-domain waveforms of demodulated output voltage and applied voltage (obtained from a reference divider) when about 70 kV_{RMS} is applied to the OVT prototype.

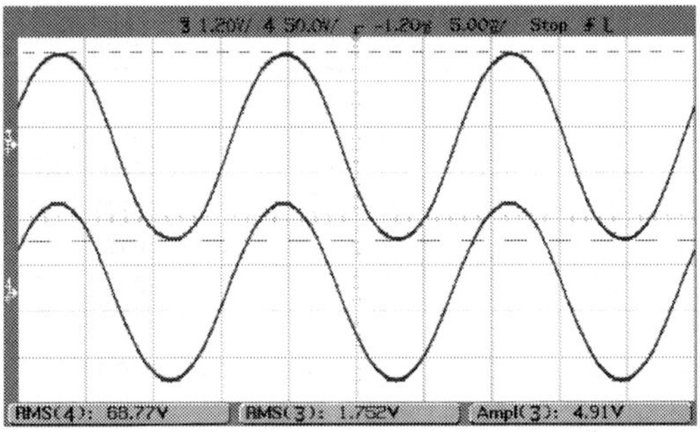

Fig. 11. Waveforms and RMS values of both the output demodulated voltage (channel 3) and ac high voltage applied to the sensor interferometer (channel 4).

In Fig. 12 frequency domain analyses are presented for both the output demodulated voltage and the ac high voltage applied to the sensor interferometer presented in Fig. 11.

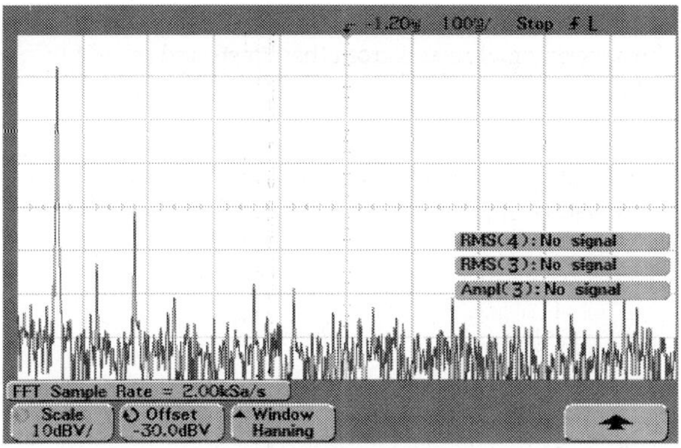

(a)

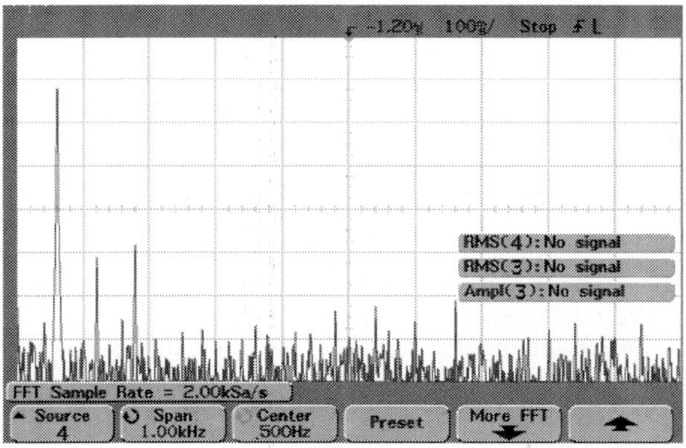

(b)

Fig. 12. Fast Fourier transforms, in the range from 0 kHz to 1 kHz: (a) of output demodulated voltage and (b) of ac high voltage applied to sensor interferometer

Unfortunately, for 70 kV_{RMS} applied to the sensor, the correspondent frequency domain analysis observed in Fig.12(b) can't be extended for frequencies higher than 300 Hz due to the bandwidth limitation of the reference divider used for measuring the amplitude of ac high voltage applied to the sensor interferometer. In spite of that, a great likeness between this spectrum and the one shown in Fig. 12(a) can be observed up to 300 Hz.

5.2 Normalization by an optical power reference sample technique

5.2.1 Principle of operation

To take advantage of WLI technique in a less expensive configuration capable to reduce the dependency of the OVT output voltage, which carries out the measurement information, $V_{ac}(t)$, on optical intensity fluctuations, an alternative encoding technique using an unmodulated recover interferometer was proposed (Santos et al., 2003), as shown in Fig. 13.

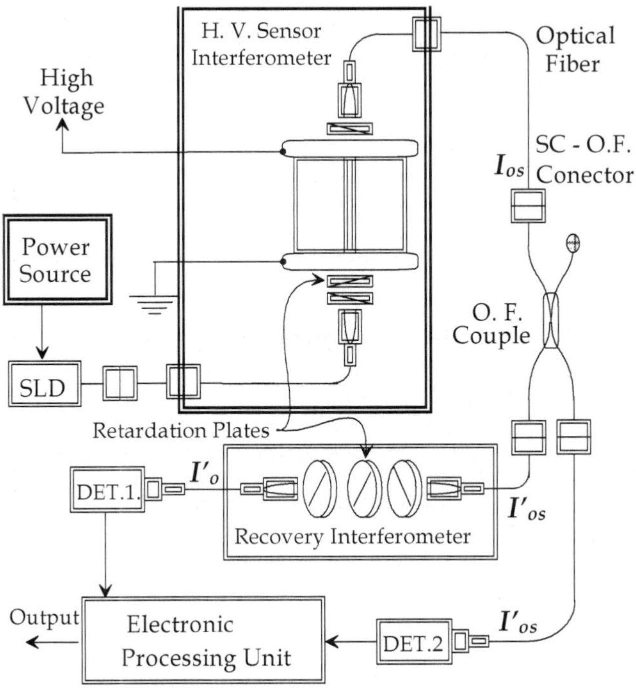

Fig. 13. OVT using WLI sensor system configuration with unmodulated recovery interferometer.

In this system, for recovering the interferometric signal a simple recovery interferometer, composed of a compensation plate inserted between two aligned polarizers, is placed in series with the optical fiber in the return path of sensing system by means of a directional coupler, which is an optical fiber beam splitter. From eq. (12), the optical intensity at the output of the recovery interferometer, I_o, is given by (Santos, 1996):

$$I_o = \frac{1}{4}\alpha_S \alpha_R T_1 T_2 I_i \left\{ 1 + \frac{1}{2} e^{-(\pi \Delta L/L_c)^2} \cos 2\pi \frac{\Delta L}{\lambda_0} \right\} \tag{38}$$

where I_i is the optical power at the input of the system (output of the SLD light source), α_S is the attenuation of light from the SLD to the input of sensor interferometer, λ_0 is the central wavelength of the light source and ΔL is the total OPD in the sensor system, given by: $\Delta L = \Delta L_s - \Delta L_c$.

The compensation plate used in the recovery interferometer is identical to the one used in the sensor interferometer. However, it is mounted in a special positioner, which can rotate around two axes to allow a fine tuning of the OPD introduced. Using this two axes rotation capability, it is possible to set the OPD of recovery interferometer in such way that: $\Delta L_{cr} = \Delta L_{cs} - \lambda_0/4$. In such condition, from eq. (38), I_o becomes:

$$I_o = \frac{1}{4}\alpha_S \alpha_R T_1 T_2 I_i \left\{ 1 + \frac{1}{2} e^{-\left[\frac{\lambda_0}{2}\left(\pi \frac{V}{V_\pi} - \frac{\pi}{2}\right)/L_c\right]^2} \cos\left(\pi \frac{V}{V_\pi} - \frac{\pi}{2}\right) \right\} \quad (39)$$

Since in practical applications L_c is much larger than λ_0, the exponential term in (39) remains close to the unity when $|V| < V_\pi$. Therefore, I_o can be approximated by:

$$I_o = I_{\bar{o}} \left(1 + \frac{1}{2} \sin \pi \frac{V}{V_\pi} \right) \quad (40)$$

where $I_{\bar{o}} = \frac{1}{4}\alpha_S \alpha_R T_1 T_2 I_i = \frac{1}{4}\alpha_t I_i$, is the average of output light intensity.

The response curve for this kind of WLI sensor system is similar to the response curve of a simple polarimetric Pockels modulator and, in this case, also there is a dependency of output light intensity, I_o, on the total attenuation of system. To take advantage of WLI technique and eliminate such dependency, an electronic signal processing scheme was developed, using as reference a sample of light intensity at output of sensor interferometer, I_{os}, which is given by:

$$I_{os} = \frac{1}{2}\alpha_s T_1 T_2 I_i \left\{ 1 + e^{-(\pi \Delta L_s / L_c)^2} \cos 2\pi \frac{\Delta L_s}{\lambda_0} \right\} \quad (41)$$

Since $\Delta L_s >> L_c$, the exponential term in eq. (40) approaches to zero. Therefore, I_{os} reduces to:

$$I_{os} = \frac{1}{2}\alpha_s T_1 T_2 I_i = \frac{2}{\alpha_R} I_{\bar{o}} \quad (42)$$

As can be seem in Fig. 13, placing a directional coupler between the output of sensor interferometer and the input of recover interferometer it is possible to obtain the signal I'_{os}(t), which is detected by the photodetector (DET. 2) and used to compensate the signal obtained in the output of recover interferometer, I'_o, for any optical power variation. Using a 3 dB directional coupler, the incoming optical intensity is divided in equal parts between the two outputs. Therefore, the optical power I'_{os}, which reaches the photodetector (DET. 2), is:

$$I'_{os} = \alpha_{ex} I_{os} / 2 \quad (43)$$

where α_{ex} is the excess loss of directional coupler and I_{os} is the optical intensity at output of sensor interferometer.

The optical intensity at output of recovery interferometer, I'_o(t), is given by:

$$I'_o(t) = \frac{1}{2}\alpha_r I'_{os} \left(1 + \frac{1}{2} \sin \pi \frac{V_{ac}(t)}{V_\pi} \right) \quad (44)$$

where α_r is the total attenuation of recovery interferometer, I'_{os} is the optical power at output of optical fiber coupler and V_π is the half-wave voltage of sensor interferometer.

A blocks diagram of electronic signal processing unit required by such system is given in Fig. 14.

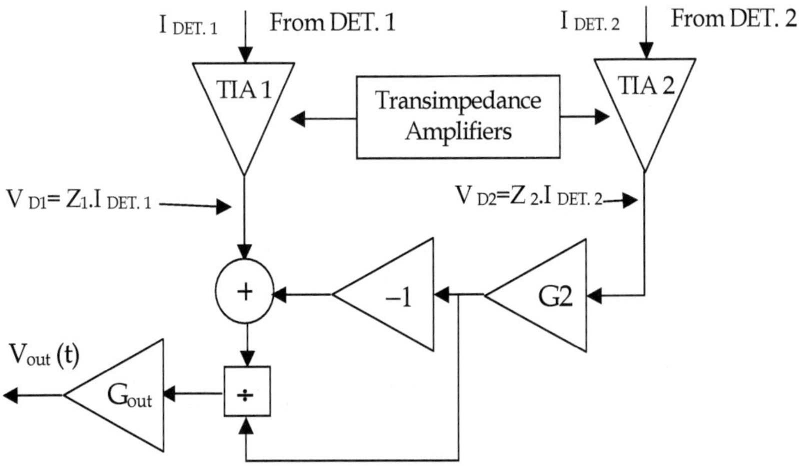

Fig. 14. Blocks diagram of electronic signal processing unit.

Applying the signals I'_{os} and I'_o(t) to the electronic signal processing unit proposed in Fig. 14, if the transimpedance gains Z_1 and Z_2 are equals and the gain $G_2 = \alpha_r/2$, the output signal, V_{out}(t), can be linearized (when $V_{ac}(t) << V_\pi$) and approximated by the following expression:

$$V_{out}(t) \cong \frac{G_{out}}{2} \cdot \pi \cdot \frac{V_{ac}(t)}{V_\pi} \tag{45}$$

where G_{out} is the gain of last amplifier stage.

From eq. (45) it is clear that V_{out} *(t)* is independent on the optical intensity at optical fiber link, as desired.

5.2.2 Experimental results

To verify the V_{out} dependency on optical power I_{os}, an experimental setup implementing the system described in Fig. 13 was built using prototypes of high voltage Pockels cell and recovery interferometer previously developed (Santos et al., 2000, 2003), a 3 dB directional coupler (commercially available), a SLD operating in a wavelength of 1.3 μm as broadband light source and an electronic signal processing unit developed according to the blocks diagram shown in Fig. 14. To introduce variations in I_{os}, an adjustable optical attenuator was inserted between the output of sensor interferometer and the input of directional coupler.

Experiments were conducted changing the attenuation imposed by the adjustable attenuator in a range of 0 to -8.66 dB. During the variation of the attenuation the voltage applied to the OVT was maintained constant in a reference value of 4.5 kV_{pp}. In Table 2, bellow, the attenuation values introduced by the attenuator, the RMS values of measured signals in the electronic processing unit output and the relative variations of correspondent V_{out} value, named ΔV_{out}, are shown.

Attenuation (dB)	V_{out} **(mV_{RMS})**	ΔV_{out} **(%)**
0	131,9	0
-2,5	131,0	-0,7
-3,46	131,0	-0,7
-4,75	131,1	-0,6
-5,54	132,0	0,1
-6,51	132,0	0,1
-7,49	133,2	1,0
-8,66	133,0	0,8

Table 2. Attenuation values introduced by the attenuator, RMS values of measured signals in the electronic processing unit output and relative variations of the V_{out} values (ΔV_{out}).

From values presented in table 2, it is shown that, for a variation of about -9 dB in average optical power carried by the optical fiber link, less than 1% of variation is observed in output of WLI sensor system.

6. Conclusions

Both alternatives of WLI based OVTs presented have adequate performances for measuring ac high voltages while keeping low sensibility to fluctuations on average light intensity in the optical fiber link.

The advantages exhibited by the first approach, compared to the second one, are the better characteristics in terms of resolution and thermal and temporal stabilities. However, the cost and complexity of that kind of solution are higher than the second one. Such characteristics make first solution suitable for applications where high performance is demanded.

The main advantages of the second approach are the lower cost and smaller complexity of the entire system. Therefore, it is recommended for applications demanding moderate performances and lower prices.

New optical encoding techniques and electronic signal processing techniques applied to WLI sensor systems are still in development, and it is expected that many improvements arise in this field in a near future.

7. Acknowledgment

The authors would like to thank to the University of São Paulo - USP - and to São Paulo Research Foundation - FAPESP - for the financial support to this work.

8. References

Almeida, J. C. J. de & Filho, O. V. A. (2003). *Demodulador de desvio de fase óptico não recíproco num sensor óptico interferométrico, via detecção dos valores dos picos da corrente detectada no fotodetector acoplado à saída do interferômetro óptico,* patent required to the Instituto Nacional da Propriedade Industrial - INPI, according to the registry number PI0303.688-0, Brazil

Almeida, J. C. J. de, (2001). *Nova Técnica de Processamento de Sinal no Domínio do Tempo de Giroscópios Interferométricos de Sagnac a Fibra Óptica,* Ph.D. Thesis, State University of Campinas, Brazil

Bull, J. D., Jaeger, N. A. F., & Rahmatian, F. (2005). A new hybrid current sensor for high-voltage applications. IEEE Transactions on Power Delivery, Vol. 20, No. 1, (Jan 2005), pp. 32-38, ISSN 0885-8977.

Burns, W. K. (1994). *Optical Fiber Rotation Sensing,* (Ed. 1), Academic Press, Inc., ISBN 0-1214.6075-4, Boston, USA

Chen, S.; Palmer, A. W.; Grattan, K. T. V.; & Meggit, B. T. (1992). Digital signal-processing techniques for electronically scanned optical-fiber white-light interferometry, *Applied Optics*, Vol. 31, No. 28, (October 1992), pp. 6003-6010, ISSN 1559-128X

Chen, X.; Ye, W.; Zhang, H. & Liu, T. (2010). Spectral domain demodulation of fibre optics position and displacement sensors by Fourier-transform spectral interferogram. *Proceedings of 9th International Conference on Optical Communications and Networks (ICOCN 2010)*, ISBN 978-1-84919-314-6, Nanjing, China, October 2010.

Crotti , G.; Sardi, A.; Kuljaca, N.; Mazza, P.; De Donà, G.; Brand, U.; Giraud, M.; A. Andersson & Weiss, S. (2006). "On-site live verification of HV instrument transformer accuracy", CIGRE General Session 41, Aug. 27 - Sep. 1, 2006, paper A3-204

Franco, M. A. R.; Passaro, A.; Sircilli, F. & Cardoso, J. R. (1999). Finite element analysis of anisotropic optical waveguide with arbitrary index profile. *IEEE Transactions on Magnetics*, Vol.35, No. 3, (May 1999), pp. 1546-1549, ISSN 0018-9464

Françon, M. (1967). *Optical Interferometry*, Academic Press, ISBN 0-1226.6350-0, New York, USA

Kim, J.H. (2008). An All Fiber White Light Interferometric Absolute Temperature Measurement System. Sensors, 8, 6825-6845. ISSN 1424-8220

Lequine, M.; Lecot, C.; Giovannini, H.; & Huard, S. (1990). A dual wavelegth passive homodyne detection unit for fiber coupled white light interferometers, In: *Fiber-Optic Sensors IV*, Vol. 1267, Editor: Ralf T. Kersten, pp. 288-293, Proceedings of SPIE, ISBN 9780819403148

Liao, H. & Yang, Y. (2010). A Linnik Scanning White-Light Interferometry System Using a MEMS Digital-to-Analog Converter, Proc. Eurosensors XXIV, September 5-8, 2010, Linz, Austria Volume 5, 2010, Pages 758-761, ISSN: 1877-7058.

Mariscotti, A. (2009). A Rogowski winding with high voltage immunity, In: EUROCON 2009 pp. 1129-1133, ISBN 978-1-4244-3860-0, St. Petersburg, RU, May 18-23 2009.

Murtaza, G. (1991). *Dual Wavelength referenced intensity modulated optical fibre sensor system*, PhD Thesys, Manchester Polytechnic, CNAA

Ning, Y. N.; Grattan, K. T. V.; & Palmer, A.W. (1995). In: *Applications of Photonic Technology*, Editors: Lampropoulos, G. A.; Chrostowsky, J. & Measures, R. M., pp. 339-342, Plenum Press, ISBN 0-3064.5011-9

Rahmatian, F. (2010). High-voltage current and voltage sensors for a smarter transmission grid and their use in live-line testing and calibration, In: *IEEE Power and Energy Society General Meeting*, Minneapolis, MN, ISSN: 1944-9925, E-ISBN: 978-1-4244-8357-0, Print ISBN: 978-1-4244-6549-1, July 2010

Rubini Jr., J.; A. Passaro, Abe, N. M. & Santos, J. C. (2004) Analysis of the electric field distribution in an electrooptic sensor for pulsed high-voltage measurements, *Proceedings of Fifth IEE International Conference on Computation in Electromagnetics - CEM2004*, pp. 71-72, ISBN 0-8634.1400-1, Stratford-upon-Avon, UK, April 19-22, 2004

Santos, J. C. (2009). Contribuições para o desenvolvimento de transformadores de potencial a fibras ópticas (TPs Ópticos) aplicáveis em sistemas elétricos de potência, L.D. Thesis, University of São Paulo, Brazil.

Santos, J. C.; Côrtes, A. L. & Hidaka, K. (1999). A New Electro-optical Method for recovering White Light Interferometric Signals, *Proceedings of 1999 International Microwave And Optoeletronics Conference - IMOC'99*, ISBN 0-7803-5807-4, Rio de Janeiro, Brazil, August 1999

Santos, J. C.; Côrtes, A. L.; Hidaka, K. & Silva, L. P. C. (2003). Improved optical sensor for high voltage measurement using white light interferometry, *Proceedings of SBMO/IEEE MTT-S IMOC 2003*, ISBN 0-7803-7824-5, Fóz do Iguacú, Paraná, Brazil, September 2003

Santos, J.C.; Hidaka, K. & Côrtes, A. L. (2002). Optical High Voltage Measurement Transformer Using White Light Interferometry, *Proceedings of 2002 IEEE/PES T&D Latin America Conference*, in CD-ROM, São Paulo, Brazil, March 2002

Santos, J.C.; Taplamacioglu, M. C. & Hidaka, K. (2000) Pockels high-voltage measurement system. *IEEE Transactions on Power Delivery*, Vol.15, No.1, (January 2000), pp. 8–13, ISSN 0885-8977

Tselikov, A.; Arruda, J. U. & Blake, J. (1998). Zero-Crossing Demodulation for Open Loop Sagnac Interferometers, *IEEE Journal of Lightwave Technology*, Vol. 16, No. 9, (September 1998), pp. 1613-1619, ISSN 0733-8724

Velosa, J.; Gouveia, C.; Frazao, O.; Jorge, P. & Baptista, J. (2011). Digital Control of a White Light System for Optical Fiber Interferometers. *IEEE Sensors Journal*, Vol. PP, No. 99, (April 2011), pp. 1 – 1, ISSN: 1530-437X

Yariv, A. & Yeh, P. (1984). *Optical waves in cristals*. (Ed. 1), Wiley-Interscience, ISBN 0-4714.3081-1, New York, USA

Yu, B.; A. Wang, Gary Pickrell & Xu, J. (2005) "Tunable-optical-filter-based white-light interferometry for sensing," Opt. Lett. 30, 1452-1454 (2005), Optics Letters, Vol. 30, Issue 12, pp. 1452-1454, ISSN: 0146-9592